Handbook of Statistical Modeling for the Social and Behavioral Sciences

AD.

Handbook of Statistical Modeling for the Social and Behavioral Sciences

Edited by

Gerhard Arminger
Bergische Universität Wuppertal
Wuppertal, Germany

Clifford C. Clogg
Late of Pennsylvania State University
University Park, Pennsylvania

and

Michael E. Sobel
University of Arizona
Tucson, Arizona

PLENUM PRESS • NEW YORK AND LONDON

Library of Congress Cataloging-in-Publication Data

Handbook of statistical modeling for the social and behavioral sciences / edited by Gerhard Arminger, Clifford C. Clogg, and Michael E. Sobel.
p. cm.
Includes bibliographical references and index.
ISBN 0-306-44805-X
1. Social sciences--Statistical methods. 2. Psychology--Statistical methods. I. Arminger, Gerhard. II. Clogg, Clifford C. III. Sobel, Michael E.
HA29.H2487 1994
300'.1'5195--dc20 94-43088
CIP

ISBN 0-306-44805-X

A Division of Plenum Publishing Corporation
233 Spring Street, New York, N.Y. 10013

10 9 8 7 6 5 4 3 2

Printed in the United States of America

To Paula Weißenbacher
To the Memory of Richard G. Clogg
To the Memory of Irvin Sobel and Peggy Sobel

Contributors

Gerhard Arminger, Department of Economics, Bergische Universität–GH Wuppertal, D-42097 Wuppertal, Germany

Michael W. Browne, Department of Psychology, Ohio State University, 142 Townshend Hall, 1885 Neil Avenue Mall, Columbus, Ohio 43210, USA

Clifford C. Clogg,† Department of Sociology and Department of Statistics, Pennsylvania State University, University Park, Pennsylvania 16802, USA

Alfred Hamerle, Lehrstuhl für Statistik, Universität Regensburg, Universitätsstr. 31, D-93053 Regensburg, Germany

Cheng Hsiao, Department of Economics, University of Southern California, Los Angeles, California 90089–0253, USA

Roderick J. A. Little, Department of Biostatistics, University of Michigan, 1420 Washington Heights, Ann Arbor, Michigan 48109-2029, USA

Nicholas T. Longford, Educational Testing Service, Princeton, New Jersey 08541, USA

Trond Petersen, Walter A. Haas School of Business, University of California, Berkeley, California 94720, USA

Gerd Ronning, Abteilung Statistik und Ökonometrie I, Department of Economics, Eberhard-Karls-Universität, Mohlstr. 36, D-72074 Tübingen, Germany

Nathaniel Schenker, Department of Biostatistics, UCLA School of Public Health, 10833 Le Conte Avenue, Los Angeles, California 90024-1772, USA

Michael E. Sobel, Department of Sociology, University of Arizona, Tucson, Arizona 85721, USA

† Deceased

Foreword

It is a pleasure to be able to contribute a foreword to this impressive handbook on quantitative methods for the analysis of data. Too often attempts such as this appear to consist of rather disconnected chapters on favorite, but possibly narrow, topics of distinguished contributors. Not so with this one! The editors have done an admirable job of blending contributions from distinguished researchers into a coherent package.

Throughout the chapters, simple but realistic examples are used to introduce fundamental ideas, and the individual authors do an especially good job of relating more advanced procedures to more basic ones, which should already be familiar to most researchers. Also, all chapters indicate, at least to some extent, the availability of software for implementing the procedures being discussed; this enterprise is always a bit hazardous in that software is constantly being born, modified, and dying, but the choice to include such references is clearly preferable to excluding them.

The selection of topics is also excellent for a researcher approaching data already collected. The general focus on conceptual parametric modeling is on-target, as such models allow the formulation of crisp scientific hypotheses, and for the natural estimation of effects and intervals for them in addition to statistical tests. As these models and their applications become more extended and developed, I expect to see more full probability (Bayesian) modelling being used so that nuisance parameters and small sample complications can be more systematically handled. Eventually, this will lead to even more discussion of computational issues, including inference by simulation, especially iterative simulation.

This is a forward-looking book with many fine contributions– congratulations to the editors and to the other authors of individual chapters.

DONALD B. RUBIN
Professor and Chairman
Department of Statistics
Harvard University

IN MEMORIAM

Clifford C. Clogg passed away on May 7, 1995. Cliff was an excellent colleague, and we shall miss working with him. But most of all, we mourn the loss of a very dear and special friend.

G.A.
M.E.S.

Preface

This is a research handbook and a reference work for quantitative methodologists, applied statisticians, empirical researchers, and graduate students in disciplines such as communications, demography, economics, education, geography, political science, psychology, statistics, and sociology. Although the focus is on models and methods for the social and behavioral sciences, this volume should also be of interest to epidemiologists and others in the health sciences, as well as to business researchers, especially those engaged in organizational or marketing research.

Researchers in these fields face a number of common modeling problems. First, much or even most of the research in these substantive fields is based on nonexperimental data, such as surveys and quasi-experiments. Second, variables to be modeled are usually measured with error. Failing to take measurement error into account typically leads to inferences that are not reliable. Third, measurements in these disciplines may be quantitative and continuous at one extreme or categorical and nominal at another. Fourth, in recent years, longitudinal data have become more common, and special methods are needed to take full advantage of the information in such data sets.

Each of the chapters in this handbook addresses one or more of the above issues. In Chapter 1, Sobel considers the difficulties that arise in attempting to use nonexperimental data to draw causal inferences, while in Chapter 2, Little and Schenker discuss modern methods for dealing with another ubiquitous problem: missing data. In modeling data, researchers usually attempt to describe how one or more specified dependent variables is or are related to independent variables or predictors, and the inferences that are made can depend heavily on auxiliary assumptions. For example, the assumption of normal homoscedastic errors in regression analysis needs to be examined. Arminger's chapter on mean structures focuses on the construction of parametric models for the relation between a set of predictor variables and one or more specified dependent variables; here the relationship is to be described by the "mean structure," and one wants to use models and methods that are valid with as few auxiliary assumptions as possible. Browne and Arminger's chapter on mean and covariance structure models discusses latent variable models for the case where both the observed and unobserved variables are metrical, and the case where one or more of the observed variables are ordered categorical and the unobserved variables are metrical. Sobel's chapter on discrete data focuses primarily on log-linear models for contingency tables, where all the observed variables in the analysis are categorical, while Clogg's chapter on the latent class model focuses on latent variable models where both the observed and unobserved variables are discrete; he also briefly takes up the case where the observed variables are discrete and the unobserved variable is metrical, as in the Rasch model.

The next three chapters focus on methods for longitudinal data. Hsiao considers models for the analysis of panel data; here measurements on the sample of respondents are taken on several occasions. He shows how to take advantage of the panel setup to answer questions that cannot be addressed with data from a cross-sectional study, or that can be answered more confidently with panel data. Panel data are also the focus of the chapter by Hamerle and Ronning; whereas Hsiao is concerned with the case where the specified dependent variable is metrical, Hamerle and Ronning take up the case where the dependent variable is discrete. Finally, Petersen considers dynamic models for the analysis of event histories, or survival models, in which a respondent can change states at arbitrary times in the study interval.

Longford's chapter considers multilevel or hierarchical models. In many ways, these models are similar to the panel-data models discussed by Hsiao. However, in panel studies (without missing data) the observations are fully crossed (time by person), whereas in multilevel models, the observations are nested within larger clusters, e.g., persons within schools within school districts. The observations within a cluster tend to be more similar than those in different clusters, and Longford shows how to estimate models that take this clustering into account.

A number of other important topics were excluded from the volume. For example, there is no chapter on model selection principles; but each chapter addresses this topic in context. Given the focus on modeling, we did not include material on the collection of data, or on sampling design. We also focused on parametric modeling, excluding, for example, consideration of nonparametric estimation of regression functions and graphical procedures. In addition, we do not discuss time series models, as there is already a large literature on the subject. Similarly, a number of other topics were excluded, including spatial models and network analysis, the latter somewhat specific to sociology. Finally, the subject of graphical modeling is not considered, despite a large statistical literature on the topic. These and other topics are surely important, but it would have been difficult to include all of them in a one-volume handbook.

In order that the handbook be accessible and useful to empirical workers and advanced graduate students, we asked our authors to follow a common format. Each chapter introduces the models in a simple context, illustrating the types of problems and data for which the models are useful. We encouraged our authors to use examples throughout their chapters and to draw upon familiar models or procedures to motivate their contributions. In addition, all the chapters include material on software that can be used to estimate the models studied. Each chapter is largely self-contained, thereby allowing a researcher who wants to study a certain type of model useful in his or her work to do so by focusing on a particular chapter, without having to study the rest of the handbook in depth. Similarly, instructors can easily organize an advanced-graduate level course around one or more of the handbook's themes by focusing attention on several of the chapters. For example, a one semester course on longitudinal analysis might take up the chapters by Hsiao, Hamerle and Ronning, and Petersen.

At the same time, the handbook is also intended for use by those interested in the more technical aspects of these subjects. Therefore, we also asked our authors to prepare reviews that represent the "state of the art" in their area. Not only did all of our authors do so, but many of the chapters also contain original material.

We are grateful to our authors for fulfilling the difficult task of writing clear papers targeted to diverse users, and for the good humor, grace, and patience they showed with our requests for rewrites and revisions. We very much appreciate the new material in a number of the chapters. We are also grateful to our editor, Eliot Werner, who has helped us at every stage of this project. Finally, we especially want to thank Daniel Enache and Ulrich Küsters, who wrote the LaTeX macros used to produce this book.

Contents

Contributors vii

Foreword *by Donald B. Rubin* ix

Preface xii

1 Causal Inference in the Social and Behavioral Sciences 1

Michael E. Sobel

1 Introduction 1
2 Deterministic Causation in Philosophy 4
3 Probabilistic Causation: Variations on a Deterministic Regularity Account 10
3.1 Philosophical Treatments 10
3.2 Granger Causation in Economics 14
4 Causation and Statistics: An Experimental Approach 17
5 Causal Inference in "Causal Models" 27
6 Discussion 32

2 Missing Data 39

Roderick J. A. Little and Nathaniel Schenker

1 Introduction 39
1.1 Examples 40
1.2 Important Concepts 42
1.3 Naive Approaches 44
1.4 More Principled Approaches 46
2 Weighting Adjustments for Unit Nonresponse 46
3 Maximum Likelihood Assuming Ignorable Nonresponse 48
3.1 Maximum-Likelihood Theory 48
3.2 The Expectation-Maximization Algorithm 49
3.3 Some Important Ignorable Maximum-Likelihood Methods 51
4 Nonignorable Nonresponse Models 55
4.1 Introduction 55
4.2 Probit Selection Model 56
4.3 Normal Pattern-Mixture Models 58
5 Multiple Imputation 59

5.1 Imputation 59
5.2 Theoretical Motivation for Multiple Imputation 62
5.3 Creating a Multiply Imputed Data Set 63
5.4 Analyzing a Multiply Imputed Data Set 65
6 Other Bayesian Simulation Methods 66
6.1 Data Augmentation 67
6.2 The Gibbs Sampler 67
6.3 The Use of Iterative Simulation to Create Multiple Imputations . . 68
7 Discussion 69

3 Specification and Estimation of Mean Structures: Regression Models 77

Gerhard Arminger

1 Introduction 77
2 The Linear Regression Model 80
2.1 Model Specification 80
2.2 Estimation of Regression Coefficients 84
2.3 Regression Diagnostics 89
2.4 Multivariate Linear Regression 97
3 Maximum Likelihood Estimation 100
3.1 Loglikelihood function 100
3.2 Properties of the ML Estimator 101
3.3 Likelihood Ratio, Wald, and Lagrange Multiplier Tests 104
3.4 Restrictions on Parameters 108
4 ML Estimation Under Misspecification 111
5 Pseudo-ML Estimation 113
5.1 Mean Structures 113
5.2 The Linear Exponential Family 114
5.3 Properties of PML Estimators 121
5.4 Computation of PML Estimators With Fisher Scoring 124
5.5 PML Wald and PML Lagrange Multiplier Tests 128
5.6 Regression Diagnostics Under PML Estimation 129
6 Quasi Generalized PML Estimation 131
6.1 Specification of Mean and Variance 131
6.2 Properties of PML Estimation With Nuisance Parameters 132
6.3 Computation of QGPML Estimators 135
6.4 QGPML Wald, Lagrange Multiplier, and Likelihood Ratio Tests . 135
6.5 Regression Diagnostics Under QGPML Estimation 136
7 Univariate Nonlinear Regression Models 139
7.1 Models for Count Data 139
7.2 Standard Nonlinear Regression Models 143
7.3 Models For Dichotomous Outcomes 146
7.4 Quantit Models for Censored Outcomes 150
7.5 Generalized Linear Models 153

8 Multivariate Nonlinear Regression Models 160
8.1 Models for Ordered Categorical Variables 160
8.2 Models for Doubly Censored and Classified Metric Outcomes . . 164
8.3 Unordered Categorical Variables 166
8.4 Generalized Estimating Equations for Mean Structures 172
9 Software 177

4 Specification and Estimation of Mean- and Covariance-Structure Models 185

Michael W. Browne and Gerhard Arminger

1 Introduction 185
1.1 Background and Notation 186
1.2 Scaling Considerations for Mean, Covariance, and Correlation Structures 187
1.3 Fitting the Moment Structure 188
2 Large Sample Properties of Estimators 194
2.1 Lack of Fit of the Model and the Assumption of Population Drift . 195
2.2 Reference Functions and Correctly Specified Discrepancy Functions 195
3 Computational Aspects 200
4 Examples of Moment Structures 203
4.1 The Factor Analysis Model 203
4.2 Structural Equation Models 205
4.3 Other Mean and Covariance Structures 216
5 Mean and Covariance Structures with Nonmetric Dependent Variables . . 220
5.1 Unconditional and Conditional Mean and Covariance Structures . 221
5.2 Inclusion of Threshold Models 223
5.3 Conditional Polyserial and Polychoric Covariance and Correlation Coefficients 226
5.4 Estimation 227
5.5 Multigroup Analysis 232
5.6 Example: Achievement in and Attitude toward High School Mathematics 232
6 Software 241

5 The Analysis of Contingency Tables 251

Michael E. Sobel

1 Introduction 251
2 Introductory Examples 253
2.1 Some Models for Univariate Distributions 253
2.2 Measuring Association in the Two-by-Two Table: The Odds Ratio 259
3 Odds Ratios for Two- and Three-Way Tables 264
3.1 Odds Ratios for Two-Way Tables 264
3.2 Odds Ratios for Three-Way Tables 265

4 Models for the Two-Way Table 266
4.1 Basic Models 266
4.2 Models for Square Tables 270
4.3 Models for Ordinal Variables 274
5 Models for the Three-Way Table 282
5.1 Basic Models 282
5.2 Collapsibility in Models for the Three-Way Table 285
5.3 Models for Tables with a One-to-One Correspondence among Categories 288
5.4 Models for Tables With Ordered Variables 289
6 Higher-Way Tables 291
7 Estimation Theory 293
8 Residual Analysis and Model-Selection Procedures 298
9 Software 300
9.1 GLIM 300
9.2 BMDP 301
9.3 SAS 301
9.4 SPSS 302
9.5 GAUSS 302
9.6 CDAS 302
9.7 S-Plus 303

6 Latent Class Models **311**

Clifford C. Clogg†

1 Introduction 311
2 Computer Programs 312
3 Latent Class Models and Latent Structure Models 313
4 Basic Concepts and Notation 315
5 The Model Defined and Alternative Forms 317
5.1 Measuring Fit 318
5.2 Alternative Forms of the Model 319
6 An Example: Latent Classes in the American Occupational Structure . . . 321
6.1 Standard Latent Class Models for Two-Way Tables 321
6.2 Some Related Models 324
7 Research Contexts Giving Rise to Latent Classes and Latent Class Models 327
7.1 Medical Diagnosis 327
7.2 Measuring Model Fit with Latent Class Evaluation Models 328
7.3 Rater Agreement 330
7.4 Latent Class Models for Missing Categories 332
8 Exploratory Latent Class Analysis and Clustering 333
9 Predicting Membership in Latent Classes 336
10 Latent Class Models in Multiple Groups: Categorical Covariates in Latent Class Analysis 340

† Deceased

11 Scaling, Measurement, and Scaling Models as Latent Class Models 343
11.1 Ordinal X 343
11.2 Classical Scaling Models 344
11.3 The Rasch Model and Related Models 348
11.4 Extending Latent Class Models to Other Scaling Contexts 351
12 Conclusion 352

7 Panel Analysis for Metric Data 361

Cheng Hsiao

1 Introduction 361
2 A General Framework 367
2.1 The Basic Model 367
2.2 A Bayes Solution 368
3 Two Extreme Cases — All Cross-Sectional Units Have the Same Behavioral Pattern versus Different Units Have Different Behavioral Patterns . . 374
3.1 A Common Model for All Cross-Sectional Units 374
3.2 Different Models for Different Cross-Sectional Units 374
4 Variable Intercept Model 375
5 Error Components Models 376
6 Random Coefficients Models 382
7 Mixed Fixed and Random Coefficients Models 384
8 Random or Fixed Effects (Parameters) 386
8.1 An Example 386
8.2 Some Basic Considerations 388
8.3 Correlations between Effects and Included Explanatory Variables . 390
8.4 Hypothesis Testing or Model Selection 393
9 Conclusion 395

8 Panel Analysis for Qualitative Variables 401

Alfred Hamerle and Gerd Ronning

1 Introduction 401
2 Some Regression Models for Binary Outcomes 402
2.1 Probit Model, Logit Model, Linear Probability Model, and Maximum Likelihood Estimation 402
2.2 Generalized Least Squares Estimation When There Are Repeated Observations 407
2.3 A Note on Interpretation 409
2.4 Models for Limited Dependent Variables 409
3 Binary Regression Models for Panel Data 411
3.1 The Fixed Effects Logit Model 413
3.2 Random Effects Models 417
3.3 Random Coefficients Models 422

3.4 Probit Models With Autocorrelated Errors 423
3.5 Autoregressive Probit Models 429
3.6 Panel Models for Ordinal Data 431
4 Markov Chain Models 433
5 Tobit Models for Panel Data 435
6 Models for Count Data 437
6.1 Poisson Distribution and Negative Binomial Distribution 437
6.2 Mixtures of Poisson Distributions 438
6.3 The Poisson Model 438
6.4 A Model with Overdispersion 439
6.5 Maximum Quasi-likelihood Estimation Under Overdispersion . . 441
6.6 An Example with Cross-Sectional Data 442
6.7 Panel Models for Count Data 444

9 Analysis of Event Histories **453**

Trond Petersen

1 Introduction 453
2 Motivation 455
3 The Hazard-Rate Framework 456
3.1 Basic Concepts 456
3.2 Discrete-Time Formulations 457
3.3 Continuous-Time Formulations 458
4 Time-Independent Covariates 465
5 Time-Dependent Covariates 469
6 Observability of the Dependent Variable 476
7 Repeated Events 478
8 Multistate Processes: Discrete State Space 481
9 Multistate Processes: Continuous State Space 483
10 Estimation Procedures 488
11 Unobserved Heterogeneity 492
12 Time-Aggregation Bias 495
13 Continuous– Versus Discrete–Time Models 498
14 Structural Models for Event Histories 500
15 Sampling Plans 501
15.1 A Conditional Likelihood for t_a, given t_b 504
15.2 Likelihood for t_b and Joint Likelihood for t_a and t_b 505
15.3 Full Likelihood in t_b, t_a, and x 508
16 Left Censoring 511
17 Conclusion 512

10 Random Coefficient Models **519**

Nicholas T. Longford

1 Introduction . . . 519
1.1 An Illustration . . . 522
1.2 Clustered Design . . . 523
2 Models With a Single Explanatory Variable . . . 524
2.1 Patterns of Variation . . . 526
2.2 Contextual Models . . . 529
2.3 Terminology: A Review . . . 530
2.4 Applications . . . 531
3 The General Two-Level Model . . . 533
3.1 Categorical Variables and Variation . . . 536
3.2 Multivariate Regression as a Random Coefficient Model . . . 536
3.3 Contextual Models . . . 537
3.4 Random Polynomials . . . 538
3.5 Fixed and Random Parts . . . 538
3.6 Model Identification . . . 539
4 Estimation . . . 540
4.1 The Fisher Scoring Algorithm . . . 544
4.2 Diagnostics . . . 546
4.3 Model Selection . . . 546
5 Multiple Levels of Nesting . . . 547
5.1 Estimation . . . 549
5.2 Proportion of Variation Explained in Multilevel Models . . . 549
6 Generalized Linear Models . . . 551
6.1 Estimation . . . 552
6.2 Quasi-likelihood . . . 553
6.3 Extensions for Dependent Data . . . 554
6.4 Estimation for Models With Dependent Data . . . 555
7 Factor Analysis and Structural Equations . . . 557
7.1 Factor Analysis . . . 557
7.2 Structural Equation Models . . . 561
8 Example: Wage Inflation in Britain . . . 562
9 Software . . . 568
9.1 ML3 . . . 569
9.2 VARCL . . . 569
9.3 HLM . . . 570
9.4 Outlook . . . 570

Index **579**

Chapter 1

Causal Inference in the Social and Behavioral Sciences

MICHAEL E. SOBEL

1 Introduction

The human propensity to think in causal terms is well known (Young 1978), and the manner in which judgments about causation are made in everyday life has been studied extensively by psychologists (Einhorn and Hogarth 1986; White 1990). No doubt this propensity contributes, for better or worse, to the persistence of causal language in scientific discourse, despite some influential attempts (for example, Russell 1913) to banish such talk to the prescientific era.

In the social and behavioral sciences, causal talk is currently abundant, and at least in some quarters in sociology, especially since the introduction of path analysis (Wright 1921) to the sociological community (Duncan 1966), the impression that explanation and causation are one and the same is given. (Duncan himself does not make such claims.) Furthermore, with a little bit of substantive theory and some data, inferences about causal relations are readily made using statistics. One simply uses selected variables to draw a path diagram that purports to correspond with some theoretical notions of interest, uses data to estimate the parameters of linear equations corresponding to said diagram, and these parameter estimates, and functions of these, give the effects of interest. In reaction to such excesses, several writers (Cliff 1983; Freedman 1987; Holland 1988; Sobel 1990, 1993, 1994) have criticized the cavalier approach to the assessment of causal relations that is often associated with the utilization of modern path models and covariance structure mod-

---

MICHAEL E. SOBEL • Department of Sociology, University of Arizona, Tucson, Arizona 85721, USA. •
For helpful comments and discussion, I am grateful to Gerhard Arminger, Peter Brantley, Henry Byerly, David R. Cox, Otis Dudley Duncan, Judea Pearl, and Herbert L. Smith. Portions of this material were presented at the August 1991 meetings of the American Sociological Association in Cincinnati, Ohio.

Handbook of Statistical Modeling for the Social and Behavioral Sciences, edited by Gerhard Arminger, Clifford C. Clogg, and Michael E. Sobel. Plenum Press, New York, 1995.

els (often called causal models). Similarly, economists (Judge, Griffiths, Hall, Lütkepohl, and Lee 1985; Leamer 1985; Zellner [1979] 1984) have expressed dissatisfaction with the concept of Granger causation, which is often used in conjunction with time series models.

This chapter critically examines the literature on causal inference in the social and behavioral sciences. For a related examination, with emphasis on epidemiology, see Cox (1992). Insofar as the term "causal inference" is vague, it is important to understand that throughout I shall use this term to refer to the act of using evidence to infer causal relations. Attention focuses on the relationship between causal inference and an account of causation to which many social scientists who make causal inferences ascribe, if only implicitly.

Several contributions are offered. First, I bring together a philosophical account of causation (Collingwood [1940] 1948; Gasking 1955; Harré and Madden 1975; von Wright 1971) that hinges on notions such as manipulability with a formalization of this account and a resulting approach to causal inference (Rubin 1974, 1977, 1978, 1980, 1990) that derives from the statistical literature on experimental design.

It is important to note that the account, its formalization, and the approach are general: the account is not limited to the case where the manipulations can actually take place, and the approach in no way hinges on the ability to actually perform an experiment, randomized or otherwise. The formalization comports well with the general idea behind the account, but the account leaves room for other formalizations as well. Some details will be given later. The close correspondence between the account, its formalization, and the resulting approach to causal inference appears to have gone largely unnoticed: philosophers (with the possible exception of Giere (1980)) have done little to attempt to formalize the account and/or implement an approach to causal inference based on this account of causation, and statisticians have done little to say how their work ties to a causal account. By noting how this formal approach to causal inference, which is applicable to both experimental and nonexperimental (observational) studies, dovetails with the account of causation, thereby bringing epistemological and conceptual aspects of the causal relationship into correspondence, I accomplish several things. First, once this account (hereafter called a manipulative account) is modified slightly, the correspondence privileges the foregoing approach to causal inference (hereafter referred to as an experimental approach) without committing the sin of verificationism. Further, if other approaches to causal inference fail to square with this approach, it is because these approaches are deficient, provided a manipulative account of causation is under consideration. Of course, if an alternative account of causation is under consideration, a different approach to inference might be privileged.

Given the foregoing relationship between causal accounts and causal inference, it is important that researchers understand the causal concepts they are using. In that vein, empirical workers almost never explicitly indicate the causal concepts they are using, but they often implicitly commit to a manipulative account when interpreting empirical results. Similar remarks apply to methodologists and statisticians when they indicate how model parameters ought to be interpreted. Unfortunately, many researchers do not appear to understand the consequences of adopting this view for the collection, analysis, and interpretation of data. Thus, I show how the experimental approach to causal inference relies on considerations external to the data and models that are typically employed in nonexperimental (observational) studies. Further, the considerations are not the usual ones identified in the social and behavioral science literatures. It is also hoped that recognition of the issues above will encourage social scientists who wish to make causal inferences (in this experimental sense) to put a higher premium on evidence from well-designed experiments (when these are possible).

I also identify a number of problems with various treatments of causation and/or causal inference in philosophy, economics, statistics, psychology, and sociology. Some of the problems identified appear fatal to the corresponding treatment (in the sense that the approach does not support the accompanying connotations that are typically imparted), suggesting that social scientists would be well advised to not hinge causal inferences on such treatments. Other treatments appear to be more promising, but these have not been fully worked through for the cases that interest social scientists. Some suggestions for further work are also given.

Finally, empirical workers in the social sciences often incorrectly equate explanation with causation. Although this chapter focuses on the causal relation and causal inference, and does not take up the more general subject of explanation, I hope to indicate, if only by elimination and suggestion, that many of the processes and phenomena that are of interest to social and behavioral scientists are not causal or at least not entirely causal. Explicit recognition of this fact should help researchers to think more clearly about what they are actually attempting to find out, and to make more appropriate inferences about the phenomenon under investigation. For example, researchers are often interested in processes that operate over time, such as human development or the intergenerational flow of status. Here, statistical modeling can help to give a parsimonious description of the relationships among successive events. In many cases, researchers simply equate such a description with "causal explanation" without even attempting to say what the terms "causation" and "explanation" mean. In doing so, they lose sight of other types of explanation that may actually be of more interest. In addition, by virtue of an implicit commitment to a manipulative account of causation, these researchers incur the additional risk of falling into the trap of thinking that the statistical analysis supports policy interventions. For further material on the types of noncausal questions that are central in scientific activity, see, for example, Bunge (1979).

Since empirical research in the social and behavioral sciences often involves the use of statistical models, attention centers on "statistical" or "probabilistic" accounts of causation. However, probabilistic accounts historically derive from corresponding deterministic accounts, and therefore it is useful to begin by examining the literature on deterministic causation. The chapter is organized as follows: Section 2 selectively reviews the philosophical literature on deterministic causation. Here attention focuses on regularity theories and manipulative accounts of causation. Section 3 considers some accounts of probabilistic causation in philosophy (for example, Suppes 1970) that may be viewed as an outgrowth of deterministic regularity theories. Previous criticisms of these accounts are discussed, and a number of new criticisms are offered. Next, the related notion of Granger causation (Granger 1969), which is used in economics, is examined. Previous criticisms of this notion are also discussed, and a number of new criticisms are offered. Section 4 formalizes the manipulative account and takes up the resulting approach to causal inference. I relate this approach to the usual model-based approaches to causal inference in the social and behavioral sciences, using single-equation models to show the dependence of the usual approaches on a number of assumptions that are usually implicit (at best). Next, some limitations and ambiguities of the account and its formalization are identified. In Section 5, the discussion is extended to the simultaneous-equation models (causal models) that are often used in social sciences to draw causal inferences, and some recent attempts in the econometric literature to redefine exogeneity and causality (though not Granger causality) are examined. In addition, the usual approach to causal inference in sociology (and some parts of psychology), which involves notions of direct, total, and indirect effects, is considered. It is concluded that these new econometric concepts do not correspond strongly to the causal notions they are intended to capture. Similar remarks apply to the effect decompositions that come from sociology and psychometrics. I conclude by examining an approach due to Holland (1988) that offers promise, if suitably generalized.

2 Deterministic Causation in Philosophy

Aristotle construed causation broadly, equating this with what would now be called explanation. Two of the types of causes identified by Aristotle, the material and formal causes, connect objects or events to their concomitant properties by supplying a linkage that is not typically viewed as causal now. In plainer terms, material and formal causes are invoked to answer questions of the form "why" by statements of the form "because" (Barnes 1982). In post-medieval science, it is usual to distinguish between causation and explanation, seeing causation as a special type of explanation. Thus, current discussions of causation focus on either efficient or final causes, the other two types identified by Aristotle. These are most readily associated with the notion that causation is or involves some form of action, and both efficient and final causes are featured in explanations in the social and behavioral sciences. In this chapter, causation is also viewed as a special kind of explanation, and only efficient causation is explicitly discussed. The exclusion of final causes is not intended to suggest that teleological causation is strictly reducible to efficient causation, as some philosophers have argued. At the same time, the causal processes that underlie teleological accounts can generally be described in terms of efficient causes (Mackie 1974), and it is this fact that warrants such an exclusion.

In this section, two deterministic causal accounts are examined. Roughly speaking, regularity theories take the cause (sometimes called the full cause or the philosophical cause) of a generic event E to be the complex of generic antecedents necessary, sufficient, or necessary and sufficient for E. It is understood here that the antecedents are "distinct" from E. For further material on the notion of "distinctness" see Mackie (1974). Under this account, a successful causal explanation is one that yields "the causes of effects," to borrow language from Mill ([1843] 1973). According to Collingwood ([1940] 1948, p. 287) explanations of this form are the goal that is sought in "the theoretical sciences of nature." By way of contrast, in a manipulative account an event C causes an event E if E occurs when the experimenter induces C, and E^c (the complement of E) occurs when the experimenter induces C^c. Here, attention focuses on the "effect of causes" and there is no presumption that C is the full cause of E. Collingwood argues that this notion of causation dominates in the applied sciences, such as medicine.

I begin by considering regularity accounts. Since Hume is often regarded as a precursor to such treatments, his analysis of causation ([1739] 1978, [1740] 1988, [1748] 1988) provides a convenient starting point for the discussion. The prototypical example that Hume considered is the case of two colliding billiard balls. Here, one ball is moving toward a second ball at rest. The two balls collide, setting the second ball in motion. In analyzing this situation, Hume notes that the motion of the first ball preceded that of the second, and that the collision between the two balls sets the second ball in motion. On the basis of these observations, he argues that temporal priority and spatio-temporal contiguity are intrinsic components of cause-effect relationships. However, Hume adds that a singular instance of the colliding billiard balls does not, in and of itself, lead to a causal inference. Such an inference is warranted only by the additional observation that every time the setup has been repeated (or adequately approximated), the same effect appeared. This is the criterion of constant conjunction. According to Hume, these three components exhaust the ontological aspect of causation.

The constant conjunction criterion means that the events C and E featured in a causal statement of the form "C caused E" are general, referring to classes of instances, as opposed to a singular instance. (This creates some problems that cannot be taken up here.) From this point of view, a causal statement of the form "C caused E", where C and E are now viewed as singular instances, means only that C both preceded and was contiguous to E, and that the instance in hand can be subsumed under a wider class of observed instances. In this sense, singular causal statements are completely derivative from the broader class. Second, insofar as the analysis purports to be exhaustive, the causal relation resides, given contiguity and temporal priority, only in the relation of constant conjunction, and not in any necessity that goes beyond constant conjunction, logical or otherwise, by which the cause suffices for the effect. Our idea of the causal relation may involve such notions of necessity, for example, productive principles (powers, forces, or in modern terminology, causal mechanisms) by which a cause brings about an effect, but according to Hume, these notions exist only in our minds, and not in the world itself. (At some points Hume ([1748] 1988), in dealing with necessity as production, appears to take the weaker position that even if such forces exist in the world, they will always be unknowable to us.) Third, insofar as causation hinges on regularities of succession that have been observed in past instances, and as the previous analysis exhausts the meaning of causation, inference to future instances is not warranted in any logical sense. Rather, such inductive inferences depend solely upon beliefs in the uniformity of nature.

Hume's analysis has been criticized extensively. First, it is not general enough to accomplish the goals of a regularity theory. Mill ([1843] 1973), who accepts Hume's critique of necessity as production and the temporal priority criterion, points out that E may follow C whenever C occurs, but it may also follow B whenever B occurs, even if C fails to occur. That is, there may be a plurality of causes. The full cause might then be $(C \cup B)$, where $\cup$ is the operator "or" in set theory. Second, he points out that the cause C might be conjunctive, that is, $C = (C_1 \cap C_2)$, where $\cap$ is the operator "and" in set theory. (Although Hume does not refer to conjunctive causes, there is nothing in his analysis that excludes such a case.)

Second, the relevance of the contiguity criterion has also been questioned. Properly understood, the criterion permits events that are not spatially proximate to be causally related, provided there is a chain of contiguous causes connecting the cause and effect (Hume [1739] 1978). As Bunge (1979) points out, this turns the criterion into a hypothesis that does not derive from experience. And as a hypothesis, some have argued that it is suspect, citing the case of quantum theory in physics. Here, there are relationships (supported by experimental evidence) that apparently feature action at a distance, and some authors have interepreted these as causal. For a brief account, see Skyrms (1988). From this point of view, although contiguity may be a feature of most real-world causal relationships, it appears that a completely general concept of causation which purports to have ontological support would not include the contiguity criterion.

Third, the temporal priority criterion has been widely debated. Philosophers such as Russell and Schopenhauer have sided with Hume, while others have argued that cause and effect can occur contemporaneously. Kant, for example, took up the case of a ball resting upon a cushion, thereby creating an indentation. And Collingwood ([1940] 1948) argues that when the full cause of an event is sought, cause and effect must be contemporaneous. Otherwise, the effect that actually happens could depend on events that occur during the time interval between cause and effect (Cook and Campbell 1979). Others, like Bunge (1979), distinguish between the concept of causation and physical causation in the world. According to Bunge, the concept of causation is independent of temporal priority, but when causes and effects are separated by a distance, physical causation involves temporal priority (at least according to 20th-century theories in physics).

Fourth, and perhaps most importantly, as Mackie (1974) points out, by denying that the causal relationship contains any form of necessity, Hume loses the ability to distinguish causal sequences from sequences, like night and day, that most philosophers and scientists do not intuitively accept as causal, and therefore wish to call noncausal. Mill, who was also aware of this, attempts to deal with the problem by incorporating a notion of necessity into his account. He argues that it is not enough for the cause to be the invariable antecedent in the Humean sense of de facto regularity. In addition, the cause must also be the unconditional invariable antecedent. By this Mill means that the effect will occur in the presence of the putative cause, even under changed circumstances in other antecedents, which may have to be imagined because they have not been observed. For example let C, E, and B be three events, with C and B antecedent to E, and suppose C is necessary and sufficient for E in the observed data. Suppose also that B always occurs in the observed data. Mill is saying that before we conclude C causes E, we must be able to say that C is necessary and sufficient for E even if the event B^c (the complement of the event B) had occurred. This event may never be observed in the actual data, which means that Mill is arguing that a causal statement must be capable of sustaining certain kinds of counterfactuals.

Modern regularity theorists often argue that legitimate causal statements should sustain counterfactual conditionals, and they use this criterion to distinguish causal sequences from sequences that merely exhibit universal de facto association. Mackie (1974), who accepts the idea that a causal statement should sustain a counterfactual conditional, nonetheless argues by way of several examples that Mill's attempt to distinguish between causal and noncausal sequences is inadequate. According to him, it is necessary to also include a notion of causal priority before causal and noncausal sequences can be distinguished. Further, as he notes, since the examples are ones in which there is a clear temporal order among the events discussed, temporal priority in and of itself cannot accomplish the job of causal priority. This is a blow to regularity theories that put forth the temporal priority criterion and then equate this with causal priority.

Philosophers who either accept Mackie's argument or simply argue that causes and effects may be contemporaneous are thus faced with the problem of introducing a suitable notion of causal priority (causal order). In the absence of such a notion, when C and E are contemporaneous, instead of saying C causes E, one might just as well say that E causes C. For that matter, one might just as well say that C and E cause each other, thereby robbing the concept of causation of its essential asymmetry. But philosophers do not typically accept either of these alternatives as meaningful.

Scientists usually argue that theory dictates the nature of causal ordering, where theory is usually understood to mean statements about causal mechanisms. In other words, causal ordering depends upon the notion of causal mechanism, whose ontological status is denied by followers of the Humean tradition. This maneuver, to be successful, requires explication of the concept of causal mechanism. For some attempts to explicate this concept, see, for example, Harré (1972). It is not clear that such attempts have been successful.

Mackie (1974) argues that the notion of a causal mechanism should hinge on the prior notion of causal order, and he identifies causal order with a notion he calls "fixity," that is, he takes the statement "C is causally prior to E" to mean that the event C is fixed prior to the time that E is fixed. Note that this does not preclude C and E from being contemporaneous, as a future event may be determined by a sufficient set of events in the past. Nor does this appear to rule out "backward" causation.

One way to obtain fixity is by human intervention. But Mackie argues that the concept of fixity is broader than the anthropomorphic notion that would result if fixity were obtainable only through intervention by a human (or human-like) agent. However, except for brief mention (p. 185) of the notion of an explanatory account (which seems like an appeal to the notion of causal mechanism, a notion Mackie already argued requires exposition in terms of the concept of causal priority, and not vice versa), Mackie does little to indicate how fixity can be obtained. These points are important, not only because of the potential circularity of Mackie's argument, but because the relationship of C to E could easily depend upon the manner in which said fixity is obtained, at least in the nondeterministic cases that are of interest in this chapter. To see this, consider the following example, in which educational level precedes subsequent earnings. Under the first scenario, persons choose their level of education, and under the second, persons are randomly assigned by an experimenter to levels of education. Will the relationship between education and earnings be the same under the two different scenarios? Of course not. Under the first scenario, if persons want to maximize lifetime earnings and have good information on how much they would earn under their various educational choices, they will choose the level of education that allows them to maximize lifetime earnings. Under the second scenario, such behavior is precluded. It is clear that under the second scenario, Mackie would take education to be causally prior to subsequent earnings. Under the first scenario, most persons might still view education as causally prior to subsequent earnings, and Mackie's notions do not seem to preclude this, especially if the good information persons have is stochastic, that is, good on average, but not necessarily with respect to any particular case.

On the logical front, Simon (1952) argues that a proposition a_1 has causal precedence over a_2 if the set of laws determining a_1 are a proper subset of the laws determining a_2. Subsequent work by Simon (1953) and Simon and Rescher (1966) carries this argument over to variables and functions embedded in systems of equations, and identifies the functions with mechanisms. Connections between the concepts of exogeneity, endogeneity, and causal order are given. Simon (1952) carefully points out that his efforts to formalize the notion of causal order are not ontological in character. By way of contrast, Bunge (1979) argues that causal priority is essentially an ontological problem; as such, syntactic treatments are inadequate. According to him, the meaning of causal priority remains an open issue.

The issues raised in the preceding discussion are important to social and behavioral scientists who use causal language. But regularity theories have also been criticized by philosophers who argue that such accounts, which are unsuccessful in any case, do not comport with the manner in which the word "cause" is used in either ordinary or scientific language (von Wright 1971). In this vein, Collingwood ([1940] 1948, p. 285) argues that scientists often use the word "cause" in his second sense, where "that which is 'caused' is an event in nature, and its 'cause' is an event or state of things by producing or preventing which we can produce or prevent that whose cause it is said to be." When the word "cause" is used in this sense, which is certainly closer to the way an experimentalist uses causal terminology, a different picture emerges.

First, under this account, there is no presumption that the cause is the full cause that is required in a regularity account. Rather, the cause is simply the state the experimenter produces by manipulating a particular variable. To see how this notion corresponds with a regularity account, let us suppose, to keep matters simple and in accord with the philosophical literature, that variable X has two states, x and x^c. Similarly, Y, W, and Z are variables with two states, y and y^c, w and w^c, and z and z^c, respectively. Suppose that y occurs if and only if the full cause $((x \cap w) \cup z)$ occurs. In light of the previous remarks concerning fixity, it may be important to know how the full cause comes about, but for the moment I suppose that no matter how it comes about, y occurs. Now, suppose for the moment that variables X, W, and Z can be manipulated, at least hypothetically. Viewing X as the manipulated variable, y occurs when the experimenter assigns the value x to X, provided $W = w$ or $Z = z$, where the values of W or Z might be viewed as background factors or standing conditions. This point was recognized by Anderson (1938), who argued that the background constitutes a field in which the cause operates. Note also that in the case at hand, had the experimenter assigned X the value x^c, y would not have occurred (unless $Z = z$); thus the manipulative account is theoretically capable of sustaining counterfactual conditionals.

Second, it is just as legitimate to view W as the experimental variable, in which case W, rather than X, is manipulated. Now the values of the variables X and Z constitute the causal field. Thus, causes are relative (Collingwood [1940] 1948). The relativity of causation is important, for it implies that alternative causal explanations of a given phenomenon are not necessarily at odds. Typically, in the social sciences, however, researchers proceed as if the particular explanation they are proposing is inimical to others. Thus, for example, sociologists often spend a great deal of time attacking psychological and economic explanations, as well as one another; similarly, researchers from these other disciplines often deny the validity of sociological explanations. From the standpoint here, many of these attacks are misdirected. To be concrete, a physiologist might attempt to explain the onset of depression by altered chemical activity in the brain while a sociologist might with equal legitimacy attempt to give an account that makes reference to dramatic life changes, such as divorce, death of a parent, etcetera. In making this last statement, I do not mean to imply that either the physiologist or the sociologist could actually manipulate the cause in the real world, only that they could hypothetically manipulate the cause. Similarly, a manipulation of social structure may produce declines in birth rates in third-world countries; so might the introduction of birth-control clinics and technologies.

Third, under this account, the issue of causal priority is neatly resolved, for the state y may be said to be causally prior to x if it is Y that is manipulated, and x is causally prior to y if it is X that is manipulated. Further, since the effect of the cause is measured subsequent to the manipulation, the causal relation apparently features temporal priority. Collingwood ([1940] 1948) claims that temporal priority is an integral feature of the causal relation in the manipulative account. Subsequently, I shall argue that this is incorrect and that to argue for temporal priority on the grounds that the effect is measured subsequent to the imposition of the cause is tantamount to verificationism.

Fourth, if the cause is manipulated independently of variables that are temporally prior to it, the problem of distinguishing noncausal from causal sequences is partially resolved, for at least the cause is not confounded with the effect of prior variables, that is, there are no prior variables that screen off the relationship between the cause and the effect. (Some philosophers would also require that there be no variables occuring in the time interval between cause and effect that screen off the relationship, a point to which I shall return.)

The causal accounts considered in this section are deterministic, and hence the discussion may appear irrelevant to current research in the social and behavioral sciences, with its emphasis on statistical relations. This is not so, for two reasons.

First, many historical and comparative sociologists and political scientists compare case studies in a deterministic framework, and the foregoing discussion is relevant to such work. A limited methodological attempt to help these workers systematize and formalize these comparisons is given by Ragin (1987), who attempted to implement Mill's methods (and variants thereof). Ragin appears to be unaware of earlier treatments, such as that in Mackie (1974, pp. 297–321). Ragin argues that statistical methods are inappropriate when researchers have few cases, and hence that other methods are needed. Especially in such instances, he argues that his approach is superior to a statistical approach. Apart from the dubious logic of the argument, which suggests that a choice between statistical and nonstatistical methods can be made on the basis of the number of cases, as opposed to the nature of the phenomenon under investigation, Ragin subsequently fails to appreciate Mill's point that casual statements should sustain counterfactuals. The implication of Mill's point is that the observed data can be used to eliminate causes, but not, per se, to establish causal relationships. And in this vein, it is evident that with more observed data (in general) more causes can be eliminated. Conversely, with fewer cases, fewer causes can be eliminated.

But Ragin fails to see the relation between elimination and the number of cases and he also equates (incorrectly) the failure to eliminate causes with the establishment of causal relationships. Thus, he ends up making the curious argument that the methods he proposes are especially well suited to establishing causal relationships with few cases.

Second, there have been numerous attempts to make causal accounts nondeterministic. As noted, historically these attempts hinge on the deterministic accounts, and as such, many of the critical issues that arise in discussing the adequacy of the nondeterministic accounts are similar to the issues just considered.

3 Probabilistic Causation: Variations on a Deterministic Regularity Account

3.1 Philosophical Treatments

A number of authors have argued the need for a nondeterministic treatment of causation (Good 1961, 1962; Reichenbach 1956; Salmon 1984; Skyrms 1988; Suppes 1970). Various grounds have been given, including: (1) the observation that in everyday life and in scientific activity, causal language is used to discuss phenomena that are not transparently deterministic; (2) the argument that the world is nondeterministic, as purportedly evidenced by quantum theory; and (3) the argument that even if the world is deterministic, matters are often too complicated to permit a deterministic account.

Much of the philosophical literature on probabilistic causation may be viewed as an attempt to use probability theory to abandon the constant conjunction criterion, while retaining many other features of a regularity account. (An exception is Giere 1980.) As such, many of the issues debated in the deterministic context carry over to the probabilistic context.

In addition, new concerns are also raised, since at least according to some, it is necessary to exposit the meaning of the term "probabilistic causation" (Fetzer 1988; Salmon 1984; Skyrms 1988). This is because the axioms of probability theory do not speak for themselves. This means that either a notion of deterministic causation from which probabilistic causation arises must be explicated (philosophers who write on the subject of probabilistic causation have not taken this route), or alternatively, the term "probability" must be explicated and tied to some suitable notion of "probabilistic cause," for example, causal propensity. The concerns above are critical to the approach of this section. However, space considerations preclude detailed discussion; hence, in this section, I will assume (without committing myself to the assumption) that the following accounts can be buttressed by a suitable notion of "probabilistic causation." Thus, the criticisms outlined in this section are "internal" to the basic account, that is, these criticisms do not question the fundamental premises of the argument. A critique on "external grounds" is a subject for a different manuscript; for some steps in this direction, see Holland (1986) and Sobel (1993).

Recall that a successful regularity account should distinguish causal from noncausal sequences. This is the key consideration that motivates most probabilistic accounts. In the deterministic context, it was seen that some writers attempted to distinguish between causal sequences and sequences that merely exhibit de facto universal association by requiring causal sequences to fulfill some additional conditions that transcend the observed data. Philosophers working in the probabilistic context have taken another tack. Instead of trying to say what causation is, most have attempted to characterize noncausal relations. In some treatments, the basic argument is that if two events are associated, but the relationship is not causal, there must be a set of prior (not necessarily in the strictly temporal sense) events that "accounts" for this association. This turns the issue of probabilistic causation into the issue of spurious association (Simon 1954), and on the surface this approach seems to have the advantage that it does not require the use of counterfactual conditionals that transcend the observed (or potentially observed) data. Instead, if spuriousness is suspected, one looks for the prior events that account for the association. In principle, by collecting the right data, that is, finding the correct prior events, the causal issue can apparently be decided empirically.

A typical approach is Suppes' (1970) attempt to modify the Humean account, which I now briefly examine. Suppes begins (p. 12) by defining an event $C_{t'}$ to be a prima facie cause of an event E_t if and only if (1) t' refers to a time point prior to time t, (2) the event $C_{t'}$ has positive probability, and (3) $C_{t'}$ is positively relevant to E_t, that is, the conditional probability of E_t given $C_{t'}$, denoted $Pr(E_t \mid C_{t'})$, satisfies $Pr(E_t \mid C_{t'}) > Pr(E_t)$. He then gives several definitions of spuriousness. For convenience, I examine his second definition (p. 25). Here, $C_{t'}$ is a spurious cause of E_t if and only if (4) $C_{t'}$ is a prima facie cause of E_t, (5) there is a time period t'' prior to t' and a partition $\Pi_{t''}$ made up of events $B_{t''}$ such that for all events in the partition (a) the event $B_{t''} \cap C_{t'}$ has positive probability, and (b) $C_{t'} \perp\!\!\!\perp E_t \mid B_{t''}$, where the symbol $\perp\!\!\!\perp$ taken from Dawid (1979) is used to denote independence. Suppes also considers indirect causes and a number of other issues, but these shall not concern us here.

Suppes' account has been criticized extensively. With respect to (1) recall that some philosophers would allow causes and effects to be contemporaneous (which raises some problems for a probabilistic theory that Suppes discussed). With respect to (2), events with probability 0 occur all the time. As for (3), Skyrms (1988) notes that independence between two events does not imply conditional independence of the events, given a third event, and thus $C_{t'}$ could be a genuine (in some sense) cause without being a prima facie cause, as required by (4). But this is not really a criticism of the definition of spuriousness, per se. However, others (Davis 1988; Salmon 1984) have criticized the positive relevance criterion in (3). With respect to (5), Davis (1988) argues that by requiring t'' to precede t', Suppes cannot assign the accidental generalization to the class of spurious causes. He gives the example of a pitcher who tips his hat at time t' before throwing his infamous fastball and striking out the batter. Finally, although Suppes intends his analysis to apply to both frequentist and subjective notions of probability, he offers no support for this intent, nor does he offer a causal account from which probability relations may be derived.

The remarks above show that Suppes' analysis at best might tell us if two associated events $C_{t'}$ and E_t are spuriously related, on the basis of prior events. But problems remain. Most philosophers want to take spuriousness to mean that $C_{t'}$ and E_t share a common cause, and many interpret Suppes in this light as well, although Suppes does not claim this. This interpretation, however, is incorrect (as illustrated by the example below).

The principle of the common cause, put forth in philosophy by Reichenbach (1956), is the idea that if two distinct events C and E are associated, it is either because they are causally related or they share a distinct common cause B. (B must occur no later than C, the putative cause, which occurs no later than E.) In the latter case, it is argued that C and E should be independent, given B. Intuitively, this idea appears to have some appeal. But philosophers like Salmon (1984) have given examples (called interactive forks) where one would want to say that two events share a common cause, but conditioning on the common cause does not render the two events independent. This is a criticism of (5b). At the same time, these examples neither suggest nor fail to suggest whether B should be called a common cause in the case where B does screen off C and E. In that vein, it is always possible to find a partition such that clause (5) in Suppes' definition is satisfied. Good (1962) points this out in commenting on similar proposals by Reichenbach (1956) and Good (1961). Evidently there is no such thing as probabilistic causation at all.

There are several ways to deal with the foregoing problem. One is to argue, following Hempel (1968), for a partition based on the current knowledge situation. A second way is to partition on the history of the world before C. This is the tactic taken by Granger (1969). At the inferential level, both ways are identical, for Granger's definition is clearly nonoperational. But even when such strategies are followed, it is not hard to produce genuine probabilistic examples (as opposed to examples which mix deterministic and probabilistic events) where the relationship between the putative cause and the effect is screened off by a prior event that one may not want to call a common cause. To the best of my knowledge, examples of this type have not been given in the literature. A simple example follows.

Let the random vector $(X_1, X_2, \varepsilon_B, \varepsilon_C, \varepsilon_E)'$ follow a normal distribution with mean vector $\mathbf{0}$ and covariance matrix:

$$\begin{pmatrix} 1 & \gamma & \mathbf{0}' \\ \gamma & 1 & \mathbf{0}' \\ \mathbf{0} & \mathbf{0} & \boldsymbol{I}_3 \end{pmatrix}$$

where $\gamma \neq 0$ and $\boldsymbol{I}_3$ denotes the identity matrix of order 3. Let

$$B = X_1 + X_2 + \varepsilon_B, \tag{1.1}$$

$$C = \alpha X_1 + \varepsilon_C, \tag{1.2}$$

$$E = \beta X_2 + \varepsilon_E, \tag{1.3}$$

where $\alpha \neq 0$, $\beta \neq 0$. To (1.2) and (1.3) I append a "causal" interpretation, namely that X_1 probabilistically causes C, and X_2 probabilistically causes E. B, however, is just a linear combination of the respective causal variables, and the error, as opposed to a genuine cause of C and/or E. Temporal subscripts could be placed on these variables, if desired. (Saying that one variable causes another is an abuse of language that is common among scientists. Provided it is understood that the actual referent is to events which can be constructed from these random variables, no special problems are created by this language, which I shall use freely.) With these assumptions, the covariance between C and E is $\alpha\beta\gamma$, which is nonzero. Now (to conform with philosophical treatments that require considerations of events with nonzero probability) let C' and E' be Borel sets with positive probability such that $Pr(C \in C', E \in E') \neq Pr(C \in C')Pr(E \in E')$. So C' is a prima facie cause of E'. (Note I have not shown C' is positively relevant to E', but the events can be chosen so that this criterion is also satisfied, if desired.) Using standard results on the normal distribution (Anderson 1984, chapter 2), the distribution of $((C, E) \mid B)$ is normal with covariance matrix:

$$\begin{pmatrix} \frac{\alpha^2(2-\gamma^2)+(2\gamma+3)}{3+2\gamma} & \frac{\alpha\beta(\gamma^2+\gamma-1)}{3+2\gamma} \\ \frac{\alpha\beta(\gamma^2+\gamma-1)}{3+2\gamma} & \frac{\beta^2(2-\gamma^2)+(2\gamma+3)}{3+2\gamma} \end{pmatrix}$$

Setting the covariance to 0, that is, making C and E independent, given B, and solving for γ yields $\gamma \doteq .618$, irrespective of the values of β and α, which are only assumed to be nonzero (as X_1 causes C and X_2 causes E). The result now follows by forming the events C' and E' from C and E.

The previous example suggests, along the lines of Mackie's (1974) argument for the deterministic case, that probability relations cannot, in the absence of a concept of causal priority, distinguish causal from noncausal sequences. Otte (1981) has also made this argument. However, every example he gives relies on deterministic causation. Thus, what Otte really shows is that Suppes' analysis cannot handle deterministic causes, contrary to Suppes' claims. No support for the more general conclusion Otte reaches is given in his paper or in the literature. However, it is easy to establish the argument in a purely stochastic framework. To that end, I show (in the simplest case), using variables B, C, and E, that it is impossible, by using probability relations alone, to distinguish between wedges (B and C cause E), forks (B causes C and B causes E), and simple causal chains (B causes C and C causes E).

Let $(B, C, E)'$ follow a normal distribution with mean vector $\mathbf{0}$ and variance covariance matrix:

$$\begin{pmatrix} 1 & \gamma_1 & \gamma_2 \\ \gamma_1 & 1 & \gamma_3 \\ \gamma_2 & \gamma_3 & 1 \end{pmatrix}$$

Linear equations corresponding to the case of a (interactive) fork are:

$$\begin{aligned} C &= \beta_{CBF}B + \varepsilon_C \\ E &= \beta_{EBF}B + \varepsilon_E, \end{aligned} \tag{1.4}$$

where ε_C and ε_E are normal random variables uncorrelated with B, with variances $\sigma^2_{\varepsilon_C}$ and $\sigma^2_{\varepsilon_E}$, respectively, and covariance $\sigma_{\varepsilon_C\varepsilon_E}$. Regarding the variance of B as given, and choosing $\beta_{CBF} = \gamma_1$, $\beta_{EBF} = \gamma_2$, $\sigma^2_{\varepsilon_C} = 1 - \gamma_1^2$, $\sigma^2_{\varepsilon_E} = 1 - \gamma_2^2$, $\sigma_{\varepsilon_C\varepsilon_E} = \gamma_3 - \gamma_1\gamma_2$, the covariance matrix above is reproduced. For the case of a wedge, a linear equation is:

$$E = \beta_{EBW}B + \beta_{ECW}C + \tilde{\varepsilon}_E, \tag{1.5}$$

where $\tilde{\varepsilon}_E$ is a normal random variable uncorrelated with B and C, with variance $\sigma^2_{\tilde{\varepsilon}_E}$. Treating the variances of B and C, and the covariance of B and C as given, the covariance matrix is reproduced by setting $\beta_{EBW} = (\gamma_2 - \gamma_1\gamma_3)/(1 - \gamma_1^2)$, $\beta_{ECW} = (\gamma_3 - \gamma_1\gamma_2)/(1 - \gamma_1^2)$, and $\sigma^2_{\tilde{\varepsilon}_E} = 1 - (\beta^2_{EBW} + \beta^2_{ECW} + 2\beta_{EBW}\beta_{ECW})$. Finally, linear equations for a chain are:

$$\begin{aligned} C &= \beta_{CBC}B + \varepsilon^*_C \\ E &= \beta_{ECC}C + \varepsilon^*_E, \end{aligned} \tag{1.6}$$

where ε^*_C and ε^*_E are normal random variables uncorrelated with B, with variances $\sigma^2_{\varepsilon^*_C}$ and $\sigma^2_{\varepsilon^*_E}$, respectively, and covariance $\sigma_{\varepsilon^*_C\varepsilon^*_E}$. Setting $\beta_{CBC} = \gamma_1$, $\sigma^2_{\varepsilon^*_C} = 1 - \gamma_1^2$, $\beta_{ECC} = \gamma_2/\gamma_1$, $\sigma_{\varepsilon^*_C\varepsilon^*_E} = \gamma_3 - \gamma_2/\gamma_1$, and $\sigma^2_{\varepsilon^*_E} = 1 + \gamma_2^2/\gamma_1^2 - 2\gamma_3\gamma_2/\gamma_1$ reproduces the covariance matrix.

The example above illustrates the more general point that causal order, viewed as a concept, is a property of a model and not of the data. Econometricians have known this since at least Basmann (1965). The example strongly suggests that philosophers would do well to abandon attempts to characterize probabilistic causality without reference to some notion of causal priority that goes beyond temporal priority. As noted, some philosophers have also argued this, and there have been attempts, using recursive structural equation models, to deal with such issues, for example, Cartwright (1989).

Another recent attempt to derive causal statements from the probability distribution of a set of observed variables is due to Pearl and Verma (1991), who, according to Cox (1992), replace the notion of causal priority by the notion of simple structure. They use recursive structural equation models to express the conditional independence relations among variables in as simple a way as possible. By requiring the class of models to be recursive from the outset (see definition 2 of a causal theory in Pearl and Verma 1991), these authors exclude from further consideration cases like (1.4) and (1.6) where ε_C and ε_E are correlated. Pearl (personal communication) has argued that it is not so much that a model with correlated errors cannot be causal, but rather, because of the principle of the common cause (applied to the errors), such a model is incomplete. Note that this argument rests on acceptance of the principle of the common cause.

In addition, Pearl and Verma (1991) give general definitions of spurious association (both with and without temporal information) and genuine causation, claiming that their definitions comport with a manipulative account of causation. For a criticism of this attempt, see Sobel (1993).

I now turn to a discussion of the notion of Granger causation.

3.2 Granger Causation in Economics

Economists have proposed several approaches to causal inference that are of interest. One approach, which is now common in empirical work in sociology and psychology and to which I shall return, stems from econometric research on simultaneous equation models in the 1940s. Here, a set of variables is partitioned into exogenous and endogenous subsets, and a stochastic model (with as many equations as endogenous variables) is proposed. The exogenous variables are viewed as being determined outside the model, and until recently the exogeneity assumption was considered untestable. Exogenous variables are also typically viewed as causes of the endogenous variables, and some endogenous variables are causally prior to others. Causal priority is a property of the model that is not testable. Dissatisfaction with this approach, in particular with the arbitrariness of assumptions about causal priority, for example, the arbitrary classification of variables as exogenous or endogenous and the assumptions used to identify the model, led econometricians to develop probabilistic accounts that are similar to those proposed by Suppes (1970). The idea was to free empirical researchers from the reliance on untestable assumptions by providing a notion of probabilistic causation and an approach to causal inference that relies on the data alone. Not surprisingly, the critiques of these ideas by economists led back to the conclusion that some notion of causal priority is indispensable if inferences that might be called causal were to be drawn.

Granger (1980), whose earlier work (1969) on causation in time series models touched off a large and technical literature on the subject (see Geweke 1984 for an excellent review), proceeds as follows. He defines Ω_n as the history of the world up to and including the discrete time n, excluding deterministic relations among components of this history. Let $\boldsymbol{Y}_n$ denote the random vector $\boldsymbol{Y}$ at time n. Granger says that $\boldsymbol{Y}_n$ causes $\boldsymbol{X}_{n+1}$ if $Pr(\boldsymbol{X}_{n+1} \in A \mid \Omega_n) \neq Pr(\boldsymbol{X}_{n+1} \in A \mid \Omega_n - \boldsymbol{Y}_n)$ for some set A. One can also consider the history of Y up to time n, that is, $\tilde{\boldsymbol{Y}}_n = \{\boldsymbol{Y}_t\}_{t \leq n}$, and define $\tilde{\boldsymbol{Y}}_n$ as a cause of $\boldsymbol{X}_{n+1}$ if $Pr(\boldsymbol{X}_{n+1} \in A \mid \Omega_n) \neq Pr(\boldsymbol{X}_{n+1} \in A \mid \Omega_n - \tilde{\boldsymbol{Y}}_n)$ for some set A, as in some other treatments. If $(\boldsymbol{X}_{n+1} \perp\!\!\!\perp \tilde{\boldsymbol{Y}}_n) \mid (\Omega_n - \tilde{\boldsymbol{Y}}_n)$, one could say $\tilde{\boldsymbol{Y}}_n$ does not cause $\boldsymbol{X}_{n+1}$ (Florens and Mouchart, 1982). Granger (1980) does not actually define noncausation for the exact case at hand, though his definition of noncausation is consistent with the above. The definition by Florens and Mouchart is also consistent with the above, except that their definition, which refers to the entire history of the sequences, requires the statement above to hold at all times. Granger (1969) also discusses feedback systems and instantaneous causation. In the present context, a feedback system refers to the case where $\boldsymbol{Y}_n$ (or $\tilde{\boldsymbol{Y}}_n$) causes $\boldsymbol{X}_{n+1}$ and $\boldsymbol{X}_n$ (or $\tilde{\boldsymbol{X}}_n$, defined analogously to $\tilde{\boldsymbol{Y}}_n$) causes $\boldsymbol{Y}_{n+1}$. Finally, $\boldsymbol{Y}_n$ causes $\boldsymbol{X}_n$ instantaneously if $Pr(\boldsymbol{X}_n \in A \mid \Omega_{n-1} \cup \boldsymbol{Y}_n) \neq Pr(\boldsymbol{X}_n \in A \mid \Omega_{n-1})$ for some set A. Granger (1980) also discusses a weaker form of causation, called causation in mean, which need not concern us further.

The foregoing definitions are not operational because the stochastic history of the world is inaccessible. Granger operationalizes these definitions by (1) replacing Ω_n with an information set J'_n that includes sequences of distinct random vectors, that is, $J'_n = (\tilde{\boldsymbol{X}}_n, \tilde{\boldsymbol{Y}}_n, \tilde{\boldsymbol{Z}}_n)$, and (2) relativizing the previous definitions with respect to J'_n. This is akin to the strategy of partitioning suggested by Hempel (1968).

Granger's account has distinct advantages over that of Suppes (1970). First, Granger does not need to define prima facie causes (in the sense of Suppes), nor does he include the criterion of positive statistical relevance. His definitions can handle conditional distributions, where the event conditioned on has probability zero. Further, partitioning on the stochastic history of the world (or in the operational case, on a particular information set) is tantamount to choosing a fixed partition, thereby avoiding the problem that one can always find a partition such that the putative cause and effect are independent, given events in the partition. In addition, Granger's account excludes deterministic causation, and thereby avoids the type of objections Otte (1981) raised against Suppes.

While Granger's account, see also related material by Chamberlain (1982) and Florens and Mouchart (1982), avoids some of the technical pitfalls in Suppes (1970), it is open to a number of criticisms previously raised. For these reasons, some economists have objected to the use of the term causation in this context. For example, Leamer (1985) argues for simply using the term "precedence." (A better term would be "predictive precedence.") Judge et al. (1985) state that the notion of causation used here is not philosophically acceptable. Zellner ([1979] 1984) notes that the definitions above merely serve to indicate whether one time series predicts another. He argues, following Feigl (1953, p. 408), that causality should be defined as "predictability according to a law" (or set of laws). Thus, Granger's account or similar accounts are causal, from Zellner's point of view, only if economic theory suggests a relation between the series in question. For criticism of the notion that the causal relation should be equated with predictability according to a law, see Bunge 1979, Byerly 1990, and Wold 1966. Briefly, these authors argue that predictability and causation are different (unless one adopts a Humean position) and there are many different types of lawfulness that one might not call causal.

The critiques above suggest, albeit in different language, that causal priority should not be reduced to temporal priority, as in the bulk of Granger's account. Sims (1977), a proponent of the approach, also recognizes this point. He attempts to provide a deeper rationale for Granger's notion of causation. In so doing, he reintroduces considerations that transcend the available data. To begin, Sims (1977) generalizes Simon's (1952) definition of causal order. He defines S as the set of possible outcomes, and A and B as subsets of S that arise from imposing restrictions. Functions P_X and P_Y map S to X and Y, respectively. He then says that the ordered pair of restrictions (A, B) determines a causal order from X to Y if and only if $P_X(A \cap B) = P_X(A)$ and $P_Y(A) = Y$. The generalization is defective, for if $P_X(A) = X$ and $P_Y(A \cap B) = P_Y(A) = Y$, there is both a causal order from X to Y and a causal order from Y to X. Since causal order is viewed as a property of the model, and not as a property of the world, Sims also proposes the concept of structure to link the model to real- world phenomena. Roughly, the idea is that if X is causally prior to Y, and one inputs a set of restrictions (in the form of the set A above) the model describes the real world if the outputs in the real world correspond with $P_Y(A \cap B)$. Recalling the previous discussion, the idea of structure is related to Mackie's idea of fixity; in short, structure produces fixity by means of intervention. Thus, in Sim's account, structure not only connects the model to the real world, it also links the notion of causation to intervention (even if these interventions cannot actually take place). This does not necessarily make the account satisfactory, but it does seems to be an improvement on previous treatments of probabilistic causation, both in philosophy and economics, where almost any variable, no matter how it comes about, is sometimes viewed as causal. For further details, the reader should consult Sims (1977) or Geweke (1984).

There is one other remark that should be made before leaving this topic. Although Granger actually believes that temporal priority is a necessary feature of the causal relationship, economists have worked with his notion of instantaneous causation. In that vein, the definition previously given is equivalent to the statement that instantaneous causation holds if $\boldsymbol{X}_n \perp\!\!\!\perp \boldsymbol{Y}_n \mid \Omega_{n-1}$ fails to hold. Thus, by the symmetry of independence, the statement $\boldsymbol{Y}_n$ causes $\boldsymbol{X}_n$ instantaneously implies the statement $\boldsymbol{X}_n$ causes $\boldsymbol{Y}_n$ instantaneously.

Hence, instantaneous causation, as operationalized by the econometricians, entails a commitment to the further conclusion that cause and effect instantaneously cause each other. But this is at odds with all previous work on the subject of causation; despite many different treatments of causation in the philosophical literature, the causal relationship is viewed as intrinsically asymmetric. Therefore (unless one wants to attempt to argue against this tradition for a causal account that is not asymmetric), the concept of instantaneous causation is inherently defective, even when concepts such as structure and the like are added to attempt to shore up the formal account.

4 Causation and Statistics: An Experimental Approach

The manipulative account of causation discussed in Section 2 suggests an alternative approach to causal inference. The idea there is to manipulate an independent variable and see how the value of a response variable Y depends upon the value of the manipulated variable. While the manipulative account has a number of positive features, philosophers have done little to indicate how this account could be implemented, and the examples discussed in the literature tend to use simple types of effects in conjunction with counterfactual conditionals that warrant strong belief. For example, the conclusion that turning the ignition switch caused the car to start (in any given instance or in general) is sustained by the belief that had the ignition not been turned, the car would not have started. Matters are much less clear, however, if we want to study the effect of exposure versus no exposure to a training program on subsequent earnings of workers (Heckman and Hotz 1989; Heckman, Hotz, and Dabos 1987). Here, a particular person can be exposed to the training program, but unlike the case of the car, it would be unreasonable to believe that we know the value of earnings that would result were that person not exposed. The same remark would hold true on average. Some headway could be made if it were reasonable to assume that an individual's pre-exposure earnings and his post-exposure earnings without training are identical, but such an assumption would not be reasonable in this substantive context; see Holland and Rubin (1983) for further material on the foregoing type of assumption.

The foregoing remarks indicate that it will generally be impossible to ascertain the effect of a manipulated variable in a particular instance, at least without imposing some strong assumptions that are not verifiable. (See Holland 1986 for further material on this issue.) That is, a singular causal statement will rely upon a counterfactual conditional, and in many cases of scientific interest, we shall not have the type of information that is needed to form (or believe) the relevant counterfactual.

Rubin (1974, 1977, 1978, 1980), drawing on work by writers such as Neyman ([1923] 1990), Kempthorne (1952), and Cox (1958), proposes a model for experimental and non-experimental studies that can be regarded as a formalization of a manipulative account. Under the model, certain quantities (parameters) are to be estimated, and estimates from the observed data either are good or poor estimates of these parameters (in a sense to be made precise).

The key idea is to begin with the effect in the singular instance. Although this quantity, sometimes called the unit causal effect, cannot typically be observed, the average of the unit causal effects (sometimes called the average causal effect) can be estimated under the right conditions. (Since an effect is by definition causal, the term "causal effect" is redundant, and therefore, despite convention, I shall simply use the term "effect"instead.) For the simplest case (which shall be extended later), Rubin proceeds as follows. To begin, he assumes a population of units, (which we shall index as $\{i : i \in I\}$), and a set of treatments $\{k : k = 1, ..., K\}$. To keep matters simple, let Y_{ik} denote the response when treatment k is applied to unit i, and suppose the Y_{ik} are drawn from the distribution of a random variable Y_k. Note that Y_{ik} is defined for every element of the population, whether or not that element is actually assigned to treatment k. Further, it is assumed that this value is unambiguously defined, in the sense that it does not depend on the treatments to which other units are asssigned. This is the stable unit treatment value assumption (SUTVA). Next, define the unit effect of treatment k versus treatment k' (which might be no treatment) as some function of Y_{ik} and $Y_{ik'}$, for example, $Y_{ik} - Y_{ik'}$. This effect is typically unobservable, but it may be possible to estimate the average (over the population) effect $E(Y_k - Y_{k'})$. In the typical experiment, an investigator assigns experimental units to one of the K treatments. Let T be the random variable denoting treatment assignment. Then the actual data observed by the investigator consist of drawings from the pairs (Y_{ik}, T_i), typically obtained from a sample (hopefully random), where T_i, which equals k, denotes the treatment to which unit i was assigned. The investigator is interested in estimating the average effects of treatments k versus alternative treatments k'. For fixed k and k', this is defined as $E(Y_k - Y_{k'}) = E(Y_k) - E(Y_{k'})$. To estimate the average effect, the investigator uses the data on units assigned to treatment k to compute an estimate of $E(Y_k)$, for example, $\bar{Y}_k$, the sample mean among units assigned to treatment k, and he uses the data on units assigned to treatment k' to estimate $E(Y_{k'})$. In actuality, however, the investigator has estimated $E(Y_k \mid T = k)$ and $E(Y_{k'} \mid T = k')$. In general, the latter quantities are not equal to the respective unconditional expectations. However, randomization makes the assumption $(Y_1, \cdots, Y_K) \underline{||} T$ plausible. In turn, this implies $E(Y_k \mid T = k) = E(Y_k)$ for all k. Randomization therefore allows the investigator to use $\bar{Y}_k - \bar{Y}_{k'}$ as an estimate of the average effect of treatment k versus treatment k'. In essence, randomization works because the units assigned to treatment k can be viewed as a random subsample from Y_k.

The approach above is also important for the analysis of data from experimental studies without randomization and nonexperimental (observational) studies, for it clearly reveals the types of assumptions that most social and behavioral scientists implicitly make when using data from such studies to make causal inferences. This is because most social scientists, without respect to the study design, use experimental language when interpreting empirical results, thereby entailing a commitment (sometimes not recognized) to a manipulative account of causation. See Sobel (1990, 1994) for more on this point in the context of covariance structure analysis.

Inferences from nonexperimental studies in the social and behavioral sciences are typically model based, and it is therefore important to relate Rubin's approach to the model based approach. I begin with the case corresponding to the simple setup previously described. For this setup, to compare the differences among the K treatments, most researchers would estimate a one-way analysis of variance model:

$$Y_{ik} = \mu_k + \varepsilon^*_{ik}, \tag{1.7}$$

where k indexes the treatment to which the unit is actually assigned, and ε^*_{ik} is a random variable with mean 0. Then, it is assumed (generally implicitly) that $\mu_k = E(Y_k)$, but this is not true in general, and in fact, $\mu_k = E(Y_k \mid T = k)$. In the absence of random assignment, the sample means $\bar{Y}_k$, which are also the ordinary least squares estimates for μ_k, are consistent (under mild conditions). Nevertheless, $\bar{Y}_k$ is not a consistent estimator of $E(Y_k)$. Therefore $\bar{Y}_k - \bar{Y}_{k'}$ does not consistently estimate the average effect of treatment k versus k'. However, if $(Y_1, \cdots, Y_K) \perp\!\!\!\perp T$, $E(Y_k|T = k) = E(Y_k)$. In this case, because $\bar{Y}_k$ is a consistent estimator of μ_k, it is also a consistent estimator of $E(Y_k)$. Therefore, $\bar{Y}_k - \bar{Y}_{k'}$ consistently estimates the average effect of treatment k versus treatment k'. In short, under random assignment, the average effect is estimated. But (as seen above) the model can also hold without sustaining the desired interpretation.

In addition to making the implicit assumption that $\mu_k = E(Y_k)$, many researchers also interpret the model to mean that unit and average effects are identical, that is, the unobserved quantity $Y_{ik} - Y_{ik'} = E(Y_k - Y_{k'})$. This is tantamount to assuming the error for unit i is invariant across potential treatments. That is, if the error for unit i under treatment k is ε_{ik}, the assumption is $\varepsilon_{ik} = \varepsilon_{ik'}$, where $k' = 1, \cdots, K$. (The reason for distinguishing between ε^*_{ik} of (1.7) and ε_{ik} is that $\varepsilon^*_{ik} = Y_{ik} - E(Y_k \mid T = k)$, whereas $\varepsilon_{ik} = Y_{ik} - E(Y_k)$, and the two errors are identical only if $E(Y_k) = E(Y_k \mid T = k)$). As before, the model given by (1.7) can hold without implying this interpretation.

To see the points above in a substantive context, consider the following (oversimplified) nonexperimental study, in which $K = 2$. A researcher wants to know the average effect of college going (versus not going to college) on subsequent earnings. To estimate the effect, he computes the mean difference in subsequent earnings between college goers and non-college goers. This is valid when T is independent of subsequent earnings (under either nonexperimental treatment). If T is not independent of subsequent earnings, as would be the case if people attend or do not attend college based on a good estimate of their subsequent earnings under the two conditions, the researcher would have perfectly good estimates of $E(Y_k \mid T = k)$, but not $E(Y_k)$. In this case, the mean difference in earnings between college goers and non-college goers estimates the actual average difference between the two groups, but this latter quantity is not the average effect. Economists are also familiar with the types of problems involved in making causal inferences from nonexperimental data. In economic parlance, this comes under the heading of selection bias (Heckman 1974, 1976). (Note, however, that until recently, economists did not describe such problems using the explicit notation suggested by Rubin's approach. This notation lends considerable clarity to the issue.)

The simple setup can be extended to the case where treatment assignment is based on a probabilistic rule that depends on a vector of observed covariates $\boldsymbol{Z}$ (Rubin 1977). The covariates usually refer to variables that are temporally prior to the cause. For exceptions, see Rosenbaum (1984b). For this case, for any given value of the covariates $\boldsymbol{Z}$, T and $\boldsymbol{Z}$ are assumed to have a joint distribution with $Pr(T = k \mid \boldsymbol{Z} = \boldsymbol{z}) > 0$ for all $\boldsymbol{z}$ and k.

Asssume now that $((Y_1, \cdots, Y_K) \| T) \mid \boldsymbol{Z}$. This is the assumption that treatment assignment is independent of the response, given a set of covariates. In experimental work, when, given a set of covariates, the investigator uses randomization to assign subjects to treatment groups, this assumption becomes plausible. (In the literature, the two assumptions above are sometimes referred to as the assumption that treatment assignment is strongly ignorable, given $\boldsymbol{Z}$.)

The importance of strongly ignorable treatment assignment for drawing causal inferences from nonexperimental studies cannot be overemphasized. In the previous setup, $E(Y_k)$ and the conditional expectation $E(Y_k \mid T = k)$ coincided when $((Y_1, \cdots, Y_K) \| T)$. In nonexperimental studies, this assumption is generally much too strong, as illustrated by the previous example. This suggests that social researchers who want to make appropriate causal inferences must either perform an experiment, which is often impossible, or find another way. Attempting to find a set of covariates that accounts for treatment assignment in nonexperimental work, while not at all trivial, nor fully testable, allows the social researcher the hope of drawing more plausible causal inferences, when treatment assignment itself is not random; details follow.

Analogously to the case previously discussed, under strongly ignorable treatment assignment, $E(Y_k \mid \boldsymbol{Z}, T) = E(Y_k \mid \boldsymbol{Z})$. Now let $\bar{Y}_{kz}$ be a consistent estimate of $E(Y_k \mid \boldsymbol{Z} = \boldsymbol{z}, T = k)$. (The notation suggests that the estimate is a sample average, but this need not be the case.) Under strongly ignorable treatment assignment, $\bar{Y}_{kz} - \bar{Y}_{k'z}$ is a consistent estimate of the average effect of treatment k versus k' when the covariates have value $\boldsymbol{z}$. An estimate of the average effect, if desired, is then obtained (at least in theory) by averaging over the distribution of $\boldsymbol{Z}$.

Several remarks appear to be in order at this point. First, in practice, an investigator may not wish to average over the distribution of $\boldsymbol{Z}$, but may be more interested in the treatment effect at particular levels of the covariates. In some instances, the distribution of Z may not be known, and a good estimate may not be available. Even if it is possible to estimate this distribution accurately, when there is heterogeneity in these effects across levels, more refined inferences can be obtained by focusing on particular levels. Of course, in general the question of whether or not to average across levels depends upon the purposes of the investigation. In some instances, the investigator may wish to average over some of the components of Z, but not others. In fact, the investigator can estimate $E(Y_k \mid \boldsymbol{Z}_1)$, where $\boldsymbol{Z} = (\boldsymbol{Z}_1', \boldsymbol{Z}_2')'$, by averaging over the conditional distribution $\boldsymbol{Z}_2 \mid \boldsymbol{Z}_1$.

Second, if one is interested in how the response varies with the covariates, say at values $\boldsymbol{z}$ and $\boldsymbol{z}^*$, one can compare the estimates of $E(Y_k \mid \boldsymbol{Z} = \boldsymbol{z})$ and $E(Y_k \mid \boldsymbol{Z} = \boldsymbol{z}^*)$. This does not mean that the comparison is to be given a causal interpretation, for $\boldsymbol{Z}$ is treated as a covariate and not as a cause. Note also (Holland 1986) that such a comparison would involve comparisons across units, which is not consistent with the manner in which the basic building blocks (the unit causal effects) are used to define average effects.

Third, as previously noted, the approach above carries over to the analysis of data from nonexperimental studies, and this is its most important application for social scientists. However, the approach appears to require measurements of the covariates. In this vein, it is important to note that when one or more of the covariates are unmeasured, it may still be possible, under certain types of assumptions, to estimate the effect of the causal variable in some cases (Heckman and Hotz 1989). When this is not the case, sensitivity analyses can be very informative; see Rosenbaum (1992) and the references therein.

Fourth, if Z is a discrete random vector and the number of possible values that Z takes is sufficiently small (relative to the sample size), under random assignment, conditional on Z, $E(Y_k \mid Z = z)$ can be estimated by averaging over units with value k of the cause, and covariates z. Estimates of average effects can then be obtained in the manner described above, provided the distribution of Z is known, or the sample studied is a random sample from the population of interest. When Z takes on many values, other techniques must be used (in either experimental or nonexperimental studies) to obtain estimates of interest. These techniques, which are not our primary concern here, go under the headings of matching, subclassification, and covariance adjustment (for example, see Rosenbaum and Rubin 1983) in the statistics literature. For an example of matching (on the propensity score) in educational research, see Rosenbaum (1986).

Fifth, the assumption of strongly ignorable treatment assignment, which is typically made (implicitly) by social researchers who estimate causal models, should not be taken lightly. By introducing other auxiliary assumptions, it is sometimes possible to test this assumption (Rosenbaum 1984a, 1987; Heckmann and Hotz 1989). More work remains to be done on this topic. However, the test is only as good as these auxiliary assumptions, and if substantive knowledge in an area is limited, the failure to reject the assumption of strongly ignorable treatment assignment should not be taken as seriously as in the case where subject matter knowledge is extensive.

To illustrate some of the foregoing material in a simple case, I elaborate on the earnings example, introducing the covariate gender, with $Z = 1$ if male, $Z = 2$ if female. Suppose that $0 < Pr(T = k \mid Z = z) < 1$ for $z = 1, 2$ and $k = 1, 2$. In this case, under the strongly ignorable treatment assignment assumption, estimates of $E(Y_k \mid Z)$ can be formed by averaging over units with identical values of Z and T. The researcher who wants an estimate of the average effect in the population can use these estimates as described above. The researcher who is interested in gender differences can compare the estimates of $E(Y_k \mid Z = 1)$ with $E(Y_k \mid Z = 2)$, for college goers and non-college goers, respectively. This will give estimates of the average gender difference in earnings among college goers and non-college goers, respectively. In light of the previous points, however, such a comparison is descriptive and these average differences should not be confused with average effects. Suppose now that the strongly ignorable treatment assignment assumption does not hold, that is, gender is not sufficient for treatment assignment. In that case, the level of the causal variable taken on by a subject also depends on covariates other than gender, and it will not be possible to estimate the average effect in the usual way. For example, suppose that Y_i measures earnings of the ith survey respondent, and $T_i = 1$ if this respondent does not go to college, $T_i = 2$ otherwise. Suppose that treatment assignment is strongly ignorable among men. Then the average effect of college going among men can be estimated by computing the mean difference between college going and non college going males. Suppose now that women who do not intend to work attend college in order to marry men who will earn high incomes. In this case, treatment assignment will not be strongly ignorable among women. This means that a comparison of mean incomes among college- and non-college-going women would not estimate the average effect of college versus lack of college on female incomes. Nor would comparing the mean difference for men with the mean difference for women tell us the difference between the effects for men and women. Finally, note that if women who attend college end up marrying men with higher incomes and subsequently do not work, independently of their reasons for going to college, treatment assignment may be strongly ignorable among women. In this case, the average effect of college going (versus non-college going) on earnings could be estimated among women.

Rubin's approach has been extended to the case where the set of treatments is not finite by Pratt and Schlaifer (1988). In this formulation, the key ingredients are: (1) a population I, as before, (2) a set of factors $\{\boldsymbol{x} : \boldsymbol{x} \in \Omega\}$, where, for convenience of exposition, $\Omega \subseteq R^K$; factors are analogous to treatments; (3) for every value $\boldsymbol{x} \in \Omega$ and for every member of I, a vector valued set of concomitants $\boldsymbol{Z}_{xi}$ and a set of disturbances $\boldsymbol{U}_{xzi}$ with a joint distribution; for any particular value $\boldsymbol{x}$, the disturbances are independent and identically distributed, for a given value of the concomitants; concomitants are analogous to covariates, and as suggested by the notation, are allowed to depend on the value of the factor; (4) a set of hypothetical random vectors $\{\boldsymbol{Y}_{xzi}\}$ that describes the value of the outcome when the factors have value $\boldsymbol{x}$ and the random variable $\boldsymbol{Z}_{xi} = \boldsymbol{z}$; (5) a function $g(\boldsymbol{x}, \boldsymbol{z}, \boldsymbol{U}_{xzi}) = \boldsymbol{Y}_{xzi}$.

The assumptions above suffice to generate the conditional distribution of the response at $\boldsymbol{x}$ when the concomitants have value $\boldsymbol{z}$, hereafter $D(\boldsymbol{Y}_{xz})$, and this is the distribution the researcher wishes to estimate. Inferences about the effect of factors, when the concomitants have value $\boldsymbol{z}$, can be obtained by comparing $D(\boldsymbol{Y}_{xz})$ across levels of the factors, provided this distribution can be estimated. The distribution $D(\boldsymbol{Y}_{xz})$ is not the distribution of the data; the data are drawn from the distribution $D(\boldsymbol{Y} \mid (\boldsymbol{X}, \boldsymbol{Z}))$, where $\boldsymbol{Y}$ denotes the observed response vector, and $\boldsymbol{X}$ is a random vector taking values in Ω. However, $D(\boldsymbol{Y}_{xz})$ can in principle be estimated from the observed data if $(\boldsymbol{Y}_{xzi} \perp\!\!\!\perp \boldsymbol{X}_i) \mid \boldsymbol{Z}_{xi}$ for every member of I. This condition, which Pratt and Schlaifer call the observability condition, is analogous to assumption of strongly ignorable treatment assignment. Note, however, it does not require conditional independence of $\boldsymbol{X}_i$ and the set of all hypothetical responses, taken jointly. Note also the possible dependence of the covariates on the cause(s).

To see how this works for the case of regression analysis, consider the case of a univariate response. The causal regression model is:

$$Y_{xzi} = \alpha + \boldsymbol{\beta}'\boldsymbol{x} + \boldsymbol{\gamma}'\boldsymbol{Z}_{xi} + \varepsilon_{xzi}, \tag{1.8}$$

where $E(\varepsilon_{xzi} \mid \boldsymbol{Z}_{xi} = \boldsymbol{z}) = 0$ for every member of I. Elements of $\boldsymbol{\beta}$ measure the unit effect of the factors, while elements of $\boldsymbol{\gamma}$ are ordinary regression coefficients.

In practice, the researcher observes only the realizations of the random variable Y_i corresponding to the one value of $\boldsymbol{X}_i$ observed and the one value of $\boldsymbol{Z}_{xi}$, $\boldsymbol{Z}_i$, that is observed. The regression model considered by the researcher is:

$$Y_i = \tilde{\alpha} + \tilde{\boldsymbol{\beta}}'\boldsymbol{X}_i + \tilde{\boldsymbol{\gamma}}'\boldsymbol{Z}_i + \varepsilon_i^*, \tag{1.9}$$

with the usual exogeneity assumption $E(\varepsilon_i^* \mid \boldsymbol{Z}_i = \boldsymbol{z}, \boldsymbol{X}_i = \boldsymbol{x}) = 0$ for every member of I. (Analogous to the discussion following (1.7), the reason for distinguishing between ε_i^* and ε_{xzi} is that $\varepsilon_i^* = Y_{xzi} - E(Y_{xzi} \mid \boldsymbol{Z}_{xi} = z, \boldsymbol{X}_i = \boldsymbol{x})$, whereas $\varepsilon_{xzi} = Y_{xzi} - E(Y_{xzi} \mid \boldsymbol{Z}_{xi} = z)$, and the two errors are identical only if $E(Y_{xzi} \mid \boldsymbol{Z}_{xi} = \boldsymbol{z}) = E(Y_{xzi} \mid \boldsymbol{Z}_{xi} = \boldsymbol{z}, \boldsymbol{X}_i = \boldsymbol{x})$.)

From the fact that $E(\varepsilon_i^* \mid \boldsymbol{Z}_i = \boldsymbol{z}, \boldsymbol{X}_i = \boldsymbol{x}) = E(\varepsilon_{xzi}^* \mid \boldsymbol{Z}_{xi} = \boldsymbol{z}, \boldsymbol{X}_i = \boldsymbol{x})$, it follows immediately that the parameters of (1.9) are equal to the parameters of (1.8) if the observability condition is satisfied (because the errors are the same when this condition is satisfied). When this condition does not hold, the parameters are not generally equal, and thus elements of $\tilde{\boldsymbol{\beta}}$ do not measure the unit effect of the factors. Further, $\gamma \neq \tilde{\gamma}$, that is, the regression coefficients in the two models are not identical. Note also that the regression model can hold without sustaining a causal interpretation. That is, if the exogeneity assumption in (1.9) is true, the population regression coefficients of (1.9) can be (under mild conditions) consistently estimated by least squares.

The results above have several implications for the manner in which sociologists and psychologists use (and should use) regressions to draw causal inferences. First, contrary to opinions that are often held in sociology and psychology, a properly specified regression need not sustain a causal interpretation, and in particular, exogeneity by itself does not suffice to permit a causal interpretation. (In making this last remark, which appears to echo the remarks of econometricians who have studied the relationship between Granger causation and exogeneity, it is important to note that here a different notion of causation is being used.) Second, most social scientists do not draw a clear distinction between factors (causes) and concomitants. Typically, all independent variables are treated as if they are factors, and all regression coefficients are interpreted as unit effects (sometimes called direct effects). Since there are no concomitants in such an analysis, the observability condition in this case is tantamount to the assumption of random assignment to levels of the factors. In most instances in social research such an assumption will not be justified. This suggests that social scientists should first consider which variables are to be treated as causes, and then attempt to measure concomitants that suffice (in the sense of making the observability condition hold) for assignment to levels of the factors. In some instances, a variable that an investigator would like to treat as a cause should be treated as a concomitant because assignment to levels of this variable has not been randomized and the investigator cannot measure or does not know the variables that are sufficient for treatment assignment. The regression coefficient corresponding to such a variable should not be interpreted as a unit effect.

The manipulative account and the resulting approach to inference, as outlined above, has several advantages over the notions of probabilistic causation treated previously. First, when randomization or conditional randomization is possible to implement, the issue of spuriousness does not arise (at least with respect to variables prior to the cause). Second, causal inference is tied to explicit counterfactual conditionals at the unit level; the notion that a causal inference supports a counterfactual conditional is typically lacking in the probabilistic versions discussed in Section 3 (though not in some of the justifications for the intuition underlying the definitions). Third, the quantity that one wants to estimate is clearly defined, irrespective of the study design; this too is lacking in much of the econometric work discussed in Section 3. Fourth, issues concerning causal priority do not tend to arise as the causes are variables that are manipulated and the effects are measured later.

The experimental approach is not without its practical difficulties. Causal inferences are sometimes difficult to make in randomized experiments because of internal validity problems such as nonrandom experimental mortality. A number of external validity problems also arise (Campbell and Stanley [1963] 1966). Some authors have raised these issues in responding to the account (for example, Granger 1986). From the point of view here, most of these issues are relevant, indeed critical to the implementation of an adequate experiment, but not particularly relevant to the fundamental perspective offered by the account itself.

Nor is randomization a panacea for all problems. Various authors (Conlisk 1985; Gail, Wieand, and Piantodosi 1984; Smith 1990) have shown that randomization does not allow consistent estimation of unit treatment effects in models where a covariate, omitted from the analysis, interacts with a treatment variable. Thus, if an investigator wants to estimate the unit treatment effect (as opposed to the average effect), randomization will not always allow this.

Other difficulties arise when the manipulative account is transferred over to the nonexperimental domain. Philosophers and statisticians have argued about the types of phenomena that could be causes. Must a cause be an event or state that can actually be induced in practice? If so, the manipulative account is irrelevant to many sociologists. At the other extreme, sociologists and psychologists who use causal models often seem to think that anything can be a cause. For example, in these disciplines the usual view is that latent variables do not correspond to any real-world entities, but are only hypothetical constructs around which theoretical work is organized. However, this does not stop anyone from speaking explicitly of the effects of these variables (for example, Jöreskog 1977) or from reporting estimates of these. Evidently, latent variables are not really real, but these not really real constructs cause real-world effects nevertheless. In some instances the response variable is a latent variable; in this case the not really real constructs cause a real-world effect that is not really real, either. For further treatment of causal inference in models with latent variables, see Sobel (1994).

Collingwood ([1940] 1948) points out (in the deterministic context) that the manipulative account of causation is both anthropocentric and anthropomorphic. G. H. von Wright (1971) argues that the manipulative account does not hinge on the actual ability to manipulate the cause, but it does hinge on the ability to potentially manipulate the cause. Holland (1986) also makes this argument. Mackie (1974) finds such accounts unacceptable, for (in part) they imply that causation is inherently linked to the activities of human beings (or entities that operate like humans, such as Nature and Prankster in Pratt and Schlaifer 1984). In the absence of such agents, causation does not exist. (But see also Section 2 for criticism of Mackie's attempt to fix the problem in a nonanthropomorphic manner.)

Pratt and Schlaifer (1988), who eschew commitment to a particular notion of causation, point out, with respect to the observability condition previously discussed, that it does not matter if the investigator selects case i with value $\boldsymbol{x}$ or sets the value of the factors on this case to $\boldsymbol{x}$. This seems to suggest that one could reduce the dependence of the manipulative account on the notion of manipulation by removing this notion and imagining, without saying how, that all observations can receive any value of the factor. This seems true in a formal sense, thereby making the account nonanthropomorphic. On the other hand, it has already been argued that the manner in which the cause is brought about may be critical for understanding the meaning of an effect. Without attempting to definitively settle this issue, when the factors are set to some value $\boldsymbol{x} \in \Omega$, either in actuality or in theory, the investigator is forced to indicate how this occurs. This clarifies the meaning of the factors, and hence of the effects. It also allows the investigator to think about the concomitants that are sufficient for assignment to levels of the factors. Finally, the content of the counterfactual conditional on which causal inferences rest is clarified (although it may still be difficult to form this counterfactual). If cases were simply selected with value $\boldsymbol{x}$, the investigator does not have to indicate how an alternative value could come about, and hence the counterfactual conditional is vague, as is the meaning of the factors and their effects. In this case, it is also virtually impossible to imagine the concomitants that are sufficient for assignment to treatment levels. A natural way to surmount the difficulties raised by selection, as opposed to manipulation, is to reintroduce manipulation by imagining the experiment in which cases are assigned to levels of the factors. Unless there is some other way of dealing with the problems raised by selection (and I do not see another way), it appears that the ability to manipulate the factors, at least hypothetically, is central, at least at the level of implementation in a real case.

Second, issues concerning causal priority do not tend to arise under a manipulative account. This is because the cause is under the control of the investigator. In addition, the effect is measured subsequent to the cause, so the causal relation apparently features temporal priority (see Section 2). Some authors (for example, Collingwood [1940] 1948; Holland 1986) view temporal priority as an inherent part of the causal relation. Sobel (1990) also requires causes to precede effects in time. In his paper, some of the results would change if instantaneous causation were allowed; as he notes, in most instances in the social sciences, the causal relation of interest features temporal priority.

It is important to note, however, that the manipulative account does not satisfactorily resolve the issue of temporal priority. The key to this account is the ability to manipulate (if only hypothetically) the cause; the fact that the effect is measured subsequently should be viewed as part of the research design, that is, as part of the experimental approach to inference. But philosophers and statisticians alike have failed to take such a view; in so doing, they have committed the sin of verificationism.

I propose to resolve this problem by distinguishing between the time at which the response is measured and the value of the response at different times, incorporating this distinction into the previous formalization of the account. Since a full treatment is beyond the scope of this chapter, matters are kept as simple as possible. Consider the case where the cause takes values $k = 0$ (say, no treatment) and $k = 1$ (treatment). Let t_0 denote the time at which treatment is initiated and let $(Y_{ik})_{t_0,t_1}$ denote the response of unit i to level k of the cause at the time t_1, where $t_1 \geq t_0$. Similarly, let $(Y_{i1} - Y_{i0})_{(t_0,t_1)}$ denote the effect for unit i. Under the usual manipulative account, t_1 denotes a fixed time at which the response is measured, and insofar as it is not possible to actually measure the response at the fixed time t_0, $t_1 > t_0$. The fact that the response is measured subsequent to the cause appears to be the basis for the claim that temporal priority is an integral feature of the causal relation under a manipulative account. This, however, confuses the approach with the ontological aspect of the causal relation.

With fixed times t_0 and t_1, as in the foregoing paragraph, there is no real reason to actually introduce the temporal subscripts into the notation for the unit effect. Suppose now we interpret the unit effect in the previous paragraph as a unit effect function, with arguments t_0 and t_1, subject as above to the restriction $t_1 \geq t_0$. For simplicity, suppose also that for all initiation times t_0 the unit effect function is a step function, that is, $(Y_{i1} - Y_{i0})_{(t_0,t_1)} = \Delta_i$ if $t_1 \geq t_0 + \lambda_i$, 0 otherwise, where λ_i is a fixed real number on the nonnegative portion of the real line. For simplicity, suppose the unit effect function is a "constant" effect function, that is, $\Delta_i = \Delta$ and $\lambda_i = \lambda$. Suppose also that $\Delta \neq 0$. (Note how the constant effect case corresponds to a treatment that is akin to the deterministic manipulative account.) In the case at hand, it seems reasonable to say that for every unit the cause is temporally prior to the effect if $\lambda > 0$, and to view the case $\lambda = 0$ as an instance of instantaneous causation. Thus, temporal priority does not appear to be an integral feature of the causal relation under a (suitably modified) manipulative account. Of course, the experimental approach will not allow for the measurement of instantaneous effects, should these exist. Thus, an experimental approach that appears to allow instantaneous causation will rely on some untestable assumptions.

A serious limitation of the usual experimental approach, especially for some types of sociological research, stems from utilization of the stable unit treatment value assumption (SUTVA) (Rubin 1978, 1980). In essence, this assumption states that the same value of the response for unit i to treatment k will occur no matter what treatments have been assigned to the other units of the population. Pratt and Schlaifer (1988) avoid making this assumption explicitly, but assume that the response $\boldsymbol{Y}_i$ that is observed when $\boldsymbol{X}_i = \boldsymbol{x}$ and the other observations have whatever values on the factors and concomitants they take on, corresponds to the value $\boldsymbol{Y}_{xzi}$ that would be observed (a counterfactual) if all observations had value $\boldsymbol{x}$ on the factors. In other words, Pratt and Schlaifer (1988) do not avoid this issue; they merely do not discuss it.

Relaxations of the SUTVA assumption for limited types of violations have been discussed in the literature (for example, Cox 1958; Holland 1987; Rosenbaum 1987). But these relaxations do not suffice to allow consideration of many questions that most philosophers and scientists would consider causal. For example, does poverty cause crime? Under the formalization of the manipulative account under consideration here, this is not even a sensible question if poverty is defined as lying below some percentile on the income distribution. If, however, poverty is defined by reference to some dollar amount, use of the SUTVA assumption (or the assumption of Pratt and Schlaifer) is tantamount to the assumption that the criminal behavior of respondent i is the same whether all persons have the same income or there is an income distribution featuring vast disparities. But this is simply not believable. Further, the problem does not lie with the question, which admits a causal interpretation under a manipulative account. Evidently, use of the SUTVA assumption precludes consideration of certain types of questions that social scientists ask all the time. In philosophy, causal questions with answers that are dependent on distributional properties of the population have been termed frequency dependent by Sober (1982), whose examples are drawn from evolutionary biology.

5 Causal Inference in "Causal Models"

Section 4 showed that the parameters of an explicitly causal regression model need not be identical with the parameters of the usual linear regression model used in social and behavioral research, even when the latter model is properly specified (at least in the mean structure). Additionally, conditions were given under which the parameters of the two models are identical. In this section, I briefly take up the use of structural equation modeling for drawing causal inferences in social and behavioral research. This approach, which is an outgrowth of econometric work on simultaneous equation models, is usually featured in conjunction with nonexperimental studies, but the interpretation of the model parameters (and parameter estimates) typically relies on notions of causation that are similar to those encountered in Section 4 (Sobel 1990, 1993, 1994). Consequently, the material on single-equation models in that section carries over to the simultaneous-equation context. In addition, these models are more complicated than the single-equation models, in particular because they feature relationships among dependent variables. As such, a number of new issues and concerns, not apparent in the single-equation case, arise with respect to the interpretation and estimation of these models.

I begin with the usual approach, as handed down from econometrics. To keep matters simple, connections with time series models will be minimized, and in the discussion of exogeneity, it will not be necessary to refer to the more general notion that a variable is predetermined. I will also assume simple random sampling from a random vector of observables $\boldsymbol{Z}$ of dimension $n + m$.

In the econometric approach, the variables in $\boldsymbol{Z}$ are first partitioned into subsets $\boldsymbol{X}$ and $\boldsymbol{Y}$, of dimension n and m, respectively. The variables in $\boldsymbol{X}$ are viewed as "exogenous," that is, determined outside the system under consideration, and these variables, which often refer to policy instruments in the economic context, are regarded as inputs into the model. Exogenous variables are causally prior to the variables in $\boldsymbol{Y}$, which are dependent or endogenous variables, and the values of the endogenous variables (which are often viewed as outputs) depend upon the values of the exogenous variables, as well as the values of other endogenous variables and stochastic disturbances. The partitioning of $\boldsymbol{Z}$ into exogenous and endogenous sets is to be decided on the basis of economic considerations (and temporal priority), and this assumption is not fully testable. For observation i from the population I, the "structural" model is:

$$\boldsymbol{Y}_i = \boldsymbol{\alpha} + \boldsymbol{B}\boldsymbol{Y}_i + \boldsymbol{\Gamma}\boldsymbol{X}_i + \boldsymbol{\varepsilon}_i, \tag{1.10}$$

where $\boldsymbol{B}$ is an $m \times m$ matrix of parameters describing relationships among the endogenous variables, with diagonal elements 0, $\boldsymbol{\Gamma}$ is an $m \times n$ matrix of parameters connecting exogenous and endogenous variables, and $\boldsymbol{\varepsilon}_i$ is a random drawing from a distribution with mean $\boldsymbol{0}$ and finite nonsingular covariance matrix $\boldsymbol{\Sigma}$. Technically, it is not necessary to assume that this covariance matrix is nonsingular. In addition, it is assumed that that matrix $(\boldsymbol{I} - \boldsymbol{B})$ is invertible, that observations are independent, and that $\boldsymbol{\varepsilon}_i$ and $\boldsymbol{X}_i$ are independent. This last assumption is the exogeneity assumption, and as stated, is stronger than required, since uncorrelatedness will suffice for deriving consistent estimates. The relationship between the exogeneity assumption and the intuitive notion that exogenous variables are determined outside the system is discussed subsequently.

The parameters of the structural model above may or may not be uniquely determined. However, the parameters $\boldsymbol{\alpha}'$, $\boldsymbol{\Pi}$, and $\boldsymbol{\Psi}$ of the reduced form equation:

$$\boldsymbol{Y}_i = \boldsymbol{\alpha}' + \boldsymbol{\Pi}\boldsymbol{X}_i + \boldsymbol{v}_i, \tag{1.11}$$

where $\boldsymbol{v}_i = (\boldsymbol{I} - \boldsymbol{B})^{-1}\varepsilon_i$, are uniquely determined, with $\boldsymbol{\alpha}' = (\boldsymbol{I} - \boldsymbol{B})^{-1}\boldsymbol{\alpha}$, $\boldsymbol{\Pi} = (\boldsymbol{I} - \boldsymbol{B})^{-1}\boldsymbol{\Gamma}$, $\boldsymbol{\Psi} = \boldsymbol{V}(\boldsymbol{v}_i) = (\boldsymbol{I} - \boldsymbol{B})^{-1}\boldsymbol{\Sigma}(\boldsymbol{I} - \boldsymbol{B}')^{-1}$ for all i. These parameters characterize the conditional mean structure $E(\boldsymbol{Y} \mid \boldsymbol{X})$ and the conditional covariance structure of $\boldsymbol{v}_i$, for all i. Using these parameters, in conjunction with restrictions on the structural parameters, it may be possible to uniquely determine (identify) the value of the unrestricted structural parameters. When this is the case, the model is said to be identified. In the following, it is assumed that the model is identified; readers who desire further information about identification can consult an econometrics textbook or Hausman (1983).

Econometric models are used to draw causal inferences in several ways. Under regularity conditions on $\boldsymbol{B}$, the reduced form parameter π_{rs} is usually interpreted as the effect of a one unit change in X_{ir} (the rth element of $\boldsymbol{X}_i$) on Y_{is} (the sth element of $\boldsymbol{Y}_s$). In economics, such effects are called equilibrium multipliers (Goldberger 1959), and in psychology and sociology these quantities are called total effects. Second, economists are typically interested in the structural form of the model. Parameters of the structural form, which are often viewed as more fundamental than the reduced form parameters, are thought to describe the behavior of agents (persons, firms, the government) in the economic system, and changes in one structural parameter may change the entire reduced form (Judge et al. 1985). The manner in which such a parameter describes this behavior is generally not stated, however.

Economists have criticized the structural-equation approach to causal inference on various grounds. First, the concept of a structural parameter is vague at best, meaningless at worst. As Lucas (1976) points out, economic agents may change their behavior when inputs into the economic system change. This would imply that the so-called structural parameters are not invariant to such changes, and in this case, the interpretations in the preceding paragraph would not hold. This is also evident from the material in Section 4. Since each reduced form equation is simply a regression, note that, as in Section 4, the regression may be properly specified, yet π_{rs} may not describe the actual change that would result under manipulation of the exogenous variable X_r. The assumption that π_{rs} does describe the effects of this manipulation is external to the model, although it is often overlooked. From Section 4, it is also clear that if all the exogenous variables are viewed as causes, this is tantamount to assuming that assignment to these variables is random.

In response to some of the concerns above, Sims (1982) argues that a parameter is structural if it is invariant under a class of modifications of the system. Similarly (but more strongly), Engle, Hendry, and Richard (1983, p. 284) define a (conditional) model to be structurally invariant "if all its parameters are invariant for any change in the distribution of the conditioning variables."

Leamer (1985) argues that these ideas about invariance should be extended to apply to the concept of exogeneity. The intuitive notion that exogenous variables are determined outside the system of interest does not correspond well with the definition that a variable is exogenous (with respect to some equation) if it is independent of the error. For example, if two random variables are bivariate normal, there are two regressions, and the right-hand-side variable in either regression is exogenous (independent of the error). Thus, exogeneity depends on the equation of interest, and it also implies a committment to certain parameter values. Leamer argues instead that a vector $\boldsymbol{X}$ should be called exogenous with respect to a dependent variable Y_s if the distribution $D(Y_s \mid \boldsymbol{X})$ is invariant to a class of interventions. Characterization of this class may, however, be difficult. This (typically untestable assumption) is similar to the concept of superexogeneity proposed by Engle, et al. (1983), except that superexogeneity entails invariance to all interventions, and it hinges on a concept of weak exogeneity that is tied to efficient statistical inference. Then, Leamer proposes to call an exogenous variable X_r causal with respect to a dependent variable Y_s if the distribution $D(Y_s \mid \boldsymbol{X})$ depends on X_r. An endogenous variable Y_1 can also be a cause of another endogenous variable Y_2 if the distribution of Y_2 depends on Z, where Z is a "surrogate exogenous variable for Y_1" (Leamer, 1985, p. 258). Leamer does little to clarify the intended meaning of this statement. It is clear, however, from Section 4, that if the observability condition holds, one can think of Y_1 as a cause, and estimate the effect of Y_1 on Y_2. It seems unlikely (given the rest of Leamer's treatment) that this is what he has in mind, however.

Psychologists and sociologists who have written on covariance structure analysis, which incorporates the model above as a special case, have tended to borrow the econometrician's approach to causal inference. But they have not been attentive to the recent concerns in the econometric literature concerning exogeneity and parameter invariance. In addition, they often focus on independent variables that are not subject to intervention (as seen earlier), and unlike much of the economic literature, they also equate the parameters relating dependent variables to other dependent variables with effects, a practice criticized below.

In that vein, the usual approach in psychology and sociology distinguishes three types of effects: total, direct, and indirect. The direct effect of a variable on another is usually thought of as the effect that is not transmitted through intervening variables, and has been defined explicitly as the outcome that results when the investigator induces a one-unit change in the independent variable, holding intervening variables constant (Alwin and Hauser 1975). Direct effects are thought to be captured by the elements of $\boldsymbol{\Gamma}$ and $\boldsymbol{B}$. Thus, γ_{rs} gives the direct effect of X_s on Y_r and β_{rs} gives the direct effect of Y_s on Y_r. Similarly, the total effect of X_r on Y_s is given by π_{rs}, and as in the economics literature, this quantity is interpreted as an equilibrium multiplier. Total effects of endogenous variables on endogenous variables have also been defined in the literature; Sobel (1990) shows that these definitions are incorrect.

Although the foregoing definitions are usually applied in nonexperimental studies, the implicit notion of causation in the literature is still manipulative (as indicated above). But investigators who use the experimental metaphor do not take it seriously. Neither do methodologists who write about such matters (as illustrated below). For example, they do not distinguish between variables that are causes and other variables that are concomitants. Any variable can be a cause, and thus one sees statements (based on a nonzero regression coefficient) that exogenous variables like father's occupation cause the endogenous variable son's intelligence (Kenny, 1979, p. 52). Do we really believe that, if we raised the father's occupation one unit on some prestige scale, the son would become more intelligent (presuming here intelligence, which is usually measured as a latent variable, is real)? Admittedly, the concept of intelligence is nebulous; nevertheless, it is doubtful whether any current notions of intelligence would allow a claim along such lines. Second, investigators attempt to consider direct effects of alleged causes, even when it is clear that one or more intervening variables the concept of an intervening variable is not even defined in the usual literature, as pointed out in Sobel (1990) cannot be held constant. Third, investigators ignore possible relations among exogenous variables when they compute total effects of exogenous variables. For example, in a status attainment model which includes father's education and father's occupation, the total effect of father's education on son's education that is reported ignores the fact that father's occupation depends on and is temporally subsequent to father's education. Duncan (1975) makes a similar point. Note also how the foregoing example illustrates that while both variables may be exogenous in the statistical sense with respect to the son's education equation, it is difficult to argue that father's occupation is determined outside the model. Fourth, investigators discuss the effects of endogenous variables on other endogenous variables, linking these to elements of the matrix B, or to functions of these elements. But if an endogenous variable is to be viewed as a cause of another endogenous variable, and causation involves at least a hypothetical manipulation, as implicit (and sometimes explicit) in the literature, how is this view to be reconciled with the treatment of the (causal) endogenous variable as a stochastic outcome of the model itself? Sobel (1990), drawing on Fisher (1970), points out that here a very strong invariance assumption is being made, namely that the parameter (or function) relating the two variables would be the same in a different system in which the so-called causal endogenous variables were exogenous. This requires a thought experiment which is difficult to perform in some instances, and the invariance assumption is also often unreasonable. Fifth, mathematical results depend on whether causation is instantaneous or whether the cause is temporally prior to the effect. In many cases, researchers view temporal priority as needed, but this is not reflected in their interpretation of the results, nor in the definition of the effects. To see how the mathematical results may hinge on explicit assumptions about such matters, see Sobel (1990). Sixth, as noted in Section 3, investigators often treat simultaneity in these models as meaning that cause and effect simultaneously cause each other. On the subject of simultaneity, see also Cox (1992), and for additional criticisms of the causal-modeling approach in psychology and sociology, see Holland (1988) and Sobel (1990, 1994).

A number of the preceding abuses would be lessened if sociologists and psychologists took the experimental metaphor seriously, as well as the associated econometric literature on exogeneity and invariance. But the econometric approach itself is not without its difficulties. To illustrate, I briefly reconsider Leamer's definition of causation. First, Leamer recognizes that the interpretation of economic data rests upon the experimental metaphor, and he espouses a notion of causation compatible with this metaphor. Thus it is reasonable to ask whether his definition of causality, which is based on the distribution of the observed data, and not on the distributions in Section 4, is in line with this metaphor.

To that end, consider the joint distribution of height and weight in an adult population, and suppose it is bivariate normal. Imagine a class of interventions in which some persons are induced to gain five pounds, others to lose five pounds. Such a set of interventions could be conducted in such a way that neither the marginal distribution of weight, nor the conditional distribution of height given weight changes. Thus, weight is exogenous (Leamer's definition) to height for such interventions. Furthermore, the distribution of height, given weight, depends on weight. Hence, using Leamer's definitions, weight causes height. However, we do not believe that inducing an adult person to lose or gain five pounds changes their height. Therefore, it is necessary to conclude that Leamer's definition of causality (in the sense he intends it) is defective. One might well object to this counterexample on the grounds that no one would consider this relationship causal in the first place or wish to consider such a class of interventions, and/or that a fuller model would uncover the spuriousness of this relationship. Such objections are beside the point. The counterexample is designed simply to show that the definition itself is defective. A better model might well produce better causal knowledge, and Leamer's definitions (to the extent these could be implemented), might work under some circumstances (which have not been spelled out), but this certainly does not vindicate the definition itself.

A more explicit approach is taken by Holland (1988), who applies the experimental approach to a recursive two-equation system with a binary independent variable, which takes values 1 in the treatment group and 0 in the control group. He distinguishes carefully between the causal model and the system of regression equations for this case. Mimicking (1.8) (Holland does not write the model in this way) the causal model (note both X and Y_1 are explicitly viewed as causes here) could be written as:

$$\begin{aligned} Y_{1xi} &= \alpha_1 + \gamma_{11}x + \varepsilon_{1xi} \\ Y_{2xy_1i} &= \alpha_2 + \gamma_{21}x + \beta_{21}y_1 + \varepsilon_{2xy_1i}. \end{aligned} \tag{1.12}$$

The first equation describes the value of the response Y_{1xi} when the independent variable is set to x, for all i, and the second describes the value of the response Y_{2xy_1i} when the independent variable is set to x, for all i, and in addition, Y_1 is set to y_1.

The corresponding simultaneous equation model is:

$$\begin{aligned} Y_{1(i)} &= \tilde{\alpha}_1 + \tilde{\gamma}_{11}X_i + \varepsilon^*_{1i} \\ Y_{2(i)} &= \tilde{\alpha}_2 + \tilde{\gamma}_{21}X_i + \tilde{\beta}_{21}Y_{1(i)} + \varepsilon^*_{2i}. \end{aligned} \tag{1.13}$$

Holland assumes random assignment to treatment and control groups. Under this assumption, consistent estimates of $\tilde{\gamma}_{11}$ are consistent for γ_{11}. Similarly, the reduced form parameter in the regression of Y_2 on X is consistent for the total effect in the causal model. But the estimates of $\tilde{\gamma}_{21}$ and $\tilde{\beta}_{21}$ are not consistent for the corresponding causal parameters. Note that Sobel's (1990) conclusion that the usual parameters of the simultaneous equation system should not be interpreted as direct effects, a conclusion reached in a different manner, reiterates this result. For further details on this case, see Holland (1988), and for other relevant material, see Angrist, Imbens, and Rubin (1993), Efron and Feldman (1991), Robins (1994), and Robins and Greenland (1992).

6 Discussion

Two deterministic accounts of causation were considered. Probabilistic notions of causation that rely primarily on the concepts of independence and conditional independence do not lead to causal inferences that comport with a manipulative account. Nor does the typical approach to causal inference in nonexperimental studies in the social and behavioral sciences yield inferences that comport with a manipulative account. In that vein, the formalization of the manipulative account discussed in Section 4 appears to offer more promise to social scientists. The formalization is identical for experimental and nonexperimental studies, and the resulting approach to causal inference reveals the types of assumptions researchers implicitly make when analyzing data in nonexperimental studies. As such, this approach provides a framework in which the nonexperimental worker can think more clearly about the types of conditions that need to be satisfied in order to make inferences in line with a manipulative account. It should also encourage sociologists, who have typically disdained experimental evidence, to pay more attention to the results from well-designed experiments (when these are possible) and/or to design experiments that will yield inferences in line with the manipulative account. This is not to say that the experimental approach is free of difficulties (some of which have been identified in Section 4), or that it can always be implemented, or that all scientific inferences should be causal in the first place. Nor am I suggesting causal inferences are always faulty in the absence of a randomized experiment. In that vein, many, if not most, scientific breakthroughs (for example, Snow lowered the death rate due to cholera by turning off water pumps in London) have been made without the benefits of a randomized experiment.

Existing formalizations of the manipulative account are not general enough to handle the types of causal questions that social scientists are often interested in addressing. Some of the limitations have already been pointed out in the text. Another problem that could benefit from a formal treatment (noted in Cox 1992) is the subject of hierarchical variation, which is usually addressed by using contextual (multilevel) models. Hopefully some of these gaps will be addressed by future workers.

REFERENCES

Alwin, D. F., and Hauser, R. M. (1975), "The Decomposition of Effects in Path Analysis," *American Sociological Review*, 40, 37–47.

Anderson, J. (1938), "The Problem of Causality," *Australasian Journal of Psychology and Philosophy*, 16, 127–142.

Anderson, T. W. (1984), *An Introduction to Multivariate Statistical Analysis* (2nd ed.), New York: John Wiley.

Angrist, J. D., Imbens, G. W., and Rubin, D. B. (1993), "Identification of Causal Effects Using Instrumental Variables," unpublished manuscript, Department of Statistics, Harvard University.

Barnes, J. (1982), *Aristotle*, Oxford: Oxford University Press.

Basmann, R. L. (1965), "A Note on the Statistical Testability of 'Explicit Causal Chains' Against the Class of 'Interdependent' Models," *Journal of the American Statistical Association*, 60, 1080–1093.

Bunge, M. (1979), *Causality and Modern Science* (3rd ed.), New York: Dover.

Byerly, H. (1990), "Causes and Laws: The Asymmetry Puzzle," in *PSA 1990, Proceedings of the 1990 Biennial Meeting of the Philosophy of Science Association*, Vol. 1, eds. A. Fine, M. Forbes, and L. Wessels, East Lansing, MI: Philosophy of Science Association.

Campbell, D. T., and Stanley, J. C. (1963) 1966, *Experimental and Quasi-Experimental Designs for Research*, Chicago: Rand McNally.

Cartwright, N. (1989), *Nature's Capacities and Their Measurement*, Oxford: Clarendon Press.

Chamberlain, G. (1982), "The General Equivalence of Granger and Sims Causality," *Econometrica*, 50, 569–581.

Cliff, N. (1983), "Some Cautions Concerning the Application of Causal Modeling," *Multivariate Behavioral Research*, 18, 115–126.

Collingwood, R. G. (1940) 1948, *An Essay on Metaphysics*, Oxford: Oxford University Press.

Conlisk, J. (1985), Comment on "Technical Problems in Social Experimentation: Cost vs. Ease of Analysis," by J. A. Hausman and D. A. Wise, in *Social Experimentation*, eds. J. A. Hausman and D. A. Wise, Chicago: University of Chicago Press, pp. 208–219.

Cook, T. D., and Campbell, D. T. (1979), *Quasi-Experimentation: Design and Analysis Issues for Field Settings*, Boston: Houghton Mifflin.

Cox, D. R. (1958), *The Planning of Experiments*, New York: John Wiley.

——— (1992), "Causality; Some Statistical Aspects," *Journal of the Royal Statistical Society*, Ser. A, 155, 291–301.

Davis, W. A. (1988), "Probabilistic Theories of Causation," in *Probability and Causality: Essays in Honor of Wesley C. Salmon*, ed. J. H. Fetzer, Dordrecht, Holland: D. Reidel, pp. 133–160.

Dawid, A. P. (1979), "Conditional Independence in Statistical Theory" (with discussion), *Journal of the Royal Statistical Society*, Ser. B, 41, 1–31.

Duncan, O. D. (1966), "Path Analysis: Sociological Examples", *American Journal of Sociology*, 72, 1–16.

———(1975) *Introduction to Simultaneous Equation Models*, New York: Academic.

Efron, B., and Feldman, D. (1991), "Compliance as an Explanatory Variable in Clinical Trials" (with discussion), *Journal of the American Statistical Association*, 86, 9–26.

Einhorn, H. J., and Hogarth, R. M. (1986), "Judging Probable Cause", *Psychological Bulletin*, 99, 3–19.

Engle, R. F., Hendry, D. F., and Richard, J. F. (1983), "Exogeneity," *Econometrica*, 51, 277–304.

Feigl, H. (1953), "Notes on Causality," in *Readings in the Philosophy of Science*, eds. H. Feigl and M. Brodbeck, New York: Appleton-Century Crofts, pp. 408–418.

Fetzer, J. H. (1988), "Probabilistic Metaphysics," in *Probability and Causality: Essays in Honor of Wesley C. Salmon*, ed. J. H. Fetzer, Dordrecht, Holland: D. Reidel, pp. 109–132.

Fisher, F. M. (1970), "A Correspondence Principle for Simultaneous Equation Models,"*Econometrica*, 38, 73–92.

Florens, J. P., and Mouchart, M. (1982), "A Note on Noncausality," *Econometrica*, 50, 583–591.

Freedman, D. A. (1987), "As Others See Us: A Case Study in Path Analysis" (with discussion), *Journal of Educational Statistics*, 12, 101–223.

Gail, H. M., Wieand, S., and Piantadosi, S. (1984), "Biased Estimates of Treatment Effects in Randomized Experiments with Nonlinear Regression and Omitted Covariates," *Biometrika*, 71, 431–444.

Gasking, D. (1955), "Causation and Recipes," *Mind*, 64, 479–487.

Geweke, J. (1984), "Inference and Causality in Economic Time Series Models," in *Handbook of Econometrics* (Vol. 2), eds. Z. Griliches and M. D. Intriligator, Amsterdam: North Holland, pp. 1101–1144.

Giere, R. (1980), "Causal Systems and Statistical Hypotheses" (with discussion), in *Applications of Inductive Logic*, eds. L. J. Cohen and M. Hesse, Oxford: Oxford University Press, pp. 251–290.

Goldberger, A. S. (1959), *Impact Multipliers and Dynamic Properties of the Klein-Goldberger Model*, Amsterdam: North Holland.

Good, I. J. (1961), "A Causal Calculus I," *British Journal of the Philosophy of Science*, 11, 305–318.

———(1962), "A Causal Calculus II," *British Journal of the Philosophy of Science*, 12, 42–51.

Granger, C. W. (1969), "Investigating Causal Relations by Econometric Models and Cross-Spectral Methods," *Econometrica*, 37, 424–438.

———(1980), "Testing for Causality: A Personal Viewpoint," *Journal of Economic Dynamics and Control*, 2, 329–352.

———(1986), Comment on "Statistics and Causal Inference," by P. W. Holland, *Journal of the American Statistical Association*, 81, 967–968.

Harré, R. (1972), *The Philosophies of Science*, Oxford: Oxford University Press.

Harré, R., and Madden, E. H. (1975), *Causal Powers: A Theory of Natural Necessity*, Oxford: Basil Blackwell.

Hausman, J. A. (1983), "Specification and Estimation of Simultaneous Equation Models," in *Handbook of Econometrics*(Vol. 1), eds. Z. Griliches and M. D. Intriligator, Amsterdam: North Holland, pp. 392–448.

Heckman, J. J. (1974), "Shadow Prices, Market Wages, and Labor Supply," *Econometrica*, 42, 679–694.

——— (1976), "The Common Structure of Statistical Models of Truncation, Sample Selection and Limited Dependent Variables and a Simple Estimator for such Models," *Annals of Economic and Social Measurement*, 5, 475–492.

Heckman, J. J., and Hotz, V. J. (1989), "Choosing Among Alternative Nonexperimental Methods for Estimating the Impact of Social Programs: The Case of Manpower Training" (with discussion), *Journal of the American Statistical Association*, 84, 862–880.

Heckman, J. J., Hotz, V. J., and Dabos, M. (1987), "Do We Need Experimental Data to Evaluate the Impact of Manpower Training on Earnings ?" *Evaluation Review*, 11, 395–427.

Hempel, C. G. (1968), "Maximal Specificity and Lawlikeness in Probabilistic Explanation," *Philosophy of Science*, 35, 116–133.

Holland, P. W. (1986), "Statistics and Causal Inference" (with discussion), *Journal of the American Statistical Association*, 81, 945–970.

——— (1987), Comment on "The Role of a Second Control Group in an Observational Study," by P. R. Rosenbaum, *Statistical Science*, 2, 306–308.

——— (1988), "Causal Inference, Path Analysis, and Recursive Structural Equation Models" (with discussion), in *Sociological Methodology, 1988*, ed. C. C. Clogg, Washington, D. C.: American Sociological Association, pp. 449–493.

Holland, P. W., and Rubin, D. B. (1983), "On Lord's Paradox," In *Principals of Modern Psychological Measurement*, eds. H. Wainer and S. Messnick, Hillsdak, NJ: Lawrence Erlbaum, pp. 3–35.

Hume, D. (1739) 1978, *A Treatise of Human Nature*, Oxford: Oxford University Press.

——— (1740) 1988, *An Abstract of a Treatise of Human Nature*, in *An Enquiry Concerning Human Understanding/David Hume: Introduction, Notes, and Editorial Arrangement by Anthony Flew*, ed. A. Flew, La Salle, IL: Open Court, pp. 29–43.

——— (1748) 1988, *An Enquiry Concerning Human Understanding*, in *An Enquiry Concerning Human Understanding/David Hume: Introduction, Notes, and Editorial Arrangement by Anthony Flew*, ed. A. Flew, La Salle, IL: Open Court, pp. 53–195.

Jöreskog, K. G. (1977), "Structural Equation Models in the Social Sciences: Specification, Estimation and Testing," in *Applications of Statistics*, ed. P. R. Krishnaiah, Amsterdam: North Holland, pp. 265–287.

Judge, G. G., Griffiths, W. E., Hall, R. C., Lütkepohl, H., and Lee, T. C. (1985), *The Theory and Practice of Econometrics* (2nd ed.), New York: John Wiley.

Kempthorne, O. (1952), *The Design and Analysis of Experiments*, New York: John Wiley.

Kenny, D. A. (1979), *Correlation and Causality*, New York: John Wiley.

Leamer, E. E. (1985), "Vector Autoregressions for Causal Inference?" (with discussion), in *Understanding Monetary Regimes*, supplement to *Journal of Monetary Economics*, eds. K. Brunner and A. Meltzer, pp. 255–318.

Lucas, R. E. (1976), "Econometric Policy Evaluation: A Critique" (with discussion), in *The Phillips Curve and Labor Markets*, supplement to *Journal of Monetary Economics*, eds. K. Brunner and A. Meltzer, pp. 19–62.

Mackie, J. L. (1974), *The Cement of the Universe*, Oxford: Oxford University Press.

Mill, J. S. (1843) 1973, *A System of Logic: Ratiocinative and Inductive*, in *The Collected Works of John Stuart Mill* (Vol. 7), ed. J. M. Robson, Toronto: University of Toronto Press.

Neyman, J. S. (1923) 1990, "On the Application of Probability Theory to Agri-Cultural Experiments. Essay on Principles. Section 9" (with discussion), *Statistical Science*, 4, 465–480.

Otte, R. (1981), "A Critique of Suppes' Theory of Probabilistic Causality," *Synthese*, 48, 167–189.

Pearl, J., and Verma, T. S. (1991), "A Theory of Inferred Causation," in *Principles of Knowledge Representation and Reasoning: Proceedings of the Second International Conference*, eds. J. A. Allen, R. Fikes, and E. Sandewall, San Mateo, CA: Morgan Kaufmann, pp. 441–452.

Pratt, J. W., and Schlaifer, R. (1984), "On the Nature and Discovery of Structure" (with discussion), *Journal of the American Statistical Association*, 79, 9–33.

——(1988), "On the Interpretation and Observation of Laws," *Journal of Econometrics*, 39, 23–52.

Ragin, C. C. (1987), *The Comparative Method*, Berkeley: University of California Press.

Reichenbach, H. (1956), *The Direction of Time*, Berkeley: University of California Press.

Robins, J. M. (1992), "Identifiability and Exchangeability for Direct and Indirect Effects," *Epidemiology*, 3, 143–155.

——(1994), "Correcting for Non-Compliance in Randomized Trials Using Structural Nested Mean Models," forthcoming in *Communications in Statistics*, Ser. A.

Rosenbaum, P. R. (1984a), "From Association to Causation in Observational Studies: The Role of Tests of Strongly Ignorable Treatment Asignment," *Journal of the American Statistical Association*, 79, 41–48.

——(1984b), "The Consequences of Adjustment for a Concomitant Variable That Has Been Affected by the Treatment," *Journal of the Royal Statistical Society*, Ser. A, 147, 656–666.

——(1986), "Dropping Out of High School in the United States: An Observational Study," *Journal of Educational Statistics*, 11, 207–224.

——(1987), "The Role of a Second Control Group in an Observational Study" (with discussion), *Statistical Science*, 2, 292–316.

——(1992), "Detecting Bias with Confidence in Observational Studies," *Biometrika*, 79, 367–374.

Rosenbaum, P. R., and Rubin, D. B. (1983), "The Central Role of the Propensity Score in Observational Studies for Causal Effects," *Biometrika*, 70, 41–55.

Rubin, D. B. (1974), "Estimating Causal Effects of Treatments in Randomized and Nonrandomized Studies," *Journal of Educational Psychology*, 66, 688–701.

———(1977), "Assignment to Treatment Groups on the Basis of a Covariate," *Journal of Educational Statistics*, 2, 1–26.

———(1978), "Bayesian Inference for Causal Effects: The Role of Randomization," *The Annals of Statistics*, 6, 34–58.

———(1980), Comment on "Randomization Analysis of Experimental Data: The Fisher Randomization Test," by D. Basu, *Journal of the American Statistical Association*, 75, 591–593.

———(1990), "Formal Modes of Statistical Inference for Causal Effects," *Journal of Statistical Planning and Inference*, 25, 279–292.

Russell, B. (1913), "On the Notion of Cause," *Proceedings of the Aristotelian Society*, New Series, 13, 1–26.

Salmon, W. C. (1984), *Scientific Explanation and the Causal Structure of the World*, Princeton, NJ: Princeton University Press.

Simon, H. A. (1952), "On the Definition of the Causal Relation," *Journal of Philosophy*, 49, 517–528.

———(1953), "Causal Ordering and Identifiability," in *Studies in Econometric Methods*, eds. W. Hood and T. Koopmans, New York: John Wiley, pp. 49–74.

———(1954), "Spurious Correlation: A Causal Interpretation," *Journal of the American Statistical Association*, 49, 467–492.

Simon, H. A., and Rescher, N. (1966), "Cause and Counterfactual," *Philosophy of Science*, 33, 323–340.

Sims, C. A. (1977), "Exogeneity and Causal Ordering in Macroeconomic Models," in *New Methods in Business Cycle Research: Proceedings from a Conference*, ed. C. A. Sims, Minneapolis: Federal Reserve Bank of Minneapolis.

———(1982), "Policy Analysis with Econometric Models," *Brookings Papers on Economic Activity*, 1, 107–164.

Skyrms, B. (1988), "Probability and Causation," *Journal of Econometrics*, 39, 53–68.

Smith, H. L. (1990), "Problems of Specification Common to Experimental and Non-experimental Social Research," in *Sociological Methodology, 1990*, ed. C. C. Clogg, Oxford: Basil Blackwell, pp. 59–91.

Sobel, M. E. (1990), "Effect Analysis and Causation in Linear Structural Equation Models," *Psychometrika*, 55, 495–515.

———(1993), "Causation, Spurious Correlation and Recursive Structural Equation Models: A Reexamination," unpublished manuscript.

———(1994), "Causal Inference in Latent Variable Models," forthcoming in *Analysis of Latent Variables in Developmental Research*, eds. A. von Eye and C. C. Clogg, Newburg Park, CA: Sage.

Sober, E. (1982), "Frequency-Dependent Causation," *Journal of Philosophy*, 79, 247–253.

Suppes, P. (1970), *A Probabilistic Theory of Causality*, Amsterdam: North Holland.

von Wright, G. H. (1971), *Explanation and Understanding*, Ithaca, N. Y.: Cornell University Press.

White, P. (1990), "Ideas About Causation in Philosophy and Psychology," *Psychological Bulletin*, 108, 3–18.

Wold, H. O. A. (1966), "On the Definition and Meaning of Causal Concepts," in *La Technique des modèles dans les sciences humaines*, ed. H. O. A. Wold, Monaco: Union Européenne d'Editions, pp. 265–275.

Wright, S. (1921), "Correlation and Causation," *Journal of Agricultural Research*, 20, 557–585.

Young, J. Z. (1978), *Programs of the Brain*, Oxford: Oxford University Press.

Zellner, A. (1979) 1984, "Causality and Econometrics," in *Basic Issues in Econometrics*, ed. A. Zellner, Chicago: University of Chicago Press, pp. 35–74.

Chapter 2

Missing Data

RODERICK J. A. LITTLE AND NATHANIEL SCHENKER

1 Introduction

Studies in the social and behavioral sciences frequently suffer from missing data. For instance, sample surveys often have some individuals who either refuse to participate or do not supply answers to certain questions, and panel studies often have incomplete data due to attrition. Recent comprehensive treatments of the subject of missing data include three volumes produced by the Panel on Incomplete Data of the Committee on National Statistics (Madow, Nisselson, and Olkin 1983; Madow and Olkin 1983; Madow, Olkin, and Rubin 1983) and Little and Rubin (1987).

There are three major problems created by missing data. First, if the units with missing values are systematically different from the units with complete data, naive analyses that ignore these differences may be biased. Second, the existence of missing data usually implies a loss of information, so that estimates will be less efficient than planned. Finally, standard statistical methods are designed for complete data sets; thus missing data often render the analyses of data from a study more complicated. In this chapter we review methods for handling missing data that seek to alleviate these problems.

This section begins with some examples that illustrate types of missing data that can occur; we will refer to these examples throughout the chapter. Some important general concepts concerning missing data are then reviewed, and some common but naive techniques for dealing with missing data are discussed. In Sections 2 through 6, we discuss less naive and more principled methods. Section 7 contains a concluding discussion.

RODERICK J.A. LITTLE • Department of Biostatistics, University of Michigan, 1420 Washington Heights, Ann Arbor, MI 48109-2029, USA. NATHANIEL SCHENKER • Department of Biostatistics, UCLA School of Public Health, 10833 Le Conte Avenue, Los Angeles, CA 90024-1772, USA. • The authors' names have been listed in alphabetical order. This research was conducted while Little was with the UCLA Department of Biomathematics. Support was provided by Grant MH37188 from the U.S. National Institute of Mental Health. The authors thank Joseph L. Schafer and Michael E. Sobel for helpful comments on the first draft.

Handbook of Statistical Modeling for the Social and Behavioral Sciences, edited by Gerhard Arminger, Clifford C. Clogg, and Michael E. Sobel. Plenum Press, New York, 1995.

1.1 Examples

Example 1: Nonresponse in the Survey of Consumer Finances

The Survey of Consumer Finances (SCF), sponsored by the Federal Reserve Board, collects information on the financial and demographic characteristics of households in the United States. The 1989 SCF used a dual-frame design, in which the two frames were (1) a multistage area frame and (2) a list frame developed from administrative files maintained by the Statistics of Income Division of the Internal Revenue Service based on 1987 tax returns (Woodburn 1991). The list frame was used to oversample wealthy individuals and thus provide a better representation of the upper tail of the wealth distribution.

Kennickell (1991) and Woodburn (1991) discussed nonresponse in the 1989 SCF and methods used to adjust for the nonresponse. Woodburn (1991) reported that the rates at which interviews were obtained were 69% in the area sample and only 34% in the list sample; thus the survey suffered from a high rate of noninterviews. The background information available, however, especially from the administrative files, provides useful information for noninterview adjustments. For example, it is clear that the interview rate decreases as wealth increases (Woodburn 1991). Kennickell (1991) discussed nonresponse on items for households that were interviewed in the SCF, and provides the nonresponse rates for a selection of these items. The extent of nonresponse varies from close to complete response on some variables to over 28% nonresponse on 1988 adjusted gross income.

Example 2: Attrition in a Panel Study on Education

Marini, Olsen, and Rubin (1980) analyzed data from a panel study of students in 10 Illinois schools. One block of variables was recorded for all of the students at the start of the study in 1957-1958. In a followup survey fifteen years later, 79% of the original students responded, resulting in a second block of variables. Some further data were collected on a subset of the students supplying the block 2 variables, resulting in a third block. The three blocks of variables form a special pattern of missing data: All students for whom block 3 variables were observed had block 2 variables observed, and all students for whom block 2 variables were observed had block 1 variables observed. In other words, block 1 is more observed than block 2, which is more observed than block 3. A fourth block of variables that violates this pattern somewhat was obtained via a questionnaire mailed to the parents of all the students in the original sample. Of these parents, 65% responded.

Example 3: Income Nonresponse in the Current Population Survey

The Current Population Survey (CPS) is a monthly survey of households conducted by the Census Bureau to gather a variety of information. In March of each year, the CPS includes a supplement in which detailed data on income are requested. Individuals are not always willing to report their incomes. As a consequence, typically about 20% of those surveyed have data missing on one or more income items (see, e.g., Madow et al. 1983, chapter 1). The CPS typically also has a small number of households for which interviews are not obtained at all.

Income nonresponse in the CPS results in reduced efficiency when the data are analyzed. More importantly, however, the nonrespondents are typically different from the respondents (David, Little, Samuhel, and Triest 1986). For example, those with high incomes tend to respond less frequently than those with middle incomes. Thus, analyzing the CPS data without adjusting for the differences between nonrespondents and respondents results in biases.

The CPS is a source of data for the government's figures on employment and income, as well as a major source of data for economists, social scientists, and others. Consequently, CPS data bases need to supply realistic answers for data analysts in spite of the problem of income nonresponse.

Example 4: Nonresponse in the Fatal Accident Reporting System

Another data base that is supplied to the public for analysis is from the Fatal Accident Reporting System (see Heitjan and Little 1991). This data base, collected for the National Highway Traffic Safety Administration, contains data on the location and time of each accident, the number and position of the vehicles involved, and the age, sex, driving record, seatbelt use, and blood alcohol content of the drivers. The last two variables are of great interest but are missing for many cases.

In handling the problem of missing data in this public-use data base, the National Highway Traffic Safety Administration has two major goals. The first is to produce a data file that has missing items filled in, so that it can be analyzed using standard methods designed for complete data. The second is to provide analysts with a means to obtain standard errors that accurately reflect the loss of information due to nonresponse.

Example 5: Lack of Comparability of Occupation Codes Over Time

In each decennial census, employment information is obtained from individuals in the form of open-ended descriptions of occupations. These descriptions are then coded into hundreds of occupations, using a scheme devised by the Census Bureau. To provide up-to-date information, the occupation classification scheme is revised for each census.

Major changes were made for the 1980 census. As a consequence, public-use data bases from the 1980 census have occupation codes that are not directly comparable to those on public-use data bases from previous censuses, in particular, the 1970 census. The lack of comparability of codes across time makes it difficult to study such topics as occupation mobility and labor force shifts by demographic characteristics. The 1970 public-use data bases contain over one million records, and therefore would be prohibitively expensive to recode according to the 1980 scheme. There exists, however, a double-coded sample of 120,000 units from the 1970 census, that is, a sample with both 1980 and 1970 occupation codes.

The occupation code problem can be viewed as a missing-data problem. A small fraction of the 1970 public-use data base contains both 1970 and 1980 occupation codes, whereas the remaining cases contain only the old codes.

1.2 Important Concepts

Unit versus Item Nonresponse

In survey settings, two forms of missing data are commonly distinguished, namely *unit nonresponse*, where entire questionnaires are missing because of inability to contact or interview a sampled individual, and *item nonresponse*, where an interview is conducted but particular questions are missing, through refusal to answer, interviewer errors, or deletion of inconsistent responses in the editing process. Example 1 concerns a survey in which both unit nonresponse and item nonresponse are present; this is a common situation in surveys. Unit nonresponse often results in a pattern where the survey variables are missing in a block for unit nonrespondents, whereas survey design variables (such as geography) are observed; it is often treated by weighting adjustments, where unit nonrespondents are dropped from the data set and respondents are assigned weights to compensate for bias arising from systematic differences between unit respondents and nonrespondents (see Section 2). Item nonresponse, on the other hand, is often treated by imputing (i.e., filling in) the missing items, or simply leaving them with a missing-value code to identify them as missing.

Patterns of Missing Data

Suppose that if the data were complete, they could be arranged in an $(N \times V)$ data matrix $\boldsymbol{X} = \{x_{ij}\}$, such that x_{ij} is the value of variable j for unit i, $i = 1, ..., N$; $j = 1, ..., V$. Let $\boldsymbol{M} = \{m_{ij}\}$ denote an $(N \times V)$ missing-data indicator matrix, such that $m_{ij} = 1$ if x_{ij} is missing and $m_{ij} = 0$ if x_{ij} is present. The matrix $\boldsymbol{M}$ describes the pattern of missing data. It is useful when discussing missing-data analysis to treat $\boldsymbol{M}$ as a stochastic matrix; for example the mean of m_{ij} then can be interpreted as the probability that x_{ij} is missing.

Some missing-data methods are designed for special patterns of missing data. Perhaps the most commonly considered pattern is *univariate nonresponse*, where (possibly after rearrangement of the rows and columns), x_{ij} is observed for $i = 1, ..., N$ and $j = 1, ..., V-1$, and x_{iV} is observed for $i = 1, ..., n_0$ and missing for $i = n_0 + 1, ..., n_0 + n_1 = N$. Thus, with univariate nonresponse, missing data are confined to variable V, say X_V, and $\boldsymbol{M}$ can be replaced by a vector $\boldsymbol{m} = (m_1, ..., m_N)'$, such that $m_i = 0$ if x_{iV} is observed and $m_i = 1$ if x_{iV} is missing. We will discuss this pattern frequently throughout the chapter.

Univariate nonresponse is a special case of *monotone* missing data, where (perhaps after rearranging columns), the variable X_j is observed whenever X_{j+1} is observed, for $j = 1, ..., V - 1$. Thus, for any i, $m_{ij} = 0$ implies that $m_{ij'} = 0$ for all $j' < j$. As will be discussed in Section 3, maximum-likelihood methods can be developed for a general pattern of missing data, but they sometimes simplify for special patterns such as monotone data. Marini et al. (1980) discussed monotone missing-data methods for the data described in example 2.

It is often useful to examine the pattern of missing data, using computer programs such as BMDP8D (Dixon 1988). For example, variables with high levels of nonresponse that are of minor interest might be discarded for some analyses. If one finds that two variables are never observed together, parameters measuring association between these variables cannot be estimated.

Missing-Data Mechanisms

The pattern of multivariate missing data frequently depends on the variables being considered; in particular, different variables often have different rates of missing values (for example, income often has a higher nonresponse rate than gender), and variables with similar subject matter tend to be observed or missing together (Anderson, Basilevsky, and Hum 1983; Glasser 1964). Such considerations concern the marginal distribution of M, and affect the information content in the data.

A different issue is whether missingness depends on the values of survey variables, that is, whether M and X are associated. For example, if income is subject to nonresponse, does missingness of income depend on the value of income or other variables in the data set? Such questions concern the missing-data mechanism, and are key in determining the extent of nonresponse bias. Missing data sometimes occur by design. For example, detailed questions on a certain topic may be confined to a random subsample of the study sample. In example 5, the small double-coded sample is essentially a random subsample of the entire public-use data base. In such cases it may be safe to assume that missing values are missing completely at random (MCAR), in the sense that missingness is unrelated to the survey variables X. Formally, let $p(M|X,\phi)$ denote the conditional distribution of M given X, and parameters ϕ that characterize response rates. Rubin (1976a) defined the missing data as MCAR if

$$p(M|X,\phi) = p(M|\phi) \text{ for all } X. \tag{2.1}$$

A rather basic requirement of a missing-data technique is that it yields estimates with reasonable statistical properties (such as consistency) when the missing data are MCAR.

The MCAR assumption can be assessed from the observed data by comparing distributions of respondents and nonrespondents on observed variables. In particular, for continuous variables, the BMDP8D program (Dixon 1988) provides univariate t-tests, and Little (1988a) developed a global multivariate test of the MCAR assumption. In many practical instances, particularly when missingness is not under the control of the investigator, the MCAR assumption is a hope rather than an expectation; the missing-data mechanism is likely to be related to X. For example, in a retrospective survey, missing values caused by difficulties of recall may be more likely in older individuals. Missing values of HIV status in a survey of AIDS may be more likely among individuals who have contracted the infection. Recall also examples 1 and 3, in which there was evidence that nonrespondents differed systematically from respondents.

An important issue if the missing data are not MCAR concerns whether differences in characteristics of nonrespondents and respondents can be captured in terms of variables measured for both respondents and nonrespondents. For example, if missingness of a variable depends solely on age, and age is recorded for nonrespondents and respondents, then nonresponse bias can be controlled by an analysis that stratifies on age or adjusts for age in some fashion. Rubin's (1976a) definition of missing at random (MAR) addresses this issue. Formally, let X_{obs} denote the observed portion of X and let X_{mis} denote the missing portion. The missing data are MAR if

$$p(M|X_{obs}, X_{mis}, \phi) = p(M|X_{obs}, \phi) \text{ for all } X_{mis}; \tag{2.2}$$

that is, the distribution of M given X depends only on variables X_{obs} that are recorded in the data set. If income is missing and age is fully observed, then the missing data are MAR if missingness of income depends only on age; however if for subjects of a given age, missingness of income depends on the value of income, then the missing data are not MAR, since missingness depends on values of a variable that is sometimes missing (namely income itself). Some authors use the term "missing at random" in an informal way that corresponds to our definition of MCAR; we shall follow Rubin's (1976a) usage, since it formalizes a distinction that is important when assessing missing-data methods.

We noted that in example 3, income nonrespondents differ systematically from respondents, so that the missing data are not MCAR. However, considerable covariate information is generally available for respondents and nonrespondents, and can be used to predict nonrespondent income values. David et al. (1986) provided evidence from external data that differences in wages and salary between nonrespondents and respondents can be accounted for in terms of these covariates; that is, missing wages and salary appear to be MAR.

1.3 Naive Approaches

Complete-Case Analysis

Complete-case (CC) analysis simply discards cases with any missing values, and is the standard treatment of missing data in statistical packages. This method is also known as *listwise deletion*. Advantages are ease of implementation, and the fact that valid (but not necessarily optimal) inferences are obtained when the missing data are MCAR. The latter follows from the fact that if the missing data are MCAR, then the complete cases are a random subsample of the original sample.

On the other hand, the rejection of incomplete cases seems an unnecessary waste of information: if the number of variables is large, then even a sparse pattern of missing X's can result in a substantial number of incomplete cases. It seems reasonable to seek ways of incorporating the incomplete cases into the analysis. One approach is to drop variables with high levels of nonresponse; in the regression context, Rubin (1976b) described measures of a covariate's predictive value that take into account degree of incompleteness. Other ways of incorporating incomplete cases are discussed below.

Aside from efficiency considerations, a serious problem with dropping incomplete cases is that the complete cases are often a biased sample, that is, the missing data are often not MCAR. The size of the resulting bias depends on the degree of association between M and X, the amount of missing data, and the specifics of the analysis. For example, consider an analysis of income measured at a number of time points, where missing data arise from attrition from the sample. If the factors attributing to attrition are related to income (for example, people move out of the area to seek other employment), then the complete-case estimate of the mean of income from the reduced sample may be seriously biased. However, comparisons of mean income levels between time points may be less vulnerable to bias than some simple alternative methods, such as comparison of the cross-sectional means (Little and Su 1989).

Bias and variance considerations aside, statisticians rightly resist attempts to analyze data selectively, and hence aim to analyze all the data to the extent possible; CC is at best a compromise method. It serves as a useful baseline method for comparisons, and it can be better than methods that use the incomplete cases in an unsatisfactory way.

Available-Case Analysis

Available-case (AC) methods use the largest sets of available cases for estimating individual parameters (Little and Rubin 1987, sect. 3.3). For example, suppose the objective is to estimate the correlation matrix of a set of continuous variables $X_1, ..., X_V$. Complete-case analysis uses the set of complete cases to estimate all the correlations; AC analysis uses the set of cases with both X_j and X_k observed to estimate the correlation of X_j and X_k, $1 \leq j, k \leq V$. Since the sample base of available cases for measuring each correlation includes the set of complete cases, the AC method appears to make better use of available information. The sample base changes from correlation to correlation, however, creating potential problems when the missing data are not MCAR or variables are highly correlated. In the presence of high correlations, there is no guarantee that the AC correlation matrix is positive definite, and hence analyses that require a positive definite covariance matrix (such as multiple regression) may be problematic.

Haitovsky's (1968) simulations concerning regression with highly correlated continuous data found AC markedly inferior to CC. On the other hand, Kim and Curry (1977) found AC superior to CC in simulations based on weakly correlated data. Simulation studies comparing AC regression estimates with maximum likelihood (ML) under normality (Section 3) suggest that ML is superior even when underlying normality assumptions are violated (Azen, Van Guilder, and Hill 1989; Little 1988c; Muthén, Kaplan, and Hollis 1987). Little and Su (1989) suggested that comparisons of estimates over time based on CC analysis may be preferable to comparisons based on AC when missing data are not MCAR.

Unlike CC analysis, standard errors of AC estimates generally require custom formulas, even when the missing data are MCAR. For example, see Van Praag, Dijkstra, and Van Velzen (1985).

Unconditional Mean Imputation

Methods that impute or fill in the missing values have the advantage that observed values in the incomplete cases are retained, unlike CC analysis. A common naive approach imputes missing X's by their unconditional sample means. Wilks (1932) and Afifi and Elashoff (1967) discussed this method in bivariate settings. Unconditional mean imputation can yield satisfactory point estimates of some parameters such as unconditional means and totals, but it yields inconsistent estimates of other parameters, even under the MCAR assumption. In particular, variances from the filled-in data are clearly understated by imputing means, and associations between variables are distorted. Thus, the method yields an inconsistent estimate of the covariance matrix (Haitovsky 1968). Obvious corrections for bias lead to an AC method (Little and Rubin 1987, chap. 3). Biases in the resulting estimated slopes can be compensated for by reductions in variance relative to CC if the fraction of complete cases is small (Afifi and Elashoff 1967). Inferences (tests and confidence intervals) based on the filled-in data are seriously distorted by bias and overstated precision, however. Thus unconditional mean imputation cannot be generally recommended. Better imputation methods are described in Section 5 below.

1.4 More Principled Approaches

We now turn to more principled approaches to the analysis of incomplete data. Section 2 discusses weighting adjustments for unit nonresponse. Section 3 reviews ML methods that assume an ignorable missing-data mechanism, and Section 4 concerns ML methods under a nonignorable missing-data mechanism. Section 5 concerns multiple-imputation methods. Finally, in Section 6, other Bayesian simulation methods for missing-data models are considered.

2 Weighting Adjustments for Unit Nonresponse

We noted in Section 1.3 that a potentially serious problem with CC analysis is bias arising from the fact the complete cases are not a random subsample of all cases. A useful modification of CC analysis is to assign a nonresponse weight to the respondents to remove or reduce nonresponse bias. In probability sampling, a sampling weight inversely proportional to the probability of selection is often used to adjust for differential selection probabilities. If nonresponse is viewed as another stage of selection, then it is natural to multiply the sampling weight by a nonresponse weight that is the inverse of the probability of response given selection. Whereas sampling weights are determined by the sample design and hence known, nonresponse weights are based on unknown nonresponse probabilities, which need to be estimated from the data.

A standard approach to weighting for nonresponse is to form adjustment cells based on background variables measured for respondents and nonrespondents; often these are based on geographical areas or groupings of similar areas based on aggregate socioeconomic data. The nonresponse weight for individuals in an adjustment cell is then the inverse of the response rate in that cell. For example, in the CPS, a noninterview weighting adjustment based on cross-classification by race and urbanicity is used (Hanson 1978).

This weighting method removes the component of nonresponse bias attributable to differential nonresponse rates across the adjustment cells, and eliminates bias if respondents can be regarded as a random subsample of the original sample within the adjustment cells. To increase the plausibility of the latter assumption, it is important to attempt to record background characteristics of nonrespondents and respondents that are predictive of nonresponse, and then use these variables to form adjustment cells. Characteristics should also be predictive of survey outcomes, to control variance.

If more than one background variable is measured, adjustment cells can be based on a joint classification. With more than say three or four background variables this approach may yield too many adjustment cells, and nonresponse weights that are unstable. The cells may then need to be collapsed or otherwise coarsened. One approach to this problem is *response propensity stratification*, where (a) the indicator for unit nonresponse is regressed on the background variables, using the combined data for respondents and nonrespondents and a method such as logistic regression appropriate for a binary outcome; (b) predicted response probabilities $\hat{P}$ are computed for respondents based on the regression in (a); and (c) adjustment cells are formed based on a categorized version of $\hat{P}$. Theory (Little 1986; Rosenbaum and Rubin 1983) suggests that this is an effective method for removing nonresponse bias attributable to the background variables. For applications to real surveys, see Czajka, Hirabayashi, Little, and Rubin (1992); Göksel, Judkins, and Mosher (1991) and Woodburn (1991).

A related adjustment method is post-stratification, which weights respondents to match the distribution of variables available from external census or survey data. For example, if a survey is based on a random sample of census blocks, and the distribution of age is available for the population in these blocks from census data, then survey respondents can be classified by age and then weighted so that the weighted distribution of age matches the census age distribution. Specifically, the weight $w_h = rP_h/r_h$ is computed for each sample case in age class h, where r_h is the number of respondents in age class h, P_h is the population proportion from a census, and r is the respondent sample size; weights are scaled so that they sum to the respondent sample size. Note that the post-stratifying variable (here age) does not need to be known for survey nonrespondents, as in the previous weighting method.

Post-stratification can improve the accuracy of survey estimates, both by reducing bias and increasing precision. In the above example, the mean age of the population might be estimated by the sample mean, given an equal probability sample. The mean weighted by the $\{w_h\}$ is much more precise, since it essentially reproduces the population mean aside from effects of grouping. Furthermore, if the unweighted mean is biased by differential nonresponse by age, the post-stratified mean corrects for this bias. These properties are of academic interest given the availability of data on age from the census, but they also apply in diluted form to other survey variables that are correlated with age. Post-stratification is very common in practice, playing an important role in many government surveys (see, for example, Hanson 1978; Harte 1982; Waterton and Lievesley 1987). For a recent discussion of post-stratification and extensions to two or more post-stratifiers, see Little (1993a).

Weighting methods can be useful for removing or reducing the bias in CC analysis. They can be applied both to continuous and categorical variables. Indeed, unlike imputation or the likelihood-based methods of the next section, weighting methods do not model the joint distribution of the data at all; models, when used, are confined to the conditional distribution of the nonresponse indicator variable given the observed covariates. As a result, weighting methods are often relatively easy to implement compared with modeling or imputation. However, weighting methods do have some serious limitations. First, weighted estimates can have unacceptably high variance, as when outlying values of a variable are given large weights. Estimators that employ sampling weights can have similar problems, as pointed out in Basu's (1971) famous circus example. The concern may be more realistic in nonresponse settings since selection by nonresponse is not under the control of the investigator. Second, variance estimation for weighted estimates is problematic. Explicit formulas are available for simple estimators such as means under simple random sampling (for example, Oh and Scheuren 1983), but methods are not well developed for more complex problems, and often ignore the component of variability from estimating the weight from the data.

3 Maximum Likelihood Assuming Ignorable Nonresponse

3.1 Maximum-Likelihood Theory

We now turn to methods that are based on explicit models for the incomplete data. The method of maximum likelihood for incomplete data, in full generality, requires specification of a model for the distribution of $\boldsymbol{X}$ and $\boldsymbol{M}$. *Selection models* specify this distribution as follows:

$$p(\boldsymbol{X}, \boldsymbol{M}|\boldsymbol{\theta}, \boldsymbol{\psi}) = p(\boldsymbol{X}|\boldsymbol{\theta})p(\boldsymbol{M}|\boldsymbol{X}, \boldsymbol{\psi}), \tag{2.3}$$

where $p(\boldsymbol{X}|\boldsymbol{\theta})$ represents a model for the distribution of the data matrix $\boldsymbol{X}$ in the absence of missing values, $p(\boldsymbol{M}|\boldsymbol{X}, \boldsymbol{\psi})$ represents a model for the missing-data mechanism, and $(\boldsymbol{\theta}, \boldsymbol{\psi})$ are unknown parameters. Usually interest is in the parameters $\boldsymbol{\theta}$, with $\boldsymbol{\psi}$ being "nuisance" parameters. For example, the rows of $\boldsymbol{X}$ may be assumed to have independent multivariate normal distributions with mean $\boldsymbol{\mu}$, covariance matrix $\boldsymbol{\Sigma}$, and $\boldsymbol{\theta} = (\boldsymbol{\mu}, \boldsymbol{\Sigma})$.

If $\boldsymbol{X}$ were completely known, the likelihood of $\boldsymbol{\theta}$ and $\boldsymbol{\psi}$ would be (2.3), treated as a function of $\boldsymbol{\theta}$ and $\boldsymbol{\psi}$ with $\boldsymbol{X}$ and $\boldsymbol{M}$ fixed at their observed values. Maximum-likelihood estimates of $\boldsymbol{\theta}$ and $\boldsymbol{\psi}$ are the values of $\boldsymbol{\theta}$ and $\boldsymbol{\psi}$ that maximize this function. Note that provided $\boldsymbol{\theta}$ and $\boldsymbol{\psi}$ are not functionally related, only the term $p(\boldsymbol{X}|\boldsymbol{\theta})$ contributes to ML estimation of $\boldsymbol{\theta}$, that is, the model for the missing-data mechanism is not relevant for inference about $\boldsymbol{\theta}$.

When values in the matrix $\boldsymbol{X}$ are missing, the likelihood of $\boldsymbol{\theta}$ and $\boldsymbol{\psi}$ is by definition proportional to the marginal density of the observed entries of $\boldsymbol{X}$ (say $\boldsymbol{X}_{obs}$) and $\boldsymbol{M}$, treated as a function of $\boldsymbol{\theta}$ and $\boldsymbol{\psi}$ with $\boldsymbol{X}_{obs}$ and $\boldsymbol{M}$ fixed. By standard distribution theory, this is obtained formally by integrating the missing entries of $\boldsymbol{X}$ (say $\boldsymbol{X}_{mis}$) out of the joint density of $\boldsymbol{X}$ and $\boldsymbol{M}$, that is:

$$L(\boldsymbol{\theta}, \boldsymbol{\psi}|\boldsymbol{X}_{obs}, \boldsymbol{M}) = \int p(\boldsymbol{X}_{obs}, \boldsymbol{X}_{mis}|\boldsymbol{\theta})p(\boldsymbol{M}|\boldsymbol{X}_{obs}, \boldsymbol{X}_{mis}, \boldsymbol{\psi})d\boldsymbol{X}_{mis}. \tag{2.4}$$

Maximum-likelihood estimates of $\boldsymbol{\theta}$ and $\boldsymbol{\psi}$ are obtained by maximizing this function with respect to $\boldsymbol{\theta}$ and $\boldsymbol{\psi}$.

Maximum-likelihood estimation for $\boldsymbol{\theta}$ is often considered without explicitly including a model for the missing-data mechanism. The likelihood *ignoring the missing-data mechanism* (Rubin 1976a) is the likelihood of $\boldsymbol{\theta}$ based on the marginal density of $\boldsymbol{X}_{obs}$, ignoring the contribution to (2.4) from the model for $\boldsymbol{M}$:

$$L(\boldsymbol{\theta}|\boldsymbol{X}_{obs}) = \int p(\boldsymbol{X}_{obs}, \boldsymbol{X}_{mis}|\boldsymbol{\theta})d\boldsymbol{X}_{mis}. \tag{2.5}$$

Maximum-likelihood ignoring the missing-data mechanism means maximizing (2.5) rather than the full likelihood (2.4). The missing-data mechanism is called *ignorable* if inference about θ based on the likelihood (2.5) is equivalent to inference about θ based on the likelihood (2.4).

Rubin (1976a) showed that the missing-data mechanism is ignorable if (a) θ and ψ are distinct parameters, that is, are not functionally related; and (b) the missing data are MAR, that is, (2.2) is satisfied. The first of these conditions is often reasonable in practical settings. Condition (b) is the important one; a key feature is that MAR is weaker than the MCAR condition. That is, ignorable ML methods allow missingness to depend on the data matrix X, provided it does so only through *observed* values X_{obs}. This is an important feature of ML estimation, since other methods (such as those considered in Section 1) often require the more stringent MCAR condition. Even in cases where MAR is violated and hence maximizing (2.5) is not strictly ML, ignorable ML can improve on naive alternatives since MAR is a weaker assumption than MCAR. For examples, see the simulations in Muthén et al. (1987) and Little (1988c).

Ignorable ML methods are desirable since (a) specifying an appropriate model for the missing data mechanism is often a difficult task; (b) even when the mechanism is nonignorable, estimates under the ignorable model can be superior to estimates from a misspecified nonignorable model; and (c) even if the missing-data model is correctly specified, information for estimating θ and ψ jointly may be very limited, and rest strongly on the untestable assumptions made about the distribution of X. We shall return to these issues in Section 4.

Once ML estimates of parameters have been obtained, standard errors, hypothesis tests for particular values, and confidence intervals can be derived by applying standard ML methods; they have appropriate statistical properties provided sample sizes are large enough for the log-likelihood function to be quadratic in the neighborhood of the ML estimate. In particular, these quantities take into account the fact that values are missing, unlike naive analyses of imputed data sets by complete-data methods. For example, the covariance matrix of the ML estimates $(\hat{\theta}, \hat{\psi})$ of (θ, ψ) can be estimated as the inverse of the information matrix, namely the negative second derivative of the logarithm of the likelihood $L(\theta, \psi | X_{obs}, M)$ in (2.4) with respect to (θ, ψ) evaluated at $(\hat{\theta}, \hat{\psi})$. If the missing-data mechanism is ignorable, the covariance matrix of the ML estimate $\hat{\theta}$ of θ can be estimated more simply as the inverse of the second derivative of the logarithm of $L(\theta | X_{obs})$ in (2.5) with respect to θ, evaluated at $\hat{\theta}$. Given large samples, difficulties with these procedures are mainly computational, particularly when the likelihoods are complicated and the parameters are high-dimensional. Other approaches are to approximate the information matrix, for example by numerical approximation of the derivatives, or to use the bootstrap to compute standard errors (Efron 1979).

Maximum-likelihood is essentially a large-sample tool. In small samples the usual normal-theory tests and confidence intervals based on the ML estimates and their covariance matrix may have unsatisfactory statistical properties, and alternatives to ML inference (such as those considered in Sections 5 and 6) may be worthwhile.

3.2 The Expectation-Maximization Algorithm

In many incomplete-data problems, the likelihood of (2.4) or (2.5) is a complicated function, and explicit expressions for the ML estimates do not exist. Standard numerical ML algorithms, such as the Newton-Raphson algorithm, can be applied, but other algorithms exploit the missing-data aspect of the problem and may have advantages. The best known of these algorithms is the expectation-maximization (EM) algorithm (Dempster, Laird, and Rubin 1977). Applications of the algorithm date back at least to McKendrick (1926), and Orchard and Woodbury (1972) first noted the general applicability of the underlying idea, calling it the "missing information principle." We describe the EM algorithm for maximizing the likelihood (2.5) ignoring the missing-data mechanism; the extension to nonignorable likelihoods is straightforward (Little and Rubin 1987, chap. 11). Let $\ell(\boldsymbol{\theta}|\boldsymbol{X}_{obs}, \boldsymbol{X}_{mis})$ denote the log-likelihood of $\boldsymbol{\theta}$ based on the hypothetical complete data $\boldsymbol{X} = (\boldsymbol{X}_{obs}, \boldsymbol{X}_{mis})$. Let $\boldsymbol{\theta}^{(t)}$ denote an estimate of $\boldsymbol{\theta}$ at iteration t of the algorithm. Iteration $t + 1$ consists of an E step and an M step. The E step consists of taking the expectation of $\ell(\boldsymbol{\theta}|\boldsymbol{X}_{obs}, \boldsymbol{X}_{mis})$ over the conditional distribution of $\boldsymbol{X}_{mis}$ given $\boldsymbol{X}_{obs}$, evaluated at $\boldsymbol{\theta} = \boldsymbol{\theta}^{(t)}$. That is, the expected log-likelihood

$$Q(\boldsymbol{\theta}|\boldsymbol{\theta}^{(t)}) = \int \ell(\boldsymbol{\theta}|\boldsymbol{X}_{obs}, \boldsymbol{X}_{mis}) p(\boldsymbol{X}_{mis}|\boldsymbol{X}_{obs}, \boldsymbol{\theta} = \boldsymbol{\theta}^{(t)}) d\boldsymbol{X}_{mis}$$

is formed. In practice, the E step often corresponds to a form of imputation of the missing data, thus providing a link between ML and imputation methods.

The M step determines $\boldsymbol{\theta}^{(t+1)}$ by maximizing this expected log-likelihood:

$$Q(\boldsymbol{\theta}^{(t+1)}|\boldsymbol{\theta}^{(t)}) \geq Q(\boldsymbol{\theta}|\boldsymbol{\theta}^{(t)}) \text{ for all } \boldsymbol{\theta}.$$

The new estimate $\boldsymbol{\theta}^{(t+1)}$ then replaces $\boldsymbol{\theta}^{(t)}$ in the next iteration. It can easily be shown that each step of EM increases the likelihood of $\boldsymbol{\theta}$ given $\boldsymbol{X}_{obs}$. Also, under quite general conditions, EM converges to the maximum of this function. In particular, if a unique finite ML estimate of $\boldsymbol{\theta}$ exists, EM will find it.

The EM algorithm is particularly useful when the M step is noniterative or available using existing software. A variant of EM when the M step is noniterative is the ECM algorithm of Meng and Rubin (1993). Note that the EM algorithm does not involve computing and inverting an information matrix at each iteration. This feature can be useful in problems with many parameters, since the information matrix is square with dimension equal to the number of parameters. Standard errors based on the inverted information matrix, however, are not an output of EM and hence, if needed, require a separate computation. An extension of EM, the SEM algorithm (Meng and Rubin 1991), uses successive iterates of EM to construct the information matrix. Although EM is reliable in that it increases the likelihood at each iteration, it can be painfully slow to converge in problems where the fraction of missing information (defined in terms of eigenvalues of the information matrix) is large.

Starting values $\boldsymbol{\theta}^{(0)}$ for EM may be based on a naive method, such as CC, AC, or unconditional mean imputation. It should be noted that the likelihood for incomplete-data problems can have multiple maxima, particularly when the fraction of missing information is large. Thus it is a useful precaution to run EM with more than one starting value, to check that it converges to the same final estimate.

3.3 Some Important Ignorable Maximum-Likelihood Methods

We now discuss some common ignorable models, which we view as providing baseline methods for exploiting the information in the incomplete cases.

Multivariate Normal Model

The multivariate normal model underlies many multivariate statistical methods, so ML for multivariate normal data with missing values is an important method. Let $\boldsymbol{x}_i$ denote the data for the ith case (assuming there were no missing values), and assume the multivariate normal model

$$\boldsymbol{x}_i \sim_{ind} N_V(\boldsymbol{\mu}, \boldsymbol{\Sigma}),\ i = 1, ..., N, \tag{2.6}$$

where $N_V(\boldsymbol{\mu}, \boldsymbol{\Sigma})$ denotes the multivariate normal distribution with $(V \times 1)$ mean vector $\boldsymbol{\mu}$ and $(V \times V)$ covariance matrix $\boldsymbol{\Sigma}$. Let $x_{obs,i}$ denote the subvector of observed variables for case i, and let $\boldsymbol{\mu}_i$ and $\boldsymbol{\Sigma}_i$ denote the corresponding subvector of means and submatrix of variances and covariances. If the missing-data mechanism is assumed ignorable, the log-likelihood of $\boldsymbol{\theta} = (\boldsymbol{\mu}, \boldsymbol{\Sigma})$ is

$$\ell(\boldsymbol{\mu}, \boldsymbol{\Sigma}|\boldsymbol{X}_{obs}) = -0.5\sum_{i=1}^{N}\{\log|\boldsymbol{\Sigma}_i| + (\boldsymbol{x}_{obs,i} - \boldsymbol{\mu}_i)'\boldsymbol{\Sigma}_i^{-1}(\boldsymbol{x}_{obs,i} - \boldsymbol{\mu}_i)\}. \tag{2.7}$$

For monotone missing-data patterns, explicit estimates of $(\boldsymbol{\mu}, \boldsymbol{\Sigma})$ that maximize (2.7) can be derived using Anderson's (1957) method of *factored likelihoods*. The parameters $(\boldsymbol{\mu}, \boldsymbol{\Sigma})$ are transformed in such a way that the likelihood factorizes into distinct factors corresponding to complete-data problems; for details, see Little and Rubin (1987, chap. 6). Marini et al. (1980) applied this method to the data described in example 2, after deleting a small number of values to create a monotone pattern.

We present here ML estimates for a simple monotone pattern, namely, univariate nonresponse, where missing values are confined to X_V. The likelihood is re-expressed as the product of two complete-data likelihoods, one for the parameters of the marginal distribution of $X_1, ..., X_{V-1}$ based on all N cases, and one for the parameters of the conditional distribution of X_V given $X_1, ..., X_{V-1}$ based on the complete cases. Under the normal model, the sets of parameters in these two factors are distinct, so the factors can be maximized separately, yielding a combination of available-case analysis for the parameters of the distribution of $X_1, ..., X_{V-1}$ and complete-case analysis for the parameters of the regression of X_V on $X_1, ..., X_{V-1}$. The former analysis yields ML estimates $\{\hat{\mu}_j, \hat{\sigma}_{jk};\ 1 \le j, k \le V-1\}$ that are the sample means and covariance matrix of $X_1, ..., X_{V-1}$ from all N cases. Maximum-likelihood estimates of the remaining parameters can be written in the following instructive form:

$$\hat{\mu}_V = N^{-1}(\sum_{i=1}^{n_0} x_{iV} + \sum_{i=n_0+1}^{N} \hat{x}_{iV}), \tag{2.8}$$

$$\hat{\sigma}_{jV} = N^{-1}(\sum_{i=1}^{n_0} x_{ij}x_{iV} + \sum_{i=n_0+1}^{N} x_{ij}\hat{x}_{iV}) - \hat{\mu}_j\hat{\mu}_V,\ 1 \le j \le V-1, \tag{2.9}$$

$$\hat{\sigma}_{VV} = N^{-1}(\sum_{i=1}^{n_0} x_{iV}^2 + \sum_{i=n_0+1}^{N} \hat{x}_{iV}^2) - \hat{\mu}_V^2 + (N - n_0)s^2/N, \quad (2.10)$$

where (i) $\hat{x}_{iV}$ is the predicted value of x_{iV} from the regression of X_V on $X_1, ..., X_{V-1}$, computed by ordinary least squares on the complete cases, and (ii) s^2 is the residual variance from that regression. The statistic $\hat{\mu}_V$ is called the *regression estimate* of μ_V.

If the term involving s^2 in (2.10) is omitted, (2.8)-(2.10) are equivalent to *regression imputation*, where missing values $\{x_{iV} : i = n_0 + 1, ..., N\}$ are replaced by $\hat{x}_{iV}$ and the mean and covariance matrix are estimated from the filled-in data. The last term in (2.10) corrects for underestimation of the variance of X_V that results from imputing predictions $\hat{x}_{iV}$ for the missing values: note that these regression imputations have an interpretation as (estimated) conditional means, and hence tend to be less variable than the observed values of X_V. Thus ML for the normal model is effectively regression imputation, with a correction for bias in the variance of X_V. Another resolution of the bias problem is given by the multiple-imputation method of Section 5, where multiple *draws* from the conditional distribution of X_V given $X_1, ..., X_{V-1}$ are imputed rather than conditional *means* $\hat{x}_{iV}$ as here.

For non-monotone patterns of missing data, maximization of (2.7) requires an iterative algorithm. The EM algorithm for this problem can be viewed as an iterative form of regression imputation. In fact the E step effectively imputes the missing values in each case by the predictions from the regression of the missing variables on the observed variables, with coefficients based on current estimates of the parameters. The M step estimates the mean and covariance matrix from the filled-in data, with corrections for the covariance matrix for imputing predicted means. Specific formulas were first presented in Orchard and Woodbury (1972). An accessible reference, with a discussion of standard errors, is Little and Rubin (1987, sect. 8.2). This algorithm is available as the BMDPAM program in the BMDP statistical package (Dixon 1988), and in a GAUSS module (Schoenberg 1988).

So far our discussion has focussed on estimation of the mean and covariance matrix of the variables, but other functions of these parameters may be of interest. If $\hat{\boldsymbol{\theta}}$ is the ML estimate of $\boldsymbol{\theta}$, then the ML estimate of a function $g(\boldsymbol{\theta})$ is simply $g(\hat{\boldsymbol{\theta}})$, that is, the function evaluated at $\hat{\boldsymbol{\theta}}$.

In particular, suppose interest centers on the regression of X_V on, say, X_1, X_2, and X_4. The parameters of this regression are well-known functions of $\boldsymbol{\mu}$ and $\boldsymbol{\Sigma}$, the mean and covariance matrix of $X_1, ..., X_V$. Hence ML estimates of the regression parameters are obtained by evaluating these functions with ML estimates $\hat{\boldsymbol{\mu}}$ and $\hat{\boldsymbol{\Sigma}}$ from the above algorithm substituted for $\boldsymbol{\mu}$ and $\boldsymbol{\Sigma}$. For a recent review on missing data in regression, see Little (1992).

Multinomial Models for Discrete Data

If variables are discrete, incomplete data can be rearranged as a V-way contingency table with supplemental margins. For example, suppose that $V = 2$, and that X_1 takes I values and X_2 takes J values. Then fully observed cases can be classified into the $(I \times J)$ contingency table defined by joint levels of X_1 and X_2. Cases with X_1 observed and X_2 missing form a supplemental X_1 margin with I entries, and cases with X_2 observed and X_1 missing form a supplemental X_2 margin with J entries.

Complete-data analysis of contingency tables is often based on a multinomial model for the cell counts. That is, if the contingency table has C cells with count n_c in cell c, $\boldsymbol{n} = \{n_1, ..., n_C\}'$ is assumed to be multinomially distributed with index $n_+ = \sum_c n_c$ and probabilities $\boldsymbol{\pi} = \{\pi_1, ..., \pi_C\}'$, where $\sum_c \pi_c = 1$. The complete-data likelihood is

$$L(\boldsymbol{\pi}|\boldsymbol{n}) = \prod_{c=1}^{C} \pi_c{}^{n_c}, \tag{2.11}$$

and the ML estimate of π_c is simply n_c/n_+, the proportion of cases in cell c.

With supplemental margins representing partially observed data, ML estimation is readily accomplished via EM. Since the logarithm of the complete-data likelihood (2.11) is linear in the cell counts $\{n_c\}$, the E step simply estimates these complete-data counts given the data and current estimates of the cell probabilities. The effect is to classify counts in the supplemental margins into the full table proportionately according to current estimates of the cell probabilities. The M step performs complete-data ML estimation on the filled-in contingency table; that is, the estimated probability in a cell is the proportion of observed and fractionally allocated supplemental counts in that cell (Chen and Fienberg 1974).

As with models for continuous data, it is often useful to modify the model by placing special structure on the cell probabilities. In particular, log-linear models decompose the logarithm of the cell probabilities as the sum of a constant, main effects and higher order terms, and then set some of the terms in the decomposition to zero (see, for example, Bishop, Fienberg, and Holland 1975). The EM algorithm can also be used to estimate log-linear models from data with supplemental margins (Dempster et al. 1977; Fuchs 1982). The E step allocates the partially classified cases using probabilities calculated from the current parameter estimates, as before. The M step computes new parameter estimates by fitting the log-linear model to the filled-in data. This step is itself iterative for log-linear models that do not have explicit ML estimates from complete data. This may not be of major concern, however, since statistical software for fitting log-linear models to complete data is readily available; moreover, maximization in the M step can be replaced by a single cycle of iterative proportional fitting to yield an ECM algorithm that converges reliably (Meng and Rubin 1993). For more details on ML for partially classified contingency tables, see Little and Rubin (1987, chap. 9).

General Location Model

Little and Schluchter (1985) discussed models for mixed continuous and categorical variables that provide relatively simple and computationally feasible EM algorithms with missing data. Write $\boldsymbol{x}_i = (\boldsymbol{y}_i', \boldsymbol{z}_i')'$, where $\boldsymbol{y}_i$ represent continuous variables and $\boldsymbol{z}_i$ represent categorical variables. The general location model (Olkin and Tate 1961) assumes a multinomial distribution for $\boldsymbol{z}_i$, and a multivariate normal distribution for $\boldsymbol{y}_i$ given $\boldsymbol{z}_i$, with a distinct mean $\boldsymbol{\mu}_c$ for each cell c defined by the distinct values of $\boldsymbol{z}_i$, and common within-cell covariance matrix $\boldsymbol{\Sigma}$. (The latter is a special case of the model that underlies discriminant analysis.) This model can be elaborated by restricting the multinomial probabilities of the distribution of z_i by a log-linear model, and by restricting the means μ_c by a multivariate analysis of variance model for y_i given z_i. The EM algorithm for this model when components of by_i and/or bz_i are missing generalizes the EM algorithms for normal and multinomial data described above. Transformations of the ML estimates of parameters of the joint distribution of y_i and z_i yield ML estimates for other important problems, including logistic regression with missing values, and normal linear regression with missing continuous and/or categorical predictors. For details, see Little and Schluchter (1985), or Little and Rubin (1987, sect. 10.2).

Normal Models for Repeated Measures

When the vector x_i consists of repeated measurements of the same variable, incomplete data arise commonly through attrition or missed visits. Extensions of the multivariate normal model have been developed that (a) expand the mean structure μ to provide for within-subject and between-subject covariates, and (b) place constraints on the covariance matrix Σ. A broad class of models, proposed by Laird and Ware (1982), has the form

$$x_i = C_i\beta + Z_i\theta_i + e_i, \tag{2.12}$$

where for each subject i, C_i and Z_i are known design matrices, with dimensions $(V \times p)$ and $(V \times q)$ respectively, β is an unknown $(p \times 1)$ vector of fixed effects, θ_i is an unknown $(q \times 1)$ vector of random effects, and e_i is an unknown $(V \times 1)$ vector of random errors. It is further assumed that independently

$$\theta_i \sim_{ind} N_q(\mathbf{0}, D) \text{ and } e_i \sim_{ind} N_V(\mathbf{0}, \Sigma(\phi)), \tag{2.13}$$

where D is the unknown $(q \times q)$ covariance matrix of the random effects, and Σ is the $(V \times V)$ covariance matrix of the errors, assumed to be a known function of a set of unknown variance parameters ϕ. Averaging over the unknown random effects θ_i shows that this model is equivalent to

$$x_i \sim_{ind} N_V(C_i\beta, \Omega_i(D, \phi)), \tag{2.14}$$

where $\Omega_i(D, \phi) = Z_i D Z_i' + \Sigma(\phi)$, with unknown parameters β, ϕ, and D. The multivariate normal model (2.6) is a special case of this model, where C_i is the identity matrix, $\mu = \beta$, $Z_i = \mathbf{0}$, and $\Sigma(\beta) = \Sigma$, an unstructured covariance matrix. Maximum-likelihood estimation for the model given by (2.12)–(2.14) when some values in x_i are missing can be accomplished using modifications of EM or Newton-type algorithms (Jennrich and Schluchter 1986; Laird 1988; Laird and Ware 1982; Lindstrom and Bates 1988; Schluchter 1988).

For practitioners, the main issues are in specifying the mean structure, as reflected in the choice of C_i, and the covariance structure, as reflected in the choice of D and $\Sigma(\phi)$. Software for fitting models of this form is becoming increasingly available. In particular, the BMDP progam 5V (Dixon 1988) and SAS Proc Mixed (SAS 1992) provide for a flexible choice of mean structure, specified in terms of between-subject and within-subject covariates, and various forms for the covariance matrix Ω_i.

Non-Normal Models for Repeated Measures

A body of recent research has focused on methods of repeated-measures analysis for data with non-normal errors. Robust ML estimation based on the multivariate t distribution is discussed in Lange, Little, and Taylor (1989). Another important case concerns repeatedly measured categorical (and in particular binary) outcomes. With independent measurements, a natural approach to this problem is to apply ML for the generalized linear model (GLIM), an extension of regression to non-normal error structures, such as Poisson errors for categorical outcomes (McCullagh and Nelder 1989). With repeated-measures data, the ML approach requires an extension of GLIM models that specifies the joint distribution of outcomes over the repeated measures. This turns out to be problematic; in particular, the natural extension for categorical outcomes is to log-linear models, but the parameters of log-linear models do not have a convenient interpretation, particularly in the presence of missing data (Liang, Zeger, and Qaqish 1992). This situation has led to the development of alternatives to ML that avoid specification of the joint distribution of the outcomes. Two possible approaches are as follows:

A. Fit a GLIM to the pooled data, ignoring the correlation structure, and then estimate standard errors from jackknife or bootstrap samples, where the entire cases are resampled (Lipsitz, Laird, and Harrington 1990; Moulton and Zeger 1989). This approach exploits the fact that the analysis ignoring correlations can yield consistent estimates even when correlation is present.

B. Specify the mean and covariance matrix of the repeated measures, and then apply the generalized estimating equations (GEE) approach to estimation (Liang et al. 1992; Prentice and Zhao 1991; Zeger and Liang 1986). This approach yields consistent estimates of location parameters even when the covariance matrix is misspecified, and robust "sandwich-type" estimators of standard errors are also available. When these techniques are refined and made more widely available, the practitioner will have a fairly comprehensive set of tools for analyzing repeated-measures data with missing values.

The methods considered here are asymptotic, and hence generally require large samples. Recent developments for small data sets have concerned Bayesian methods for normal models. Bayesian methods have been made more practical by the availability of new computational tools such as Laplace approximations (Kass and Steffey 1989) and the Gibbs sampler outlined in Section 6 (Gelfand, Hills, Racine-Poon, and Smith 1990).

4 Nonignorable Nonresponse Models

4.1 Introduction

We now consider ML estimation for nonignorable nonresponse models, where terms representing the missing-data mechanism need to be included in the likelihood. Two ways of specifying the joint distribution of X and M can be contrasted (Little and Rubin 1987, chap. 10). Selection models, as introduced in Section 3.1, specify

$$p(\boldsymbol{X}, \boldsymbol{M}|\boldsymbol{\theta}, \boldsymbol{\psi}) = p(\boldsymbol{X}|\boldsymbol{\theta})p(\boldsymbol{M}|\boldsymbol{X}, \boldsymbol{\psi}), \tag{2.15}$$

where $p(\boldsymbol{X}|\boldsymbol{\theta})$ represents the complete-data model for $\boldsymbol{X}$, $p(\boldsymbol{M}|\boldsymbol{X}, \boldsymbol{\psi})$ represents the model for the missing-data mechanism, and $(\boldsymbol{\theta}, \boldsymbol{\psi})$ are unknown parameters. On the other hand, *pattern-mixture models* (Glynn, Laird, and Rubin 1986; Little 1993) specify

$$p(\boldsymbol{X}, \boldsymbol{M}|\boldsymbol{\phi}, \boldsymbol{\pi}) = p(\boldsymbol{X}|\boldsymbol{M}, \boldsymbol{\phi})p(\boldsymbol{M}|\boldsymbol{\pi}), \tag{2.16}$$

where the distribution of $\boldsymbol{X}$ is conditioned on the missing data pattern $\boldsymbol{M}$. The terminology "pattern-mixture" reflects the fact that the resulting marginal distribution of $\boldsymbol{X}$ is a mixture of distributions.

Equations (2.15) and (2.16) are simply two different ways of factoring the joint distribution of $\boldsymbol{X}$ and $\boldsymbol{M}$. When $\boldsymbol{M}$ is independent of $\boldsymbol{X}$, the missing data are MCAR, and the two specifications are equivalent as long as $\boldsymbol{\theta} = \boldsymbol{\phi}$ and $\boldsymbol{\psi} = \boldsymbol{\pi}$. When the missing data are not MCAR and distributional assumptions are added, (2.15) and (2.16) can yield different models.

For nonignorable models for contingency tables, see Little (1985a), Fay (1986), Little and Rubin (1987, sect. 11.6), Baker and Laird (1988), or Rubin, Schafer, and Schenker (1988). We confine attention here to normal models for continuous outcomes. In particular, the next section outlines a nonignorable selection model commonly used in econometrics, and Section 4.3 discusses an analogous normal pattern-mixture model.

4.2 Probit Selection Model

Consider the problem of univariate nonresponse defined in Section 1.2. There are two missing-data patterns, complete cases ($m_i = 0$) and cases with x_{iV} missing ($m_i = 1$). A probit selection model for these data assumes that:

a) $(x_{i1}, x_{i2}, ..., x_{iV})$ is normal with mean $\boldsymbol{\mu}$ and covariance matrix $\boldsymbol{\Sigma}$, and

b) m_i given $(x_{i1}, x_{i2}, ..., x_{iV})$ is Bernoulli with probability

$$p(m_i = 1|x_{i1}, x_{i2}, ..., x_{iV}, \boldsymbol{\psi}) = \Phi(\psi_0 + \sum_{j=1}^{V} \psi_j x_{ij}) \tag{2.17}$$

where Φ denotes the standard normal cumulative distribution (that is, the probit) function. If interest centers on the regression of X_V on $X_1, ..., X_{V-1}$, then *a*) can be replaced by

a') x_{iV} given $(x_{i1}, ..., x_{i,V-1})$ is normal with mean $\beta_0 + \sum_{j=1}^{V-1} \beta_j x_{ij}$ and variance σ^2.

This specification is less restrictive than *a*) since it does not make any distributional assumptions about the regressors. The probit selection model considered by Heckman (1976) is defined by *a'*) and *b*). The probit form (2.17) of selection can be interpreted as arising when an unobserved variable Z with a joint normal distribution with X_V crosses a threshold; for this reason the model is sometimes called a *stochastic censoring* model. Heckman's application concerned female labor force participation, with X_V representing wages and salary, the selection (or missing-data) mechanism arising since women not in the labor force did not have a value of X_V, and Z being interpreted as a theoretical construct ("reservation wage") that represented the net value to the woman of participating in the labor force rather than staying at home. Lillard, Smith, and Welch (1986) applied this model to the problem discussed in example 3, that is, missing income data in the CPS, with an additional parameter for a power transformation of X_V to normality. A key issue in the selection model (2.17) is whether $\psi_V = 0$, since in that case missingness of X_V depends only on variables that are fully observed, so the missing data are MAR; the probit selection model then reduces to a special case of the ignorable multivariate normal model discussed in Section 3.3.

Two approaches to estimating the parameters of this model can be distinguished, least squares and ML. Maximum-likelihood requires an iterative algorithm such as EM (e.g., Little and Rubin 1987, sect. 11.4). Least squares, the method proposed in Heckman (1976), is popular since it does not require specialized software to implement. It can be shown that the model implies that for respondents ($m_i = 0$):

$$E(x_{iV}|x_{i1}, ..., x_{i,V-1}, m_i = 0) = \beta_0 + \sum_{j=1}^{V-1} \beta_j x_{ij} + \delta\varphi(q_i)/(1 - \Phi(q_i)), \tag{2.18}$$

where φ is the standard normal density, and $\Phi(q_i) \equiv \psi_0^* + \sum_{j=1}^{V-1} \psi_j^* x_{ij} = \text{pr}(m_i = 0|x_{i1}, \ldots, x_{i,V-1})$, the probit of the probability of response given $x_{i1}, \ldots, x_{i,V-1}$. Note that q_i differs from the right side of (2.17), since the latter also conditions on the value of x_{iV}. Hence, consistent estimates of the coefficients are obtained by a) estimating the probit regression of the response indicator m_i on the regressors $x_{i1}, \ldots, x_{i,V-1}$; b) estimating q_i by $\hat{q}_i$, the linear predictor from the probit regression in a); and c) regressing x_{iV} on $x_{i1}, \ldots, x_{i,V-1}$ and $\varphi(\hat{q}_i)/(1 - \Phi(\hat{q}_i))$, by least squares.

Although the probit selection model is ingenious, its value in practice is controversial. The basic issue is whether the data really provide enough information to allow all of the parameters to be simultaneously estimated in a satisfactory way. For example, consider the least-squares estimation procedure based on the regression model (2.18). If $\varphi(\hat{q}_i)/(1 - \Phi(\hat{q}_i))$ were in fact linear in $\hat{q}_i$, then the coefficient δ, which represents the adjustment for nonignorable nonresponse, would be inestimable since it multiplies a covariate which is a linear combination of variables $(x_{i1}, ..., x_{i,V-1})$ already included in the regression. It is only the fact that $\varphi(\hat{q}_i)/(1 - \Phi(\hat{q}_i))$ is a nonlinear function that allows δ to be estimated. Hence the estimation method can be very unstable and yield contradictory results, such as negative predictions for outcomes known to be positive. Even when stability is not a problem, the method relies crucially on the correctness of both the linear model specification a'), and the validity of the transformation $\varphi(q_i)/(1 - \Phi(q_i))$ of q_i in (2.18), which is based on the normality assumption for the unobserved variable Z. These assumptions are often questionable in practice, and results may be sensitive to minor departures from them. For these reasons the approach cannot be generally recommended.

The ML method often gives more stable results, but it has its own set of problems. In particular, it is highly vulnerable to the assumption that the residuals from the regression a')

are normal in the unselected population, an assumption which cannot be checked because this distribution can only be examined in the selected population. See Little and Rubin (1987, chap. 11) for more discussion of this point.

There is one circumstance where the methods discussed here may work reasonably well, if one or more of the regression coefficients $\{\beta_j\}$ or $\{\psi_j^*\}$ in the models for X_V or m can be assumed to be zero, that is, predictors that are included in one of the equations are excluded from the other. In our view, however, knowledge of such identifying restrictions is usually too limited and tentative to form a basis for selection bias adjustments (Little 1985b). Simulations (Nelson 1984; Stolzenberg and Relles 1990) suggest that even when such specifications are made correctly, selection models can yield unstable results.

Variations and extensions of the probit selection model have been proposed (Greenlees, Reece, and Zieschang 1982; Lee 1982; Olsen 1982), some of which relax particular normal assumptions. In our view, however, all of these models require the existence of identifying restrictions as described in the previous paragraph to have much hope of success.

4.3 Normal Pattern-Mixture Models

If the factorization (2.16) is applied to the example of univariate nonresponse, with normal specifications, we obtain the following normal pattern-mixture model:

$a)$ given $m_i = r$, $(x_{i1}, x_{i2}, ..., x_{iV})$ is normal with mean and covariance matrix $\phi^{(r)} = \{\boldsymbol{\mu}^{(r)}, \boldsymbol{\Sigma}^{(r)}\}$, $r = 0, 1$;

$b)$ m_i is marginally Bernoulli with $\text{pr}(m_i = 1) = \pi$.

This model implies that marginally $(X_1, X_2, ..., X_V)$ is a mixture of the multivariate normal distributions for each pattern, rather than multivariate normal as in the multivariate normal selection model. The two models are equivalent when the missing data are MCAR, since then missingness of X_V is independent of $X_1, X_2, ..., X_V$, $\psi_j = 0$ for all j, $\Phi(\psi_0) = \pi$, and $\boldsymbol{\mu}^{(r)} = \boldsymbol{\mu}$, $\boldsymbol{\Sigma}^{(r)} = \boldsymbol{\Sigma}$ for $r = 0, 1$.

Pattern-mixture models are often underidentified. In particular, the model above implies that conditional on $X_1, ..., X_{V-1}$:

$a')$ x_{iV} given $(x_{i1}, ..., x_{i,V-1})$ and $m_i = r$ is normal with mean $\beta_0^{(r)} + \sum_{j=1}^{V-1} \beta_j^{(r)} x_{ij}$ and variance $\sigma^{(r)2}$, for $r = 0, 1$.

Clearly the data supply no information about the parameters $\{\beta_j^{(1)}\}$ and $\sigma_1^{(1)}$, since by definition x_{iV} is missing when $m_i = 1$.

One approach to this problem is to assume identifying restrictions on the unidentified parameters. In the above example, one might assume

$$\beta_j^{(1)} = \beta_j^{(0)},\ j = 0, ..., V-1;\ \text{and}\ \sigma_1^{(1)} = \sigma_1^{(0)}. \quad (2.19)$$

This implies that X_V and $\boldsymbol{M}$ are conditionally independent given $X_1, ..., X_{V-1}$, that is, the missing data are MAR. Resulting ML estimates are the same as those for the ignorable selection model. Little (1993b,c) showed that for general patterns of missing data, the class of potential identifying restrictions is rich, yielding a variety of interesting estimation methods.

Rubin (1977) used a Bayesian prior distribution to relate the unidentified parameters to identified analogs. Specifically, write $\boldsymbol{\beta}^{(r)} = (\beta_1^{(r)}, ..., \beta_{V-1}^{(r)})$ for the vector of slopes for each pattern r, and $\phi^{(r)} = \beta_0^{(r)} + \sum_{j=1}^{V-1} \beta_j^{(r)}\bar{x}_j^{(0)}$ for the means of X_V for each pattern, evaluated at the sample means $\{\bar{x}_j^{(0)}\}$ of $\{X_j\}$ for pattern $m_i = 0$. Rubin assumed noninformative Jeffreys priors for $\boldsymbol{\beta}^{(0)}$, $\phi^{(0)}$, and $\sigma_1^{(0)}$, $\sigma_1^{(1)} = \sigma_1^{(0)}$, and

$$\boldsymbol{\beta}^{(1)} \sim N(\boldsymbol{\beta}^{(0)}, \psi_1^2\boldsymbol{\beta}^{(0)}\boldsymbol{\beta}^{(0)\prime}),\ \phi^{(1)} \sim N(\phi^{(0)}, \psi_2^2\phi^{(0)2})$$

for the nonrespondent parameters. Thus the priors for $\boldsymbol{\beta}^{(1)}$ and $\phi^{(1)}$ are centered at their complete-case analogs, but have a priori uncertainty measured by the parameters ψ_1 and ψ_2, which measure the degree to which the slopes and predicted means may differ between respondents and nonrespondents. The identifying restrictions (2.19) are obtained when $\psi_1 = \psi_2 = 0$. Rubin (1977) considered Bayesian inference for parameters of interest, for a range of plausible choices of ψ_1 and ψ_2. A summary of Rubin's analysis can be found in Little and Rubin (1987, Section 11.5).

The key message underlying this approach may be more important than the specifics of the model: The data have essentially nothing to say about the conditional distribution of X_V given $X_1, ..., X_{V-1}$ for incomplete cases, so a sensitivity analysis is needed to assess the impact on inferences about parameters of interest of differences in this distribution from the analogous distribution for respondents. If the missing-data mechanism is not well understood and follow-up data on nonrespondents are not available, a sensitivity analysis may be the most rational approach to the problem of nonignorable nonresponse, although it is certainly less satisfying than a single answer.

5 Multiple Imputation

We now turn to imputation methods for dealing with nonresponse in surveys. In Section 5.1, we discuss several issues that arise in developing imputation procedures and in the process describe some common methods; we also discuss why multiple imputation is needed. Sections 5.2 through 5.4 deal with various aspects of multiple imputation.

5.1 Imputation

A common technique for handling item nonresponse in a survey is to impute a value for each missing datum. This procedure is especially well suited to data bases distributed to the public, such as the FARS, CPS, and occupation data bases described in Section 1.1, for several reasons. First, imputation restores the rectangular form of the data matrix, so that users can apply standard methods of analysis designed for complete data. Second, imputation is often carried out by the data producer, who typically knows more about the reasons for nonresponse and has more information available than the subsequent data analysts. For example, the data producer may have detailed geographic information available that cannot be passed on to the public but that can be used in creating imputations. Finally, when imputation is performed by the data producer, the nonresponse problem is fixed in the same way for all subsequent data analysts, which ensures consistency of analyses across data users.

Some Important Principles

Little (1988b) gave a detailed discussion of principles of imputation. We discuss two such principles here. First, imputations should be based on the predictive distribution of the missing values given the observed values. Ideally then, all observed items should be taken into account in creating imputations. Of course, it is often infeasible to base an imputation model on all variables collected in a survey. Conditioning the imputations on as many variables as possible, and using subject-matter knowledge to help select variables, does help to reduce bias due to nonresponse, however. This is one argument against the naive technique of imputing unconditional means that was mentioned in Section 1.3.

An improvement over unconditional mean imputation is *conditional mean imputation*, in which each missing value is replaced by an estimate of its conditional mean given the observed values. For example, in the case of univariate nonresponse (see Section 1.2), one approach is to classify cases into cells based on similar values of $X_1, ..., X_{V-1}$, and then to impute the within-cell mean of X_V calculated from the complete cases for each missing value of X_V. Another more general approach is regression imputation (see Section 3.3), in which the regression of X_V on $X_1, ..., X_{V-1}$ is estimated from the complete cases, and the resulting prediction equation is used to impute the estimated conditional mean for each missing value of X_V.

Although conditional mean imputation incorporates information from the observed variables, it can lead to distorted estimates of quantities that are not linear in the data, such as percentiles, measures of association, and measures of variability. Therefore, a second important principle for creating imputations is to use random draws, not best predictions, to preserve the distribution of variables in the filled-in data set. An example is *stochastic regression imputation*, in which each missing value is replaced by its regression prediction with a random error added on, and the random error has variance equal to the estimated residual variance.

Explicit versus Implicit Models

Imputation procedures can be based on models that are explicit, implicit, or a combination of the two. Explicit models are those usually discussed in mathematical statistics: normal linear regression, generalized linear models, and so forth; thus stochastic regression imputation is based on an explicit model. Implicit models are those that underlie procedures for fixing up data structures in practice; they often have a nonparametric flavor to them.

A common class of procedures based on implicit models are *hot-deck* procedures, in which imputations for nonrespondents are drawn from the respondents' values. An example is the hot-deck method used by the Census Bureau for imputing income in the CPS (example 3). For each nonrespondent on one or more income items, the CPS hot-deck finds a matching respondent based on the variables that are observed for both; the missing items for the nonrespondent are then replaced by the respondent's values. For matching purposes, all variables are grouped into categories, and the number of variables used to define matches is large. When no match can be found for a nonrespondent based on all of the variables, the CPS hot-deck searches for a match at a lower level of detail, by omitting some variables and collapsing the categories of others.

The CPS hot-deck is essentially based on an implicit saturated analysis of variance model (Lillard et al. 1986). A natural alternative examined by David et al. (1986) is to create income imputations using a more parsimonious regression model for income. David et al. (1986) found both the CPS hot-deck and stochastic regression imputation methods effective. The main advantage of the regression method is that it does not necessarily include all interactions. Thus it has the potential to work better when fewer data are available, as well as to allow the inclusion of more predictors. The CPS hot-deck method, on the other hand, has the advantage that it always imputes values that are observed in the data set. In addition, hot-deck methods can produce multivariate imputations when many variables are missing more easily than methods based on explicit models.

An alternative to the technique of finding exact matches on categorized variables in hot-deck imputation is to define a distance function based on the variables that are observed for both nonrespondents and respondents. The missing values for each nonrespondent are then imputed from a respondent that is close to the nonrespondent in terms of the distance function. One such method, proposed in a different context in Rubin (1986), was termed *predictive mean matching* by Little (1988b). Consider univariate nonresponse, and suppose that a model predicting X_V from the other variables, $X_1, ..., X_{V-1}$, has been estimated using the complete cases. For each nonrespondent, predictive mean matching finds a respondent whose predicted X_V value is close to that of the nonrespondent. The respondent's observed X_V value is then imputed to the nonrespondent. Thus, predictive mean matching imputation combines features of an explicit method and an implicit method.

An extension of predictive mean matching to multivariate missing values was used by Heitjan and Little (1991) to impute missing values in the FARS data (example 4). Lazzeroni, Schenker, and Taylor (1990) showed in a simulation study that because predictive mean matching uses observed data values for imputation, it has some robustness to misspecification of the model used for matching.

Ignorable versus Nonignorable Models

The distinction between ignorable and nonignorable models for nonresponse was made in Sections 3 and 4. The basic idea is conveyed by the simple example of univariate nonresponse. Ignorable models assert that a nonrespondent's X_V value is only randomly different from that of a respondent having the same values of $X_1, ..., X_{V-1}$. In contrast, nonignorable models assert that even if a respondent and nonrespondent appear identical with respect to $X_1, ..., X_{V-1}$, their X_V values differ systematically.

Either an ignorable or a nonignorable model can be used as the basis for an imputation procedure. For example, the CPS hot-deck and the stochastic regression imputation methods studied by David et al. (1986) are based on ignorable models. Greenlees et al. (1982) and Lillard et al. (1986) discussed how imputations for missing CPS data could be based on nonignorable models such as those discussed in Section 4. It would also be possible to create a nonignorable hot-deck procedure; for example, respondents' values that are to be imputed to nonrespondents could be multiplied by an inflation or deflation factor that depends on the variables that are observed.

A crucial point about the use of nonignorable models is that there is no direct evidence in the data to address the validity of their underlying assumptions. Thus, whenever possible, it is prudent to consider several nonignorable models and explore the sensitivity of analyses to the choice of model.

Introduction to Multiple Imputation

Despite the advantages discussed at the beginning of this section, there is a major disadvantage to imputing just one value for each missing value in a survey. Because a single imputed value cannot represent all of the uncertainty about which value to impute, analyses that treat imputed values just like observed values generally underestimate uncertainty, even if nonresponse is modeled correctly and random imputations are created. For example, large-sample results in Rubin and Schenker (1986) show that for simple situations with 30% of the data missing, single imputation under the correct model results in nominal 90% confidence intervals having actual coverages below 80%. The inaccuracy of nominal levels is even more extreme in multiparameter testing problems (Rubin 1988).

Multiple imputation (Rubin 1978, 1987) alleviates the above drawback by replacing each missing datum with two or more values drawn from an appropriate distribution for the missing values under the posited assumptions about nonresponse. The result is two or more completed data sets, each of which is analyzed using the same standard method. The analyses are then combined in a way that reflects the extra variability due to missing data. Another benefit of multiple imputation is that it generally results in more efficient point estimates than does single random imputation. Multiple imputation can also be carried out under several different models for nonresponse to display sensitivity to the choice of model. For a recent review of multiple imputation, see Rubin and Schenker (1991).

A large-scale application of multiple imputation was carried out during the 1980s for the missing occupation code problem of example 5. Five sets of 1980 codes were imputed for a 1970 public-use data base of 1.6 million cases using models estimated from the double-coded sample of 120,000 cases. Publications on various aspects of the project include Rubin and Schenker (1987), Treiman, Bielby, and Cheng (1988), Weidman (1989), Clogg, Rubin, Schenker, Schultz, and Weidman (1991), and Schenker, Treiman, and Weidman (1993). The rest of this section outlines the theoretical motivation for multiple imputation as well as how to create and analyze a multiply imputed data set.

5.2 Theoretical Motivation for Multiple Imputation

Multiple imputation can be motivated most easily from the Bayesian perspective. For simplicity, we discuss the development for the case of ignorable nonresponse; Rubin (1987, chaps. 3, 6) discussed issues that arise when nonresponse is nonignorable. Letting θ now denote the estimand in a survey, we can write its posterior density as

$$p(\boldsymbol{\theta}|\boldsymbol{X}_{obs}) = \int p(\boldsymbol{\theta}|\boldsymbol{X}_{obs}, \boldsymbol{X}_{mis})p(\boldsymbol{X}_{mis}|\boldsymbol{X}_{obs})d\boldsymbol{X}_{mis}. \tag{2.20}$$

Equation (2.20) shows that the posterior density of $\boldsymbol{\theta}$ can be obtained by averaging the density for $\boldsymbol{\theta}$ obtained with complete data over the predictive distribution of the missing values. Thus, one way to compute $p(\boldsymbol{\theta}|\boldsymbol{X}_{obs})$ (say, at a particular value $\boldsymbol{\theta}_0$) would be to repeatedly draw values of $\boldsymbol{X}_{mis}$ from $p(\boldsymbol{X}_{mis}|\boldsymbol{X}_{obs})$, calculate $p(\boldsymbol{\theta}_0|\boldsymbol{X}_{obs}, \boldsymbol{X}_{mis})$ separately for each draw, and then average the values. In principle, multiple imputations are repeated draws from the predictive distribution of $\boldsymbol{X}_{mis}$ under the posited model for the data and the missing-data mechanism. Hence, multiple imputation allows the data analyst to approximate (2.20) by analyzing the completed data set under each draw and then combining the analyses.

Although multiple imputation was developed within the Bayesian framework (Rubin 1978), it has been shown to possess good frequentist properties. Theoretical and empirical work reported in Herzog and Rubin (1983), Rubin and Schenker (1986, 1987), Rubin (1987, chap. 4), Heitjan and Little (1991), Li, Meng, Raghunathan, and Rubin (1991), and Li, Raghunathan, and Rubin (1991), has shown that multiple imputation is far superior to single imputation with regard to validity of interval estimates and significance levels, even when the procedures used are only approximate.

5.3 Creating a Multiply Imputed Data Set

In principle, multiple imputations should be $K > 1$ independent draws from the posterior predictive distribution of X_{mis} under appropriate models for the data and the posited nonresponse mechanism. In practice, several important issues arise. These include the problems discussed in Section 5.1 for single imputation, that is, choosing between explicit and implicit models and choosing between ignorable and nonignorable models. In addition, for multiple imputation, there are issues concerning how to incorporate proper variability and how to choose K, as we now discuss.

Proper Imputation Methods

Imputation procedures that incorporate appropriate variability among the K sets of imputations within a model are called "proper," a term that is defined precisely in Rubin (1987, chap. 4). Proper imputation methods correctly reflect the sampling variability when creating imputations under a model, and as a result lead to valid inferences.

Consider again the example of univariate nonresponse. Suppose that nonresponse is ignorable, so that respondents and nonrespondents with common values of $X_1, ..., X_{V-1}$ differ only randomly in their X_V values. The simple hot-deck procedure that randomly draws imputations for nonrespondents from matching respondents' X_V values is not proper because it ignores the sampling variability due to the fact that the population distribution of X_V values given $X_1, ..., X_{V-1}$ is not known, but is estimated from the respondent values in the sample. Rubin and Schenker (1986, 1991) discussed the use of the bootstrap

(Efron 1979) to make the hot-deck procedure proper, and called the resulting procedure an "approximate Bayesian bootstrap." Consider a collection of units with the same values of $X_1, ..., X_{V-1}$, n_0 of which are respondents and n_1 of which are nonrespondents. For each of the K sets of imputations, the bootstrapped hot-deck first draws n_0 possible values of X_V at random with replacement from the n_0 observed values of X_V, and then imputes the n_1 missing values of X_V by drawing randomly with replacement from the set of n_0 possible values. Rubin and Schenker (1986) showed that for simple random samples with a single variable measured, the drawing of imputations from a possible sample of n_0 values rather than from the observed sample of n_0 values generates appropriate between-imputation variability. The bootstrap has been used in conjunction with other types of multiple-imputation methods by Heitjan and Little (1991) and Dorey, Little, and Schenker (1993).

The bootstrapped hot-deck has two stages: the first involves drawing a bootstrap sample and the second involves drawing imputations from the bootstrap sample. A two-stage procedure is often necessary to create proper imputations when explicit modeling is used as well. In the context of ignorable nonresponse, suppose that the predictive distribution of the missing values has been parameterized using a parameter $\boldsymbol{\beta}$. Then it can be expressed as

$$p(\boldsymbol{X}_{mis}|\boldsymbol{X}_{obs}) = \int p(\boldsymbol{X}_{mis}|\boldsymbol{X}_{obs}, \boldsymbol{\beta})p(\boldsymbol{\beta}|\boldsymbol{X}_{obs})d\boldsymbol{\beta}, \tag{2.21}$$

where $p(\boldsymbol{\beta}|\boldsymbol{X}_{obs})$ is the posterior distribution of $\boldsymbol{\beta}$. Equation (2.21) shows that drawing a value from the predictive distribution of $\boldsymbol{X}_{mis}$ requires first drawing a value of $\boldsymbol{\beta}$ from its posterior distribution and then drawing a value of $\boldsymbol{X}_{mis}$ conditional on the drawn value of $\boldsymbol{\beta}$. This parametric two-stage process was used with logistic regression models in creating imputations for the occupation code problem of example 5; see Rubin and Schenker (1987) and Clogg et al. (1991) for details. Bootstrapping a hot-deck imputation procedure as above may be thought of as a nonparametric analogue to drawing values of $\boldsymbol{X}_{mis}$ from its predictive distribution.

Choosing the Number of Imputations

The larger the number of draws (K) of $\boldsymbol{X}_{mis}$ used in multiple imputation, the more precisely (2.20) can be simulated. Since a distinct analysis is required for each of the K completed data sets (as discussed in Section 5.4), however, the computing time for analysis increases directly with K.

The effect of the size of K on precision depends in part on the fraction of information about $\boldsymbol{\theta}$ that is missing due to nonresponse, γ (Rubin 1987, chaps. 3, 4). With ignorable nonresponse and no covariates, γ equals that fraction of data values that are missing. When there are many variables in a survey, however, γ is typically smaller than this fraction because of the dependence between variables and the resulting ability to improve prediction of missing values from observed ones.

Rubin (1987, chap. 4) showed that for the estimation procedure to be described in Section 5.4, the efficiency of the point estimate for $\boldsymbol{\theta}$ based on K imputations relative to the estimate based on an infinite number of imputations is $(1+\gamma/K)^{-1}$. Thus, for the moderate fractions of missing information that occur in most surveys, a small number of imputations results in nearly fully efficient estimates.

In addition, for typical fractions of missing information, Rubin and Schenker (1986, 1987) and Rubin (1987, chap. 4) have shown that if the imputation procedure follows the Bayesian paradigm of drawing $\boldsymbol{X}_{mis}$ from its posterior predictive distribution as suggested by (2.20), then multiple-imputation inferences (as described in Section 5.4) generally have close to their nominal coverage or significance levels, even when only a few imputations (say $K = 3$) are used. In fact, results in Rubin and Schenker (1986) suggest that improvements in the coverage of multiple-imputation intervals due to increasing K are linear in $1/(K-1)$. Thus the incremental gains from increasing K diminish as K increases.

5.4 Analyzing a Multiply Imputed Data Set

The exact computation of (2.20) by simulation would require (a) an infinite number of draws of $\boldsymbol{X}_{mis}$ from its predictive distribution and (b) the calculation of $p(\boldsymbol{\theta}_0|\boldsymbol{X}_{obs}, \boldsymbol{X}_{mis})$ for every value $\boldsymbol{\theta}_0$. This section describes an approximation to (2.20) for scalar θ given by Rubin and Schenker (1986) that can be used even when only a small number of imputations of $\boldsymbol{X}_{mis}$ are available. Approximations that are useful for multiparameter testing problems were developed in Rubin (1987, chap. 3), Schenker and Welsh (1988), Li, Meng, Raghunathan, and Rubin (1991), Li, Raghunathan, and Rubin (1991), and Meng and Rubin (1992).

Given large samples, inference for θ with complete data is typically based on a point estimate $\hat{\theta}$, a variance U, and a normal reference distribution; for example, the typical interval estimate has the form

$$\hat{\theta} \pm \Phi^{-1}(1-\alpha/2)U^{1/2}. \tag{2.22}$$

In the presence of nonresponse, however, with K imputations of the missing values X_{mis} under the posited model for the missing data, there are K completed data sets and hence K sets of completed-data statistics, say $\hat{\theta}_k$ and U_k, $k = 1, \ldots, K$.

The K sets of completed-data statistics are combined to create one multiple-imputation inference as follows. The estimate of θ is

$$\bar{\theta} = K^{-1}\sum_{k=1}^{K}\hat{\theta}_k,$$

the average of the K completed-data estimates of θ. Also, let

$$\bar{U} = K^{-1}\sum_{k=1}^{K}U_k$$

be the average of the K completed-data variances, and

$$B = (K-1)^{-1}\sum_{k=1}^{K}(\hat{\theta}_k - \bar{\theta})^2$$

be the between-imputation variance of the completed-data estimates of θ. Then the total variance of $(\theta - \bar{\theta})$ is given by the sum of the within-imputation component $(\bar{U})$ and the between-imputation component (B) multiplied by a finite-K correction $(1 + K^{-1})$:

$$T = \bar{U} + (1 + K^{-1})B.$$

Interval estimates and significance levels are obtained using a t distribution with center $\bar{\theta}$, scale $T^{1/2}$, and degrees of freedom

$$\nu = (K - 1)(1 + r^{-1})^2,$$

where

$$r = (1 + K^{-1})B/\bar{U}$$

is the ratio of the between-imputation component of variance to the within-imputation component. Thus, for example, a $100(1 - \alpha)\%$ interval estimate for θ is

$$\bar{\theta} \pm t_\nu(1 - \alpha/2)T^{1/2},$$

where $t_\nu(1 - \alpha/2)$ is the $1 - \alpha/2$ quantile of the t distribution with ν degrees of freedom.

When r is small, indicating that there is little extra variability due to missing data, ν is large and the multiple-imputation analysis is based on a nearly normal reference distribution, as used in the standard complete-data interval (2.22). When r is large, however, ν is close to the degrees of freedom $(K - 1)$ associated with B. In general, r estimates the quantity $\gamma/(1 - \gamma)$, where γ is the fraction of information about θ that is missing due to nonresponse.

6 Other Bayesian Simulation Methods

Maximum-likelihood techniques such as those discussed in Sections 3 and 4 are most useful when sample sizes are large, since then the log-likelihood is nearly quadratic and can be summarized well using the ML estimate $\hat{\boldsymbol{\theta}}$ and its standard error. In other words, with large samples, the usual normal approximations are suitable. Even when ML methods are appropriate, however, they may be intractable when there are missing data; for instance, the E step of the EM algorithm can be quite difficult, even though the M step is often constructed to be straightforward.

When sample sizes are small or ML techniques are intractable, it can be useful to resort to simulation methods. Such methods are often easier to implement than analytic methods. Moreover, simulating a distribution for the parameters of interest, $\boldsymbol{\theta}$, avoids the restriction to the normal distributions that are often used as approximations with ML methods.

Multiple imputation, discussed in Section 5, is one such simulation method. In Section 5.2, we described how multiple imputation could in principle be used to simulate the exact posterior distribution of $\boldsymbol{\theta}$. Usually, however, multiple imputation is used jointly with the approximations of Section 5.4, in which case it becomes a large-sample technique. In the next two sections, we describe iterative simulation-based methods for approximating the posterior distribution of $\boldsymbol{\theta}$ that do not require large samples for their efficacy. Although our development is in the context of missing-data problems, the methods described can be used for many other types of problems and can be motivated from other points of view; see, for example, Gelfand and Smith (1990) and Casella and George (1992).

6.1 Data Augmentation

Data augmentation (Tanner and Wong 1987) is an iterative method of simulating the posterior distribution of $\boldsymbol{\theta}$ that combines features of the EM algorithm and multiple imputation as described below. Let $p^{(t)}(\boldsymbol{\theta}|\boldsymbol{X}_{obs})$ denote the approximation to the posterior density of $\boldsymbol{\theta}$ at iteration t. The algorithm cycles through the following three steps until convergence:

1. Draw K values, say $\boldsymbol{\theta}_1^{(t)}, ..., \boldsymbol{\theta}_K^{(t)}$, as independently and identically distributed (iid) with density $p^{(t)}(\boldsymbol{\theta}|\boldsymbol{X}_{obs})$.

2. For $k = 1, ..., K$, draw $\boldsymbol{X}_{mis,k}^{(t+1)}$ with density $p(\boldsymbol{X}_{mis}|\boldsymbol{X}_{obs}, \boldsymbol{\theta}_k^{(t)})$.

3. Update the approximate posterior density of $\boldsymbol{\theta}$ to

$$p^{(t+1)}(\boldsymbol{\theta}|\boldsymbol{X}_{obs}) = K^{-1}\sum_{k=1}^{K} p(\boldsymbol{\theta}|\boldsymbol{X}_{obs}, \boldsymbol{X}_{mis,k}^{(t+1)}).$$

Steps 1 and 2, which together are termed the "imputation step" by Tanner and Wong (1987), multiply impute $\boldsymbol{X}_{mis}$ from the current approximation to the predictive distribution $p(\boldsymbol{X}_{mis}|\boldsymbol{X}_{obs})$. Step 3, which updates the posterior distribution of $\boldsymbol{\theta}$, is termed the "posterior step" by Tanner and Wong (1987). The data-augmentation algorithm can be thought of as a small-sample refinement of the EM algorithm using simulation, with the imputation step corresponding to the E step and the posterior step corresponding to the M step. Note that the data-augmentation algorithm allows K to vary across iterations. To obtain a reasonable approximation to the entire posterior distribution of $\boldsymbol{\theta}$, a large value of K should be used at the final iteration.

Tanner and Wong (1987) showed that under regularity conditions, the data-augmentation algorithm converges to the posterior distribution of $\boldsymbol{\theta}$. For the example of bivariate normal data with missing values, Tanner and Wong (1987) illustrated how the algorithm can uncover the bimodality of the posterior distribution of the correlation coefficient.

6.2 The Gibbs Sampler

Rather than drawing $K > 1$ values at each iteration of the data-augmentation algorithm, we could set $K = 1$. Given a value of θ at iteration t, say $\boldsymbol{\theta}^{(t)}$, the algorithm would then cycle through the following steps until convergence:

1. Draw $\boldsymbol{X}_{mis}^{(t+1)}$ with density $p(\boldsymbol{X}_{mis}|\boldsymbol{X}_{obs}, \boldsymbol{\theta}^{(t)})$.

2. Draw $\boldsymbol{\theta}^{(t+1)}$ with density $p(\boldsymbol{\theta}|\boldsymbol{X}_{obs}, \boldsymbol{X}_{mis}^{(t+1)})$.

This iterative procedure was termed "chained data augmentation" by Tanner (1991).

Chained data augmentation is a special case of the "Gibbs sampler" algorithm (Gelfand and Smith 1990; Geman and Geman 1984; Li 1988), which has also been called "Markov imputation" and "stochastic relaxation." Just as the E step of the EM algorithm is sometimes intractable, it can be difficult to perform the sampling required in step one of the chained data augmentation algorithm. This difficulty can sometimes be alleviated by partitioning $\boldsymbol{X}_{mis}$ into L parts, say $(\boldsymbol{X}_{mis(1)}, ..., \boldsymbol{X}_{mis(L)})$, such that it is easy to draw each part given all of the other parts as well as $\boldsymbol{X}_{obs}$ and $\boldsymbol{\theta}$. Given the current values at iteration t, say $\boldsymbol{\theta}^{(t)}, \boldsymbol{X}^{(t)}_{mis(1)}, \boldsymbol{X}^{(t)}_{mis(2)}, \ldots, \boldsymbol{X}^{(t)}_{mis(L)}$, the algorithm cycles through the following steps until convergence:

1. Draw $\boldsymbol{X}^{(t+1)}_{mis(1)}$ with density $p(\boldsymbol{X}_{mis}|\boldsymbol{X}_{obs}, \boldsymbol{X}^{(t)}_{mis(2)}, \boldsymbol{X}^{(t)}_{mis(3)}, ..., \boldsymbol{X}^{(t)}_{mis(L)}, \boldsymbol{\theta}^{(t)})$.

2. Draw $\boldsymbol{X}^{(t+1)}_{mis(2)}$ with density $p(\boldsymbol{X}_{mis}|\boldsymbol{X}_{obs}, \boldsymbol{X}^{(t+1)}_{mis(1)}, \boldsymbol{X}^{(t)}_{mis(3)}, \ldots, \boldsymbol{X}^{(t)}_{mis(L)}, \boldsymbol{\theta}^{(t)})$.

$\vdots$

L. Draw $\boldsymbol{X}^{(t+1)}_{mis(L)}$ with density $p(\boldsymbol{X}_{mis}|\boldsymbol{X}_{obs}, \boldsymbol{X}^{(t+1)}_{mis(1)}, \boldsymbol{X}^{(t+1)}_{mis(2)}, \ldots, \boldsymbol{X}^{(t+1)}_{mis(L-1)}, \boldsymbol{\theta}^{(t)})$.

L+1. Draw $\boldsymbol{\theta}^{(t+1)}$ with density $p(\boldsymbol{\theta}|\boldsymbol{X}_{obs}, \boldsymbol{X}^{(t+1)}_{mis(1)}, \boldsymbol{X}^{(t+1)}_{mis(2)}, \ldots, \boldsymbol{X}^{(t+1)}_{mis(L)})$.

Geman and Geman (1984) showed that under regularity conditions, the sequence $(\boldsymbol{\theta}^{(t)}, \boldsymbol{X}^{(t)}_{mis(1)}, \boldsymbol{X}^{(t)}_{mis(2)}, \ldots, \boldsymbol{X}^{(t)}_{mis(L)})$ converges in distribution to a draw from the posterior distribution of $(\boldsymbol{\theta}, \boldsymbol{X}_{mis})$ as $t \to \infty$.

Both chained data augmentation and the Gibbs sampler can be run independently K times to generate K iid draws from the approximate joint posterior distribution of $\boldsymbol{\theta}$ and $\boldsymbol{X}_{mis}$. Starting values for the algorithms can be drawn from initial approximations to the joint posterior distribution. Tanner and Wong (1987), Gelfand et al. (1990), Gelman and Rubin (1992), and Geyer (1992) discussed techniques for monitoring the convergence of the algorithms.

6.3 The Use of Iterative Simulation to Create Multiple Imputations

Our introduction to data augmentation and the Gibbs sampler described them as methods for simulating the posterior distribution of $\boldsymbol{\theta}$ when large-sample methods are inadequate or intractable. The developments of Sections 6.1 and 6.2 showed that these methods can be used to simulate draws of $\boldsymbol{X}_{mis}$ from its posterior predictive distribution as well. Thus iterative Bayesian simulation can be used as the basis for multiple-imputation algorithms. For the Survey of Consumer Finances described in example 1, a variation of the Gibbs sampler is used to multiply impute missing items, as described in Kennickell (1991).

Schafer (1991) developed algorithms that use iterative Bayesian simulation to multiply impute rectangular data sets with arbitrary patterns of missing values when the missing-data mechanism is ignorable. The methods are applicable when the rows of the complete-data matrix can be modeled as iid observations from the following multivariate models discussed in Section 3.3: multivariate normal, multinomial (including log-linear models), and the general location model. In addition, Schafer (1991) proposed an extension to a model for census households, so that missing items for households and persons in the census can be imputed.

7 Discussion

As discussed in Madow, Nisselson, and Olkin (1983), the most important step in dealing with missing data is to try to eliminate it as much as possible during the data-collection stage. Given that some missing data are bound to exist after data collection, however, it is also useful to try to collect covariates that are likely to be predictive of the missing values, so that an adequate adjustment can be made. In addition, the process that leads to missing values should be noted during the collection of data. This will assist in appropriate modeling of the missing-data mechanism when an adjustment for the missing values is performed.

The methods that we have discussed in this chapter may be classified into three broad categories: weighting adjustments (for example, Section 2), imputation (for example, Sections 5 and 6), and direct analysis of the observed data (for example, Sections 3, 4, and 6). Weighting methods are useful for handling unit nonresponse. As mentioned in Section 2, however, methods for calculating standard errors from weighted data need to be developed more fully. Imputation is useful mainly for data bases that are to be distributed to the public by data-collection organizations, for the reasons discussed in Section 5. It can also be useful in the context of individual studies, however, since it can simplify the analysis of an incomplete data set. If missing values are to be imputed, multiple imputation should be employed to reflect the variability due to missing data. Direct analysis of the observed data is useful for individual studies. Computational methods such as the EM algorithm, data augmentation, and the Gibbs sampler can be very useful in performing direct analysis. Continued development of packaged software, such as mentioned in Section 3.3, is needed; such software should make the use of direct analysis more widespread.

REFERENCES

Afifi, A. A., and Elashoff, R. M. (1967), "Missing observations in multivariate statistics II: point estimation in simple linear regression," *Journal of the American Statistical Association*, 62, 595–604.

Anderson, A. B., Basilevsky, A., and Hum, D. P. J. (1983), "Missing data: a review of the literature," in P.H. Rossi, J. D. Wright, and A. B. Anderson (eds.), *Handbook of Survey Research*, pp. 415–494, New York: Academic Press.

Anderson, T. W. (1957), "Maximum likelihood estimation for the multivariate normal distribution when some observations are missing," *Journal of the American Statistical Association*, 52, 200–203.

Azen, S. P., Van Guilder, M., and Hill, M. A. (1989), "Estimation of parameters and missing values under a regression model with non-normally distributed and non-randomly incomplete data," *Statistics in Medicine*, 8, 217–228.

Baker, S. G., and Laird, N. M. (1988), "Regression analysis for categorical variables with outcome subject to nonignorable nonresponse," *Journal of the American Statistical Association*, 81, 29–41.

Basu, D. (1971), "An essay on the logical foundations of survey sampling, Part 1," in V. P. Godambe and D. A. Sprott (eds.), *Foundations of Statistical Inference*, pp. 203–242. Toronto: Holt, Rinehart, and Winston.

Bishop, Y. M. M., Fienberg, S. E., and Holland, P. W. (1975), *Discrete Multivariate Analysis: Theory and Practice*, Cambridge, MA: MIT Press.

Casella, G., and George, E. I. (1992), "Explaining the Gibbs sampler," *The American Statistician*, 46, 167–174.

Chen, T., and Fienberg, S. E. (1974), "Two-dimensional contingency tables with both completely and partially classified data," *Biometrics*, 30, 629–642.

Clogg, C. C., Rubin, D. B., Schenker, N., Schultz, B., and Weidman, L. (1991), "Multiple imputation of industry and occupation codes in census public-use samples using Bayesian logistic regression," *Journal of the American Statistical Association*, 86, 68–78.

Czajka, J. L., Hirabayashi, S. M., Little, R. J. A., and Rubin, D. B. (1992), "Projecting from advance data using propensity modeling: an application to income and tax statistics," *Journal of Business and Economic Statistics*, 10, 117.-132.

David, M. H., Little, R. J. A., Samuhel, M. E., and Triest, R. K. (1986), "Alternative methods for CPS income imputation," *Journal of the American Statistical Association*, 81, 29–41.

Dempster, A. P., Laird, N. M., and Rubin, D. B. (1977), "Maximum likelihood from incomplete data via the EM algorithm," *Journal of the Royal Statistical Society*, Ser. B, 39, 1–38.

Dixon, W. J., ed. (1988), *BMDP Statistical Software*, Los Angeles: University of California Press.

Dorey, F. J., Little, R. J. A., and Schenker, N. (1993), "Multiple imputation for threshold-crossing data with interval censoring," *UCLA Statistics Series*, No. 81. To appear in *Statistics in Medicine*.

Efron, B. (1979), "Bootstrap methods: Another look at the jackknife," *The Annals of Statistics*, 7, 1–26.

Fay, R. E. (1986), "Causal models for patterns of nonresponse," *Journal of the American Statistical Association*, 81, 354–365.

Fuchs, C. (1982), "Maximum likelihood estimation and model selection in contingency tables with missing data," *Journal of the American Statistical Association*, 77, 270–278.

Gelfand, A. E., Hills, S. E., Racine-Poon, A., and Smith, A. F. M. (1990), "Illustration of Bayesian inference in normal data models using Gibbs sampling," *Journal of the American Statistical Association*, 85, 972–985.

Gelfand, A. E., and Smith, A. F. M. (1990), "Sampling-based approaches to calculating marginal densities," *Journal of the American Statistical Association*, 85, 398–409.

Gelman, A., and Rubin, D. B. (1992), "Inference from iterative simulation using multiple sequences (with discussion)," *Statistical Science*, 4, 457–511.

Geman, S., and Geman, D. (1984), "Stochastic relaxation, Gibbs distributions, and the Bayesian restoration of images," *IEEE Transactions on Pattern Analysis and Machine Intelligence*, 6, 721–741.

Geyer, C. J. (1992), "Practical Markov chain Monte Carlo (with discussion)," *Statistical Science*, 4, 473–511.

Glasser, M. (1964), "Linear regression analysis with missing observations among the independent variables," *Journal of the American Statistical Association*, 59, 834–844.

Glynn, R., Laird, N. M., and Rubin, D. B. (1986), "Selection modeling versus mixture modeling with nonignorable nonresponse," in H. Wainer (ed.), *Drawing Inferences from Self-Selected Samples*, pp. 119–146, New York: Springer-Verlag.

Göksel, H., Judkins, D. R., and Mosher, W. D. (1991), "Nonresponse adjustments for a telephone follow-up to a national in-person survey," in American Statistical Association, *1991, Proceedings of the section on Survey Research Methods*, pp. 581-586.

Greenlees, J. S., Reece, W. S., and Zieschang, K. O. (1982), "Imputation of missing values when the probability of response depends on the variable being imputed," *Journal of the American Statistical Association*, 77, 251–261.

Haitovsky, Y. (1968), "Missing data in regression analysis," *Journal of the Royal Statistical Society*, Ser. B, 30, 67–81.

Hanson, R. H. (1978), "The current population survey: design and methodology," Technical Paper No. 40, Washington, DC: U.S. Bureau of the Census.

Harte, J. M. (1982), "Post-stratification approaches in the Corporation Statistics of Income Program," in American Statistical Association, *1982, Proceedings of the section on Survey Research Methods*, 250–253.

Heckman, J. (1976), "The common structure of statistical models of truncation, sample selection and limited dependent variables, and a simple estimator for such models," *Annals of Economic and Social Measurement*, 5, 475–492.

Heitjan, D. F., and Little, R. J. A. (1991), "Multiple imputation for the fatal accident reporting system," *Applied Statistics*, 40, 13–29.

Herzog, T. N., and Rubin, D. B. (1983), "Using multiple imputations to handle nonresponse in sample surveys," in W. G. Madow, I. Olkin, and D. B. Rubin (eds.), *Incomplete Data in Sample Surveys*, Volume 2: *Theory and Bibliographies*, pp. 209-245, New York: Academic Press.

Jennrich, R. I., and Schluchter, M. D. (1986), "Unbalanced repeated-measures models with structured covariance matrices," *Biometrics*, 42, 805–820.

Kass, R. E., and Steffey, D. (1989), "Approximate Bayesian inference in conditionally independent hierarchical models (parametric empirical Bayes models)," *Journal of the American Statistical Association*, 84, 717–726.

Kennickell, A. B. (1991), "Imputation of the 1989 Survey of Consumer Finances: stochastic relaxation and multiple imputation," in American Statistical Association, *1991, Proceedings of the section on Survey Research Methods*, pp. 1–10.

Kim, J. O., and Curry, J. (1977), "The treatment of missing data in multivariate analysis," *Sociological Methods and Research*, 6, 215–240.

Laird, N. M. (1988), "Missing data in longitudinal studies," *Statistics in Medicine*, 7, 305–315.

Laird, N. M., and Ware, J. H. (1982), "Random-effects models for longitudinal data," *Biometrics*, 38, 963–974.

Lange, K. L., Little, R. J. A., and Taylor, J. M. G. (1989), "Robust statistical modeling using the t distribution," *Journal of the American Statistical Association*, 84, 881–896.

Lazzeroni, L. C., Schenker, N., and Taylor, J. M. G. (1990), "Robustness of multiple imputation techniques to model specification," in American Statistical Association, *1990, Proceedings of the section on Survey Research Methods*, pp. 260–265.

Lee, L. F. (1982), "Some approaches to the correction of selectivity bias," *Review of Economic Studies*, 49, 355–372.

Li, K. H. (1988), "Imputation using Markov chains," *Journal of Statistical Computation and Simulation*, 30, 57–79.

Li, K. H., Meng, X. L., Raghunathan, T. E., and Rubin, D. B. (1991), "Significance levels from repeated p-values with multiply imputed data," *Statistica Sinica*, 1, 65–92.

Li, K. H., Raghunathan, T. E., and Rubin, D. B. (1991), "Large-sample significance levels from multiply imputed data using moment-based statistics and an F reference distribution," *Journal of the American Statistical Association*, 86, 1065–1073.

Liang, K. Y., Zeger, S. L., and Qaqish, B. (1992), "Multivariate regression analysis for categorical data," *Journal of the Royal Statistical Society*, Ser. B, 54, 3–40.

Lillard, L., Smith, J. P., and Welch, F. (1986), "What do we really know about wages: the importance of nonreporting and census imputation," *Journal of Political Economy*, 94, 489–506.

Lindstrom, M. J., and Bates, D. M. (1988), "Newton-Raphson and EM algorithms for linear mixed-effects models for repeated-measures data," *Journal of the American Statistical Association*, 88, 1014–1022.

Lipsitz, S. R., Laird, N. M., and Harrington, D. P. (1990), "Using the jackknife to estimate the variance of regression estimators from repeated measures studies," *Communications in Statistics*, Ser. A, 19, 821–845.

Little, R. J. A. (1985a), "Nonresponse adjustments in longitudinal surveys: models for categorical data," *Bulletin of the International Statistical Institute*, 15.1, 1–15.

———(1985b), "A note about models for selectivity bias," *Econometrica*, 53, 1469–1474.

———(1986), "Survey nonresponse adjustments," *International Statistical Review*, 54, 139–157.

———(1988a), "A test of missing completely at random for multivariate data with missing values," *Journal of the American Statistical Association*, 83, 1198–1202.

———(1988b), "Missing data adjustments in large surveys," *Journal of Business and Economic Statistics*, 6, 287–301.

———(1988c), "Robust estimation of the mean and covariance matrix from data with missing values," *Applied Statistics*, 37, 23–38.

———(1992), "Regression with incomplete X's; a review," *Journal of the American Statistical Association*, 87, 1227–1237.

———(1993a), "Post-stratification: a modeler's perspective," *Journal of the American Statistical Association*, 88, 1001–1012.

———(1993b), "Pattern-mixture models for multivariate incomplete data," *Journal of the American Statistical Association*, 88, 125–134.

———(1993c), "A class of pattern-mixture models for normal incomplete data," To appear in *Biometrika*.

Little, R. J. A., and Rubin, D. B. (1987), *Statistical Analysis with Missing Data*, New York: Wiley.

Little, R. J. A., and Schluchter, M. D. (1985), "Maximum likelihood estimation for mixed continuous and categorical data with missing values," *Biometrika*, 72, 497–512.

Little, R. J. A., and Su, H. L. (1989), "Item nonresponse in panel surveys," in D. Kasprzyk, G. Duncan, G. Kalton, and M. P. Singh (eds.), *Panel Surveys*, pp. 400–425, New York: Wiley.

Madow, W. G., Nisselson, H., and Olkin, I. (eds.) (1983), *Incomplete Data in Sample Surveys*, Volume 1: *Report and Case Studies*. Academic Press, New York.

Madow, W. G., and Olkin, I. (eds.) (1983), *Incomplete Data in Sample Surveys*, Volume 3: *Proceedings of the Symposium*. Academic Press, New York.

Madow, W. G., Olkin, I., and Rubin, D. B. (eds.) (1983), *Incomplete Data in Sample Surveys*, Volume 2: *Theory and Bibliographies*. Academic Press, New York.

Marini, M. M., Olsen, A. R., and Rubin, D. B. (1980), "Maximum-likelihood estimation in panel studies with missing data," *Sociological Methodology*, 11, 314–357.

McCullagh, P., and Nelder, J. A. (1989), *Generalized Linear Models*, second edition, London: Chapman and Hall.

McKendrick, A. G. (1926), "Applications of mathematics to medical problems," *Proceedings of the Edinburgh Mathematics Society*, 44, 98–130.

Meng, X. L., and Rubin, D. B. (1991), "Using EM to obtain aysmptotic variance-covariance matrices: the SEM algorithm," *Journal of the American Statistical Association* 86, 899–909.

———(1992), "Performing likelihood ratio tests with multiply-imputed data sets," *Biometrika*, 79, 103–111.

———(1993), "Maximum likelihood estimation via the ECM algorithm: a general framework," *Biometrika*, 80, 267–278.

Moulton, L. H., and Zeger, S. L. (1989), "Analyzing repeated measures on generalized linear models via the bootstrap," *Biometrics*, 45, 381–394.

Muthén, B., Kaplan, D., and Hollis, M. (1987), "On structural equation modeling with data that are not missing completely at random," *Psychometrika*, 52, 431–462.

Nelson, F. D. (1984), "Efficiency of the two-step estimator for models with endogenous sample selection," *Journal of Econometrics*, 24, 181–196.

Oh, H. L., and Scheuren, F. S. (1983), "Weighting adjustments for unit nonresponse," in W. G. Madow, I. Olkin, and D. B. Rubin (eds.), *Incomplete Data in Sample Surveys*, Volume 2: *Theory and Bibliographies*, pp. 143–184, New York: Academic Press.

Olkin, I., and Tate, R. F. (1961), "Multivariate correlation models with mixed discrete and continuous variables," *Biometrika*, 72, 448–465.

Olsen, R. J. (1982), "Distributional tests for selectivity bias and a more robust likelihood estimator," *International Economic Review*, 23, 223–240.

Orchard, T., and Woodbury, M. A. (1972), "A missing information principle: theory and applications," *Proceedings of the Sixth Berkeley Symposium on Mathematical Statistics and Probability*, 1, 697–715.

Prentice, R. L., and Zhao, L. P. (1991), "Estimating equations for parameters in means and covariances of multivariate discrete and continuous responses," *Biometrics*, 47, 825–839.

Rosenbaum, P. R., and Rubin, D. B. (1983), "The central role of the propensity score in observational studies for causal effects," *Biometrika*, 70, 41–55.

Rubin, D. B. (1976a), "Inference and missing data," *Biometrika*, 63, 581–592.

———(1976b), "Comparing regressions when some predictor values are missing," *Technometrics*, 18, 201–205.

———(1977), "Formalizing subjective notions about the effect of nonrespondents in sample surveys," *Journal of the American Statistical Association*, 72, 538–543.

———(1978), "Multiple imputations in sample surveys – A phenomenological Bayesian approach to nonresponse," in American Statistical Association, *1978, Proceedings of the section on Survey Research Methods*, pp. 20–34.

———(1986), "Statistical matching and file concatenation with adjusted weights and multiple imputations," *Journal of Business and Economic Statistics*, 4, 87–94.

———(1987), *Multiple Imputation for Nonresponse in Surveys*, New York: Wiley.

———(1988), "An overview of multiple imputation," in American Statistical Association, *1988, Proceedings of the section on Survey Research Methods*, pp. 79–84.

Rubin, D. B., Schafer, J. L. and Schenker, N. (1988), "Imputation strategies for missing values in post-enumeration surveys," *Survey Methodology*, 14, 209–221.

Rubin, D. B. and Schenker, N. (1986), "Multiple imputation for interval estimation from simple random samples with ignorable nonresponse," *Journal of the American Statistical Association*, 81, 366–374.

———(1987), "Interval estimation from multiply-imputed data: A case study using census agriculture industry codes," *Journal of Official Statistics*, 3, 375–387.

———(1991), "Multiple imputation in health-care databases: An overview and some applications," *Statistics in Medicine*, 10, 585–598.

SAS (1992), "The MIXED Procedure," chapter 16 in: *SAS/STAT Software: Changes and Enhancements, Release 6.07*. Technical Report P-229, SAS Institute, Inc., Cary, NC.

Schafer, J. L. (1991), *Algorithms for Multiple Imputation and Posterior Simulation from Incomplete Multivariate Data with Ignorable Nonresponse*. Ph.D. Thesis, Department of Statistics, Harvard University.

Schenker, N., Treiman, D.J., and Weidman, L. (1993), "Analyses of public-use data with multiply-imputed industry and occupation codes," *Applied Statistics*, 42, 545–556.

Schenker, N., and Welsh, A. H. (1988), "Asymptotic results for multiple imputation," *The Annals of Statistics*, 16, 1550–1566.

Schluchter, M. D. (1988), "Analysis of incomplete multivariate data using linear models with structured covariance matrices," *Statistics in Medicine*, 7, 317–324.

Schoenberg, R. S. (1988), "MISS: a program for missing data," in *GAUSS Programming Language*, Aptech Systems Inc., P.O. Box 6487, Kent, WA 98064.

Stolzenberg, R. M. and Relles, D. A. (1990), "Theory testing in a world of constrained research design — The significance of Heckman's censored sampling bias correction for nonexperimental research," *Sociological Methods and Research*, 18, 395–415.

Tanner, M. A. (1991), *Tools for Statistical Inference: Observed Data and Data Augmentation Methods*, New York: Springer-Verlag.

Tanner, M. A., and Wong, W. H. (1987), "The calculation of posterior distributions by data augmentation," *Journal of the American Statistical Association*, 82, 528–550.

Treiman, D. J., Bielby, W. T., and Cheng, M. T. (1988), "Evaluating a multiple-imputation method for recalibrating 1970 U.S. census detailed industry codes to the 1980 standard," *Sociological Methodology*, 18, 309–345.

Van Praag, B. M. S., Dijkstra, T. K., and Van Velzen, J. (1985), "Least-squares theory based on general distributional assumptions with an application to the incomplete observations problem," *Psychometrika*, 50, 25–36.

Waterton, J., and Lievesley, D. (1987), "Attrition in a panel study of attitudes," *Journal of Official Statistics*, 3, 267–282.

Weidman, L. (1989), "Final report: industry and occupation imputation," Statistical Research Division Report Number Census/SRD/89/03, Washington, DC: U.S. Bureau of the Census.

Wilks, S. S. (1932), "Moments and distribution of estimates of population parameters from fragmentary samples," *The Annals of Mathematical Statistics* 3, 163–195.

Woodburn, L. (1991), "Using auxiliary information to investigate nonresponse bias," in American Statistical Association, *1991, Proceedings of the section on Survey Research Methods*, pp. 278–283.

Zeger, S. L., and Liang, K. Y. (1986), "Longitudinal data analysis for discrete and continuous outcomes," *Biometrics*, 42, 121–130.

Chapter 3

Specification and Estimation of Mean Structures: Regression Models

GERHARD ARMINGER

1 Introduction

A major activity in the social sciences is modeling the dependence of one or more outcome or dependent variables on some explanatory or predictor variables. As Hastie and Tibshirani (1990, chap. 4, sec. 2) point out, the goals of modeling this dependence are *description* (to find out more about the process by which the dependent variable is generated), *inference* (to detect the contribution of each explanatory variable for prediction), and *prediction* (to forecast the value of a dependent variable for a specific combination of the values of the explanatory variables).

Modeling dependence in the social sciences has to take into account circumstances that differ substantially from those encountered in the natural sciences. First, experimentation is usually not feasible and is replaced by survey research, implying that the explanatory variables cannot be manipulated and fixed by the researcher. Therefore, the explanatory variables are usually random rather than fixed. The lack of experimental research may preclude causal interpretation of the research findings (cf. Sobel, Chapter 1). Second, the number of possible explanatory variables is often quite large, unlike the small number of carefully chosen treatment variables frequently found in the natural sciences. Third, the measurement level of a dependent variable is not always continuous like income or discrete like number of visits to a doctor. It may also be censored metric (household expenses for durable goods within a year), classified metric (grouped income), dichotomous (ques-

GERHARD ARMINGER • Department of Economics, Bergische Universität–GH Wuppertal, 42097 Wuppertal, Germany. • I wish to acknowledge the reviews and comments of C. Clogg, U. Küsters, M. Sobel, and A. Ziegler on earlier versions of this chapter. I am indebted to A. Ziegler for his help in the preparation of examples and to D. Enache for his help in typesetting the manuscript.

Handbook of Statistical Modeling for the Social and Behavioral Sciences, edited by Gerhard Arminger, Clifford C. Clogg, and Michael E. Sobel. Plenum Press, New York, 1995.

tionnaire items with categories *yes* and *no*), ordered categorical (five-point Likert scales in attitude research), or unordered categorical (choice of occupation). Fourth, scientists will not always specify a detailed dependence model by making assumptions about the functional form of the error variance or about the distribution of the errors. Assumptions will often only concern the average value of the dependent variable as a function of the explanatory variables.

As a consequence, this chapter deals with the specification and estimation of models for the conditional mean of a dependent variable given a set of explanatory variables. The treatment is restricted to standard parametric models with a fixed number of parameters that characterize the mean structure. These standard models can deal adequately with dependent variables of different measurement levels and with inference and prediction even when the number of explanatory variables is large. They are weak in approximating the true conditional mean structure if the model for the mean structure is not properly specified. Misspecification may easily occur by overlooking nonlinear dependencies or interactions between explanatory variables.

To detect the specific form of the conditional mean structure, the reader is advised to look into the semiparametric estimation of mean structures where the mean of a dependent variable is approximated by computing a smooth local average of the outcome variable for a given point in the explanatory variables. However, this estimation is restricted to the support of the explanatory variables in the data set at hand. It may therefore be difficult to make reliable predictions. In addition, these direct estimation methods are riddled with the *curse of dimensionality* (cf. Hastie and Tibshirani 1990, chap. 4, sec. 2) implying that estimation becomes progressively more difficult and less reliable when the number of explanatory variables increases. The treatment of these important topics is outside the scope of this chapter. The reader is therefore referred to Härdle (1990), Hastie and Tibshirani (1990), and Manski (1988, 1991, 1993).

Since social scientists cannot always make completely valid assumptions about the distribution of the error terms, it is important to be able to estimate the parameters of the mean structure with as few assumptions as possible. Consequently, parameters that characterize the distribution of the error term will be considered as nuisance parameters in contrast to the parameters that determine the mean structure.

In Section 2 of this chapter the univariate linear regression model is reviewed, as it is a natural starting point for all models that follow. Specification is mainly concerned with the construction of the matrix of explanatory variables. In estimation, the focus is on the maximum likelihood (ML) estimation of the parameters of the mean structure under different assumptions about the error term. The appropriate asymptotic confidence intervals, confidence regions, and test statistics are given. Coefficients of determination and regression diagnostics such as standardized and studentized residuals, leverage, and influential points are discussed. The multivariate linear regression model, which extends the univariate model, is briefly introduced.

In Section 3, ML estimation is discussed under the assumption that the conditional density of a dependent variable given the explanatory variables has been specified correctly up to an unknown parameter vector. Asymptotic confidence intervals, confidence regions and test statistics such as the Likelihood Ratio (LR), Lagrange Multiplier (LM) or score and Wald (W) statistics are given. The multivariate delta method and minimum distance estimation are introduced to deal with restrictions on the parameters of a model. Section 4 discusses the properties of the ML estimator if the conditional density of the dependent variable is misspecified. Then, the ML estimator can only be interpreted as an estimate of the parameter vector that minimizes the discrepancy between the true density of the dependent variable and the density assumed under the model chosen by the researcher. In this case, ML estimation has been called quasi-maximum-likelihood (QML) estimation (White 1982) because estimation is based on a misspecified density.

In Section 5, it is assumed that the density of the dependent variable is misspecified except for the first moment. Then, ML estimation using a misspecified density can still be used to compute strongly consistent estimates of the parameters of the mean structure. The misspecified density function chosen by the researcher must be a member of the univariate or multivariate linear exponential family that includes as special cases the normal, Poisson, binomial, and multinomial distributions. For this case, the ML method has been called pseudo-maximum-likelihood (PML) estimation by Gourieroux, Monfort and Trognon (1984). The assumed and the true density functions need not be the same except for the first moment. Although the ML estimates of the parameters of the mean structure are consistent, the asymptotic covariance matrix of these parameter estimates has to be estimated in a different way than in standard ML estimation. If not only the first moment, but the first and the second moment are equal for the true and the misspecified density, the efficiency of the PML estimator may be increased by using the information from the second moment. In this case, the ML method is called the quasi-generalized, pseudo-maximum-likelihood (QGPML) method. This method is discussed in Section 6.

In Section 7, univariate nonlinear regression models are specified. In particular, models for count data, dichotomous outcomes, and censored metric variables are derived from threshold models. Univariate generalized linear models (GLMs) are considered as special cases of nonlinear regression models, and quasi-likelihood models are considered as special mean structures where the parameters are estimated using a univariate QGPML method. In Section 8, multivariate nonlinear regression models are discussed. Models for doubly censored, grouped metric variables and ordered categorical outcomes are derived from threshold models. The multinomial logit and the multinomial probit model are considered as random utility maximization (RUM) models. The relationship between the PML and QGPML approach and the approach of generalized estimating equations (GEE) of Liang and Zeger (1986) for parameter estimation in mean structures is discussed. Finally, a short overview of statistical programs to compute parameter estimates is given.

Throughout the chapter it is assumed that the data $\{\boldsymbol{y}_i, \boldsymbol{x}_i\}, i = 1, \ldots, n$ that are analyzed have been generated by independent identical random experiments. This is equivalent to a cross-sectional random sample of size n often found in behavioral research. The sample size n is assumed to be large in comparison to the number of estimated parameters so that the asymptotic statistical inference from ML, PML, QGPML, or GEE estimation is justified. Repeated observations on sample points from a cross-sectional sample with a fixed number T of time points (panel data) can often be treated as cross-sectional data coming from n individuals with T dependent variables. Models for time series data are not treated here.

2 The Linear Regression Model

2.1 Model Specification

Let y_i be a metric observed dependent variable such as income, for which the distances between outcomes are known. The most common way in the behaviorial sciences to analyze the dependence of y_i on explanatory variables is the univariate linear regression model

$$y_i = \boldsymbol{x}_i'\boldsymbol{\beta} + \epsilon_i \ , \quad i = 1, \ldots, n \tag{3.1}$$

where $\boldsymbol{x}_i$ is a $q \times 1$ vector of observed explanatory variables, $\boldsymbol{\beta}$ is a $q \times 1$ vector of regression coefficients, and ϵ_i is an unobserved random variable called the error or disturbance. The first element of $\boldsymbol{x}_i$ is usually 1 so that β_1 is the regression constant or intercept. If $E(\epsilon_i|\boldsymbol{x}_i) = 0$ then $E(y_i|\boldsymbol{x}_i) = \mu_i = \boldsymbol{x}_i'\boldsymbol{\beta}$, which constitutes the systematic part of the model. More generally, one may consider $E(y_i|\boldsymbol{x}_i) = \mu(\boldsymbol{x}_i, \boldsymbol{\vartheta}) = \mu_i$ as a function of $\boldsymbol{x}_i$ and a parameter vector $\boldsymbol{\vartheta}$. The function μ_i is called the mean structure of y_i given $\boldsymbol{x}_i$. If the whole mean structure or a part of it is linear in the parameters, then the parameter vector $\boldsymbol{\vartheta}$ or a part of it are often denoted as $\boldsymbol{\beta}$. The linear regression model for the whole sample may be written in matrix notation as

$$\boldsymbol{y} = \boldsymbol{X}\boldsymbol{\beta} + \boldsymbol{\epsilon} \tag{3.2}$$

with $\boldsymbol{X} \sim n \times q$ as regressor matrix. The matrix $\boldsymbol{X}$ is also called the model matrix because its form plays an important role in the specification of the mean structure, $\mu_i = \boldsymbol{x}_i'\boldsymbol{\beta}$. In experimental work, $\boldsymbol{X}$ is also called design matrix.

Possible applications of the linear regression model follow directly from the form of the mean structure. If the jth variable is continuous then the regression coefficient β_j is the partial derivative, that is the rate of change, of μ_i with respect to x_{ij}

$$\beta_j = \frac{\partial \mu_i}{\partial x_{ij}} \quad . \tag{3.3}$$

Characteristic for the linear model is that the rate of change is constant for the whole range of $E(y_i|\boldsymbol{x}_i)$. If the dependence of μ_i on x_{ij} is roughly linear, the coefficient β_j provides a simple description of this dependence. Since β_j is a partial coefficient, it summarizes the whole contribution of x_{ij} to the mean structure. If $\boldsymbol{\beta}$ is estimated by $\hat{\boldsymbol{\beta}}$, prediction with a new value $\boldsymbol{x}_p$ is easily computed with $\hat{\mu}_p = \boldsymbol{x}_p'\hat{\boldsymbol{\beta}}$. If one or more variables x_{ij} are continuous, then μ_i can take any value in $(-\infty, +\infty)$. This fact makes the use of linear models for mean structures with a limited range such as probabilities where $\mu_i \in (0, 1)$ given $\boldsymbol{x}_i$ unattractive. For this purpose, the models of Sections 7 and 8 should be used.

To compare the regression coefficients across different explanatory variables the coefficients are standardized by multiplication with the standard deviations σ_j of x_{ij}.

$$\beta_j^{(s)} = \beta_j \sigma_j \quad \text{regression coefficient standardized for } x_{ij} \tag{3.4}$$

Division of $\beta_j^{(s)}$ by the standard deviation σ_y of the dependent variable yields the standardized beta coefficient of path analysis.

The interpretation of the regression coefficients is different if the x_{ij}'s are not continuous. Let A denote a categorical explanatory variable like region or occupation with K disjoint categories $A_1, \ldots, A_k, \ldots A_K$. If the expected values of y_i are different for each category of A and depend only on A, the mean structure can be written as

$$\mu_i = \beta_k^{(A)} \iff \quad \text{case } i \text{ is in category } A_k \ (i \in A_k) \quad . \tag{3.5}$$

This formulation is embedded into the linear regression model by defining dummy variables

$$x_{ik}^{(A)} = \begin{cases} 1 & \text{if } i \in A_k \\ 0 & \text{otherwise} \end{cases} \quad \text{and} \quad \mu_i = x_{i1}^{(A)}\beta_1^{(A)} + x_{i2}^{(A)}\beta_2^{(A)} + \ldots + x_{iK}^{(A)}\beta_K^{(A)} \, . \tag{3.6}$$

In this formulation, there is no regression constant. K dummy variables are used to represent the K categories of A. The parameter $\beta_k^{(A)}$ is the expected value $E(y_i|i \in A_k)$. If the model matrix includes a regression constant, that is $x_{i1} = 1$ for all $i = 1, \ldots, n$, then the mean structure can be written as

$$\mu_i = \beta_1 + x_{i2}\beta_2^{(A)} + \ldots + x_{iK}^{(A)}\beta_K^{(A)} \tag{3.7}$$

using the first category as a *reference* or *baseline*. The regression constant β_1 represents the expected value in the first category. If β_1 is included in the mean structure, $\beta_k^{(A)}$ represents the difference between $E(y_i|i \in A_k)$ and $E(y_i|i \in A_1)$. This is immediately seen from

$$\mu_i = \beta_1 \quad \text{if } i \in A_1, \quad \text{and} \quad \mu_i = \beta_1 + \beta_k^{(A)} \quad \text{if } i \in A_k, k = 2, \ldots, K \, . \tag{3.8}$$

In this formulation, A_1 is the reference category to which all other categories $A_2, \ldots, A_K$ are compared. If a regression constant is included, only $K - 1$ dummy variables

$$x_{ik}^{(A)} = \begin{cases} 1 & \text{if } i \in A_k, k = 2, \ldots, K \\ 0 & \text{otherwise} \end{cases} \tag{3.9}$$

are necessary to represent all K categories of A. The choice of the reference category is arbitrary if the categories of A do not follow a natural order.

A different coding scheme for categorical explanatory variables is effect coding in the analysis of variance

$$\mu_i = \beta_1 + \beta_k^{(A)} \quad \text{if} \quad i \in A_k, \ k = 1, \ldots, K \tag{3.10}$$

with the restriction

$$\sum_{k=1}^{K} \beta_k^{(A)} = 0 \quad \Rightarrow \quad \beta_K^{(A)} = -\sum_{k=1}^{K-1} \beta_k^{(A)} \quad . \tag{3.11}$$

β_1 is the unconditional mean and $\beta_k^{(A)}$ are the deviations in category A_k from the general (grand) mean. The corresponding $K-1$ variables $x_{ik}, k = 1, \ldots, K-1$ are

$$x_{ik} = \begin{cases} 1 & \text{if } i \in A_k, k = 1, \ldots, K-1 \\ 0 & \text{if } i \text{ is not in } A_k \\ -1 & \text{if } i \in A_K \end{cases} \quad . \tag{3.12}$$

Although this coding scheme has a long tradition in the analysis of variance to test differences between treatments and is used in many computer programs to describe categorical explanatory variables, it is not very useful for data from observational studies (rather than experimental studies). The value of the restriction (3.11) breaks down if additional explanatory categorical variables are introduced and the number of observations is not equal for each cell of the cross-classification of two or more categorical variables. The coding scheme of (3.9) is used in the theory of generalized linear models (McCullagh and Nelder 1989) and the computer program GLIM (1993). Other coding schemes and their possible uses are described in Bock (1975).

The specification of a linear model often includes both categorical and continuous explanatory variables. Let A be a categorical variable with categories $A_1, \ldots, A_K$ and let z be a continuous variable. A is represented by $K-1$ dummy variables $x_{ik}, k = 2, \ldots, K$. Typical specifications are the following four models:

$$\text{M1: } \mu_i = \beta_1 + x_{i2}^{(A)}\beta_2^{(A)} + \ldots + x_{iK}^{(A)}\beta_K^{(A)}$$

$$\text{M2: } \mu_i = \beta_1 + z_i\gamma_1$$

$$\text{M3: } \mu_i = \beta_1 + x_{i2}^{(A)}\beta_2^{(A)} + \ldots + x_{iK}^{(A)}\beta_K^{(A)} + z_i\gamma_1$$

$$\text{M4: } \mu_i = \beta_1 + x_{i2}^{(A)}\beta_2^{(A)} + \ldots + x_{iK}^{(A)}\beta_K^{(A)} + z_i\gamma_1 + x_{i2}^{(A)}z_i\gamma_2 + \ldots + x_{iK}^{(A)}z_i\gamma_K \quad .$$

In model M1, one considers only the expected values of y_i for each category of A. The regression constant β_1 is the expected value of y_i given that case i is in the first category of A. The expected value of the dependent variable is not conditioned on the values of z. In model M2 the expected value is only conditioned on z. The regression constant β_1 is the expected value of y_i given that z_i is 0. Model M3 specifies a joint effect of A and z on the conditional expected value of y_i. Here, β_1 is the regression constant of the first group and $\beta_k^{(A)}$ represents the difference between the regression constant of category A_k and category A_1. The rate of change in μ_i with respect to z_i is assumed to be equal across all categories of A. Finally, in model M4 one assumes that the intercepts [denoted by $(\beta_1 + \beta_k^{(A)})$] vary across categories of A, and that the regression coefficients of z_i vary across the categories of A. The regression coefficient of z_i for the first category is given by γ_1, the regression coefficient of z_i for the kth category is $\gamma_1 + \gamma_k$ for $k = 2, \ldots, K$. These different regression models for each category A_k are specified by introducing the interaction terms $x_{ik}^{(A)} z_i$, which are defined as products of the dummy variables and the continuous explanatory variable. Hypotheses about equalities of intercepts and/or regression coefficients of z_i can easily be specified as differences between parameters $\beta_k^{(A)}$ and/or $\gamma_k^{(A)}$.

The notion of interaction can be extended to two or more continuous variables. In the case of continuous variables $z_1, z_2, \ldots, z_p$, the additional variables $z_1^2, z_1 z_2, z_2^2, \ldots, z_p^2$ are defined as products of the corresponding variables. Note however, the model for μ_i becomes nonlinear in the variables which may change the interpretation of the parameters. This is seen by looking at the rate of change in μ_i. If one considers for instance the model

$$\mu_i = \beta_1 + z_i\beta_2 + z_i^2\beta_3 \quad \text{with} \quad \frac{\partial \mu_i}{\partial z_i} = \beta_2 + 2z_i\beta_3 \quad , \tag{3.13}$$

one finds that the rate of change is no longer constant but depends on the level of the explanatory variable. Hence the inclusion of such interaction terms should be considered with great care. Preferably, there should exist a substantive theory that dictates the inclusion of such interaction terms, for example, synergy between variables z_{i1} and z_{i2}.

In the case of three categorical variables A, B, C, interactions are also defined by simple multiplication. If the variables A, B, C are represented by dummy variables $x_{ik}^{(A)}, k = 2, \ldots, K, x_{il}^{(B)}, l = 2, \ldots, L$, and $x_{im}^{(C)}, m = 2, \ldots, M$, then the interaction between A and B is defined as

$$A.B = \{x_{i2}^{(A)} x_{i2}^{(B)}, \ldots, x_{i2}^{(A)} x_{iL}^{(B)}, \ldots, x_{iK}^{(A)} x_{iL}^{(B)}\} \quad . \tag{3.14}$$

These additional $(K-1) \cdot (L-1)$ variables are then included as regressors in the linear model. In a similar way, one can define the interaction among A, B, and C as

$$A.B.C = \{x_{i2}^{(A)} x_{i2}^{(B)} x_{i2}^{(C)}, \ldots, x_{i2}^{(A)} x_{i2}^{(B)} x_{iM}^{(C)}, \ldots, x_{iK}^{(A)} x_{iL}^{(B)} x_{iM}^{(C)}\} \tag{3.15}$$

While the interaction of a continuous variable with itself is meaningful if one uses multiplication, i.e. $z_{i2} = z_{i1}^2$, the interaction of a categorical variable with itself reproduces the dummy variables of the categorical variable. Therefore, the convention $A.A = A$ is used.

If categorical variables are denoted by $A, B, C, \ldots$ and continuous variables are denoted by $X, Y, Z \ldots$, one can employ the model formulas developed by Wilkinson and Rogers (1973) to specify in a simple and concise way the mean structure of linear and generalized linear models. The four models M1–M4 may be written as:

$$\text{M1} = A, \quad \text{M2} = Z, \quad \text{M3} = A + Z, \quad \text{M3} = A + Z + A.Z \quad .$$

The operator + denotes the addition of model terms in a model formula with the convention $A + A = A$. If a continuous variable Z is added, only the variable itself enters the model. If a categorical variable A is added, the corresponding dummy variables $x_{ik}^{(A)}, k = 2, \ldots, K$ are generated by the program and entered as additional variables. The dot operator in the term $A.Z$ performs the multiplications between the dummy variables of A and the variable Z to generate the interaction terms. The + and the . operator are used to generate additional symbolic operations that are defined and described by McCullagh and Nelder (1989) and in the GLIM program manual. Similar operators are available in SAS and in GAUSS. These operators and related symbolic notation are often used to specify models of various kinds.

In all programs, the model formula generates a vector $\boldsymbol{x}_i$ of dummy variables for categorical explanatory variables and interaction terms for categorical and/or continuous variables. If the model formula generates many variables, for instance the dummy variables for three or higher-order interactions of categorical variables, the model matrix may contain dummy variables for interactions which are not realized in the sample. Typical examples are explanatory variables like type of schooling and occupation for the dependent variable income where certain combinations of type of schooling and occupation are not observed, but the model formula automatically creates dummy variables for all interactions. These dummy variables will always be 0 yielding one or more columns of zeros in $\boldsymbol{X}$. These columns must be eliminated from the $\boldsymbol{X}$ matrix. In GLIM, the message that the corresponding parameter is *aliased* is switched on. Technically, this type of aliasing is called extrinsic aliasing because it has been caused by a lack of data.

2.2 Estimation of Regression Coefficients

The classical assumptions for the estimation of the parameters $\boldsymbol{\beta}$ in a linear regression model are:

- $E(\epsilon_i) = 0$ (correct specification of mean structure) (3.16)
- $\boldsymbol{x}_i$ is nonstochastic, $\boldsymbol{X}$ is of full column rank (3.17)
- $\lim_{n \to \infty} \left(\frac{1}{n} \boldsymbol{X}'\boldsymbol{X} \right) = \boldsymbol{S}$ is positive definite (3.18)
- $V(\epsilon) = \sigma^2$ (homoscedasticity) (3.19)
- $\epsilon_i \sim \mathcal{N}(0, \sigma^2)$ (normal distribution of errors). (3.20)

These assumptions do not suffice to prove the properties of the estimators discussed below rigorously. Additional regularity conditions must be introduced that vary for specific models and sampling designs [cf. for instance Schmidt (1976) and White (1980)]. These classical assumptions are considered as a starting point and the consequences of relaxing these assumptions are discussed below.

The ordinary least squares (OLS) estimator of $\boldsymbol{\beta}$ is obtained by minimizing the objective function

$$Q(\boldsymbol{\beta}) = (\boldsymbol{y} - \boldsymbol{X}\boldsymbol{\beta})'(\boldsymbol{y} - \boldsymbol{X}\boldsymbol{\beta}) \tag{3.21}$$

giving

$$\hat{\boldsymbol{\beta}}_{OLS} = (\boldsymbol{X}'\boldsymbol{X})^{-1}\boldsymbol{X}'\boldsymbol{y} \quad . \tag{3.22}$$

The estimator of σ^2 that is usually associated with $\hat{\boldsymbol{\beta}}_{OLS}$ is

$$\hat{\sigma}^2_{OLS} = \frac{1}{n-q}\sum_{i=1}^{n}(y_i - \boldsymbol{x}_i'\hat{\boldsymbol{\beta}}_{OLS})^2 \quad . \tag{3.23}$$

The inverse of $(\boldsymbol{X}'\boldsymbol{X})$ is assumed to exist. If this is not the case, then one or more parameters cannot be estimated. To avoid numerical complications, one can use a generalized inverse of $(\boldsymbol{X}'\boldsymbol{X})$. For most applications, the sweep inverse (Goodnight 1979), which is a special case of a generalized inverse, can be used (Rao 1973, chap. 1, pp. 24). Each column $\boldsymbol{x}_j \sim n \times 1,\ j = 2, \ldots, q$ is checked to ascertain whether it is linearly dependent on the columns $\boldsymbol{x}_k, \quad k = 1, \ldots, j-1$. If linear dependence exists, the jth column and jth row in $(\boldsymbol{X}'\boldsymbol{X})$ are set to $\boldsymbol{0}$. The corresponding parameter is then aliased.

If the classical assumptions are fulfilled, the small sample properties of $\hat{\boldsymbol{\beta}}_{OLS}$ and $\hat{\sigma}^2_{OLS}$ are well known (cf. for instance Greene 1990, chapter 6). In particular, $\hat{\boldsymbol{\beta}}_{OLS}$ and $\hat{\sigma}^2_{OLS}$ are unbiased estimators of $\boldsymbol{\beta}$ and σ^2. In addition,

$$\sqrt{n}(\hat{\boldsymbol{\beta}}_{OLS} - \boldsymbol{\beta}) \sim \mathcal{N}\left(\boldsymbol{0}, \sigma^2(\frac{1}{n}\boldsymbol{X}'\boldsymbol{X})^{-1}\right) \tag{3.24}$$

holds, that is $\hat{\boldsymbol{\beta}}_{OLS}$ follows a multivariate normal distribution even in finite samples. Within the class of linear unbiased estimators, $\hat{\boldsymbol{\beta}}_{OLS}$ has minimum variance.

To derive confidence intervals and test statistics for significance tests for individual parameters $\beta_j, j = 1, \ldots, q$, σ^2 is replaced by $\hat{\sigma}^2_{OLS}$ and the jth diagonal element of $(\boldsymbol{X}'\boldsymbol{X})^{-1}$ is denoted by c_{jj}. The test statistic

$$t_j^\star = \frac{\hat{\beta}_j - \beta_{j0}}{\sqrt{\hat{\sigma}^2_{OLS}c_{jj}}} \tag{3.25}$$

follows a central t distribution with $n-q$ degrees of freedom. A two-sided confidence interval for a given probability $1-\alpha$ is then given by

$$P\left(\hat{\beta}_j - t_{n-q;1-\alpha/2}\sqrt{\hat{\sigma}^2_{OLS}c_{jj}} \le \beta_j \le \hat{\beta}_j + t_{n-q;1-\alpha/2}\sqrt{\hat{\sigma}^2_{OLS}c_{jj}}\right) = 1-\alpha. \tag{3.26}$$

The null hypothesis of the two-sided hypothesis $H_0 : \beta_j = \beta_{j0}$ against $H_1 : \beta_j \neq \beta_{j0}$ is rejected with significance level α if

$$\left|t_j^\star\right| > t_{n-q;1-\alpha/2} \quad . \tag{3.27}$$

In addition to computing confidence intervals for test statistics and single regression coefficients β_j, one may be interested in computing a confidence region for the whole parameter vector $\boldsymbol{\beta}$ or a subset of $\boldsymbol{\beta}$ and performing a significance test for linear restrictions of the form

$$\boldsymbol{R\beta} - \mathbf{c} = \boldsymbol{0} \tag{3.28}$$

where $\boldsymbol{R}$ is a $s \times q$ matrix of constants and $\boldsymbol{c}$ is a $s \times 1$ vector of constants with $s \leq q$ and $\boldsymbol{R}$ is of rank s. The most common of these restrictions is $\beta_2 = \beta_3 = \ldots = \beta_q = 0$. Confidence regions and tests of $H_0 : \boldsymbol{R\beta} - \boldsymbol{c} = \boldsymbol{0}$ against $H_1 : \boldsymbol{R\beta} - \boldsymbol{c} \neq \boldsymbol{0}$ are based on the F statistic (cf. Greene 1990, chap. 6)

$$F^\star = (\boldsymbol{R}\hat{\boldsymbol{\beta}}_{OLS} - \boldsymbol{c})'(\hat{\sigma}^2_{OLS}\boldsymbol{R}(\boldsymbol{X}'\boldsymbol{X})^{-1}\boldsymbol{R}')^{-1}(\boldsymbol{R}\hat{\boldsymbol{\beta}}_{OLS} - \boldsymbol{c})/s \quad . \tag{3.29}$$

which follows a central F distribution with $(s, n - q)$ degrees of freedom under the null hypothesis.

The joint confidence region for the parameter vector $\boldsymbol{\beta}$ is therefore given by the confidence ellipsoid with confidence probability $1 - \alpha$.

$$CE(1 - \alpha) = \{\boldsymbol{\beta} : (\boldsymbol{\beta} - \hat{\boldsymbol{\beta}}_{OLS})'(\boldsymbol{X}'\boldsymbol{X})(\boldsymbol{\beta} - \hat{\boldsymbol{\beta}}_{OLS})/\hat{\sigma}^2_{OLS} \leq qF_{1-\alpha}(q, n - q)\} \tag{3.30}$$

If only a subset of $\boldsymbol{\beta}$ is chosen, then the corresponding subset of the covariance matrix of $(\hat{\boldsymbol{\beta}}_{OLS} - \boldsymbol{\beta})$ given by $\hat{\boldsymbol{V}}(\hat{\boldsymbol{\beta}}_{OLS}) = \hat{\sigma}^2_{OLS}(\boldsymbol{X}'\boldsymbol{X})^{-1}$ must be inverted to form $CE(1 - \alpha)$. The null hypothesis $\boldsymbol{R\beta} - \boldsymbol{c} = \boldsymbol{0}$ is rejected in favor of the hypothesis $\boldsymbol{R\beta} - \boldsymbol{c} \neq \boldsymbol{0}$ if the test statistic $F^\star > F_{1-\alpha}(s, n - q)$.

What happens if the classical assumptions for the estimation of $\boldsymbol{\beta}$ are eliminated? First, the explanatory variables $\boldsymbol{x}_i$ may be stochastic. Additionally, it is assumed that the parameters of the density of $\boldsymbol{x}_i$ are distinct from the parameters of the conditional density of y_i given $\boldsymbol{x}_i$. The assumptions $E(\epsilon_i) = 0$ and $V(\epsilon_i) = \sigma^2$ are replaced by $E(\epsilon_i|\boldsymbol{x}_i) = 0$ and $V(\epsilon_i|\boldsymbol{x}_i) = \sigma^2$. The assumption that $E(\epsilon_i|\boldsymbol{x}_i) = 0$ implies $E(\epsilon_i\boldsymbol{x}_i) = \boldsymbol{0}$, that is, the error term and the explanatory variables are uncorrelated. If $\boldsymbol{x}_i$ is stochastic, the finite sample properties of $\hat{\boldsymbol{\beta}}_{OLS}$ remain valid as shown in Greene (1990, chap. 10). Hence, conditional on $\boldsymbol{X}$, one can use the $t^\star$ and $F^\star$ statistics as in the case of fixed regressors. However, these results hold only given the model matrix $\boldsymbol{X}$ of the sample analyzed. Unconditionally, the results change as is seen immediately by looking at the asymptotic covariance of $\hat{\boldsymbol{\beta}}_{OLS}$. The conditional asymptotic covariance matrix of $\sqrt{n}(\hat{\boldsymbol{\beta}} - \boldsymbol{\beta})$ given $\boldsymbol{X}$ is

$$\boldsymbol{V}(\sqrt{n}\hat{\boldsymbol{\beta}}_{OLS}|\boldsymbol{X}) = \sigma^2(\frac{1}{n}\boldsymbol{X}'\boldsymbol{X})^{-1} \quad . \tag{3.31}$$

Taking the expected value over $\boldsymbol{x}$ yields the unconditional asymptotic covariance of $\sqrt{n}(\hat{\boldsymbol{\beta}}_{OLS} - \boldsymbol{\beta})$:

$$\boldsymbol{V}(\sqrt{n}\hat{\boldsymbol{\beta}}_{OLS}) = \sigma^2\boldsymbol{M}_2^{-1} \tag{3.32}$$

where $\boldsymbol{M}_2$ is the second-order moment matrix of $\boldsymbol{x}$. A consistent estimator of $\boldsymbol{M}_2$ is given by $(\frac{1}{n}\boldsymbol{X}'\boldsymbol{X})$. Hence $\boldsymbol{V}(\sqrt{n}\hat{\boldsymbol{\beta}}_{OLS}|\boldsymbol{X})$ can also be considered as a consistent estimator of $\boldsymbol{V}(\sqrt{n}\hat{\boldsymbol{\beta}}_{OLS})$.

Under the assumptions of homoscedasticity and normality of ϵ_i, the estimation of $\boldsymbol{\beta}$ and σ^2 may be performed using the ML methods discussed in Section 3 of this chapter. The density of y_i given $\boldsymbol{x}_i$, $\boldsymbol{\beta}$, and σ^2 and the corresponding log-likelihood function divided by the sample size n are given by

$$f(y_i|\boldsymbol{x}_i,\boldsymbol{\beta},\sigma^2) = \frac{1}{\sqrt{2\pi\sigma^2}}\exp\left(-\frac{1}{2\sigma^2}(y_i - \boldsymbol{x}_i'\boldsymbol{\beta})'(y_i - \boldsymbol{x}_i'\boldsymbol{\beta})\right), \tag{3.33}$$

$$l(\boldsymbol{\beta},\sigma^2) = \frac{1}{n}\sum_{i=1}^{n}\left(-\frac{1}{2\sigma^2}(y_i - \boldsymbol{x}_i'\boldsymbol{\beta})'(y_i - \boldsymbol{x}_i'\boldsymbol{\beta})\right) - \ln\sqrt{2\pi\sigma^2}\,. \tag{3.34}$$

Maximization of $l(\boldsymbol{\beta},\sigma^2)$ by differentiating $l(\boldsymbol{\beta},\sigma^2)$ with respect to $\boldsymbol{\beta}$ and σ^2 and solving the likelihood equations $\partial l(\boldsymbol{\beta},\sigma^2)/\partial\boldsymbol{\beta} = \mathbf{0}$ and $\partial l(\boldsymbol{\beta},\sigma^2)/\partial\sigma^2 = 0$ yields the ML estimates

$$\hat{\boldsymbol{\beta}}_{ML} = (\boldsymbol{X}'\boldsymbol{X})^{-1}\boldsymbol{X}'\boldsymbol{y} \quad , \tag{3.35}$$

$$\hat{\sigma}^2_{ML} = \frac{1}{n}\sum_{i=1}^{n}(y_i - \boldsymbol{x}_i'\hat{\boldsymbol{\beta}}_{ML})^2 \quad . \tag{3.36}$$

Application of standard ML theory shows that $\hat{\boldsymbol{\beta}}_{ML}$ is a strongly consistent and best asymptotically normal (BAN) estimator of $\boldsymbol{\beta}$:

$$\sqrt{n}(\hat{\boldsymbol{\beta}}_{ML} - \boldsymbol{\beta}) \overset{A}{\sim} \mathcal{N}\left(\mathbf{0}, \sigma^2(\frac{1}{n}\boldsymbol{X}'\boldsymbol{X})^{-1}\right). \tag{3.37}$$

Of course, $\hat{\boldsymbol{\beta}}_{ML}$ is the same as the OLS estimator $\hat{\boldsymbol{\beta}}_{OLS}$ and therefore has the same finite sample properties. However, $\hat{\sigma}^2_{ML}$ is biased in small samples, but is an asymptotically unbiased and a strongly consistent estimator of σ^2. Since $\hat{\sigma}^2_{ML}$ is consistent it may be used to replace σ^2 in the asymptotic covariance matrix of $\sqrt{n}(\hat{\boldsymbol{\beta}}_{ML})$. Therefore, the usual z statistic may be derived,

$$z_j^\star = \frac{\hat{\beta}_j - \beta_j}{\sqrt{\hat{\sigma}^2 c_{jj}}} \quad , \tag{3.38}$$

which follows a univariate standard normal distribution asymptotically. An asymptotic two-sided confidence interval with probability $1-\alpha$ for β_j is given by

$$P\left(\hat{\beta}_j - z_{1-\alpha/2}\sqrt{\hat{\sigma}^2_{ML}c_{jj}} \le \beta_j \le \hat{\beta}_j + z_{1-\alpha/2}\sqrt{\hat{\sigma}^2_{ML}c_{jj}}\right) = 1-\alpha. \tag{3.39}$$

The $t_j^\star$ statistic of the small sample case is replaced by the $z_j^\star$ statistic. The null hypothesis $\beta_j = \beta_{j0}$ is rejected in favor of $\beta_j \ne \beta_{j0}$ for a given significance level α if

$$|z^\star| > z_{1-\alpha/2} \quad . \tag{3.40}$$

To test the null hypothesis $\boldsymbol{R\beta} - \boldsymbol{c} = \mathbf{0}$ against $\boldsymbol{R\beta} - \boldsymbol{c} \ne \mathbf{0}$ with rank$(\boldsymbol{R}) = s$, one uses the fact that

$$\sqrt{n}(\boldsymbol{R}\hat{\boldsymbol{\beta}}_{ML} - \mathbf{c}) \overset{A}{\sim} \mathcal{N}\left(\mathbf{0}, \sigma^2\boldsymbol{R}(\frac{1}{n}\boldsymbol{X}'\boldsymbol{X})^{-1}\boldsymbol{R}'\right), \tag{3.41}$$

which follows immediately from the asymptotic normality of $\sqrt{n}(\hat{\boldsymbol{\beta}}_{ML} - \boldsymbol{\beta})$. Standard results for quadratic forms of multivariate normal random variables (cf. Rao 1973, chap. 3, pp. 185) show that the statistic

$$W = (\boldsymbol{R}\hat{\boldsymbol{\beta}}_{ML} - \boldsymbol{c})'(\boldsymbol{R}(\boldsymbol{X}'\boldsymbol{X})^{-1}\boldsymbol{R}')^{-1}(\boldsymbol{R}\hat{\boldsymbol{\beta}}_{ML} - \boldsymbol{c})/\hat{\sigma}^2_{ML} \tag{3.42}$$

is distributed asymptotically as a χ^2 distributed random variable with s degrees of freedom if the null hypothesis is correct. The null hypothesis is rejected at significance level α if $W > \chi^2_{1-\alpha,s}$. The statistic W is called a Wald statistic. Other statistics like the LR and the LM statistic are discussed in Section 3. The Wald statistic is very similar to the $F^\star$ statistic of equation (3.29) except that $\hat{\sigma}^2_{OLS}$ has been replaced by $\hat{\sigma}^2_{ML}$ and the $F(s, n-q)$ distribution function is replaced by the χ^2_s distribution function divided by s. Consequently, the confidence ellipsoid with probability $1 - \alpha$ is for large samples given by

$$CE(1-\alpha) = \{\boldsymbol{\beta} : (\boldsymbol{\beta} - \hat{\boldsymbol{\beta}}_{ML})'(\boldsymbol{X}'\boldsymbol{X})(\boldsymbol{\beta} - \hat{\boldsymbol{\beta}}_{ML})/\hat{\sigma}^2_{ML} \leq \chi^2_{1-\alpha,q}\}. \tag{3.43}$$

What happens if the assumption of normality of ϵ_i given $\boldsymbol{x}_i$, is abandoned but the assumption of homoscedasticity is retained? In this case there is a model for the mean structure $\mu_i = \boldsymbol{x}_i'\boldsymbol{\beta}$ and the variance $V(\epsilon_i|\boldsymbol{x}_i) = \sigma^2$ but there is no model for the density $f(y_i|\boldsymbol{x}_i)$ of y_i given $\boldsymbol{x}_i$. Hence, only the first two moments of the unknown conditional density of y_i given $\boldsymbol{x}_i$ are specified. The parameter σ^2 is mathematically independent of $\boldsymbol{\beta}$, that is σ^2 can vary independently of the value of $\boldsymbol{\beta}$. It is—in relation to the parameter of interest, that is $\boldsymbol{\beta}$—a nuisance parameter of no interest by itself. The usual way to estimate $\boldsymbol{\beta}$ without the normality assumption is to apply OLS estimation yielding

$$\tilde{\boldsymbol{\beta}} = (\boldsymbol{X}'\boldsymbol{X})^{-1}\boldsymbol{X}'\boldsymbol{y} \tag{3.44}$$

and to rely on the large sample properties of $\tilde{\boldsymbol{\beta}}$ given $\boldsymbol{X}$.

$$\sqrt{n}(\tilde{\boldsymbol{\beta}} - \boldsymbol{\beta}) \overset{A}{\sim} \mathcal{N}\left(\boldsymbol{0}, \sigma^2(\frac{1}{n}\boldsymbol{X}'\boldsymbol{X})^{-1}\right) \tag{3.45}$$

The unknown parameter σ^2 is replaced either by the estimator $\hat{\sigma}^2_{OLS}$ or by $\hat{\sigma}^2_{ML}$ to derive the usual z and χ^2 statistics for statistical inference in large samples.

Instead of using the OLS principle one might as well use the ML method to derive the estimator $\tilde{\boldsymbol{\beta}}$. The question is whether the ML method for the estimation of mean structures may be used in general even if the model is not linear and if the true density $f_\star(y_i|\boldsymbol{x}_i)$ is not normal. The answer is yes. In fact the OLS principle may be viewed as a special case of the QGPML method of Section 6; thus, a linear mean structure with a nuisance parameter may be estimated using ML based on conditional normal density even though the true density $f_\star(y_i|\boldsymbol{x}_i)$ is not normal.

Until now homoscedasticity has been assumed. In many applications one may not be willing to assume that the variance of the error term is constant for all observations but may not know how to specify $V(\epsilon_i|\boldsymbol{x}_i)$ as a function of $\boldsymbol{x}_i$. In this case only the mean structure $\mu_i = \boldsymbol{x}_i'\boldsymbol{\beta}$ is specified but the variance of ϵ_i given $\boldsymbol{x}_i$ remains unspecified and may vary across the observations in an unknown way. Also, normality of ϵ_i cannot be assumed. Even in this case a consistent estimator of $\boldsymbol{\beta}$ can be derived using the ML method based on the wrong assumption that ϵ_i given $\boldsymbol{x}_i$ is normal with a constant variance of 1. The corresponding PML estimator is again $\hat{\boldsymbol{\beta}}_{PML} = (\boldsymbol{X}'\boldsymbol{X})^{-1}\boldsymbol{X}'\boldsymbol{y}$. However, the asymptotic covariance matrix of $\sqrt{n}(\hat{\boldsymbol{\beta}}_{PML} - \boldsymbol{\beta})$ is different from the ML covariance matrix and the estimator $\hat{\boldsymbol{\beta}}_{PML}$ is usually inefficient. The PML estimator $\hat{\boldsymbol{\beta}}_{PML}$ is consistent and asymptotically normal. The asymptotic covariance matrix $\boldsymbol{V}(\sqrt{n}\hat{\boldsymbol{\beta}}_{PML})$ of $\sqrt{n}\hat{\boldsymbol{\beta}}_{PML}$ is consistently estimated by

$$\hat{\boldsymbol{V}}(\sqrt{n}\hat{\boldsymbol{\beta}}_{PML}) = (\frac{1}{n}\boldsymbol{X}'\boldsymbol{X})^{-1}(\frac{1}{n}\boldsymbol{X}'\hat{\boldsymbol{\Omega}}\boldsymbol{X})(\frac{1}{n}\boldsymbol{X}'\boldsymbol{X})^{-1} \tag{3.46}$$

where $\hat{\boldsymbol{\Omega}} = \text{diag}\{e_1^2, \ldots, e_n^2\}$ is the diagonal matrix of squared residuals $e_i = y_i - \boldsymbol{x}_i'\hat{\boldsymbol{\beta}}$. This is the heteroscedastic consistent (HC) covariance matrix of White (1980), which is a special case of the so-called information sandwich. Confidence intervals and regions and test statistics for $\boldsymbol{\beta}$ may be derived from the asymptotic normality of $\hat{\boldsymbol{\beta}}_{PML}$ with $\hat{\boldsymbol{V}}(\sqrt{n}\hat{\boldsymbol{\beta}}_{PML})$ of equation (3.46). Corrections of $\hat{\boldsymbol{V}}(\sqrt{n}\hat{\boldsymbol{\beta}}_{PML})$ for small samples are given in MacKinnon and White (1985).

Finally, the violation of the assumption that the explanatory variables and the errors are uncorrelated is considered. If $E(\epsilon_i \boldsymbol{x}_i) \neq \boldsymbol{0}$ then the PML estimate $\hat{\boldsymbol{\beta}}_{PML}$ of $\boldsymbol{\beta}$ will be inconsistent for $\boldsymbol{\beta}_0$, as is immediately seen from

$$\hat{\boldsymbol{\beta}}_{PML} = (\boldsymbol{X}'\boldsymbol{X})^{-1}\boldsymbol{X}'\boldsymbol{y} = \boldsymbol{\beta} + \left(\frac{1}{n}\boldsymbol{X}'\boldsymbol{X}\right)^{-1}\left(\frac{1}{n}\boldsymbol{X}'\boldsymbol{\epsilon}\right). \tag{3.47}$$

If $\boldsymbol{x}_i$ and ϵ_i are correlated then $\frac{1}{n}\sum_{i=1}^{n}\boldsymbol{x}_i\epsilon_i$ will converge to $E(\boldsymbol{x}_i\epsilon_i) \neq \boldsymbol{0}$. This inconsistency of $\hat{\boldsymbol{\beta}}_{PML}$ typically occurs if μ_i is nonlinearly dependent on $\boldsymbol{x}_i$ or if errors correlated with $\boldsymbol{x}_i$ are introduced by choosing a "wrong" estimation method. However, $\hat{\boldsymbol{\beta}}_{PML}$ still minimizes the predicted error, which may be thought of as a discrepancy measure between the true distribution of y given $\boldsymbol{x}$ and the wrongly assumed normal distribution. Since misspecification of the mean structure cannot be remedied by using specific estimators, while all other violations of the classic assumptions can be remedied by using the PML estimator with the HC covariance matrix, only a test for misspecification of the mean structure will be given in the next subsection. The general topic of misspecification is vast and cannot be treated within this chapter. For a general survey of misspecification tests the reader is referred to Godfrey (1988) and MacKinnon (1992). A special treatment of the linear model is found in Long and Trivedi (1993).

2.3 Regression Diagnostics

Coefficients of Determination

The contributions of individual independent variables to predicting the dependent variable are usually quantified by computing standardized regression coefficients and the $t_j^\star$ or $z_j^\star$

statistics of the last subsection. The joint contribution of all variables in the linear regression model is quantified by the coefficient of determination

$$R^2 = \frac{\sum_{i=1}^{n}(y_i - \bar{y})^2 - \sum_{i=1}^{n}(y_i - \hat{y}_i)^2}{\sum_{i=1}^{n}(y_i - \bar{y})^2} = \frac{\sum_{i=1}^{n}(\hat{y}_i - \bar{y})^2}{\sum_{i=1}^{n}(y_i - \bar{y})^2} \tag{3.48}$$

where $\hat{y}_i = \hat{\mu}_i = \boldsymbol{x}_i'\hat{\boldsymbol{\beta}}$ and $\bar{y}$ is the arithmetic mean of the y_i's. This coefficient can be interpreted as the proportion of variability of y that is explained by the explanatory variables. It may also be interpreted as a coefficient of the proportional reduction of error (PRE) generated by the explanatory variables. As Greene (1990, chap. 6) notes, there are two problems with R^2. First, the value of R^2 is normed to lie in the interval $[0, 1]$ only if a column of ones is included in the regressor matrix $\boldsymbol{X}$. Second, R^2 always increases if variables are added to the model, even if the variable is not relevant and only uses up degrees of freedom. To alleviate the second shortcoming, an adjusted coefficient of determination is computed (Theil 1961):

$$R^2_{adj} = 1 - \frac{n}{n-q}\left(1 - R^2\right). \tag{3.49}$$

R^2 or R^2_{adj} are valuable guides for judging and discriminating between different regression models. If the R^2 of a regression model is rather low (as is often the case in social science applications), the estimates of the model parameters may be very unstable, and one should be not surprised to get quite different results if other samples are used or if other explanatory variables are included in the model. The coefficients of determination are derived from the sum of squared errors of a null model (without regressors) and the fitted model (with regressors), which may be considered as measures of the discrepancy between the data and the given model that are not normed. Additional discrepancy measures are discussed in Section 7.

Ordinary Residuals and a Test for Misspecification

The classical assumptions of correct specification, homoscedasticity, and normality can be checked informally by using residual plots, as discussed by Cook and Weisberg (1982), McCullagh and Nelder (1989, chap. 12) and more formally by specification tests (cf. Godfrey 1988; Long and Trivedi 1993; and MacKinnon 1992).

The first and most important question is whether the hypothesis $E(y|\boldsymbol{x}) = \boldsymbol{x}'\boldsymbol{\beta}$ correctly specifies the mean structure. One wants to know if there is a lack of fit and to check the alternative hypothesis that there does not exist a vector $\boldsymbol{b}$ such that $E(y|\boldsymbol{x}) = \boldsymbol{x}'\boldsymbol{b}$, for all $\boldsymbol{x}$ in the domain of the explanatory variables in $\boldsymbol{x}$. Correct specification of the mean structure is often confused with the consistent estimation of all or a subset of parameters in a statistical model. To clarify this issue, a model is considered where y is a function of $\boldsymbol{x}$ and an additional variable z (cf. Clogg, Petkova, and Shidaheh 1992):

$$y = \boldsymbol{x}'\boldsymbol{\beta} + z\gamma + \epsilon \quad \text{with} \quad E(\epsilon|\boldsymbol{x}, z) = 0 \quad \text{and} \quad E(y|\boldsymbol{x}, z) = \boldsymbol{x}'\boldsymbol{\beta} + z\gamma\,. \tag{3.50}$$

The expected value of y given only $\boldsymbol{x}$ is found by taking the expected value over z given $\boldsymbol{x}$. The model is correctly specified if there exists a vector $\boldsymbol{b}$ such that $E(y|\boldsymbol{x}) = \boldsymbol{x}'\boldsymbol{b}$ is a linear combination of the variables in $\boldsymbol{x}$ with weights $\boldsymbol{b}$. Without assumptions about z given $\boldsymbol{x}$, $E(y|\boldsymbol{x})$ is given by

$$E(y|\boldsymbol{x}) = \boldsymbol{x}'\boldsymbol{\beta} + E(z|\boldsymbol{x})\gamma \quad . \tag{3.51}$$

Therefore, $E(y|\boldsymbol{x})$ is in general not a linear combination of the variables in $\boldsymbol{x}$. However, if $\gamma = 0$, then $E(y|\boldsymbol{x})$ is linear in $\boldsymbol{x}$ and only a function of $\boldsymbol{\beta}$. $\boldsymbol{\beta}$ can be estimated consistently only from $\boldsymbol{x}$ without z. If $E(z|\boldsymbol{x}) = E(z)$, then $\boldsymbol{\beta}$ can be estimated from $\boldsymbol{x}$ alone with the exception of the regression constant. However, if $E(z|\boldsymbol{x})$ is a linear function of $\boldsymbol{x}$, say $E(z|\boldsymbol{x}) = \boldsymbol{x}'\boldsymbol{\delta}$ with parameter vector $\boldsymbol{\delta}$, then $E(y|\boldsymbol{x})$ is given by

$$E(y|\boldsymbol{x}) = \boldsymbol{x}'\boldsymbol{\beta} + \boldsymbol{x}'\boldsymbol{\delta}\gamma = \boldsymbol{x}'(\boldsymbol{\beta} + \boldsymbol{\delta}\gamma) = \boldsymbol{x}'\boldsymbol{b} \quad \text{with} \quad \boldsymbol{b} = (\boldsymbol{\beta} + \boldsymbol{\delta}\gamma). \tag{3.52}$$

In this case, the mean structure $E(y|\boldsymbol{x})$ is correctly specified in the sense that $E(y|\boldsymbol{x}) = \boldsymbol{x}'\boldsymbol{b}$ is a linear function of $\boldsymbol{x}$. On the other hand, the OLS estimator $\hat{\boldsymbol{b}}$ based only on $\boldsymbol{x}$ is a consistent estimator for $\boldsymbol{b} = (\boldsymbol{\beta} + \boldsymbol{\delta}\gamma)$, but is not a consistent estimator for $\boldsymbol{\beta}$, which may be of primary substantive interest. A misspecification test for the mean structure will therefore be a lack-of-fit test that may detect nonlinearities in the dependence of y on $\boldsymbol{x}$, but it cannot test for inconsistently estimated parameters induced by omitted variables.

Nonlinearities may be detected by using ordinary residuals. The ordinary residual in a linear model is defined by

$$e_i = y_i - \hat{\mu}_i = y_i - \boldsymbol{x}_i'\hat{\boldsymbol{\beta}} \tag{3.53}$$

where $\hat{\boldsymbol{\beta}}$ is the OLS, ML, or PML estimator. If the e_i's are plotted against the predicted values $\hat{\mu}_i$ or the explanatory variables $x_{ij}, j = 1, \ldots, q$, there should be no local or global trends. If there is systematic variation of the e_i values with $\hat{\mu}_i$ or x_{ij}, the mean structure may be misspecified. Even if there is no more systematic variation of the residuals about the variables in the model, they should also be plotted against explanatory variables that have not been included in the model until now. If there is additional systematic variation, these variables should be included to increase the predictive power of the model.

A formal lack-of-fit test for the specification of $E(y|\boldsymbol{x})$ based on the greatest partial sum of residuals has been proposed by Su and Wei (1991) in the context of generalized linear models. A simple formal test is the RESET test of Ramsey (1969), which checks for nonlinearities in the mean structure by adding p additional variables in the regression equation. Let $\hat{\mu}_i$ be the predicted value from the model $y_i = \boldsymbol{x}_i'\boldsymbol{\beta} + \epsilon$. The errors are assumed to be normally distributed with $\mathcal{N}(0, \sigma^2)$. The regression equation for the specification test is given by

$$y_i = \boldsymbol{x}_i'\boldsymbol{\beta} + \sum_{j=1}^{p} \hat{\mu}_i^{j+1}\alpha_j + \epsilon_i^\star \quad . \tag{3.54}$$

Since the predicted values are functions of $\boldsymbol{x}_i$, this equation introduces nonlinearities into the model–albeit in a restricted way. OLS regression yields the estimates $\hat{\boldsymbol{\beta}}$ and $\hat{\boldsymbol{\alpha}}$ as well as their estimated joint covariance matrix $\hat{\boldsymbol{V}}(\hat{\boldsymbol{\beta}}, \hat{\boldsymbol{\alpha}})$. The covariance matrix of $\hat{\boldsymbol{\alpha}}$ is denoted by $\hat{\boldsymbol{V}}(\boldsymbol{\alpha})$. The test stastic for a RESET test of order p is then given by

$$F^{\star} = \hat{\boldsymbol{\alpha}}' \hat{\boldsymbol{V}}(\boldsymbol{\alpha})^{-1} \hat{\boldsymbol{\alpha}}/p \quad , \tag{3.55}$$

which follows a $F(p, n-q-p)$ distribution under the null hypothesis of correct specification. If the assumptions of homoscedasticity and normality are not maintained, a modified version of the RESET test using the HC covariance matrix should be used (Davidon and MacKinnon 1985; Long and Trivedi 1993).

Standardized and Studentized Residuals

If the assumption of homoscedasticity $V(\epsilon_i | \boldsymbol{x}_i) = \sigma^2$ is maintained and σ^2 is estimated by $\hat{\sigma}^2_{OLS}$ or $\hat{\sigma}^2_{ML}$ denoted by $\hat{\sigma}^2$, one may use standardized and studentized residuals to check the assumption of constant variance. The standardized residual is defined by

$$r_i = \frac{e_i}{\hat{\sigma}} \quad . \tag{3.56}$$

The standardized residuals are supposed to have a variance of 1. This is only approximately true since the variance covariance matrix of $\boldsymbol{e} = (e_i, \ldots, e_n)'$ is, conditional on $\boldsymbol{X}$, given by

$$\boldsymbol{V}(\boldsymbol{e}) = E_{\epsilon}(\boldsymbol{y} - \boldsymbol{X}\hat{\boldsymbol{\beta}})(\boldsymbol{y} - \boldsymbol{X}\hat{\boldsymbol{\beta}})' = \sigma^2(\boldsymbol{I} - \boldsymbol{X}(\boldsymbol{X}'\boldsymbol{X})^{-1}\boldsymbol{X}'). \tag{3.57}$$

The matrix $\boldsymbol{H} = \boldsymbol{X}(\boldsymbol{X}'\boldsymbol{X})^{-1}\boldsymbol{X}'$ with elements $H_{il} = \boldsymbol{x}_i'(\boldsymbol{X}'\boldsymbol{X})^{-1}\boldsymbol{x}_l$ is called the hat matrix, since $\hat{\boldsymbol{y}} = \boldsymbol{H}\boldsymbol{y}$. $\boldsymbol{H}$ is idempotent and of rank q (cf. Rao 1973, chap. 1, p. 46), the matrix $\boldsymbol{I} - \boldsymbol{H}$ is idempotent with rank $n - q$. The estimated variance of e_i is therefore

$$\hat{V}(e_i) = \hat{\sigma}^2(1 - H_{ii}) \tag{3.58}$$

and varies across observations. Replacing σ^2 by $\hat{V}(e_i)$ yields the studentized residual:

$$s_i = \frac{e_i}{\hat{\sigma}(1 - H_{ii})^{1/2}} \quad . \tag{3.59}$$

The difference between the standardized and the studentized residuals will be relevant if the model matrix $\boldsymbol{X}$ contains many variables, relative to the number of observations. This is the case for loglinear models in the analysis of contingency tables, where similar residuals are used. Both residuals are used to check for homoscedasticity by plotting r_i or s_i against $\hat{\mu}_i$ and individual regressors x_{ij}. If homoscedasticity holds, the spread of the residuals should be the same for all values of $\hat{\mu}_i$ and/or x_{ij}.

These residuals can also be used to check the assumption of normality of errors by plotting the ordered residuals $r_{(i)}$ or $s_{(i)}$ on the y axis against the expected order statistics of a normal sample (full normal plot) on the x axis. The expected order statistics are approximated by $\Phi^{-1}[(i - 3/8)/(n + 1/4)]$ for $i = 1, \ldots, n$. If the errors are approximately normal, the plot yields approximately a straight line with a slope of 1 through the origin.

New approaches to judge the effect of adding explanatory variables on the residuals in linear regression by making use of fast computer graphics for animation and rotation have been proposed by Cook and Weisberg (1989). For a discussion of formal tests for homoscedasticity and normality the reader is referred to Godfrey (1988) and Long and Trivedi (1993).

Leverage and Influential Points

In addition to checking the residuals one should also check whether any point $\{y_i, \boldsymbol{x}_i\}$ of the sample has an overly large influence on the prediction of y_i and/or on the estimation of $\boldsymbol{\beta}$. The first check is to look for leverage points. Note that $\hat{\mu}_i$ is defined by

$$\hat{\mu}_i = \boldsymbol{x}_i'\hat{\boldsymbol{\beta}} = \boldsymbol{x}_i'(\boldsymbol{X}'\boldsymbol{X})^{-1}\boldsymbol{X}'\boldsymbol{y} = \sum_{l=1}^{n} H_{il}y_l = H_{ii}y_i + \sum_{l\neq i} H_{il}y_l \quad . \tag{3.60}$$

This equation implies that the predicted value for the ith observation depends on the contribution by the ith observation itself and by all other observations. If H_{ii} is relatively large in comparison to $H_{il}, l \neq i$, then the observation should be checked whether it is a outlier in $\boldsymbol{X}$ or whether it has a great effect on the estimation of $\boldsymbol{\beta}$. Since $\boldsymbol{H}$ is idempotent with rank q, its trace is q. Hence, the average value of H_{ii} should be q/n. If H_{ii} is much greater than q/n, the ith observation should be checked because the predicted value $\hat{\mu}_i$ is mainly based on y_i itself and not on a weighted average of all observations. Belsley, Kuh, and Welsch (1980) propose $2q/n$, and Bollen and Jackman (1985) propose $3q/n$ as critical values for H_{ii}. In large samples, the value $3q/n$ seems appropriate. To detect leverage points, one usually plots H_{ii} on the y axis against i on the x axis.

The second check is to look for the influence of a data point on the estimation of $\boldsymbol{\beta}$ by re-estimating $\boldsymbol{\beta}$ from $\boldsymbol{X}_{(i)}$, where $\boldsymbol{X}_{(i)}$ denotes the model matrix $\boldsymbol{X}$ without the observation i. The estimate $\hat{\boldsymbol{\beta}}_{(i)}$ without observation i is given by (cf. Cook and Weisberg 1982, chap. 3)

$$\hat{\boldsymbol{\beta}}_{(i)} = (\boldsymbol{X}_{(i)}'\boldsymbol{X}_{(i)})^{-1}\boldsymbol{X}_{(i)}'\boldsymbol{y}_{(i)} = \hat{\boldsymbol{\beta}} - (\boldsymbol{X}'\boldsymbol{X})\boldsymbol{x}_i(1 - H_{ii})^{-1}e_i \quad . \tag{3.61}$$

The influence of the ith data point is a mixture of the effect of possible outliers in $\boldsymbol{X}$ given in the values H_{ii} and of outliers in $\boldsymbol{y}$ given in e_i. To judge whether the distance between the estimates $\hat{\boldsymbol{\beta}}_{(i)}$ and of $\hat{\boldsymbol{\beta}}$ is relevant, one uses—for large samples—the definition of the asymptotic confidence ellipsoid

$$CE(1-\alpha) = \{\boldsymbol{\beta} : (\boldsymbol{\beta} - \hat{\boldsymbol{\beta}})'(\boldsymbol{X}'\boldsymbol{X})(\boldsymbol{\beta} - \hat{\boldsymbol{\beta}})/\hat{\sigma}^2 \leq \chi^2_{1-\alpha,q}\}, \tag{3.62}$$

which is centered at $\hat{\boldsymbol{\beta}}$. Cook (1977) therefore proposes the statistic

$$C_i = (\hat{\boldsymbol{\beta}}_{(i)} - \hat{\boldsymbol{\beta}})'(\boldsymbol{X}'\boldsymbol{X})(\hat{\boldsymbol{\beta}}_{(i)} - \hat{\boldsymbol{\beta}})/\hat{\sigma}^2 \quad . \tag{3.63}$$

which is compared to the critical value $\chi^2_{1-\alpha,q}$ for a given value of $1-\alpha$. The statistic C_i is called the Cook statistic. If $C_i > \chi^2_{1-\alpha,q}$ then the ith observation appears to have an unduly large influence on the estimation of $\boldsymbol{\beta}$ from the sample. Note that C_i is not a test statistic since the common asymptotic distribution of $\hat{\boldsymbol{\beta}}_{(i)}$ and $\hat{\boldsymbol{\beta}}$ has not been derived.

Example: Dependence of Income on Schooling and Professional Experience

To illustrate the previous material, the dependence of monthly income on schooling and professional experience is considered. This dependence has often been discussed in human capital theory (Becker 1975; Mincer 1958). The population consists of male employees working full time in West Germany in 1984. The data come from the first wave (1984) of the German Socio–Economic panel (Wagner, Schupp, and Rendtel 1991). The analysis is performed on a random subset of size 300 of the original data. Three different models are considered. The models and their estimated coefficients are given in Table 1.

Table 1. Three Alternative Models for the Explanation of Log Income

Explanatory variables	Model 1	Model 2	Model 3
Regression constant	7.249	6.825	7.213
	(0.096) [0.095]	(0.117) [0.106]	(0.120) [0.112]
Years of schooling (S)	0.058	0.062	0.029
	(0.007) [0.007]	(0.007) [0.007]	(0.008) [0.008]
Professional experience (P)	0.058	0.445	0.373
in steps of ten years	(0.014) [0.014]	(0.068) [0.062]	(0.065) [0.059]
Professional experience	—	-0.082	-0.070
squared (PS)	—	(0.014) [0.013]	(0.014) [0.012]
Qualified employee (Q)	—	—	0.142
	—-	—	(0.034) [0.033]
Highly qualified employee (HQ)	—	—	0.332
	—	—	(0.046) [0.050]
R^2	0.186	0.265	0.376
R^2_{adj}	0.180	0.258	0.365
df	297	296	294

NOTE: The variables Q and HQ are dummy variables. $HQ = 1$ if the employee is a highly qualified white collar worker. The values in parentheses are the ML standard errors, the values in brackets are the PML standard errors which are heteroscedasticity consistent. All coefficients are significant using an $\alpha = 0.05$ test level using t_j^* or z_j^* statistics.

In human capital theory, interest centers on the effect of S on log income Y. The coefficient $\beta_S \times 100$ as the derivative of the expected value of log income with respect to years of schooling gives approximately the percent change of income for an additional year of schooling. In the first model, only S and P are entered as explanatory variables. The value of $\hat{\beta}_S \approx 0.06$ indicates a 6% increase of income for each additional year of schooling. Approximately the same effect is achieved for every 10 years of professional experience. The $R^2_{\text{adj}} = 0.18$ shows that 18% of the variance of log income has been explained by the linear model $E(Y|S,P) = \beta_1 + S\beta_2 + P\beta_3$. The standard errors of the ML and PML estimators are practically equal, indicating that the assumption of homoscedasticity may be maintained. Hence, superficially, this model looks quite good. However, to detect possible nonlinearities, the ordinary residuals e_i are plotted against the predicted values $\hat{\mu}_i$ and against the explanatory variables. The plot of the residuals e_i against professional experience is shown in Figure 1.

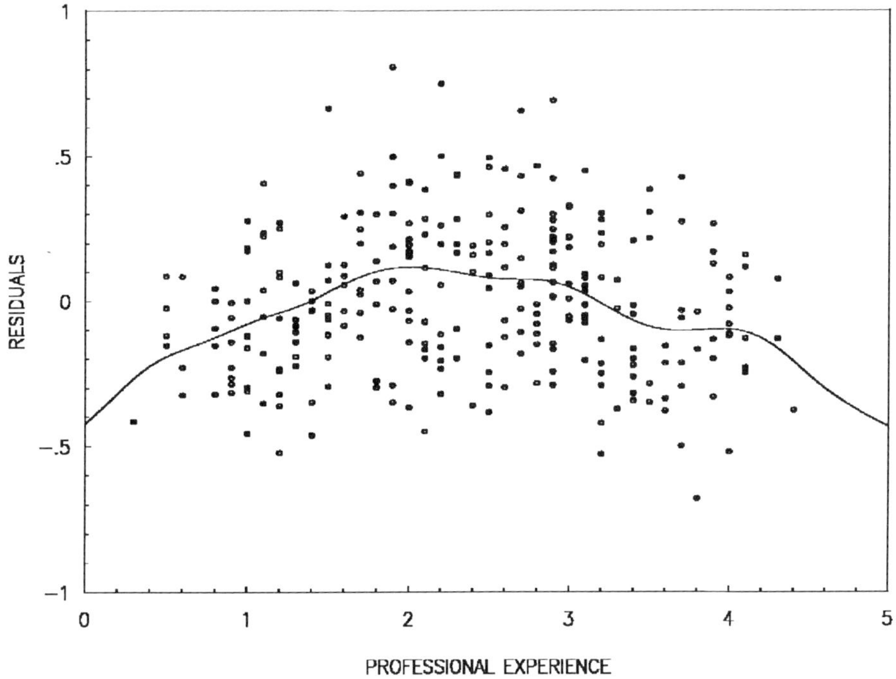

FIGURE 1. Standardized Residuals of Model 1 Versus Professional Experience. The fitted line is a local second-order polynomial fitted with OLS in the variable P for Y where the weights decrease with distance to the point for which the estimate $\hat{Y}$ is computed (cf. McLain 1974).

The residuals show a quadratic trend with professional experience indicating a marked deviation from linearity. A RESET test of order 3 for misspecification yields a test statistic of $F^\star = 2.336$ with a computed $\alpha = 0.074$, which is not significant at the $\alpha = 0.05$ test level but is significant at the $\alpha = 0.1$ test level. Taking both results into account, the inclusion of a quadratic term for P appears warranted. Model 2 includes the variable $PS = P$ squared. The results show a marked increase in R^2_{adj}. Plots of the residuals against the predicted values and the explanatory variables show no local or global trends. The expected value $E(Y|S,P)$ can safely be described as a linear function of the form $E(Y|S,P) = \beta_1 + S\beta_2 + P\beta_3 + PS\beta_4$.

The RESET test of order 3 gives now an $F^\star = 1.60$ with a computed test level of $\alpha = 0.189$ indicating that there is no more lack to fit. The frequency plot of the standardized residuals in Figure 2 shows no outliers and no deviance from normality.

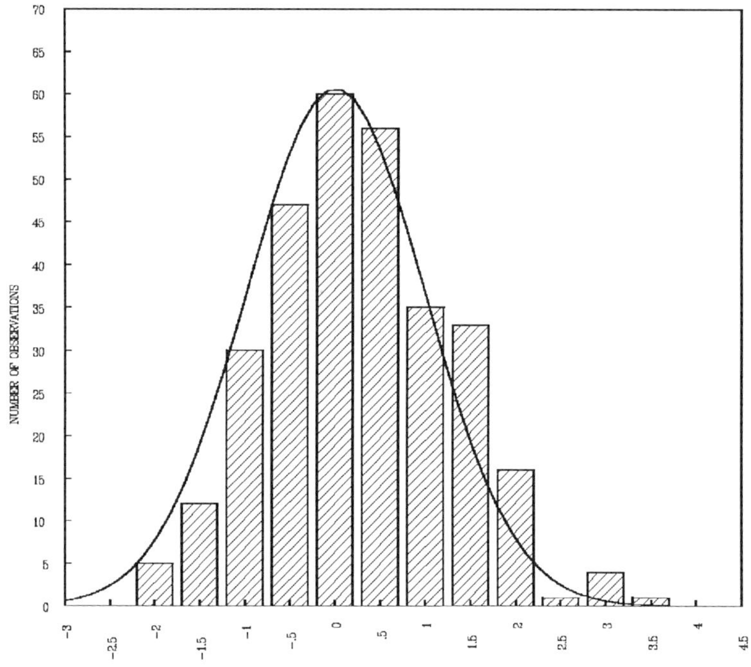

FIGURE 2. Histogram of Standardized Residuals of Model 2. The fitted line is the expected normal curve.

Only 15 values of the hat matrix exceed the value $3q/n = 0.04$, the highest of which has value 0.058. The highest value of the Cook distance is 0.0318, which is much lower than the comparison value $\chi^2_{0.95,6} = 12.6$. Hence, no important influential points are detected. There seems to be no evidence for model misspecification.

Does this result mean that this model is correctly specified in the sense that the estimate $\hat{\beta}_S$ of the parameter β is stable? This is not the case, as seen by looking at model 3, where two occupational dummy variables are included. The R^2_{adj} is pushed up further to a value of 0.364, which is quite high for a social science model. However, the estimate of β_S drops from 0.063 in model 2 to 0.029 in model 3, which is a drop of over 50%. The reason for this drop is that the occupational variables are correlated with schooling and act as moderator variables between schooling and income. The important lesson to be learned from this example is therefore that instabilities of the regression coefficients can only be detected by actually including additional explanatory variables.

2.4 Multivariate Linear Regression

In the multivariate linear regression model one collects G univariate linear regressions of the form $y_{ig} = \boldsymbol{x}'_{ig}\boldsymbol{\beta}_g + \epsilon_{ig}$ for each component y_{ig} of the ith dependent multivariate observation $\boldsymbol{y}_i$ of dimension G. The regressors $\boldsymbol{x}_{ig}$ for the univariate regressions are collected in a matrix $\boldsymbol{X}_i$ of dimension $G \times q$, the regression coefficients are collected in the vector $\boldsymbol{\beta}$ of length q. The error terms are collected in the $G \times 1$ vector ϵ_i. The model may then be written as

$$\boldsymbol{y}_i = \boldsymbol{X}_i\boldsymbol{\beta} + \epsilon_i, \quad i = 1, \ldots, n. \tag{3.64}$$

Important special cases are illustrated with two components of $\boldsymbol{y}_i$. In the first case, the explanatory variables are equal for each $g = 1, 2$, but each dependent variable is described by different coefficient vectors.

$$\begin{pmatrix} y_1 \\ y_2 \end{pmatrix} = \begin{pmatrix} \boldsymbol{x}'\boldsymbol{\beta}_1 \\ \boldsymbol{x}'\boldsymbol{\beta}_2 \end{pmatrix} + \begin{pmatrix} \epsilon_1 \\ \epsilon_2 \end{pmatrix} = \begin{pmatrix} \boldsymbol{x}' & \boldsymbol{0}' \\ \boldsymbol{0}' & \boldsymbol{x}' \end{pmatrix} \begin{pmatrix} \boldsymbol{\beta}_1 \\ \boldsymbol{\beta}_2 \end{pmatrix} + \begin{pmatrix} \epsilon_1 \\ \epsilon_2 \end{pmatrix} \tag{3.65}$$

If the regressors for all dependent variables are the same, then $\boldsymbol{X}_i$ may be written as $\boldsymbol{I} \otimes \boldsymbol{x}'_i$, where $\boldsymbol{I}$ is the identity matrix and $\otimes$ is the Kronecker product. (The Kronecker product $\boldsymbol{C} = \boldsymbol{A} \otimes \boldsymbol{B}$ of the matrices $\boldsymbol{A} \sim I \times J$, $\boldsymbol{B} \sim K \times L$ is the $IK \times JL$ matrix $\boldsymbol{C} = (a_{ij}\boldsymbol{B})$, $i = 1, \ldots, I; j = 1, \ldots, J$.)

In the second case, the explanatory variables vary with g, but the regression coefficients do not change.

$$\begin{pmatrix} y_1 \\ y_2 \end{pmatrix} = \begin{pmatrix} \boldsymbol{x}'_1\boldsymbol{\beta} \\ \boldsymbol{x}'_2\boldsymbol{\beta} \end{pmatrix} + \begin{pmatrix} \epsilon_1 \\ \epsilon_2 \end{pmatrix} = \begin{pmatrix} \boldsymbol{x}'_1 \\ \boldsymbol{x}'_2 \end{pmatrix} \boldsymbol{\beta} + \begin{pmatrix} \epsilon_1 \\ \epsilon_2 \end{pmatrix} \tag{3.66}$$

This model is typically used for analyzing panel data where y_1 and y_2 denote the same variable observed at two time points and the explanatory variables vary over time.

In the third case, the two models are merged.

$$\begin{pmatrix} y_1 \\ y_2 \end{pmatrix} = \begin{pmatrix} \boldsymbol{x}'_1\boldsymbol{\beta} + \boldsymbol{z}'\boldsymbol{\gamma}_1 \\ \boldsymbol{x}'_2\boldsymbol{\beta} + \boldsymbol{z}'\boldsymbol{\gamma}_2 \end{pmatrix} + \begin{pmatrix} \epsilon_1 \\ \epsilon_2 \end{pmatrix} = \begin{pmatrix} \boldsymbol{x}'_1 & \boldsymbol{z}' & \boldsymbol{0}' \\ \boldsymbol{x}'_2 & \boldsymbol{0}' & \boldsymbol{z}' \end{pmatrix} \begin{pmatrix} \boldsymbol{\beta} \\ \boldsymbol{\gamma}_1 \\ \boldsymbol{\gamma}_2 \end{pmatrix} + \begin{pmatrix} \epsilon_1 \\ \epsilon_2 \end{pmatrix} \tag{3.67}$$

Again, this model is often used in the analysis of panel data where the explanatory variables $\boldsymbol{x}_1$ and $\boldsymbol{x}_2$ vary over time but have constant coefficients, and the coefficients associated with the time constant variable $\boldsymbol{z}$ may change over time.

To estimate the parameter vector $\boldsymbol{\beta}$ of the linear mean structure, the assumptions of the univariate case are adapted to the multivariate case. Furthermore, it is assumed that $\boldsymbol{X}_i$ is stochastic.

$$E(\epsilon_i|\boldsymbol{X}_i) = \boldsymbol{0} \quad \text{(correct specification of mean structure)} \tag{3.68}$$

$$V(\epsilon_i|\boldsymbol{X}_i) = \boldsymbol{\Sigma} \quad \text{(multivariate homoscedasticity)} \tag{3.69}$$

$$\epsilon_i \sim \mathcal{N}(\boldsymbol{0}, \boldsymbol{\Sigma}) \quad \text{(multivariate normality)} \tag{3.70}$$

The matrix $\boldsymbol{\Sigma}$ is the covariance matrix of the errors of the linear models for the dependent variables given the explanatory variables. If the dependent variables are observations of one variable over G time points, the elements of $\boldsymbol{\Sigma}$ are serial covariances. If the structure of $\boldsymbol{\Sigma}$ is known, more efficient estimates of $\boldsymbol{\beta}$ may be found by estimating the parameters of $\boldsymbol{\Sigma}$ and taking these parameters into account during the estimation of $\boldsymbol{\beta}$. For the moment, no specification of $\boldsymbol{\Sigma}$ is assumed.

Under the assumption of multivariate normality, the ML method may be used to estimate $\boldsymbol{\beta}$ based on the multivariate normal density:

$$f(\boldsymbol{y}_i|\boldsymbol{X}_i, \boldsymbol{\beta}, \boldsymbol{\Sigma}) = (2\pi)^{-G/2}|\boldsymbol{\Sigma}|^{-1/2}\exp\{-\frac{1}{2}(\boldsymbol{y}_i - \boldsymbol{X}_i\boldsymbol{\beta})'\boldsymbol{\Sigma}^{-1}(\boldsymbol{y}_i - \boldsymbol{X}_i\boldsymbol{\beta})\} \tag{3.71}$$

Using the property that the trace of the product of two matrices $\boldsymbol{A}, \boldsymbol{B}$ is commutative, that is, $\mathrm{tr}\boldsymbol{A}\cdot\boldsymbol{B} = \mathrm{tr}\boldsymbol{B}\cdot\boldsymbol{A}$ when $\boldsymbol{B}\cdot\boldsymbol{A}$ is well defined, one finds

$$(\boldsymbol{y}_i - \boldsymbol{X}_i\boldsymbol{\beta})'\boldsymbol{\Sigma}^{-1}(\boldsymbol{y}_i - \boldsymbol{X}_i\boldsymbol{\beta}) = \mathrm{tr}\left\{(\boldsymbol{y}_i - \boldsymbol{X}_i\boldsymbol{\beta})(\boldsymbol{y}_i - \boldsymbol{X}_i\boldsymbol{\beta})'\boldsymbol{\Sigma}^{-1}\right\}. \tag{3.72}$$

Hence, the kernel of the log-likelihood function of the sample is (after division by n):

$$l(\boldsymbol{\beta}, \boldsymbol{\Sigma}) = -\frac{1}{2}\left[\ln|\boldsymbol{\Sigma}| - \mathrm{tr}\left\{\left(\frac{1}{n}\sum_{i=1}^{n}(\boldsymbol{y}_i - \boldsymbol{X}_i\boldsymbol{\beta})(\boldsymbol{y}_i - \boldsymbol{X}_i\boldsymbol{\beta})'\right)\boldsymbol{\Sigma}^{-1}\right\}\right] \tag{3.73}$$

The log-likelihood function as a function of $\boldsymbol{\beta}$ and $\boldsymbol{\Sigma}$ can be maximized with respect to $\boldsymbol{\beta}$ and $\boldsymbol{\Sigma}$ simultaneously. This problem is taken up in Chapter 4 (Browne and Arminger), where the mean as well as the covariance structure are parameterized in a model. Here, the parameters of $\boldsymbol{\Sigma}$ are considered as nuisance parameters and a two-step strategy is used to estimate $\boldsymbol{\beta}$ and $\boldsymbol{\Sigma}$ (cf. Schmidt 1976, chap. 2, sec. 6).

If $\boldsymbol{\Sigma}$ is known, then maximization of $l(\boldsymbol{\beta}, \boldsymbol{\Sigma})$ yields the ML estimate

$$\hat{\boldsymbol{\beta}} = \left(\sum_{i=1}^{n}\boldsymbol{X}_i'\boldsymbol{\Sigma}^{-1}\boldsymbol{X}_i\right)^{-1}\sum_{i=1}^{n}\boldsymbol{X}_i'\boldsymbol{\Sigma}^{-1}\boldsymbol{y}_i = (\boldsymbol{X}'\boldsymbol{\Sigma}_{BD}^{-1}\boldsymbol{X})^{-1}\boldsymbol{X}'\boldsymbol{\Sigma}_{BD}^{-1}\boldsymbol{y}\,. \tag{3.74}$$

where $\boldsymbol{X}$ is the $nG \times q$ matrix of individual $\boldsymbol{X}_i$'s, $\boldsymbol{y}$ is the $nG \times 1$ vector of $\boldsymbol{y}_i$'s and $\boldsymbol{\Sigma}_{BD}$ is the $nG \times nG$ block-diagonal matrix $\boldsymbol{I}_n \otimes \boldsymbol{\Sigma}$, where $\boldsymbol{I}_n$ is the $n \times n$ identity matrix.

If Σ is unknown, Σ in (3.74) may be replaced by a consistent estimator $\tilde{\Sigma}$ which may be computed using the OLS estimator $\tilde{\beta}$ for β in a first estimation step:

$$\tilde{\beta} = (X'X)^{-1}X'y \quad . \tag{3.75}$$

$\tilde{\beta}$ minimizes the objective function

$$Q(\beta) = \text{tr}\left(\frac{1}{n}\sum_{i=1}^{n}(y_i - X_i\beta)(y_i - X_i\beta)'\right), \tag{3.76}$$

which is equivalent to ML estimation of β under the assumption that $\Sigma = I_G$. The OLS estimator will usually be inefficient because the covariances between the error terms are not taken into account. A consistent estimator of Σ is given by

$$\tilde{\Sigma} = \frac{1}{n}\sum_{i=1}^{n}(y_i - X_i\tilde{\beta})(y_i - X_i\tilde{\beta})' \quad . \tag{3.77}$$

Finally, the second-step estimator $\hat{\beta}$ may be computed as a weighted least squares estimator:

$$\hat{\beta} = \left(\sum_{i=1}^{n} X_i'\tilde{\Sigma}^{-1}X_i\right)^{-1}\sum_{i=1}^{n} X_i'\tilde{\Sigma}^{-1}y_i \quad . \tag{3.78}$$

$\sqrt{n}(\hat{\beta} - \beta)$ is asymptotically normal distributed with $\mathcal{N}(\mathbf{0}, V(\sqrt{n}\hat{\beta}))$. $V(\sqrt{n}\hat{\beta})$ is consistently estimated by

$$\hat{V}(\sqrt{n}\hat{\beta}) = \frac{1}{n}\sum_{i=1}^{n}(X_i'\tilde{\Sigma}^{-1}X_i)^{-1} \quad . \tag{3.79}$$

The asymptotic normality of $\hat{\beta}$ allows the construction of confidence intervals for individual parameters β_j, confidence ellipsoids for subsets of β, and z and χ^2 statistics for significance tests analogously to the univariate case.

If the assumption of normality of ϵ_i is given up, but the hypothesis of homoscedasticity is retained, then ML estimation of β from (3.78) still yields consistent and asymptotically normal estimators as discussed in Section 6. The estimated asymptotic covariance matrix is again given by (3.79).

Finally, the assumption of homoscedasticity is given up and replaced by general heteroscedasticity, that is $V(\epsilon_i) = \Omega_i, i = 1, \ldots, n$. If one uses ML estimation with the (possibly wrong) assumed conditional density of y_i given $X_i \sim \mathcal{N}(X_i\beta, I_G)$, one still gets the consistent estimator $\tilde{\beta}$ of (3.75), which is asymptotically normal. However, the asymptotic covariance matrix of $\sqrt{n}(\tilde{\beta} - \beta)$ must take the general heteroscedasticity into account. A consistent estimator of the asymptotic covariance matrix is given by

$$\hat{V}(\sqrt{n}\tilde{\beta}) = \left(\frac{1}{n}\sum_{i=1}^{n} X_iX_i'\right)^{-1}\left(\frac{1}{n}\sum_{i=1}^{n} X_i\tilde{\Omega}_iX_i'\right)\left(\frac{1}{n}\sum_{i=1}^{n} X_iX_i'\right)^{-1} \tag{3.80}$$

where $\tilde{\Omega}_i$ is $G \times G$ outer product matrix $(y_i - X_i\tilde{\beta})(y_i - X_i\tilde{\beta})'$ of residuals which corresponds to the squared residuals of the asymptotic covariance matrix of $\sqrt{n}(\hat{\beta} - \beta)$ in the univariate regression model.

3 Maximum Likelihood Estimation

3.1 Loglikelihood function

In this section the principle of maximum likelihood (ML) estimation, the related likelihood ratio (LR), Wald (W), and Lagrange multiplier (LM) test statistics and estimation under restriction of parameters are discussed. It is assumed that the data $\{\boldsymbol{y}_i, \boldsymbol{x}_i\}$, $i = 1, \ldots, n$ are independently and identically distributed (iid). $\boldsymbol{y}$ is a dependent variable of dimension $G \times 1$. The joint density of $\boldsymbol{y}$ and $\boldsymbol{x}$ is given by

$$f(\boldsymbol{y}, \boldsymbol{x}) = f_\star(\boldsymbol{y}|\boldsymbol{x})g(\boldsymbol{x}) \tag{3.81}$$

where $f_\star(\boldsymbol{y}|\boldsymbol{x})$ denotes the true conditional density of $\boldsymbol{y}$ given $\boldsymbol{x}$ and $g(\boldsymbol{x})$ denotes the marginal density of $\boldsymbol{x}$. If $\boldsymbol{y}$ consists of discrete random variables, the conditional densities are conditional probabilities.

In ML estimation, it is assumed that $f_\star(\boldsymbol{y}|\boldsymbol{x})$ is known up to an unknown parameter vector $\boldsymbol{\vartheta}$ of dimension $q \times 1$. The assumed density function indexed by $\boldsymbol{\vartheta}$ is denoted by $f(\boldsymbol{y}|\boldsymbol{x}, \boldsymbol{\vartheta})$, and there exists a parameter vector $\boldsymbol{\vartheta}_0$ such that $f_\star(\boldsymbol{y}|\boldsymbol{x}) = f(\boldsymbol{y}|\boldsymbol{x}, \boldsymbol{\vartheta}_0)$. The marginal density $g(\boldsymbol{x})$ is assumed to not depend on $\boldsymbol{\vartheta}$. For given $\boldsymbol{\vartheta}$, the density of the iid sample $\{\boldsymbol{y}_i, \boldsymbol{x}_i\}$, $i = 1, \ldots, n$, is

$$L(\boldsymbol{\vartheta}) = \prod_{i=1}^{n} f(\boldsymbol{y}_i|\boldsymbol{x}_i, \boldsymbol{\vartheta})g(\boldsymbol{x}_i). \tag{3.82}$$

If the values of the sample $\{\boldsymbol{y}_i, \boldsymbol{x}_i\}$ are considered fixed and the parameter vector variable, then $L(\boldsymbol{\vartheta})$ is a function of $\boldsymbol{\vartheta}$ and is called the likelihood function. The domain of admissible values of $\boldsymbol{\vartheta}$ is the parameter space Θ. If a value $\hat{\boldsymbol{\vartheta}} \in \Theta$ globally maximizes $L(\boldsymbol{\vartheta})$ in the given sample, it is called the ML estimator of $\boldsymbol{\vartheta}_0$.

Instead of maximizing $L(\boldsymbol{\vartheta})$ to find $\hat{\boldsymbol{\vartheta}}$, one may maximize the logarithm of $L(\boldsymbol{\vartheta})$ divided by n (normed log-likelihood function):

$$\tilde{l}(\boldsymbol{\vartheta}) = \frac{1}{n}\sum_{i=1}^{n} \ln f(\boldsymbol{y}_i|\boldsymbol{x}_i, \boldsymbol{\vartheta}) + \frac{1}{n}\sum_{i=1}^{n} \ln g(\boldsymbol{x}_i). \tag{3.83}$$

The maximization of $\tilde{l}(\boldsymbol{\vartheta})$ does not involve $g(\boldsymbol{x}_i)$. Hence, the second term can be omitted. It is only necessary to maximize the kernel of the normed log-likelihood function:

$$l(\boldsymbol{\vartheta}) = \frac{1}{n}\sum_{i=1}^{n} \ln f(\boldsymbol{y}_i|\boldsymbol{x}_i, \boldsymbol{\vartheta}) \tag{3.84}$$

Note that the same kernel is obtained if the explanatory variables are nonstochastic. However, asymptotic theory is different for the stochastic and the nonstochastic case. The kernel $l(\boldsymbol{\vartheta})$ and $\hat{\boldsymbol{\vartheta}}$ depend on the sample size n and may therefore be indexed by n.

3.2 Properties of the ML Estimator

To derive the properties of the ML estimator, the regularity assumptions in White (1982, 1983) are assumed to hold. Here, only those arguments necessary to derive the asymptotic properties of the ML estimator are sketched. Detailed technical proofs are found in Amemiya (1985), Gallant (1987), and White (1982, 1983). Here, the arguments of White are followed.

First, strong consistency of the ML estimator $\hat{\boldsymbol{\vartheta}}$ is considered. The kernel $l(\boldsymbol{\vartheta})$ of (3.84) is a normed sum of independent random variables. As $n \to \infty$, this sum, using a strong law of large numbers, converges strongly for all $\boldsymbol{\vartheta} \in \boldsymbol{\Theta}$ to its expected value, where the expectation is first taken over $\boldsymbol{y}$ given $\boldsymbol{x}$ and then over $\boldsymbol{x}$ (Amemiya 1973; Jennrich 1969; White 1982):

$$E_g[E_f \ln f(\boldsymbol{y}|\boldsymbol{x}, \boldsymbol{\vartheta})] = \int\int \ln f(\boldsymbol{y}|\boldsymbol{x}, \boldsymbol{\vartheta}) f(\boldsymbol{y}|\boldsymbol{x}, \boldsymbol{\vartheta}_0) d\boldsymbol{y} g(\boldsymbol{x}) d\boldsymbol{x} \quad . \tag{3.85}$$

E_g and E_f denote the expectation operators with respect to the marginal density $g(\boldsymbol{x})$ and the conditional density $f(\boldsymbol{y}|\boldsymbol{x}, \boldsymbol{\vartheta}_0)$ at the true parameter value $\boldsymbol{\vartheta}_0$.

The sequence of maximized kernel functions $l_n(\hat{\boldsymbol{\vartheta}}_n)$ converges almost surely (a.s.) to the maximum of $E_g[E_f \ln f(\boldsymbol{y}|\boldsymbol{x}, \boldsymbol{\vartheta})]$ as a function of $\boldsymbol{\vartheta}$. To find the maximum of $E_g[E_f \ln f(\boldsymbol{y}|\boldsymbol{x}, \boldsymbol{\vartheta})]$, one considers the following function:

$$\begin{aligned} I\left[f(\boldsymbol{y}|\boldsymbol{x}, \boldsymbol{\vartheta}_0), f(\boldsymbol{y}|\boldsymbol{x}, \boldsymbol{\vartheta})\right]) &= E_g\left[E_f \ln \frac{f(\boldsymbol{y}|\boldsymbol{x}, \boldsymbol{\vartheta}_0)}{f(\boldsymbol{y}|\boldsymbol{x}, \boldsymbol{\vartheta})}\right] \\ = E_g[E_f \ln f(\boldsymbol{y}|\boldsymbol{x}, \boldsymbol{\vartheta}_0)] &- E_g[E_f \ln f(\boldsymbol{y}|\boldsymbol{x}, \boldsymbol{\vartheta})] \quad . \end{aligned} \tag{3.86}$$

This function is the Kullback-Leibler Information Criterion (KLIC), which measures the average discrepancy between the density $f(\boldsymbol{y}|\boldsymbol{x}, \boldsymbol{\vartheta}_0)$ and $f(\boldsymbol{y}|\boldsymbol{x}, \boldsymbol{\vartheta})$. For any two density functions $f(\boldsymbol{y}|\boldsymbol{x}, \boldsymbol{\vartheta}_0)$ and $f(\boldsymbol{y}|\boldsymbol{x}, \boldsymbol{\vartheta})$ it may be shown with Jensen's inequality (Rao 1973, chap. 1 p. 59) that $I(\cdot, \cdot) \geq 0$. The value 0 is taken if and only if $f(\boldsymbol{y}|\boldsymbol{x}, \boldsymbol{\vartheta}_0) = f(\boldsymbol{y}|\boldsymbol{x}, \boldsymbol{\vartheta})$ a.s. (Almost surely is dropped in the sequel.) The term in the second line of (3.86) is the expected value of the log-likelihood function. It takes a maximum if the corresponding KLIC is 0, that is if $f(\boldsymbol{y}|\boldsymbol{x}, \boldsymbol{\vartheta}) = f(\boldsymbol{y}|\boldsymbol{x}, \boldsymbol{\vartheta}_0)$. If the model is identified, that is, $f(\boldsymbol{y}|\boldsymbol{x}, \boldsymbol{\vartheta}_0^{(1)}) = f(\boldsymbol{y}|\boldsymbol{x}, \boldsymbol{\vartheta}_0^{(2)}) \Rightarrow \boldsymbol{\vartheta}_0^{(1)} = \boldsymbol{\vartheta}_0^{(2)}$, then there exists a unique maximum $\boldsymbol{\vartheta}_0$ to which the sequence of ML estimators converges strongly. Hence, the ML estimator $\hat{\boldsymbol{\vartheta}}$ is strongly consistent. The critical point is here the identification of the parameters in the density. This must be analyzed for each specific model separately. Often, these expressions are not tractable analytically, and hence one must resort to checking local identifiability at the ML estimate in the sample. This is performed by checking whether the matrix of second derivatives of $l(\boldsymbol{\vartheta})$ or the estimated expected matrix of second derivatives is of full rank at $\hat{\boldsymbol{\vartheta}}$ (Rothenberg 1971).

The ML estimator is computed by first defining the score vector $\boldsymbol{s}(\boldsymbol{\vartheta})$ of the log-likelihood function

$$\boldsymbol{s}(\boldsymbol{\vartheta}) = \frac{\partial l(\boldsymbol{\vartheta})}{\partial \boldsymbol{\vartheta}} = \frac{1}{n} \sum_{i=1}^{n} \frac{\partial \ln f(\boldsymbol{y}_i|\boldsymbol{x}_i, \boldsymbol{\vartheta})}{\partial \boldsymbol{\vartheta}} \tag{3.87}$$

and then solving the likelihood equation

$$\boldsymbol{s}(\boldsymbol{\vartheta}) = \mathbf{0} \tag{3.88}$$

for $\boldsymbol{\vartheta}$. Solving the likelihood equations usually gives only a local maximum. Finding the global maximum cannot be insured.

Now, the asymptotic distribution of $\sqrt{n}(\hat{\boldsymbol{\vartheta}} - \boldsymbol{\vartheta}_0)$ is considered. The $q \times 1$ vector of first derivatives of the individual log-likelihood function $l_i(\boldsymbol{\vartheta}) = \ln f(\boldsymbol{y}_i|\boldsymbol{x}_i, \boldsymbol{\vartheta})$ at the value $\boldsymbol{\vartheta}$ is denoted by $\boldsymbol{q}_i(\boldsymbol{\vartheta})$ (the individual score vectors), and the $q \times q$ matrix of second derivatives is denoted by $\boldsymbol{Q}_i(\boldsymbol{\vartheta})$. If the index i is not used then the derivatives are denoted by $\boldsymbol{q}(\boldsymbol{\vartheta})$ and $\boldsymbol{Q}(\boldsymbol{\vartheta})$:

$$\boldsymbol{q}_i(\boldsymbol{\vartheta}) = \frac{\partial l_i(\boldsymbol{\vartheta})}{\partial \boldsymbol{\vartheta}} = \frac{\partial \ln f(\boldsymbol{y}_i|\boldsymbol{x}_i, \boldsymbol{\vartheta})}{\partial \boldsymbol{\vartheta}}, \quad \boldsymbol{q}(\boldsymbol{\vartheta}) = \frac{\partial \ln f(\boldsymbol{y}|\boldsymbol{x}, \boldsymbol{\vartheta})}{\partial \boldsymbol{\vartheta}}, \tag{3.89}$$

$$\boldsymbol{Q}_i(\boldsymbol{\vartheta}) = \frac{\partial^2 l_i(\boldsymbol{\vartheta})}{\partial \boldsymbol{\vartheta} \partial \boldsymbol{\vartheta}'} = \frac{\partial^2 \ln f(\boldsymbol{y}_i|\boldsymbol{x}_i, \boldsymbol{\vartheta})}{\partial \boldsymbol{\vartheta} \partial \boldsymbol{\vartheta}'}, \quad \boldsymbol{Q}(\boldsymbol{\vartheta}) = \frac{\partial^2 \ln f(\boldsymbol{y}|\boldsymbol{x}, \boldsymbol{\vartheta})}{\partial \boldsymbol{\vartheta} \partial \boldsymbol{\vartheta}'}. \tag{3.90}$$

A first-order Taylor expansion of (3.88) about $\boldsymbol{\vartheta}_0$ gives

$$\mathbf{0} = \frac{1}{n} \sum_{i=1}^{n} \boldsymbol{q}_i(\boldsymbol{\vartheta}_0) + \left(\frac{1}{n} \sum_{i=1}^{n} \boldsymbol{Q}_i(\tilde{\boldsymbol{\vartheta}}) \right) (\hat{\boldsymbol{\vartheta}} - \boldsymbol{\vartheta}_0) \tag{3.91}$$

where $\tilde{\boldsymbol{\vartheta}}$ lies on the line segment between $\hat{\boldsymbol{\vartheta}}$ and $\boldsymbol{\vartheta}_0$. Premultiplication by $\sqrt{n}$ and assuming that $\left(\frac{1}{n} \sum_{i=1}^{n} \boldsymbol{Q}_i(\tilde{\boldsymbol{\vartheta}})\right)$ is regular yields

$$\sqrt{n}(\hat{\boldsymbol{\vartheta}} - \boldsymbol{\vartheta}_0) = -\left(\frac{1}{n} \sum_{i=1}^{n} \boldsymbol{Q}_i(\tilde{\boldsymbol{\vartheta}}) \right)^{-1} \frac{1}{\sqrt{n}} \sum_{i=1}^{n} \boldsymbol{q}_i(\boldsymbol{\vartheta}_0). \tag{3.92}$$

The matrix of second derivatives converges to its expected value by a strong law of large numbers, that is,

$$\left(\frac{1}{n} \sum_{i=1}^{n} \boldsymbol{Q}_i(\tilde{\boldsymbol{\vartheta}}) \right) \xrightarrow{a.s.} E_g[E_f \boldsymbol{Q}(\boldsymbol{\vartheta}_0)] \equiv \boldsymbol{A}(\boldsymbol{\vartheta}_0). \tag{3.93}$$

The matrix $-\boldsymbol{A}(\boldsymbol{\vartheta}_0)$ is the Fisher information matrix.

The expected value of the first derivatives at $\boldsymbol{\vartheta}_0$ is $\mathbf{0}$ using the assumption that differentiation and integration can be interchanged:

$$\begin{aligned} E_g[E_f \frac{1}{\sqrt{n}} \sum_{i=1}^{n} \boldsymbol{q}_i(\boldsymbol{\vartheta}_0)] &= \sqrt{n} E_g \left[\int \frac{\partial \ln f(\boldsymbol{y}|\boldsymbol{x}, \boldsymbol{\vartheta}_0)}{\partial \boldsymbol{\vartheta}} f(\boldsymbol{y}|\boldsymbol{x}, \boldsymbol{\vartheta}_0) d\boldsymbol{y} \right] \\ &= \sqrt{n} E_g \left[\frac{\partial}{\partial \boldsymbol{\vartheta}} \int f(\boldsymbol{y}|\boldsymbol{x}, \boldsymbol{\vartheta}_0) d\boldsymbol{y} \right] = \mathbf{0} \quad . \end{aligned} \tag{3.94}$$

The notation $\partial \ln f(\boldsymbol{y}|\boldsymbol{x}, \boldsymbol{\vartheta}_0)/\partial\boldsymbol{\vartheta}$ denotes that the first derivative of $\ln f(\boldsymbol{y}|\boldsymbol{x}, \boldsymbol{\vartheta}_0)$ is evaluated at the point $\boldsymbol{\vartheta}_0$.

The covariance matrix of the first derivatives is derived by using the iid assumption and $E_g[E_f(q_i(\boldsymbol{\vartheta}_0))] = \mathbf{0}$:

$$\begin{aligned} \boldsymbol{V}\left(\frac{1}{\sqrt{n}}\sum_{i=1}^{n}\boldsymbol{q}_i(\boldsymbol{\vartheta}_0)\right) &= \frac{1}{n}\sum_{i=1}^{n}\boldsymbol{V}\left(\boldsymbol{q}_i(\boldsymbol{\vartheta}_0)\right) \\ = \frac{1}{n}\sum_{i=1}^{n}\boldsymbol{V}\left(\boldsymbol{q}(\boldsymbol{\vartheta}_0)\right) &= E_g E_f\left[\boldsymbol{q}(\boldsymbol{\vartheta}_0)\boldsymbol{q}'(\boldsymbol{\vartheta}_0)\right] \equiv \boldsymbol{B}(\boldsymbol{\vartheta}_0). \end{aligned} \tag{3.95}$$

Using a multivariate central limit theorem (Rao 1973, chap. 2, p. 128), the asymptotic distribution of the first derivatives is

$$\frac{1}{\sqrt{n}}\sum_{i=1}^{n}\boldsymbol{q}_i(\boldsymbol{\vartheta}_0) \overset{A}{\sim} \mathcal{N}(\mathbf{0}, \boldsymbol{B}(\boldsymbol{\vartheta}_0)). \tag{3.96}$$

Using Slutsky's theorem (Rao 1973, chap. 2, p. 122), premultiplication of the left side of (3.96) with $(-\boldsymbol{A}(\boldsymbol{\vartheta}_0))^{-1}$ yields the asymptotic distribution of $\sqrt{n}(\hat{\boldsymbol{\vartheta}} - \boldsymbol{\vartheta}_0)$, that is,

$$\sqrt{n}(\hat{\boldsymbol{\vartheta}} - \boldsymbol{\vartheta}_0) \overset{A}{\sim} \mathcal{N}\left(\mathbf{0}, (-\boldsymbol{A}(\boldsymbol{\vartheta}_0))^{-1}\boldsymbol{B}(\boldsymbol{\vartheta}_0)(-\boldsymbol{A}(\boldsymbol{\vartheta}_0))^{-1}\right). \tag{3.97}$$

This formulation of the asymptotic distribution of the parameter estimates $\hat{\boldsymbol{\vartheta}}$ involves the expected value of the second derivatives as well as the covariance matrix of the first derivatives.

An important simplification occurs if the central assumption of ML estimation, that is the equality of the assumed and the true conditional density function, is correct. In this case, one finds that

$$-\boldsymbol{A}(\boldsymbol{\vartheta}_0) = \boldsymbol{B}(\boldsymbol{\vartheta}_0). \tag{3.98}$$

This last equality is derived by using the differentiation rule for quotients:

$$\begin{aligned} -E_f\boldsymbol{Q}(\boldsymbol{\vartheta}_0) &= -\int\frac{\partial^2 \ln f(\boldsymbol{y}|\boldsymbol{x}, \boldsymbol{\vartheta}_0)}{\partial\boldsymbol{\vartheta}\partial\boldsymbol{\vartheta}'}f(\boldsymbol{y}|\boldsymbol{x}, \boldsymbol{\vartheta}_0)d\boldsymbol{y} \\ &= -\int\frac{\partial}{\partial\boldsymbol{\vartheta}}\left(\frac{\partial f(\boldsymbol{y}|\boldsymbol{x}, \boldsymbol{\vartheta}_0)}{\partial\boldsymbol{\vartheta}'}\cdot\frac{1}{f(\boldsymbol{y}|\boldsymbol{x}, \boldsymbol{\vartheta}_0)}\right)f(\boldsymbol{y}|\boldsymbol{x}, \boldsymbol{\vartheta}_0)d\boldsymbol{y} \\ &= -\int\frac{\partial^2 f(\boldsymbol{y}|\boldsymbol{x}, \boldsymbol{\vartheta}_0)}{\partial\boldsymbol{\vartheta}\partial\boldsymbol{\vartheta}'}d\boldsymbol{y} + E_f\left[\boldsymbol{q}(\boldsymbol{\vartheta}_0)\boldsymbol{q}'(\boldsymbol{\vartheta}_0)\right]. \end{aligned} \tag{3.99}$$

Interchanging integration and differentiation shows that the first term of the right hand side of the last equation is $\mathbf{0}$. Therefore, a simplified formula for the limiting distribution of the ML estimates $\hat{\boldsymbol{\vartheta}}$ is given by

$$\sqrt{n}(\hat{\boldsymbol{\vartheta}} - \boldsymbol{\vartheta}_0) \overset{A}{\sim} \mathcal{N}\left(\mathbf{0}, \boldsymbol{B}(\boldsymbol{\vartheta}_0)^{-1}\right) = \mathcal{N}\left(\mathbf{0}, (-\boldsymbol{A}(\boldsymbol{\vartheta}_0))^{-1}\right). \tag{3.100}$$

The implication of this result is that the asymptotic covariance matrix of $\hat{\boldsymbol{\vartheta}}$ can be computed either using the matrix of the cross-products of the first derivatives or using the matrix of second derivatives. Consistent estimates of the Fisher Information matrix $-\boldsymbol{A}(\boldsymbol{\vartheta}_0)$ and of $\boldsymbol{B}(\boldsymbol{\vartheta}_0)$ are found by replacing $\boldsymbol{\vartheta}_0$ by $\hat{\boldsymbol{\vartheta}}$ in $-\boldsymbol{A}(\boldsymbol{\vartheta}_0)$ and $\boldsymbol{B}(\boldsymbol{\vartheta}_0)$ if the integrals can be solved analytically. Otherwise, one may use the moment estimators

$$-\hat{\boldsymbol{A}}(\hat{\boldsymbol{\vartheta}}) = -\frac{1}{n}\sum_{i=1}^{n} \boldsymbol{Q}_i(\hat{\boldsymbol{\vartheta}}) \quad , \tag{3.101}$$

$$\hat{\boldsymbol{B}}(\hat{\boldsymbol{\vartheta}}) = \frac{1}{n}\sum_{i=1}^{n} \boldsymbol{q}_i(\hat{\boldsymbol{\vartheta}})\boldsymbol{q}_i(\hat{\boldsymbol{\vartheta}})' \quad . \tag{3.102}$$

$-\hat{\boldsymbol{A}}(\hat{\boldsymbol{\vartheta}})$ is called the observed Fisher information matrix since the expectation is approximated by taking the mean. $\hat{\boldsymbol{B}}(\hat{\boldsymbol{\vartheta}})$ is the outer product gradient form of the Fisher information matrix (cf. Berndt, Hall, Hall, and Hausman 1974). This form is particularly useful if one uses the Lagrange multiplier test considered subsequently. As a consequence, different, but under correct specification, asymptotically equivalent estimators of the asymptotic covariance matrix are given by

$$\hat{\boldsymbol{V}}_1(\sqrt{n}\hat{\boldsymbol{\vartheta}}) = -\left(\frac{1}{n}\sum_{i=1}^{n} \boldsymbol{Q}_i(\hat{\boldsymbol{\vartheta}})\right)^{-1} \Rightarrow \hat{\boldsymbol{V}}_1(\hat{\boldsymbol{\vartheta}}) = -\left(\sum_{i=1}^{n} \boldsymbol{Q}_i(\hat{\boldsymbol{\vartheta}})\right)^{-1} \tag{3.103}$$

$$\hat{\boldsymbol{V}}_2(\sqrt{n}\hat{\boldsymbol{\vartheta}}) = \left(\frac{1}{n}\sum_{i=1}^{n} \boldsymbol{q}_i(\hat{\boldsymbol{\vartheta}})\boldsymbol{q}_i(\hat{\boldsymbol{\vartheta}})'\right)^{-1} \Rightarrow \hat{\boldsymbol{V}}_2(\hat{\boldsymbol{\vartheta}}) = \left(\sum_{i=1}^{n} \boldsymbol{q}_i(\hat{\boldsymbol{\vartheta}})\boldsymbol{q}_i(\hat{\boldsymbol{\vartheta}})'\right)^{-1} . \tag{3.104}$$

These estimated covariance matrices, $\hat{\boldsymbol{V}}_1(\sqrt{n}\hat{\boldsymbol{\vartheta}})$ and $\hat{\boldsymbol{V}}_2(\sqrt{n}\hat{\boldsymbol{\vartheta}})$, converge to their expected value, where the expectation is not only taken over the $\boldsymbol{y}$ variables given $\boldsymbol{x}$ but also over $\boldsymbol{x}$.

Third, the ML estimator is asymptotically efficient. Its asymptotic covariance matrix is equal to the Rao-Cramér bound for consistent estimators (cf. Rao 1973, chap. 5, p. 350).

Summing up these results, we find that the ML estimator $\hat{\boldsymbol{\vartheta}}$ is a consistent, asymptotically normal, and efficient estimator for $\boldsymbol{\vartheta}_0$ under the assumption of correct specification of the conditional density function. The properties sketched above only apply for large samples. The small sample properties may be quite different. The estimates of the asymptotic covariance matrix may be used to derive confidence intervals for individual parameters and elliptic confidence regions for sets of parameters as in Section 2. The subject of tests is taken up next.

3.3 Likelihood Ratio, Wald, and Lagrange Multiplier Tests

Hypotheses about the parameters $\boldsymbol{\vartheta}_0$ of a specific model are usually cast in the form of restrictions:

$$\boldsymbol{r}(\boldsymbol{\vartheta}_0) = \boldsymbol{0} \tag{3.105}$$

where $\boldsymbol{r}(\cdot)$ is a $s \times 1$ vector of linear or nonlinear restrictions with $s \leq q$, which is assumed to be a continously differentiable function of the parameter vector $\boldsymbol{\vartheta}$. The $s \times q$ matrix of first derivatives at $\boldsymbol{\vartheta}$ is denoted by $\boldsymbol{R}(\boldsymbol{\vartheta})$ and has full rank at $\boldsymbol{\vartheta}_0$. This condition insures that there are no restrictions which are dependent on each other and therefore superfluous. The most important example are linear restrictions of the form

$$\boldsymbol{R}\boldsymbol{\vartheta}_0 - \boldsymbol{c} = \boldsymbol{0} \quad . \tag{3.106}$$

Restrictions like setting parameters equal to fixed values such as 0 or setting parameters equal to each other or writing parameters as linear functions of other parameters may be formulated in this way.

The ML estimate of $\boldsymbol{\vartheta}_0$ has been denoted by $\hat{\boldsymbol{\vartheta}}$. To test the null hypothesis $H_0 : \boldsymbol{r}(\boldsymbol{\vartheta}_0) = \boldsymbol{0}$ against the alternative hypothesis $H_1 : \boldsymbol{r}(\boldsymbol{\vartheta}_0) \neq \boldsymbol{0}$, one estimates $\boldsymbol{\vartheta}_0$ under H_0. In general, one maximizes the log-likelihood function under the restriction using the Lagrange multiplier method

$$l(\boldsymbol{\vartheta}, \boldsymbol{\lambda}) = l(\boldsymbol{\vartheta}) + \boldsymbol{\lambda}'\boldsymbol{r}(\boldsymbol{\vartheta}) \tag{3.107}$$

where $\boldsymbol{\lambda}$ is the $s \times 1$ vector of Lagrange multipliers. $\boldsymbol{\lambda}'\boldsymbol{r}(\boldsymbol{\vartheta}) = \boldsymbol{0}$ for $\boldsymbol{\vartheta} = \boldsymbol{\vartheta}_0$ and $l(\boldsymbol{\vartheta}, \boldsymbol{\lambda})$ is maximized with respect to $\boldsymbol{\vartheta}$ and $\boldsymbol{\lambda}$ (cf. Goldberger 1964). The estimate of $\boldsymbol{\vartheta}_0$ under H_0 is denoted by $\tilde{\boldsymbol{\vartheta}}$. To test the null hypothesis against the unrestricted alternative, one computes one of the three classical asymptotic test statistics, that is the likelihood ratio, the Wald, or the Lagrange multiplier statistic. A thorough discussion of the derivation and properties of these statistics is found in Godfrey (1988).

The first test statistic is the likelihood ratio statistic

$$LR = 2n(l(\hat{\boldsymbol{\vartheta}}) - l(\tilde{\boldsymbol{\vartheta}})). \tag{3.108}$$

The difference $(l(\hat{\boldsymbol{\vartheta}}) - l(\tilde{\boldsymbol{\vartheta}}))$ is multiplied by n because the normed log-likelihood function has been denoted by $l(.)$. If the null hypothesis is true, then LR is asymptotically distributed as a central χ^2 variable with s degrees of freedom. The calculation of LR involves the knowledge of the conditional density function of $\boldsymbol{y}$ given $\boldsymbol{x}$ and computing the ML estimate $\hat{\boldsymbol{\vartheta}}$ as well as the estimate $\tilde{\boldsymbol{\vartheta}}$; $g(\boldsymbol{x})$ does not depend on $\boldsymbol{\vartheta}$ by assumption and therefore drops out. The ML estimation must be performed twice, which may be time consuming when the models are complicated, the data sets are large, or the numerical procedures are cumbersome.

The second test statistic is the Wald statistic:

$$W = n(\hat{\boldsymbol{\vartheta}} - \boldsymbol{\vartheta}_0)'\hat{\boldsymbol{R}}'(\hat{\boldsymbol{R}}\hat{\boldsymbol{V}}\hat{\boldsymbol{R}}')^{-1}\hat{\boldsymbol{R}}(\hat{\boldsymbol{\vartheta}} - \boldsymbol{\vartheta}_0) \tag{3.109}$$

where $\hat{\boldsymbol{R}} = \boldsymbol{R}(\hat{\boldsymbol{\vartheta}})$ and $\hat{\boldsymbol{V}}$ is a consistent estimator of the asymptotic covariance of $\sqrt{n}\hat{\boldsymbol{\vartheta}}$. W is also distributed as a central χ^2 under H_0 with s degrees of freedom. Calculation of the Wald statistic involves only the computation of the unrestricted ML estimate $\hat{\boldsymbol{\vartheta}}$. Additionally, it relies on the asymptotic distribution of $\hat{\boldsymbol{\vartheta}}$ and not necessarily on the knowledge of the conditional density function of $\boldsymbol{y}$ given $\boldsymbol{x}$. This property will be taken up in the context of PML estimation.

The third test statistic is Rao's efficient score (also called the Lagrange multiplier statistic):

$$LM = n\left(\frac{\partial l(\tilde{\boldsymbol{\vartheta}})}{\partial \boldsymbol{\vartheta}}\right)' \tilde{\boldsymbol{V}} \left(\frac{\partial l(\tilde{\boldsymbol{\vartheta}})}{\partial \boldsymbol{\vartheta}}\right) \tag{3.110}$$

where $\tilde{\boldsymbol{V}}$ is a consistent estimator of the asymptotic covariance matrix of $\sqrt{n}\tilde{\boldsymbol{\vartheta}}$. LM is also distributed as a central χ^2 under the H_0 with s degrees of freedom. Calculation of the LM statistic involves only the computation of the restricted ML estimate $\tilde{\boldsymbol{\vartheta}}$. Although only the computation of the restricted estimate $\tilde{\boldsymbol{\vartheta}}$ is necessary for the LM statistic, the first and possibly second derivatives with respect to *all* parameters including the restricted parameters must be obtained. The derivation of the LM statistic is sketched in Section 5 in the context of PML estimation.

If the outer product moment estimator of the asymptotic covariance matrix is used, then the LM statistic can be written in a very simple form. Let $\tilde{\boldsymbol{W}}$ denote the $n \times q$ matrix of first derivatives $\boldsymbol{q}_i(\tilde{\boldsymbol{\vartheta}})$ and let $\mathbf{1}_n$ denote the $n \times 1$ vector of ones. Then the score vector is

$$\frac{\partial l(\tilde{\boldsymbol{\vartheta}})}{\partial \boldsymbol{\vartheta}} = \frac{1}{n}\tilde{\boldsymbol{W}}'\mathbf{1}_n \tag{3.111}$$

and the outer product estimate of the asymptotic covariance matrix of $\sqrt{n}\tilde{\boldsymbol{\vartheta}}$ is given by

$$\tilde{\boldsymbol{V}}(\sqrt{n}\tilde{\boldsymbol{\vartheta}}) = (\frac{1}{n}\tilde{\boldsymbol{W}}'\tilde{\boldsymbol{W}})^{-1} \quad . \tag{3.112}$$

As a result, the LM statistic may be written as

$$LM = \mathbf{1}_n'\tilde{\boldsymbol{W}}(\tilde{\boldsymbol{W}}'\tilde{\boldsymbol{W}})^{-1}\tilde{\boldsymbol{W}}'\mathbf{1}_n \quad . \tag{3.113}$$

The null hypothesis is rejected at a given significance level α if the test statistic chosen is greater than $\chi^2_{1-\alpha,s}$. The three test statistics are asymptotically equivalent under the null, that is the rejection rate of H_0 is the same for LR, W, and LM if the sample size is large. However, for small sample sizes, the test statistics may differ considerably and yield different conclusions. For tests of linear restrictions on the parameters of the classical linear regression model with normally distributed errors, there exists the following inequality relation:

$$W \geq LR \geq LM \tag{3.114}$$

A proof of this inequality is found in Godfrey (1988, chap. 2). However, in the case of the classical linear regression model with normal errors, exact t and F statistics instead of the asymptotic statistics are used to test restrictions in the first place.

The Wald test should be used with caution for testing nonlinear restrictions. Hypotheses can often be expressed with different but equivalent restrictions. Lafontaine and White (1986) consider the hypothesis H_0: $\beta = 1$ in a simple linear model $y_i = \alpha + x_i\beta + z_i\gamma + \epsilon_i$, $i = 1, \dots, 17$ with data from Theil (1971, p. 102). Equivalent expressions for H_0 are $\beta^2 = 1$, $\beta^3 = 1$, $\beta^{-1} = 1$ and so forth. The authors show that the corresponding test statistics of the Wald test are not equal and lead to rather different conclusions about the rejection of the original null hypothesis. Further enquiries into the behavior of the Wald tests of non-linear restrictions are found in Phillips and Park (1988) and Dagenais and Dufour (1991).

Finally, an important special case is considered. There are no restrictions on the first $q-s$ parameters which are collected in the vector $\boldsymbol{\vartheta}_0^{(1)}$, and the s restrictions on the remaining parameters collected in $\boldsymbol{\vartheta}_0^{(2)}$ are of the form

$$H_0 : \boldsymbol{\vartheta}_0^{(2)} = \boldsymbol{c}, \quad \text{for instance } \boldsymbol{c} = \boldsymbol{0} \quad . \tag{3.115}$$

$\boldsymbol{\vartheta}_0^{(2)}$ may not be a function of $\boldsymbol{\vartheta}_0^{(1)}$. Further, it is assumed that the score vector and the consistent estimator $\hat{\boldsymbol{V}}$ of the asymptotic covariance matrix of $\sqrt{n}\hat{\boldsymbol{\vartheta}}$ are partitioned analogously:

$$\left(\frac{\partial l(\boldsymbol{\vartheta})}{\partial \boldsymbol{\vartheta}}\right)' = \left[\left(\frac{\partial l(\boldsymbol{\vartheta})}{\partial \boldsymbol{\vartheta}^{(1)}}\right)', \left(\frac{\partial l(\boldsymbol{\vartheta})}{\partial \boldsymbol{\vartheta}^{(2)}}\right)'\right] \quad , \quad \hat{\boldsymbol{V}} = \begin{pmatrix} \hat{\boldsymbol{V}}_{11} & \hat{\boldsymbol{V}}_{12} \\ \hat{\boldsymbol{V}}_{21} & \hat{\boldsymbol{V}}_{22} \end{pmatrix} . \tag{3.116}$$

Using the following notation for the inverse of the partitioned matrix

$$\hat{\boldsymbol{V}}^{-1} = \begin{pmatrix} \hat{\boldsymbol{V}}^{11} & \hat{\boldsymbol{V}}^{12} \\ \hat{\boldsymbol{V}}^{21} & \hat{\boldsymbol{V}}^{22} \end{pmatrix} , \tag{3.117}$$

one obtains simplified formulas for the test statistics

$$LR = 2n\left[l(\hat{\boldsymbol{\vartheta}}^{(1)}, \hat{\boldsymbol{\vartheta}}^{(2)}) - l(\tilde{\boldsymbol{\vartheta}}^{(1)}, \boldsymbol{c})\right] , \tag{3.118}$$

$$W = n\hat{\boldsymbol{\vartheta}}^{(2)\prime}\hat{\boldsymbol{V}}^{22}\hat{\boldsymbol{\vartheta}}^{(2)} \tag{3.119}$$

where $\hat{\boldsymbol{V}}^{22} = \left(\hat{\boldsymbol{V}}_{22}^{-1} - \hat{\boldsymbol{V}}_{21}\hat{\boldsymbol{V}}_{11}^{-1}\hat{\boldsymbol{V}}_{12}\right)^{-1}$ (Rao 1973, chap. 1, p. 33). If $l(\boldsymbol{\vartheta})$ evaluated at the point $\boldsymbol{c}$ is denoted by $l(\boldsymbol{c})$, then LM may be written as

$$LM = n\left(\frac{\partial l(\boldsymbol{c})}{\partial \boldsymbol{\vartheta}^{(2)}}\right)' \tilde{\boldsymbol{V}}_{22}\left(\frac{\partial l(\boldsymbol{c})}{\partial \boldsymbol{\vartheta}^{(2)}}\right) . \tag{3.120}$$

These simplifications occur because the first $q - s$ parameters are not restricted. Hence, the score vector is $\boldsymbol{0}$ for these parameters.

3.4 Restrictions on Parameters

The Multivariate Delta Method

In the last subsection the Lagrange multiplier method was used to estimate $\boldsymbol{\vartheta}_0$ under the restriction $\boldsymbol{r}(\boldsymbol{\vartheta}_0) = \mathbf{0}$. As an alternative to constrained optimization, one can use unconstrained optimization methods and take the restrictions into account by using reparameterizations of $\boldsymbol{\vartheta}$. Fundamental parameters (collected in a vector $\boldsymbol{\tau} \sim p \times 1$ with $p \leq q$ [Küsters 1987; McDonald 1980]) are useful for this purpose. Assume that $\boldsymbol{\vartheta}$ is a continuously differentiable function of $\boldsymbol{\tau}$ with $\boldsymbol{\vartheta}_0 = \boldsymbol{\vartheta}(\boldsymbol{\tau}_0)$. The first and second derivatives of the log-likelihood function then are taken with respect to $\boldsymbol{\tau}$ rather than $\boldsymbol{\vartheta}$. If $\hat{\boldsymbol{\tau}}$, the ML estimator of $\boldsymbol{\tau}_0$ is asymptotically normal,

$$\hat{\boldsymbol{\tau}} \overset{A}{\sim} \mathcal{N}(\boldsymbol{\tau}_0, \boldsymbol{V}(\hat{\boldsymbol{\tau}})), \tag{3.121}$$

then one can use the multivariate delta method (Rao 1973, chap. 6, p. 388) to derive the asymptotic distribution of $\hat{\boldsymbol{\vartheta}}$:

$$\hat{\boldsymbol{\vartheta}} \overset{A}{\sim} \mathcal{N}\left(\boldsymbol{\vartheta}(\boldsymbol{\tau}_0), \left[\frac{\partial \boldsymbol{\vartheta}'(\boldsymbol{\tau}_0)}{\partial \boldsymbol{\tau}}\right]' \boldsymbol{V}(\hat{\boldsymbol{\tau}}) \left[\frac{\partial \boldsymbol{\vartheta}'(\boldsymbol{\tau}_0)}{\partial \boldsymbol{\tau}}\right]\right) \tag{3.122}$$

The asymptotic covariance matrix of $\hat{\boldsymbol{\vartheta}}$ is estimated by replacing $\boldsymbol{\tau}_0$ with $\hat{\boldsymbol{\tau}}$.

Important examples of reparameterizations are (cf. Küsters 1987):

- Equality restrictions. Setting $\vartheta_j = \vartheta_h$ is done by setting $\vartheta_j = \tau_k$ and $\vartheta_h = \tau_k$.

- Linear restrictions of the form

$$\sum_{j=1}^{J} c_j \vartheta_j = d \tag{3.123}$$

with c_j and d as known constants may be written as functions of the unrestricted parameters $\tau_1, \ldots, \tau_{J-1}$:

$$\vartheta_j = \tau_j,\ j = 1, \ldots, J-1; \ \ \vartheta_J = \frac{1}{c_J}\left[d - \sum_{j=1}^{J-1} c_j \tau_j\right]. \tag{3.124}$$

- Inequality restrictions of the form $0 \leq \vartheta_1 \leq \vartheta_2 \leq \vartheta_3$. To take these inequalities into account, one writes

$$\vartheta_1 = \tau_1^2, \quad \vartheta_2 = \tau_1^2 + \tau_2^2, \quad \vartheta_3 = \tau_1^2 + \tau_2^2 + \tau_3^2 \quad . \tag{3.125}$$

The parameter vector $\boldsymbol{\tau}$ is unrestricted.

- Domain restrictions. Substantive reasons may suggest that the domain of a parameter ϑ_j is restricted to the interval (c, d) with known constants c and d. The logit transformation may be used for reparameterization:

$$\vartheta_j = c + (d - c)\frac{\exp(\tau_j)}{1 + \exp(\tau_j)} \quad . \tag{3.126}$$

$l(\boldsymbol{\vartheta}(\boldsymbol{\tau}))$ is maximized with respect to $\boldsymbol{\tau}$ instead of $\boldsymbol{\vartheta}$ by deriving the first and second derivatives of $l(\boldsymbol{\vartheta}(\boldsymbol{\tau}))$ with respect to $\boldsymbol{\vartheta}$ and then to $\boldsymbol{\tau}$:

$$\frac{\partial \ln f(\boldsymbol{y}|\boldsymbol{x}, \boldsymbol{\vartheta}(\boldsymbol{\tau}))}{\partial \boldsymbol{\tau}} = \frac{\partial \boldsymbol{\vartheta}'(\boldsymbol{\tau})}{\partial \boldsymbol{\tau}} \frac{\partial \ln f(\boldsymbol{y}|\boldsymbol{x}, \boldsymbol{\vartheta}(\boldsymbol{\tau}))}{\partial \boldsymbol{\vartheta}} \tag{3.127}$$

$$\begin{aligned} \frac{\partial^2 \ln f(\boldsymbol{y}|\boldsymbol{x}, \boldsymbol{\vartheta}(\boldsymbol{\tau}))}{\partial \boldsymbol{\tau} \partial \boldsymbol{\tau}'} &= \left(\frac{\partial \boldsymbol{\vartheta}'(\boldsymbol{\tau})}{\partial \boldsymbol{\tau}}\right) \frac{\partial^2 \ln f(\boldsymbol{y}|\boldsymbol{x}, \boldsymbol{\vartheta}(\boldsymbol{\tau}))}{\partial \boldsymbol{\vartheta} \partial \boldsymbol{\vartheta}'} \left(\frac{\partial \boldsymbol{\vartheta}'(\boldsymbol{\tau})}{\partial \boldsymbol{\tau}}\right)' \\ &+ \sum_{j=1}^{q} \frac{\partial \ln f(\boldsymbol{y}|\boldsymbol{x}, \boldsymbol{\vartheta}(\boldsymbol{\tau}))}{\partial \vartheta_j} \frac{\partial^2 \vartheta_j(\boldsymbol{\tau})}{\partial \boldsymbol{\tau} \partial \boldsymbol{\tau}'} \quad . \end{aligned} \tag{3.128}$$

In the computation of the LM statistic the restricted estimates of $\boldsymbol{\vartheta}_0$ are then easily calculated as functions $\boldsymbol{\vartheta}$ of the unrestricted estimator $\hat{\boldsymbol{\tau}}$.

Minimum Distance Estimation

Minimum distance estimation (MDE) is a general tool to estimate parameters of restricted models from parameters of unrestricted models, when the asymptotic distribution of the unrestricted parameters is known.

Let $\boldsymbol{\vartheta}$ be a $q \times 1$ parameter vector which may be written as a continuously differentiable function of fundamental parameters collected in a $p \times 1$ vector $\boldsymbol{\tau}$ with $p \leq q$, that is, $\boldsymbol{\vartheta} = \boldsymbol{\vartheta}(\boldsymbol{\tau})$. The true values of $\boldsymbol{\vartheta}$ and $\boldsymbol{\tau}$ are denoted by $\boldsymbol{\vartheta}_0$ and $\boldsymbol{\tau}_0$. The estimate $\hat{\boldsymbol{\vartheta}}$ is a CAN estimator, that is, $\sqrt{n}(\hat{\boldsymbol{\vartheta}} - \boldsymbol{\vartheta}_0) \overset{A}{\sim} \mathcal{N}(\mathbf{0}, \boldsymbol{V}(\sqrt{n}\hat{\boldsymbol{\vartheta}}))$ where $\boldsymbol{V}(\sqrt{n}\hat{\boldsymbol{\vartheta}})$ is the asymptotic covariance matrix of $\sqrt{n}(\hat{\boldsymbol{\vartheta}} - \boldsymbol{\vartheta}_0)$. Furthermore, there exists a consistent estimator $\hat{\boldsymbol{V}}$ of the asymptotic covariance matrix of $\sqrt{n}\hat{\boldsymbol{\vartheta}}$. The parameter vector $\boldsymbol{\tau}$ is assumed to be first order identified, that is, $\boldsymbol{\vartheta}(\boldsymbol{\tau}_1) = \boldsymbol{\vartheta}(\boldsymbol{\tau}_2) \Rightarrow \boldsymbol{\tau}_1 = \boldsymbol{\tau}_2$ a.s. The MDE $\hat{\boldsymbol{\tau}}$ of $\boldsymbol{\tau}_0$ is defined as the minimum of the Mahalanobis distance

$$Q(\boldsymbol{\tau}) = n[\hat{\boldsymbol{\vartheta}} - \boldsymbol{\vartheta}(\boldsymbol{\tau})]'\hat{\boldsymbol{V}}^{-1}[\hat{\boldsymbol{\vartheta}} - \boldsymbol{\vartheta}(\boldsymbol{\tau})]. \tag{3.129}$$

A minimum is found by setting the vector of first derivatives of $Q(\boldsymbol{\tau})$ with respect to $\boldsymbol{\tau}$ to $\mathbf{0}$ and using matrix differentiation rules.

$$\mathbf{0} = \frac{\partial Q(\hat{\boldsymbol{\tau}})}{\partial \boldsymbol{\tau}} = -\sqrt{n}\, \frac{\partial \boldsymbol{\vartheta}'(\hat{\boldsymbol{\tau}})}{\partial \boldsymbol{\tau}} \hat{\boldsymbol{V}}^{-1} \sqrt{n}[\hat{\boldsymbol{\vartheta}} - \boldsymbol{\vartheta}(\hat{\boldsymbol{\tau}})] \tag{3.130}$$

The asymptotic distribution of $\sqrt{n}(\hat{\tau} - \tau_0)$ is found as follows (cf. Küsters 1987, chap. 5, sec. 2). First, set $\hat{\vartheta} - \vartheta(\hat{\tau}) = \hat{\vartheta} - \vartheta(\tau_0) - (\vartheta(\hat{\tau}) - \vartheta(\tau_0))$. Second, a first-order Taylor expansion of $\vartheta(\hat{\tau})$ about τ_0 yields

$$\vartheta(\hat{\tau}) - \vartheta(\tau_0) = \left(\frac{\partial \vartheta'(\tilde{\tau})}{\partial \tau}\right)' (\hat{\tau} - \tau_0), \tag{3.131}$$

where $\tilde{\tau}$ lies on the line segment between $\hat{\tau}$ and τ_0. Third, inserting (3.131) into (3.130) and dividing by $\sqrt{n}$ gives

$$\left(\frac{\partial \vartheta'(\hat{\tau})}{\partial \tau}\right) \hat{V}^{-1} \left(\frac{\partial \vartheta'(\tilde{\tau})}{\partial \tau}\right)' \sqrt{n}(\hat{\tau} - \tau_0) = \left(\frac{\partial \vartheta'(\hat{\tau})}{\partial \tau}\right) \hat{V}^{-1} \sqrt{n}(\hat{\vartheta} - \vartheta(\tau_0)). \tag{3.132}$$

If $Q(\tau)$ has a global minimum at τ_0, then $\hat{\tau}$ is a (strongly) consistent estimate of τ_0. Since $\partial \vartheta'(\tau)/\partial \tau$ is a continuous function of τ, $\partial \vartheta'(\hat{\tau})/\partial \tau$ and $\partial \vartheta'(\tilde{\tau})/\partial \tau$ converge to $\vartheta'(\tau_0)/\partial \tau$. $\hat{V}$ converges to $V(\sqrt{n}\hat{\vartheta})$ denoted by V. Hence, $\sqrt{n}(\hat{\tau} - \tau_0)$ is asymptotically distributed as

$$\left[\left(\frac{\partial \vartheta'(\tau_0)}{\partial \tau}\right) V^{-1} \left(\frac{\partial \vartheta'(\tau_0)}{\partial \tau}\right)'\right]^{-1} \left(\frac{\partial \vartheta'(\tau_0)}{\partial \tau}\right) V^{-1} \sqrt{n}(\hat{\vartheta} - \vartheta(\tau_0)), \tag{3.133}$$

where all matrices are constant and $\sqrt{n}(\hat{\vartheta} - \vartheta(\tau_0))$ converges in distribution to $\mathcal{N}(\mathbf{0}, V)$. Hence,

$$\sqrt{n}(\hat{\tau} - \tau_0) \overset{A}{\sim} \mathcal{N}\left(\mathbf{0}, \left[\left(\frac{\partial \vartheta'(\tau_0)}{\partial \tau}\right) V^{-1} \left(\frac{\partial \vartheta'(\tau_0)}{\partial \tau}\right)'\right]^{-1}\right). \tag{3.134}$$

The asymptotic covariance matrix $V(\sqrt{n}\hat{\tau})$ is estimated by replacing τ_0 with $\hat{\tau}$.

$$\hat{V}(\sqrt{n}\hat{\tau}) = \left[\left(\frac{\partial \vartheta'(\hat{\tau})}{\partial \tau}\right) \left(\hat{V}(\sqrt{n}\hat{\vartheta})\right)^{-1} \left(\frac{\partial \vartheta'(\hat{\tau})}{\partial \tau}\right)'\right]^{-1} \tag{3.135}$$

For finding the minimum of $Q(\tau)$ and for computing the consistent estimator $\hat{V}(\sqrt{n}\hat{\tau})$, only the first-order derivatives of ϑ with respect to τ are needed. These derivatives can either be derived analytically or computed numerically. If one uses numerical differentiation routines provided in program systems like GAUSS, one can routinely estimate quite complicated fundamental parameters τ and their asymptotic covariance matrix.

A typical example of MDE is the estimation of structural parameters in simultaneous-equation models from the parameters of the reduced form which have been estimated in a multivariate linear regression model. The simultaneous equation system is given by

$$y_i = A y_i + \Gamma x_i + \delta_i\,, \quad i = 1, \ldots, n\,. \tag{3.136}$$

y_i is the vector of endogenous variables, x_i is the vector of exogenous variables, and δ_i is a vector of disturbances with $E(\delta_i|x_i) = \mathbf{0}$ and a covariance matrix $\Sigma = V(\delta_i|X_i)$ that is unrestricted. $\boldsymbol{A}$ is the matrix of endogenous regression coefficients with the property that the inverse of $(\boldsymbol{I} - \boldsymbol{A})$ exists. Unlike the multivariate regression model, the dependent variables in $\boldsymbol{y}_i$ appear on the left and the right side of the equation, allowing the endogenous variables to depend not only on $\boldsymbol{x}_i$ but also on the other components of $\boldsymbol{y}_i$.

The reduced form of the simultaneous equation model is given by

$$\boldsymbol{y}_i = (\boldsymbol{I} - \boldsymbol{A})^{-1}\boldsymbol{\Gamma}\boldsymbol{x}_i + (\boldsymbol{I} - \boldsymbol{A})^{-1}\boldsymbol{\delta}_i = \boldsymbol{B}\boldsymbol{x}_i + \boldsymbol{\epsilon}_i \quad \text{with} \tag{3.137}$$

$$\boldsymbol{B} = (\boldsymbol{I} - \boldsymbol{A})^{-1}\boldsymbol{\Gamma} \quad \text{and} \quad V(\boldsymbol{\epsilon}_i) = (\boldsymbol{I} - \boldsymbol{A})^{-1}\boldsymbol{\Sigma}(\boldsymbol{I} - \boldsymbol{A})^{-1'} \quad . \tag{3.138}$$

The matrix $\boldsymbol{B}$ of regression coefficients of the reduced form can be estimated like the vector $\boldsymbol{\beta}$ in the multivariate linear regression model. In fact, $\boldsymbol{\beta}$ is the vectorized form of $\boldsymbol{B}$, that is, $\boldsymbol{\beta} = (\boldsymbol{B}'_{1\cdot}, \boldsymbol{B}'_{2\cdot}, \ldots, \boldsymbol{B}'_{G\cdot})'$ where the rows of $\boldsymbol{B}$ are transposed and stacked above each other. $\boldsymbol{B}$ and therefore $\boldsymbol{\beta}$ is a continuously differentiable function of the parameters in $\boldsymbol{A}$ and $\boldsymbol{\Gamma}$ that are collected in a $p \times 1$ vector of fundamental parameters $\boldsymbol{\tau}$, that is, $\boldsymbol{\beta} = \boldsymbol{\beta}(\boldsymbol{\tau})$. It is assumed that the condition of first-order identifiability is fulfilled. Under this condition, the parameter vector $\boldsymbol{\tau}$ can be estimated by minimum-distance estimation. Other examples are the estimation methods for covariance structure analysis discussed by Browne and Arminger in Chapter 4.

4 ML Estimation Under Misspecification

The key assumption of ML estimation is that the true conditional density $f_\star(\boldsymbol{y}|\boldsymbol{x})$ is correctly specified as a density $f(\boldsymbol{y}|\boldsymbol{x}, \boldsymbol{\vartheta})$ except for the unknown parameter vector $\boldsymbol{\vartheta}_0$. White (1982, 1983) discusses in detail what happens if the researcher chooses a density $f(\boldsymbol{y}|\boldsymbol{x}, \boldsymbol{\vartheta})$ that is misspecified. Misspecification means that there does not exist a vector $\boldsymbol{\vartheta}_0$ of $f(\boldsymbol{y}|\boldsymbol{x}, \boldsymbol{\vartheta})$ such that $f_\star(\boldsymbol{y}|\boldsymbol{x}) = f(\boldsymbol{y}|\boldsymbol{x}, \boldsymbol{\vartheta}_0)$. Generally, misspecification implies that the conditional moments of $\boldsymbol{y}$ given $\boldsymbol{x}$ under $f_\star$ do not agree with the conditional moments of $\boldsymbol{y}$ given $\boldsymbol{x}$ under $f(\boldsymbol{y}|\boldsymbol{x}, \boldsymbol{\vartheta})$ for any $\boldsymbol{\vartheta}$. It may not even be possible to parameterize $f_\star$ partially in the vector $\boldsymbol{\vartheta}$ that characterizes $f(\boldsymbol{y}|\boldsymbol{x}, \boldsymbol{\vartheta})$. If the density is misspecified, White (1982) uses the expression quasi ML (QML) estimation instead of ML estimation, although the computations do not change. Here, a short summary of his findings is given. First, it may be shown that the kernel

$$l(\boldsymbol{\vartheta}) = \frac{1}{n}\sum_{i=1}^{n} \ln f(\boldsymbol{y}_i|\boldsymbol{x}_i, \boldsymbol{\vartheta})$$

computed from the misspecified densities converges (uniformly) for all $\boldsymbol{\vartheta} \in \boldsymbol{\Theta}$ to its expected value

$$E_g[E_{f_\star} \ln f(\boldsymbol{y}|\boldsymbol{x}, \boldsymbol{\vartheta})] = \int \left(\int \ln f(\boldsymbol{y}|\boldsymbol{x}, \boldsymbol{\vartheta}) f_\star(\boldsymbol{y}|\boldsymbol{x}) d\boldsymbol{y}\right) g(\boldsymbol{x}) d\boldsymbol{x} \quad . \tag{3.139}$$

The expected value for $\ln f(\boldsymbol{y}|\boldsymbol{x}, \boldsymbol{\vartheta})$ is taken now over the true density $f_\star(\boldsymbol{y}|\boldsymbol{x})$ and not the density $f(\boldsymbol{y}|\boldsymbol{x}, \boldsymbol{\vartheta})$ assumed by the researcher.

The sequence of maximized kernel functions $l_n(\hat{\boldsymbol{\vartheta}}_n)$ now converges to the maximum of $E_g[E_{f_\star} \ln f(\boldsymbol{y}|\boldsymbol{x}, \boldsymbol{\vartheta})]$ as a function of $\boldsymbol{\vartheta}$. To interpret the maximum of $E_g[E_{f_\star} \ln f(\boldsymbol{y}|\boldsymbol{x}, \boldsymbol{\vartheta})]$, one considers again the KLIC as a measure of the average discrepancy between the assumed density $f(\boldsymbol{y}|\boldsymbol{x}, \boldsymbol{\vartheta})$ and the true density $f_\star(\boldsymbol{y}|\boldsymbol{x})$.

$$I\left[f_\star(\boldsymbol{y}|\boldsymbol{x}), f(\boldsymbol{y}|\boldsymbol{x}, \boldsymbol{\vartheta})\right] = E_g[E_{f_\star} \ln f_\star(\boldsymbol{y}|\boldsymbol{x})] - E_g[E_{f_\star} \ln f(\boldsymbol{y}|\boldsymbol{x}, \boldsymbol{\vartheta})] \tag{3.140}$$

Since the first term on the right-hand side of (3.140) is independent of $\boldsymbol{\vartheta}$ and therefore a constant, the KLIC is minimized if the second term is maximized in $\boldsymbol{\vartheta}$. If there exists a unique global maximum which is denoted by $\boldsymbol{\vartheta}_\star$, the ML estimator $\hat{\boldsymbol{\vartheta}}$ of the parameter vector of the assumed density $f(\boldsymbol{y}|\boldsymbol{x}, \boldsymbol{\vartheta})$ converges to this value $\boldsymbol{\vartheta}_\star$. As a consequence, the ML estimate $\hat{\boldsymbol{\vartheta}}$ can still be interpreted in a meaningful way even if the conditional density is misspecified completely. In this case $\hat{\boldsymbol{\vartheta}}$ estimates the parameter value $\boldsymbol{\vartheta}_\star$, which minimizes the discrepancy between the model assumed and the data. Therefore, it is called the *minimum ignorance estimator* by White (1982). It is important to note that $f(\boldsymbol{y}|\boldsymbol{x}, \boldsymbol{\vartheta})$ may be completely misspecified so that the conditional moments of the true and the assumed density do not agree at all. The parameter vector $\boldsymbol{\vartheta}$ may then not have any meaningful substantive interpretation. Nevertheless, the parameter $\boldsymbol{\vartheta}_\star$ is meaningful statistically since it produces the best possible approximation to the true density (in the sense of the KLIC) by the assumed density.

The QML estimates obtained from the quasi-log-likelihood function are asymptotically normally distributed about the parameter $\boldsymbol{\vartheta}_\star$. The simplification (3.98) which has been found if the assumed and the true density are equal up to the parameter vector to be estimated no longer holds. That is, here

$$\sqrt{n}(\hat{\boldsymbol{\vartheta}} - \boldsymbol{\vartheta}_\star) \overset{A}{\sim} \mathcal{N}\left(\mathbf{0}, (-\boldsymbol{A}(\boldsymbol{\vartheta}_\star))^{-1}\boldsymbol{B}(\boldsymbol{\vartheta}_\star)(-\boldsymbol{A}(\boldsymbol{\vartheta}_\star))^{-1}\right). \tag{3.141}$$

If one keeps in mind that most researchers misspecify the model for $f_\star(\boldsymbol{y}|\boldsymbol{x})$ and do not incorporate the first and second derivatives of the quasi-log-likelihood function into the estimation of the asymptotic covariance matrix of the estimated parameters, it is obvious that their estimated parameters can usually be interpreted only as minimum ignorance estimators and that the standard errors and test statistics may be far away from the correct asymptotic values, depending on the discrepancy between the assumed density and the actual density that generated the data.

An important special case is the linear regression model for a univariate dependent variable

$$y_i = \boldsymbol{x}_i'\boldsymbol{\beta} + \epsilon_i \quad , \tag{3.142}$$

in which ϵ_i is assumed to be distributed $\mathcal{N}(0, 1)$. The corresponding assumed density $f(y|\boldsymbol{x}, \boldsymbol{\beta}) \sim \mathcal{N}(\boldsymbol{x}'\boldsymbol{\beta}, 1)$ may be completely misspecified not only with regard to the distributional form but also with regard to the variance (homoscedasticity may not hold) and the conditional expected value (the linear form may be wrong). The QML estimate of $\boldsymbol{\beta}$ is, using matrix notation,

$$\hat{\boldsymbol{\beta}} = (\boldsymbol{X}'\boldsymbol{X})^{-1}\boldsymbol{X}'\boldsymbol{y} \quad .$$

$\hat{\boldsymbol{\beta}}$ converges to $\boldsymbol{\beta}_\star$ which maximizes the expression

$$E_g[E_{f_\star} \ln f(y|\boldsymbol{x}, \boldsymbol{\beta})] \propto -\int\int (y - \boldsymbol{x}'\boldsymbol{\beta})^2 f_\star(y|\boldsymbol{x}) dy g(\boldsymbol{x}) d\boldsymbol{x} \quad . \tag{3.143}$$

The right-hand side is maximized if the integral is minimized with respect to $\boldsymbol{\beta}$. This integral can be interpreted as the mean prediction error. $\boldsymbol{\beta}_\star$ is therefore the value which minimizes the squared prediction error over the population of y and $\boldsymbol{x}$, and may be a good model for forecasts in spite of the misspecification of the true density.

5 Pseudo-ML Estimation

5.1 Mean Structures

In Section 4 it was shown that misspecification of the conditional density function of $\boldsymbol{y}$ given $\boldsymbol{x}$ yields in general ML estimates $\hat{\boldsymbol{\vartheta}}$ that can only be meaningfully interpreted as minimum-ignorance estimators. In this section, only partial misspecification of the density is assumed.

The true but unknown density of $\boldsymbol{y}$ given $\boldsymbol{x}$ is denoted by $f_\star(\boldsymbol{y}|\boldsymbol{x})$ with conditional expected value $E_{f_\star}(\boldsymbol{y}|\boldsymbol{x})$. Furthermore, the researcher has specified – up to an unknown parameter vector $\boldsymbol{\vartheta}$ of dimension $q \times 1$ – a density $f(\boldsymbol{y}|\boldsymbol{x}, \boldsymbol{\vartheta})$ where $\boldsymbol{\vartheta}$ is a parameterization of $E_f(\boldsymbol{y}|\boldsymbol{x}, \boldsymbol{\vartheta}) = \boldsymbol{\mu}(\boldsymbol{x}, \boldsymbol{\vartheta})$. If there exists a vector $\boldsymbol{\vartheta}_0$ such that

$$E_{f_\star}(\boldsymbol{y}|\boldsymbol{x}) = E_f(\boldsymbol{y}|\boldsymbol{x}, \boldsymbol{\vartheta}_0), \tag{3.144}$$

then $\boldsymbol{\vartheta}_0$ is also a parameterization of $E_{f_\star}(\boldsymbol{y}|\boldsymbol{x})$.

Gourieroux et al. (1984) have analyzed the properties of pseudo-ML (PML) estimators of $\boldsymbol{\vartheta}_0$ which are based on the assumed density $f(\boldsymbol{y}|\boldsymbol{x}, \boldsymbol{\vartheta})$ [see also Gourieroux and Monfort (1993)]. They show that PML estimation of $\boldsymbol{\vartheta}_0$ based on the assumed density yields a consistent estimator $\hat{\boldsymbol{\vartheta}}$ of $\boldsymbol{\vartheta}_0$ if and only if the assumed density is a member of the univariate or multivariate linear exponential family. Since the assumed density may be misspecified—except for the mean structure—this estimation method is called pseudo-ML estimation. Regularity conditions and technical proofs are found in their paper.

The stochastic model for $\boldsymbol{y}$ given $\boldsymbol{x}$ under the true density $f_\star(\boldsymbol{y}|\boldsymbol{x})$ is given by

$$\begin{aligned} &\boldsymbol{y} = \boldsymbol{\mu}(\boldsymbol{x}, \boldsymbol{\vartheta}_0) + \boldsymbol{\epsilon} \quad \text{with} \quad E(\boldsymbol{\epsilon}|\boldsymbol{x}) = \boldsymbol{0} \quad , \\ &E_{f_\star}(\boldsymbol{y}|\boldsymbol{x}) = E_f(\boldsymbol{y}|\boldsymbol{x}, \boldsymbol{\vartheta}_0) = \boldsymbol{\mu}(\boldsymbol{x}, \boldsymbol{\vartheta}_0) \quad , \text{and} \\ &V_{f_\star}(\boldsymbol{y}|\boldsymbol{x}) = \boldsymbol{V}(\boldsymbol{\epsilon}|\boldsymbol{x}) = \boldsymbol{\Omega}_\star(\boldsymbol{x}) \quad . \end{aligned} \tag{3.145}$$

Only the mean structure of $\boldsymbol{y}$ given $\boldsymbol{x}$ is specified in this model in the parameter vector $\boldsymbol{\vartheta}_0$. $E(\boldsymbol{\epsilon}|\boldsymbol{x}) = \boldsymbol{0}$ implies that the error term and the regressor variables $\boldsymbol{x}$ are uncorrelated. The covariance matrix $\boldsymbol{\Omega}_\star(\boldsymbol{x})$ of $\boldsymbol{y}$ given $\boldsymbol{x}$ is not specified in any way. It is an arbitrary function indexed by $\boldsymbol{x}$ allowing for general multivariate heteroscedasticity.

The stochastic model for $\boldsymbol{y}$ given $\boldsymbol{x}$ under the assumed distribution $f(\boldsymbol{y}|\boldsymbol{x}, \boldsymbol{\vartheta})$ is

$$\begin{aligned} \boldsymbol{y} &= \boldsymbol{\mu}(\boldsymbol{x}, \boldsymbol{\vartheta}_0) + \epsilon^\star \quad \text{with} \quad E(\epsilon^\star|\boldsymbol{x}) = \mathbf{0} \quad , \\ E_{f_\star}(\boldsymbol{y}|\boldsymbol{x}) &= E_f(\boldsymbol{y}|\boldsymbol{x}, \boldsymbol{\vartheta}_0) = \boldsymbol{\mu}(\boldsymbol{x}, \boldsymbol{\vartheta}_0) \quad , \text{ and} \\ V_f(\boldsymbol{y}|\boldsymbol{x}) &= \boldsymbol{V}(\epsilon^\star|\boldsymbol{x}) = \boldsymbol{\Sigma}(\boldsymbol{x}, \boldsymbol{\vartheta}_0, \boldsymbol{\Phi}) \quad . \end{aligned} \tag{3.146}$$

The mean structures for $\boldsymbol{y}$ given $\boldsymbol{x}$ are equal under the true and the assumed density. However, while the conditional covariance matrix has not been specified under the true density, the choice of a specific member of the linear exponential family for estimation implies the choice of a conditional covariance matrix that may depend on $\boldsymbol{x}$, the mean structure parameter $\boldsymbol{\vartheta}_0$, and a nuisance parameter $\boldsymbol{\Phi}$. An example is the choice of a Poisson density where the conditional variance of y given $\boldsymbol{x}$ is equal to the expected value $\mu(\boldsymbol{x}, \boldsymbol{\vartheta}_0)$, that is, $\Sigma(\boldsymbol{x}, \boldsymbol{\vartheta}_0, \Phi) = \mu(\boldsymbol{x}, \boldsymbol{\vartheta}_0)$. Throughout this section it is assumed that the parameters $\boldsymbol{\vartheta}_0$ are first-order identifiable, that is, $\boldsymbol{\mu}(\boldsymbol{x}, \boldsymbol{\vartheta}_1) = \boldsymbol{\mu}(\boldsymbol{x}, \boldsymbol{\vartheta}_2) \Rightarrow \boldsymbol{\vartheta}_1 = \boldsymbol{\vartheta}_2$.

5.2 The Linear Exponential Family

In this subsection, parameterizations, univariate and multivariate examples, and properties of the linear exponential family with a nuisance parameter are reviewed (cf. Barndorff-Nielsen 1978; Gourieroux et al. 1984; McCullagh and Nelder 1989). For the discussion of these properties, the conditioning variable $\boldsymbol{x}$ is dropped.

Canonical and Expected-Value Parameterization

The density for a $G \times 1$ random vector $\boldsymbol{y}$ from the linear exponential family is given by

$$f_\kappa(\boldsymbol{y}|\boldsymbol{\kappa}, \boldsymbol{\Phi}) = \exp\{\boldsymbol{y}'\boldsymbol{\kappa} + u(\boldsymbol{y}, \boldsymbol{\Phi}) - v(\boldsymbol{\kappa}, \boldsymbol{\Phi})\}. \tag{3.147}$$

The exponent is linear in $\boldsymbol{y}$ with parameter vector $\boldsymbol{\kappa} \in \mathcal{K} \subset \mathcal{R}^G$. $\boldsymbol{\kappa}$ is therefore called the canonical or natural parameter of the linear exponential family. $\boldsymbol{\Phi}$ is a $G \times G$ matrix of nuisance parameters that are fixed, $u(.,.)$ is a scalar function of $\boldsymbol{y}$ and $\boldsymbol{\Phi}$, and $v(\boldsymbol{\kappa}, \boldsymbol{\Phi})$ is a scalar function of $\boldsymbol{\kappa}$ and $\boldsymbol{\Phi}$ which may be written as a normalizing constant, that is,

$$0 < \exp\{v(\boldsymbol{\kappa}, \boldsymbol{\Phi})\} = \int \exp\{\boldsymbol{y}'\boldsymbol{\kappa} + u(\boldsymbol{y}, \boldsymbol{\Phi})\} d\boldsymbol{y} < \infty \quad . \tag{3.148}$$

Statistically, $v(\boldsymbol{\kappa}, \boldsymbol{\Phi})$ is the cumulant generating function of $f(\boldsymbol{y}|\boldsymbol{\kappa}, \boldsymbol{\Phi})$, that is, the logarithm of the moment-generating function (Kendall and Stuart 1976, chap. 3, p. 69). It is assumed that the covariance matrix is positive definite. All moments exist, integration and differentation may be exchanged. The expected value $\boldsymbol{\mu}$ and the covariance matrix $\boldsymbol{\Sigma}$ may be written as

$$\boldsymbol{\mu} = \frac{\partial v(\boldsymbol{\kappa}, \boldsymbol{\Phi})}{\partial \boldsymbol{\kappa}} \quad , \quad \boldsymbol{\Sigma} = \frac{\partial^2 v(\boldsymbol{\kappa}, \boldsymbol{\Phi})}{\partial \boldsymbol{\kappa} \partial \boldsymbol{\kappa}'} = \frac{\partial \boldsymbol{\mu}'}{\partial \boldsymbol{\kappa}} \quad . \tag{3.149}$$

These results may be derived immediately by differentiating $\int f_\kappa(\boldsymbol{y}|\boldsymbol{\kappa},\boldsymbol{\Phi})d\boldsymbol{y} = 1$ with respect to $\boldsymbol{\kappa}$ on both sides and exchanging differentiation and integration.

$$\int \left(\boldsymbol{y} - \frac{\partial v(\boldsymbol{\kappa},\boldsymbol{\Phi})}{\partial \boldsymbol{\kappa}}\right) \exp\{\boldsymbol{y}'\boldsymbol{\kappa} + u(\boldsymbol{y},\boldsymbol{\Phi}) - v(\boldsymbol{\kappa},\boldsymbol{\Phi})\}d\boldsymbol{y} = \mathbf{0} \tag{3.150}$$

Since $\boldsymbol{\Sigma} = \partial\boldsymbol{\mu}'/\partial\boldsymbol{\kappa}$ and $\boldsymbol{\Sigma}$ is assumed to be positive definite, there exists (locally) a one-to-one function that maps $\boldsymbol{\kappa}$ onto $\boldsymbol{\mu}$ and $\boldsymbol{\mu}$ onto $\boldsymbol{\kappa}$ (inverse-function theorem, Apostol 1979, chap. 13, sec. 3).

$$\boldsymbol{\mu} = \frac{\partial v(\boldsymbol{\kappa},\boldsymbol{\Phi})}{\partial \boldsymbol{\kappa}} = \boldsymbol{c}^{-1}(\boldsymbol{\kappa},\boldsymbol{\Phi}) \quad \text{and} \quad \boldsymbol{\kappa} = \boldsymbol{c}(\boldsymbol{\mu},\boldsymbol{\Phi}) \tag{3.151}$$

Consequently, the density of the exponential family may be reparameterized using the expected value $\boldsymbol{\mu}$ and $\boldsymbol{\Phi}$ instead of the canonical parameters $\boldsymbol{\kappa}$ and $\boldsymbol{\Phi}$.Hypotheses about the mean structure $\boldsymbol{\mu}$ can then formulated by writing $\boldsymbol{\mu}$ as a function of the parameter vector $\boldsymbol{\vartheta}$ as in Section 5.1.

$$f(\boldsymbol{y}|\boldsymbol{\mu},\boldsymbol{\Phi}) = \exp\{a(\boldsymbol{\mu},\boldsymbol{\Phi}) + b(\boldsymbol{y},\boldsymbol{\Phi}) + \boldsymbol{y}'\boldsymbol{c}(\boldsymbol{\mu},\boldsymbol{\Phi})\} \tag{3.152}$$

$\boldsymbol{\kappa}$ is replaced by $\boldsymbol{c}(\boldsymbol{\mu},\boldsymbol{\Phi})$ and $-v(\boldsymbol{\kappa},\boldsymbol{\Phi})$ is replaced by $a(\boldsymbol{\mu},\boldsymbol{\Phi})$. $u(\boldsymbol{y},\boldsymbol{\Phi})$ equals $b(\boldsymbol{y},\boldsymbol{\Phi})$. $\mathcal{M} \subset \mathcal{R}^G$ is the domain of $\boldsymbol{\mu}$. This reparameterization is used by Gourieroux et al. (1984).

Examples of the Univariate Linear Exponential Family

Poisson distribution: The density function of a Poisson distributed variable $y \in \{0, 1, \ldots\}$ with expected value $\mu \in (0,\infty)$ is

$$f(y|\mu,\Phi) = \frac{\mu^y}{y!}\exp(-\mu). \tag{3.153}$$

Since the density function is not a function of Φ, Φ may be chosen arbitrarily and is set to 1. The parameterization in μ is

$$a(\mu,\Phi) = -\mu\ , \quad b(y,\Phi) = -\ln y!\ , \quad c(\mu,\Phi) = \ln\mu \quad . \tag{3.154}$$

The canonical parameter is $\kappa = \ln\mu$. The parameterization in κ is

$$u(y,\Phi) = -\ln y!\ , \quad v(\kappa,\Phi) = \exp(\kappa) \quad . \tag{3.155}$$

The variance as a function of μ is

$$\Sigma = \mu \quad . \tag{3.156}$$

The Poisson density is typically used to model count variables including frequencies in cells of contingency tables. The loglinear models for contingency tables based on the canonical parameterization of the Poisson density are treated by Sobel in Chapter 5.

Bernoulli distribution: The density function of a Bernoulli distributed variable $y \in \{0,1\}$ with expected value $\mu \in (0,1)$ is

$$f(y|\mu,\Phi) = \mu^y(1-\mu)^{1-y} \quad , \tag{3.157}$$

where μ is the probability of success, that is $y = 1$. Φ is set to 1. The parameterization in μ is

$$a(\mu,\Phi) = \ln(1-\mu)\,, \quad b(y,\Phi) = 0\,, \quad c(\mu,\Phi) = \ln\frac{\mu}{1-\mu} \quad . \tag{3.158}$$

The canonical parameter $\kappa = \ln[\mu/(1-\mu)]$ which is the logit transformation of the probability of success. The parameterization in κ is

$$u(y,\Phi) = 0\,, \quad v(\kappa,\Phi) = \ln(1+\exp(\kappa)). \tag{3.159}$$

The variance as a function of μ is

$$\Sigma = \mu(1-\mu). \tag{3.160}$$

The Bernoulli density is used to model dichotomous outcomes. The logit model as a regression model for dichotomous outcomes is treated in Section 7 of this chapter.

Binomial distribution: If a Bernoulli experiment is repeated m times, the number $y_i \in \{0,1\ldots,m\}$ of successes follows a binomial distribution with expected value $\mu \in (0,m)$ and density

$$f(y|\mu,\Phi) = \binom{m}{y}\left(\frac{\mu}{m}\right)^y\left(1-\frac{\mu}{m}\right)^{m-y}. \tag{3.161}$$

Φ is set to 1. The parameterization in μ is

$$a(\mu,\Phi) = m\ln\left(1-\frac{\mu}{m}\right), \quad b(y,\Phi) = \ln\binom{m}{y}, \quad c(\mu,\Phi) = \ln\left(\frac{\mu}{m-\mu}\right). \tag{3.162}$$

The canonical parameter $\kappa = \ln[\mu/(m-\mu)]$ is again the logit of μ/m. The parameterization in κ is

$$u(y,\Phi) = \ln\binom{m}{y}, \quad v(\kappa,\Phi) = m\ln(1+\exp(\kappa)). \tag{3.163}$$

The variance as a function of μ is

$$\Sigma = \mu\left(1-\frac{\mu}{m}\right). \tag{3.164}$$

The binomial density is used to model dichotomous outcomes that are grouped. Models for grouped dichotomous variables can always be written as models for individual dichotomous variables. However, approximate methods for logit models such as using empirical logits $\ln[(y+1/2)/(m-y-1/2)]$ only work if grouped data are available.

Negative binomial distribution: The density function of a negative binomial distributed variable $y \in \{0, 1, \ldots\}$ with expected value $\mu \in (0, \infty)$ is

$$f(y|\pi, \Phi) = \binom{\Phi + y - 1}{y} \pi^{\Phi}(1 - \pi)^{y} \tag{3.165}$$

where π is the probability of success in a Bernoulli experiment. If Φ is set to 1, then

$$f(y|\pi, 1) = \pi(1 - \pi)^{y} \tag{3.166}$$

is the density of the Pascal (or geometric) distribution. $f(y|\pi, 1)$ is the probability that in a series of Bernoulli experiments the number of failures before the first success is equal to y. The density explicitly depends on Φ. The parameterization in $\mu = \Phi[(1 - \pi)/\pi]$ is:

$$a(\mu, \Phi) = \Phi \ln\left(\frac{\Phi}{\Phi + \mu}\right), \; b(y, \Phi) = \ln\binom{\Phi + y - 1}{y}, \; c(\mu, \Phi) = \ln\left(\frac{\mu}{\Phi + \mu}\right) \tag{3.167}$$

The canonical parameter is $\kappa = \ln\,[\mu/(\Phi + \mu)]$. The parameterization in κ is

$$u(y, \Phi) = \ln\binom{\Phi + y - 1}{y}, \quad v(\kappa, \Phi) = -\Phi \ln\,(1 - \exp(\kappa)). \tag{3.168}$$

The variance as a function of μ and Φ is

$$\Sigma = \mu + \frac{\mu^2}{\Phi} \quad \Rightarrow \quad \Phi = \frac{\mu^2}{(\Sigma - \mu)} \quad . \tag{3.169}$$

The negative binomial density is used as an alternative to the Poisson density for modeling counts and frequencies. The variance is given by $\mu + \mu^2/\Phi$ and is therefore greater than the variance of a Poisson variable if $\Phi > 0$.

Normal (Gaussian) distribution: The density of a normally distributed variable $y \in (-\infty, \infty)$ is

$$f(y|\mu, \Phi) = \frac{1}{\sqrt{2\pi\Phi}} \exp\left\{-\frac{1}{2}\frac{(y - \mu)^2}{\Phi}\right\} \quad . \tag{3.170}$$

A parameterization in μ is

$$a(\mu, \Phi) = -\frac{\mu^2}{2\Phi}, \quad b(y, \Phi) = -\frac{1}{2}\ln\,(2\pi\Phi) - \frac{y^2}{2\Phi}, \quad c(\mu, \Phi) = \frac{\mu}{\Phi} \quad . \tag{3.171}$$

The canonical parameter $\kappa = \mu/\Phi$. The parameterization in κ is

$$u(y, \Phi) = -\frac{1}{2}\ln\,(2\pi\Phi) - \frac{y^2}{2\Phi}, \quad v(\kappa, \Phi) = -\frac{1}{2}\kappa^2\Phi \quad . \tag{3.172}$$

The variance as a function of μ and Φ is

$$\Sigma = \Phi \quad . \tag{3.173}$$

In contrast to all other examples given here for the univariate linear exponential family, the variance Σ of y depends only on Φ and not on μ.

Gamma distribution: The density of a gamma distributed variable $y \in (0, \infty)$ with expected value $\mu \in (0, \infty)$ is

$$f(y|\mu, \Phi) = \left(\frac{\Phi}{\mu}\right)^{\Phi} y^{\Phi-1} \frac{\exp(-\Phi y/\mu)}{\Gamma(\Phi)} \quad . \tag{3.174}$$

The parameterization in μ is

$$a(\mu, \Phi) = \Phi \ln\frac{\Phi}{\mu} \, , \quad b(y, \Phi) = (\Phi - 1)\ln y - \ln\Gamma(\Phi) \, , \quad c(y, \Phi) = -\frac{\Phi}{\mu} . \tag{3.175}$$

The variance as a function of μ and Φ is

$$\Sigma = \frac{\mu^2}{\Phi} \quad \Rightarrow \Phi = \frac{\mu^2}{\Sigma} \quad . \tag{3.176}$$

Since the range of y and μ is restricted to $(0, \infty)$, the gamma density is used to model continuous outcomes that are positive. Typical examples are squared deviations from a mean or from a predicted value. If $\Phi = 1$, then κ is the inverse of the mean and is the failure rate in the exponential distribution. Φ is the inverse of the squared coefficient of variation.

Parameterization of Generalized Linear Models

With the exception of the negative binomial density, all the examples of the univariate linear exponential family given here can also be written in the form

$$f_G(y|\bar{\kappa}, \Phi) = \exp\,[y\bar{\kappa} - b^\star(\bar{\kappa})]/a^\star(\Phi) + c^\star(y, \Phi)\} \quad . \tag{3.177}$$

This is the form of the density f_κ that is used by McCullagh and Nelder (1989) in the theory of generalized linear models. $\bar{\kappa}$ is again the canonical parameter, which is proportional to κ of (3.147). Φ is again the nuisance parameter. The notation $a^\star, b^\star$, and $c^\star$ has been chosen to stay as closely to the notation of McCullagh and Nelder (1989) as possible without confusion with the notation in (3.152). $a^\star(\Phi)$ usually has the form $a^\star \cdot \Phi$ where $a^\star$ is a known weight. If $a^\star(\Phi) = \Phi$, the relationship between (3.177) and the parameterization in κ of (3.147) is given by

$$\bar{\kappa} = \kappa\Phi \, , \quad b^\star(\bar{\kappa}) = v(\kappa, \Phi) \cdot \Phi \, , \text{ and } c^\star(y, \Phi) = u(y, \Phi). \tag{3.178}$$

The relationship between (3.177) and the parameterization in μ and Φ of (3.152) may be written as

$$\bar{\kappa} = c(\mu, \Phi) \cdot \Phi \, , \quad b^\star(\bar{\kappa}) = -a(\mu, \Phi) \cdot \Phi \, , \text{ and } c^\star(y, \Phi) = b(y, \Phi). \tag{3.179}$$

The expected value and the variance are

$$\mu = \frac{\partial b^\star(\bar{\kappa})}{\partial \bar{\kappa}} \, , \quad \Sigma = \Phi \frac{\partial^2 b^\star(\bar{\kappa})}{\partial \kappa} = \Phi \frac{\partial \mu}{\partial \bar{\kappa}} = \Phi \cdot \tau(\mu). \tag{3.180}$$

The variance is written in this parameterization as the product of the nuisance parameter Φ and the function $\tau(\mu)$. Therefore, Φ is called the dispersion parameter in generalized linear models and $\tau(\mu)$ is the variance function which models the dependence of $V(y)$ on μ under the linear exponential family. Φ is replaced by $a^\star(\Phi)$ if a weight function is used.

Examples of the Multivariate Linear Exponential Family

Multinomial distribution: The case of a multinomial variable $\boldsymbol{y} \sim G \times 1$ with $y_g \in \{0,1\}$, $g = 1,\ldots,G$, $\sum_{g=1}^{G} y_g = 1$, and $E(y_g) = \mu_g \in (0,1)$, $\sum_{g=1}^{G} \mu_g = 1$ is considered. This corresponds to a sample size of 1. The expected values are the probabilities that a categorical variable A with G categories takes on category g. Only $G - 1$ elements of $\boldsymbol{y}$ are free; the last element is given by the restriction $\sum_{g=1}^{G} y_g = 1$. The density function is

$$f(\boldsymbol{y}|\boldsymbol{\mu}, \boldsymbol{\Phi}) = \prod_{g=1}^{G} \mu_g^{y_g} \quad . \tag{3.181}$$

The parameterization in $\boldsymbol{\mu}$ depends only on μ_g, $g = 1,\ldots,G-1$ because $\mu_G = 1 - \sum_{g=1}^{G-1} \mu_g$.

$$a(\boldsymbol{\mu}, \boldsymbol{\Phi}) = \ln\left(1 - \sum_{g=1}^{G-1} \mu_g\right), \quad b(\boldsymbol{y}, \boldsymbol{\Phi}) = 0 \tag{3.182}$$

$$\boldsymbol{c}(\boldsymbol{\mu}, \boldsymbol{\Phi}) = \left(\ln\frac{\mu_1}{\mu_G}, \ln\frac{\mu_2}{\mu_G}, \ldots, \ln\frac{\mu_{G-1}}{\mu_G}\right)' \tag{3.183}$$

The density does not depend on $\boldsymbol{\Phi}$, $\boldsymbol{\Phi}$ is set to $\boldsymbol{I}_G$. The canonical parameter $\kappa_g = \ln(\mu_g/\mu_G)$ is the logit of the gth category with reference category G. μ_g is given as

$$\mu_g = \frac{\exp(\kappa_g)}{1 + \sum_{g=1}^{G-1} \exp(\kappa_g)}, \quad g = 1,\ldots,G-1 \, . \tag{3.184}$$

The parameterization in $\boldsymbol{\kappa}$ is

$$u(\boldsymbol{y}, \boldsymbol{\Phi}) = 0 \, , \quad v(\boldsymbol{\kappa}, \boldsymbol{\Phi}) = \ln\left(1 + \sum_{g=1}^{G-1} \exp(\kappa_g)\right) . \tag{3.185}$$

The covariance matrix $\boldsymbol{V}(\boldsymbol{y})$ has the elements

$$v_{gg} = \mu_g(1 - \mu_g) \, , \quad v_{gh} = -\mu_g \mu_h \, , \text{ and } g \neq h = 1,\ldots,G \, . \tag{3.186}$$

$\boldsymbol{V}(y)$ is only positive semidefinite because $\sum_{g=1}^{G} y_g = 1$. The multinomial density is used to model polytomous outcomes. Examples are the regression models for censored and classified metric variables, ordered categorical variables and unordered categorical variables that are discussed in Section 8 of this chapter.

Multivariate normal distribution: The density of a multivariate normal vector $\boldsymbol{y} \sim G \times 1 \in \mathcal{R}^G$ with expected value $\boldsymbol{\mu} \in \mathcal{R}^G$ and positive definite covariance matrix $\boldsymbol{\Phi}$ is

$$f(y|\boldsymbol{\mu}, \boldsymbol{\Phi}) = (2\pi)^{G/2}|\boldsymbol{\Phi}|^{-1/2} \exp\left\{-\frac{1}{2}(\boldsymbol{y} - \boldsymbol{\mu})'\boldsymbol{\Phi}^{-1}(\boldsymbol{y} - \boldsymbol{\mu})\right\} \quad . \tag{3.187}$$

A parameterization in $\boldsymbol{\mu}$ is

$$a(\boldsymbol{\mu}, \boldsymbol{\Phi}) = -\frac{1}{2}\boldsymbol{\mu}'\boldsymbol{\Phi}^{-1}\boldsymbol{\mu} ,$$

$$b(\boldsymbol{y}, \boldsymbol{\Phi}) = -\frac{G}{2}\ln(2\pi) - \frac{1}{2}\boldsymbol{y}'\boldsymbol{\Phi}^{-1}\boldsymbol{y}' - \frac{1}{2}\ln|\boldsymbol{\Phi}| \text{ , and} \tag{3.188}$$

$$\boldsymbol{c}(\boldsymbol{\mu}, \boldsymbol{\Phi}) = \boldsymbol{\Phi}^{-1}\boldsymbol{\mu} \quad . \tag{3.189}$$

The canonical parameter is $\boldsymbol{\kappa} = \boldsymbol{\Phi}^{-1}\boldsymbol{\mu}$. A parameterization in $\boldsymbol{\kappa}$ is

$$u(\boldsymbol{y}, \boldsymbol{\Phi}) = -\frac{G}{2}\ln(2\pi) - \frac{1}{2}\boldsymbol{y}'\boldsymbol{\Phi}^{-1}\boldsymbol{y}' - \frac{1}{2}\ln|\boldsymbol{\Phi}| \text{ , and} \tag{3.190}$$

$$v(\boldsymbol{\kappa}, \boldsymbol{\Phi}) = \frac{1}{2}\boldsymbol{\kappa}'\boldsymbol{\Phi}\boldsymbol{\kappa} \quad . \tag{3.191}$$

The covariance matrix as a function of $\boldsymbol{\mu}$ and $\boldsymbol{\Phi}$ is

$$\boldsymbol{V}(\boldsymbol{y}) = \boldsymbol{\Sigma} = \boldsymbol{\Phi} , \quad \Rightarrow \quad \boldsymbol{\Phi} = \boldsymbol{\Sigma} \quad . \tag{3.192}$$

The multivariate normal distribution is the most often used distribution in multivariate statistical analysis of G variate metric outcomes. However, it may also be used to derive estimators for vectors of multivariate counts or dichotomous outcomes, as is shown for multivariate nonlinear regression models in Section 8 of this chapter.

Properties of the Linear Exponential Family Parameterized in the Mean

Important properties of the linear exponential family are:

1. $$\frac{\partial \boldsymbol{c}'(\boldsymbol{\mu}, \boldsymbol{\Phi})}{\partial \boldsymbol{\mu}} = \frac{\partial \boldsymbol{c}(\boldsymbol{\mu}, \boldsymbol{\Phi})}{\partial \boldsymbol{\mu}'} = \boldsymbol{\Sigma}^{-1} \tag{3.193}$$

   This property follows from the fact that the canonical parameter $\boldsymbol{\kappa} = \boldsymbol{c}(\boldsymbol{\mu}, \boldsymbol{\Phi})$, equation (3.149), and the fact that the covariance matrix $\boldsymbol{\Sigma}$ is symmetric and positive definite.

2. $$\frac{\partial a(\boldsymbol{\mu}, \boldsymbol{\Phi})}{\partial \boldsymbol{\mu}} + \frac{\partial \boldsymbol{c}'(\boldsymbol{\mu}, \boldsymbol{\Phi})}{\partial \boldsymbol{\mu}}\boldsymbol{\mu} = \boldsymbol{0} \tag{3.194}$$

   This property follows immediately from $\int f(\boldsymbol{y})d\boldsymbol{y} = 1$ by differentiating both sides with regard to $\boldsymbol{\mu}$ and exchanging differentiation and integration.

   A useful variant of this property is the equation

$$\frac{\partial a(\boldsymbol{\mu}, \boldsymbol{\Phi})}{\partial \boldsymbol{\mu}} + \frac{\partial \boldsymbol{c}'(\boldsymbol{\mu}, \boldsymbol{\Phi})}{\partial \boldsymbol{\mu}}\boldsymbol{y} = -\frac{\partial \boldsymbol{c}'(\boldsymbol{\mu}, \boldsymbol{\Phi})}{\partial \boldsymbol{\mu}}\boldsymbol{\mu} + \frac{\partial \boldsymbol{c}'(\boldsymbol{\mu}, \boldsymbol{\Phi})}{\partial \boldsymbol{\mu}}\boldsymbol{y} \tag{3.195}$$

3. $$\frac{\partial^2 a(\boldsymbol{\mu},\boldsymbol{\Phi})}{\partial\boldsymbol{\mu}\partial\boldsymbol{\mu}'}+\sum_{g=1}^{G}\frac{\partial^2 c_g(\boldsymbol{\mu},\boldsymbol{\Phi})}{\partial\boldsymbol{\mu}\partial\boldsymbol{\mu}'}\mu_g+\frac{\partial \boldsymbol{c}'(\boldsymbol{\mu},\boldsymbol{\Phi})}{\partial\boldsymbol{\mu}}=\boldsymbol{0} \tag{3.196}$$

$c_g(\boldsymbol{\mu},\boldsymbol{\Phi})$ and μ_g are the gth elements of $\boldsymbol{c}(\boldsymbol{\mu},\boldsymbol{\Phi})$ and $\boldsymbol{\mu}$. This property follows from differentiating (3.194) with respect to $\boldsymbol{\mu}'$ and (3.193).

4. For any $\boldsymbol{\mu}\in\mathcal{M}$ and fixed true value $\boldsymbol{\mu}_0\in\mathcal{M}$, the following inequality holds:

$$a(\boldsymbol{\mu},\boldsymbol{\Phi})+\boldsymbol{c}'(\boldsymbol{\mu},\boldsymbol{\Phi})\boldsymbol{\mu}_0\leq a(\boldsymbol{\mu}_0,\boldsymbol{\Phi})+\boldsymbol{c}'(\boldsymbol{\mu}_0,\boldsymbol{\Phi})\boldsymbol{\mu}_0 \tag{3.197}$$

with equality if and only if (iff) $\boldsymbol{\mu}=\boldsymbol{\mu}_0$ almost surely. This inequality follows directly from Kullback's inequality (Rao 1973, chap. 1, p. 59) for arbitrary densities $g^\star(\boldsymbol{y})$ and $f^\star(\boldsymbol{y})$.

$$\int \ln\frac{g^\star(\boldsymbol{y})}{f^\star(\boldsymbol{y})}g^\star(\boldsymbol{y})d\boldsymbol{y}\geq 0 \tag{3.198}$$

The value 0 is taken iff $g^\star(\boldsymbol{y})=f^\star(\boldsymbol{y})$ almost surely. If one sets $g^\star(\boldsymbol{y})=f(\boldsymbol{y}|\boldsymbol{\mu}_0,\boldsymbol{\Phi})$ and $f^\star(\boldsymbol{y})=f(\boldsymbol{y}|\boldsymbol{\mu},\boldsymbol{\Phi})$, one finds

$$\int \ln\frac{f(\boldsymbol{y}|\boldsymbol{\mu}_0,\boldsymbol{\Phi})}{f(\boldsymbol{y}|\boldsymbol{\mu},\boldsymbol{\Phi})}f(\boldsymbol{y}|\boldsymbol{\mu}_0,\boldsymbol{\Phi})d\boldsymbol{y}\geq 0 \quad . \tag{3.199}$$

Substitution of (3.152) and evaluation of the integral gives property 4.

5.3 Properties of PML Estimators

In a mean-structure model it is assumed that the expected value $E_{f_\star}(\boldsymbol{y}|\boldsymbol{x})=\boldsymbol{\mu}(\boldsymbol{x},\boldsymbol{\vartheta}_0)$ is parameterized in $\boldsymbol{\vartheta}_0$. The covariance matrix $\boldsymbol{V}(\boldsymbol{y}|\boldsymbol{x})=\boldsymbol{\Omega}_\star(\boldsymbol{x})$ exists but is not specified. To estimate $\boldsymbol{\vartheta}_0$ one uses the assumed density of the linear exponential family with fixed or previously determined nuisance parameter $\boldsymbol{\Phi}$. These fixed nuisance parameters may vary across sample elements as a function $\boldsymbol{\Phi}(\boldsymbol{x})$, but this additional complication does not change the argument for consistency given below. The normed log-likelihood function of the assumed density is

$$l(\boldsymbol{\vartheta})=\frac{1}{n}\sum_{i=1}^{n}[a(\boldsymbol{\mu}(\boldsymbol{x}_i,\boldsymbol{\vartheta}),\boldsymbol{\Phi})+b(\boldsymbol{y}_i,\boldsymbol{\Phi})+\boldsymbol{y}_i'\boldsymbol{c}(\boldsymbol{\mu}(\boldsymbol{x}_i,\boldsymbol{\vartheta}),\boldsymbol{\Phi})]. \tag{3.200}$$

The linear exponential family contains the univariate and the multivariate normal distribution as special cases given a fixed nuisance parameter. Hence, maximization of a quasi-log-likelihood function based on the normal distribution is equal to minimizing the normed sum of least-squares in a linear or nonlinear regression model with no specific assumptions about the second moments of the error. The PML estimation based on the normal distribution as the assumed density is therefore equivalent to least squares estimation and to minimum distance estimation in nonlinear models.

Any member of the exponential family can be used to estimate $\boldsymbol{\vartheta}_0$. Some regularity conditions have to be fulfilled such as the requirement that the domain space for $\boldsymbol{y}$ in the true density $f_\star(\boldsymbol{y}|\boldsymbol{x})$ is a subset of the domain space for $\boldsymbol{y}$ of the assumed density $f(\boldsymbol{y}|\boldsymbol{x},\boldsymbol{\vartheta})$. This implies for instance, that the Poisson density can only be used for positive, discrete variables, while the normal density can be used for continuous as well as for discrete variables of any range.

If one assumes that $l(\boldsymbol{\vartheta})$ converges to the expected value $E_g E_{f_\star} \ln f(\boldsymbol{y}|\boldsymbol{x},\boldsymbol{\vartheta},\boldsymbol{\Phi})$, strong consistency of the estimator $\hat{\boldsymbol{\vartheta}}$ that maximizes $l(\boldsymbol{\vartheta})$ is established if the global maximum of $E_g E_{f_\star} \ln f(\boldsymbol{y}|\boldsymbol{x},\boldsymbol{\vartheta},\boldsymbol{\Phi})$ is $\boldsymbol{\vartheta}_0$. To show this, one first writes the expected value of the logarithm of the assumed density:

$$E_g E_{f_\star} \ln f(\boldsymbol{y}|\boldsymbol{x},\boldsymbol{\vartheta},\boldsymbol{\Phi}) =$$

$$\int\!\!\int \left[a(\boldsymbol{\mu}(\boldsymbol{x},\boldsymbol{\vartheta}),\boldsymbol{\Phi})) + b(\boldsymbol{y},\boldsymbol{\Phi}) + \boldsymbol{y}'\boldsymbol{c}(\boldsymbol{\mu}(\boldsymbol{x},\boldsymbol{\vartheta}),\boldsymbol{\Phi}))\right] f_\star(\boldsymbol{y}|\boldsymbol{x}) d\boldsymbol{y} g(\boldsymbol{x}) d\boldsymbol{x}\,. \qquad (3.201)$$

Integration over the domain of $\boldsymbol{y}$, using the specification $E_{f_\star}(\boldsymbol{y}|\boldsymbol{x}) = \int \boldsymbol{y} f_\star(\boldsymbol{y}|\boldsymbol{x}) d\boldsymbol{y} = \boldsymbol{\mu}(\boldsymbol{x},\boldsymbol{\vartheta}_0)$, gives

$$\int \left[a(\boldsymbol{\mu}(\boldsymbol{x},\boldsymbol{\vartheta}),\boldsymbol{\Phi})) + b(\boldsymbol{y},\boldsymbol{\Phi}) + \boldsymbol{\mu}_0'\boldsymbol{c}(\boldsymbol{\mu}(\boldsymbol{x},\boldsymbol{\vartheta}),\boldsymbol{\Phi})\right] g(\boldsymbol{x}) d\boldsymbol{x} \quad . \qquad (3.202)$$

Application of property 4 of the linear exponential family yields the inequality

$$\int \left[a(\boldsymbol{\mu}(\boldsymbol{x},\boldsymbol{\vartheta}),\boldsymbol{\Phi}) + \boldsymbol{\mu}(\boldsymbol{x},\boldsymbol{\vartheta}_0)'\boldsymbol{c}(\boldsymbol{\mu}(\boldsymbol{x},\boldsymbol{\vartheta}),\boldsymbol{\Phi})\right] g(\boldsymbol{x}) d\boldsymbol{x} \leq$$

$$\int \left[a(\boldsymbol{\mu}(\boldsymbol{x},\boldsymbol{\vartheta}_0),\boldsymbol{\Phi}) + \boldsymbol{\mu}(\boldsymbol{x},\boldsymbol{\vartheta}_0)'\boldsymbol{c}(\boldsymbol{\mu}(\boldsymbol{x},\boldsymbol{\vartheta}_0),\boldsymbol{\Phi})\right] g(\boldsymbol{x}) d\boldsymbol{x}\,.$$

Equality and therefore the maximum of $E_g E_{f_\star} \ln f(\boldsymbol{y}|\boldsymbol{x},\boldsymbol{\vartheta},\boldsymbol{\Phi})$ is attained if $\boldsymbol{\mu}(\boldsymbol{x},\boldsymbol{\vartheta}) = \boldsymbol{\mu}(\boldsymbol{x},\boldsymbol{\vartheta}_0)$. From the assumption of first-order identifiability follows the equality of $\boldsymbol{\vartheta}$ and $\boldsymbol{\vartheta}_0$. Consequently, the ML estimator from the assumed density is strongly consistent for $\boldsymbol{\vartheta}_0$ if the mean structure is specified correctly. Note, that in any other respect, the assumed density need not agree with the true density. This consistent estimator $\hat{\boldsymbol{\vartheta}}$ is called the PML estimator of $\boldsymbol{\vartheta}_0$. One important consequence of this result is that whenever the conditional mean structure is correctly specified one can use linear or nonlinear least squares to get consistent estimates of $\boldsymbol{\vartheta}_0$ regardless of $\boldsymbol{\Omega}_\star(\boldsymbol{x})$ or any other properties of the true distribution. For example, the parameters of a loglinear or a logit model can be estimated consistently by using the mean structure of a loglinear or a logit model but by choosing the normal density for the log-likelihood function or equivalently nonlinear least squares.

Furthermore, Gourieroux et al. (1984) show that the linear exponential family is the only family of distributions with the property that the parameters of the mean structure are estimated consistently with ML regardless of the true density. This result implies that the use of assumed densities that do not belong to the exponential families such as the uniform distribution yield inconsistent estimators of the parameters of the mean structure.

The asymptotic covariance matrix of $\sqrt{n}(\hat{\boldsymbol{\vartheta}} - \boldsymbol{\vartheta}_0)$ follows immediately from the results in Section 3 on ML estimation. Since the assumed and the true density need not be equal, one finds:

$$\sqrt{n}(\hat{\boldsymbol{\vartheta}} - \boldsymbol{\vartheta}_0) \overset{A}{\sim} \mathcal{N}\left(\mathbf{0}, (-\boldsymbol{A}(\boldsymbol{\vartheta}_0))^{-1}\boldsymbol{B}(\boldsymbol{\vartheta}_0)(-\boldsymbol{A}(\boldsymbol{\vartheta}_0))^{-1}\right). \tag{3.203}$$

The computation of the first and second derivatives of the quasi-log-likelihood function is simplified using properties of the linear exponential family (cf. Gourieroux et al. 1984). The abbreviation $\boldsymbol{\mu}$ is used for $\boldsymbol{\mu}(\boldsymbol{x}, \boldsymbol{\vartheta})$. The arguments $\boldsymbol{\mu}$ and $\boldsymbol{\Phi}$ are left out in the functions $a(\boldsymbol{\mu}, \boldsymbol{\Phi})$, $b(\boldsymbol{y}, \boldsymbol{\Phi})$ and $\boldsymbol{c}(\boldsymbol{\mu}, \boldsymbol{\Phi})$ for notational simplicity. The vector of first order derivatives of the exponential family density with respect to $\boldsymbol{\vartheta}$ is given by [cf. (3.89)]

$$\boldsymbol{q}(\boldsymbol{y}|\boldsymbol{x}, \boldsymbol{\vartheta}) = \frac{\partial}{\partial \boldsymbol{\vartheta}}(a + b + \boldsymbol{y}'\boldsymbol{c}) = \frac{\partial \boldsymbol{\mu}'}{\partial \boldsymbol{\vartheta}}\frac{\partial \boldsymbol{c}'}{\partial \boldsymbol{\mu}}(\boldsymbol{y} - \boldsymbol{\mu}), \tag{3.204}$$

where $\partial \boldsymbol{\mu}'/\partial \boldsymbol{\vartheta}$ is the $q \times G$ matrix of first derivatives of $\boldsymbol{\mu}'$ with respect to $\boldsymbol{\vartheta}$. The last equality sign follows from property 2, equation (3.194), of the exponential family.

Consequently, the matrix $\boldsymbol{B}(\boldsymbol{\vartheta}_0)$ of expected crossproducts of first derivatives is

$$\begin{aligned} \boldsymbol{B}(\boldsymbol{\vartheta}_0) &= E_g E_{f_\star}\left[\boldsymbol{q}(\boldsymbol{y}|\boldsymbol{x}, \boldsymbol{\vartheta}_0)\boldsymbol{q}(\boldsymbol{y}|\boldsymbol{x}, \boldsymbol{\vartheta}_0)'\right] \\ &= E_g\left[\frac{\partial \boldsymbol{\mu}'}{\partial \boldsymbol{\vartheta}}\frac{\partial \boldsymbol{c}'}{\partial \boldsymbol{\mu}}\left(E_{f_\star}(\boldsymbol{y} - \boldsymbol{\mu})(\boldsymbol{y} - \boldsymbol{\mu})'\right)\left(\frac{\partial \boldsymbol{c}'}{\partial \boldsymbol{\mu}}\right)'\left(\frac{\partial \boldsymbol{\mu}'}{\partial \boldsymbol{\vartheta}}\right)'\right] \\ &= E_g\left[\frac{\partial \boldsymbol{\mu}'}{\partial \boldsymbol{\vartheta}}\frac{\partial \boldsymbol{c}'}{\partial \boldsymbol{\mu}}\boldsymbol{\Omega}_\star(\boldsymbol{x})\left(\frac{\partial \boldsymbol{c}'}{\partial \boldsymbol{\mu}}\right)'\left(\frac{\partial \boldsymbol{\mu}'}{\partial \boldsymbol{\vartheta}}\right)'\right]. \end{aligned} \tag{3.205}$$

The derivatives of $\boldsymbol{\mu}$ with respect to $\boldsymbol{\vartheta}$ are evaluated at $\boldsymbol{\vartheta}_0$.

The matrix of second-order derivatives [cf. (3.90)] of the exponential family density with respect to $\boldsymbol{\vartheta}$ is computed by using the product rule (3.195) of property 2 and (3.196) of property 3 of the linear exponential family:

$$\begin{aligned} \boldsymbol{Q}(\boldsymbol{y}|\boldsymbol{x}, \boldsymbol{\vartheta}) &= \frac{\partial^2}{\partial \boldsymbol{\vartheta}\partial \boldsymbol{\vartheta}'}(a + b + \boldsymbol{y}'\boldsymbol{c}) = \frac{\partial}{\partial \boldsymbol{\vartheta}}\left[\sum_g \frac{\partial \mu_g}{\partial \boldsymbol{\vartheta}'}\left(\frac{\partial a}{\partial \mu_g} + \frac{\partial \boldsymbol{c}'}{\partial \mu_g}\boldsymbol{y}\right)\right] \\ &= \sum_g \frac{\partial^2 \mu_g}{\partial \boldsymbol{\vartheta}\partial \boldsymbol{\vartheta}'}\left(\frac{\partial a}{\partial \mu_g} + \frac{\partial \boldsymbol{c}'}{\partial \mu_g}\boldsymbol{y}\right) + \frac{\partial \boldsymbol{\mu}'}{\partial \boldsymbol{\vartheta}}\left[\frac{\partial^2 a}{\partial \boldsymbol{\mu}\partial \boldsymbol{\mu}'} + \sum_g \frac{\partial^2 c_g}{\partial \boldsymbol{\mu}\partial \boldsymbol{\mu}'}y_g\right]\frac{\partial \boldsymbol{\mu}}{\partial \boldsymbol{\vartheta}'} \\ &= \sum_g \frac{\partial^2 \mu_g}{\partial \boldsymbol{\vartheta}\partial \boldsymbol{\vartheta}'}\frac{\partial \boldsymbol{c}'}{\partial \mu_g}(\boldsymbol{y} - \boldsymbol{\mu}) + \frac{\partial \boldsymbol{\mu}'}{\partial \boldsymbol{\vartheta}}\left[\sum_g \frac{\partial^2 c_g}{\partial \boldsymbol{\mu}\partial \boldsymbol{\mu}'}(y_g - \mu_g) - \frac{\partial \boldsymbol{c}'}{\partial \boldsymbol{\mu}}\right]\frac{\partial \boldsymbol{\mu}}{\partial \boldsymbol{\vartheta}'}. \end{aligned} \tag{3.206}$$

Since $E_{f_\star}(\boldsymbol{y} - \boldsymbol{\mu}(\boldsymbol{x}, \boldsymbol{\vartheta}_0)) = \mathbf{0}$, by (3.206) the matrix $-\boldsymbol{A}(\boldsymbol{\vartheta}_0)$ of expected second derivatives may be written as

$$-\boldsymbol{A}(\boldsymbol{\vartheta}_0) = -E_g E_{f_\star}\left[\boldsymbol{Q}(\boldsymbol{y}|\boldsymbol{x}, \boldsymbol{\vartheta}_0)\right] = E_g\left[\frac{\partial \boldsymbol{\mu}'}{\partial \boldsymbol{\vartheta}}\frac{\partial \boldsymbol{c}'}{\partial \boldsymbol{\mu}}\frac{\partial \boldsymbol{\mu}}{\partial \boldsymbol{\vartheta}'}\right]. \tag{3.207}$$

Again, the derivatives with respect to $\boldsymbol{\vartheta}$ are evaluated at $\boldsymbol{\vartheta}_0$.

With $\partial \boldsymbol{c}'/\partial\boldsymbol{\mu} = \boldsymbol{\Sigma}^{-1}$, one can write the expressions for $\boldsymbol{B}(\boldsymbol{\vartheta}_0)$ and $-\boldsymbol{A}(\boldsymbol{\vartheta}_0)$ as follows:

$$\boldsymbol{B}(\boldsymbol{\vartheta}_0) = E_g\left[\frac{\partial\boldsymbol{\mu}'}{\partial\boldsymbol{\vartheta}}\boldsymbol{\Sigma}_0^{-1}\boldsymbol{\Omega}_\star(\boldsymbol{x})\boldsymbol{\Sigma}_0^{-1}\frac{\partial\boldsymbol{\mu}}{\partial\boldsymbol{\vartheta}'}\right] \tag{3.208}$$

$$-\boldsymbol{A}(\boldsymbol{\vartheta}_0) = E_g\left[\frac{\partial\boldsymbol{\mu}'}{\partial\boldsymbol{\vartheta}}\boldsymbol{\Sigma}_0^{-1}\frac{\partial\boldsymbol{\mu}}{\partial\boldsymbol{\vartheta}'}\right]. \tag{3.209}$$

These expressions are inserted in (3.203).

The asymptotic covariance matrix of the PML estimator $\hat{\boldsymbol{\vartheta}}$ depends on the covariance matrix $\boldsymbol{\Sigma}_0 = \boldsymbol{\Sigma}(\boldsymbol{x}, \boldsymbol{\vartheta}_0, \boldsymbol{\Phi})$ of the chosen element of the linear exponential family which has been used as the assumed density for the PML estimation, as well as on the covariance matrix $\boldsymbol{\Omega}_\star(\boldsymbol{x})$ of $\boldsymbol{y}$ given $\boldsymbol{x}$ under the true but unknown density $f_\star(\boldsymbol{y}|\boldsymbol{x})$. Additionally, $\boldsymbol{V}(\sqrt{n}\hat{\boldsymbol{\vartheta}}) = -\boldsymbol{A}(\boldsymbol{\vartheta}_0)^{-1}$ if the assumed and the true conditional density agree not only in the first but also in the second moment.

Consistent estimators $\hat{\boldsymbol{B}}$ and $\hat{\boldsymbol{A}}$ are found by replacing the expectation E_g by $\frac{1}{n}\sum_{i=1}^n$, $g(\boldsymbol{x})$ by $\frac{1}{n}$ and $\boldsymbol{\vartheta}_0$ by $\hat{\boldsymbol{\vartheta}}$. $\boldsymbol{\Phi}$ is assumed to be given or predetermined. $\boldsymbol{\mu}(\boldsymbol{x}_i, \hat{\boldsymbol{\vartheta}}, \boldsymbol{\Phi})$ and $\boldsymbol{\Sigma}(\boldsymbol{x}_i, \hat{\boldsymbol{\vartheta}}, \boldsymbol{\Phi})$ are denoted by $\hat{\boldsymbol{\mu}}_i$ and $\hat{\boldsymbol{\Sigma}}_i$.

$$\hat{\boldsymbol{B}} = \frac{1}{n}\sum_{i=1}^n\left[\frac{\partial\hat{\boldsymbol{\mu}}_i'}{\partial\boldsymbol{\vartheta}}\hat{\boldsymbol{\Sigma}}_i^{-1}(\boldsymbol{y}_i - \hat{\boldsymbol{\mu}}_i)(\boldsymbol{y}_i - \hat{\boldsymbol{\mu}}_i)'\hat{\boldsymbol{\Sigma}}_i^{-1}\left(\frac{\partial\hat{\boldsymbol{\mu}}_i'}{\partial\boldsymbol{\vartheta}}\right)'\right] \tag{3.210}$$

$$-\hat{\boldsymbol{A}} = \frac{1}{n}\sum_{i=1}^n\left[\frac{\partial\hat{\boldsymbol{\mu}}_i'}{\partial\boldsymbol{\vartheta}}\hat{\boldsymbol{\Sigma}}_i^{-1}\left(\frac{\partial\hat{\boldsymbol{\mu}}_i'}{\partial\boldsymbol{\vartheta}}\right)'\right] \tag{3.211}$$

The estimator $-\hat{\boldsymbol{A}}$ in the last equation is a consistent estimator of the Fisher information matrix, that is of the expected value of the second derivatives of the quasi-log-likelihood function. A consistent estimator of the asymptotic covariance matrix of $\sqrt{n}\hat{\boldsymbol{\vartheta}}$ is therefore given by

$$\hat{\boldsymbol{V}}(\sqrt{n}\hat{\boldsymbol{\vartheta}}) = (-\hat{\boldsymbol{A}})^{-1}\hat{\boldsymbol{B}}(-\hat{\boldsymbol{A}})^{-1} \quad . \tag{3.212}$$

This estimator may be used to construct asymptotic confidence intervals for individual parameters, confidence regions for sets of parameters, and test statistics for hypotheses about parameters.

5.4 Computation of PML Estimators With Fisher Scoring

The PML estimator $\hat{\boldsymbol{\vartheta}}$ is computed by maximizing the quasi-log-likelihood function $l(\boldsymbol{\vartheta})$ of the assumed density of the sample. Maximization usually involves an iterative procedure. The rationale for the gradient methods of optimization is a first-order Taylor series expansion of the first derivatives of $l(\boldsymbol{\vartheta})$. Let $\boldsymbol{\vartheta}^{(q)}$ be the parameter vector of iteration step q, $q = 0, \ldots$. The relevant approximation is

$$\frac{\partial l(\boldsymbol{\vartheta}^{(q+1)})}{\partial \boldsymbol{\vartheta}} \approx \frac{\partial l(\boldsymbol{\vartheta}^{(q)})}{\partial \boldsymbol{\vartheta}} + \frac{\partial^2 l(\boldsymbol{\vartheta}^{(q)})}{\partial \boldsymbol{\vartheta} \partial \boldsymbol{\vartheta}'}(\boldsymbol{\vartheta}^{(q+1)} - \boldsymbol{\vartheta}^{(q)}). \tag{3.213}$$

If the matrix of second derivatives can be inverted, this approximation may be solved for $\boldsymbol{\vartheta}^{(q+1)}$ to yield

$$\boldsymbol{\vartheta}^{(q+1)} = \boldsymbol{\vartheta}^{(q)} - \left(\frac{\partial^2 l(\boldsymbol{\vartheta}^{(q)})}{\partial \boldsymbol{\vartheta} \partial \boldsymbol{\vartheta}'} \right)^{-1} \frac{\partial l(\boldsymbol{\vartheta}^{(q)})}{\partial \boldsymbol{\vartheta}} \quad . \tag{3.214}$$

The last equation describes one step in the multivariate Newton-Raphson (NR) procedure to maximize $l(\boldsymbol{\vartheta})$. To avoid the cumbersome computation of the matrix of second derivatives, modifications of the NR procedure have been proposed. They are of the form

$$\boldsymbol{\vartheta}^{(q+1)} = \boldsymbol{\vartheta}^{(q)} - \gamma^{(q)} \left(\boldsymbol{K}(\boldsymbol{\vartheta}^{(q)}) \right)^{-1} \frac{\partial l(\boldsymbol{\vartheta}^{(q)})}{\partial \boldsymbol{\vartheta}} \tag{3.215}$$

where the scalar $\gamma(q)$ is the step length in the qth iteration step and the matrix $\boldsymbol{K}(\boldsymbol{\vartheta}^{(q)})$ is an approximation to the Hessian matrix of second derivatives of $l(\boldsymbol{\vartheta})$.

The Fisher Scoring procedure (Fisher 1925; Kale 1961, 1962) is defined by setting $\gamma(q) = 1$ in every step and approximating the Hessian matrix by its expected value, that is, the negative Fisher information matrix. This matrix itself may be replaced by a consistent estimator. In PML estimation, this is the matrix $\boldsymbol{K}(\boldsymbol{\vartheta}^{(q)}) = \boldsymbol{A}(\boldsymbol{\vartheta}^{(q)})$ from (3.211):

$$\boldsymbol{\vartheta}^{(q+1)} = \boldsymbol{\vartheta}^{(q)} - \left(\boldsymbol{A}(\boldsymbol{\vartheta}^{(q)}) \right)^{-1} \frac{\partial l(\boldsymbol{\vartheta}^{(q)})}{\partial \boldsymbol{\vartheta}} \quad . \tag{3.216}$$

Optimization procedures with different approximations to the Hessian matrix such as the Davidon-Fletcher-Powell (DFP) or the Broyden-Fletcher-Goldfarb-Shanno (BFGS) procedure may also be used (cf. for instance, Luenberger 1984).

The derivation of Fisher scoring for the estimation of mean structures is simplified by introducing the following notation:

$$\boldsymbol{M}_i = \frac{\partial \boldsymbol{\mu}(\boldsymbol{x}_i, \boldsymbol{\vartheta})}{\partial \boldsymbol{\vartheta}'} \tag{3.217}$$

is the $G \times q$ matrix of first derivatives of $\boldsymbol{\mu}$ with regard to $\boldsymbol{\vartheta}$ evaluated at $\boldsymbol{\vartheta}$ for the ith sample element. These matrices are stacked on top of each other to form the $nG \times q$ matrix $\boldsymbol{M}$, which corresponds to the regressor matrix in linear multivariate regression. Note that restrictions on the parameter $\boldsymbol{\vartheta}$ may be introduced by reparameterizing $\boldsymbol{\vartheta}$ as a continuously differentiable function of the fundamental parameter vector $\boldsymbol{\tau}$, as discussed in Section 3 and applying the chain rule:

$$\frac{\partial \boldsymbol{\mu}(\boldsymbol{x}, \boldsymbol{\vartheta}(\boldsymbol{\tau}))}{\partial \boldsymbol{\tau}'} = \frac{\partial \boldsymbol{\mu}(\boldsymbol{x}, \boldsymbol{\vartheta}(\boldsymbol{\tau}))}{\partial \boldsymbol{\vartheta}'} \frac{\partial \boldsymbol{\vartheta}(\boldsymbol{\tau})}{\partial \boldsymbol{\tau}'} \quad . \tag{3.218}$$

The ordinary residuals $\boldsymbol{e}_i \sim G \times 1$ evaluated at $\boldsymbol{\vartheta}$ are defined by

$$\boldsymbol{e}_i = \boldsymbol{y}_i - \boldsymbol{\mu}(\boldsymbol{x}_i, \boldsymbol{\vartheta}) \tag{3.219}$$

and are collected in the $nG \times 1$ vector $\boldsymbol{e}$ in the sample.

The $G \times G$ covariance matrices $\boldsymbol{\Sigma}_i = \boldsymbol{\Sigma}(\boldsymbol{x}_i, \boldsymbol{\vartheta}, \boldsymbol{\Phi})$ of the assumed density are collected in the $nG \times nG$ block diagonal matrix $\boldsymbol{\Sigma}$:

$$\Sigma = \begin{pmatrix} \Sigma_1 & \dots & \dots & \mathbf{0} \\ \mathbf{0} & \Sigma_2 & \dots & \mathbf{0} \\ \vdots & \vdots & \ddots & \vdots \\ \mathbf{0} & \dots & \dots & \Sigma_n \end{pmatrix} \tag{3.220}$$

The cross-products of the individual residual vectors at $\boldsymbol{\vartheta}$ are denoted by $\boldsymbol{\Omega}_i = \boldsymbol{e}_i\boldsymbol{e}_i' \sim G \times G$. Collecting the individual cross-products yields the $nG \times nG$ matrix:

$$\Omega = \begin{pmatrix} \Omega_1 & \dots & \dots & \mathbf{0} \\ \mathbf{0} & \Omega_2 & \dots & \mathbf{0} \\ \vdots & \vdots & \ddots & \vdots \\ \mathbf{0} & \dots & \dots & \Omega_n \end{pmatrix} \tag{3.221}$$

The superscripts hat $\hat{}$, tilde $\tilde{}$, and bar $\bar{}$ are put on top of the vector e and the matrices M, Σ, and $\boldsymbol{\Omega}$ if $\boldsymbol{\vartheta}$ is replaced by $\hat{\boldsymbol{\vartheta}}$, $\tilde{\boldsymbol{\vartheta}}$, and $\bar{\boldsymbol{\vartheta}}$.

Substituting $\boldsymbol{\vartheta}^{(q)}$ for $\hat{\boldsymbol{\vartheta}}$ in (3.211) and computation of $\frac{\partial l(\boldsymbol{\vartheta}^{(q)})}{\partial \boldsymbol{\vartheta}}$ with (3.204) gives:

$$-\boldsymbol{A}^{(q)}(\boldsymbol{\vartheta}^{(q)}) = \frac{1}{n}\sum_{i=1}^{n} \boldsymbol{M}_i'^{(q)}(\boldsymbol{\Sigma}_i^{(q)})^{-1}\boldsymbol{M}_i^{(q)} \quad , \tag{3.222}$$

$$\frac{\partial l(\boldsymbol{\vartheta}^{(q)})}{\partial \boldsymbol{\vartheta}} = \frac{1}{n}\sum_{i=1}^{n} \boldsymbol{M}_i'^{(q)}(\boldsymbol{\Sigma}_i^{(q)})^{-1}\boldsymbol{e}_i^{(q)} \quad . \tag{3.223}$$

Iteration step $q + 1$ of the Fisher scoring procedure is therefore given by

$$\boldsymbol{\vartheta}^{(q+1)} = \boldsymbol{\vartheta}^{(q)} + \left(\sum_{i=1}^{n} \boldsymbol{M}_i'^{(q)}(\boldsymbol{\Sigma}_i^{(q)})^{-1}\boldsymbol{M}_i^{(q)}\right)^{-1} \left(\sum_{i=1}^{n} \boldsymbol{M}_i'^{(q)}(\boldsymbol{\Sigma}_i^{(q)})^{-1}\boldsymbol{e}_i^{(q)}\right) . \tag{3.224}$$

In matrix notation, equation (3.224) may be written as

$$\boldsymbol{\vartheta}^{(q+1)} = \boldsymbol{\vartheta}^{(q)} + \left(\boldsymbol{M}'^{(q)}(\boldsymbol{\Sigma}^{(q)})^{-1}\boldsymbol{M}^{(q)}\right)^{-1} \boldsymbol{M}'^{(q)}(\boldsymbol{\Sigma}^{(q)})^{-1}\boldsymbol{e}^{(q)} \quad . \tag{3.225}$$

With the notation $\boldsymbol{z}^{(q)} = \boldsymbol{M}^{(q)}\boldsymbol{\vartheta}^{(q)} + \boldsymbol{e}^{(q)}$ and premultiplication of the last equation with $\left(\boldsymbol{M}'^{(q)}(\boldsymbol{\Sigma}^{(q)})^{-1}\boldsymbol{M}^{(q)}\right)$, one finds that $\boldsymbol{\vartheta}^{(q+1)}$ is the result of a weighted least squares regression of the vector $\boldsymbol{z}^{(q)}$ on $\boldsymbol{M}^{(q)}$.

$$\boldsymbol{\vartheta}^{(q+1)} = \left(\boldsymbol{M}'^{(q)}(\boldsymbol{\Sigma}^{(q)})^{-1}\boldsymbol{M}^{(q)}\right)^{-1} \boldsymbol{M}'^{(q)}(\boldsymbol{\Sigma}^{(q)})^{-1}\boldsymbol{z}^{(q)} \tag{3.226}$$

The vector $\boldsymbol{z}^{(q)}$ is the multivariate analogue of the "working vector" of generalized linear models (McCullagh and Nelder 1989). The iteration is finished if a suitable termination criterion is fulfilled (cf. Dennis and Schnabel 1983). The whole procedure is therefore called iterative weighted least squares (IWLS). A critical issue concerns the existence of the PML estimator and the choice of starting values. These questions cannot be answered in general but have to be discussed separately for specific models. Note that the inverse of $\boldsymbol{M}'^{(q)}(\boldsymbol{\Sigma}^{(q)})^{-1}\boldsymbol{M}^{(q)}$ may not exist. In this case a generalized inverse such as the sweep inverse should be used. The corresponding parameter will be aliased.

The estimated asymptotic covariance matrix of $\sqrt{n}(\hat{\boldsymbol{\vartheta}} - \boldsymbol{\vartheta}_0)$ may be calculated from (3.210) and (3.211) as

$$\hat{\boldsymbol{V}}(\sqrt{n}\hat{\boldsymbol{\vartheta}}) = (-\hat{\boldsymbol{A}})^{-1}\hat{\boldsymbol{B}}(-\hat{\boldsymbol{A}})^{-1} \text{ with} \tag{3.227}$$

$$-\hat{\boldsymbol{A}} = \frac{1}{n}\hat{\boldsymbol{M}}'\hat{\boldsymbol{\Sigma}}^{-1}\hat{\boldsymbol{M}} = \frac{1}{n}\sum_{i=1}^{n}\hat{\boldsymbol{M}}_i'\hat{\boldsymbol{\Sigma}}_i^{-1}\hat{\boldsymbol{M}}_i\text{ , and} \tag{3.228}$$

$$\hat{\boldsymbol{B}} = \frac{1}{n}\hat{\boldsymbol{M}}'\hat{\boldsymbol{\Sigma}}^{-1}\hat{\boldsymbol{\Omega}}\hat{\boldsymbol{\Sigma}}^{-1}\hat{\boldsymbol{M}} = \frac{1}{n}\sum_{i=1}^{n}\hat{\boldsymbol{M}}_i'\hat{\boldsymbol{\Sigma}}_i^{-1}\hat{\boldsymbol{\Omega}}_i\hat{\boldsymbol{\Sigma}}_i^{-1}\hat{\boldsymbol{M}}_i \quad . \tag{3.229}$$

Example: Arbitrary Heteroscedasticity in Linear Regression

As a simple example of PML estimation, the linear regression model for a univariate dependent variable $y_i = \boldsymbol{x}_i'\boldsymbol{\vartheta}_0 + \epsilon_i$ is considered where the variance $V(\epsilon_i) = \omega_i^2$ remains unspecified and the conditional density $f_\star(y_i|\boldsymbol{x}_i)$ is unknown. The assumed density for y_i given $\boldsymbol{x}$ is $\mathcal{N}(\boldsymbol{x}_i'\boldsymbol{\vartheta}_0, 1)$. The variance of the assumed density is set to 1. In this case one finds:

$$\begin{aligned} \boldsymbol{M}_i &= \boldsymbol{x}_i', \quad \boldsymbol{M} = \boldsymbol{X}, \quad \boldsymbol{\Sigma}_i = 1, \quad \boldsymbol{\Sigma} = \boldsymbol{I}, \\ e_i &= y_i - \boldsymbol{x}_i'\boldsymbol{\vartheta}, \quad \boldsymbol{e} = \boldsymbol{y} - \boldsymbol{X}\boldsymbol{\vartheta}, \\ \boldsymbol{\Omega}_i &= e_i^2, \quad \boldsymbol{\Omega} = \text{diag}\{e_1^2, \ldots, e_n^2\} \quad . \end{aligned} \tag{3.230}$$

The PML estimator of $\boldsymbol{\vartheta}_0$ is given by the first step of the IWLS procedure and is equal to the OLS estimator

$$\hat{\boldsymbol{\vartheta}} = (\boldsymbol{X}'\boldsymbol{X})^{-1}\boldsymbol{X}'\boldsymbol{y} \quad . \tag{3.231}$$

The estimators of the expected matrices of first and second derivatives are given by

$$-\hat{\boldsymbol{A}} = \frac{1}{n}\boldsymbol{X}'\boldsymbol{X} \quad , \tag{3.232}$$

$$\hat{\boldsymbol{B}} = \frac{1}{n}\boldsymbol{X}'\hat{\boldsymbol{\Omega}}\boldsymbol{X} \quad . \tag{3.233}$$

Hence, the estimate of the asymptotic covariance matrix of $\hat{\boldsymbol{\vartheta}}$ is

$$\hat{\boldsymbol{V}}(\hat{\boldsymbol{\vartheta}}) = \frac{1}{n}(-\hat{\boldsymbol{A}})^{-1}\hat{\boldsymbol{B}}(-\hat{\boldsymbol{A}})^{-1} = (\boldsymbol{X}'\boldsymbol{X})^{-1}\boldsymbol{X}'\hat{\boldsymbol{\Omega}}\boldsymbol{X}(\boldsymbol{X}'\boldsymbol{X})^{-1} \quad . \tag{3.234}$$

This is the estimator of White (1980) for $\boldsymbol{V}(\hat{\boldsymbol{\vartheta}})$ for linear regression models given in Section 2.2; this estimator of the asymptotic covariance matrix of $\hat{\boldsymbol{\vartheta}}$ is robust under any type of heteroscedasticity. As previously noted, this result holds only for large samples.

5.5 PML Wald and PML Lagrange Multiplier Tests

In the context of mean structures, testing is usually restricted to hypotheses about the parameter vector $\boldsymbol{\vartheta}_0$ of $\boldsymbol{\mu}(\boldsymbol{x}, \boldsymbol{\vartheta}_0)$. As in general ML estimation, hypotheses are formulated as restrictions of the form

$$\boldsymbol{r}(\boldsymbol{\vartheta}_0) = \mathbf{0} \tag{3.235}$$

where $\boldsymbol{r}(.)$ is an $s \times 1$ vector ($s \leq q$) of restrictions (possibly nonlinear), assumed to be continously differentiable with respect to $\boldsymbol{\vartheta}$. The $s \times q$ matrix of first derivatives at $\boldsymbol{\vartheta}$ is again denoted by $\boldsymbol{R}(\boldsymbol{\vartheta})$ and has full rank at $\boldsymbol{\vartheta}_0$. The PML estimate of $\boldsymbol{\vartheta}_0$ has been denoted by $\hat{\boldsymbol{\vartheta}}$. To test the null hypothesis $H_0 : \boldsymbol{r}(\boldsymbol{\vartheta}_0) = \mathbf{0}$ against the alternative hypothesis $H_1 : \boldsymbol{r}(\boldsymbol{\vartheta}_0) \neq \mathbf{0}$, one can use either the Wald test statistic or the Lagrange multiplier test statistic. If only the mean structure is specified, the asymptotic distribution of the pseudo-log-likelihood ratio statistic

$$LR_{PML} = 2n[l(\hat{\boldsymbol{\vartheta}}) - l(\tilde{\boldsymbol{\vartheta}})] \tag{3.236}$$

is unknown. Therefore, the log-likelihood ratio statistic cannot be used under PML estimation.

First, the Wald statistic is considered. The formulation of the Wald statistic given in Section 3 depends only on the matrix $\boldsymbol{R}(\hat{\boldsymbol{\vartheta}}) = \hat{\boldsymbol{R}}$ and the estimate of the asymptotic covariance matrix of $\sqrt{n}\hat{\boldsymbol{\vartheta}}$. This asymptotic covariance matrix has been derived for PML estimation. Consequently, the corresponding Wald statistic can be written as follows

$$W_{PML} = n(\hat{\boldsymbol{\vartheta}} - \boldsymbol{\vartheta}_0)'\hat{\boldsymbol{R}}' \left[\hat{\boldsymbol{R}}(-\hat{\boldsymbol{A}})^{-1}\hat{\boldsymbol{B}}(-\hat{\boldsymbol{A}})^{-1}\hat{\boldsymbol{R}}'\right]^{-1} \hat{\boldsymbol{R}}(\hat{\boldsymbol{\vartheta}} - \boldsymbol{\vartheta}_0). \tag{3.237}$$

The Wald statistic is distributed as a central χ^2 variable under H_0 with s degrees of freedom. Only the unrestricted estimator $\hat{\boldsymbol{\vartheta}}$ of $\boldsymbol{\vartheta}_0$ is calculated. The test of nonlinear restrictions using the PML Wald statistic involves similar problems as the use of the ML Wald statistic.

The second statistic is a version of Rao's efficient score or the Lagrange multiplier statistic. To derive this test statistic, one maximizes the quasi-log-likelihood function of PML estimation under the restriction that $\boldsymbol{r}(\boldsymbol{\vartheta}_0) = \mathbf{0}$. Setting the first derivatives of the Lagrangean [cf. (3.107)] with respect to $\boldsymbol{\vartheta}$ and $\boldsymbol{\lambda}$ equal to $\mathbf{0}$ yields the system

$$\begin{pmatrix} \dfrac{\partial l(\boldsymbol{\vartheta})}{\partial \boldsymbol{\vartheta}} + \boldsymbol{\lambda}'\boldsymbol{R}(\boldsymbol{\vartheta}) \\ \boldsymbol{r}(\boldsymbol{\vartheta}) \end{pmatrix} = \begin{pmatrix} \mathbf{0} \\ \mathbf{0} \end{pmatrix} . \tag{3.238}$$

The solution of this equation system is denoted by $(\tilde{\boldsymbol{\vartheta}}', \tilde{\boldsymbol{\lambda}}')'$. With a first-order Taylor expansion of $\partial l(\tilde{\boldsymbol{\vartheta}})/\partial \boldsymbol{\vartheta}$ and $\boldsymbol{r}(\tilde{\boldsymbol{\vartheta}})$ about $\boldsymbol{\vartheta}_0$, one may show that $\sqrt{n}\tilde{\boldsymbol{\lambda}}$ follows an asymptotic normal distribution [cf. for instance Küsters (1987, chap. 5, p. 60-63)]:

$$\sqrt{n}\tilde{\boldsymbol{\lambda}}_{PML} \overset{A}{\sim} \mathcal{N}(\mathbf{0}, \tilde{\boldsymbol{V}}_{PML}) \quad \text{where} \tag{3.239}$$

$$\tilde{\boldsymbol{V}}_{PML} = \left(\boldsymbol{R}'\boldsymbol{A}^{-1}\boldsymbol{R}\right)^{-1} \left(\boldsymbol{R}'\boldsymbol{A}^{-1}\boldsymbol{B}\boldsymbol{A}^{-1}\boldsymbol{R}\right) \left(\boldsymbol{R}'\boldsymbol{A}^{-1}\boldsymbol{R}\right)^{-1} \tag{3.240}$$

Here, $\boldsymbol{R} = \boldsymbol{R}(\boldsymbol{\vartheta}_0)$, $\boldsymbol{A} = \boldsymbol{A}(\boldsymbol{\vartheta}_0)$, and $\boldsymbol{B} = \boldsymbol{B}(\boldsymbol{\vartheta}_0)$. The asymptotic normality of $\sqrt{n}\tilde{\boldsymbol{\lambda}}_{PML}$ and $\sqrt{n}\tilde{\boldsymbol{\lambda}}_{PML}$ implies that the corresponding quadratic form

$$Q_1 = n\tilde{\boldsymbol{\lambda}}'_{PML}\tilde{\boldsymbol{V}}^{-1}_{PML}\tilde{\boldsymbol{\lambda}}_{PML} \qquad (3.241)$$

is asymptotically centrally χ^2 distributed with s degrees of freedom. Substituting $\partial l(\tilde{\boldsymbol{\vartheta}})/\partial\boldsymbol{\vartheta}$ for $\boldsymbol{R}(\tilde{\boldsymbol{\vartheta}})'\tilde{\boldsymbol{\lambda}}$, $\tilde{\boldsymbol{A}}$ for $\boldsymbol{A}$, $\tilde{\boldsymbol{B}}$ for $\boldsymbol{B}$, and $\tilde{\boldsymbol{R}} = \boldsymbol{R}(\tilde{\boldsymbol{\vartheta}})$ for $\boldsymbol{R}$ yields the corresponding Lagrange multiplier test statistic:

$$LM_{PML} = n\frac{\partial l(\tilde{\boldsymbol{\vartheta}})}{\partial\boldsymbol{\vartheta}'}\tilde{\boldsymbol{A}}^{-1}\tilde{\boldsymbol{R}}\left(\tilde{\boldsymbol{R}}'\tilde{\boldsymbol{A}}^{-1}\tilde{\boldsymbol{B}}\tilde{\boldsymbol{A}}^{-1}\tilde{\boldsymbol{R}}\right)^{-1}\tilde{\boldsymbol{R}}'\tilde{\boldsymbol{A}}^{-1}\frac{\partial l(\tilde{\boldsymbol{\vartheta}})}{\partial\boldsymbol{\vartheta}} \quad . \qquad (3.242)$$

For computation, the following substitutions (using property 1 of the exponential family and the notation of Section 5.4) are useful:

$$\frac{\partial l(\tilde{\boldsymbol{\vartheta}})}{\partial\boldsymbol{\vartheta}'} = \frac{1}{n}\sum_{i=1}^{n}\frac{\partial l_i(\tilde{\boldsymbol{\vartheta}})}{\partial\boldsymbol{\vartheta}'} = \frac{1}{n}\sum_{i=1}^{n}\tilde{\boldsymbol{M}}_i\left[\frac{\partial a(\tilde{\boldsymbol{\mu}}_i,\boldsymbol{\Phi})}{\partial\boldsymbol{\mu}} + \tilde{\boldsymbol{\Sigma}}_i^{-1}\boldsymbol{y}_i\right] \qquad (3.243)$$

$$-\tilde{\boldsymbol{A}} = \frac{1}{n}\tilde{\boldsymbol{M}}'\tilde{\boldsymbol{\Sigma}}\tilde{\boldsymbol{M}} \qquad (3.244)$$

$$\tilde{\boldsymbol{B}} = \frac{1}{n}\tilde{\boldsymbol{M}}'\tilde{\boldsymbol{\Sigma}}^{-1}\tilde{\boldsymbol{\Omega}}\tilde{\boldsymbol{\Sigma}}^{-1}\tilde{\boldsymbol{M}} \quad . \qquad (3.245)$$

$a(\tilde{\boldsymbol{\mu}}_i,\boldsymbol{\Phi})$ and $\tilde{\boldsymbol{\Sigma}}_i^{-1}$ depend on the chosen member of the exponential family.

5.6 Regression Diagnostics Under PML Estimation

The asymptotic test statistics discussed in the last subsection are mainly used for testing a restricted model under the hypothesis that the unrestricted model captures adequately the mean structure of $\boldsymbol{y}$ given $\boldsymbol{x}$. A critical assessment of this hypothesis is often facilitated by using the extensions of the regression diagnostics of the linear model to the case when the parameters of the mean structure are estimated with PML methods.

Ordinary Residuals

The ordinary residual $\hat{\boldsymbol{e}}_i$ under the model $\boldsymbol{\vartheta}_0$ is defined just as the ordinary residual in linear regression:

$$\hat{\boldsymbol{e}}_i = \boldsymbol{y}_i - \boldsymbol{\mu}(\boldsymbol{x}_i,\hat{\boldsymbol{\vartheta}}). \qquad (3.246)$$

If $\hat{\boldsymbol{e}}_i$ is multivariate, one can use the quadratic form

$$\hat{q}_i = \hat{\boldsymbol{e}}_i'\hat{\boldsymbol{e}}_i \quad . \qquad (3.247)$$

If there is no model for $\boldsymbol{\Omega}_*(\boldsymbol{x}_i)$, then the ordinary residuals cannot be standardized. Hence, these residuals can only be used to check whether there is a systematic variation of the residuals with one or more of the regressors in $\boldsymbol{x}$. For univariate models one plots the residuals against the predicted values $\mu(\boldsymbol{x}_i,\hat{\boldsymbol{\vartheta}})$ or the values of specific regressors. If there is systematic variation, the mean structure may be misspecified.

Influential Points and Modified Cook Statistics

To judge the stability of a parameter estimate with regard to individual points, the Cook statistic has been used in the linear model. As a general approach to measure the influence of individual points, Cook and Weisberg (1982, chap. 5) propose the likelihood distance

$$LD_i = 2\left[nl(\hat{\boldsymbol{\vartheta}}) - (n-1)l(\hat{\boldsymbol{\vartheta}}_{(i)})\right] \qquad (3.248)$$

to judge the distance $\hat{\boldsymbol{\vartheta}} - \hat{\boldsymbol{\vartheta}}_{(i)}$. $\hat{\boldsymbol{\vartheta}}_{(i)}$ is the ML estimate of $\boldsymbol{\vartheta}_0$ after eliminating the ith sample point. If $LD_i > \chi^2_{1-\alpha;q}$ then i is considered to be an influential point which might be dropped from the analysis. The reason for this proposal is the definition of the asymptotic confidence region of the LR statistic (Cox and Hinkley 1974, chap. 9):

$$\{\boldsymbol{\vartheta} : 2n\left[l(\hat{\boldsymbol{\vartheta}}) - l(\boldsymbol{\vartheta})\right] \le \chi^2_{1-\alpha;q}\} \qquad (3.249)$$

for given $1-\alpha$. The distribution of the LD statistic is not known. To find the asymptotic distribution of LD_i, one would have to consider the joint asymptotic distribution of both estimators.

The LD statistic depends on the existence of a properly chosen log-likelihood function. If the likelihood function is replaced by the pseudo-likelihood function, the likelihood distance must be replaced by a statistic that is asymptotically equivalent to LD_i if $\boldsymbol{A}(\boldsymbol{\vartheta}_0) = -\boldsymbol{B}(\boldsymbol{\vartheta}_0)$ but is robust against violations of this condition. Since the PML estimator $\hat{\boldsymbol{\vartheta}}$ is asymptotically normal with expected value $\boldsymbol{\vartheta}_0$ and estimated covariance matrix $\hat{\boldsymbol{V}}(\sqrt{n}\hat{\boldsymbol{\vartheta}}) = (-\hat{\boldsymbol{A}})^{-1}\hat{\boldsymbol{B}}(-\hat{\boldsymbol{A}})^{-1}$, the asymptotic confidence region is given by

$$\{\boldsymbol{\vartheta} : n(\hat{\boldsymbol{\vartheta}} - \boldsymbol{\vartheta})'\left((-\hat{\boldsymbol{A}})^{-1}\hat{\boldsymbol{B}}(-\hat{\boldsymbol{A}})^{-1}\right)^{-1}(\hat{\boldsymbol{\vartheta}} - \boldsymbol{\vartheta}) \le \chi^2_{1-\alpha;q}\} \quad . \qquad (3.250)$$

This confidence region yields a modified likelihood distance which has been proposed by Cook and Weisberg (1982, chap. 5) for ML estimation if the contour of the log-likelihood function is approximately elliptic. This is equivalent to the assumption that the log-likelihood function can approximated by a quadratic term. Since this formulation depends not on the true log-likelihood function, but only on the expected first and second derivatives of the assumed liglikelihood function, one can define the modified Cook statistic for PML estimation:

$$\hat{C}_i = n(\hat{\boldsymbol{\vartheta}} - \hat{\boldsymbol{\vartheta}}_{(i)})'\left((-\hat{\boldsymbol{A}})^{-1}\hat{\boldsymbol{B}}(-\hat{\boldsymbol{A}})^{-1}\right)^{-1}(\hat{\boldsymbol{\vartheta}} - \hat{\boldsymbol{\vartheta}}_{(i)}) \qquad (3.251)$$

Again, for given $1-\alpha$ one compares the modified Cook statistic with the critical value $\chi^2_{1-\alpha;q}$. If $\hat{C}_i$ is greater than the critical value, then the data point is considered to be highly influential for the estimation of the parameters.

The computation of the modified Cook statistic is rather cumbersome because PML estimation must be performed $n+1$ times. To decrease computational effort considerably, one can approximate the modified Cook statistic with the formula (Arminger 1992)

$$\tilde{C}_i = \hat{e}_i'\hat{\boldsymbol{\Sigma}}_i^{-1/2\prime}\boldsymbol{K}_i\boldsymbol{Z}_i\left(\sum_{i=1}^{n}\boldsymbol{Z}_i'\boldsymbol{D}_i\boldsymbol{Z}_i\right)^{-1}\boldsymbol{Z}_i'\boldsymbol{K}_i\hat{\boldsymbol{\Sigma}}_i^{-1/2}\hat{e}_i \quad . \qquad (3.252)$$

The matrices $\boldsymbol{Z}_i$, $\boldsymbol{D}_i$, and $\boldsymbol{K}_i$ are defined as

$$\boldsymbol{Z}_i = \hat{\boldsymbol{\Sigma}}_i^{-1/2} \hat{\boldsymbol{M}}_i \sim G \times q \,, \tag{3.253}$$

where $\hat{\boldsymbol{\Sigma}}_i^{1/2}$ is the Cholesky decomposition $\hat{\boldsymbol{\Sigma}}_i = \hat{\boldsymbol{\Sigma}}_i^{1/2} \hat{\boldsymbol{\Sigma}}_i^{1/2\prime}$ of $\hat{\boldsymbol{\Sigma}}_i$.

$$\boldsymbol{D}_i = \hat{\boldsymbol{\Sigma}}_i^{-1/2\prime} \hat{e}_i \hat{e}_i' \hat{\boldsymbol{\Sigma}}_i^{-1/2} \tag{3.254}$$

$\boldsymbol{I}$ is the $G \times G$ identity matrix.

$$\boldsymbol{K}_i = \boldsymbol{I} + \left[\boldsymbol{I} - \boldsymbol{Z}_i \left(\sum_{i=1}^{n} \boldsymbol{Z}_i' \boldsymbol{Z}_i \right)^{-1} \boldsymbol{Z}_i' \right]^{-1} \boldsymbol{Z}_i \left(\sum_{i=1}^{n} \boldsymbol{Z}_i' \boldsymbol{Z}_i \right)^{-1} \boldsymbol{Z}_i' \tag{3.255}$$

In the univariate case, (3.252) may be simplified. The matrices $\hat{\boldsymbol{\Sigma}}_i$ and $\boldsymbol{K}_i$ now are scalar values so that the notation $v_i = \boldsymbol{Z}_i (\sum_{i=1}^{n} \boldsymbol{Z}_i' \boldsymbol{Z}_i)^{-1} \boldsymbol{Z}_i'$ yields $K_i = 1/(1 - v_i)$ and therefore,

$$\tilde{C}_i = \hat{e}_i^2 \boldsymbol{Z}_i (\sum_{i=1}^{n} \boldsymbol{Z}_i' \boldsymbol{D}_i \boldsymbol{Z}_i)^{-1} \boldsymbol{Z}_i' / \hat{\Sigma}^{-1} (1 - v_i)^2 \quad . \tag{3.256}$$

The approximate value of the modified Cook statistic is then compared with the critical value $\chi^2_{1-\alpha;G}$ for a given level α. If the Cook statistic exceeds this critical value the observation i should possibly be dropped from the analysis since it unduly influences the estimator $\hat{\boldsymbol{\vartheta}}$.

6 Quasi Generalized PML Estimation

6.1 Specification of Mean and Variance

In Section 5, only a model for the expected value of $\boldsymbol{y}$ given $\boldsymbol{x}$ parameterized in $\boldsymbol{\vartheta}_0$ has been assumed. In this section, it is additionally assumed that the covariance matrix $\boldsymbol{V}_{f_\star}(\boldsymbol{y}|\boldsymbol{x})$ can be correctly specified with a parametric model. The mean *and* the covariance structure under the true density are assumed to be equal to the mean and covariance structure of the linear exponential family with nuisance parameter $\boldsymbol{\Phi}$. The third and higher-order moments may differ for the true and the assumed density. The covariance matrix is given by

$$\boldsymbol{V}_{f_\star}(\boldsymbol{y}|\boldsymbol{x}) = \boldsymbol{V}_f(\boldsymbol{y}|\boldsymbol{x}, \boldsymbol{\vartheta}_0, \boldsymbol{\alpha}_0) = \boldsymbol{\Omega}(\boldsymbol{x}, \boldsymbol{\vartheta}_0, \boldsymbol{\alpha}_0) \tag{3.257}$$

where the function $\boldsymbol{\Omega}$ may depend on $\boldsymbol{x}$, the parameter vector $\boldsymbol{\vartheta}_0$ of the mean structure, and a $p \times 1$ vector $\boldsymbol{\alpha}_0$. The vector $\boldsymbol{\alpha}_0$ is independent of $\boldsymbol{\vartheta}_0$, that is, the values of $\boldsymbol{\alpha}_0$ can vary without being restricted by specific values of $\boldsymbol{\vartheta}_0$. The information contained in $\boldsymbol{\Omega}(\boldsymbol{x}, \boldsymbol{\vartheta}_0, \boldsymbol{\alpha})$ is used only for computing a more efficient estimate of $\boldsymbol{\vartheta}_0$ than the PML estimate. Therefore, the parameters in $\boldsymbol{\alpha}_0$ are considered as nuisance parameters and there is no genuine substantive interest in them. The more efficient estimates are usually computed in two steps and are called quasi-generalized PML estimates by Gourieroux et al. (1984). Again, the derivation and the properties of QGPML estimation have been given by Gourieroux et al. (1984), where regularity conditions and proofs are found.

If the parameters of the conditional covariance matrix parameters are of substantive interest and possibly depend on the parameters of the mean structure, simultaneous mean and covariance structures have to be formulated. The parameters of joint mean and covariance structures may be estimated with PML estimation using the quadratic instead of the linear exponential family. (This subject is taken up by Browne and Arminger in Chapter 4.)

With regard to $\boldsymbol{\alpha}$, conditional identifiability of second order given $\boldsymbol{\vartheta}$ is assumed, that is $\boldsymbol{\Omega}(\boldsymbol{x}, \boldsymbol{\vartheta}, \boldsymbol{\alpha}_1) = \boldsymbol{\Omega}(\boldsymbol{x}, \boldsymbol{\vartheta}, \boldsymbol{\alpha}_2) \Rightarrow \boldsymbol{\alpha}_1 = \boldsymbol{\alpha}_2$ holds for all fixed $\boldsymbol{x}$ and $\boldsymbol{\vartheta}$.

Important special cases of mean structures with nuisance parameters are the linear regression model with homoscedasticity

$$\mu(\boldsymbol{x}, \boldsymbol{\vartheta}_0) = \boldsymbol{x}'\boldsymbol{\vartheta}_0, \quad \Omega(\boldsymbol{x}, \boldsymbol{\vartheta}_0, \boldsymbol{\alpha}_0) = \Omega_0 \tag{3.258}$$

and the family of generalized linear models with overdispersion, but without assumptions about the higher moments of the true density $f_\star(y|\boldsymbol{x})$:

$$\mu(\boldsymbol{x}, \boldsymbol{\vartheta}_0) = g(\boldsymbol{x}'\boldsymbol{\vartheta}_0), \quad \Omega(\boldsymbol{x}, \boldsymbol{\vartheta}_0, \boldsymbol{\alpha}_0) = \Phi_0 \cdot \tau(\mu(\boldsymbol{x}, \boldsymbol{\vartheta}_0)). \tag{3.259}$$

The functional forms $g(.)$ and $\tau(.)$ are assumed to be known, the parameters $\boldsymbol{\vartheta}_0$ and $\boldsymbol{\alpha}_0 = \Phi_0$ are unknown. This class of models comprises as special cases the generalized linear models based on the univariate exponential family (cf. Section 5.2) and the quasi-likelihood models of Wedderburn (1974). In these models, the function $g(.)$ is the inverse link function and $\tau(.)$ is the variance function. The parameter Φ_0 is called the dispersion parameter.

6.2 Properties of PML Estimation With Nuisance Parameters

As before, the basis for estimation is again the linear exponential family with nuisance parameter $\boldsymbol{\Phi}$. The density of the ith observation is

$$f(\boldsymbol{y}_i|\boldsymbol{\mu}_i, \boldsymbol{\Phi}_i) = \exp\{a(\boldsymbol{\mu}_i, \boldsymbol{\Phi}_i) + b(\boldsymbol{y}_i, \boldsymbol{\Phi}_i) + \boldsymbol{y}_i'\boldsymbol{c}(\boldsymbol{\mu}_i, \boldsymbol{\Phi}_i)\} \tag{3.260}$$

where the nuisance parameter $\boldsymbol{\Phi}_i$ can vary over individual observations. The inverse covariance matrix of the ith observation under the assumed density of the linear exponential family is therefore given by

$$\boldsymbol{\Sigma}_i^{-1} = \frac{\partial \boldsymbol{c}'(\boldsymbol{\mu}_i, \boldsymbol{\Phi}_i)}{\partial \boldsymbol{\mu}_i} \quad . \tag{3.261}$$

In the examples of the linear exponential family discussed in Section 5.2, $\boldsymbol{\Phi}_i$ may be written as a differentiable function of $\boldsymbol{\mu}_i$ and $\boldsymbol{\Sigma}_i$, that is,

$$\boldsymbol{\Phi}_i = \boldsymbol{H}(\boldsymbol{\mu}_i, \boldsymbol{\Sigma}_i). \tag{3.262}$$

Since it is now assumed that not only the mean structure $\boldsymbol{\mu}_i = \boldsymbol{\mu}(\boldsymbol{x}_i, \boldsymbol{\vartheta}_0)$ but also the covariance structure $\boldsymbol{\Omega}(\boldsymbol{x}_i, \boldsymbol{\vartheta}_0, \boldsymbol{\alpha}_0)$ is equal for the assumed and the true density, the covariance matrix $\boldsymbol{\Sigma}_i$ of the assumed density is set equal to $\boldsymbol{\Omega}(\boldsymbol{x}_i, \boldsymbol{\vartheta}_0, \boldsymbol{\alpha}_0)$. Hence, $\boldsymbol{\Phi}_i$ can be written as a function $\boldsymbol{H}$ of $\boldsymbol{\mu}_i$ and $\boldsymbol{\Omega}_i$.

$$\boldsymbol{\Phi}_i = \boldsymbol{H}\left[\boldsymbol{\mu}(\boldsymbol{x}_i, \boldsymbol{\vartheta}_0), \boldsymbol{\Omega}(\boldsymbol{x}_i, \boldsymbol{\vartheta}_0, \boldsymbol{\alpha}_0)\right] \tag{3.263}$$

Estimation proceeds in two steps. In the first step a consistent estimator $\tilde{\boldsymbol{\vartheta}}$ is computed for $\boldsymbol{\vartheta}_0$ using the PML method of the last subsection. Additionally, a consistent estimator $\tilde{\boldsymbol{\alpha}}$ is computed for $\boldsymbol{\alpha}_0$; this will depend on the chosen model. Often one takes simple moment or least-squares estimators that yield consistent, but not necessarily efficient estimates of $\boldsymbol{\alpha}_0$. The corresponding nuisance parameter is computed as

$$\tilde{\boldsymbol{\Phi}}_i = \boldsymbol{H}\left[\boldsymbol{\mu}(\boldsymbol{x}_i, \tilde{\boldsymbol{\vartheta}}), \boldsymbol{\Omega}(\boldsymbol{x}_i, \tilde{\boldsymbol{\vartheta}}, \tilde{\boldsymbol{\alpha}})\right]. \tag{3.264}$$

In the second step, the estimator $\hat{\boldsymbol{\vartheta}}$ for $\boldsymbol{\vartheta}_0$ is estimated by maximizing the quasi-log-likelihood function

$$\tilde{l}(\boldsymbol{\vartheta}) = \frac{1}{n}\sum_{i=1}^{n} \ln f\left(\boldsymbol{y}_i | \boldsymbol{\mu}(\boldsymbol{x}_i, \boldsymbol{\vartheta}), \tilde{\boldsymbol{\Phi}}_i\right) \tag{3.265}$$

with regard to $\boldsymbol{\vartheta}$. This two-step estimator is called the quasi-generalized PML (QGPML) estimator by Gourieroux et al. (1984). It is a generalization of the GLS estimator under parameterized heteroscedasticity discussed by Jobson and Fuller (1980) and Amemiya (1985, chap. 6, p. 202).

The quasi-log-likelihood function $\tilde{l}(\boldsymbol{\vartheta})$ converges to its expected value:

$$E_g E_{f_\star}\left[\ln f\left(\boldsymbol{y} | \boldsymbol{x}, \boldsymbol{\mu}(\boldsymbol{x}, \boldsymbol{\vartheta}), \boldsymbol{\Phi}(\boldsymbol{x}, \boldsymbol{\vartheta}_0, \boldsymbol{\alpha}_0)\right)\right]. \tag{3.266}$$

Maximization of (3.266) with respect to $\boldsymbol{\vartheta}$ is identical to maximizing the expected value of a quasi-log-likelihood function based on the linear exponental family with known nuisance parameter $\boldsymbol{\Phi}(\boldsymbol{x}, \boldsymbol{\vartheta}_0, \boldsymbol{\alpha}_0) = \boldsymbol{H}\left[\boldsymbol{\mu}(\boldsymbol{x}, \boldsymbol{\vartheta}_0), \boldsymbol{\Omega}(\boldsymbol{x}, \boldsymbol{\vartheta}_0, \boldsymbol{\alpha}_0)\right]$. The covariance matrix $\boldsymbol{\Sigma}(\boldsymbol{x}, \boldsymbol{\vartheta})$ of the linear exponential family with fixed $\boldsymbol{\Phi}(\boldsymbol{x}, \boldsymbol{\vartheta}_0, \boldsymbol{\alpha}_0)$ at $\boldsymbol{\vartheta}_0$ is given by

$$\boldsymbol{\Sigma}(\boldsymbol{x}, \boldsymbol{\vartheta}_0, \boldsymbol{\Phi})^{-1} = \frac{\partial}{\partial \boldsymbol{\mu}} \boldsymbol{c}'\left[\boldsymbol{\mu}(\boldsymbol{x}, \boldsymbol{\vartheta}_0), \boldsymbol{\Phi}(\boldsymbol{x}, \boldsymbol{\vartheta}_0, \boldsymbol{\alpha}_0)\right] = \boldsymbol{\Omega}(\boldsymbol{x}, \boldsymbol{\vartheta}_0, \boldsymbol{\alpha}_0)^{-1}. \tag{3.267}$$

Hence, the asymptotic covariance matrix of $\sqrt{n}(\hat{\boldsymbol{\vartheta}} - \boldsymbol{\vartheta}_0)$ is found by substituting $\boldsymbol{\Omega}(\boldsymbol{x}, \boldsymbol{\vartheta}_0, \boldsymbol{\alpha}_0)$ in $\boldsymbol{\Sigma}_0(\boldsymbol{x})$ in (3.208) and (3.209), yielding $-\boldsymbol{A}(\boldsymbol{\vartheta}_0) = \boldsymbol{B}(\boldsymbol{\vartheta}_0)$:

$$\boldsymbol{V}(\sqrt{n}(\hat{\boldsymbol{\vartheta}}) = \left[E_g\left(\frac{\partial \boldsymbol{\mu}'}{\partial \boldsymbol{\vartheta}} \boldsymbol{\Omega}(\boldsymbol{x}, \boldsymbol{\vartheta}_0, \boldsymbol{\alpha}_0)\left(\frac{\partial \boldsymbol{\mu}'}{\partial \boldsymbol{\vartheta}}\right)'\right)\right]^{-1} \tag{3.268}$$

The asymptotic covariance matrix is consistently estimated by

$$\hat{\boldsymbol{V}}(\sqrt{n}\hat{\boldsymbol{\vartheta}}) = \left[\frac{1}{n}\sum_{i=1}^{n}\left(\frac{\partial \hat{\boldsymbol{\mu}}_i'}{\partial \boldsymbol{\vartheta}} \boldsymbol{\Omega}(\boldsymbol{x}_i, \hat{\boldsymbol{\vartheta}}, \tilde{\boldsymbol{\alpha}})\left(\frac{\partial \hat{\boldsymbol{\mu}}_i'}{\partial \boldsymbol{\vartheta}}\right)'\right)\right]^{-1}. \tag{3.269}$$

Computation of the QGPML estimator requires a consistent estimator $\tilde{\alpha}$ for α_0. If y is a scalar, then $\tilde{\alpha}$ may be computed by minimizing the unweighted least-squares function divided by n:

$$Q(\alpha) = \frac{1}{n}\sum_{i=1}^{n}\left[(y_i - \mu(\boldsymbol{x}_i, \tilde{\boldsymbol{\vartheta}}))^2 - \Omega(\boldsymbol{x}_i, \tilde{\boldsymbol{\vartheta}}, \boldsymbol{\alpha})\right]^2 \tag{3.270}$$

with respect to $\boldsymbol{\alpha}$. The minimization of the objective function in (3.270) corresponds again to PML estimation under the assumption that the squared residual e_i^2 has expected value $\Omega(\boldsymbol{x}_i, \tilde{\boldsymbol{\vartheta}}, \boldsymbol{\alpha})$ and that the density function of e_i^2 given $\boldsymbol{x}_i$ and $\tilde{\boldsymbol{\vartheta}}$ is unknown. The assumed density is the Gaussian density with expected value $\Omega(\boldsymbol{x}_i, \tilde{\boldsymbol{\vartheta}}, \boldsymbol{\alpha})$ and variance 1.

This procedure may be extended to the multivariate case by replacing the squared residual $(y_i - \mu(\boldsymbol{x}_i, \tilde{\boldsymbol{\vartheta}}))^2$ with the vector $\boldsymbol{s}(\boldsymbol{x}_i, \tilde{\boldsymbol{\vartheta}})$ that consists of the elements of the lower triangular matrix of $(\boldsymbol{y}_i - \boldsymbol{\mu}(\boldsymbol{x}_i, \tilde{\boldsymbol{\vartheta}}) \cdot (\boldsymbol{y}_i - \boldsymbol{\mu}(\boldsymbol{x}_i, \tilde{\boldsymbol{\vartheta}}))'$ including the diagonal. $\Omega(\boldsymbol{x}_i, \tilde{\boldsymbol{\vartheta}}, \boldsymbol{\alpha})$ is replaced by the vector $\omega(\boldsymbol{x}_i, \tilde{\boldsymbol{\vartheta}}, \boldsymbol{\alpha})$ that consists of the elements of the lower triangular matrix of $\boldsymbol{\Omega}(\boldsymbol{x}_i, \tilde{\boldsymbol{\vartheta}}, \boldsymbol{\alpha})$ including the diagonal. The consistent estimator $\tilde{\boldsymbol{\alpha}}$ is then found by minimizing the function

$$Q(\boldsymbol{\alpha}) = \sum_{i=1}^{n}[\boldsymbol{s}(\boldsymbol{x}_i, \tilde{\boldsymbol{\vartheta}}) - \omega(\boldsymbol{x}_i, \tilde{\boldsymbol{\vartheta}}, \boldsymbol{\alpha})]'[\boldsymbol{s}(\boldsymbol{x}_i, \tilde{\boldsymbol{\vartheta}}) - \omega(\boldsymbol{x}_i, \tilde{\boldsymbol{\vartheta}}, \boldsymbol{\alpha})] \tag{3.271}$$

with respect to $\boldsymbol{\alpha}$.

Carroll, Wu, and Ruppert (1988) consider the heteroscedastic linear regression model

$$y_i = \boldsymbol{x}_i'\boldsymbol{\vartheta}_0 + \delta_i \quad \text{with} \quad E(\delta_i|\boldsymbol{x}_i) = 0 \quad \text{and} \quad \delta_i = \sigma\Omega(\boldsymbol{x}_i'\boldsymbol{\vartheta}_0, \boldsymbol{x}_i, \boldsymbol{\alpha})\epsilon_i \tag{3.272}$$

where the random variable ϵ_i has expected value 0 and variance 1 and Ω is the variance function multiplied with a scalar σ. These authors show that the number of iterations c of the unweighted least-squares two-step estimator that is necessary for the stabilization of the asymptotic covariance matrix of $\sqrt{n}(\hat{\boldsymbol{\vartheta}} - \boldsymbol{\vartheta}_0)$ is $c \geq 1$, if the variance of δ_i is independent of the mean and $c \geq 2$ if the variance is dependent on the mean. If a different estimation method is used for $\boldsymbol{\alpha}_0$, then c has to be increased by 1. It may be surmised that similar results hold for general mean structures. Therefore, the two-step procedure given above should be iterated until covariance stabilization is reached.

Until now, only the mean structure $E(\boldsymbol{y}|\boldsymbol{x}) = \boldsymbol{\mu}(\boldsymbol{x}, \boldsymbol{\vartheta})$ and the covariance $\boldsymbol{V}(\boldsymbol{y}|\boldsymbol{x}) = \boldsymbol{\Omega}(\boldsymbol{x}, \boldsymbol{\vartheta}, \boldsymbol{\alpha})$ have been specified. If it may be assumed that the true density of $\boldsymbol{y}$ given $\boldsymbol{x}$ is itself a member of the exponential family with expected value $\boldsymbol{\mu}(\boldsymbol{x}, \boldsymbol{\vartheta})$ and covariance matrix $\boldsymbol{\Omega}(\boldsymbol{x}, \boldsymbol{\vartheta}, \boldsymbol{\alpha})$, then Gourieroux et al. (1984) show that the QGPML estimator $\hat{\boldsymbol{\vartheta}}$ given above is asymptotically equivalent to the ML estimator of $\boldsymbol{\vartheta}_0$ that is based on the density of the exponential family, where $\boldsymbol{\vartheta}_0$ and $\boldsymbol{\alpha}_0$ are estimated jointly.

6.3 Computation of QGPML Estimators

In the first stage the parameter $\boldsymbol{\vartheta}_0$ is estimated consistently by $\tilde{\boldsymbol{\vartheta}}$, using Fisher scoring. For given $\tilde{\boldsymbol{\vartheta}}$, a consistent estimator $\tilde{\boldsymbol{\alpha}}$ for $\boldsymbol{\alpha}_0$ is computed using (3.270) or (3.271). In the second step, $\tilde{\boldsymbol{\alpha}}$ is considered as fixed and the PML estimation of $\boldsymbol{\vartheta}_0$ is carried out as in Section 5.4 using the matrix $\boldsymbol{\Omega}(\boldsymbol{x}_i, \tilde{\boldsymbol{\vartheta}}, \tilde{\boldsymbol{\alpha}})$ as weight matrix in the IWLS algorithm. Let $\boldsymbol{\vartheta}^{(q)}$ be the parameter vector in the qth iteration step of the IWLS procedure. Then the computation of the parameter vector $\boldsymbol{\vartheta}^{(q+1)}$ for step $q+1$ is given by

$$\boldsymbol{\vartheta}^{(q+1)} = \left(\boldsymbol{M}^{(q)\prime}(\boldsymbol{\Omega}^{(q)})^{-1}\boldsymbol{M}^{(q)}\right)^{-1} \boldsymbol{M}^{(q)\prime}(\boldsymbol{\Omega}^{(q)})^{-1}\boldsymbol{z}^{(q)} \quad , \tag{3.273}$$

$$\boldsymbol{z}^{(q)} = \boldsymbol{M}^{(q)}\boldsymbol{\vartheta}^{(q)} + \boldsymbol{e}^{(q)} \quad . \tag{3.274}$$

The matrix $\boldsymbol{\Omega}^{(q)}$ is again the blockdiagonal matrix of matrices $\boldsymbol{\Omega}_i^{(q)} = \boldsymbol{\Omega}(\boldsymbol{x}_i, \boldsymbol{\vartheta}^{(q)}, \tilde{\boldsymbol{\alpha}})$. Instead of setting $\boldsymbol{\Omega}_i^{(q)} = \boldsymbol{\Omega}(\boldsymbol{x}_i, \boldsymbol{\vartheta}^{(q)}, \tilde{\boldsymbol{\alpha}})$, one can also iterate with $\boldsymbol{\Omega}_i^{(q)} = \boldsymbol{\Omega}(\boldsymbol{x}_i, \tilde{\boldsymbol{\vartheta}}, \tilde{\boldsymbol{\alpha}})$ and replace $\tilde{\boldsymbol{\vartheta}}$ by $\hat{\boldsymbol{\vartheta}}$ after the procedure has converged. In the QGPML procedure, the matrix $\boldsymbol{\Sigma}^{(q)}$ is replaced by $\boldsymbol{\Omega}^{(q)}$ since the covariance structure of $\boldsymbol{y}$ given $\boldsymbol{x}$ is assumed to be known. After convergence of the IWLS procedure, the estimate of the asymptotic covariance matrix of $\sqrt{n}(\hat{\boldsymbol{\vartheta}} - \boldsymbol{\vartheta}_0)$ is computed as

$$\hat{\boldsymbol{V}}(\sqrt{n}\hat{\boldsymbol{\vartheta}}) = \left(\frac{1}{n}\hat{\boldsymbol{M}}'\hat{\boldsymbol{\Omega}}^{-1}\hat{\boldsymbol{M}}\right)^{-1} \quad . \tag{3.275}$$

This is the usual form of the estimated asymptotic covariance matrix of parameters for non-linear weighted least squares. For many models the two-stage procedure can be simplified to a one-stage procedure. An important example is the class of univariate quasi-likelihood models of Wedderburn (1974), briefly discussed in Section 7.

6.4 QGPML Wald, Lagrange Multiplier, and Likelihood Ratio Tests

As in PML estimation, hypotheses are formulated as restrictions about $\boldsymbol{\vartheta}_0$ of the form $\boldsymbol{r}(\boldsymbol{\vartheta}_0) = \boldsymbol{0}$, with the same notation and conditions as in Section 5.5. The QGPML estimate of $\boldsymbol{\vartheta}_0$ is denoted by $\hat{\boldsymbol{\vartheta}}$. To test the null hypothesis $H_0 : \boldsymbol{r}(\boldsymbol{\vartheta}_0) = \boldsymbol{0}$ against the alternative hypothesis $H_1 : \boldsymbol{r}(\boldsymbol{\vartheta}_0) \neq \boldsymbol{0}$, one can use the Wald test statistic, the Lagrange multiplier test statistic, or the likelihood ratio statistic under the assumed density.

The Wald statistic (W) can be derived from the asymptotic covariance matrix $\boldsymbol{V}(\sqrt{n}\hat{\boldsymbol{\vartheta}})$ and is given by

$$W_{QGPML} = n(\hat{\boldsymbol{\vartheta}} - \boldsymbol{\vartheta}_0)'\hat{\boldsymbol{R}}'\left[\hat{\boldsymbol{R}}(-\hat{\boldsymbol{A}})^{-1}\hat{\boldsymbol{R}}'\right]^{-1}\hat{\boldsymbol{R}}(\hat{\boldsymbol{\vartheta}} - \boldsymbol{\vartheta}_0) \quad \text{with} \tag{3.276}$$

$$-\hat{\boldsymbol{A}} = \frac{1}{n}\hat{\boldsymbol{M}}'\hat{\boldsymbol{\Omega}}^{-1}\hat{\boldsymbol{M}} \quad .$$

The Wald statistic is distributed as a central χ^2 variable under the H_0 with s degrees of freedom. Only the unrestricted estimator $\hat{\boldsymbol{\vartheta}}$ of $\boldsymbol{\vartheta}_0$ is calculated.

The second statistic is Rao's efficient score or the Lagrange multiplier statistic (LM). The derivation is similar to the derivation of LM_{PML} in Section 5. The estimator of $\boldsymbol{\vartheta}_0$ under the restriction $\boldsymbol{r}(\boldsymbol{\vartheta}_0) = \mathbf{0}$ is denoted by $\tilde{\boldsymbol{\vartheta}}$. The values of the matrices $\boldsymbol{M}$, $\boldsymbol{A}$, and $\boldsymbol{\Omega}$ are also calculated at $\tilde{\boldsymbol{\vartheta}}$. The LM test statistic is given by

$$LM_{QGPML} = n\frac{\partial l(\tilde{\boldsymbol{\vartheta}})}{\partial \boldsymbol{\vartheta}'}(-\tilde{\boldsymbol{A}})^{-1}\frac{\partial l(\tilde{\boldsymbol{\vartheta}})}{\partial \boldsymbol{\vartheta}} \quad , \tag{3.277}$$

$$\frac{\partial l(\tilde{\boldsymbol{\vartheta}})}{\partial \boldsymbol{\vartheta}'} = \frac{1}{n}\sum_{i=1}^{n}\frac{\partial l_i(\tilde{\boldsymbol{\vartheta}})}{\partial \boldsymbol{\vartheta}'} = \frac{1}{n}\sum_{i=1}^{n}\tilde{\boldsymbol{M}}_i'\left[\frac{\partial a(\tilde{\boldsymbol{\mu}}_i)}{\partial \boldsymbol{\mu}} + \left(\boldsymbol{\Omega}_i(\boldsymbol{x}_i, \tilde{\boldsymbol{\vartheta}}, \bar{\boldsymbol{\alpha}})\right)^{-1}\boldsymbol{y}_i\right] . \tag{3.278}$$

The LM statistic under QGPML estimation is equivalent to the LM statistic in ML estimation. However, it applies only to the parameter vector $\boldsymbol{\vartheta}_0$ of the mean structure and not to the parametric specification of the entire density function of $\boldsymbol{y}$ given $\boldsymbol{x}$ as in ML estimation.

Finally, the likelihood ratio statistic (LR) is considered under QGPML estimation. If the covariance matrix $\boldsymbol{\Omega}_*(\boldsymbol{x})$ of the true density is specified, then $-\boldsymbol{A}(\boldsymbol{\vartheta}_0) = \boldsymbol{B}(\boldsymbol{\vartheta}_0)$. Under this assumption, a second-order Taylor expansion of $l(\hat{\boldsymbol{\vartheta}}, \boldsymbol{\alpha}_0) - l(\boldsymbol{\vartheta}_0, \boldsymbol{\alpha}_0)$ and $l(\tilde{\boldsymbol{\vartheta}}, \boldsymbol{\alpha}_0) - l(\boldsymbol{\vartheta}_0, \boldsymbol{\alpha}_0)$ about $\boldsymbol{\vartheta}_0$ shows that the pseudo-LR statistic is asymptotically χ^2 distributed with s degrees of freedom under H_0. The computation of the pseudo-LR statistic generally involves the following steps. First, a consistent estimator $\bar{\boldsymbol{\vartheta}}$ of $\boldsymbol{\vartheta}_0$ is computed without restrictions using PML estimation. Given $\bar{\boldsymbol{\vartheta}}$, a consistent estimator $\bar{\boldsymbol{\alpha}}$ for $\boldsymbol{\alpha}_0$ is computed. Given $\bar{\boldsymbol{\alpha}}$, the second-stage estimators $\hat{\boldsymbol{\vartheta}}$ (without restrictions) and $\tilde{\boldsymbol{\vartheta}}$ (with restrictions) are calculated. Hence, the pseudo-LR statistic is computed as

$$LR_{QGPML} = 2n[l(\hat{\boldsymbol{\vartheta}}, \bar{\boldsymbol{\alpha}}) - l(\tilde{\boldsymbol{\vartheta}}, \bar{\boldsymbol{\alpha}})], \tag{3.279}$$

where $l(\hat{\boldsymbol{\vartheta}}, \bar{\boldsymbol{\alpha}})$ and $l(\tilde{\boldsymbol{\vartheta}}, \bar{\boldsymbol{\alpha}})$ are based on the linear exponential family with covariance matrix $\boldsymbol{\Omega}(\boldsymbol{x}, \boldsymbol{\vartheta}_0, \boldsymbol{\alpha}_0)$, estimated by $\boldsymbol{\Omega}(\boldsymbol{x}, \bar{\boldsymbol{\vartheta}}, \bar{\boldsymbol{\alpha}})$.

6.5 Regression Diagnostics Under QGPML Estimation

Standardized and Studentized Residuals

The ordinary residual is $\hat{e}_i = \boldsymbol{y}_i - \boldsymbol{\mu}(\boldsymbol{x}_i, \hat{\boldsymbol{\vartheta}})$ as before. If the covariance matrix of $\epsilon_i = \boldsymbol{y}_i - \boldsymbol{\mu}(\boldsymbol{x}_i, \boldsymbol{\vartheta}_0)$ is estimated by $\hat{\boldsymbol{\Omega}}_i = \boldsymbol{\Omega}(\boldsymbol{x}_i, \hat{\boldsymbol{\vartheta}}, \hat{\boldsymbol{\alpha}})$ using the QGPML method, the standardized residual for variable y_{ig} in $\boldsymbol{y}_i$ is defined in analogy to the linear regression model as the residual divided by the square root of the estimated variance of the error term:

$$r_{ig} = \frac{\hat{e}_{ig}}{\sqrt{\hat{\Omega}_{igg}}} \quad . \tag{3.280}$$

In this definition, one assumes that the covariance matrix $\hat{\boldsymbol{\Omega}}_i$ of ϵ_i is practically the same as the covariance matrix of the residual $\hat{e}_i$, which is denoted by $\hat{\boldsymbol{V}}(\hat{e}_i)$. This will be correct if the sample size is large. The quadratic form of the standardized residual for vector residuals is given by

$$q_i^r = \hat{e}_i' \hat{\Omega}_i^{-1} \hat{e}_i \quad . \tag{3.281}$$

If one assumes that the univariate standardized residuals are approximately distributed as $\mathcal{N}(\mathbf{0}, \boldsymbol{I}_G)$, then one can use the usual $1-\alpha$ quantile of the univariate standard normal distribution to detect outliers in the dependent variables. If the vector of standardized residuals is used, an outlier is found if $q_i^r \geq \chi^2_{1-\alpha,G}$.

To ensure proper standardization of the residuals one computes their covariance matrix $\hat{\boldsymbol{V}}(\hat{e}_i)$ for each residual. The estimator of $\hat{\boldsymbol{V}}(\hat{e}_i)$ given below is an extension of the results of Pregibon (1981) for GLMs to multivariate mean structure models. The derivation and technical details are given in Arminger (1992). First, the QGPML analogue of the multivariate hat matrix is given

$$\hat{\boldsymbol{H}} = \hat{\boldsymbol{\Omega}}^{-1/2} \hat{\boldsymbol{M}} (\hat{\boldsymbol{M}}' \hat{\boldsymbol{\Omega}}^{-1} \hat{\boldsymbol{M}})^{-1} \hat{\boldsymbol{M}}' \hat{\boldsymbol{\Omega}}^{-1/2\prime} \quad . \tag{3.282}$$

$\hat{\boldsymbol{\Omega}}^{1/2}$ is the Cholesky decomposition of $\hat{\boldsymbol{\Omega}} = \hat{\boldsymbol{\Omega}}^{1/2} \cdot (\hat{\boldsymbol{\Omega}}^{1/2})'$. The $G \times G$ hat matrix for the ith observation is

$$\hat{\boldsymbol{H}}_{ii} = \hat{\boldsymbol{\Omega}}_i^{-1/2} \hat{\boldsymbol{M}}_i (\hat{\boldsymbol{M}}' \hat{\boldsymbol{\Omega}}^{-1} \hat{\boldsymbol{M}})^{-1} \hat{\boldsymbol{M}}_i' \hat{\boldsymbol{\Omega}}_i^{-1/2\prime} \quad . \tag{3.283}$$

The $nG \times nG$ covariance matrix of all residuals is estimated by

$$\hat{\boldsymbol{V}}(\hat{e}) = \hat{\boldsymbol{\Omega}}^{1/2} (\boldsymbol{I} - \hat{\boldsymbol{H}}) \hat{\boldsymbol{\Omega}}^{1/2\prime} \quad . \tag{3.284}$$

The estimated covariance matrix of the ith residual is

$$\hat{\boldsymbol{V}}(\hat{e}_i) = \hat{\boldsymbol{\Omega}}_i^{1/2} (\boldsymbol{I} - \hat{\boldsymbol{H}}_{ii}) \hat{\boldsymbol{\Omega}}_i^{1/2\prime} \quad . \tag{3.285}$$

Using these results one can write the studentized residual for variable y_{ig} as

$$s_{ig} = \frac{\hat{e}_{ig}}{\sqrt{\hat{\Omega}_{igg}(1 - \hat{H}_{igg})}} \tag{3.286}$$

where $\hat{H}_{igg}$ is the gth diagonal element of $\hat{\boldsymbol{H}}_{ii}$. Special cases of the last equation are the studentized residuals of linear regression (cf. Cook and Weisberg 1982, chap. 2, sec. 2) and of generalized linear models (cf. McCullagh and Nelder 1989, chap. 12).

The quadratic form of a multivariate studentized residual is given by

$$q_i^s = \hat{e}_i' \left[\hat{\boldsymbol{\Omega}}_i^{1/2} (\boldsymbol{I} - \hat{\boldsymbol{H}}_i) \hat{\boldsymbol{\Omega}}_i^{1/2\prime} \right]^{-1} \hat{e}_i \quad . \tag{3.287}$$

The studentized residual is used in the same way as the standardized residual to detect outliers in the dependent variables. Discrepancies between r_{ig} and s_{ig} are often found when the number q of estimated parameters is relatively large in comparison to the sample size n. This is the case for aggregated data such as in the analysis of contingency tables using loglinear or logit models. In this case, studentized residuals should be used.

Leverage and Influential Points

The matrix $\boldsymbol{H} \sim nG \times nG$ corresponds to the hat matrix $\boldsymbol{X}(\boldsymbol{X}'\boldsymbol{X})^{-1}\boldsymbol{X}'$ of linear regression in Section 2.3 and has similar properties. It is idempotent with rank q and trace tr$\boldsymbol{H} = q$. Consequently, the matrix $(\boldsymbol{I} - \boldsymbol{H})$ is idempotent with rank $nG - q$ and trace tr$(\boldsymbol{I} - \boldsymbol{H}) = nG - q$. The estimated covariance matrix of the error terms $\hat{\boldsymbol{\Omega}}$ can therefore be decomposed into the estimated covariance matrix of the residuals and the estimated covariance matrix of the estimated conditional means $\boldsymbol{\mu}(\boldsymbol{x}, \hat{\boldsymbol{\vartheta}})$.

$$\hat{\boldsymbol{\Omega}} = \hat{\boldsymbol{\Omega}}^{1/2}(\boldsymbol{I} - \hat{\boldsymbol{H}})\hat{\boldsymbol{\Omega}}^{1/2\prime} + \hat{\boldsymbol{\Omega}}^{1/2}\hat{\boldsymbol{H}}\hat{\boldsymbol{\Omega}}^{1/2\prime} \tag{3.288}$$

In the univariate case $\hat{H}_{ii}$ may be interpreted as the proportion of the estimated variance of $\mu(\boldsymbol{x}_i, \hat{\boldsymbol{\vartheta}})$ of the estimated error variance $\hat{\Omega}_i$. Since the tr$\boldsymbol{H} = q$, the average proportion of $\hat{V}(\hat{\mu}_i)$ of $\hat{\Omega}_i$ is given by q/n. The upper bound of the proportion is 1. Hence, the proportion $\hat{H}_{ii}$ can be interpreted as a measure of leverage of the ith observation on the fit of the regression function to the observed value y_i. Therefore, H_{ii} can be used in the same way to detect leverage points as the diagonal elements of the hat matrix in linear regression.

To detect influential points, the modified Cook statistic of Section 5.6 is adapted to QGPML estimation by replacing $\hat{\boldsymbol{\Sigma}}_i$ in (3.253) by $\hat{\boldsymbol{\Omega}}_i$ such that $\boldsymbol{Z}_i = \hat{\boldsymbol{\Omega}}_i^{-1/2}\hat{\boldsymbol{M}}_i$. $\boldsymbol{K}_i$ is defined in (3.255). With this notation, (3.252) of Section 5.6 yields the modified Cook statistic for mean structures with a specified covariance matrix:

$$\tilde{C}_i = \hat{\mathbf{e}}_i'\hat{\boldsymbol{\Omega}}_i^{-1/2\prime}\boldsymbol{K}_i\boldsymbol{Z}_i\left(\sum_{i=1}^{n}\boldsymbol{Z}_i'\boldsymbol{Z}_i\right)^{-1}\boldsymbol{Z}_i'\boldsymbol{K}_i\hat{\boldsymbol{\Omega}}_i^{-1/2}\hat{\mathbf{e}}_i \tag{3.289}$$

With $v_i = \boldsymbol{Z}_i(\sum_{i=1}^{n}\boldsymbol{Z}_i'\boldsymbol{Z}_i)^{-1}\boldsymbol{Z}_i'$, the Cook statistic for a univariate dependent variable may be written as

$$\tilde{C}_i = \frac{\hat{e}_i^2 v_i}{\hat{\Omega}_i^{-1}(1 - v_i)^2} \quad . \tag{3.290}$$

Here, v_i is equal to the diagonal element $\hat{H}_{ii}$ of the hat matrix $\hat{\boldsymbol{H}}$. Therefore, the expression $\hat{e}_i^2/[\hat{\Omega}_i^{-1}(1 - v_i)^2]$ is equal to the squared of the studentized residual and the modified Cook statistic can be computed directly from the studentized residuals:

$$\tilde{C}_i = \frac{s_i^2 H_{ii}}{(1 - H_{ii})} \tag{3.291}$$

$\tilde{C}_i$ is the modified Cook statistic that has been derived by Cook and Weisberg (1982, chap. 5, p. 187) for the case of ML estimation.

7 Univariate Nonlinear Regression Models

In nonlinear regression models, the mean structure $\mu_i = E(y_i|x_i)$ of a random variable y_i may depend not only on the parameters but also on the variables in a nonlinear way. A general model is therefore written in the form

$$y_i = \mu_i + \epsilon_i, \quad E(\epsilon_i|\boldsymbol{x}_i) = 0, \quad i = 1, \ldots, n \quad \text{and} \quad \mu_i = \mu(\boldsymbol{x}_i, \boldsymbol{\vartheta}). \tag{3.292}$$

$\boldsymbol{\vartheta} \sim q \times 1$ is a vector of parameters that defines the mean structure μ_i, which is a function $\mu(\cdot, \cdot)$ of the explanatory variables $\boldsymbol{x}_i$ and of $\boldsymbol{\vartheta}$. The true value of $\boldsymbol{\vartheta}$ is denoted by $\boldsymbol{\vartheta}_0$. Depending on the model, regularity assumptions such as differentiability of order k are imposed on $\mu(\cdot, \cdot)$ as a function of $\boldsymbol{\vartheta}$.

Most of the nonlinear models in the behaviorial sciences typically arise when nonmetric data have to be analyzed where linear models violate range restrictions for the predictor of the dependent variable. These models are often extensions of linear models devised to deal with nonmetric variables. However, they still have an important linear component. Examples are threshold models such as the logit and probit models for dichotomous outcomes and the class of generalized linear models. Genuinely nonlinear models are mainly found in the natural sciences. Interesting models and applications are discussed by Bates and Watts (1988).

The models below may have different specifications with regard to the density $f(y|\boldsymbol{x})$ given $\boldsymbol{x}$. If only the mean structure μ_i is specified, then the variance $V(\epsilon_i|\boldsymbol{x}_i) = \Omega_i$ is allowed to vary freely over $i = 1, \ldots, n$ and the density is not specified except for the first moment. If there is additional knowledge about the second moment, the variance $V(\epsilon_i|\boldsymbol{x})$ may be specified as a function of μ_i and/or of $\boldsymbol{x}_i$ and additional parameters collected in a vector $\boldsymbol{\alpha}$ of parameters. Finally, the density itself may be specified. For each model, these choices will be briefly discussed.

The estimation procedures described in Sections 3–6 are only asymptotically valid and therefore yield valid results only for large data sets. For the correct construction of confidence intervals, confidence regions for small samples for $\boldsymbol{\vartheta}$ are estimated by ML methods based on conditional normality of y given x [see Bates and Watts (1988)]. Otherwise resampling methods such as bootstrapping (Efron 1979) must be used. Regression diagnostics such as coefficients of determination, residuals, leverage points, and Cook statistics are given in the last subsection.

7.1 Models for Count Data

An important class of dependent variables are count data $y_i = 0, 1, 2, \ldots$ taking on any natural number between 0 and ∞. Examples are number of members of households, number of patents held by a firm, the number of traffic accidents within a three-hour period in a given area, and the frequency of a cell in a contingency table. If $y_i \in 0, 1, 2, \ldots$, then the expected value is restricted to the positive real numbers, that is $E(y_i|\boldsymbol{x}_i) = \mu_i \in (0, \infty)$. Therefore, a linear model of the form $\mu_i = \boldsymbol{x}_i'\boldsymbol{\beta}$ may not be a sensible model for the mean structure because variables in $\boldsymbol{x}$ that are unrestricted may yield predictions $\hat{\mu}_i$ that are below

0 and therefore outside the admissible range of values for $\hat{\mu}_i$. This mistake may be avoided with the choice of a nonlinear regression function. Two typical choices are the loglinear model with

$$\eta_i = \boldsymbol{x}_i'\boldsymbol{\beta} \quad \text{and} \quad \mu_i = \exp(\eta_i) = \exp(\boldsymbol{x}_i'\boldsymbol{\beta}) \tag{3.293}$$

and the square root linear model with

$$\eta_i = \boldsymbol{x}_i'\boldsymbol{\beta} \quad \text{and} \quad \mu_i = \left(\frac{\eta_i}{2}\right)^2 = \left(\frac{\boldsymbol{x}_i'\boldsymbol{\beta}}{2}\right)^2 \quad . \tag{3.294}$$

The linear component of the model $\eta_i = \boldsymbol{x}_i'\boldsymbol{\beta}$ is called a linear predictor in the GLM language. The function μ_i as a function of η_i is called an inverse link function, which is a one-to-one function of η_i. On the other hand, η_i as a function of μ_i is called the link function in GLM language. In the loglinear model, $\eta_i = \ln \mu_i$, and in the square root linear model, $\eta_i = 2\sqrt{\mu_i}$.

The interpretation of the parameter β_j is found by looking at the first derivative of μ_i with respect to x_{ij} if μ_i is differentiable with respect to x_{ij}. The first derivative in the loglinear model is

$$\frac{\partial \mu_i}{\partial x_{ij}} = \frac{\partial \mu_i}{\partial \eta_i}\frac{\partial \eta_i}{\partial x_{ij}} = \mu_i \beta_j \Rightarrow \beta_j = \frac{1}{\mu_i}\frac{\partial \mu_i}{\partial x_{ij}} \quad . \tag{3.295}$$

The parameter β_j in the loglinear model is therefore the relative change of μ_i that corresponds to a unit change in x_{ij}.

The interpretation of β_j in the square-root linear model is less natural since the first derivative is given by

$$\frac{\partial \mu_i}{\partial x_{ij}} = \frac{\partial \mu_i}{\partial \eta_i}\frac{\partial \eta_i}{\partial x_{ij}} = \sqrt{\mu_i}\beta_j \Rightarrow \beta_j = \frac{1}{\sqrt{\mu_i}}\frac{\partial \mu_i}{\partial x_{ij}} \quad . \tag{3.296}$$

The first justification for the choice of these two nonlinear models is obviously the property that any predicted value $\hat{\mu}_i$ will be positive. A second justification may be found from the assumption that y_i given $\boldsymbol{x}_i$ follows a Poisson distribution with expected value μ_i.

The Poisson distribution is a member of the linear exponential family treated in section 5.2. The link function $\eta_i = \ln \mu_i$ corresponds therefore to the canonical parameter $\kappa_i = \ln \mu_i$ of the Poisson distribution.

The square root transformation may be justified by noting that the expected value and the variance of the square root z_i of a Poisson variable y_i may be approximated by a first-order Taylor expansion about μ_i:

$$z_i = 2\sqrt{y_i} \approx 2\sqrt{\mu_i} + \frac{1}{\sqrt{\mu_i}}(y_i - \mu_i) \Rightarrow E(z_i) \approx 2\sqrt{\mu_i} \text{ and } V(z_i) \approx 1 \tag{3.297}$$

Hence, $z_i = 2\sqrt{y_i}$ is a transformation that stabilizes the variance.

The parameter vector $\boldsymbol{\vartheta}$ is estimated with the ML method based on the Poisson density using the method of iterated weighted least squares as outlined in Section 5.4. The log-linear model specification for the analysis of count data in contingency tables and the ML estimation of $\boldsymbol{\vartheta}$ are considered in detail by Sobel in Chapter 5.

Neither the log transformation nor the square root transformation depend upon or imply that the data y_i given $\boldsymbol{x}_i$ actually follow a Poisson distribution. If this assumption is dubious, one may consider only partial specification of the density y_i given $\boldsymbol{x}_i$. For example, suppose the mean structure is given as $\mu_i = \exp(\boldsymbol{x}_i'\boldsymbol{\beta})$, but that $V(y_i|\boldsymbol{x}_i) = \Omega_i$ is unspecified. Then PML estimation based either on the Poisson distribution or on the normal distribution with $y_i \sim \mathcal{N}(\mu_i, 1)$ or any other member of the exponential family may be used (except for the binomial, where the range of μ is bounded from above by the sample size). The nuisance parameter Φ is set to 1.

A more restrictive assumption is that—although the density $f(y_i|\boldsymbol{x}_i)$ is not known—at least $V(y_i|\boldsymbol{x}_i)$ may be specified. A first example is a model with constant variance $V(y_i|\boldsymbol{x}_i) = \Phi_0$. A second example is a model with over- or underdispersion where $V(y_i|\boldsymbol{x}_i) = \Phi_0\mu_i$. If $\Phi_0 = 1$ then the variance is equal to the variance of a Poisson distributed variable. If $\Phi_0 > 1$ then the variable is overdispersed by comparison with the Poisson variance. If $\Phi_0 < 1$ then the variable is underdispersed. The parameter vector $\boldsymbol{\vartheta}_0$ is estimated with PML methods. The nuisance parameter Φ_0 is estimated consistently by

$$\hat{\Phi} = \frac{1}{n}\sum_{i=1}^{n}(e_i - \mu_i(\boldsymbol{x}_i, \hat{\boldsymbol{\vartheta}}))^2, \quad \tilde{\Phi} = \frac{1}{n-q}\sum_{i=1}^{n}(e_i - \mu_i(\boldsymbol{x}_i, \hat{\boldsymbol{\vartheta}}))^2 \quad . \tag{3.298}$$

To compute the ML, PML, or QGPML estimates of $\boldsymbol{\vartheta}_0$, only the first derivatives of μ_i with respect to $\boldsymbol{\vartheta}_0$ are needed. In the loglinear model one has $\partial\mu_i/\partial\boldsymbol{\vartheta} = \mu_i\boldsymbol{x}_i$, and in the square root linear model, $\partial\mu_i/\partial\boldsymbol{\vartheta} = \sqrt{\mu_i}\boldsymbol{x}_i$.

Example: Frequency of Traffic Accidents

As an example of the PML estimation based on the Poisson density, the number of traffic accidents in three-hour periods, in the city of Hamburg, Germany, is considered. The dependent variable is the number of accidents observed during the first six months of 1985 ($n = 1365$ observations) ranging from 0 to 45 in a three-hour period. The explanatory variables of interest are dryness of the road (0 if wet, 1 if dry) and severity of weather condition coded from 1 ("normal" weather without fog and/or precipitation) to 9 (thunderstorms with heavy precipitation). For simplicity's sake this variable is considered as metric with equal distances, although the correct measurement level is ordered categorical. Of course, many other variables (such as month, day of the week, time of the day,) that proxy the amount of traffic are important predictors of the number of traffic accidents. However, the weather's effect remains stable whether the proxies are included in the model or not. Two simple loglinear models are considered.

As Table 2 presents, the number of traffic accidents increases with deteriorating weather conditions and decreases when the road is dry. Although the Pseudo R^2 (cf. sec. 7.5) is low for both models, one should note that the parameter estimates for weather and dryness

Table 2. PML Estimation Under the Poisson Assumption

Explanatory variables	Model 1	Model 2
Regression constant	2.112	2.0833
	(0.0369)	(0.0285) [0.0709]
Tuesday	0.0593	-
	(0.0343)	
Wednesday	0.0057	-
	0.0348	
Thursday	0.0652	-
	(0.0344)	-
Friday	0.0171	-
	(0.0335)	
Saturday	0.0051	-
	0.0349)	
Sunday	-0.3878	-
	(0.0394)	
Weather	0.0526	0.0634
	(0.0054)	(0.0053) [0.0135]
Dryness	-0.1425	-0.1389
	(0.0238)	(0.0236) [0.0563]
Pseudo R^2	0.0854	0.0490
df	1356	1362

NOTE: The variables Tuesday–Sunday and Dryness are dummy variables. The values in parentheses are the ML standard errors, the values in brackets for the second model are the PML standard errors. The coefficients for weather and dryness are significant on an $\alpha = 0.05$ test level using $z_j^\star$ statistics. The Pseudo R^2 is computed from the logarithm of the likelihood ratio statistic of model 1 and 2 in comparision with a null model that only includes the regression constant.

change very little if day of the week is excluded from the model, indicating that day of the week and weather conditions are uncorrelated. The PML estimates of the standard errors are somewhat larger than the ML estimates, indicating that the assumption of a Poisson distribution with $V(y_i|\boldsymbol{x}_i) = \mu_i$ may be incorrect.

7.2 Standard Nonlinear Regression Models

The standard nonlinear regression model is the nonlinear analogue of the linear regression model with the classical assumptions that the error term in the model (3.292) is homoscedastic and follows a normal distribution with $\epsilon_i \sim \mathcal{N}(0, \sigma^2)$. A typical model may again be generated as a combination of a linear model $\eta_i = \boldsymbol{x}_i'\boldsymbol{\beta}$ for a linear predictor and an inverse link function which is parameterized in a flexible way. Such parametric functions are the simple Box-Cox transformation (Box and Cox 1964)

$$\eta = \frac{\mu^{\lambda_1} - 1}{\lambda_1} \tag{3.299}$$

and the Box-Cox transformation with a shift parameter λ_2 [proposed by Pregibon (1980)]:

$$\eta = \frac{(\mu + \lambda_2)^{\lambda_1} - 1}{\lambda_1} \tag{3.300}$$

In the first case, $\lambda_1 = 0$ yields the log-linear link function $\eta = \ln \mu$. In the second case, $\lambda_1 = 1, \lambda_2 = 1$ yields the identity link function $\eta = \mu$.

The parameters of the inverse link function are collected in the vector $\boldsymbol{\lambda}$. $\boldsymbol{\lambda}$ and $\boldsymbol{\beta}$ together form the parameter vector $\boldsymbol{\vartheta}$ such that $\mu_i = \mu(\boldsymbol{x}_i, \boldsymbol{\vartheta}) = \mu(\boldsymbol{x}_i, (\boldsymbol{\lambda}', \boldsymbol{\beta}')')$. Since $\boldsymbol{\lambda}$ consists of parameters that are estimated jointly with the parameters for the linear model, the fit to the data will be improved in comparison to a fixed link function such as the logarithmic or square root link. In fact, the estimates of λ_1 and λ_2 can be used to specify the link function. For example, if the data analyzed are count data, an estimate $\hat{\lambda} \approx 0.5$ in (3.299) may suggest a square root transformation of μ_i rather than a log transformation.

If the error terms in (3.292) follow a normal distribution with $\epsilon_i \sim \mathcal{N}(0, \sigma^2)$, one can use the ML method based on the normal density of y given $\boldsymbol{x}$.

$$f(y_i|\boldsymbol{x}_i, \boldsymbol{\vartheta}, \sigma^2) = \frac{1}{\sqrt{2\pi}\sigma} \exp\{-\frac{1}{2\sigma^2}(y_i - \mu(\boldsymbol{x}_i, \boldsymbol{\vartheta}))^2\} \tag{3.301}$$

Hence, the normed log-likelihood function of the sample is, except for a constant,

$$l(\boldsymbol{\vartheta}, \sigma^2) = -\frac{1}{2}\ln\sigma^2 - \frac{1}{2\sigma^2}\sum_{i=1}^{n}(y_i - \mu(\boldsymbol{x}_i, \boldsymbol{\vartheta}))^2 \quad . \tag{3.302}$$

The corresponding iterated least squares method (cf. Section 5.4) yields the ML estimator $\hat{\boldsymbol{\vartheta}}$:

$$\hat{\boldsymbol{\vartheta}} = \left(\hat{\boldsymbol{M}}'\hat{\boldsymbol{M}}\right)^{-1}\hat{\boldsymbol{M}}'\hat{\boldsymbol{z}} \quad . \tag{3.303}$$

Here, $\hat{M}$ is the $n \times q$ matrix of first derivatives $\partial\mu(\boldsymbol{x}_i, \boldsymbol{\vartheta})/\partial\boldsymbol{\vartheta}'$ evaluated at $\hat{\boldsymbol{\vartheta}}$ and $\boldsymbol{z} = \hat{M}\hat{\boldsymbol{\vartheta}} + \boldsymbol{e}$ where $\boldsymbol{e}$ is the vector of residuals $e_i = y_i - \mu(\boldsymbol{x}_i, \hat{\boldsymbol{\vartheta}})$. In the nonlinear model, the model matrix $\boldsymbol{X}$ of the linear model has beeen replaced by the matrix $\hat{M}$ of first derivatives at $\hat{\boldsymbol{\vartheta}}$. Consistent estimators of σ^2 are given by

$$\hat{\sigma}^2 = \frac{1}{n}\sum_{i=1}^{n}(y_i - \mu(\boldsymbol{x}_i, \hat{\boldsymbol{\vartheta}}))^2 \text{ and} \tag{3.304}$$

$$\tilde{\sigma}^2 = \frac{1}{n-q}\sum_{i=1}^{n}(y_i - \mu(\boldsymbol{x}_i, \hat{\boldsymbol{\vartheta}}))^2 \quad . \tag{3.305}$$

The estimator $\hat{\boldsymbol{\vartheta}}$ is consistent for $\boldsymbol{\vartheta}$ and best asymptotically normal with asymptotic covariance matrix $\boldsymbol{V}(\sqrt{n}\hat{\boldsymbol{\vartheta}})$, which is estimated consistently by

$$\hat{\boldsymbol{V}}(\sqrt{n}\hat{\boldsymbol{\vartheta}}) = \hat{\sigma}^2(\frac{1}{n}\hat{M}'\hat{M})^{-1} \quad . \tag{3.306}$$

The confidence intervals and elliptic confidence regions based on the asymptotic normality of $\hat{\boldsymbol{\vartheta}}$ may—unlike the linear case—be quite misleading for nonlinear models, especially if the sample size is small. This is shown in detail by Bates and Watts (1988, chaps. 6 and 7), who recommend the use of profile t plots and profile traces for computing marginal likelihood intervals for parameters in nonlinear models estimated by the ML method.

If the assumption that ϵ_i is normal is dropped, the parameter vector $\boldsymbol{\vartheta}$ may still be estimated using nonlinear least-squares methods. However, the nonlinear least squares method minimizes the negative of the log-likelihood function of the sample based on the possibly incorrect normality assumption. As in the linear case, the least-squares method can be interpreted as a special case of the QGPML method under the (possibly incorrect) assumption of normality. A consistent estimator of the asymptotic covariance matrix of $\sqrt{n}\hat{\boldsymbol{\vartheta}}$ is again given by $\hat{\boldsymbol{V}}(\sqrt{n}\hat{\boldsymbol{\vartheta}})$ of equation (3.306).

If the assumption of homoscedasticity is also dropped, the ML estimator $\hat{\boldsymbol{\vartheta}}$ is still a consistent estimator of the mean structure $\boldsymbol{\vartheta}_0$. However, the estimate of the asymptotic covariance matrix of $\hat{\boldsymbol{\vartheta}}$ must be adjusted to take the possible heteroscedasticity of ϵ_i into account. Using the PML theory of Section 5.3, a consistent estimator of the asymptotic covariance matrix of $\sqrt{n}(\hat{\boldsymbol{\vartheta}} - \boldsymbol{\vartheta})$ is given by

$$\hat{\boldsymbol{V}}(\sqrt{n}\hat{\boldsymbol{\vartheta}}) = (\frac{1}{n}\hat{M}'\hat{M})^{-1}(\frac{1}{n}\hat{M}'\hat{\Omega}\hat{M})(\frac{1}{n}\hat{M}'\hat{M})^{-1} \tag{3.307}$$

where $\boldsymbol{M}$ is again the $n \times q$ matrix of first derivatives of μ_i with respect to $\boldsymbol{\vartheta}$ at $\hat{\boldsymbol{\vartheta}}$, and $\hat{\Omega}$ is the diagonal matrix of squared residuals $e_i^2 = (y_i - \mu(\boldsymbol{x}_i, \hat{\boldsymbol{\vartheta}}))^2$.

Table 3. PML Estimation Under the Normality Assumption

Explanatory variables	Model 1	Model 2
Regression constant	2.109	2.065
	(0.0781)	(0.0599) [0.0701]
Tuesday	0.0560	-
	(0.0733)	
Wednesday	- 0.0128	-
	0.0756	
Thursday	0.0378	-
	(0.0742)	-
Friday	0.1502	-
	(0.0706)	
Saturday	0.0108	-
	(0.0754)	
Sunday	- 0.3911	-
	(0.0994)	
Weather	0.0575	0.0697
	(0.0105)	(0.0105)[0.0143]
Dryness	- 0.1436	- 0.1406
	(0.0495)	(0.0501) [0.0544]
Pseudo R^2	0.0842	0.0551
df	1356	1362

NOTE: The values in parentheses are ML standard errors, the values in brackets for the second model are the PML standard errors.

Example: Frequency of Traffic Accidents

As an example of PML estimation based on the normal density with constant variance, the dependence of traffic accidents on weather and road conditions is reconsidered. The results are given in Table 3.

The results for the regression coefficients shown in Table 2 and Table 3 are identical, illustrating that both estimation methods yield the same estimates for the parameters in the mean structure of both models. However, the ML estimates for the standard deviations of the parameter estimates in model 2 are quite different under the Poisson and the normality assumption. On the other hand, the PML estimates of the standard errors are identical, illustrating the robustness of PML estimation.

7.3 Models For Dichotomous Outcomes

Specification of Threshold Models

Many outcome variables in the behavioral sciences are dichotomous, taking on one of two categories. Such a variable is denoted by z, and the categories are coded as 1 and 2. Typical examples are individual test items which an individual may or may not solve, college education (yes or no), employment (yes or no), and the wish to have children (yes or no). The random variable that corresponds to the variable z_i is a vector $\boldsymbol{y}_i$ with two components $y_{i1} \in \{0,1\}$ and $y_{i2} \in \{0,1\}$ which can be interpreted as frequencies that z_i falls into class 1 [implying $(y_{i1} = 1, y_{i2} = 0)$] or 2 [implying $(y_{i1} = 0, y_{i2} = 1)$]. Since these frequencies sum to 1, it is sufficient to consider only the random variable y_{i2}, which is now denoted y_i. The probability that z_i falls into category 2 is then given by

$$P(z_i = 2|\boldsymbol{x}_i) = P(y_i = 1|\boldsymbol{x}_i). \tag{3.308}$$

Since the random variable y_i can only take on the values 0 and 1, the expected value is

$$E(y_i|\boldsymbol{x}_i) = \mu_i = 0 \cdot P(y_i = 0|\boldsymbol{x}_i) + 1 \cdot P(y_i = 1|\boldsymbol{x}_i) = P(y_i = 1|\boldsymbol{x}_i). \tag{3.309}$$

y_i can therefore be written as

$$y_i = \mu_i + \varepsilon_i = P(y_i = 1|\boldsymbol{x}_i) + \varepsilon_i \quad . \tag{3.310}$$

Therefore, the probability $P(z_i = 2|\boldsymbol{x}_i)$ may be interpreted as the expected value of y_i given $\boldsymbol{x}_i$.

A simple model for $E(y_i|\boldsymbol{x}_i)$ is the linear probability model $E(y_i|\boldsymbol{x}_i) = \boldsymbol{x}_i'\boldsymbol{\beta}$. However, the range of μ_i is restricted to (0, 1) since μ_i is a probability. If there are continuous components in $\boldsymbol{x}_i$, then the predicted values may lie outside the admissible interval (0, 1) . Therefore, the linear probability model should not be used.

To derive a better model for the mean structure $E(y_i|\boldsymbol{x}_i)$, one may start with the assumption that there is a continous dependent univariate unobserved (latent) variable $y_i^\star$ for which a linear model holds:

$$y_i^\star = \boldsymbol{x}_i'\boldsymbol{\beta} + \delta_i, \quad i = 1, \ldots, n \tag{3.311}$$

The $\boldsymbol{x}_i$'s are again the regressors with $x_{i1} = 1$, $\boldsymbol{\beta}$ is the parameter vector of the mean structure, and δ_i is the error term with $E(\delta_i|\boldsymbol{x}_i) = 0, V(\delta_i|\boldsymbol{x}_i) = \sigma_\delta^2$ and distribution function $F(\delta_i)$, which, for the moment, is assumed to be unknown. The error term δ_i is not connected with the error term ϵ_i for the discrete frequency variable y_i. There exists an observation rule which links $y_i^\star$ to the observed outcome z_i and therefore also to y_i:

$$z_i = \begin{cases} 1 & \text{if} \quad y_i^\star \leq \tau \\ 2 & \text{if} \quad y_i^\star > \tau \end{cases} \quad . \tag{3.312}$$

The value τ is an additional unknown parameter which may be interpreted as a threshold yielding a very rough measurement model with $z_i = 1$ if $y_i^\star$ stays below the threshold and $z_i = 2$ if $y_i^\star$ crosses it. Since $P(z_i = 1|\boldsymbol{x}_i) = 1 - P(z_i = 2|\boldsymbol{x}_i)$, only $P(z_i = 2|\boldsymbol{x}_i)$ need be computed.

$$\begin{aligned} P(z_i = 2|\boldsymbol{x}_i) &= P(y_i^\star > \tau|\boldsymbol{x}_i) = P(\boldsymbol{x}_i'\boldsymbol{\beta} + \delta_i > \tau) = 1 - P(\boldsymbol{x}_i'\boldsymbol{\beta} + \delta_i \leq \tau) \\ &= 1 - P(\delta_i \leq \tau - \boldsymbol{x}_i'\boldsymbol{\beta}) = 1 - F(\tau - \boldsymbol{x}_i'\boldsymbol{\beta}) \end{aligned} \tag{3.313}$$

The probability that $z_i = 2$ given $\boldsymbol{x}_i$ depends on the distribution function of the error term δ at the value $\tau - \boldsymbol{x}_i'\boldsymbol{\beta}$. The same value of F is obtained by $F(\tau + \alpha - (\alpha + \boldsymbol{x}_i'\boldsymbol{\beta}))$ and by $F((\tau - \boldsymbol{x}_i'\boldsymbol{\beta})\gamma/\gamma)$; that is, only the difference between the threshold τ and the regression constant is identified, and the regression coefficients are identified only up to a scale (cf. Nelson 1975). This is a consequence of the fact that the arbitrary coding of z has neither an absolute point nor a scale. Therefore, the mean and the standard deviation of $y_i^\star$ are not defined via the observed variable z_i. Hence, the threshold value and the scale are arbitrarily restricted by setting τ and σ_δ^2 to fixed values such as $\tau = 0$ and $\sigma_\delta^2 = 1$. However, the problem with non-identified scales becomes much more complicated if one looks at a simultaneous equation system for dichotomous dependent variables (cf. Sobel and Arminger 1992).

In addition to fixing τ and σ_δ^2 one may choose the distribution function for δ_i. If $\tau = 0$ and $\delta_i \sim \mathcal{N}(0, 1)$, then one obtains the probit model

$$P(z_i = 2|\boldsymbol{x}_i) = 1 - F(\tau - \boldsymbol{x}_i'\boldsymbol{\beta}) = 1 - \Phi(0 - \boldsymbol{x}_i'\boldsymbol{\beta}) = \Phi(\boldsymbol{x}_i'\boldsymbol{\beta}). \tag{3.314}$$

Here, $\Phi(.)$ denotes the distribution function of a univariate normal random variable. Equivalently, the mean structure of the variable y_i given $\boldsymbol{x}_i$ is given by the nonlinear model

$$E(y_i|\boldsymbol{x}_i) = \Phi(\boldsymbol{x}_i'\boldsymbol{\beta}). \tag{3.315}$$

The probit model can be written as a combination of a linear predictor $\eta_i = \boldsymbol{x}_i'\boldsymbol{\beta}$ with inverse link function $\mu_i = \Phi(\eta_i)$ and link function $\eta_i = \Phi^{-1}(\mu_i)$.

Alternatively, one can assume that δ_i follows a standard logistic distribution with $E(\delta_i|\boldsymbol{x}_i) = 0$ and $\sigma^2 = \pi^2/3$. This specification with $\tau = 0$ yields the logit model with

$$P(z_i = 2|\boldsymbol{x}_i) = 1 - F(\tau - \boldsymbol{x}_i'\boldsymbol{\beta}) = \frac{\exp(\boldsymbol{x}_i'\boldsymbol{\beta})}{1 + \exp(\boldsymbol{x}_i'\boldsymbol{\beta})} \quad . \tag{3.316}$$

As in the probit model, the mean structure may be written as a function of a linear predictor $\eta_i = \boldsymbol{x}_i'\boldsymbol{\beta}$, an inverse link function $\mu_i = \exp(\boldsymbol{x}_i'\boldsymbol{\beta})/(1 + \exp(\boldsymbol{x}_i'\boldsymbol{\beta}))$, and a link function

$\eta_i = \ln[\mu_i/(1-\mu_i)]$. Since the standard deviation of δ_i in the probit model is $\sigma_{\delta P} = 1$ and in the logit model is $\sigma_{\delta L} = \pi/\sqrt{3}$, the absolute values of parameters in the logit model will always be greater than in the probit model by a factor of about 1.6. However, since the distribution functions of δ_i for the probit and logit models are similar, the ratios for each pair of regression coefficients should be the same in both models.

As an alternative to probit and logit models, one can use distributions functions $F(\delta_i)$ that are parameterized in a flexible way so that the parameters of $F(\delta_i)$ are estimated simultaneously with the parameters of the linear predictor. These models are called quantit models (Copenhaver and Mielke 1977). Semiparametric approaches are found in Gallant and Nychka (1987) and Gabler, Laisney, and Lechner (1993)

The interpretation of the coefficients β_j of a general quantit model is quite simple if one looks at (3.311). In this equation, β_j is, as in the linear model the amount of change in the expected value $E(y_i^\star|\boldsymbol{x}_i) = \mu_i^\star$ corresponding to a unit change in x_j.

However, the rate of change in the probability $P(z_i = 2|\boldsymbol{x}_i)$ with respect to x_{ij} will be quite different. In the probit model, $P(z_i = 2|\boldsymbol{x}_i) = \Phi(\boldsymbol{x}_i'\boldsymbol{\beta})$. Hence, one finds a nonconstant first derivative of $P(z_i = 2|\boldsymbol{x}_i)$ with respect to x_{ij},

$$\frac{\partial P(z_i = 2|\boldsymbol{x}_i)}{\partial x_{ij}} = \frac{\partial \Phi(\boldsymbol{x}_i'\boldsymbol{\beta})}{\partial x_{ij}} = \frac{\partial \Phi(\boldsymbol{x}_i\boldsymbol{\beta})}{\partial \boldsymbol{x}_i'\boldsymbol{\beta}}\beta_j = \varphi(\boldsymbol{x}_i\boldsymbol{\beta})\beta_j \quad , \tag{3.317}$$

where $\varphi(\boldsymbol{x}_i'\boldsymbol{\beta})$ is the standard normal density at point $\boldsymbol{x}_i'\boldsymbol{\beta}$. This implies that the instantaneous rate of change of $\mu_i^\star$ with respect to x_{ij}, given by β_j, is dampened by the value of the density function at the point $\boldsymbol{x}_i'\boldsymbol{\beta}$. If $\boldsymbol{x}_i'\boldsymbol{\beta}$ is 0 then $\varphi(\boldsymbol{x}_i'\boldsymbol{\beta})$ takes on its maximum, $P(z = 2|\boldsymbol{x}_i) = 0.5$, and the instantaneous rate of change is very high. However, if $|\boldsymbol{x}_i'\boldsymbol{\beta}| > 3$ then $\varphi(\boldsymbol{x}_i'\boldsymbol{\beta})$ is near 0, and the instantaneous rate of change will be rather small.

The scale of the variable $y^\star$ depends on the value chosen for the variance σ_δ^2 of δ_i. The unconditional variance of $y^\star$ is given by

$$V(y^\star) = \boldsymbol{\beta}'\boldsymbol{V}(\boldsymbol{x})\boldsymbol{\beta} + V(\delta) = \boldsymbol{\beta}'\boldsymbol{V}(\boldsymbol{x})\boldsymbol{\beta} + \sigma_\delta^2 \quad . \tag{3.318}$$

If σ_δ^2 is set to 1 then $V(y^\star) = \boldsymbol{\beta}'\boldsymbol{V}(\boldsymbol{x})\boldsymbol{\beta} + 1$.

Quantit models are not restricted to models with a linear structure for $E(y_i^\star|\boldsymbol{x}_i) = h(\boldsymbol{x}_i, \boldsymbol{\vartheta})$. In general, one considers nonlinear quantit models where $h(\boldsymbol{x}_i, \boldsymbol{\vartheta})$ is the mean structure for $y_i^\star$ given $\boldsymbol{x}_i$ and the error distribution function $F(\delta_i)$ is known.

To estimate the parameter vector $\boldsymbol{\vartheta}_0$ in the mean structure $E(y_i|\boldsymbol{x}_i)$, one may specify the conditional distribution of y_i given $\boldsymbol{x}_i$ either completely or partially. The usual complete specification is that $y_i \in \{0, 1\}$ follows a Bernoulli distribution with expected value $\mu_i = P(y_i = 1|\boldsymbol{x}_i)$. From the fact that the Bernoulli distribution is a member of the linear exponential family, it follows immediately that the logit transformation $\eta_i = \ln[\mu_i/(1-\mu_i)]$ is the canonical link function for the Bernoulli distribution. If the same Bernoulli trial is repeated m_i times for each $i = 1, \ldots, n$, then the number y_i of successes in m_i trials follows the binomial distribution with expected value $E(y_i|\boldsymbol{x}_i) = \mu_i$, $V(y_i|\boldsymbol{x}_i) = \mu_i(m_i - \mu_i)/m_i$ and natural link function $\eta_i = \ln[\mu_i/(m_i - \mu_i)]$. The log-likelihood function for ML estimation of the parameters is based on the Bernoulli distribution.

The assumption of a Bernoulli distribution for y_i given $\boldsymbol{x}_i$ is not necessary to obtain consistent estimates of the parameter vector $\boldsymbol{\vartheta}$ for the mean structure. If it is only assumed that $V(\epsilon_i|\boldsymbol{x}_i) = \Omega_i$ where Ω_i is unspecified, a consistent estimate of $\boldsymbol{\vartheta}$ may be found by using PML estimation based either on the Bernoulli distribution, or the normal distribution with the assumption $\epsilon_i \sim \mathcal{N}(0, 1)$, or the normal distribution with the assumption $\epsilon_i \sim \mathcal{N}(0, \sigma_i^2 = \mu_i(1 - \mu_i))$.

If assumptions are made about the conditional variance $V(\epsilon_i|\boldsymbol{x}_i)$ as a function of some parameters collected in $\boldsymbol{\alpha}$ and/or the mean structure, then QGPML estimation may be employed. A typical example is again overdispersion (or underdispersion), where $V(\epsilon|\boldsymbol{x}_i) = \Phi_0\mu_i(1 - \mu_i)$; here Φ_0 is an additional dispersion parameter.

To compute the ML, PML, or QGPML estimates of $\boldsymbol{\vartheta}_0$, only the first derivatives of μ_i with respect to $\boldsymbol{\vartheta}_0$ are needed. In the probit model, one has $\partial\mu_i/\partial\boldsymbol{\vartheta} = \varphi(\boldsymbol{x}'\boldsymbol{\vartheta})\boldsymbol{x}_i$, and in the logit model, $\partial\mu_i/\partial\boldsymbol{\vartheta} = \mu_i(1 - \mu_i)\boldsymbol{x}_i$.

Coefficients of Determination

Because of the special importance of dichotomous dependent variables, the more common coefficients of determination proposed in the literature (cf. Amemiya 1981) are given here. The arithmetic mean of $y_i \in \{0, 1\}$ is denoted by $\bar{y}$, $\hat{\mu}_i$ denotes the estimated expected value of y_i under the chosen model, and $\bar{\mu}$ the arithmetic mean of the $\hat{\mu}_i$'s.

Efron's R^2_{EF} considers the dependent variable y_i just as a metric variable. Therefore R^2_{EF} is equivalent to the coefficient of determination in the linear regression model.

$$R^2_{EF} = \frac{\sum_{i=1}^{n}(y_i - \bar{y})^2 - \sum_{i=1}^{n}(y_i - \hat{\mu}_i)^2}{\sum_{i=1}^{n}(y_i - \bar{y})^2} \qquad \text{Efron's } R^2 \tag{3.319}$$

Efron's R^2_{EF} may severely underestimate the strength of the relationship between y and $\boldsymbol{x}$ as discussed by Cox and Wermuth (1992).

Goldberger's R^2_{GO} is the squared correlation coefficient between $\boldsymbol{y}$ and $\hat{\boldsymbol{\mu}}$. It is equal to R^2_{EF} for continuous variables that are normally distributed.

$$R^2_{GO} = \frac{\left(\sum_{i=1}^{n}(y_i - \bar{y})\hat{\mu}_i\right)^2}{\sum_{i=1}^{n}(y_i - \bar{y})^2\sum_{i=1}^{n}(\hat{\mu}_i - \bar{\mu})^2} \qquad \text{Goldberger's } R^2 \tag{3.320}$$

Instead of considering the correlation between observed values of y_i and expected values μ_i, one may also consider the correlation between the values of $y_i^\star$ of the threshold model and its expected values $\mu_i^\star$ using (3.318) (McKelvey and Zavoina 1975). For the probit and logit models, R^2_{MZ} is defined by

$$R^2_{MZ} = \frac{\boldsymbol{\beta}'\hat{\boldsymbol{V}}(\boldsymbol{x})\boldsymbol{\beta}}{\boldsymbol{\beta}'\hat{\boldsymbol{V}}(\boldsymbol{x})\boldsymbol{\beta} + \sigma_\delta^2} \quad . \tag{3.321}$$

$\hat{V}(\boldsymbol{x})$ is the empirical covariance matrix of the explanatory variables. $\sigma_\delta^2 = 1$ in the probit model and is equal to $\pi^2/3$ in the logit model. Since the variance σ_δ^2 is constant for $y_i^\star$, the statistic R^2_{MZ} may be considered as the proportion of explained variance under the model.

If the observed variables y_i given $\boldsymbol{x}_i$ follow a Bernoulli distribution, a Pseudo R^2 defined by McFadden (1974) may be used as a coefficient of determination.

$$R^2_{MF} = \frac{l(\hat{\boldsymbol{\mu}}|M_0) - l(\hat{\boldsymbol{\mu}}|M_1)}{l(\hat{\boldsymbol{\mu}}|M_0)} \tag{3.322}$$

where $l(\hat{\boldsymbol{\mu}}|M_1)$ denotes the log-likelihood function under the current model and $l(\hat{\boldsymbol{\mu}}|M_0)$ under the model that includes only the regression constant.

Example: Employment Status in the German Socioeconomic Panel

The employment status of a sample of 1246 men of the German Socio-Economic Panel (hereafter GSOEP) from 1985 is considered as the dependent variable where 0 denotes employed and 1 denotes unemployed. A simple model for the probability to be unemployed is given here. Flaig, Licht, and Steiner (1993) discuss the relevant variables in the GSOEP data set in detail and specify more complicated models.

Parameter estimates for probit and logit models based on ML estimation using the Bernoulli distribution, the ML standard errors, and coefficients of determination are given in Table 4.

The probit and logit parameter estimates appear quite different at first glance. However, recall this is only a result of the different definition of σ_δ^2 for probit and logit models. Division by the standard errors shows that the $z^\star$ values are about the same yielding equivalent substantive conclusions. The variables ALH, BB1, BB2, and FST are not significantly different from 0 at an $\alpha = 0.05$ level test in this analysis. The number of months of unemployment from 1974 to 1984 has a positive linear and a negative quadratic coefficient. The regression coefficients for ALT have opposite signs. Being severely handicapped (GZ) increases the probability of being unemployed greatly, while being a white collar worker (BST2) decreases it greatly. The coefficients of determination for the probit and the logit models are about the same. The first three coefficients take on equivalent values which indicate that about 30 percent of the variation in $y_i^\star$ is explained by the explanatory variables. R^2_{MF} is much higher; it cannot be interpreted as a measure of the proportion of explained variance.

7.4 Quantit Models for Censored Outcomes

Threshold models are useful not only for analyzing dichotomous variables but also for analyzing a dependent metric variable that has been observed only partially. A typical example is the Tobit model (Tobin 1958), where $\boldsymbol{x}_i$ consists of variables like income upon which the amount of money $z_i^\star$ spent on durable goods may depend. However, since durable goods are not bought by all individuals in a cross-sectional sample in a given observation

Table 4. Probit and Logit Models for Unemployment Status

Explanatory variables	Probit-parameters	Probit-s.e.	Logit-parameters	Logit-s.e.
CONST	3.085	(1.746)	6.481	(3.539)
ALD	0.108	(0.017)	0.206	(0.032)
ALDSQ	-0.086	(0.040)	-0.166	(0.074)
ALH	-0.045	(0.055)	-0.110	(0.097)
ALT	-0.246	(0.097)	-0.492	(0.198)
ALTSQ	0.306	(0.128)	0.612	(0.262)
GZ	0.713	(0.209)	1.345	(0.404)
BB1	-0.208	(0.157)	-0.387	(0.324)
BB2	-0.010	(0.269)	0.042	(0.562)
BST2	-0.396	(0.195)	-0.894	(0.410)
FST	-0.230	(0.174)	-0.431	(0.347)
R^2_{EF}	0.278		0.280	
R^2_{GO}	0.278		0.280	
R^2_{MZ}	0.295		0.317	
R^2_{MF}	0.462		0.458	

NOTE: ALD = duration of unemployment between 1974 and 1984 in months, ALDSQ = $\text{ALD}^2/100$, ALH = number of unemployment spells between 1974 and 1984, ALT = age in years, ALTSQ = $\text{ALT}^2/100$, GZ = 1 if a person is severely handicapped and 0 otherwise, BB1 = 1 if a person has finished some professional education and 0 otherwise, BB2 = 1 if a person has a university degree and 0 otherwise, BST2 = 1 if a person is a white collar employee and 0 otherwise, FST = 1 if a person is married and 0 otherwise.

period, observations $(z_i^\star, \boldsymbol{x}_i)$ are only available for a subset of the sample. Let the model for $z_i^\star$ be a linear model of the form

$$z_i^\star = \boldsymbol{x}_i'\boldsymbol{\beta} + \epsilon_i^\star \quad \text{with} \quad \epsilon_i^\star \sim \mathcal{N}(0, \sigma^2). \tag{3.323}$$

The following observation rule holds where τ is a known threshold,

$$z_i = \begin{cases} z_i^\star & \text{if } z_i^\star > \tau \\ \tau & \text{if } z_i^\star \leq \tau \end{cases}, \tag{3.324}$$

for instance 0 as in the example by Tobin. If z_i instead of $z_i^\star$ is regressed on $\boldsymbol{x}_i$, the ML estimates $\hat{\beta}$ may be biased. The amount of bias depends on the proportion of elements in the sample where $z_i = \tau$. If the estimation is based only on the observations where $z_i = z_i^\star$, the estimate is unbiased, but it is inefficient because it is only based on a part of the sample.

This model is characterized by two random variables that can be observed. The first random variable, y_i , is—as in the quantit models for dichotomous outcomes—an indicator that takes on the value 0 if the dependent variable is observed and 1 if it is censored. The variable $z_i^\star$ can only be observed if $y_i = 0$. Otherwise, it is only known that $z_i^\star$ is less than or equal to τ. The expected value of interest is $E(z_i^\star|\boldsymbol{x}_i)$. However, $E(z_i^\star|\boldsymbol{x}_i)$ cannot be written as a simple parametric function of $\boldsymbol{x}_i$ and $\boldsymbol{\beta}$.

Instead of first defining a mean structure and then finding an estimation method corresponding to different assumptions about the error term, a direct ML approach is taken to construct a consistent and efficient estimator for $\boldsymbol{\beta}$ under the strong assumption that $f(\epsilon_i^\star)$ is known. To define the log-likelihood function, one has to distinguish between the contribution $z_i = z_i^\star$ to the likelihood function and of $z_i = \tau$. Let $F(\epsilon_i^\star)$ be the distribution of $\epsilon_i^\star$ and $f(\epsilon_i^\star)$ be the corresponding density function parameterized in a vector $\boldsymbol{\alpha}$. The individual likelihood functions are:

$$L_i(\boldsymbol{\beta}, \boldsymbol{\alpha}) = f(z_i - \boldsymbol{x}_i'\boldsymbol{\beta}) \quad \text{if} \quad z_i = z_i^\star \quad , \tag{3.325}$$

$$L_i(\boldsymbol{\beta}, \boldsymbol{\alpha}) = F(\tau - \boldsymbol{x}_i'\boldsymbol{\beta}) \quad \text{if} \quad z_i = \tau \quad . \tag{3.326}$$

The log-likelihood function of the sample divided by n is therefore given by

$$l(\boldsymbol{\beta}, \boldsymbol{\alpha}) = \frac{1}{n}\left[\sum_{i=1}^{n} y_i \ln F(\tau - \boldsymbol{x}_i'\boldsymbol{\beta}) + \sum_{i=1}^{n}(1 - y_i) \ln f(z_i - \boldsymbol{x}_i'\boldsymbol{\beta})\right] . \tag{3.327}$$

Substituting the density and the distribution function of the normal distribution with $\boldsymbol{\alpha} = \sigma^2$ yields the log-likelihood function

$$l(\boldsymbol{\beta}, \sigma^2) = \frac{1}{n}\left[\sum_{i=1}^{n} y_i \ln \Phi\left(\frac{\tau - \boldsymbol{x}_i'\boldsymbol{\beta}}{\sigma}\right) + \sum_{i=1}^{n}(1 - y_i) \ln \varphi\left(\frac{z_i - \boldsymbol{x}_i'\boldsymbol{\beta}}{\sigma}\right)\right] \tag{3.328}$$

where φ is the standard normal density. This is the specification chosen by Tobin (1958). Since τ is known and σ^2 is identified in the model that generates the observations with $z_i = z_i^\star$, there is no identification problem with σ^2 as in the dichotomous case. The first sum contains the information from the censored observations and the second sum the information from the uncensored observations. The components $\Phi([\tau - \boldsymbol{x}_i]/\sigma)$ are the expected values of y_i that is the probability of censoring under the probit model. Hence, the Tobit model may be considered as a hybrid between a probit model for dichotomous outcomes and a regression model with normality assumptions for a metric outcome. The linear models for $E(y_i^\star|\boldsymbol{x}_i)$ may be replaced by nonlinear models of the form $E(z_i^\star|\boldsymbol{x}_i) = \mu(\boldsymbol{x}_i, \boldsymbol{\vartheta})$.

The above model has been extended in many ways. A first variant is censoring from above with the following observation rule:

$$z_i = \begin{cases} \tau & \text{if } z_i^\star \geq \tau \\ z_i^\star & \text{if } z_i^\star < \tau \end{cases} \tag{3.329}$$

The log-likelihood function divided by n is then given by

$$l(\boldsymbol{\beta}, \sigma^2) = \frac{1}{n}\sum_{i=1}^{n} y_i \ln(1 - F(\tau - \boldsymbol{x}_i\boldsymbol{\beta})) + \frac{1}{n}\sum_{i=1}^{n}(1 - y_i) \ln f(y_i - \boldsymbol{x}_i\boldsymbol{\beta}) \tag{3.330}$$

Another variant is censoring from below and above; this will be treated in the context of multivariate nonlinear regression models.

7.5 Generalized Linear Models

Model Specification

The univariate linear model with the assumption of normal errors, the loglinear model with the assumption of a dependent variable that follows a Poisson distribution, and the logit and probit model for a Bernoulli distributed dependent variable are special instances of the class of generalized linear models introduced by Nelder and Wedderburn (1972). A thorough and extensive treatment of this topic is found in McCullagh and Nelder (1989). In this subsection, these models are placed in the context of nonlinear models estimated with PML and QGPML methods.

The classic GLM is defined by three components: the linear predictor η, the link function $g(.)$, and the density of the univariate dependent variable y_i given explanatory variables. Let $y_i = \mu_i + \varepsilon_i$ with $E(\varepsilon_i|\boldsymbol{x}_i) = 0$. The components are given by

$$\eta_i = \boldsymbol{x}_i'\boldsymbol{\beta}\,, \quad \text{(linear predictor)} \tag{3.331}$$

$$\mu_i = g(\eta_i)\,, \quad \text{(link funktion)} \tag{3.332}$$

$$\eta_i = g^{-1}(\mu_i)\,, \quad \text{(inverse link function)} \tag{3.333}$$

$$f_G(y_i|\bar{\kappa}_i, \Phi) = \exp\left\{\frac{y_i\bar{\kappa}_i - b^\star(\bar{\kappa}_i)}{a_i^\star(\Phi)} - c^\star(y_i, \Phi)\right\}\,. \quad \text{(linear exponential family)} \tag{3.334}$$

$\bar{\kappa}_i$ is the canonical parameter of the linear exponential family in the parameterization of generalized linear models of Section 5.2. $a_i^\star(\Phi)$ is a known function which is usually set to $1, \Phi$ or Φ/w_i where w_i is a fixed weight that may vary for each element of the sample. The relationship between the parameter $\boldsymbol{\beta}$ in the linear predictor $\eta_i = \boldsymbol{x}_i'\boldsymbol{\beta}$, $\bar{\kappa}_i$ and Φ is given through the expected value and the variance as in Section 5.2.

$$E(y_i|\boldsymbol{x}_i) = \mu_i = \frac{\partial b^\star(\bar{\kappa}_i)}{\partial \bar{\kappa}_i} \tag{3.335}$$

$$V(y_i|\boldsymbol{x}_i) = \Sigma_i = a_i^\star(\Phi)\frac{\partial^2 b^\star(\bar{\kappa}_i)}{\partial \bar{\kappa}_i^2} = a_i^\star(\Phi)\frac{\partial \mu_i}{\partial \bar{\kappa}_i} = a_i^\star(\Phi)\tau(\mu_i) \tag{3.336}$$

Table 5. Canonical Parameters, Expected Values, and Variances in GLMs

Density	$\bar{\kappa}$	μ	$V(y)$
Poisson	$\ln \mu$	$\exp(\bar{\kappa})$	μ
Bernoulli	$\ln [\mu/(1-\mu)]$	$\exp(\bar{\kappa})/(1+\exp(\bar{\kappa}))$	$\mu(1-\mu)$
Binomial with sample size m	$\ln [\mu/(n-\mu)]$	$m\,[\exp(\bar{\kappa})/(1+\exp(\bar{\kappa})]$	$\mu(m-\mu)/m$
Gaussian	μ	$\bar{\kappa}$	Φ
Inverse Gaussian	$1/\mu^2$	$1/\sqrt{\bar{\kappa}}$	μ^3/Φ
Gamma	μ^{-1}	$\bar{\kappa}^{-1}$	μ^2/Φ

where $\tau(\mu_i)$ is again the variance function.

The canonical parameters $\bar{\kappa}$ are given as a function of the expected values μ and vice versa in Table 5 (cf. McCullagh and Nelder 1989, chap. 2, sec. 1). Additionally, the variances are given as a function of μ and Φ.

The natural choice for a link function is therefore $\eta_i = \bar{\kappa}_i = \boldsymbol{x}_i'\boldsymbol{\beta}$, yielding the loglinear model for the Poisson, the logit model for the Bernoulli and the binomial, the linear model for the Gaussian and the reciprocal linear model for the gamma density. However, the choice of a canonical link function $\eta_i = \bar{\kappa}_i$ is taken here purely on statistical grounds. From a substantive viewpoint, a different choice for the link, such as the square root for count data or the probit for dichotomous outcomes, may be preferred.

ML and QGPML Estimation

ML estimation involves joint estimation of the true values $\boldsymbol{\beta}_0$ and Φ_0. Alternatively, one can use QGPML estimation of $\boldsymbol{\beta}_0$. The variance Σ_i of the assumed density and the variance Ω_i of the true density are identical since correct specification of the density is assumed. If

$a_i^\star(\Phi) = \Phi$, then the components of the IWLS algorithm in Section 5.4 are given by

$$M_i^{(q)} = \frac{\partial \mu_i^{(q)}}{\partial \boldsymbol{\beta}'} = \frac{\partial \mu_i^{(q)}}{\partial \eta_i}\frac{\partial \eta_i}{\partial \boldsymbol{\beta}'} = \frac{\partial \mu_i^{(q)}}{\partial \eta_i}\boldsymbol{x}_i' \tag{3.337}$$

$$\Sigma_i^{(q)} = \Phi\frac{\partial \mu_i^{(q)}}{\partial \bar{\kappa}_i} = \Phi\tau\left(\mu_i^{(q)}\right) \Rightarrow \Sigma_i^{(q)^{-1}} = \Phi^{-1}\left(\frac{\partial \bar{\kappa}_i^{(q)}}{\partial \mu_i}\right) \tag{3.338}$$

$$z_i^{(q)} = \frac{\partial \mu_i^{(q)}}{\partial \eta_i}\eta_i^{(q)} + \left(y_i - \mu_i^{(q)}\right). \tag{3.339}$$

Iteration step $q + 1$ of Fisher scoring is therefore

$$\boldsymbol{\beta}^{(q+1)} = \left(\sum_{i=1}^{n} M_i^{(q)\prime}\Sigma_i^{-1(q)}M_i^{(q)}\right)^{-1}\sum_{i=1}^{n} M_i^{(q)\prime}\Sigma_i^{-1(q)}z_i^{(q)} \quad . \tag{3.340}$$

Since $\Sigma_i = \Phi\tau\left(\mu_i^{(q)}\right)$, the dispersion parameter Φ cancels out so that the QGPML estimation can be performed in one step of the QGPML procedure in Section 6.3.

$$\boldsymbol{\beta}^{(q+1)} = \left(\sum_{i=1}^{n} M_i^{(q)\prime}\tau\left(\mu_i^{(q)}\right)^{-1)}M_i^{(q)}\right)^{-1}\sum_{i=1}^{n} M_i^{(q)\prime}\left(\tau\left(\mu_i^{(q)}\right)\right)^{(-1)}z_i^{(q)} \tag{3.341}$$

The QGPML estimator is denoted by $\hat{\boldsymbol{\beta}}$. The corresponding linear predictor and the expected values are denoted by $\hat{\eta}_i$ and $\hat{\mu}_i$. Consistent estimators of the dispersion parameter Φ_0 are given by

$$\hat{\Phi} = \frac{1}{n}\sum_{i=1}^{n}\frac{(y_i - \hat{\mu}_i)^2}{\tau(\hat{\mu}_i)} \qquad \text{and} \qquad \tilde{\Phi} = \frac{1}{n-q}\sum_{i=1}^{n}\frac{(y_i - \hat{\mu}_i)^2}{\tau(\hat{\mu}_i)} \quad . \tag{3.342}$$

The asymptotic covariance matrix of $\hat{\boldsymbol{\beta}}$ is given by

$$\hat{V}(\boldsymbol{\beta}) = \hat{\Phi}\left(\sum_{i=1}^{n}\hat{M}_i'\tau(\hat{\mu}_i)^{-1}\hat{M}_i\right)^{-1} \quad . \tag{3.343}$$

The IWLS algorithm and the estimator of the asymptotic covariance matrix are simplified if the canonical link is used. Then, the variance function $\tau(\mu_i) = \partial\mu_i/\partial\bar{\kappa}_i = \partial\mu_i/\partial\eta_i$ yields

$$\boldsymbol{\beta}^{(q+1)} = \left(\sum_{i=1}^{n}\boldsymbol{x}_i\frac{\partial \eta_i^{(q)}}{\partial \mu_i^{(q)}}\boldsymbol{x}_i'\right)^{-1}\sum_{i=1}^{n}\boldsymbol{x}_i\left(\eta_i^{(q)} + \frac{\partial \eta_i^{(q)}}{\partial \mu_i^{(q)}}(y_i - \mu_i^{(q)})\right), \quad \text{and} \tag{3.344}$$

$$\hat{V}(\hat{\boldsymbol{\beta}}) = \hat{\Phi}\left(\sum_{i=1}^{n}\boldsymbol{x}_i\frac{\partial \eta_i^{(q)}}{\partial \mu_i^{(q)}}\boldsymbol{x}_i'\right)^{-1} \quad . \tag{3.345}$$

Quasi-Likelihood and QGPML Estimation

Generalized linear models may be extended to models where the variance $V(y_i|\boldsymbol{x}_i)$ is written as a function of the dispersion parameter Φ and an arbitrary variance function $\tau(\mu_i)$; $\tau(\mu_i)$ may be specified as any continuous function of μ_i while $\tau(\mu_i)$ is restricted by the density of the linear exponential family in classic generalized linear models. The density is unspecified except for the mean and the variance. The variance function may in principle differ for each member of the sample but is usually taken to be the same for all individuals. Models with this specification have been denoted as quasi likelihood (QL) models by Wedderburn (1974). They may be considered as special cases of models where, in addition to the expected value, the variance is specified as a function of $\boldsymbol{x}_i$, the parameter $\boldsymbol{\vartheta}$ of the mean structure and a vector $\boldsymbol{\alpha}$ of nuisance parameters, that is $V(y_i|\boldsymbol{x}_i) = \Omega(\boldsymbol{x}_i, \boldsymbol{\vartheta}, \boldsymbol{\alpha})$. The estimation of such models has already been treated in the context of QGPML estimation, therefore, the subject of QL models is not pursued further.

Residuals and Influential Points

Ordinary, standardized, and studentized residuals as well as leverage and influential points may be derived immediately from 6.5. The QGPML estimates of the expected values μ_i and the variances Ω_i are estimated by $\hat{\mu}_i$ and $\hat{\Omega}_i$ under the model. If one uses the notation of Section 5.4 with $\hat{\boldsymbol{\Omega}}$ as the $n \times n$ matrix of variances $\hat{\Omega}_i$, $\hat{\boldsymbol{M}}$ as the $n \times q$ matrix of row vectors $\hat{\boldsymbol{M}}_i = (\partial\hat{\mu}_i/\partial\eta_i)\boldsymbol{x}_i'$, and the $n \times n$ hat matrix $\hat{\boldsymbol{H}} = \hat{\boldsymbol{\Omega}}^{-1/2}\hat{\boldsymbol{M}}(\hat{\boldsymbol{M}}'\hat{\boldsymbol{\Omega}}\hat{\boldsymbol{M}})^{-1}\hat{\boldsymbol{M}}'\hat{\boldsymbol{\Omega}}^{-1/2}$, one finds:

$$e_i = y_i - \hat{\mu}_i\,, \quad \text{(ordinary residual)} \tag{3.346}$$

$$r_i = \frac{y_i - \hat{\mu}_i}{\sqrt{\hat{\Omega}_i}}\,, \quad \text{(standardized residual)} \tag{3.347}$$

$$\hat{H}_{ii} = i\text{th diagonal element of } \hat{\boldsymbol{H}}\,, \quad \text{(measure of leverage)} \tag{3.348}$$

$$s_i = \frac{y_i - \hat{\mu}_i}{\sqrt{\hat{\Omega}_i(1 - \hat{H}_{ii})}}\,, \quad \text{(studentized residual) and} \tag{3.349}$$

$$\tilde{C}_i = \frac{s_i^2\hat{H}_{ii}}{(1 - \hat{H}_{ii})}\,. \quad \text{(Cook statistic as measure of influence)} \tag{3.350}$$

If the density of a GLM is specified fully through the exponential family, an additional type of residuals, deviance residuals, may be defined (as discussed subsequently).

Discrepancy Measures and Coefficients of Determination

To judge the fit not only of generalized linear models but of nonlinear regression models in general, one can use (as in the linear model) the sum of squared errors of the model that includes only a regression constant

$$SST = \sum_{i=1}^{n}(y_i - \bar{y})^2 \quad \text{with} \quad \bar{y} = \frac{1}{n}\sum_{i=1}^{n} y_i \tag{3.351}$$

and of the fitted model

$$SSE = \sum_{i=1}^{n}(y_i - \hat{\mu}_i)^2 \quad \text{with} \quad \hat{\mu}_i = \mu(\boldsymbol{x}_i, \hat{\boldsymbol{\vartheta}}) \tag{3.352}$$

to form a coefficient of determination

$$R^2 = \frac{SST - SSE}{SST} \quad . \tag{3.353}$$

As in a linear model, this coefficient may be considered as the proportion of explained variance. The assumption that one implicitly makes by using this coefficient is that the variation of the errors ϵ_i in a nonlinear regression model is constant and may be estimated consistently by SSE/n. If the variance is not constant, SSE/n may be considered as an estimator of the average variation of ϵ_i and R^2 may be considered as a descriptive measure of the discrepancy between the fitted and observed values of the sample. As in Section 2.3, R^2 may be adjusted to take into account the degrees of freedom.

A different approach the derivation of discrepancy measures hinges on the likelihood ratio statistic:

$$LR = -2\ln\frac{L(\hat{\boldsymbol{\vartheta}})}{L_s} \tag{3.354}$$

where $L(\hat{\boldsymbol{\vartheta}})$ is the likelihood function of the data under the model parameterized in $\boldsymbol{\vartheta}$ and L_s is the likelihood function of the data when the data y_i are fitted perfectly, that is $y_i = \hat{\mu}_i$. The use of the likelihood ratio statistic requires that the distribution of y_i given $\boldsymbol{x}_i$ has been specified completely. The assumption of GLMs is that y_i given $\boldsymbol{x}_i$ is generated by a member of the linear exponential family with nuisance parameter Φ. The terms $\ln L(\hat{\boldsymbol{\vartheta}})$ and $\ln L_s$ may be written as the sums of individual elements $l_i(\hat{\boldsymbol{\vartheta}})$ and $l_i(s)$, which are the contributions of the ith sample element to the log-likelihood functions $\ln L(\hat{\boldsymbol{\vartheta}})$ and $\ln L_s$.

$$LR = -2\left[\ln L(\hat{\boldsymbol{\vartheta}}) - \ln L_s\right] = 2\sum_{i=1}^{n}\left[l_i(s) - l_i(\hat{\boldsymbol{\vartheta}})\right] \tag{3.355}$$

The likelihood contribution $l_i(\hat{\boldsymbol{\vartheta}})$ is given by setting $\hat{\mu} = \mu(\boldsymbol{x}_i, \hat{\boldsymbol{\vartheta}})$:

$$l_i(\hat{\boldsymbol{\vartheta}}) = \left[y_i\bar{\kappa}(\hat{\mu}_i) - b^\star(\bar{\kappa}(\hat{\mu}_i))\right]/a_i^\star(\Phi) - c^\star(y_i, \Phi) \tag{3.356}$$

In the saturated model, $\hat{\mu}_i$ is set to y_i such that

$$l_i(s) = \left[y_i\bar{\kappa}(y_i) - b^\star(\bar{\kappa}(y_i))\right]/a_i^\star(\Phi) - c^\star(y_i, \Phi). \tag{3.357}$$

Therefore, the LR statistic for the linear exponential family with nuisance parameter Φ is given by

$$LR = 2\sum_{i}^{n} \frac{1}{a_i^\star(\Phi)} \left[(y_i \bar{\kappa}(y_i) - b^\star(\bar{\kappa}(y_i))) - (y_i \bar{\kappa}(\hat{\mu}_i) - b^\star(\bar{\kappa}(\hat{\mu}_i)))\right] . \quad (3.358)$$

In most applications, $a_i^\star(\Phi) = \Phi / w_i$ where w_i is a fixed weight. If the weight $w_i = 1$, then the LR statistic may be written as

$$LR = 2\sum_{i}^{n} \frac{1}{\Phi} \left[((y_i \bar{\kappa}(y_i) - b^\star(\bar{\kappa}(y_i))) - ((y_i \bar{\kappa}(\hat{\mu}_i) - b^\star(\bar{\kappa}(\hat{\mu}_i)))\right] = D^\star(\boldsymbol{y}; \boldsymbol{\mu}), \quad (3.359)$$

which is called the scaled deviance in McCullagh and Nelder (1989, chap. 2, sec. 3). The term $D(\boldsymbol{y}; \boldsymbol{\mu}) = \Phi D^\star(\boldsymbol{y}; \boldsymbol{\mu})$ is called the unscaled deviance.

From the definition of $b^\star(\bar{\kappa})$ in Section 5.2 and the relationship between the canonical parameter $\bar{\kappa}$ and the mean μ_i, the scaled deviances in Table 6 may be derived (McCullagh and Nelder 1989, chap. 2, sec. 3).

Table 6. Scaled Deviances in the Linear Exponential Family

Distribution	Scaled deviance
Normal	$\frac{1}{\Phi} \sum_{i=1}^{n} (y_i - \hat{\mu}_i)^2$
Poisson	$2 \sum_{i=1}^{n} y_i \ln \frac{y_i}{\hat{\mu}_i} - (\hat{y}_i - \hat{\mu}_i)$
Binomial	$2 \sum_{i=1}^{n} \left[y_i \ln \frac{y_i}{\hat{\mu}_i} + (m_i - y_i) \ln \left(\frac{m_i - y_i}{m_i - \hat{\mu}_i} \right) \right]$
Gamma	$\frac{2}{\Phi} \sum_{i=1}^{n} \left(- \ln \frac{y_i}{\hat{\mu}_i} + \frac{y_i - \hat{\mu}_i}{\hat{\mu}_i} \right)$
Inverse Gaussian	$\frac{1}{\Phi} \sum_{i=1}^{n} \frac{(y_i - \hat{\mu}_i)^2}{\hat{\mu}_i^2 y_i}$

NOTE: In the binomial distribution, m_i denotes the number of Bernoulli trials and y_i the number of successes. If $m_i = 1$ then $y_i \in \{0, 1\}$. If $y_i = 0$ then $y_i \ln[y_i/\hat{\mu}_i]$ is set to 0. If $y_i = 1$ then $(m_i - y_i) \ln[(m_i - y_i)/(m_i - \hat{\mu}_i)]$ is set to 0.

The unscaled deviance $D(\boldsymbol{y};\hat{\boldsymbol{\mu}}_i)$ for the normal distribution is equivalent to the sum of squared errors $SSE = \sum_{i=1}^{n}(y_i - \hat{\mu}_i)^2$ in the linear model. In the Poisson and in the binomial distribution, the nuisance parameter $\Phi = 1$ and $D^{\star}(\boldsymbol{y};\boldsymbol{\mu}) = D(\boldsymbol{y};\boldsymbol{\mu})$. The unscaled deviance $D(\boldsymbol{y};\boldsymbol{\mu})$ is equal to the G^2 statistic of loglinear models (cf. Sobel, Chapter 5).

Since the likelihood ratio statistic LR is equal to the scaled deviance rather than the unscaled deviance, the unscaled deviance can only be used for testing hypotheses about $\boldsymbol{\vartheta}_0$, if $\Phi = 1$. In this case, $D(\boldsymbol{y};\hat{\boldsymbol{\mu}}_i)$ follows asymptotically a central χ^2 distribution with $n - q$ degrees of freedom if the model which has been specified is correct and the number of parameters in $\boldsymbol{\vartheta}$ is q. If the dispersion parameter must be estimated from the data, the saturated model where $\hat{\mu}_i = y_i$ cannot be used by definition since Φ is always estimated as 0. Hence, only the differences between deviances can be used for model comparisons. Let M_0 denote a model with parameter vector $\boldsymbol{\vartheta}^{(M_0)} \sim p \times 1$ which is nested within a model M_1 with parameter vector $\boldsymbol{\vartheta}^{(M_1)} \sim q \times 1$, that is $\boldsymbol{\vartheta}^{(M_0)} \subset \boldsymbol{\vartheta}^{(M_1)}$ and $p < q$. The corresponding estimates of the expected values are denoted by $\hat{\boldsymbol{\mu}}_0$ and $\hat{\boldsymbol{\mu}}_1$. The difference in the scaled deviances

$$D^{\star}(\boldsymbol{y};\hat{\boldsymbol{\mu}}_0) - D^{\star}(\boldsymbol{y};\hat{\boldsymbol{\mu}}_1) \tag{3.360}$$

is asymptotically χ^2 distributed with $q - p$ of freedom if M_0 holds true. To test hypotheses about $\boldsymbol{\vartheta}$, a model M_1 must be chosen from which Φ is estimated using (3.342).

The difference between deviances is easily calculated by using the general formula for the deviance of (3.359) and replacing $\bar{\kappa}(y_i)$ by $\bar{\kappa}(\hat{\mu}_{i1})$ and $\bar{\kappa}(\hat{\mu}_i)$ by $\bar{\kappa}(\hat{\mu}_{i0})$. This additive property of the deviance allows us to define a coefficient of determination as the proportion of reduced deviance (PED) which – for the normal distribution – is equal to the proportion of reduced error (PRE) coefficient of Section 2.3. Let M_0 be the model that contains only a regression constant and M_1 the model under consideration.

$$\text{PED}(M_1|M_0) = \frac{D(\boldsymbol{y};\hat{\boldsymbol{\mu}}_0) - D(\boldsymbol{y};\hat{\boldsymbol{\mu}}_1)}{D(\boldsymbol{y};\hat{\boldsymbol{\mu}}_0)} \tag{3.361}$$

Since Φ cancels out, PED$(M_1|M_0)$ may be computed with the unscaled deviances.

A word of caution is in order if one uses the deviances and PED coefficients for grouped data as it is often done in the analysis of contingency tables using logit models for dichotomous or polytomous dependent variables. The number y_i is then the number of successful trials in m_i trials within a given constellation $\boldsymbol{x}_i$ of explanatory variables. Consequently, the model with a perfect fit only gives a perfect fit for the grouped and not for the individual data. Hence, the fit usually deteriorates if more explanatory variables are used, which leads to a larger number of groups and smaller values of m_i. For a baseline, the models M_0 and M_1 and their deviances should be calculated for the individual data with $m_i = 1$ and $y_i \in \{0, 1\}$.

Another measure of discrepancy is the sum of standardized residuals denoted as Pearson's X^2 statistic by McCullagh and Nelder (1989, chap. 2, sec. 3).

$$X^2 = \sum_{i=1}^{n} \frac{(y_i - \hat{\mu}_i)^2}{V(y_i|\boldsymbol{x}_i)} \quad , \tag{3.362}$$

which in the Poisson case is the familiar χ^2 statistic for testing hypotheses about models for contingency tables. This statistic may be used as a discrepancy measure for any GLM or more generally for any mean structure model with a specified variance. While it is easily computed, it does not have the additive property of the deviance. Therefore, differences of X^2 statistics for nested models $M_0 \subset M_1$ do not follow a χ^2 distribution asymptotically if M_0 holds. Because of lack of additivity, a coefficient of determination cannot be defined with this statistic.

Pearson's X^2 statistic consists of the squared standardized errors. $D^\star(\boldsymbol{y}; \hat{\boldsymbol{\mu}})$ consists of the individual contributions to the likelihood ratio statistic. Since $D^\star(\boldsymbol{y}; \hat{\boldsymbol{\mu}})$ is χ^2_{n-q} distributed if the model is correctly specified, it is surmised that the individual contributions

$$d_i = \frac{2}{\Phi}\left[((y_i\bar{\kappa}(y_i) - b^\star(\bar{\kappa}(y_i))) - ((y_i\bar{\kappa}(\hat{\mu}_i) - b^\star(\bar{\kappa}(\hat{\boldsymbol{\mu}}_i)))\right] \tag{3.363}$$

follow approximately a χ^2_1 and $r_i^D = \text{sign}(y_i - \hat{\mu}_i)\sqrt{d_i}$ approximately a $\mathcal{N}(0, 1)$ distribution. $\sqrt{d_i}$ is a measure of discrepancy between y_i and $\hat{\boldsymbol{\mu}}_i$ and the sign function shows the direction of the discrepancy. r_i^D is called the deviance residual in McCullagh and Nelder (1989, chap. 2, sec. 4) and is used alternatively to the Pearson residual r_i, especially in the analysis of contingency tables.

8 Multivariate Nonlinear Regression Models

8.1 Models for Ordered Categorical Variables

In Section 7.3, threshold models for dichotomous outcomes were introduced. These models have been extended to ordered categorical data by McKelvey and Zavoina (1975). Again, an unobserved variable $y_i^\star$ is considered:

$$y_i^\star = \mu^\star(\boldsymbol{x}_i, \boldsymbol{\vartheta}) + \delta_i \tag{3.364}$$

Here, $E(y_i^\star|\boldsymbol{x}_i) = \mu^\star(\boldsymbol{x}_i, \boldsymbol{\vartheta}) = \mu_i^\star$ is the mean structure with true parameter $\boldsymbol{\vartheta}_0$, δ_i is an error term with $E(\delta_i) = 0, V(\delta_i) = \sigma_\delta^2, E(\boldsymbol{x}\delta_i) = \boldsymbol{0}$ and known distribution function $F_\delta(.)$. The most common models are linear models with $\mu_i^\star = \boldsymbol{x}_i'\boldsymbol{\beta}$ where $\boldsymbol{x}_i$ has the first element 1 and δ_i follows the normal or logistic distribution. The dichotomous observation rule is replaced by an observation rule for ordered categorical data:

$$z_i = k \iff y_i^\star \in (\tau_{k-1}, \tau_k], \quad k = 1, \ldots, K \tag{3.365}$$

$\tau_0 = -\infty, \tau_1, \ldots, \tau_K = +\infty$ are $K+1$ threshold values yielding K categories of the observed ordered categorical variable z_i. $\tau_0 = -\infty$ and $\tau_K + \infty$ are fixed. The observed variable z_i takes on the category $k = 1, \ldots, K$ if $y_i^\star$ lies in the interval $(\tau_{k-1}, \tau_k]$ for $k = 1, \ldots, K-1$ and in the interval (τ_{k-1}, τ_k) if $k = K$. If the model for $y_i^\star$ is linear with $y_i^\star = \boldsymbol{x}_i'\boldsymbol{\beta} + \delta_i$, then the probability that $z_i = k$ is given as

$$P(z_i = k|\boldsymbol{x}_i) = P(\tau_{k-1} < y_i^\star \le \tau_k) = P(\tau_{k-1} < \boldsymbol{x}_i'\boldsymbol{\beta} + \delta_i \le \tau_k) =$$

$$P(\delta_i \le \tau_k - \boldsymbol{x}_i'\boldsymbol{\beta}) - P(\delta_i \le \tau_{k-1} - \boldsymbol{x}_i'\boldsymbol{\beta}) = F(\tau_k - \boldsymbol{x}_i'\boldsymbol{\beta}) - F(\tau_{k-1} - \boldsymbol{x}_i'\boldsymbol{\beta}). \quad (3.366)$$

The threshold values τ_k, $k = 1, \ldots, K-1$, the parameter vector $\boldsymbol{\beta}$, and σ_δ^2 are unknown. As in the dichotomous threshold model, only the difference $\tau_k - \beta_1$ is identified and $\boldsymbol{\beta}$ is only identified up to scale. Therefore, τ_1 can be set to 0. If δ_i follows a standard normal distribution with $\sigma_\delta^2 = 1$, then the model for z_i is called an ordinal probit model. If δ_i follows a standard logistic distribution with $\sigma_\delta^2 = \pi^{2/3}$, then the model for z_i is called an ordinal logit model. Similar identification restrictions have to be taken into consideration if the linear model $\mu_i^\star = \boldsymbol{x}_i'\boldsymbol{\beta}$ is replaced by the general nonlinear model $\mu_i^\star = \mu(\boldsymbol{x}_i, \boldsymbol{\vartheta})$ for $E(y_i^\star|\boldsymbol{x}_i)$. The free threshold parameters $\tau_2, \ldots, \tau_{K-1}$ are collected in the vector $\boldsymbol{\tau}$.

Maximum likelihood estimation of $\boldsymbol{\tau}$ and $\boldsymbol{\vartheta}$ is based on the assumption that there exists a vector $y_i \sim K \times 1$ with $y_{ik} \in \{0, 1\}$ and $\sum_{k=1}^K y_{ik} = 1$ that follows a multinomial distribution. The expected values of y_{ik} are given by

$$E(y_{ik}|\boldsymbol{x}_i) = \mu_{ik} = P(z_i = k|\boldsymbol{x}_i) = F(\tau_k - \mu(\boldsymbol{x}_i, \boldsymbol{\vartheta})) - F(\tau_{k-1} - \mu(\boldsymbol{x}_i, \boldsymbol{\vartheta})). \quad (3.367)$$

This corresponds to the specification of a mean structure for the vector $\boldsymbol{y}_i$ of dependent variables with

$$E(\boldsymbol{y}_i|\boldsymbol{x}_i) = \boldsymbol{\mu}_i + \epsilon_i \quad (3.368)$$

where $\boldsymbol{\mu}_i' = (P(z_i = 1|\boldsymbol{x}_i), \ldots, P(z_i = K|\boldsymbol{x}_i))$. Additionally, the covariance matrix $\boldsymbol{V}(\epsilon_i|\boldsymbol{x}_i)$ is specified with elements (cf. Section 5.2)

$$v_{ikl} = \delta_{ikl}\mu_{ik} - \mu_{ik}\mu_{il}, \quad k, l = 1, \ldots, K \quad (3.369)$$

where δ_{ikl} is the Kronecker delta with $\delta_{ikl} = 1$ if $k = l$ and $\delta_{ikl} = 0$ otherwise. The covariance matrix is not positive definite because $y_{iK} = 1 - \sum_{k=1}^K y_{ik}$. Therefore, only the first $K-1$ rows and columns need to be considered.

The individual likelihood function from the multinomial distribution is

$$L_i(\boldsymbol{\tau}, \boldsymbol{\vartheta}) = P(z_i = 1|\boldsymbol{x}_i)^{y_{i1}} P(z_i = 2|\boldsymbol{x}_i)^{y_{i2}} \ldots P(z_i = K|\boldsymbol{x}_i)^{y_{iK}} \quad . \quad (3.370)$$

The log-likelihood function divided by n is then

$$l(\boldsymbol{\tau}, \boldsymbol{\vartheta}) = \frac{1}{n} \sum_{i=1}^n \sum_{k=1}^K y_{ik} \ln P(z_i = k|\boldsymbol{x}_i) \quad . \quad (3.371)$$

The components of the individual log-likelihood function for the ordinal probit model may be written with the restriction $\tau_1 = 0, \sigma_\delta^2 = 1$ as follows:

$$\begin{aligned} P(z_i = 1|\boldsymbol{x}_i) &= \Phi(-\boldsymbol{x}_i'\boldsymbol{\beta}) \\ P(z_i = k|\boldsymbol{x}_i) &= \Phi(\tau_k - \boldsymbol{x}_i'\boldsymbol{\beta}) - \Phi(\tau_{k-1} - \boldsymbol{x}_i'\boldsymbol{\beta}), \quad k = 2, \ldots, K-1 \\ P(z_i = K|\boldsymbol{x}_i) &= 1 - \Phi(\tau_{K-1} - \boldsymbol{x}_i'\boldsymbol{\beta}) \end{aligned} \quad (3.372)$$

where $\Phi(.)$ denotes the standard normal distribution function.

Instead of specifying the distribution function of $\boldsymbol{y}_i$ completely with the multinomial distribution, one might also use PML estimation for the parameters in the first $K-1$ components of $E(\boldsymbol{y}_i|\boldsymbol{x}_i) = \boldsymbol{\mu}(\boldsymbol{x}_i, \boldsymbol{\vartheta})$. PML estimation could be based either on the multinomial or on the multivariate normal distribution with covariance matrix $\boldsymbol{\Sigma}_i = \boldsymbol{I}$ for the assumed density. QGPML estimation based on the multinomial or multivariate normal density would proceed with the covariance matrix defined in (3.369) as assumed covariance matrix.

The adequacy of a threshold model to ordered categorical data can be judged by using either the R^2_{MZ} of McKelvey and Zavoina or the R^2_{MF} of McFadden defined in Section 7.3. The coefficients of determination of Efron and Goldberger cannot be carried over to ordered categorical variables because the dependent variable $\boldsymbol{y}_i$ is now multivariate rather than univariate.

Threshold models are of course not the only models that have been proposed for analyzing ordered categorical data. However, they seem to be closest to the thinking of social and behaviorial scientists and can be extended to construct mean and covariance structure models for nonmetric dependent variables, as is shown by Browne and Arminger in Chapter 4. Cumulative and sequential models are treated in McCullagh and Nelder (1989, chap. 5) and in Tutz (1990).

Example: Attitude Toward Guest Workers

To illustrate the foregoing material, the attitude of Germans toward guest workers is considered. The data come from the German general social survey test-retest study in 1984 that is described in detail in Porst and Zeifang (1987). Here, a subsample of 1320 Germans from the complete sample of 3004 persons is analyzed. The dependent variable *repatriation of guest workers* is the statement *guest workers should be sent back home when there is a shortage of jobs in the Federal Republic*, where the responses are scaled as a seven-point Likert scale with 1 as *strongly disagree* to 7 as *strongly agree*. The explanatory variables are age (between 18 and 95 years), gender (0 is male and 1 is female), and the Treiman occupational prestige scale. The data are analyzed with three different assumptions about the dependent variable. Under the first assumption, the dependent variable is assumed to be metric with equal distance between the seven categories. Under the second assumption, the dependent variable is analyzed as an ordered categorical variable with seven categories using a probit model. Under the third assumption, categories 1, 2 and 3, 4 and 5, and 6 and 7 have been collapsed to form a new trichotomous variable, again analyzed with a probit model. The results are given in Table 7.

The distances between the estimated thresholds in model 2 are about the same, implying that the use of the first model with the assumption of equal distances between the categories may be justified. In the first model, the error variance is estimated, but in the second and third models, the error variances are set to 1. As a consequence, the parameters of Age, Gender, and Prestige are only identified up to scale. Therefore, the parameter estimates in the second and third model are different from the first model, but are about the same as in the first model if they are multiplied by the estimated standard deviation of the errors in the first model. This can be seen immediately from checking the z^* values (which are

Table 7. Three Models for an Ordered Categorical Variable

Variable	Model 1 (Metric)		Model 2 (7 Categories)		Model 3 (3 Categories)	
Constant	4.892		1.390		0.698	
	(0.303)	(16.148)	(0.158)	(8.784)	(0.170)	(4.118)
τ_2			0.379		0.778	
			(0.029)	(13.241)	0.036	(21.846)
τ_3			0.740			
			(0.036)	(20.310)		
τ_4			1.131			
			(0.042)	(27.023)		
τ_5			1.516			
			(0.047)	(32.351)		
τ_6			1.826			
			(0.052)	(35.305)		
Age	0.021		0.011		0.010	
	(0.005)	(4.056)	(0.003)	(4.112)	(0.003)	(3.675)
Gender	-0.030		-0.001		-0.045	
	(0.119)	(-0.252)	(0.061)	(-0.021)	(0.068)	(-0.663)
Prestige	-0.043		-0.022		-0.022	
	(0.005)	(-8.377)	(0.002)	(-8.939)	0.003	(-8.260)
Variance	4.176		1.0		1.0	
R^2	0.073		.082		.081	

NOTE: The first number in parentheses is the standard error, the second number in parentheses is the z^* value. In all three models, the parameter estimates for Age, Gender, and Prestige have the same sign. The corresponding z^* values indicate that the coefficients for Age and Prestige are significantly different from 0 on a test level $\alpha = 0.05$. The coefficient for Gender is not significant. Agreement with the statement that guestworkers hould be repatriated in times of job shortage increases with age and decreases with occupational prestige. The coefficient of determination is the squared multiple correlation coefficient in the first model and the R^2 of McKelvey and Zavoina in the second and third model. The R^2's are quite low indicating that there are additional important explanatory variables and a possible misspecification of the model.

equivalent in all three models). The difference between the second and the third model is negligible, which shows that collapsing the seven categories of the dependent variable into three categories would not have effected the parameter estimates and the substantive results. However, this is not a foregone conclusion. Before doing an ordered probit or logit analysis, the researcher should check whether there are enough observations in each category of the dependent variable (at least 5% of the sample) to avoid numerical instability of the parameter estimates.

8.2 Models for Doubly Censored and Classified Metric Outcomes

Tobin's (1958) model, discussed in Section 7.4, has been extended by Rosett and Nelson (1975) to the case where the dependent variable $y_i^\star = \boldsymbol{x}_i'\boldsymbol{\beta} + \delta_i$ with $\delta_i \sim \mathcal{N}(0, \sigma^2)$ is censored from below and above. One may consider a nonlinear model with $y_i^\star = \mu^\star(\boldsymbol{x}_i, \boldsymbol{\vartheta})$ with true value $\boldsymbol{\vartheta}_0$ and a known distribution function $F(.)$ of δ_i which is parameterized in $\boldsymbol{\vartheta}$ and in a vector $\boldsymbol{\alpha}$ of nuisance parameters. Under the assumption of normality and homoscedasticity of δ_i, only the nuisance parameter $\alpha = \sigma^2$ is used. The observed variable z_i follows a threshold model with observation rule

$$z_i = \begin{cases} \tau_1 & \text{if} \quad y_i^\star \le \tau_1 \\ y_i^\star & \text{if} \quad \tau_1 < y_i^\star < \tau_2 \\ \tau_2 & \text{if} \quad y_i^\star \ge \tau_2 \end{cases} \quad . \tag{3.373}$$

The thresholds τ_1 and τ_2 are known. Corresponding to z_i, a multivariate dependent variable $\boldsymbol{y}_i \sim 3 \times 1$ is constructed where $y_{ik} \in \{0, 1\}$ and $\sum_{k=1}^{3} y_{ik} = 1$. $y_{i1} = 1$ if $z_i = \tau_1$, $y_{i2} = 1$ if $z_i = y_i^\star$ and $y_{i3} = 1$ if $z_i = \tau_2$. The individual log-likelihood function is then given by

$$L_i(\boldsymbol{\vartheta}, \boldsymbol{\alpha}) = P(z_i = \tau_1|\boldsymbol{x}_1)^{y_{i1}} P(z_i = y_i^\star|\boldsymbol{x}_1)^{y_{i2}} P(z_i = \tau_2|\boldsymbol{x}_1)^{y_{i3}} \quad . \tag{3.374}$$

Therefore, the log-likelihood of the sample divided by n may be written as

$$\begin{aligned} l(\boldsymbol{\vartheta}, \boldsymbol{\alpha}) &= \frac{1}{n}\sum_{i=1}^{n} y_{i1} \ln F(\tau_1 - \mu^\star(\boldsymbol{x}_i, \boldsymbol{\vartheta})) + \frac{1}{n}\sum_{i=1}^{n} y_{i3} \ln\left[1 - F(\tau_2 - \mu^\star(\boldsymbol{x}_i, \boldsymbol{\vartheta}))\right] \\ &\quad + \frac{1}{n}\sum_{i=1}^{n} y_{i2} \ln f(y_i^\star - \mu^\star(\boldsymbol{x}_i, \boldsymbol{\vartheta})). \end{aligned} \tag{3.375}$$

As in the simple Tobit model, the log-likelihood consists of a mixture of quantit models for the censored observations and a fully specified model for the density of $y_i^\star$ given $\boldsymbol{x}_i$. As in the Tobit model, the variance σ^2 may be identified from the uncensored observations.

In some applications, a metric variable $y_i^\star$ can be observed only as a grouped or classified metric variable. A typical example is the monthly income of employees, observed in income brackets.

Let $y_i^\star = \mu^\star(\boldsymbol{x}_i, \boldsymbol{\vartheta}) + \delta_i$ with $E(\delta_i) = 0$ where the distribution function $F(.)$ of δ_i is known and parameterized in $\boldsymbol{\vartheta}$ and $\boldsymbol{\alpha}$. The true values are denoted by $\boldsymbol{\vartheta}_0$ and $\boldsymbol{\alpha}_0$. Usually $\mu^\star(\boldsymbol{x}_i, \boldsymbol{\vartheta}) = \boldsymbol{x}_i'\boldsymbol{\beta}$ and $\delta_i \sim \mathcal{N}(0, \sigma^2)$ so that the nuisance parameter $\boldsymbol{\alpha}$ consists only of the variance σ^2. This case has been considered by Stewart (1983). The observation rule is equivalent to the ordered categorical case with

$$z_i = k \Longleftrightarrow y_i^\star \in (\tau_{k-1}, \tau_k] \quad k = 1, \ldots, K \,. \tag{3.376}$$

In contrast to the ordered categorical case previously considered, the thresholds are known and need not be estimated. As in the ordered categorical case, $\tau_0 = -\infty$ and $\tau_K = +\infty$. The parameters are fully identified in a linear model for $E(y_i^\star|\boldsymbol{x}_i)$ and not only identified up to scale. The variance is identified from the known thresholds. The probability that $z_i = k$ is

$$P(z_i = k|\boldsymbol{x}_i) = F(\tau_k - \mu^\star(\boldsymbol{x}_i, \boldsymbol{\vartheta})) - F(\tau_{k-1} - \mu^\star(\boldsymbol{x}_i, \boldsymbol{\vartheta})). \tag{3.377}$$

Again, a vector of dependent variables $y_i \sim K \times 1$ is constructed with $y_{ik} \in \{0, 1\}$ and $\sum_{k=1}^K y_{ik} = 1$, with expected values

$$E(y_{ik}|\boldsymbol{x}_i) = \mu_{ik} = P(z_i = k|\boldsymbol{x}_i). \tag{3.378}$$

Under the assumption of a multinomial distribution for $\boldsymbol{y}_i$ given $\boldsymbol{x}_i$, the individual contribution to the likelihood function is

$$L_i(\boldsymbol{\vartheta}, \boldsymbol{\alpha}) = P(z_i = 1|\boldsymbol{x}_1)^{y_{i1}} P(z_i = 2|\boldsymbol{x}_1)^{y_{i2}} \ldots P(z_i = \kappa|\boldsymbol{x}_1)^{y_{i\kappa}} \quad . \tag{3.379}$$

The log-likelihood of the sample divided by n is therefore

$$l(\boldsymbol{\vartheta}, \boldsymbol{\alpha}) = \frac{1}{n} \sum_{i=1}^{n} \sum_{k=1}^{K} y_{ik} \ln P(z_i = k|\boldsymbol{x}_i). \tag{3.380}$$

Example: Income as a Classified Metric Dependent Variable

As an example of a regression models for a classified metric dependent variable, the dependence of monthly income on schooling and professional experience in the German Socioeconomic Panel (considered in Section 2.3) is reanalyzed. The same data are used as in Section 2.3. The logarithm of the monthly income has been coded into five classes with boundaries $(-\infty, 7.7]$, $(7.7, 8.0]$, $(8.0, 8.2]$, $(8.2, 8.5]$, $(8.5, \infty)$. The parameters estimated here correspond to the parameters in model 2 in Table 1. The results are given in Table 8.

In contrast to the models for metric dependent variables, the threshold models for dichotomous, ordered categorical, censored, and classified metric outcomes contain rather restrictive assumptions about the error term. In most applications, the error δ_i of the model for $y_i^\star$ is assumed to follow either a normal or a logistic distribution, although any distribution function $F(\delta)$ could have generated the data. Obviously, the wrong choice of the error distribution is a source of misspecification, possibly yielding inconsistent estimators of the parameter vector $\boldsymbol{\vartheta}$ of the mean structure. On the other hand, parameter estimates of $\boldsymbol{\beta}$ in a linear model $\mu_i^\star = \boldsymbol{x}_i'\boldsymbol{\beta}$ computed from linear regression of dichotomous or ordered categorical outcomes z_i on continuous regressors $\boldsymbol{x}_i$ are often very similar (up to scale) to parameter estimates from probit or logit regression. An explanation of the robust behavior of OLS regression often observed is found in Ruud (1983, 1986) and Li and Duan (1989).

Table 8. Regression for Log Income as a Metric and as a Classified Metric Variable

Explanatory variables	Metric outcome	Classified metric outcome
Regression constant	6.825 (0.117)	6.745 (0.134)
Years of schooling	0.062 (0.007)	0.068 (0.008)
Professional experience	0.445 (0.068)	0.442 (0.078)
Professional experience squared	-0.082 (0.014)	-0.079 (0.016)
R^2	0.265	0.274

NOTE: The values in parentheses are ML standard errors. The results for analyzing income as a classified metric dependent variable are equivalent to the analysis as a metric variable. The parameter estimates are very similar, the estimates of the standard errors are on the average 10% greater for the classified metric outcome than in the same model for the metric outcome, reflecting the loss of information if one uses the classified rather than the original outcome.

8.3 Unordered Categorical Variables

Let z_i denote an unordered categorical variable with categories $k = 1, \ldots, K$. Associated with z_i is a vector $\boldsymbol{y}_i$ of random variables $y_{ik} \in \{0, 1\}$ with $\sum_{k=1}^{K} y_{ik} = 1$. $y_{ik} = 1$ if $z_i = k$, otherwise $y_{ik} = 0$. Under the assumption that $\boldsymbol{y}_i$ is multinomially distributed, the expected values of y_{ik} are given by $\mu_{ik} = P(z_i = k)$. The elements of the covariance matrix have been given in (3.369). Typical examples of unordered categorical variables are consumer choice of a certain brand, choice of a college by a student, and choice of mode of transportation. Since the categories of z_i are now unordered, a threshold model cannot be used to specify the expected value μ_{ik}. However, latent-variable models for the specification of μ_{ik} have been proposed by Thurstone (1927), Bock and Jones (1968), and McFadden (1974). McFadden (1974, 1981) has introduced the concept of random utility maximization (RUM), which is first explained for a dichotomous dependent variable.

A RUM Model for Dichotomous Outcomes

Let $y_1^\star$ and $y_2^\star$ be two unobserved random variables with

$$y_1^\star = \nu_1 + \delta_1 \text{ and} \tag{3.381}$$

$$y_2^\star = \nu_2 + \delta_2 \tag{3.382}$$

where ν_1 and ν_2 form the systematic part and δ_1 and δ_2 the random part of the model with $E(\delta_1) = 0$ and $E(\delta_2) = 0$. The element index i is eliminated in this subsection for simplicity's sake. The unobserved variables can be interpreted as the utilities of the alternatives 1 and 2 of variable z. The second alternative is chosen if $y_2^\star > y_1^\star$, that is,

$$z = 2 \Longleftrightarrow y_2^\star > y_1^\star \Longleftrightarrow \nu_1 + \delta_1 \leq \nu_2 + \delta_2 \Longleftrightarrow \delta_1 - \delta_2 \leq \nu_2 - \nu_1 \quad . \tag{3.383}$$

Consequently, $P(z = 2)$ may be written as

$$P(z = 2) = P(\delta_1 - \delta_2 \leq \nu_2 - \nu_1). \tag{3.384}$$

Note that $P(z = 1) = 1 - P(z = 2)$. The probabilities $P(z = k)$ and therefore the expected values of y_k depend on the specification of the distribution function $F(.)$ of the difference $\delta_1 - \delta_2$ of the error terms δ_1 and δ_2. Further, only the differences $\nu_2 - \nu_1$ and $\delta_1 - \delta_2$ enter the specification of $P(z = 2)$. From a substantive viewpoint, this implies that only the difference between the utilities and not their absolute values are identified. Let $\epsilon = \delta_1 - \delta_2$ and $\mu = \nu_2 - \nu_1$. Then one finds

$$P(z = 2) = P(\epsilon \leq \mu) = F(\mu) \tag{3.385}$$

where $F(.)$ is the distribution function of ϵ.

As in the threshold models, the unobservable variables $y_k^\star, k = 1, 2$ have no natural scale. Hence, the variances of δ_1 and δ_2 are set to an arbitrary value. As an example that includes explanatory variables, one may consider the model

$$y_1^\star = \nu_1 + \delta_1 \quad \text{with} \quad \nu_1 = \boldsymbol{z}_1'\boldsymbol{\alpha} + \boldsymbol{x}'\boldsymbol{\gamma}_1 \quad \text{and} \quad \delta_1 \sim \mathcal{N}(0, 0.5) \tag{3.386}$$

$$y_2^\star = \nu_2 + \delta_2 \quad \text{with} \quad \nu_2 = \boldsymbol{z}_2'\boldsymbol{\alpha} + \boldsymbol{x}'\boldsymbol{\gamma}_2 \quad \text{and} \quad \delta_2 \sim \mathcal{N}(0, 0.5) \tag{3.387}$$

which implies that regression coefficients are identified only up to scale. The variables in $\boldsymbol{z}_1$ and $\boldsymbol{z}_2$ may be interpreted as properties of the alternatives 1 and 2. The parameter vector $\boldsymbol{\alpha}$ contains the coefficients effect of these alternative specific regressors. The variables in $\boldsymbol{x}$ are characteristics of the decision-making unit. As an example, consider the choice between two modes of transport, private car or public transportation. Variables in $\boldsymbol{z}_1$ and $\boldsymbol{z}_2$ are price and transportation time, which vary across alternatives, while variables in $\boldsymbol{x}$ are occupation and income, which do not vary across alternatives. However, $\boldsymbol{\gamma}_1$ and $\boldsymbol{\gamma}_2$ may be different for each alternative. For identification the variances are set to $1/2$. Consequently $E(\epsilon) = 0, V(\epsilon) = 1, \mu = (\boldsymbol{z}_2 - \boldsymbol{z}_1)'\boldsymbol{\alpha} + \boldsymbol{x}'(\boldsymbol{\gamma}_2 - \boldsymbol{\gamma}_1)$. In this model $\boldsymbol{\alpha}$ and $\boldsymbol{\beta} = \boldsymbol{\gamma}_2 - \boldsymbol{\gamma}_1$ are identified, which corresponds to the concept that utilities are relative. The probability for choosing the second alternative is chosen is then given

$$P(z = 2) = P(\delta_1 - \delta_2 \leq \nu_2 - \nu_1) = P(\epsilon \leq \mu) = \Phi((\boldsymbol{z}_2 - \boldsymbol{z}_1)'\boldsymbol{\alpha} + \boldsymbol{x}'\boldsymbol{\beta}) \tag{3.388}$$

The expected value μ_2 is therefore defined in the same way as in the threshold model for dichotomous probits. Since there are only two alternatives, there are no variables that vary across alternatives.

The Multinomial Logit Model as a RUM Model

The RUM model may be extended from two categories to K categories by introducing K unobserved utilities $y_k^\star$ with expected values ν_k and error term δ_k.

$$y_k^\star = \nu_k + \delta_k, k = 1, \ldots, K \tag{3.389}$$

Furthermore, it is assumed that the error terms δ_k, $k = 1, \ldots, K$, are independent of each other and identically distributed following the standard extreme value distribution with density

$$f(\delta) = \exp(-\delta)\exp(-\exp(-\delta)) \tag{3.390}$$

and cumulative distribution function

$$F(\delta) = \exp(-\exp(-\delta)). \tag{3.391}$$

The expected value of δ is Euler's constant $E(\delta) = 0.577$, which must be subtracted from δ to yield an expected value of 0 for the error term. In Madalla (1983) it is shown that this specification yields the multinomial logit model. Category k of y is chosen if the utility $y_k^\star = \max\{y_1^\star, \ldots, y_K^\star\}$, that is $y_k^\star > y_1^\star$ and $y_k^\star > y_2^\star \ldots$ and $y_k^\star > y_K^\star$.

$$P(z = k) = \frac{\exp(\nu_k)}{\sum_{j=1}^{K} \exp(\nu_j)}, \quad k = 1, \ldots, K \tag{3.392}$$

As in the dichotomous RUM model, only the differences $\nu_k - \nu_1$, $k = 1, \ldots, K$ are identified. If $\nu_1 = 0$, one has the familiar multinomial logit model with the first category as the reference category:

$$P(z = k) = \frac{\exp(\nu_k)}{1 + \sum_{j=2}^{K} \exp(\nu_j)}, \quad k = 2, \ldots, K, \ P(z = 1) = 1 - \sum_{k=2}^{K} P(z = k). \tag{3.393}$$

Explanatory variables may be introduced by specifying different expected values ν_k, $k = 1, \ldots, K$ for each observation $i = 1, \ldots, n$. The explanatory variables can vary as a function of the ith observation but also as a function of the possible alternatives. A linear model for the utilities $y_{ik}^\star$, $k = 1, \ldots, K$ may be written as

$$y_{ik}^\star = \boldsymbol{z}_{ik}^{\star\prime}\boldsymbol{\alpha} + \boldsymbol{x}_i'\boldsymbol{\gamma}_k + \delta_{ik} \quad . \tag{3.394}$$

Since only the differences $\nu_{ik} - \nu_{ij}$ are identified, the values of $\boldsymbol{z}_{ik}^\star$ are set to $\mathbf{0}$ for the first alternative by subtracting $\boldsymbol{z}_{i1}^\star$ from all $\boldsymbol{z}_{ik}^\star$, $k = 2, \ldots, K$. Additionally, $\boldsymbol{\gamma}_1 = \mathbf{0}$. Hence, the mean structure for the individual alternatives is defined as

$$\mu_{ik} = (\boldsymbol{z}_{ik}^\star - \boldsymbol{z}_{i1}^\star)\boldsymbol{\alpha} + \boldsymbol{x}_i\boldsymbol{\gamma}_k, \quad k = 2, \ldots, K \quad . \tag{3.395}$$

The values of μ_{ik}, $k = 2, \ldots, K$ are collected in a vector $\boldsymbol{\mu}_i$ which may be written as

$$\boldsymbol{\mu}_i = \boldsymbol{X}_i \boldsymbol{\beta} \quad \text{with} \tag{3.396}$$

$$\boldsymbol{X}_i = \begin{pmatrix} \boldsymbol{z}^{\star}_{i2} - \boldsymbol{z}^{\star}_{i1}, & \boldsymbol{x}_i & \boldsymbol{0} & \ldots & \boldsymbol{0} \\ \boldsymbol{z}^{\star}_{i3} - \boldsymbol{z}^{\star}_{i1}, & \boldsymbol{0} & \boldsymbol{x}_i & \ldots & \boldsymbol{0} \\ \vdots & \vdots & \vdots & \ddots & \vdots \\ \boldsymbol{z}^{\star}_{ik} - \boldsymbol{z}^{\star}_{i1}, & \boldsymbol{0} & \boldsymbol{0} & \ldots & \boldsymbol{x}_i \end{pmatrix} \quad \text{and} \quad \boldsymbol{\beta} = \begin{pmatrix} \alpha \\ \gamma_2 \\ \vdots \\ \gamma_K \end{pmatrix}. \tag{3.397}$$

The value of μ_{i1} has been set to 0. The multinomial logit model may then be written as

$$P(z_i = k|\boldsymbol{X}_i) = \frac{\exp(\mu_{ik})}{1 + \sum_{j=2}^{K} \exp(\mu_{ij})}, \quad k = 2, \ldots, K \text{ and}$$

$$P(z_i = 1|\boldsymbol{X}_i) = \frac{1}{1 + \sum_{j=2}^{K} \exp(\mu_{ij})}. \tag{3.398}$$

ML estimation of the parameters is based on the log-likelihood function for the multinomial distribution for y_{ik}, that is

$$l(\boldsymbol{\beta}) = \frac{1}{n} \sum_{i=1}^{n} \sum_{k=1}^{K} y_{ik} \ln P(z_i = k|\boldsymbol{X}_i). \tag{3.399}$$

The multinomial logit model may also be seen as a special case of a multivariate generalized linear model for the vector $\boldsymbol{y}_i$ where the mean structure and covariance structure is specified as in equations (3.368)–(3.369) with $\mu_{ik} = P(z_i = k|\boldsymbol{X}_i)$.

Although many computer programs already exist to actually calculate the ML estimate of $\boldsymbol{\beta}$, a word of caution is in order for researchers getting their first experience with the multinomial logit. Throughout this subsection it is tacitly assumed that the full set $\{1, \ldots, K\}$ of alternatives (which is also called choice set) is available for all observations $i = 1, \ldots, n$. However, the choice set may be restricted and may be different for subgroups in the sample (cf. Kohn, Manski, and Mundel 1976). To deal with this problem it may be judicious to split the K dimensional multinomial logit into $K-1$ dichotomous logits where the pairs $\{1,2\}$, $\{1,3\}, \cdots \{1,K\}$ are considered successively rather than simultaneously, as proposed by Begg and Gray (1984). The parameter estimates $\tilde{\boldsymbol{\beta}}$ obtained in this way are consistent for $\boldsymbol{\beta}$, but the estimate of the asymptotic covariance matrix $\boldsymbol{V}(\hat{\boldsymbol{\beta}})$ will be inconsistent.

The Assumption of the Independence of Irrelevant Alternatives

Although the multinomial logit model is widely used, its properties make its use too restricted for certain types of application. By using the multinomial logit model one assumes that the odds ratio for every pair of alternatives is independent of all other alternatives. This property is called independence of irrelevant alternatives (IIA). By writing the selection probabilities $P(z = k)$ as a function of the expected utilities ν_k, $k = 1, \ldots, K$,

$$P(z = k) = \frac{\exp(\nu_k)}{\sum_{k=1}^{K} \exp(\nu_j)} \qquad k = 1, \cdots, K \quad ,$$

one can see immediately that the ratio

$$\frac{P(z = k)}{P(z = j)} = \frac{\exp(\nu_k)}{\exp(\nu_j)} = \exp(\nu_k - \nu_j)$$

depends only on the difference between these two utilities and not on any other alternatives. The drawback of this approach is best illustrated by the famous red bus–blue bus example of transportation research (Debreu 1960; cf. Train 1986, chap. 2, sec. 3). It is assumed that somebody can go to work either by taking a car (c) or a blue bus (bb). If the utilities of these two alternatives are equal, then $P(z = c) = P(z = bb) = 0.5$. The ratio $P(z = c)/P(z = bb) = 1$. Now a third alternative, that is, a red bus (rb), is introduced which is in every respect the same as the blue bus exept for the color. Since the utilities for the red bus and the blue bus are the same, the ratio of $P(z = bb)/P(z = rb) = 1$. On the other hand, the ratio $P(z = c)/P(z = bb) = 1$ since it is independent of the red bus in the multinomial logit model. The two ratios can only be 1 if $P(z = c) = P(z = bb) = P(z = rb) = 1/3$. In reality, however, this assumption seems unrealistic. The alternatives blue or red bus would be considered to be the same with $P(z = bb) + P(z = rb) = 0.5$ and $P(z = c) = 0.5$. This specification, which is due to a high correlation in the utilities between the alternatives blue bus or red bus, cannot be captured by the multinomial logit model.

This counterintuitive behavior of the multinomial logit model also occurs in other RUM models whenever the errors δ_j, $j = 1, \ldots, K$ are assumed to be independent of each other, but are in fact dependent on each other. To overcome the restrictive IIA assumption, generalizations of the multinomial logit model such as the tree-extreme-value (TEV) model (McFadden 1981) have been proposed.

The Multinomial Probit Model

The most general and therefore most useful alternative to the multinomial logit model is the multinomial probit model, first discussed in detail by Daganzo (1979). Starting from the model $y_k^\star = \nu_k + \delta_k$, $k = 1, \cdots, K$ for the utility $y_k^\star$ of alternative k, it is now assumed that the error vector $\boldsymbol{\delta} \sim \mathcal{N}(\mathbf{0}, \boldsymbol{\Sigma})$. Note that the scale of $y_k^\star$ is not defined, the diagonal elements $\Sigma_{kk'}$, $k = 1, \cdots, K$ are therefore set equal to an arbitrary value, for instance $\Sigma_{kk} = 1$. In addition to the parameters $\boldsymbol{\beta}$ that define the linear mean structure $\nu_k = \boldsymbol{x}_k \boldsymbol{\beta}$ for the kth utility, the free parameters in $\boldsymbol{\Sigma}$ describing possible correlations between error term of the alternatives have to be estimated.

From the multinomial distribution for the vector $\boldsymbol{y}_i$, the log-likelihood function is given by

$$l(\boldsymbol{\beta}, \boldsymbol{\Sigma}) = \frac{1}{n} \sum_{i=1}^{n} \sum_{k=1}^{K} y_{ik} \ln P(z_i = k | \boldsymbol{X}_i)$$

where $\boldsymbol{X}_i$ is given as in equation (3.397), and $P(z_i = k)$ is the selection probability for the kth alternative. However, the assumption of independence no longer holds. Therefore, the probability

$$P(z = k) = P(y_k^\star > y_1^\star, \cdots y_k^\star > y_j^\star, \cdots y_k > y_k^\star) \tag{3.400}$$

cannot be written as a function of the product of one-dimensional densities as in the multinomial logit model.

Let $\boldsymbol{v}_k^\star = (y_k^\star - y_1^\star, \cdots, y_k^\star - y_K^\star)'$ be the $(K - 1) \times 1$ vector of differences in the utility function excluding the difference $y_k^\star - y_k^\star = 0$. The vector $\boldsymbol{v}_k^\star$ then follows a multivariate normal distribution

$$\boldsymbol{v}_k^\star \sim \mathcal{N}(\boldsymbol{C}_k \boldsymbol{\nu}, \boldsymbol{C}_k \boldsymbol{\Sigma} \boldsymbol{C}_k')$$

where $\boldsymbol{C}_k$ is the $(K - 1) \times K$ matrix

$$\begin{pmatrix} -1 & 0 & \cdots & 1 & \cdots & 0 \\ 0 & -1 & \cdots & 1 & \cdots & 0 \\ \vdots & & & & & \\ 0 & 0 & \cdots & 1 & \cdots & -1 \end{pmatrix}$$

with a column of ones in the kth column that maps $\boldsymbol{y}^\star$ into $\boldsymbol{v}_k^\star$. Consequently, the selection probability is given by the $(K - 1)$ dimensional integral

$$P(z = k) = \int_0^\infty \cdots \int_0^\infty \varphi(v_{k1}^\star, \cdots, v_{kj}^\star, \cdots v_{k(K-1)}^\star) dv_{k1}^\star \cdots dv_{k(K-1)}^\star \tag{3.401}$$

of the multivariate normal density $\varphi(\boldsymbol{v}_k^\star)$ within the boundaries 0 and $+\infty$.

This integral has to be computed for each observation i for different values of $\boldsymbol{X}_i$. The numerical computation of $P(z_i = k|\boldsymbol{X}_i)$ is feasible only for $K = 4$ (or fewer) alternatives. Recently, there have been attempts to improve the simulation attempts of Lerman and Manski (1981) to simulate the probabilities $P(z_i = k|\boldsymbol{X}_i)$ and to estimate $\boldsymbol{\beta}$ and $\boldsymbol{\Sigma}$ from these simulated probabilities. The interested reader is referred to the papers of McFadden (1989), Pakes and Pollard (1989), and Börsch-Supan and Vassilis (1990). Since the estimation of a multinomial probit is extremely cumbersome, one should try to avoid the correlation of errors in the utilities $y_k^\star$ by including explanatory variables that might account for the correlation and return to the model of independence, for which a multinomial logit model is appropriate (cf. Train 1986, chap. 2, sec. 2).

8.4 Generalized Estimating Equations for Mean Structures

In Section 5.3, PML estimation of the parameter vector $\boldsymbol{\vartheta}$ of the mean structure

$$\boldsymbol{y}_i = \boldsymbol{\mu}(\boldsymbol{x}_i, \boldsymbol{\vartheta}) + \epsilon_i \tag{3.402}$$

with $E(\epsilon_i) = 0$ and unspecified covariance matrix $\boldsymbol{V}(\epsilon_i) = \boldsymbol{\Omega}_i$ was considered. If the assumed density of $\boldsymbol{y}_i$ given $\boldsymbol{x}_i$ is the multivariate normal density with fixed covariance matrix $\boldsymbol{I}_G$, the kernel of the pseudo-log-likelihood function divided by n may be written as

$$l(\boldsymbol{\vartheta}) = -\frac{1}{n}\sum_{i=1}^{n}(\boldsymbol{y}_i - \boldsymbol{\mu}(\boldsymbol{x}_i, \boldsymbol{\vartheta}))'(\boldsymbol{y}_i - \boldsymbol{\mu}(\boldsymbol{x}_i, \boldsymbol{\vartheta})). \tag{3.403}$$

Differentiation of $l(\boldsymbol{\vartheta})$ with respect to $\boldsymbol{\vartheta}$ and setting the first derivatives to $\mathbf{0}$ yields the (pseudo) log-likelihood equations

$$\boldsymbol{s}(\boldsymbol{\vartheta}) = \frac{1}{n}\sum_{i=1}^{n}\boldsymbol{M}_i'(\boldsymbol{y}_i - \boldsymbol{\mu}(\boldsymbol{x}_i, \boldsymbol{\vartheta})) = \mathbf{0} \tag{3.404}$$

where $\boldsymbol{M}_i = \partial\boldsymbol{\mu}_i/\partial\boldsymbol{\vartheta}'$ is the $G \times q$ matrix of first derivatives of $\mu_{ig}, g = 1, \ldots, G$ with $\boldsymbol{\vartheta}_j = 1, \ldots, q$. The PML estimator of the true value $\boldsymbol{\vartheta}_0$ of $\boldsymbol{\vartheta}$ is denoted by $\hat{\boldsymbol{\vartheta}}$. If $\boldsymbol{M}_i$ is evaluated at $\hat{\boldsymbol{\vartheta}}$, it is denoted by $\hat{\boldsymbol{M}}_i$. The asymptotic covariance matrix of $\sqrt{n}\hat{\boldsymbol{\vartheta}}$ is estimated by the information sandwich

$$\hat{\boldsymbol{V}} = \hat{\boldsymbol{A}}^{-1}\hat{\boldsymbol{B}}\hat{\boldsymbol{A}}^{-1} \tag{3.405}$$

$$\hat{\boldsymbol{A}} = \frac{1}{n}\sum_{i=1}^{n}\hat{\boldsymbol{M}}_i'\hat{\boldsymbol{M}}_i \tag{3.406}$$

$$\hat{\boldsymbol{B}} = \frac{1}{n}\sum_{i=1}^{n}\hat{\boldsymbol{M}}_i'(\boldsymbol{y}_i - \hat{\boldsymbol{\mu}}_i)(\boldsymbol{y}_i - \hat{\boldsymbol{\mu}}_i)'\hat{\boldsymbol{M}}_i \quad . \tag{3.407}$$

Instead of assuming that the covariance matrix of the assumed normal density is equal to $\boldsymbol{I}_G$, it may be set to a fixed value $\boldsymbol{\Sigma}_i \sim G \times G$. Then the kernel of the pseudo-log-likelihood divided by n is given by

$$l(\boldsymbol{\vartheta}) = -\frac{1}{n}\sum_{i=1}^{n}\left((\boldsymbol{y}_i - \boldsymbol{\mu}(\boldsymbol{x}_i\boldsymbol{\vartheta}))'\boldsymbol{\Sigma}_i^{-1}(\boldsymbol{y} - \boldsymbol{\mu}(\boldsymbol{x}_i, \boldsymbol{\vartheta})\right). \tag{3.408}$$

The corresponding pseudo-likelihood equations are

$$\boldsymbol{s}(\boldsymbol{\vartheta}) = \frac{1}{n}\sum_{i=1}^{n}\boldsymbol{M}_i'\boldsymbol{\Sigma}_i^{-1}(\boldsymbol{y}_i - \boldsymbol{\mu}(\boldsymbol{x}_i, \boldsymbol{\vartheta})) = \mathbf{0} \quad . \tag{3.409}$$

These equations are called generalized estimating equations for the mean structure by Liang and Zeger (1986) (GEE1). Generalized estimating equations for the first and second moments (GEE2) have been proposed by Zhao and Prentice (1991) and are discussed in detail in Liang, Zeger, and Qaqish (1992).

The information sandwich for the PML estimator $\hat{\boldsymbol{\vartheta}}$ of $\boldsymbol{\vartheta}_0$ computed from (3.409) has to take $\boldsymbol{\Sigma}_i$ into account. The asymptotic covariance matrix of $\sqrt{n}(\hat{\boldsymbol{\vartheta}} - \boldsymbol{\vartheta}_0)$ is estimated by

$$\hat{\boldsymbol{V}}(\sqrt{n}\hat{\boldsymbol{\vartheta}}) = \hat{\boldsymbol{A}}^{-1}\hat{\boldsymbol{B}}\hat{\boldsymbol{A}}^{-1} \tag{3.410}$$

$$\hat{\boldsymbol{A}} = \frac{1}{n}\sum_{i=1}^{n}\hat{\boldsymbol{M}}_i\boldsymbol{\Sigma}_i^{-1}\hat{\boldsymbol{M}}_i' \tag{3.411}$$

$$\hat{\boldsymbol{B}} = \frac{1}{n}\sum_{i=1}^{n}\hat{\boldsymbol{M}}_i\boldsymbol{\Sigma}_i^{-1}(\boldsymbol{y}_i - \hat{\boldsymbol{\mu}}_i)(\boldsymbol{y}_i - \hat{\boldsymbol{\mu}}_i)(\boldsymbol{y}_i - \hat{\boldsymbol{\mu}}_i)\boldsymbol{\Sigma}_i^{-1}\hat{\boldsymbol{M}}_i' \quad . \tag{3.412}$$

If $\boldsymbol{\Sigma}_i$ has been chosen such that $\boldsymbol{\Sigma}_i = \boldsymbol{\Omega}_i$, then the asymptotic covariance matrix of $\sqrt{n}\hat{\boldsymbol{\vartheta}}$ is consistently estimated by

$$\hat{\boldsymbol{V}}(\sqrt{n}\hat{\boldsymbol{\vartheta}}) = \hat{\boldsymbol{A}}^{-1} \tag{3.413}$$

as discussed in the subsection of QGPML estimation. If $\boldsymbol{\Sigma}_i$ is not equal to $\boldsymbol{\Omega}_i$ then the general form (3.412) must be used to estimate $\boldsymbol{V}(\sqrt{n}\hat{\boldsymbol{\vartheta}})$ consistently. However, if $\boldsymbol{\Sigma}_i$ is prudently chosen so that $\boldsymbol{\Sigma}_i$ is close to $\boldsymbol{\Omega}_i$, the estimator $\hat{\boldsymbol{\vartheta}}$ will be more efficient than the estimator computed with the identity matrix as weight matrix. The weight matrix $\boldsymbol{\Sigma}_i$ is called the working covariance in GEE1 estimation.

Until now it has been assumed that $\boldsymbol{\Sigma}_i$ is fixed. In this case, the GEE1 approach may be considered as a special case of PML estimation for mean structures. Liang and Zeger (1986) consider a slightly more general form. First, the length of the vector of dependent variables $\boldsymbol{y}_i$ may differ for each element of the sample such that $\boldsymbol{y}_i \sim G_i \times 1$, $\boldsymbol{M}_i' \sim q \times G_i$ and $\boldsymbol{\Sigma}_i \sim G_i \times G_i$. This feature allows the researcher to treat vectors where data are missing completely at random, for instance missing by design. Examples are the incomes of family members where families have a different number of members, and the employment status in a panel study where observations are missing completely at random. Second, the working covariance matrix $\boldsymbol{\Sigma}_i$ itself may be parameterized as a function of $\boldsymbol{x}_i$, $\boldsymbol{\vartheta}$, and a vector $\boldsymbol{\alpha}$ of additional parameters with true value $\boldsymbol{\alpha}_0$ which are also estimated from the data. Estimation of $\boldsymbol{\vartheta}_0$ then proceeds iteratively by solving the generalized estimating equations

$$\boldsymbol{s}(\boldsymbol{\vartheta}) = \frac{1}{n}\sum_{i=1}^{n}\boldsymbol{M}_i'\boldsymbol{\Sigma}_i^{-1}(\boldsymbol{x}_i, \boldsymbol{\vartheta}, \boldsymbol{\alpha})(\boldsymbol{y}_i - \boldsymbol{\mu}(\boldsymbol{x}_i, \boldsymbol{\vartheta})) = \boldsymbol{0} \quad . \tag{3.414}$$

In the first step, $\boldsymbol{\Sigma}_i$ is fixed by setting it to $\boldsymbol{I}_{G_i}$ or by fixing $\boldsymbol{\vartheta}$ and $\boldsymbol{\alpha}$. Then $\boldsymbol{\vartheta}$ is estimated by solving $\boldsymbol{s}(\boldsymbol{\vartheta}) = \boldsymbol{0}$ yielding $\boldsymbol{\vartheta}^{(1)}$. From $\boldsymbol{\vartheta}^{(1)}$, the residuals $\boldsymbol{e}_i^{(1)} = \boldsymbol{y}_i - \boldsymbol{\mu}(\boldsymbol{x}_i, \boldsymbol{\vartheta}^{(1)})$ may be computed, which allows the estimation of $\boldsymbol{\alpha}_0$ by a consistent estimator $\hat{\boldsymbol{\alpha}}$; $\hat{\boldsymbol{\alpha}}$ may be computed by minimizing

$$Q(\boldsymbol{\alpha}) = \frac{1}{n} \sum_{i=1}^{n} \sum_{j=1}^{G_i} \sum_{k=1}^{j} (\hat{e}_{ijk}^{(1)} - \Sigma_{ijk}(\boldsymbol{x}_i, \boldsymbol{\vartheta}^{(1)} \boldsymbol{\alpha}))^2 \quad . \tag{3.415}$$

The consistent estimator $\hat{\boldsymbol{\alpha}}$ is then used to compute $\boldsymbol{\Sigma}(\boldsymbol{x}_i, \boldsymbol{\vartheta}^{(1)}, \hat{\boldsymbol{\alpha}})$ and to start the next iteration. This procedure is similar to QGPML estimation except that it is not assumed that $\boldsymbol{\Omega}_i$ is consistently estimated by the working covariance matrix $\boldsymbol{\Sigma}(\boldsymbol{x}_i, \hat{\boldsymbol{\vartheta}}, \hat{\boldsymbol{\alpha}})$ where $\hat{\boldsymbol{\vartheta}}$ denotes the GEE1 estimator of $\boldsymbol{\vartheta}_0$. Therefore, the information sandwich (3.412) is used to estimate $\boldsymbol{V}(\sqrt{n}\hat{\boldsymbol{\vartheta}})$ where $\boldsymbol{\Sigma}_i$ is replaced by $\boldsymbol{\Sigma}(\boldsymbol{x}_i, \hat{\boldsymbol{\vartheta}}, \hat{\boldsymbol{\alpha}})$.

Multivariate Dichotomous Outcomes

The most common application of the GEE1 estimator is the estimation of the parameters of a mean structure for a vector $\boldsymbol{y}_i$ of dichotomous outcomes $y_{ig} \in \{0, 1\}, g = 1, \ldots, G_i$.

$$y_{ig} = \mu(\boldsymbol{x}_{ig}, \boldsymbol{\vartheta}) + \epsilon_{ig}, \quad g = 1, \ldots, G_i \tag{3.416}$$

Each y_{ig} is supposed to be marginally Bernoulli distributed with $E(y_{ig}|\boldsymbol{x}_{ig}) = \mu(\boldsymbol{x}_{ig}, \boldsymbol{\vartheta}) = \mu_{ig}$ and $V(y_{ig}|\boldsymbol{x}_{ig}) = \mu_{ig}(1 - \mu_{ig})$. The probability μ_{ig} is typically specified with a logit or a probit model. However, the full distribution of $\boldsymbol{y}_i$ given $\boldsymbol{x}_i$ is not specified and the covariance matrix $\boldsymbol{\Omega}_i$ of $\boldsymbol{y}_i$ given $\boldsymbol{x}_i$ is not specified except for the diagonal elements.

The covariances between the errors ϵ_{ig} should be taken into account to achieve a more efficient estimation of $\boldsymbol{\vartheta}_0$. Since the variances are specified, it is enough to choose a working correlation matrix $\boldsymbol{R}(\boldsymbol{x}_i, \boldsymbol{\vartheta}, \boldsymbol{\alpha})$ instead of $\boldsymbol{\Sigma}(\boldsymbol{x}_i, \boldsymbol{\vartheta}, \boldsymbol{\alpha})$. A typical specification for $\boldsymbol{R}(\boldsymbol{x}_i, \boldsymbol{\vartheta}, \boldsymbol{\alpha})$ is to set $\boldsymbol{R}(\boldsymbol{x}_i, \boldsymbol{\vartheta}, \boldsymbol{\alpha}) = \boldsymbol{R}$ such that the correlations are equal for all observations $i = 1, \ldots, n$, but no specific structure is assumed for the elements of the working correlation matrix. If the observations come from panel data with the same dichotomous variable observed over G_i time points, the models of exchangeability, that is $\rho_{gg'} = \alpha$, and of autocorrelation, that is $\rho_{gg'} = \alpha^{|g-g'|}, g \neq g'$, for an $AR(1)$ process in the error terms may be sensible.

Example: Employment Status in the GSOEP

As an example, employment status in the GSOEP is reconsidered. Instead of using only the first wave, four waves from 1985 to 1988 are considered. Again, the sample consists of 1320 German men observed over all four years. A probit model is specified for each wave with

$$y_{ig} = \Phi(\gamma_g + \boldsymbol{x}_{ig}'\boldsymbol{\beta}) + \epsilon_{ig}, \quad g = 1, \ldots, 4 \tag{3.417}$$

where $\Phi(.)$ is distribution function of a standard normal random variable.

The coefficients of the explanatory variables are supposed to be constant over time. The different situation in the German economy each year is captured by time-varying intercepts

γ_g. First, the parameters of four marginal probit models are estimated, not taking into account the restriction that $\boldsymbol{\beta}$ is constant over time and the fact that the errors ϵ_{ig} may be correlated.

The parameter estimates do not vary greatly over the four waves. The variables that are significant in almost every wave are duration and duration squared of former unemployment, age and age squared, being handicapped and being a white collar worker. The interpretation is the same as for the parameters of the same example in Table 5.

Table 9. Marginal Probit Models for Unemployment Status

Explanatory variables	Wave 1	Wave 2	Wave 3	Wave 4
CONST	3.088 (1.764)	3.249 (1.923)	3.499 (2.003)	2.947 (1.548)
ALD	0.108 (6.259)	0.091 (5.088)	0.065 (3.835)	0.084 (4.910)
ALDSQ	-0.086 (-2.168)	-0.074 (-1.797)	-0.057 (-1.579)	-0.057 (-1.463)
ALH	-0.045 (-0.831)	0.050 (0.892)	0.071 (1.320)	0.009 (0.168)
ALT	-0.246 (-2.526)	-0.266 (-2.948)	-0.259 (-2.850)	-0.240 (-2.517)
ALTSQ	0.306 (2.384)	0.350 (3.049)	0.328 (2.901)	0.318 (2.777)
GZ	0.713 (3.408)	0.652 (3.407)	0.776 (4.193)	0.724 (3.860)
BB1	-0.208 (-1.323)	-0.113 (-0.813)	-0.101 (-0.733)	-0.330 (-2.435)
BB2	-0.010 (-0.038)	-0.299 (-1.076)	-0.120 (-0.455)	-0.194 (-0.761)
BST2	-0.397 (-2.026)	-0.391 (-2.268)	-0.396 (-2.275)	-0.374 (-2.112)
FST	-0.230 (-1.319)	-0.115 (-0.692)	-0.309 (-1.993)	-0.253 (-1.522)
	Correlations of y_{it}, given x_{it}			
Wave 1	1.000			
Wave 2	0.354	1.000		
Wave 3	0.215	0.230	1.000	
Wave 4	0.146	0.216	0.495	1.000

NOTE: The values in parentheses are $z^{\star}$ values that are computed without taking the positive correlation of y_{ig} given $\boldsymbol{x}_{ig}$ into account.

Table 10. Restricted Probit Models Estimated With GEE1

Explanatory variables	Unspecified $\boldsymbol{R}$	Exchangeability	AR(1) process
CONST (Wave 1)	3.726	3.722	3.619
	(3.162) [3.516]	(3.203) [3.615]	(3.056) [3.419]
CONST (Wave 2)	3.623	3.604	3.519
	(3.130) [3.421]	(3.147) [3.503]	(3.021) [3.327]
CONST (Wave 3)	3.347	3.349	3.249
	(2.916) [3.160]	(2.947)[3.254]	(2.818)[3.071]
CONST (Wave 4)	3.421	3.408	3.304
	(2.907) [3.200]	(2.920)[3.281]	(2.791) [3.094]
ALD	0.092	0.089	0.091
	(7.999) [7.958]	(7.744) [7.723]	(7.881) [8.045]
ALDSQ	-0.079	-0.077	-0.077
	(-3.445)[-3.043]	(-3.430) [-2.971]	(-3.396)[-3.043]
ALH	0.014	0.022	0.014
	(0.375) [0.395]	(0.563) [0.610]	(0.358) [0.382]
ALT	-0.273	-0.273	-0.266
	(-4.642) [-4.888]	(-4.681) [-5.048]	(-4.519)[-4.793]
ALTSQ	0.356	0.356	0.348
	(4.834) [5.069]	(4.877) [5.250]	(4.714) [4.986]
GZ	0.683	0.686	0.687
	(4.080) [5.287]	(4.099) [5.340]	(4.092) [5.417]
BB1	-0.268	-0.256	-0.284
	(-2.427)[-2.713]	(-2.301) [-2.608]	(-2.571)[-2.915]
BB2	-0.213	-0.195	-0.223
	(-0.972)[-1.225]	(-0.877) [-1.132]	(-1.032)[-1.306]
BST2	-0.439	-0.442	-0.424
	(-3.386)[-3.953]	(-3.172) [-4.006]	(-3.417)[-3.875]
FST	-0.188	-0.174	-0.198
	(-1.519) [-1.773]	(-1.419) [-1.645]	(-1.629)[-1.879]

NOTE: The values in parentheses are the $z^\star$ values computed using the information sandwich with a specified working correlation matrix. The values in brackets are the naive $z^\star$ values.

Second, three different GEE1 estimations are performed, restricting the effects of the explanatory variables to be constant over time and specifying the working correlation matrix $\boldsymbol{R}_i$ (a) to be equal for all members of the sample but without further restrictions on the structure of $\boldsymbol{R}$, (b) to have the structure of exchangeability, and (c) to have an AR(1) structure. The standard deviations are calculated using both; the information sandwich of (3.412) and the naive estimates, where it is assumed that the working covariance matrix is equal to the true covariance matrix.

The unspecified working correlation matrix $\boldsymbol{R}$ is estimated as

$$\boldsymbol{R} = \begin{pmatrix} 1.000 & & & \\ 0.319 & 1.000 & & \\ 0.182 & 0.300 & 1.000 & \\ 0.116 & 0.217 & 0.519 & 1.000 \end{pmatrix}.$$

The correlation of the errors between two waves under the assumption of exchangeability is estimated as $\hat{\rho}_{EX} = 0.270$ and the correlation of the errors between two adjacent waves under the AR(1) assumption is estimated as $\hat{\rho}_{AR} = 0.396$.

The $z^\star$ values for the different assumptions about the working correlation matrix do not differ greatly from each other, indicating that all of these assumptions work equally well. However, all of them are lower than the naive $z^\star$ values so that the hypothesis that a parameter is equal to 0 will not be rejected as often as in the naive estimation of standard errors. Hence, if one is not sure about the covariance structure of a multivariate dichotomous variable given the explanatory variables, one should always use the adjusted standard error from the information sandwich.

9 Software

The standard models discussed in this chapter such as linear models, loglinear models for count data, and logit and probit models for dichotomous outcomes can be estimated using standard statistical software such as SPSS (1990), BMDP (1981), SAS (1988), and SYSTAT (1990). If one has to work often with complicated factorial and/or nested designs with complex interaction structures, one should use GLIM (1993). However, GLIM is restricted to univariate linear and nonlinear regression models and cannot handle the models discussed in Section 8. If one has to estimate one of the multivariate nonlinear regression models discussed in Section 8 or wants to estimate more complicated versions of these models, one should use LIMDEP 6 (1992), which has excellent facilities to deal with limited dependent and discrete variables (except for the GEE1 estimation of models for multivariate dichotomous outcomes). However, standard software (with the exception of LIMDEP 6 and SAS for the special case of linear models) will not suffice if PML, QGPML and GEE1 estimation are to be used. The GEE1 estimator for dichotomous multivariate dependent variables is available as a program in the SAS matrix language IMS and is also available in S (Becker, Chambers, and Wilks 1988). For many special models, special software must be written using a matrix programming language such as GAUSS (1993) or a powerful statistical development tool such as S.

REFERENCES

Amemiya, T. (1973), "Regression Analysis When the Dependent Variable Is Truncated Normal," *Econometrica*, 41, 997–1016.

Amemiya, T. (1981), "Qualitative Response Models: A Survey," *Journal of Economic Literature*, XIX, 1483–1536.

Amemiya, T. (1985), *Advanced Econometrics*, Cambridge: Harvard University Press.

Apostol, T. M. (1974), *Mathematical Analysis*, 2nd ed., Reading, Massachusetts.

Arminger, G. (1992), "Residuals and Influential Points in Mean Structures Estimated with Pseudo Maximum Likelihood Methods," in L. Fahrmeir, B. Francis, R. Gilchrist, and G. Tutz (eds.), *Advances in GLIM and Statistical Modelling*, New York, NY: Springer, pp. 20–26.

Barndorff-Nielsen, O. (1978) *Information and Exponential Families*, New York: Wiley.

Bates, D. M. and Watts, D. G. (1988), *Nonlinear Regression and Its Applications*, New York: John Wiley.

Becker, G. S. (1975), *Human Capital*, 2nd ed., New York: University of Chicago Press.

Becker, R. A., Chambers, J. M., and Wilks, A. R. (1988), *The New S Language*, Pacific Grove, CA: Wadsworth & Brooks/Cole Advanced Books & Software.

Begg, C. B., and Gray, R. (1984), "Calculation of Polychotomous Logistic Regression Parameters Using Individualized Regressions," *Biometrika*, 71, 11–18.

Belsley, D., Kuh, E., and Welsh, R. (1980), *Regression Diagnostics, Identifying Influential Data and Sources of Collinearity*, New York: Wiley.

Berndt, E. R., Hall, B. H., Hall, R. E., and Hausman, J. A. (1974), "Estimation and Inference in Nonlinear Structural Models," *Annals of Economic and Social Measurement*, 3, 653–666.

BMDP (1981), *BMDP Statistical Software 1981*, Los Angeles: University of California Press.

Bock, R. D. (1975), *Multivariate Statistical Methods in Behavioral Research*, New York: McGraw-Hill.

Bock, R., and Jones, L. (1968), *The Measurement and Prediction of Judgement and Choice*, San Francisco: Holden Day.

Bollen, K. A., and Jackman, R. W. (1985), "Regression Diagnostics: An Expository Treatment of Outliers and Influential Cases," *Sociological Methods and Research*, 13, 510–542.

Börsch-Supan, A., and Vassilis, H. (1990), "Smooth Unbiased Multivariate Probability Simulators for Maximum Likelihood Estimation of Limited Dependent Variable Models," *Cowles Foundation Discussion Paper*, No. 960, Yale University.

Box, G. E. P, and Cox, D. R. (1964), "An Analysis of Transformations," *Journal of the Royal Statistical Society*, 26, 211–252.

Burguete, J. F., Gallant, R., and Souza, G. (1987), "On Unification of the Asymptotic Theory of Nonlinear Econometric Models," *Econometric Reviews*, 2, 150–190.

Clogg, C. C., Petkova, E., and Shidaheh, E. S. (1992), "Statistical Methods for Analyzing Collapsibility in Regression Models," *Journal of Educational Statistics*, 17, 51–74.

Cook, R. D. (1977), "Detection of Influential Observations in Linear Regression," *Technometrics*, 19, 15–18.

Cook, R. D., and Weisberg, S. (1982), *Residuals and Influence in Regression*, New York: Chapman and Hall.

———(1989), "Regression Diagnostics with Dynamic Graphics," *Technometrics*, 31, 277–291.

Copenhaver, T. W., and Mielke, P. W. (1977), "Quantit Analysis: A Quantal Assay Refinement," *Biometrics*, 33, 175–186.

Cox, D. R., and Hinkley, D. V. (1974), *Theoretical Statistics*, London: Chapman and Hall.

Cox, D. R., and Wermuth, N. (1992), "A Comment on the Coefficient of Determination for Binary Responses," *The American Statistician*, 46, 1–4.

Daganzo, C. (1979), *Multinomial Probit – The Theory and Its Application to Demand Forecasting*, New York: Academic Press.

Dagenais, M. G., and Dufour, J. M. (1991), "Invariance, nonlinear models, and asymptotic tests," *Econometrica*, 59, No. 6, 1601–1615.

Davidon, R., and MacKinnon, J. (1985), "Heteroskedasticity-Robust Tests in Regression Directions," *Annales de l'INSEE*, 59/60, 183–218.

Debren, G. (1960), "Review of R. D. Luce, *Individual Choice Behavior*," *American Economic Review*, 50, 186–188.

Dennis, J. E., and Schnabel, R. B. (1983), *Numerical Methods for Unconstrained Optimization and Nonlinear Equations*, Englewood Cliffs: Prentice-Hall.

Efron, B. (1979), "Bootstrap Methods: Another Look at the Jackknife," *The Annals of Statistics*, 7, 1–26.

Fahrmeir, L., and Kaufman, H. (1985), "Consistency and Asymptotic Normality of the Maximum Likelihood Estimator in Generalized Linear Models," *The Annals of Statistics*, 13, 342–368.

Fisher, R. A. (1925), "Theory of Statistical Estimation," *Proceedings of the Cambridge Philosophical Society*, 22, 700.

Flaig, G., Licht, G., and Steiner, V. (1993), "Testing for State Dependence Effects in a Dynamic Model of Male Unemployment Behaviour," *ZEW Discussion Paper*, No. 93-07.

Gabler, S., Laisney, F., and Lechner, M. (1993), Seminonparametric Estimation of Binary Choice Models With an Application to Labor-Force Participation," *Journal of Business & Economic Statistics*, 11, No. 1, 61–80.

Gallant, A. R. (1987), *Nonlinear Statistical Models*, New York: Wiley.

GAUSS, Version 3.0 (1992), *System and Graphics Manual*, Kent, WA: Aptech Systems Inc.

GLIM (1993), *The GLIM System, Release 4 Manual*, Oxford: Clarendon Press.

Godfrey, L. G. (1988), *Misspecification Tests in Econometrics*, New York: Cambridge University Press.

Goldberger, A. S. (1964), *Econometric Theory*, New York: Wiley.

Goodnight, J. H. (1979), "A Tutorial on the Sweep Operator," *The American Statistician*, 33, No. 3, 149–158.

Gourieroux, C., Monfort, A., and Trognon, A. (1984), "Pseudo Maximum Likelihood Methods: Theory," *Econometrica*, 52, 681–700.

Gourieroux, C., and Monfort, A. (1993), "Pseudo-Likelihood Methods," in Maddala, G. S., Rao, C. R., and Vinod, H. D.: *Handbook of Statistics*, Vol. 11, Amsterdam: Elsevier.

Greene, W. H. (1990), *Econometric Analysis*, New York: Macmillan.

Härdle, W. (1990), *Applied Nonparametric Regression*, Cambridge: Cambridge University Press.

Hastie, T. J., Tibshirani, R. J. (1990), *Generalized Additive Models*, London: Chapman and Hall.

Jennrich, R. I. (1969), "Asymptotic properties of nonlinear least squares estimators," *Annals of Mathematical Statistics*, 40, 633–643.

Jobson, J. D., and Fuller, W. A. (1980), "Least Squares Estimation When the Covariance Matrix and Parameter Vector Are Functionally Related," *Journal of the American Statistical Association*, 75, 176–181.

Kale, B. K. (1961), "On the Solution of the Likelihood Equation by Iteration Processes," *Biometrika*, 48, 452–456.

——— (1962), "On the Solution of the Likelihood Equation by Iteration Processes – The Multiparametric Case", *Biometrika*, 48, 452–456.

Kendall, M. and Stuart, A. (1977), *The advanced Theory of Statistics*, 4th ed., Vol. 1, London: Charles Griffin.

Kohn, M., Manski, C., and Mundel, D. (1976), "An Empirical Investigation of Factors Influencing College-Going Behavior," *Annals of Economic and Social Measurement*, 5, 391–419.

Küsters, U. (1987), *Hierarchische Mittelwert- und Kovarianzstrukturmodelle mit nichtmetrischen endogenen Variablen*, Heidelberg: Physica Verlag.

Lafontaine, F., and White, K. J. (1986), "Obtaining Any Wald Statistic You Want," *Economics Letters*, 21, 35–40.

Lerman, S. R., and Manski, C. F. (1981), "On the Use of Simulated Frequencies to Approximate Choice Probabilities," in eds. C. F. Manski, and D. McFadden, *Structural Analysis of Discrete Data With Econometric Applications*, Cambridge, MA, pp. 305–319.

Li, K. C., and Duan, N. (1989), "Regression Analysis Under Link Violation," *The Annals of Statistics*, 17, 1009–1052.

Liang, K. Y., and Zeger, S. L. (1986), "Longitudinal Data Analysis Using Generalized Linear Models," *Biometrika*, 73, 13–22.

LIMDEP (1992), *Version 6.0 – User's Manual and Reference Guide*, Bellport, NY: Econometric Software, Inc.

Long, J. S., and Trivedi, P. K. (1993), "Some specification tests for the linear regression model," in eds. K. A. Bollen, and J. S. Long, *Testing Structural Equation Models*, Newbury Park, CA: Sage Publications, pp. 66–110.

MacKinnon, J. G. (1992), "Model Specification Tests and Artificial Regression," *Journal of Economic Literature*, 30, 102–146.

MacKinnon, J. G., and White, H. (1985), "Some Heteroskedasticity- Consistent Covariance Matrix Estimators with Improved Finite Sample Properties," *Journal of Econometrics*, 29, 305–325.

Maddala, G. S. (1983), *Limited Dependent and Qualitative Variables in Econometrics*, Cambridge: Cambridge University Press.

Maddala, G. S., and Lee, L. F. (1976), "Recursive Models With Qualitative Endogenous Variables," *Annals of Economic and Social Measurement*, 5/4, 525–545.

Manski, C. F. (1988), *Analog Estimation Methods in Econometrics*, New York: Chapman and Hall.

——— (1991), "Regression," *Journal of Economic Literature*, 29, 34–50.

——— (1993), "Identification Problems in the social sciences," in ed. P. V. Marsden, *Sociological Methodology*, 23, 1–56.

McCullagh, P., and Nelder, J. A. (1989), *Generalized Linear Models*, 2nd ed., New York: Chapman and Hall.

McDonald, R. P. (1978), "A Simple Comprehensive Model for the Analysis of Covariance Structures," *British Journal of Mathematical and Statistical Psychology*, 31, 59–72.

McDonald, R. P. (1980), "A Simple Comprehensive Model for the Analysis of Covariance Structures: Some Remarks on Applications," *British Journal of Mathematical and Statistical Psychology*, 33, 161–183.

McFadden, D. (1974), "Conditional Logit Analysis of Qualitative Choice Behavior," in Zarembka, P.: *Frontiers in Econometrics*, New York: Academic Press, pp. 105–142.

——— (1981), "Econometric Models of Probabilistic Choice," in C. F. Manski, and D. McFadden, *Structural Analysis of Discrete Data with Econometric Applications*, Cambridge, MA: Harvard University Press, pp. 198–272.

——— (1989), "A Method of Simulated Moments for Estimation of Discrete Response Models Without Numerical Integration," *Econometrica*, 57, 995–1026.

McKelvey, R. D, and Zavoina, W. (1975), "A Statistical Model for the Analysis of Ordinal Level Dependent Variables," *Journal of Mathematical Sociology*, 4, 103–120.

McLain, D. H. (1974), "Drawing Contours From Arbitrary Data Points," *The Computer Journal*, 17, 318–324.

Mincer, J. (1958), "Investment in Human Capital and Personal Income Distribution," *Journal of Political Economy*, 66, 281–302.

Nelder, J. A., and Wedderburn, R. W. M. (1972), "Generalized Linear Models," *Journal of the Royal Statistical Society*, Ser. A, 135, 370–384.

Nelson, F. D. (1976), "On a General Computer Algorithm for the Analysis of Models with Limited Dependent Variables," *Annals of Economic and Social Measurement*, 5, 493–509.

Pakes, A., and Pollard, D. (1989), "The Asymptotic Distribution of Simulation Experiments," *Econometrica*, 57, 1027–1057.

Phillips, P. C. B., and Park, J. Y. (1988), "On the Formulation of Wald Tests of Nonlinear Restrictions," *Econometrica*, 56, No. 5, 1065–1083.

Porst, R., and Zeifang, K. (1987), "A Description of the German General Social Survey Test-Retest Study and a Report on the Stability of the Sciodemographic Variables," *Sociological Methods & Research*, 15, No. 3, 177–218.

Pregibon, D. (1980), "Goodness of Link Tests for Generalized Linear Models," *Applied Statistics*, 29, 15–24.

——— (1981), "Logistic Regression Diagnostics," *Annals of Statistics*, 9, 705–724.

Ramsey, J. B. (1969), "Tests for Specification Errors in Classical Linear Least-Squares Regression Analysis," *Journal of the Royal Statistical Society*, Ser. B, 31, 350–371.

Rao, C. R. (1973), *Linear Statistical Inference and Its Applications*, 2nd ed., New York: John Wiley.

Rosett, R. N., and Nelson, F. D. (1976), "Estimation of the Two-Limit Probit Regression Model," *Econometrica*, 43, 141–146.

Rothenberg, T. J. (1971), "Identification in Parametric Models," *Econometrica*, 39, 577–591.

Ruud, P. A. (1983), "Sufficient Conditions for the Consistency of Maximum Likelihood Estimation Despite Misspecification of Distribution in Multinomial Discrete Choice Models," *Econometrica*, 51, 225–225.

———(1986), "Consistent Estimation of Limited Dependent Variable Models Despite Misspecification of Distribution," *Journal of Econometrics*, 32, 157–187.

SAS (1988), *SAS/STAT User's Guide*, Release 6.03, Cary, NC: SAS Institute.

Schmidt, P. (1976), *Econometrics*, New York: Marcel Dekker.

SPSS (1990), *SPSS Reference Guide*, Chicago: SPSS.

Sobel, M., and Arminger, G. (1992), "Modelling Household Fertility Decisions: A Nonlinear Simultaneous Probit Model," *Journal of the American Statistical Association*, 87, 38–47.

Stewart, M. B. (1983), "On Least Squares Estimation When the Dependent Variable Is Grouped," *Review of Economic Studies*, 50, 737–753.

Su, J. Q., and Wei, L. J. (1991), "A Lack-of-Fit Test for the Mean Function in a Generalized Linear Model," *Journal of the American Statistical Association*, 86, 420–426.

SYSTAT (1990), *SYSTAT: The System for Statistics*, Evanston, IL: SYSTAT, Inc.

Theil, H. (1961), *Economic Forecasts and Policy*, Amsterdam: North Holland.

———(1971), *Principles of Econometrics*, New York: Wiley.

Thurstone, L. (1927), "A Law of Comparative Judgement," *Psychological Review*, 34, 273–286.

Tobin, J. (1958), "Estimation of Relationships for Limited Dependent Variables," *Econometrica*, 26, 24–36.

Train, K. (1986), *Qualitative Choice Analysis: Theory, Econometrics and an Application to Automobile Demand*, Cambridge, MA: MIT Press.

Tutz, G. (1990), *Modelle für kategoriale Daten mit ordinalem Skalenniveau*, Göttingen: Vandenhoeck & Ruprecht.

Wagner, G. G., Schupp, J., and Rendtel, U. (1991), "The Socio-Economic Panel (SOEP) for Germany – Methods of Production and Management of Longitudinal Data," *Discussion Paper* No. 31a, Deutsches Institut für Wirtschaftsforschung, Berlin.

Wedderburn, R. W. M. (1974), "Quasi-Likelihood-Functions, Generalized Linear Models, and the Gauss-Newton-Method," *Biometrika*, 61, 439–447.

White, H. (1980), "A Heteroscedasticity-Consistent Covariance Matrix Estimator and a Direct Test for Heteroscedasticity," *Econometrica*, 48, 817–838.

———(1982), "Maximum Likelihood Estimation of Misspecified Models," *Econometrica*, 50, 1–25.

———(1983), "Corrigendum," *Econometrica*, 51, No. 2, 513.

Wilkinson, G. N., and Rogers, C. E. (1973), "Symbolic Description of Factorial Models for Analysis of Variance," *Applied Statistics*, 22, 392–399.

Chapter 4

Specification and Estimation of Mean- and Covariance-Structure Models

MICHAEL W. BROWNE AND GERHARD ARMINGER

1 Introduction

The analysis of moment structures originated with the factor analysis model and with some simple pattern hypotheses concerning equality of elements of mean vectors and covariance matrices. They have more recently received considerable attention and been expanded to incorporate a variety of additional models. Covariance structures, some with associated mean structures, occur in psychology, economics, education, marketing, sociology, biometrics, and other disciplines. Most models involving covariance structures that are in current use are related to the factor analysis model in some way, either by being special cases with restrictions on parameters or, more commonly, extensions incorporating additional assumptions.

This chapter will give an overview of the analysis of covariance structures and associated mean structures. Structures for covariance matrices alone, structures for covariance matrices and mean vectors simultaneously, and structures for correlation matrices will be considered. Structures generated by continuous latent variables alone will be treated, although both continuous and discrete manifest variables will be considered. Discrete latent variables are considered in Chapter 6. The estimation of parameters and assessment of fit will be considered in general and then applied to specific structures.

MICHAEL W. BROWNE • Department of Psychology, Ohio State University, 142 Townshend Hall, 1885 Neil Avenue Mall, Columbus, OH 43210, USA. GERHARD ARMINGER • Department of Economics, Bergische Universität–GH Wuppertal, 42097 Wuppertal, Germany. • We wish to acknowledge the reviews of C. C. Clogg, B. Muthén, and M. E. Sobel of an earlier version of the chapter. We are indebted to Ginger Nelson Goff for preparing the LSAY data analyzed in this chapter.

Handbook of Statistical Modeling for the Social and Behavioral Sciences, edited by Gerhard Arminger, Clifford C. Clogg, and Michael E. Sobel. Plenum Press, New York, 1995.

1.1 Background and Notation

Moment structures for manifest (or observable) variables often arise from data models that involve latent (or unobservable) variables. If the latent variables follow a multivariate normal distribution and if the model is linear, the distribution of the manifest variables is also multivariate normal. Since the only unknown parameters for a multivariate normal distribution are elements of mean vectors and covariance matrices, data models generate structures for population mean vectors and covariance matrices alone. The sample mean vector and covariance matrix are jointly sufficient statistics, and maximum likelihood estimation reduces to fitting structural models to sample mean vectors and covariance matrices.

As an illustrative example, consider the model for parallel psychological tests (e.g. Lord and Novick 1968), where each element of a $p \times 1$ vector variate $\boldsymbol{y}$ is regarded as a measure of the same true score variate and the errors are homoscedastic:

$$\boldsymbol{y} = \mathbf{1}_p z + \boldsymbol{e} \tag{4.1}$$

where $\mathbf{1}_p$ is a $p \times 1$ vector with all elements equal to one, z is a scalar latent variate representing a "true score" with expectation $E(z) = \zeta$ and variance $Var(z) = \varphi$, and $\boldsymbol{e}$ is a $p \times 1$ unobservable vector variate representing errors with a null expected value, $E(\boldsymbol{e}) = \mathbf{0}_p$, and covariance matrix $\psi \boldsymbol{I}_p$. All errors are uncorrelated with the true score, $Cov(\boldsymbol{e}\,,\, z) = \mathbf{0}_p$.

The mean vector and covariance matrix of the manifest vector variate $\boldsymbol{y}$ will be represented throughout by $E(\boldsymbol{y}) = \boldsymbol{\mu}$ and $Cov(\boldsymbol{y}\,,\, \boldsymbol{y}') = \boldsymbol{\Sigma}$ respectively. A consequence of the model (4.1) is that the manifest variable mean vector and covariance matrix have the structures

$$\boldsymbol{\mu} = \mathbf{1}\zeta \tag{4.2}$$

and

$$\boldsymbol{\Sigma} = \mathbf{1}\varphi\mathbf{1}' + \psi\boldsymbol{I} \tag{4.3}$$

respectively. Thus the model (4.1) generates the moment structures (4.2) and (4.3). If the latent variates in z and $\boldsymbol{e}$ have a joint multivariate normal distribution the manifest vector variate $\boldsymbol{y}$ will have a multivariate normal distribution with mean vector and covariance matrix as specified in (4.2) and (4.3).

In this chapter a mean structure will be denoted by $\boldsymbol{\mu}(\boldsymbol{\vartheta})$ and a covariance structure by $\boldsymbol{\Sigma}(\boldsymbol{\vartheta})$, where the $q \times 1$ vector $\boldsymbol{\vartheta}$ ($\boldsymbol{\vartheta} \in \mathcal{G} \subseteq \mathcal{R}^q$) represents all parameters in the model. For the structure in (4.2) and (4.3), $\boldsymbol{\vartheta} = (\zeta\,,\, \varphi\,,\, \psi)'$, $q = 3$, $\boldsymbol{\mu}(\boldsymbol{\vartheta}) = \mathbf{1}\zeta$, and $\boldsymbol{\Sigma}(\boldsymbol{\vartheta}) = \mathbf{1}\varphi\mathbf{1} + \psi\boldsymbol{I}$. Here the parameters involved in the mean structure and covariance structure are distinct; models do arise where some parameters are involved in both the mean and covariance structures.

If the observed variables y_j, $j = 1, \ldots, p$ of (4.1) are dichotomous rather than continuous, then threshold models for dichotomous outcomes must be added to the mean and covariance structure model of (4.2) and (4.3). In this case, the mean and covariance structure is formulated for a $p \times 1$ vector of unobserved variables $\boldsymbol{y}^\star \sim \mathcal{N}(\boldsymbol{\mu}(\boldsymbol{\theta}), \boldsymbol{\Sigma}(\boldsymbol{\theta}))$ that is normally distributed. Multivariate normality implies that $y_j^\star$, $j = 1, \ldots, p$ is univariate normal. Hence, a univariate probit model can be written for each manifest y_j. This topic is treated in detail in Section 5.

In subsequent sections of this chapter, a number of results, concerning the estimation of parameters and the assessment of fit of structural models for means and covariances, are given. In the interests of readability, regularity conditions will not be dealt with but may be found in articles referred to. In general the regularity conditions are quite mild and may reasonably be assumed to hold in many practical situations.

A regularity condition that is of some importance, however, is that population parameters should not lie on the boundary of the parameter space. For example, parameters representing variances should not be zero in the population. If this assumption is violated, asymptotic distributions of test statistics and parameter estimates are affected (Shapiro 1985b; Dijkstra 1992).

1.2 Scaling Considerations for Mean, Covariance, and Correlation Structures

We have considered a model where a structure was imposed on both the mean vector and covariance matrix. In other models a structure may be imposed on the mean vector alone, $\boldsymbol{\mu} = \boldsymbol{\mu}(\boldsymbol{\vartheta}_\mu)$, and $\boldsymbol{\Sigma}$ may be chosen to be any $p \times p$ positive definite matrix. In such situations the parameters for the covariance matrix form a vector $\boldsymbol{\sigma}$ of the $p^* = \frac{1}{2}p(p+1)$ nonduplicated elements of $\boldsymbol{\Sigma}$,

$$\boldsymbol{\sigma} = \text{vecs}(\boldsymbol{\Sigma})$$

where the operator vecs$(\cdot)$ yields a column vector formed from nonduplicated elements of a symmetric matrix,

$$\text{vecs}(\boldsymbol{\Sigma}) = (\sigma_{11}, \sigma_{21}, \sigma_{22}, \sigma_{31}, \sigma_{32}, \sigma_{33}, \ldots, \sigma_{pp})' \tag{4.4}$$

The vector $\boldsymbol{\vartheta}$ of all parameters in the model may then be partitioned as $\boldsymbol{\vartheta} = (\boldsymbol{\vartheta}'_\mu, \boldsymbol{\sigma}')'$.

In many models, for example the factor analysis model, a structure is imposed on the covariance matrix only, $\boldsymbol{\Sigma} = \boldsymbol{\Sigma}(\boldsymbol{\vartheta}_\sigma)$, $\boldsymbol{\mu}$ may be any $p \times 1$ vector, and $\boldsymbol{\vartheta} = (\boldsymbol{\mu}', \boldsymbol{\vartheta}'_\sigma)'$. A special type of covariance structure is the correlation structure where the diagonal elements of $\boldsymbol{\Sigma}$ are free and a structure is imposed on the correlation matrix $\boldsymbol{P} = \boldsymbol{P}(\boldsymbol{\vartheta}_\rho)$. In these situations $\boldsymbol{\vartheta}_\sigma = (\boldsymbol{\sigma}'_d, \boldsymbol{\vartheta}'_\rho)'$, where $\boldsymbol{\sigma}_d = (\sigma_{11}^{1/2}, \sigma_{22}^{1/2}, \ldots, \sigma_{pp}^{1/2})'$ is a $p \times 1$ vector of standard deviations. The covariance structure is represented as

$$\boldsymbol{\Sigma}(\boldsymbol{\sigma}_d, \boldsymbol{\vartheta}_\rho) = \boldsymbol{D}_{\sigma_d} \boldsymbol{P}(\boldsymbol{\vartheta}_\rho) \boldsymbol{D}_{\sigma_d} \tag{4.5}$$

where the diagonal elements of $\boldsymbol{P}(\boldsymbol{\vartheta}_\rho)$ are all equal to one.

Correlation structures are of primary interest in situations when the different variables under consideration have arbitrary origins and scales. This frequently happens in the social sciences. Structures for the mean vector and covariance matrix simultaneously are appropriate in situations where the same location and scale applies to all variables. This happens in longitudinal studies and learning experiments where the same measurement process is applied repeatedly over time. Structures for the covariance matrix alone require that all variables be on the same scale while the location parameters are arbitrary. This does not happen frequently although structures for the covariance matrix alone are sometimes used in longitudinal studies. In these situations there are sometimes trends in means which are unnecessarily disregarded.

1.3 Fitting the Moment Structure

Let the $N \times p$ matrix

$$\boldsymbol{Y} = \begin{bmatrix} \boldsymbol{y}_1' \\ \boldsymbol{y}_2' \\ \vdots \\ \boldsymbol{y}_N' \end{bmatrix}$$

represent a sample of N independent observations on the p variables of interest. If the $\boldsymbol{y}_i, i = 1, \ldots, N$, are independently and identically distributed according to a multivariate normal distribution with the mean vector, $\boldsymbol{\mu}_0 = \boldsymbol{\mu}(\boldsymbol{\vartheta}_0)$, and covariance matrix, $\boldsymbol{\Sigma}_0 = \boldsymbol{\Sigma}(\boldsymbol{\vartheta}_0)$, the sample mean vector

$$\bar{\boldsymbol{y}} = N^{-1}\boldsymbol{Y}'\boldsymbol{1}_N$$

and sample covariance matrix

$$\boldsymbol{W} = N^{-1}(\boldsymbol{Y} - \boldsymbol{1}\bar{\boldsymbol{y}}')'(\boldsymbol{Y} - \boldsymbol{1}\bar{\boldsymbol{y}}')$$

are jointly sufficient statistics. One may then obtain an asymptotically efficient estimator of $\boldsymbol{\vartheta}_0$ using $\bar{\boldsymbol{y}}$ and $\boldsymbol{W}$ instead of the full data matrix $\boldsymbol{Y}$.

Minimum Discrepancy Estimation for the Mean Vector and Covariance Matrix

Consider a discrepancy function, $F(\bar{\boldsymbol{y}}, \boldsymbol{W}; \boldsymbol{\mu}, \boldsymbol{\Sigma})$. This is a nonnegative, twice continuously differentiable function of $\bar{\boldsymbol{y}}$, $\boldsymbol{W}$, $\boldsymbol{\mu}$, and $\boldsymbol{\Sigma}$ with the property that $F(\bar{\boldsymbol{y}}, \boldsymbol{W}; \boldsymbol{\mu}, \boldsymbol{\Sigma})$ is zero if and only if $\bar{\boldsymbol{y}} = \boldsymbol{\mu}$ and $\boldsymbol{W} = \boldsymbol{\Sigma}$. If $\boldsymbol{\vartheta}$ is identified at $\boldsymbol{\vartheta}_0$, the minimizer, $\hat{\boldsymbol{\vartheta}}$, of the discrepancy function, $F(\bar{\boldsymbol{y}}, \boldsymbol{W}; \boldsymbol{\mu}(\boldsymbol{\vartheta}), \boldsymbol{\Sigma}(\boldsymbol{\vartheta}))$, will in general be a consistent estimator of $\boldsymbol{\vartheta}_0$ (Shapiro 1984; Kano 1986). Under certain additional conditions on the discrepancy function, $\hat{\boldsymbol{\vartheta}}$ will also be asymptotically efficient for $\boldsymbol{\vartheta}_0$. These conditions will be considered subsequently in Section 3. The minimum of the discrepancy function will be represented by $\hat{F} = F\left(\bar{\boldsymbol{y}}, \boldsymbol{W}; \boldsymbol{\mu}(\hat{\boldsymbol{\vartheta}}), \boldsymbol{\Sigma}(\hat{\boldsymbol{\vartheta}})\right)$. If $\hat{\boldsymbol{\vartheta}}$ is asymptotically efficient for $\boldsymbol{\vartheta}_0$, then $N\hat{F}$ will have an asymptotic chi-squared distribution with $\frac{1}{2}p(p+3)$ degrees of freedom and may be used to assess the fit of the model.

A well-known discrepancy function is

$$F_{NML}(\bar{\boldsymbol{y}}, \boldsymbol{W}; \boldsymbol{\mu}, \boldsymbol{\Sigma}) = (\bar{\boldsymbol{y}} - \boldsymbol{\mu})'\boldsymbol{\Sigma}^{-1}(\bar{\boldsymbol{y}} - \boldsymbol{\mu}) + \ln|\boldsymbol{\Sigma}| - \ln|\boldsymbol{W}| + \operatorname{tr}\boldsymbol{\Sigma}^{-1}\boldsymbol{W} - p \quad (4.6)$$

the normal theory maximum likelihood discrepancy function. Since $F_{NML}(\bar{\boldsymbol{y}}, \boldsymbol{W}; \boldsymbol{\mu}, \boldsymbol{\Sigma}) = -2N^{-1}\ln\lambda$, where λ is the likelihood ratio for testing the model against the general alternative, the minimizer, $\hat{\vartheta}_{NML}$, is the maximum likelihood estimator and $N\hat{F}_{NML}$ is the usual $-2\cdot$ (loglikelihood ratio) test statistic.

In addition to the normal theory maximum likelihood discrepancy function, we consider a general class of generalized least squares discrepancy functions. Let

$$\boldsymbol{u} = \begin{bmatrix} \bar{\boldsymbol{y}} \\ \boldsymbol{w} \end{bmatrix} \tag{4.7}$$

where $\boldsymbol{w}$ = vecs($\boldsymbol{W}$), and

$$\boldsymbol{\eta} = \begin{bmatrix} \boldsymbol{\mu} \\ \boldsymbol{\sigma} \end{bmatrix}$$

so that $\boldsymbol{u}$ and $\boldsymbol{\eta}$ are $\frac{1}{2}p(p+3) \times 1$ vectors. A class of generalized least squares discrepancy function is defined as

$$F_{GLS}(\boldsymbol{u}, \boldsymbol{\eta}) = (\boldsymbol{u} - \boldsymbol{\eta})'\boldsymbol{V}^{-1}(\boldsymbol{u} - \boldsymbol{\eta}) \tag{4.8}$$

where $\boldsymbol{V}$ is some positive definite weight matrix with $\frac{1}{2}p(p+3)$ rows and columns. If $\boldsymbol{V}$ is a consistent estimator of the covariance matrix $\boldsymbol{\Gamma}_0$ of the limiting distribution of $N^{1/2}(\boldsymbol{u}-\boldsymbol{\eta}_0)$, then the minimizer, $\hat{\boldsymbol{\vartheta}}$, of $F_{GLS}(\boldsymbol{u}, \boldsymbol{\eta}(\boldsymbol{\vartheta}))$ is asymptotically efficient within the class of minimum discrepancy estimators based on $\bar{\boldsymbol{y}}$ and $\boldsymbol{W}$ (Browne 1984a). Also the test statistic $N\hat{F}_{GLS}$, where $\hat{F}_{GLS} = F_{GLS}\left(\boldsymbol{u}, \boldsymbol{\eta}(\hat{\boldsymbol{\vartheta}})\right)$, has an asymptotic chi-squared distribution with $\frac{1}{2}p(p+3) - q$ degrees of freedom when the model holds.

In the case where $\boldsymbol{y}_i$ is multivariate normal, the limiting covariance matrix of $N^{1/2}(\boldsymbol{u} - \boldsymbol{\eta}_0)$ may be expressed as a function of $\boldsymbol{\Sigma}_0$ alone:

$$\boldsymbol{\Gamma}_{0_N} = \begin{bmatrix} \boldsymbol{\Sigma}_0 & \boldsymbol{0} \\ \boldsymbol{0}' & 2\boldsymbol{K}_p'(\boldsymbol{\Sigma}_0 \otimes \boldsymbol{\Sigma}_0)\boldsymbol{K}_p \end{bmatrix} \tag{4.9}$$

where $\boldsymbol{0}$ is a null matrix of order $p \times \frac{1}{2}p(p+1)$ and $\boldsymbol{K}_p$ is the $p^2 \times \frac{1}{2}p(p+1)$ transition matrix for a symmetric matrix (Browne 1974, sec. 2; Magnus and Neudecker 1988, p. 49). This is a matrix that provides the transition between $vec(\boldsymbol{W})$, the $p^2 \times 1$ vector formed by stacking the columns of a symmetric matrix $\boldsymbol{W}$, and the $\frac{1}{2}p(p+1) \times 1$ vector $vecs(\boldsymbol{W})$, defined in (4.4):

$$vecs(\boldsymbol{W}) = \boldsymbol{K}_p' vec(\boldsymbol{W}) \tag{4.10}$$

Elements of $\boldsymbol{K}_p$ are either 0, $\frac{1}{2}$, or 1. For example, if $p = 3$ then (4.10) becomes

$$\begin{bmatrix} s_{11} \\ s_{12} \\ s_{22} \\ s_{13} \\ s_{23} \\ s_{33} \end{bmatrix} = \begin{bmatrix} 1 & 0 & 0 & 0 & 0 & 0 & 0 & 0 & 0 \\ 0 & \frac{1}{2} & 0 & \frac{1}{2} & 0 & 0 & 0 & 0 & 0 \\ 0 & 0 & 0 & 0 & 1 & 0 & 0 & 0 & 0 \\ 0 & 0 & \frac{1}{2} & 0 & 0 & 0 & \frac{1}{2} & 0 & 0 \\ 0 & 0 & 0 & 0 & 0 & \frac{1}{2} & 0 & \frac{1}{2} & 0 \\ 0 & 0 & 0 & 0 & 0 & 0 & 0 & 0 & 1 \end{bmatrix} \begin{bmatrix} s_{11} \\ s_{21} \\ s_{31} \\ s_{12} \\ s_{22} \\ s_{32} \\ s_{13} \\ s_{23} \\ s_{33} \end{bmatrix}$$

If $\boldsymbol{V}$ in (4.8) is obtained by substituting a consistent estimate of $\boldsymbol{\Sigma}_0$ for $\boldsymbol{\Sigma}_0$ in (4.9), the resulting estimator is asymptocally efficient under normality assumptions and the test statistic $N\hat{F}_{GLS}$ has an asymptotic chi-squared distribution under the null hypothesis. Substituting $\boldsymbol{W}$ for $\boldsymbol{\Sigma}_0$ in (4.9), the resulting discrepancy function may be expressed in the alternative form to (4.8) of

$$F_{NGLS}(\bar{\boldsymbol{y}}, \boldsymbol{W}; \boldsymbol{\mu}, \boldsymbol{\Sigma}) = (\bar{\boldsymbol{y}} - \boldsymbol{\mu})'\boldsymbol{W}^{-1}(\bar{\boldsymbol{y}} - \boldsymbol{\mu}) + \frac{1}{2}\mathrm{tr}\left[\boldsymbol{W}^{-1}(\boldsymbol{W} - \boldsymbol{\Sigma})\right]^2 \tag{4.11}$$

This is known as the normal theory generalized least squares discrepancy function. Its minimizer, $\hat{\boldsymbol{\vartheta}}_{NGLS}$, is asymptotically equivalent to $\hat{\boldsymbol{\vartheta}}_{NML}$ in that $N^{1/2}(\hat{\boldsymbol{\vartheta}}_{NGLS} - \hat{\boldsymbol{\vartheta}}_{NML})$ converges to $\mathbf{0}$ in probability as $N \to \infty$.

If $\boldsymbol{V}$ in (4.8) is defined by substituting the identity matrix for $\boldsymbol{\Sigma}_0$ in (4.9), the resulting discrepancy function may be expressed in the equivalent form of

$$F_{OLS}(\bar{\boldsymbol{y}}, \boldsymbol{W}; \boldsymbol{\mu}, \boldsymbol{\Sigma}) = (\bar{\boldsymbol{y}} - \boldsymbol{\mu})'(\bar{\boldsymbol{y}} - \boldsymbol{\mu}) + \frac{1}{2}\mathrm{tr}[(\boldsymbol{W} - \boldsymbol{\Sigma})]^2 \tag{4.12}$$

This is known as the ordinary least squares discrepancy function. It is sometimes used but does not yield estimators that are asymptotically efficient if the observations are multivariate normal.

A general expression for $\boldsymbol{\Gamma}_0$ that holds whenever the $\boldsymbol{y}_i$ have a distribution with finite fourth order moments is given (e.g. Browne 1992) by

$$\boldsymbol{\Gamma}_0 = \begin{bmatrix} \boldsymbol{\Sigma}_0 & \boldsymbol{\Gamma}_{0_{\bar{y}w}} \\ \boldsymbol{\Gamma}_{0_{w\bar{y}}} & \boldsymbol{\Gamma}_{0_{ww}} \end{bmatrix}$$

where

$$\boldsymbol{\Gamma}_{0_{\bar{y}w}} = \mathrm{E}\left[(\boldsymbol{y} - \boldsymbol{\mu}_0)\, vecs'\left\{(\boldsymbol{y} - \boldsymbol{\mu}_0)(\boldsymbol{y} - \boldsymbol{\mu}_0)'\right\}\right]$$

with typical element

$$[\boldsymbol{\Gamma}_{0\bar{y}w}]_{i,jk} = \mathrm{E}(y_i - \mu_{0i})(y_j - \mu_{0j})(y_k - \mu_{0k})$$

and

$$\boldsymbol{\Gamma}_{0_{ww}} = \mathrm{E}\left[vecs\left\{(\boldsymbol{y} - \boldsymbol{\mu}_0)(\boldsymbol{y} - \boldsymbol{\mu}_0)'\right\}\, vecs'\left\{(\boldsymbol{y} - \boldsymbol{\mu}_0)(\boldsymbol{y} - \boldsymbol{\mu}_0)'\right\}\right] - \boldsymbol{\sigma}_0\boldsymbol{\sigma}_0'$$

with typical element

$$[\boldsymbol{\Gamma}_{0_{ww}}]_{ij,kl} = \mathrm{E}(y_i - \mu_{0i})(y_j - \mu_{0j})(y_k - \mu_{0k})(y_l - \mu_{0l}) - \sigma_{0ij}\sigma_{0kl}$$

so that $\boldsymbol{\Gamma}_0$ depends on second, third, and fourth order central moments of the distribution of $\boldsymbol{y}$.

A consistent estimate of this matrix is given by

$$\boldsymbol{V} = \begin{bmatrix} \boldsymbol{W} & \boldsymbol{V}_{\bar{y}w} \\ \boldsymbol{V}_{w\bar{y}} & \boldsymbol{V}_{ww} \end{bmatrix} \tag{4.13}$$

where typical elements of $\boldsymbol{V}_{\bar{y}w}$ and $\boldsymbol{V}_{ww}$ are given by

$$[\boldsymbol{V}_{\bar{y}w}]_{i,jk} = \frac{1}{N}\sum_{r=1}^{N}(y_{ri}-\bar{y}_i)(y_{rj}-\bar{y}_j)(y_{rk}-\bar{y}_k) \tag{4.14}$$

$$[\boldsymbol{V}_{ww}]_{ij,kl} = \frac{1}{N}\left\{\sum_{r=1}^{N}(y_{ri}-\bar{y}_i)(y_{rj}-\bar{y}_j)(y_{rk}-\bar{y}_k)(y_{rl}-\bar{y}_l)\right\} - w_{ij}w_{kl} \tag{4.15}$$

This estimator will be consistent if the distribution of the $\boldsymbol{y}_r$ has finite eighth-order moments. Substitution of the weight matrix $\boldsymbol{V}$ defined in (4.13) into (4.8) yields a discrepancy function F_{ADF} that yields the asymptotically distribution-free (ADF) estimator $\hat{\boldsymbol{\vartheta}}_{ADF}$. This is asymptotically efficient within the class of minimum discrepancy estimators based on $\boldsymbol{u}$ and the associated test statistic $N\hat{F}_{ADF}$ has an asymptotic chi-squared distribution under the null hypothesis for any distribution of the $\boldsymbol{y}_r$ with finite eighth-order moments.

Minimum Discrepancy Estimation for a Structured Covariance Matrix

If no structure is imposed on the mean vector $\boldsymbol{\mu}$ so that the parameter vector is of the form $\boldsymbol{\vartheta} = (\boldsymbol{\mu}', \boldsymbol{\vartheta}_\sigma)$, it turns out that for F_{NML} and for any F_{GLS} with a weight matrix $\boldsymbol{V}$, partitioned as in (4.13) with $\boldsymbol{V}_{\bar{y}w} = \boldsymbol{0}$, the minimum discrepancy estimator of $\boldsymbol{\mu}_0$ is $\hat{\boldsymbol{\mu}} = \bar{\boldsymbol{y}}$. This applies to F_{NGLS} where $\boldsymbol{V}$ is given by (4.9), with $\boldsymbol{\Sigma}_0$ replaced by $\boldsymbol{W}$ and to F_{OLS}, where $\boldsymbol{V} = \boldsymbol{I}$. The discrepancy functions may then be simplified by omitting the quadratic forms in $(\boldsymbol{y} - \boldsymbol{\mu})$ from (4.6), (4.11), and (4.12) to yield functions of $\boldsymbol{W}$ and $\boldsymbol{\Sigma}(\boldsymbol{\vartheta}_\sigma)$ alone.

Consider the normal theory maximum likelihood function F_{NML} in (4.6). This becomes

$$F_{NML}(\bar{\boldsymbol{y}}, \boldsymbol{W}; \bar{\boldsymbol{y}}, \boldsymbol{\Sigma}) = \ln|\boldsymbol{\Sigma}| - \ln|\boldsymbol{W}| + \mathrm{tr}\,\boldsymbol{\Sigma}^{-1}\boldsymbol{W} - p \tag{4.16}$$

However, $\boldsymbol{W}$ is a biased estimator of $\boldsymbol{\Sigma}_0$ and a preference has developed historically for use of the unbiased estimator

$$\boldsymbol{S} = \frac{N}{N-1}\boldsymbol{W} = \frac{1}{N-1}(\boldsymbol{Y} - \boldsymbol{1}\bar{\boldsymbol{y}}')'(\boldsymbol{Y} - \boldsymbol{1}\bar{\boldsymbol{y}}')$$

The use of $\boldsymbol{S}$ instead of $\boldsymbol{W}$ can be justified by the method of maximum likelihood if the Wishart distribution for $\boldsymbol{S}$ is employed. This yields the discrepancy function

$$F_{WML}(\boldsymbol{S}; \boldsymbol{\Sigma}) = \ln|\boldsymbol{\Sigma}| - \ln|\boldsymbol{S}| + \mathrm{tr}\boldsymbol{\Sigma}^{-1}\boldsymbol{S} - p \tag{4.17}$$

The likelihood ratio test of fit based on the Wishart distribution for $\boldsymbol{S}$ is

$$\begin{aligned} -2\ln\lambda_W &= (N-1)F_{WML}\left(\boldsymbol{S}; \boldsymbol{\Sigma}(\hat{\boldsymbol{\vartheta}}_{WML})\right) \\ &= (N-1)\left\{\ln\left|\boldsymbol{\Sigma}(\hat{\boldsymbol{\vartheta}}_{WML})\right| - \ln|\boldsymbol{S}| + \mathrm{tr}\left[\boldsymbol{\Sigma}^{-1}(\hat{\boldsymbol{\vartheta}}_{WML})\boldsymbol{S}\right] - p\right\} \end{aligned} \tag{4.18}$$

in contrast to the corresponding likelihood ratio test based on the normal distribution for $\boldsymbol{Y}$:

$$\begin{aligned} -2\ln\lambda_N &= NF_{NML}\left(\boldsymbol{W}; \boldsymbol{\Sigma}(\hat{\boldsymbol{\vartheta}}_{NML})\right) \\ &= N\left\{\ln\left|\boldsymbol{\Sigma}(\hat{\boldsymbol{\vartheta}}_{NML})\right| - \ln|\boldsymbol{W}| + \mathrm{tr}\left[\boldsymbol{\Sigma}^{-1}(\hat{\boldsymbol{\vartheta}}_{NML})\boldsymbol{W}\right] - p\right\} \end{aligned} \tag{4.19}$$

An approach based on the Wishart distribution for $\boldsymbol{S}$ for obtaining maximum likelihood estimates for the factor analysis model was used by Lawley (1940). The corresponding likelihood ratio test was of the form (4.18). Later Anderson and Rubin (1956) maximized the normal likelihood function for $\boldsymbol{Y}$ to obtain likelihood equations for the factor analysis model. The corresponding likelihood ratio test was of the form (4.19). Lawley's approach to maximum likelihood based on the Wishart likelihood function is now more commonly employed for covariance structures but the normal likelihood approach is still employed when structures on both $\boldsymbol{\mu}$ and $\boldsymbol{\Sigma}$ are considered.

When a covariance structure alone is under consideration it is common practice to use $\boldsymbol{S}$ rather than $\boldsymbol{W}$ in conjunction with the other discrepancy functions so that $F_{NGLS}(\bar{\boldsymbol{y}}, \boldsymbol{W}; \boldsymbol{\mu}, \boldsymbol{\Sigma})$ is replaced by

$$F_{WGLS}(\boldsymbol{S};\, \boldsymbol{\Sigma}) = \frac{1}{2}\,\mathrm{tr}\left[\boldsymbol{S}^{-1}(\boldsymbol{S} - \boldsymbol{\Sigma})\right]^2 \tag{4.20}$$

and $F_{OLS}(\bar{\boldsymbol{y}}, \boldsymbol{W}; \boldsymbol{\mu}, \boldsymbol{\Sigma})$ is replaced by

$$F_{OLS}(\boldsymbol{S};\, \boldsymbol{\Sigma}) = \frac{1}{2}\,\mathrm{tr}[(\boldsymbol{S} - \boldsymbol{\Sigma})]^2 \tag{4.21}$$

We shall consider a class of covariance structures where the two approaches are equivalent in the sense that, for the NML/WML and $NGLS/WGLS$ discrepancy functions,

$$F\left(\bar{\boldsymbol{y}}, \boldsymbol{W};\, \bar{\boldsymbol{y}}, \boldsymbol{\Sigma}(\hat{\boldsymbol{\vartheta}}_W)\right) = F\left(\boldsymbol{S};\, \boldsymbol{\Sigma}(\hat{\boldsymbol{\vartheta}}_S)\right) \tag{4.22}$$

where $\hat{\boldsymbol{\vartheta}}_W$ is the minimum discrepancy estimate obtained from $\boldsymbol{W}$ and $\hat{\boldsymbol{\vartheta}}_S$ is the minimum discrepancy estimate obtained from $\boldsymbol{S}$. A covariance structure $\boldsymbol{\Sigma}(\boldsymbol{\vartheta})$ is said to be invariant under a constant scaling factor (ICSF) if given any $\boldsymbol{\vartheta} \in \mathcal{G}$ and any positive scalar, α, there exists a $\boldsymbol{\vartheta}^* \in \mathcal{G}$ such that $\boldsymbol{\Sigma}(\boldsymbol{\vartheta}^*) = \alpha\boldsymbol{\Sigma}(\boldsymbol{\vartheta})$ (Browne 1984a). Nearly all covariance structures that are useful in practice have this property. If $\boldsymbol{\Sigma}(\boldsymbol{\vartheta})$ is ICSF then, given $\hat{\boldsymbol{\vartheta}}_W$, it is possible to choose $\hat{\boldsymbol{\vartheta}}_S$ such that $\boldsymbol{\Sigma}(\hat{\boldsymbol{\vartheta}}_S) = \frac{N}{N-1}\,\boldsymbol{\Sigma}(\hat{\boldsymbol{\vartheta}}_W)$ and (4.22) follows. In the case of maximum likelihood estimates the test statistics simplify when the model is ICSF, since it is then true that (e.g. Browne 1974)

$$\mathrm{tr}\left[\boldsymbol{\Sigma}^{-1}(\hat{\boldsymbol{\vartheta}}_{WML})\boldsymbol{S}\right] - p = \mathrm{tr}\left[\boldsymbol{\Sigma}^{-1}(\hat{\boldsymbol{\vartheta}}_{NML})\boldsymbol{W}\right] - p = 0$$

so that (4.18) simplifies to

$$-2\ln\lambda_W = (N-1)\left\{\ln\left|\boldsymbol{\Sigma}(\hat{\boldsymbol{\vartheta}}_{WML})\right| - \ln|\boldsymbol{S}|\right\}$$

and (4.19) simplifies to

$$-2\ln\lambda_N = N\left\{\ln\left|\boldsymbol{\Sigma}(\hat{\boldsymbol{\vartheta}}_{NML})\right| - \ln|\boldsymbol{W}|\right\}$$

The two likelihood ratio statistics differ slightly because of the different multipliers,

$$-2\ln\lambda_N = \frac{N}{N-1}(-2\ln\lambda_W)$$

but the difference is negligible if N is large.

The asymptotically distribution free (ADF) discrepancy function (Browne 1982, 1984a) generally employed when a covariance structure alone is under consideration is

$$F_{ADF}(\boldsymbol{s};\, \boldsymbol{\sigma}) = (\boldsymbol{s} - \boldsymbol{\sigma})'\boldsymbol{V}_{ww}^{-1}(\boldsymbol{s} - \boldsymbol{\sigma}) \tag{4.23}$$

where the elements of $\boldsymbol{V}_{ww}$ are defined in (4.15).

If a model for means and covariances is under consideration, the discrepancy function is obtained by substituting $\boldsymbol{V}$, given by (4.13), into (4.8). It will *not* in general be true that minimization of (4.8) with $\boldsymbol{V}$ given by (4.13) will yield $\hat{\boldsymbol{\mu}} = \bar{\boldsymbol{y}}$ or that the minimum is the same as the minimum of (4.23).

Caution has to be exercised in the application of asymptotic theory for asymptotically distribution-free procedures. The sample size required for the chi-squared distribution to provide a reasonable approximation to the distribution of the test statistic can be extremely large in models with many degrees of freedom, although smaller sample sizes seem adequate for models with fewer degrees of freedom (Muthén and Kaplan 1992).

Fitting a Structure for μ and Σ as a Covariance Structure

More attention has been given to covariance structures alone than to structures for the mean vector and covariance matrix simultaneously, and computer programs for minimizing $F_{WML}(\boldsymbol{S};\, \boldsymbol{\Sigma})$ in (4.6) are more readily available than those for minimizing $F_{NML}(\bar{\boldsymbol{y}}, \boldsymbol{W};\, \boldsymbol{\mu}, \boldsymbol{\Sigma})$ in (4.17). It is therefore convenient that appropriate substitutions of bordered matrices for $\boldsymbol{S}$ and $\boldsymbol{\Sigma}$ in $F_{WML}(\boldsymbol{S};\, \boldsymbol{\Sigma})$ yield an alternative but equivalent form of $F_{NML}(\bar{\boldsymbol{y}}, \boldsymbol{W};\, \boldsymbol{\mu}, \boldsymbol{\Sigma})$ as shown by McDonald (1980). If the bordered matrices are given by

$$\boldsymbol{S}^* = \begin{bmatrix} \boldsymbol{W} + \bar{\boldsymbol{y}}\,\bar{\boldsymbol{y}}' & \bar{\boldsymbol{y}} \\ \bar{\boldsymbol{y}}' & 1 \end{bmatrix} \tag{4.24}$$

and

$$\boldsymbol{\Sigma}^* = \begin{bmatrix} \boldsymbol{\Sigma} + \boldsymbol{\mu}\,\boldsymbol{\mu}' & \boldsymbol{\mu} \\ \boldsymbol{\mu}' & 1 \end{bmatrix} \tag{4.25}$$

then

$$F_{WML}(\boldsymbol{S}^*;\, \boldsymbol{\Sigma}^*) = F_{NML}(\bar{\boldsymbol{y}}, \boldsymbol{W};\, \boldsymbol{\mu}, \boldsymbol{\Sigma}) \tag{4.26}$$

Consequently computer programs for minimizing $F_{WML}(\boldsymbol{S};\, \boldsymbol{\Sigma})$ can also be used to minimize $F_{NML}(\bar{\boldsymbol{y}}, \boldsymbol{W};\, \boldsymbol{\mu}, \boldsymbol{\Sigma})$ by inputting a matrix of the form of (4.24) and formulating the model following (4.25).

Minimum Discrepancy Estimation of Correlation Structures

If a correlation structure $\boldsymbol{P}(\boldsymbol{\vartheta}_\rho)$ is formulated as a covariance structure,

$$\boldsymbol{\Sigma}(\boldsymbol{\sigma}_d, \boldsymbol{\vartheta}) = \boldsymbol{D}_{\sigma_d}\, \boldsymbol{P}(\boldsymbol{\vartheta}_\rho)\, \boldsymbol{D}_{\sigma_d}$$

as in (4.5), then any of the discrepancy functions appropriate for covariance structures may be employed. The standard deviations that form the diagonal elements of $\boldsymbol{D}_{\sigma_d}$ are nuisance parameters and are not interpreted. For some models, for example the unrestricted factor analysis model, it turns out conveniently that the maximum likelihood estimate of $\boldsymbol{D}_{\sigma_d}$ is

$$\hat{\boldsymbol{D}}_{\sigma_d} = \mathrm{Diag}^{1/2}(\boldsymbol{S}) \tag{4.27}$$

In such situations, the maximum likelihood estimate $\hat{\boldsymbol{\vartheta}}_\rho$ minimizes $F_{WML}(\boldsymbol{R}, \boldsymbol{P}(\boldsymbol{\vartheta}_\rho))$ and the likelihood ratio test statistic is $(N-1)F_{WML}(\boldsymbol{R}, \boldsymbol{P}(\boldsymbol{\vartheta}_\rho))$, where

$$\boldsymbol{R} = \mathrm{Diag}^{-\frac{1}{2}}(\boldsymbol{S})\,\boldsymbol{S}\,\mathrm{Diag}^{-\frac{1}{2}}(\boldsymbol{S}) \tag{4.28}$$

is the sample correlation matrix. Although the analysis of the correlation matrix, $\boldsymbol{R}$, as if it were a covariance matrix yields estimates and a test statistic that are appropriate when (4.27) is known to hold, the standard errors of the estimators appropriate for the analysis of $\boldsymbol{S}$ are not appropriate for the analysis of $\boldsymbol{R}$. Also, for some models, and for discrepancy functions other than F_{WML}, (4.27) does not generally hold. This can lead to incorrect analyses. A careful discussion of the difficulties associated with the analysis of $\boldsymbol{R}$ as a covariance matrix is provided in Cudeck (1989).

Discrepancy functions, $F(\boldsymbol{R};\,\boldsymbol{P})$, between $\boldsymbol{R}$ and $\boldsymbol{P}$ would avoid the necessity for the nuisance parameters $\boldsymbol{\sigma}_d$. Maximum likelihood estimation is not practical since the distribution of $\boldsymbol{R}$ under normality assumptions cannot be expressed in closed form. An appropriate normal theory GLS discrepancy function, however, is (Jennrich 1970; Shapiro and Browne 1990, p. 585)

$$\begin{aligned} F_{GLSR}(\boldsymbol{R};\,\boldsymbol{P}) &= \frac{1}{2}\,\mathrm{tr}\left[\boldsymbol{P}^{-1}(\boldsymbol{R}-\boldsymbol{P})\right]^2 \\ &\quad + \mathrm{diag}'[\boldsymbol{P}^{-1}(\boldsymbol{R}-\boldsymbol{P})]\left(\boldsymbol{I}+\boldsymbol{P}*\boldsymbol{P}^{-1}\right)^{-1}\mathrm{diag}[\boldsymbol{P}^{-1}(\boldsymbol{R}-\boldsymbol{P})] \end{aligned} \tag{4.29}$$

where diag$[\boldsymbol{A}]$ is a column vector formed from the diagonal elements of the square matrix $\boldsymbol{A}$ and $\boldsymbol{A} * \boldsymbol{B}$ is the Hadamard, or term by term, product of the matrices $\boldsymbol{A}$ and $\boldsymbol{B}$, with elements in the ith row and jth column related by

$$[\boldsymbol{A} * \boldsymbol{B}]_{ij} = [\boldsymbol{A}]_{ij} \cdot [\boldsymbol{B}]_{ij}$$

No practical implementation of this discrepancy function in a computer program has as yet been reported.

2 Large Sample Properties of Estimators

In order to deal with mean, covariance, and correlation structures under a single framework we consider a $p^* \times 1$ vector, $\boldsymbol{u}$, of sample moments with a limiting multivariate normal distribution with a mean vector, $\boldsymbol{\eta}_0$, and covariance matrix, $\boldsymbol{\Gamma}_0$. The vector variate $\boldsymbol{u}$ may consist of elements of $\bar{\boldsymbol{y}}$ and of $\boldsymbol{W}$ as in (4.7), of distinct elements of $\boldsymbol{S}$ alone, or of distinct free elements $\boldsymbol{R}$ alone. If $\boldsymbol{\eta}(\boldsymbol{\vartheta})$ represents the moment structure under consideration, the model is fitted by minimizing a discrepancy function (cf. Subsection 1.3.1), $F(\boldsymbol{u}, \boldsymbol{\eta}(\boldsymbol{\vartheta}))$, with respect to $\boldsymbol{\vartheta}$.

2.1 Lack of Fit of the Model and the Assumption of Population Drift

In general, small sample properties of the minimum discrepancy estimators are not available and one has to rely on approximations. If the model is assumed to fit perfectly in the population $\boldsymbol{\eta}_0 = \boldsymbol{\eta}(\boldsymbol{\vartheta}_0)$, then large sample approximations to the distributions of estimators and the minimum discrepancy test statistic may be obtained by considering limiting distributions as $N \to \infty$.

It is more realistic, however, to assume that the model yields a fairly good approximation but does not fit exactly in the population, that is $\boldsymbol{\eta}_0 \approx \boldsymbol{\eta}(\boldsymbol{\vartheta}_0)$. No parameter vector $\boldsymbol{\vartheta}_0$, such that $\boldsymbol{\eta}_0 = \boldsymbol{\eta}(\boldsymbol{\vartheta}_0)$, exists. We then adopt a broader definition of $\boldsymbol{\vartheta}_0$ as the minimizer of $F(\boldsymbol{\eta}_0, \boldsymbol{\eta}(\boldsymbol{\vartheta}))$. Then $\boldsymbol{\eta}_0$ and $\boldsymbol{\eta}(\boldsymbol{\vartheta}_0)$ need not be equal and the minimum $F(\boldsymbol{\eta}_0, \boldsymbol{\eta}(\boldsymbol{\vartheta}_0))$ may be greater than zero. Also the definition of $\boldsymbol{\vartheta}_0$ depends on the particular discrepancy function employed and different discrepancy functions will generally yield different values of $\boldsymbol{\vartheta}_0$ if the model does not fit exactly. If the model does fit exactly and $\boldsymbol{\vartheta}_0$ is identified, different discrepancy functions will have the same minimizer, $\boldsymbol{\vartheta}_0$, and a minimum of zero.

If $F(\boldsymbol{\eta}_0, \boldsymbol{\eta}(\boldsymbol{\vartheta}_0)) > 0$ and no further assumptions concerning goodness of fit are made, the distribution of the test statistic, $N\hat{F}$, will not converge to a limit as $N \to \infty$. In such situations the assumption of population drift (Wald 1943) has often been employed (e.g. Bentler and Dijkstra 1985; Browne and Shapiro 1988; Shapiro and Browne 1989) to obtain an asymptotic distribution for $N\hat{F}$. The population value of $\boldsymbol{\eta}$ is regarded as being a function of sample size, $\boldsymbol{\eta}_{0,N}$, that converges at a rate of $O\left(N^{-1/2}\right)$ to a point, $\boldsymbol{\eta}_0 = \boldsymbol{\eta}(\boldsymbol{\vartheta}_0)$, where the model is satisfied. When the asymptotic distribution of $\hat{\boldsymbol{\vartheta}}$ is considered, one also assumes (Shapiro and Browne 1989) that $F\left(\boldsymbol{\eta}_{0,N}, \boldsymbol{\eta}(\boldsymbol{\vartheta})\right)$ has the same minimizer, $\boldsymbol{\vartheta}_0$, for all N. Under the assumption of parameter drift the asymptotic distribution of $N\hat{F}$ is noncentral chi-squared (section 2.3).

At first sight this assumption of parameter drift seems to be of little relevance to the real world as the goodness of fit of the model in the population is assumed to be related to the size of the sample drawn from it. When one bears in mind that the ultimate aim is to derive an approximation for the distribution of the test statistic, the assumption becomes more acceptable. It is interpreted loosely as implying that the noncentral chi-squared distribution may be regarded as an approximation to the distribution of $N\hat{F}$ in situations where N is large *and* systematic errors due to lack of fit of the model in the population are not of a larger order of magnitude than random sampling errors.

Asymptotic noncentral distributions of test statistics for comparing nested models have been investigated in Steiger, Shapiro, and Browne (1985) and Satorra (1989). The assumption of parameter drift is then modified to account for the simultaneous parameter drift of two models (Steiger et al. 1985, p. 256).

2.2 Reference Functions and Correctly Specified Discrepancy Functions

In many applications $\boldsymbol{\eta}(\boldsymbol{\vartheta})$ is a nonlinear function of $\boldsymbol{\vartheta}$ and the discrepancy function $F(\boldsymbol{u}, \boldsymbol{\eta}(\boldsymbol{\vartheta}))$ is not a quadratic function, for example F_{NML} in (4.16). It is then not easy to obtain exact distributions for the minimizer $\hat{\boldsymbol{\vartheta}}$ or the minimum $\hat{F} = F\left(\boldsymbol{u}, \boldsymbol{\eta}(\hat{\boldsymbol{\vartheta}})\right)$. Asymptotic distributions for $\hat{\boldsymbol{\vartheta}}$ and $N\hat{F}$ may be obtained by approximating $F(\boldsymbol{u},\boldsymbol{\eta}(\boldsymbol{\vartheta}))$ by a quadratic function.

Let

$$\boldsymbol{\Delta}_0 = \frac{\partial}{\partial \boldsymbol{\vartheta}'} \boldsymbol{\eta}(\boldsymbol{\vartheta}_0) \tag{4.30}$$

be the Jacobian of the model evaluated at $\boldsymbol{\vartheta} = \boldsymbol{\vartheta}_0$ and let $\boldsymbol{H}_{\eta\eta}$ be the Hessian of $\frac{1}{2}F(\boldsymbol{u}, \boldsymbol{\eta})$ evaluated at the point $(\boldsymbol{u}, \boldsymbol{\eta}) = (\boldsymbol{\eta}_0, \boldsymbol{\eta}_0)$,

$$\boldsymbol{H}_{\eta\eta} = \frac{1}{2} \frac{\partial}{\partial \boldsymbol{\eta} \partial \boldsymbol{\eta}'} F(\boldsymbol{\eta}_0, \boldsymbol{\eta}_0) \tag{4.31}$$

It is assumed that $\boldsymbol{\vartheta}$ is identified at $\boldsymbol{\vartheta}_0$ so that $\boldsymbol{\Delta}_0$ is of full column rank q.

Lemma (cf. Shapiro 1985a; Browne 1990): Let $\tilde{\boldsymbol{\vartheta}}$ minimize the quadratic function

$$Q(\boldsymbol{\vartheta}) = \{\boldsymbol{u} - \boldsymbol{\eta}_0 - \boldsymbol{\Delta}_0 (\boldsymbol{\vartheta} - \boldsymbol{\vartheta}_0)\}' \boldsymbol{H}_{\eta\eta} \{\boldsymbol{u} - \boldsymbol{\eta}_0 - \boldsymbol{\Delta}_0 (\boldsymbol{\vartheta} - \boldsymbol{\vartheta}_0)\}$$

and let $\tilde{Q} = Q(\tilde{\boldsymbol{\vartheta}})$. Then $N^{1/2}(\hat{\boldsymbol{\vartheta}} - \boldsymbol{\vartheta}_0)$ has the same asymptotic distribution as $N^{1/2}(\tilde{\boldsymbol{\vartheta}} - \boldsymbol{\vartheta}_0)$ and $N\hat{F}$ has the same asymptotic distribution as $N\tilde{Q}$. □

The quadratic function $Q(\boldsymbol{\vartheta})$ will be referred to as a reference function for $F(\boldsymbol{u}, \boldsymbol{\eta}(\boldsymbol{\vartheta}))$ since the asymptotic distributions of $N^{1/2}(\tilde{\boldsymbol{\vartheta}} - \boldsymbol{\vartheta}_0)$ and $N\tilde{Q}$ may be obtained using standard theory of the General Linear Model (Browne 1990). The asymptotic distributions of $N^{1/2}(\hat{\boldsymbol{\vartheta}} - \boldsymbol{\vartheta}_0)$ and $N\hat{F}$ are obtained from the lemma.

Under regularity conditions the limiting distribution of $N^{1/2}(\boldsymbol{u} - \boldsymbol{\eta}_0)$ is multivariate normal with a mean vector

$$\lim_{N\to\infty} N^{1/2}(\boldsymbol{\eta}_{0,N} - \boldsymbol{\eta}_0) = \boldsymbol{\delta}$$

and a covariance matrix $\boldsymbol{\Gamma}_0$ that depends on the distribution of the data (cf. Section 1.3). Because of the assumption that $F\left(\boldsymbol{\eta}_{0,N}, \boldsymbol{\eta}(\boldsymbol{\vartheta})\right)$ has the same minimizer, $\boldsymbol{\vartheta}_0$, for all N, $\boldsymbol{\delta}$ satisfies (Shapiro and Browne 1989) $\boldsymbol{\Delta}_0'\boldsymbol{\delta} = \boldsymbol{0}$. It follows (e.g. Browne 1984a) that the limiting distribution of $N^{1/2}(\hat{\boldsymbol{\vartheta}} - \boldsymbol{\vartheta}_0)$ is multivariate normal with a null mean vector and covariance matrix

$$\boldsymbol{\Pi} = (\boldsymbol{\Delta}_0'\boldsymbol{H}_{\eta\eta}\boldsymbol{\Delta}_0)^{-1} \boldsymbol{\Delta}_0'\boldsymbol{H}_{\eta\eta}\boldsymbol{\Gamma}_0\boldsymbol{H}_{\eta\eta}\boldsymbol{\Delta}_0 (\boldsymbol{\Delta}_0'\boldsymbol{H}_{\eta\eta}\boldsymbol{\Delta}_0)^{-1} \tag{4.32}$$

In particular, if the discrepancy function has the property that

$$\boldsymbol{H}_{\eta\eta} = \boldsymbol{\Gamma}_0^{-1} \tag{4.33}$$

then the covariance matrix in (4.32) simplifies substantially:

$$\boldsymbol{\Pi} = (\boldsymbol{\Delta}_0'\boldsymbol{H}_{\eta\eta}\boldsymbol{\Delta}_0)^{-1} \tag{4.34}$$

A relationship between an attribute $\boldsymbol{H}_{\eta\eta}$ of the discrepancy function and an attribute $\boldsymbol{\Gamma}_0$ of the distribution of $\boldsymbol{u}$ is given in (4.33). When it holds, the estimator $\hat{\boldsymbol{\vartheta}}$ yielded by the discrepancy function is asymptotically efficient within the class of minimum discrepancy estimators based on $\boldsymbol{u}$. Therefore, if (4.33) holds, the discrepancy function is said to be correctly specified for the distribution of the data.

When (4.33) holds, also, it follows from the lemma that the asymptotic distribution of $N\hat{F}$ is chi-squared with $p^* - q$ degrees of freedom and noncentrality parameter

$$\lambda = \lim_{N\to\infty} NF\left(\boldsymbol{\eta}_{0,N}, \boldsymbol{\eta}(\boldsymbol{\vartheta}_0)\right) = \boldsymbol{\delta}'\boldsymbol{\Gamma}_0^{-1}\boldsymbol{\delta} \tag{4.35}$$

If the distribution of $\boldsymbol{y}$ is multivariate normal and $\boldsymbol{u}$ consists of the elements of $\bar{\boldsymbol{y}}$ and the distinct elements of $\boldsymbol{W}$, a discrepancy function is correctly specified if the Hessian matrix in (4.31), $\boldsymbol{H}_{\eta\eta}$, satisfies

$$\boldsymbol{H}_{\eta\eta}^{-1} = \boldsymbol{\Gamma}_{0\mathcal{N}}$$

where $\boldsymbol{\Gamma}_{0\mathcal{N}}$ is defined by (4.9). When $\boldsymbol{u}$ consists of the distinct elements of $\boldsymbol{S}$, and the distribution of $\boldsymbol{y}$ is multivariate normal, a discrepancy function is correctly specified if

$$\boldsymbol{H}_{\eta\eta}^{-1} = \boldsymbol{\Gamma}_{0_{\mathcal{W}}} = 2\boldsymbol{K}_p'(\Sigma_0 \otimes \Sigma_0)\boldsymbol{K}_p \tag{4.36}$$

In general, there is a whole class of discrepancy functions with the same reference function $Q(\boldsymbol{\vartheta})$ consequently yielding estimators $\hat{\boldsymbol{\vartheta}}$ with the same asymptotic distribution and test statistics $N\hat{F}$ with the same asymptotic distribution. For example, the maximum likelihood discrepancy function $F_{WML}(\boldsymbol{S}; \boldsymbol{\Sigma})$ in (4.17) and the generalized least squares discrepancy function $F_{WGLS}(\boldsymbol{S}; \boldsymbol{\Sigma})$ in (4.20), yield the same $\boldsymbol{H}_{\eta\eta}$ from (4.31) and therefore have the same reference function. This $\boldsymbol{H}_{\eta\eta}$ satisfies (4.36) so that both are correctly specified for a multivariate normal distribution. A class of discrepancy functions, $F(\boldsymbol{S}; \boldsymbol{\Sigma})$, that are correctly specified for a multivariate normal distribution has been given by Swain (1975a).

The $\boldsymbol{H}_{\eta\eta}$ yielded by $F_{OLS}(\boldsymbol{S}; \boldsymbol{\Sigma})$ does not satisfy (4.36) so that this discrepancy function is not correctly specified for a multivariate normal distribution. Also discrepancy functions that are correctly specified for a normal distribution will not be correctly specified for other distributions. The asymptotic covariance matrix of estimators yielded by an incorrectly specified discrepancy function is given by (4.32); a test statistic that has an asymptotic chi-squared distribution, whether or not the discrepancy function is correctly specified, is given in Browne (1984a, Proposition 4).

It is of interest that correct specification of the discrepancy function is a sufficient, but not necessary, condition for asymptotic efficiency of the estimator and an asymptotic chi-squared distribution for the test statistic. An incorrectly specified reference function

$$Q^*(\boldsymbol{\vartheta}) = \{\boldsymbol{u} - \boldsymbol{\eta}_0 - \boldsymbol{\Delta}_0(\boldsymbol{\vartheta} - \boldsymbol{\vartheta}_0)\}'\boldsymbol{H}_{\eta\eta}\{\boldsymbol{u} - \boldsymbol{\eta}_0 - \boldsymbol{\Delta}_0(\boldsymbol{\vartheta} - \boldsymbol{\vartheta}_0)\}, \boldsymbol{H}_{\eta\eta} \neq \boldsymbol{\Gamma}_0^{-1}$$

will have the same minimizer, $\hat{\boldsymbol{\vartheta}}$, and minimum, $Q^*(\hat{\boldsymbol{\vartheta}}) = Q(\hat{\boldsymbol{\vartheta}})$, as the correctly specified reference function

$$Q(\boldsymbol{\vartheta}) = \{\boldsymbol{u} - \boldsymbol{\eta}_0 - \boldsymbol{\Delta}_0(\boldsymbol{\vartheta} - \boldsymbol{\vartheta}_0)\}'\boldsymbol{\Gamma}_0^{-1}\{\boldsymbol{u} - \boldsymbol{\eta}_0 - \boldsymbol{\Delta}_0(\boldsymbol{\vartheta} - \boldsymbol{\vartheta}_0)\}$$

if and only if (cf. Shapiro 1987; Browne 1990) there exists a symmetric $q \times q$ matrix $\boldsymbol{C}$ such that

$$\boldsymbol{H}_{\eta\eta}^{-1} = \boldsymbol{\Gamma}_0 + \boldsymbol{\Delta}_0 \boldsymbol{C} \boldsymbol{\Delta}_0' \tag{4.37}$$

where $\boldsymbol{\Delta}_0$ is the Jacobian of the model, as defined in (4.30). When (4.37) is true an incorrectly specified discrepancy function will still yield asymptotically efficient estimators (within the class of minimum discrepancy estimators based on $\boldsymbol{u}$) and a test statistic with an asymptotic chi-squared distribution, and the covariance matrix of the estimator (cf. 4.34) may now be obtained from

$$\boldsymbol{\Pi} = (\boldsymbol{\Delta}_0' \boldsymbol{H}_{\eta\eta} \boldsymbol{\Delta}_0)^{-1} + \boldsymbol{C} \tag{4.38}$$

Since the Jacobian matrix, $\boldsymbol{\Delta}_0$, depends on the model, an incorrectly specified discrepancy function will only retain the asymptotic properties discussed earlier for some models. The incorrectly specified discrepancy function can lose these desirable asymptotic properties if minor modifications, such as constraining some parameters to equality, are made to the model.

Details of this approach to the investigation of robustness of minimum discrepancy estimators and test statistics under misspecification of the discrepancy function have been provided by Shapiro (1987). It has been applied in the investigation of robustness of maximum likelihood and other normal theory minimum discrepancy estimators in the analysis of covariance structures by Browne and Shapiro (1988), in the analysis of mean and covariance structures by Browne (1990), and in the situation where there are several groups by Satorra (1993). A different approach for the investigation of the robustness of estimators and test statistics has been provided by Amemiya and Anderson (1990) and Anderson and Amemiya (1988).

The assumption of population drift is necessary for deriving the expression in (4.32) for the estimator covariance matrix, $\boldsymbol{\Pi}$. All discrepancy functions with the same reference function yield the same $\boldsymbol{\Pi}$. A more general but less simple expression for $\boldsymbol{\Pi}$ holds without any assumptions of population drift. Let

$$\boldsymbol{H}_{\eta\vartheta} = \left. \frac{\partial^2 F(\boldsymbol{u}, \boldsymbol{\eta}(\boldsymbol{\vartheta}))}{\partial \boldsymbol{u} \partial \boldsymbol{\vartheta}'} \right|_{\boldsymbol{u}=\boldsymbol{\eta}_0, \boldsymbol{\vartheta}=\boldsymbol{\vartheta}_0} \tag{4.39}$$

$$\boldsymbol{H}_{\vartheta\vartheta} = \left. \frac{\partial^2 F(\boldsymbol{u}, \boldsymbol{\eta}(\boldsymbol{\vartheta}))}{\partial \boldsymbol{\vartheta} \partial \boldsymbol{\vartheta}'} \right|_{\boldsymbol{\vartheta}=\boldsymbol{\vartheta}_0} \tag{4.40}$$

Under regularity assumptions it may be shown (Dijkstra 1983, Theorem 3; Shapiro 1983, Theorem 5.4) that the limiting distribution of $N^{1/2}(\hat{\boldsymbol{\vartheta}} - \boldsymbol{\vartheta}_0)$ as $N \to \infty$ is multivariate normal with a null mean vector and covariance matrix

$$\boldsymbol{\Pi} = \boldsymbol{H}_{\vartheta\vartheta}^{-1} \boldsymbol{H}_{\eta\vartheta}' \boldsymbol{\Gamma}_0 \boldsymbol{H}_{\eta\vartheta} \boldsymbol{H}_{\vartheta\vartheta}^{-1} \tag{4.41}$$

The covariance matrix of the estimator, $\boldsymbol{\Pi}$, in (4.41) depends on the distribution of the data through $\boldsymbol{\Gamma}_0$, and on the discrepancy function employed through the Hessian matrices,

$\boldsymbol{H}_{\eta\vartheta}$ and $\boldsymbol{H}_{\vartheta\vartheta}$, defined in (4.39) and (4.40). Unlike (4.32) the expression in (4.41) yields different values of $\boldsymbol{\Pi}$ for different discrepancy functions with the same reference function if the model does not hold, $\boldsymbol{\eta}_0 \neq \boldsymbol{\eta}(\boldsymbol{\vartheta}_0)$. When the model does hold (4.41) reduces to (4.32). Expression (4.41) yields equivalent results to the pseudo-maximum likelihood approach used by Arminger and Schoenberg (1989) when $F_{WGLS}(\boldsymbol{S}, \boldsymbol{\Sigma})$ is employed.

It is instructive to compare formulas for $\boldsymbol{H}_{\vartheta\vartheta}$ and $\boldsymbol{H}_{\eta\vartheta}$ in (4.41) when the discrepancy functions $F_{WML}(\boldsymbol{S}; \boldsymbol{\Sigma})$ and $F_{WGLS}(\boldsymbol{S}, \boldsymbol{\Sigma})$ are employed. Let vec($\boldsymbol{\Sigma}$) represent the $p^2 \times 1$ vector formed by stacking the columns of $\boldsymbol{\Sigma}$, and let

$$\tilde{\boldsymbol{\Sigma}}_0 = \boldsymbol{\Sigma}(\boldsymbol{\vartheta}_0)\,, \quad \boldsymbol{\Delta}_\sigma = \frac{\partial}{\partial \boldsymbol{\vartheta}'}\text{vec}\{\boldsymbol{\Sigma}(\boldsymbol{\vartheta}_0)\}$$

$$\dot{\boldsymbol{\Sigma}}_k = \frac{\partial}{\partial \vartheta_k}\boldsymbol{\Sigma}(\boldsymbol{\vartheta}_0)\,, \quad \ddot{\boldsymbol{\Sigma}}_{kl} = \frac{\partial^2}{\partial\vartheta_k\partial\vartheta_l}\boldsymbol{\Sigma}(\boldsymbol{\vartheta}_0) \tag{4.42}$$

For notational brevity, the particular discrepancy function employed is not explicitly specified in the notation given in (4.42). In each situation $\boldsymbol{\vartheta}_0$ is regarded as the minimizer of the particular discrepancy function, $F_{WML}\,(\boldsymbol{\Sigma}_0, \boldsymbol{\Sigma}(\boldsymbol{\vartheta}))$ or $F_{WGLS}\,(\boldsymbol{\Sigma}_0, \boldsymbol{\Sigma}(\boldsymbol{\vartheta}))$, under consideration. An element in the kth row and lth column of matrix $\boldsymbol{X}$ will be represented by $[\boldsymbol{X}]_{kl}$. The Kronecker product of two matrices $\boldsymbol{A}$ and $\boldsymbol{B}$ will be represented by $\boldsymbol{A} \otimes \boldsymbol{B}$.

Formulas follow for the different Hessian matrices involved in (4.41).

Wishart Maximum Likelihood: $F\,(\boldsymbol{u}, \boldsymbol{\eta}(\boldsymbol{\vartheta})) = F_{WML}\,(\boldsymbol{S}, \boldsymbol{\Sigma}(\boldsymbol{\vartheta}))$

$$\boldsymbol{H}_{\eta\vartheta} = -\boldsymbol{K}_p^-\left(\tilde{\boldsymbol{\Sigma}}_0^{-1} \otimes \tilde{\boldsymbol{\Sigma}}_0^{-1}\right)\boldsymbol{\Delta}_\sigma \tag{4.43}$$

where

$$\boldsymbol{K}_p^- = (\boldsymbol{K}_p'\boldsymbol{K}_p)^{-1}\boldsymbol{K}_p'$$

and

$$\boldsymbol{H}_{\vartheta\vartheta} = \boldsymbol{\Delta}_\sigma'\left(\tilde{\boldsymbol{\Sigma}}_0^{-1} \otimes \tilde{\boldsymbol{\Sigma}}_0^{-1}\right)\boldsymbol{\Delta}_\sigma + \boldsymbol{\Omega}_{\vartheta\vartheta} \tag{4.44}$$

where

$$[\boldsymbol{\Omega}_{\vartheta\vartheta}]_{kl} = -\text{tr}\left[\tilde{\boldsymbol{\Sigma}}_0^{-1}(\boldsymbol{\Sigma}_0 - \tilde{\boldsymbol{\Sigma}}_0)\tilde{\boldsymbol{\Sigma}}_0^{-1}\left\{\ddot{\boldsymbol{\Sigma}}_{kl} - 2\dot{\boldsymbol{\Sigma}}_k\tilde{\boldsymbol{\Sigma}}_0^{-1}\dot{\boldsymbol{\Sigma}}_l\right\}\right]$$

Wishart Generalized Least Squares: $F\,(\boldsymbol{u}, \boldsymbol{\eta}(\boldsymbol{\vartheta})) = F_{WGLS}\,(\boldsymbol{S}, \boldsymbol{\Sigma}(\boldsymbol{\vartheta}))$

$$\boldsymbol{H}_{\eta\vartheta} = -\boldsymbol{K}_p^-\left(\boldsymbol{\Sigma}_0^{-1} \otimes \boldsymbol{\Sigma}_0^{-1}\right)\boldsymbol{\Delta}_\sigma + \boldsymbol{\Omega}_{\eta\vartheta} \tag{4.45}$$

where

$$\begin{aligned}\boldsymbol{\Omega}_{\eta\vartheta} &= \boldsymbol{K}_p^-\left\{\boldsymbol{\Sigma}_0^{-1} \otimes \boldsymbol{\Sigma}_0^{-1}\left(\boldsymbol{\Sigma}_0 - \tilde{\boldsymbol{\Sigma}}_0\right)\boldsymbol{\Sigma}_0^{-1}\right\}\boldsymbol{\Delta}_\sigma \\ \boldsymbol{H}_{\vartheta\vartheta} &= \boldsymbol{\Delta}_\sigma'\left(\boldsymbol{\Sigma}_0^{-1} \otimes \boldsymbol{\Sigma}_0^{-1}\right)\boldsymbol{\Delta}_\sigma + \boldsymbol{\Omega}_{\vartheta\vartheta}\end{aligned} \tag{4.46}$$

where

$$[\boldsymbol{\Omega}_{\vartheta\vartheta}]_{kl} = -\text{tr}\left[\boldsymbol{\Sigma}_0^{-1}(\boldsymbol{\Sigma}_0 - \tilde{\boldsymbol{\Sigma}}_0)\boldsymbol{\Sigma}_0^{-1}\ddot{\boldsymbol{\Sigma}}_{kl}\right]$$

If the structural model holds so that $\tilde{\boldsymbol{\Sigma}}_0 = \boldsymbol{\Sigma}_0$ (or the difference between $\tilde{\boldsymbol{\Sigma}}_0$ and $\boldsymbol{\Sigma}_0$ is small enough to be disregarded as implied by the population drift assumption) the expressions for $\boldsymbol{H}_{\eta\vartheta}$ in (4.43) and (4.45) are the same, the expressions for $\boldsymbol{H}_{\vartheta\vartheta}$ in (4.44) and (4.46) are the same, and the expression for $\boldsymbol{\Pi}$ in (4.41) can be shown to be the same as that in (4.32). Since both discrepancy functions define the same $\boldsymbol{\vartheta}_0$ when $\tilde{\boldsymbol{\Sigma}}_0 = \boldsymbol{\Sigma}_0$, the asymptotic distributions of $n^{1/2}(\hat{\boldsymbol{\vartheta}}_{WGLS} - \boldsymbol{\vartheta}_0)$ and $n^{1/2}(\hat{\boldsymbol{\vartheta}}_{WML} - \boldsymbol{\vartheta}_0)$ are the same. If the structural model does not apply to $\boldsymbol{\Sigma}_0$ so that $(\boldsymbol{\Sigma}_0 - \tilde{\boldsymbol{\Sigma}}_0) \neq \mathbf{0}$ the formulas for the two different methods differ and the two expressions for $\boldsymbol{\Pi}$ yielded by (4.41) and (4.32) will not be the same. A computationally efficient method for computing an estimate of (4.41), using (4.43) and (4.44), has been provided by Arminger and Schoenberg (1989).

No assumption of normality has been made. If the structural model holds but the distribution of the $\boldsymbol{y}_i$ is not multivariate normal, the asymptotic distributions of the WML and $WGLS$ estimators are the same but the estimators need not be asymptotically efficient. If the structural model holds and the distribution of the $\boldsymbol{y}_i$ is multivariate normal, the asymptotic distributions of the WML and $WGLS$ estimators are the same and the estimators will be asymptotically efficient.

3 Computational Aspects

We shall consider numerical aspects of fitting a structure $\boldsymbol{\mu}(\boldsymbol{\vartheta})$, $\boldsymbol{\Sigma}(\boldsymbol{\vartheta})$ for the mean vector and covariance matrix simultaneously by minimizing (cf. Section 1.3) an appropriate discrepancy function. Suitable discrepancy functions specified earlier and reproduced here for convenience are

$$F_{GLS}(\bar{\boldsymbol{y}}, \boldsymbol{W}; \boldsymbol{\mu}, \boldsymbol{\Sigma}) = (\boldsymbol{u} - \boldsymbol{\eta})'\boldsymbol{V}^{-1}(\boldsymbol{u} - \boldsymbol{\eta})$$

where

$$\boldsymbol{u} = \begin{bmatrix} \bar{\boldsymbol{y}} \\ \boldsymbol{w} \end{bmatrix}, \quad \boldsymbol{\eta} = \begin{bmatrix} \boldsymbol{\mu} \\ \boldsymbol{\sigma} \end{bmatrix}$$

$$F_{NML}(\bar{\boldsymbol{y}}, \boldsymbol{W}; \boldsymbol{\mu}, \boldsymbol{\Sigma}) = (\bar{\boldsymbol{y}} - \boldsymbol{\mu})'\boldsymbol{\Sigma}^{-1}(\bar{\boldsymbol{y}} - \boldsymbol{\mu}) + \ln|\boldsymbol{\Sigma}| - \ln|\boldsymbol{W}| + \text{tr}\left[\boldsymbol{\Sigma}^{-1}\boldsymbol{W}\right] - p$$

or

$$F_{NGLS}(\bar{\boldsymbol{y}}, \boldsymbol{W}; \boldsymbol{\mu}, \boldsymbol{\Sigma}) = (\bar{\boldsymbol{y}} - \boldsymbol{\mu})'\boldsymbol{W}^{-1}(\bar{\boldsymbol{y}} - \boldsymbol{\mu}) + \frac{1}{2}\text{tr}\left[\boldsymbol{W}^{-1}(\boldsymbol{W} - \boldsymbol{\Sigma})\right]^2$$

An iterative procedure is generally required for minimizing the discrepancy function. We shall first consider the generalized least squares discrepancy function in (4.8). A suitable minimization procedure is the Gauss-Newton method. Let $\hat{\boldsymbol{\vartheta}}$ be the minimizer of the discrepancy function

$$F(\boldsymbol{\vartheta}) = F_{GLS}(\bar{\boldsymbol{y}}, \boldsymbol{W}; \boldsymbol{\mu}(\boldsymbol{\vartheta}), \boldsymbol{\Sigma}(\boldsymbol{\vartheta}))$$

An initial approximation $\hat{\boldsymbol{\vartheta}}_0$ is required and a sequence $\{\hat{\boldsymbol{\vartheta}}_t\}$ of successive approximations to $\hat{\boldsymbol{\vartheta}}$ is obtained. Let

$$\boldsymbol{\Delta}_t = \frac{\partial}{\partial \boldsymbol{\vartheta}'} \boldsymbol{\eta}(\hat{\boldsymbol{\vartheta}}_t)$$

represent the Jacobian of the model evaluated at $\boldsymbol{\vartheta} = \hat{\boldsymbol{\vartheta}}_t$. The sequence of successive approximations to $\hat{\boldsymbol{\vartheta}}$ is defined by

$$\hat{\boldsymbol{\vartheta}}_{t+1} = \hat{\boldsymbol{\vartheta}}_t - \alpha_t \boldsymbol{\delta}_t \tag{4.47}$$

where

$$\boldsymbol{\delta}_t = \boldsymbol{H}_t^{-1} \boldsymbol{g}_t \tag{4.48}$$

$$\boldsymbol{g}_t = \boldsymbol{g}(\hat{\boldsymbol{\vartheta}}_t) = \frac{1}{2}\frac{\partial}{\partial \boldsymbol{\vartheta}} F(\boldsymbol{\vartheta}_t) = -\boldsymbol{\Delta}_t' \boldsymbol{V}^{-1}(\boldsymbol{u} - \boldsymbol{\eta}_t) \tag{4.49}$$

is the gradient of $\frac{1}{2}F(\boldsymbol{\vartheta})$ at $\boldsymbol{\vartheta} = \hat{\boldsymbol{\vartheta}}_t$,

$$\boldsymbol{H}_t = \boldsymbol{H}(\hat{\boldsymbol{\vartheta}}_t) = \boldsymbol{\Delta}_t' \boldsymbol{V}^{-1} \boldsymbol{\Delta}_t \approx \frac{1}{2}\frac{\partial^2}{\partial \boldsymbol{\vartheta} \partial \boldsymbol{\vartheta}'} F(\boldsymbol{\vartheta}_t) \tag{4.50}$$

is an approximation to the Hessian of $\frac{1}{2}F(\boldsymbol{\vartheta})$ at $\boldsymbol{\vartheta} = \hat{\boldsymbol{\vartheta}}_t$ obtained by disregarding terms involving second order derivatives of elements of $\boldsymbol{\eta}(\boldsymbol{\vartheta})$, and α_t $(0 \leq \alpha_t \leq 1)$ is a step size chosen to ensure that $F(\boldsymbol{\vartheta}_{t+1}) < F(\boldsymbol{\vartheta}_t)$. In (4.49) and (4.50) it is convenient to consider the gradient and Hessian, $\boldsymbol{H}$, of $\frac{1}{2}F(\boldsymbol{\vartheta})$ because $\boldsymbol{H}(\hat{\boldsymbol{\vartheta}})^{-1}$ yields a consistent estimate of the asymptotic covariance matrix of $N^{1/2}(\hat{\boldsymbol{\vartheta}} - \boldsymbol{\vartheta}_0)$ when the discrepancy function is correctly specified (cf. Section 2.2) and the model holds. Clearly the multipliers of $\frac{1}{2}$ in (4.49) and (4.50) cancel out in (4.48).

In the normal theory special case of generalized least squares given in (4.11), $F(\boldsymbol{\vartheta}) = F_{NGLS}(\bar{\boldsymbol{y}}, \boldsymbol{W}; \boldsymbol{\mu}(\boldsymbol{\vartheta}), \boldsymbol{\Sigma}(\boldsymbol{\vartheta}))$, the large matrix $\boldsymbol{V}$ of (4.8) is eliminated and elements of the gradient and approximate Hessian become

$$\left[g(\hat{\boldsymbol{\vartheta}}_t)\right]_i = \frac{1}{2}\frac{\partial}{\partial \vartheta_i} F(\hat{\boldsymbol{\vartheta}}_t) = -\left\{(\bar{\boldsymbol{y}} - \hat{\boldsymbol{\mu}})' \boldsymbol{W}^{-1} \dot{\boldsymbol{\mu}}_i + \frac{1}{2}\text{tr}\left[\boldsymbol{W}^{-1}(\boldsymbol{W} - \hat{\boldsymbol{\Sigma}})\boldsymbol{W}^{-1}\dot{\boldsymbol{\Sigma}}_i\right]\right\} \tag{4.51}$$

and

$$\left[\boldsymbol{H}(\hat{\boldsymbol{\vartheta}}_t)\right]_{ij} = \dot{\boldsymbol{\mu}}_i' \boldsymbol{W}^{-1} \dot{\boldsymbol{\mu}}_j + \frac{1}{2}\text{tr}\left[\boldsymbol{W}^{-1}\dot{\boldsymbol{\Sigma}}_i \boldsymbol{W}^{-1}\dot{\boldsymbol{\Sigma}}_j\right] \tag{4.52}$$

where

$$\dot{\boldsymbol{\mu}}_i = \dot{\boldsymbol{\mu}}_i(\hat{\boldsymbol{\vartheta}}_t) = \frac{\partial}{\partial \vartheta_i} \boldsymbol{\mu}(\hat{\boldsymbol{\vartheta}}_t) \tag{4.53}$$

$$\dot{\boldsymbol{\Sigma}}_i = \dot{\boldsymbol{\Sigma}}_i(\hat{\boldsymbol{\vartheta}}_t) = \frac{\partial}{\partial \vartheta_i} \boldsymbol{\Sigma}(\hat{\boldsymbol{\vartheta}}_t) \tag{4.54}$$

and $\hat{\mu} = \mu(\hat{\vartheta}_t)$, $\hat{\Sigma} = \Sigma(\hat{\vartheta}_t)$.

Corresponding expressions for the gradient and approximation for the Hessian of the maximum likelihood discrepancy function in (4.6), $F(\vartheta) = F_{NML}(\hat{y}, W; \mu(\vartheta), \Sigma(\vartheta))$, are:

$$\left[g(\hat{\vartheta}_t)\right]_i = -\left\{(\bar{y} - \hat{\mu})'\hat{\Sigma}^{-1}\dot{\mu}_i + \frac{1}{2}\mathrm{tr}\left[\hat{\Sigma}^{-1}(\tilde{W} - \hat{\Sigma})\hat{\Sigma}^{-1}\dot{\Sigma}_i\right]\right\} \quad (4.55)$$

$$\left[H(\hat{\vartheta}_t)\right]_{ij} = \dot{\mu}_i'\hat{\Sigma}^{-1}\dot{\mu}_j + \frac{1}{2}\mathrm{tr}\left[\hat{\Sigma}^{-1}\dot{\Sigma}_i\hat{\Sigma}^{-1}\dot{\Sigma}_j\right] \quad (4.56)$$

where

$$\tilde{W} = \tilde{W}(\hat{\vartheta}_t) = W + (\bar{y} - \hat{\mu})(\bar{y} - \hat{\mu})'$$

The matrix defined in (4.56) is the information matrix under the assumption of a multivariate normal distribution. The recurrence relation in (4.47), together with (4.55) and (4.56), define the Fisher scoring procedure for obtaining maximum likelihood estimates. Relationships between the Fisher scoring procedure and an iteratively reweighted Gauss-Newton procedure have been pointed out in Lee and Jennrich (1979) and in Browne and Du Toit (1992).

The expressions for the gradient and information matrix given above are general and may be adapted to specific models by the substitution of appropriate expressions for $\hat{\mu}$ and $\hat{\Sigma}$ and for derivatives of the model with respect to the parameters, $\dot{\mu}_i$ and $\dot{\Sigma}_i$, as defined by (4.53) and (4.54). Forward difference approximations for derivatives,

$$\dot{\mu}_i \approx \frac{1}{\epsilon}\mu(\hat{\vartheta} + \epsilon e_i)\,, \qquad \dot{\sigma}_i \approx \frac{1}{\epsilon}\sigma(\hat{\vartheta} + \epsilon e_i) \quad (4.57)$$

where ϵ is a small scalar and e_i is a vector with all elements equal to zero except for the ith element, which is one, are often adequate (Browne and Du Toit 1992).

While the expressions in (4.51), (4.52), (4.55), and (4.56) apply whenever the derivatives involved exist, more computationally efficient expressions exist for certain models. Some simplifications for structural equation and factor analysis models, where the $\dot{\Sigma}_i$ are of rank 2, are described in Browne (1982).

It is sometimes convenient to be able to impose constraints on the parameters in the model. An effective extension of the Fisher scoring method to enable the computation of maximum likelihood estimates subject to constraints was proposed by Aitchison and Silvey (1960). This was shown to be an iteratively reweighted constrained Gauss-Newton procedure under normality assumptions, and adapted to other discrepancy functions in Browne and Du Toit (1992).

Suppose that the parameters are required to satisfy the equality constraints,

$$c(\vartheta) = \mathbf{0} \quad (4.58)$$

where $c(\vartheta)$ is a continuously differentiable $r \times 1$ vector valued function of ϑ and let

$$L(\vartheta) = \frac{\partial}{\partial \vartheta'} c(\vartheta) \tag{4.59}$$

be the $r \times q$ constraint function Jacobian matrix. An iterative procedure, similar to that of (4.47),

$$\hat{\vartheta}_{t+1} = \hat{\vartheta}_t - \alpha_t \delta_t$$

is employed but δ_t is now defined by

$$\begin{bmatrix} \delta_t \\ \lambda_t \end{bmatrix} = \begin{bmatrix} (H_t + L_t' D_t L_t) & L_t' \\ L_t & 0 \end{bmatrix}^{-1} \begin{bmatrix} (g_t + L_t' D_t c_t) \\ c_t \end{bmatrix} \tag{4.60}$$

where $c_t = c(\hat{\vartheta}_t)$, $L_t = L(\hat{\vartheta}_t)$, D_t is a nonnegative definite scaling matrix, and λ_t is a $r \times 1$ vector of Lagrange multipliers. A discrepancy function that is penalized by constraint violations is used to choose the step size α_t.

The scaling matrix D_t in (4.60) has no effect on δ_t and λ_t and is often taken to be the null matrix (Gill, Murray, and Wright 1981, sec. 5.4). It may be chosen to be a positive definite diagonal matrix in situations where parameters are not identified and H_t is singular, in order to facilitate use of Gauss-Jordan sweeps (for example, Jennrich and Sampson 1986) to invert the indefinite bordered matrix in (4.60). After convergence the leading $q \times q$ submatrix of the matrix inverted in (4.60) yields an estimate of the asymptotic covariance matrix of $N^{1/2}(\hat{\vartheta} - \vartheta_0)$. Further details are given in Browne and Du Toit (1992).

4 Examples of Moment Structures

4.1 The Factor Analysis Model

The factor analysis model is one of the earliest examples of a covariance structure and has had a substantial influence on the development of later models, in particular the structural equation models with latent variables discussed in Section 4.2.

The model is of the form

$$x = \mu + \Lambda z + u \tag{4.61}$$

where the $p \times 1$ vector variate x represents manifest variables, the $m \times 1$ vector variate z represents common factors, and the $p \times 1$ vector variate u represents unique variables. The common factors, z, and unique variables, u, are examples of latent variables. They cannot be observed but their existence is hypothesized to explain relationships between latent variables. The elements of the $p \times m$ factor matrix Λ are partial regression weights of the manifest variables on the factors and are usually referred to as factor loadings. Elements of μ are corresponding intercepts. Under the assumptions that $E(z) = 0$ and $E(u) = 0$, (4.61) implies that the manifest variable mean vector is $E(x) = \mu$. We assume also that

$Cov(\boldsymbol{z}, \boldsymbol{u}') = \mathbf{0}$, and represent the $m \times m$ factor covariance matrix by $Cov(\boldsymbol{u}, \boldsymbol{u}') = \boldsymbol{\Psi}$, where $\boldsymbol{\Psi}$ is a diagonal matrix. The covariance structure generated by the model (4.61) is then

$$\boldsymbol{\Sigma} = Cov(\boldsymbol{x}, \boldsymbol{x}') = \boldsymbol{\Lambda}\boldsymbol{\Phi}\boldsymbol{\Lambda}' + \boldsymbol{\Psi} \tag{4.62}$$

In unrestricted factor analysis no elements of $\boldsymbol{\Lambda}$ or of $\boldsymbol{\Psi}$ are assumed to have prespecified values or to be equal to other elements. It is common practice to set

$$\boldsymbol{\Phi} = \boldsymbol{I} \tag{4.63}$$

initially, allthough this restriction may be relaxed subsequently. Under assumption (4.63) we have that $\boldsymbol{\Sigma} - \boldsymbol{\Psi} = \boldsymbol{\Lambda}\boldsymbol{\Lambda}' = \boldsymbol{\Lambda}^{\star}\boldsymbol{\Lambda}^{\star\prime}$ where $\boldsymbol{\Lambda}^{\star} = \boldsymbol{\Lambda}\boldsymbol{T}$ and $\boldsymbol{T}$ is any $m \times m$ orthogonal matrix. Consequently identification conditions must be imposed on $\boldsymbol{\Lambda}$ to identify its elements.

The conditional minimum of the normal theory maximum likelihood discrepancy function F_{WML} with respect to $\boldsymbol{\Lambda}$ given some nonnegative definite approximation $\hat{\boldsymbol{\Psi}}$ to $\boldsymbol{\Psi}$ subject to the identification conditions

$$\hat{\boldsymbol{\Lambda}}'\boldsymbol{S}^{-1}\hat{\boldsymbol{\Lambda}} = \text{Diag}(\hat{\boldsymbol{\Lambda}}'\boldsymbol{S}^{-1}\hat{\boldsymbol{\Lambda}}) \tag{4.64}$$

is given by (Jennrich and Robinson 1969)

$$\hat{\boldsymbol{\Lambda}}_{\hat{\Psi}} = \boldsymbol{S}^{1/2}\boldsymbol{U}(\boldsymbol{I} - \boldsymbol{D}_{\gamma})^{1/2} \tag{4.65}$$

where $\boldsymbol{S}^{1/2}$ is a square root of $\boldsymbol{S}$, $\boldsymbol{D}_{\gamma}$ is a $m \times m$ diagonal matrix with the m smallest eigenvalues of $\boldsymbol{S}^{-1/2}\hat{\boldsymbol{\Psi}}\boldsymbol{S}^{-1/2'}$ as diagonal elements, and the columns of the $p \times m$ matrix $\boldsymbol{U}$ consist of the corresponding standardized eigenvectors. Equation (4.65) simultaneously (Swain 1975a) provides the conditional minimum, given $\hat{\boldsymbol{\Psi}}$, of all members of the Swain family of discrepancy functions, including F_{GLS}. This result is convenient in that it provides an easily computed estimate of $\boldsymbol{\Lambda}$ when an estimate of $\boldsymbol{\Psi}$ is available. An alternative to (4.65) which uses eigenvalues and eigenvectors of $\hat{\boldsymbol{\Psi}}^{-1/2}\boldsymbol{S}\hat{\boldsymbol{\Psi}}^{-1/2'}$ is frequently employed (e.g. Jöreskog 1967). The two methods for obtaining $\hat{\boldsymbol{\Lambda}}_{\hat{\Psi}}$ are equivalent when $\hat{\boldsymbol{\Psi}}$ has positive diagonal elements, but the method that employs $\hat{\boldsymbol{\Psi}}^{-1/2}\boldsymbol{S}\hat{\boldsymbol{\Psi}}^{-1/2'}$ breaks down when some diagonal elements of $\hat{\boldsymbol{\Psi}}$ are zero, a situation that occurs fairly frequently in practice.

If $\hat{\boldsymbol{\Psi}}$ is the maximum likelihood estimate of $\boldsymbol{\Psi}$ then (4.65) yields the maximum likelihood estimate of $\boldsymbol{\Lambda}$. Similary, if $\hat{\boldsymbol{\Psi}}$ is a particular minimum discrepancy estimate of $\boldsymbol{\Psi}$ in the Swain family, then (4.65) yields the corresponding minimum discrepancy estimate of $\boldsymbol{\Lambda}$. A Newton-Raphson procedure may be used (Jöreskog 1967; Jennrich and Robinson 1969; Swain 1975a) to minimize the concentrated function

$$f(\boldsymbol{\Psi}) = F(\boldsymbol{S}; \boldsymbol{\Psi}, \hat{\boldsymbol{\Lambda}}_{\Psi}) \tag{4.66}$$

with respect to the p diagonal elements of $\boldsymbol{\Psi}$ to obtain $\hat{\boldsymbol{\Psi}}$. The estimate of $\boldsymbol{\Lambda}$ is then obtained from (4.65). This approach is considerably more efficient computationally than minimizing

$F(\boldsymbol{S}; \boldsymbol{\Lambda}, \boldsymbol{\Psi})$ simultaneously with respect to elements of $\boldsymbol{\Psi}$ and $\boldsymbol{\Lambda}$. Maximum likelihood estimates of parameters in the unrestricted factor analysis model satisfy (4.27) and it is common practice to replace $\boldsymbol{S}$ by $\boldsymbol{R}$ (cf. Section 1.3) when applying the procedure.

If a consistent closed form estimate of $\boldsymbol{\Psi}$ is available, a corresponding consistent estimate of $\boldsymbol{\Lambda}$ may be obtained from (4.65). A result due to Albert (1944) was applied by Ihara and Kano (1986) to provide consistent closed form estimates of the elements of $\boldsymbol{\Psi}$. Their approach involves the choice of partitions of the covariance matrix $\boldsymbol{S}$. Effective stepwise approaches for the choice of suitable partitions have been described by Kano (1990) and Cudeck (1991).

The identification conditions in (4.64) are applied for computational convenience and in general do not provide interpretable solutions. Prior to interpretation it is convenient to apply a transformation, $\hat{\boldsymbol{\Lambda}}^* = \hat{\boldsymbol{\Lambda}}\boldsymbol{T}$, so that elements of $\hat{\boldsymbol{\Lambda}}^*$ exhibit a simple structure with a substantial number of elements close to zero, and the remainder relatively large in absolute value. In orthogonal rotation, $\boldsymbol{T}$ is orthogonal and the assumption of uncorrelated factors in (4.63) is retained. In oblique rotation, $\boldsymbol{T}$ only satisfies $\mathrm{Diag}(\boldsymbol{T}\boldsymbol{T}')^{-1} = \boldsymbol{I}$ and factors are permitted to become correlated. A commonly used orthogonal rotation procedure is Kaiser's (1958) Varimax method, and Jennrich and Sampson's (1966) Direct Quartimin procedure is frequently used for oblique rotation. Methods for estimating standard errors of rotated factor loadings have been given by Jennrich and coauthors in a sequence of papers starting with Archer and Jennrich (1973) and culminating with Jennrich and Clarkson (1980). A recent review and discussion of the application of these methods for estimating standard errors of rotated factor loading has been given by Cudeck and O'Dell (1994).

In restricted factor analysis (Jöreskog 1969) linear restrictions are imposed on elements of $\boldsymbol{\Lambda}$, $\boldsymbol{\Phi}$, or $\boldsymbol{\Psi}$ in (4.62). Typically these restrictions involve the specification of zero elements of $\boldsymbol{\Lambda}$ or $\boldsymbol{\Phi}$ or the specification of equal elements of $\boldsymbol{\Lambda}$, $\boldsymbol{\Phi}$, or $\boldsymbol{\Psi}$. Restrictions are chosen so that the remaining parameters are identified, and the discrepancy function is minimized with respect to all free parameters simultaneously using methods of the type described in Section 3.

4.2 Structural Equation Models

Structural equation modeling with latent variables had its foundations in the work of Jöreskog (1973, 1977), Keesling (1972), and Wiley (1973). Much of the subsequent development of the model was due to Jöreskog who gave the model and corresponding computer programs the name of LISREL (LInear Structural RELations). There has been a very widely used sequence of LISREL programs, the current version being LISREL 8 (Jöreskog and Sörbom 1993).

The LISREL model is a extension of the factor analysis model (4.61) to incorporate linear structural relations among factors. The vector of common factors $\boldsymbol{z}$ is subdivided into two subvectors,

$$\boldsymbol{z} = \begin{pmatrix} \boldsymbol{z}_y \\ \boldsymbol{z}_x \end{pmatrix} \tag{4.67}$$

and the $m_y \times 1$ vector $\boldsymbol{z}_y$ and $m_x \times 1$ vector $\boldsymbol{z}_x$ are required to satisfy the linear structural relations

$$\boldsymbol{z}_y = \boldsymbol{B}_y \boldsymbol{z}_y + \boldsymbol{\Gamma} \boldsymbol{z}_x + \boldsymbol{e} \tag{4.68}$$

where $E(\boldsymbol{z}_x) = \boldsymbol{0}$, $E(\boldsymbol{e}) = \boldsymbol{0}$ and $Cov(\boldsymbol{z}_x, \boldsymbol{e}') = \boldsymbol{0}$. Elements of the $m_y \times m_y$ matrix $\boldsymbol{B}_y$ and the $m_y \times m_x$ matrix $\boldsymbol{\Gamma}$ are regression weights, also known as path coefficients. Diagonal elements of $\boldsymbol{B}_y$ are zero. The elements of $\boldsymbol{z}_x$ do not depend on any other variables and are said to be exogenous (outside the system). Elements of $\boldsymbol{z}_y$ depend on $\boldsymbol{z}_x$ through $\boldsymbol{\Gamma}$ and also on other elements of $\boldsymbol{z}_y$ through $\boldsymbol{B}_y$. They are said to be endogenous (inside the system). The $m_y \times 1$ vector variate $\boldsymbol{e}$ represents regression errors. The exogenous factor covariance matrix will be represented by $Cov(\boldsymbol{z}_x, \boldsymbol{z}_x') = \boldsymbol{\Phi}_x$ and the error covariance matrix by $\boldsymbol{\Psi} = Cov(\boldsymbol{e}, \boldsymbol{e}')$.

Manifest variables that measure the endogenous factors are represented by $\boldsymbol{y}$ and those that measure exogenous factors by $\boldsymbol{x}$. The corresponding factor analysis models are:

$$\boldsymbol{y} = \boldsymbol{\mu}_y + \boldsymbol{\Lambda}_y \boldsymbol{z}_y + \boldsymbol{u}_y \tag{4.69}$$
$$\boldsymbol{x} = \boldsymbol{\mu}_x + \boldsymbol{\Lambda}_x \boldsymbol{z}_x + \boldsymbol{u}_x \tag{4.70}$$

where the unique vector variates $\boldsymbol{u}_y$ and $\boldsymbol{u}_x$ have null expected values and covariance matrices $Cov(\boldsymbol{u}_y, \boldsymbol{u}_y') = \boldsymbol{\Theta}_{u_y}$ and $Cov(\boldsymbol{u}_x, \boldsymbol{u}_x') = \boldsymbol{\Theta}_{u_x}$. It is assumed that the vector variates $\boldsymbol{z}_x$, $\boldsymbol{e}$, $\boldsymbol{u}_x$, and $\boldsymbol{u}_y$ are mutually independently distributed. In the majority of applications, $\boldsymbol{\Psi}$, $\boldsymbol{\Theta}_{u_y}$, and $\boldsymbol{\Theta}_{u_x}$ are diagonal matrices, although they can be allowed to be nondiagonal to handle correlated errors. Typically, also, many elements of $\boldsymbol{\Lambda}_y$, $\boldsymbol{\Lambda}_x$, $\boldsymbol{B}_y$, and $\boldsymbol{\Gamma}$ are zero.

The model in (4.69), (4.70), and (4.68) generates the covariance structure

$$\boldsymbol{\Sigma} = \begin{pmatrix} \boldsymbol{\Sigma}_{yy} & \boldsymbol{\Sigma}_{yx} \\ \boldsymbol{\Sigma}_{xy} & \boldsymbol{\Sigma}_{xx} \end{pmatrix}$$

where

$$\boldsymbol{\Sigma}_{yy} = \boldsymbol{\Lambda}_y(\boldsymbol{I} - \boldsymbol{B}_y)^{-1}(\boldsymbol{\Gamma}\boldsymbol{\Phi}_x\boldsymbol{\Gamma}' + \boldsymbol{\Psi})(\boldsymbol{I} - \boldsymbol{B}_y')^{-1}\boldsymbol{\Lambda}_y' + \boldsymbol{\Theta}_{u_y} \tag{4.71}$$
$$\boldsymbol{\Sigma}_{yx} = \boldsymbol{\Lambda}_y(\boldsymbol{I} - \boldsymbol{B}_y)^{-1}\boldsymbol{\Gamma}\boldsymbol{\Phi}_x\boldsymbol{\Lambda}_x' = \boldsymbol{\Sigma}_{xy}' \tag{4.72}$$
$$\boldsymbol{\Sigma}_{xx} = \boldsymbol{\Lambda}_x\boldsymbol{\Phi}_x\boldsymbol{\Lambda}_x' + \boldsymbol{\Theta}_{u_x} \tag{4.73}$$

Parameter estimates and goodness of fit tests may be obtained using methods described in Section 3.

The LISREL model is rigidly compartmentalized. There are four different matrices of regression weights: $\boldsymbol{\Lambda}_y$ contains regression weights of manifest variables on endogenous factors, $\boldsymbol{\Lambda}_x$ contains regression weights of manifest variables on exogenous factors, $\boldsymbol{B}_y$ contains regression weights of endogenous factors on other endogenous factors, and $\boldsymbol{\Gamma}$ contains regression weights of endogenous factors on exogenous factors. There are an additional four covariance matrices: $\boldsymbol{\Phi}_x$ is the covariance matrix for exogenous factors, $\boldsymbol{\Psi}$ is the covariance matrix for regression errors, $\boldsymbol{\Theta}_y$ is the covariance matrix for unique variances corresponding to endogenous factors, and $\boldsymbol{\Theta}_x$ is the covariance matrix for unique variances corresponding to exogenous factors.

Another characteristic of the LISREL model is that all manifest variables are endogenous if considered in the system of all variables. It can be seen from (4.69) and (4.70) that both $\boldsymbol{y}$ and $\boldsymbol{x}$ are dependent variables in regression equations on the facors $\boldsymbol{z}_y$ and $\boldsymbol{z}_x$ and never act as explanatory variables. An exogenous manifest variable can, however, be handled circuitously in the LISREL framework by introducing a latent variable that duplicates the manifest variable:

$$x_i - \mu_{x_i} = 1 \times z_{x_j}$$

Here the ith row of $\boldsymbol{\Lambda}_x$ has its jth element equal to one and all other elements equal to zero and the jth diagonal element of $\boldsymbol{\Theta}_{u_x}$ is null (cf. Jöreskog 1977, Section 3.2, $\boldsymbol{\Lambda}_x = \boldsymbol{I}$, $\boldsymbol{\Theta}_{u_x} = \boldsymbol{\Theta}_\delta = \boldsymbol{0}$). Although x_i is formally endogenous, it is in fact equal to an exogenous variable z_{x_j} that is treated as a latent variable.

The LISREL model first arose as an amalgamation of the factor analysis model and the structural equations model (4.68). Subsequently it was found (McArdle 1979) that the LISREL program could be employed with many of the parameter matrices fixed (at $\boldsymbol{I}$ or $\boldsymbol{0}$) and the rest treated in an unconventional manner to handle models that did not appear to fit the original framework. Other simpler models that appear to be special cases of LISREL but also include LISREL as a special case were then proposed (Bentler and Weeks 1980; McArdle and McDonald 1984; Kiiveri 1987).

The Reticular Action Model (RAM) was proposed by McArdle and McDonald (1984) and essentially the same model, but with emphasis on the structure of $\boldsymbol{\Sigma}^{-1}$, was suggested independently by Kiiveri (1987). It involves a single matrix of regression weights and a single covariance matrix. An equivalent model, but with two matrices of regression weights instead of one, has been suggested by Bentler and Weeks (1980). The RAM considers a single set of variables that includes manifest variables and latent variables that influence each other in a linear manner. The treatment of the model given here differs slightly from that of McArdle and McDonald (1984) and Kiiveri (1987) in that error variates are included in the single set of variables under consideration. McArdle and McDonald's (1984) original RAM will be refered to as the "compact RAM."

Let t be the total number of variables in the system including manifest variables, factors, and error variables. Suppose that $\boldsymbol{v}$ is a $t \times 1$ vector consisting of *all* variables in the system. All elements of $\boldsymbol{v}$ are centered so that $E(\boldsymbol{v}) = \boldsymbol{0}$. All regression weights of some elements of $\boldsymbol{v}$ on other elements are contained in a single $t \times t$ square nonsymmetric matrix $\boldsymbol{B}_v$. Diagonal elements of $\boldsymbol{B}_v$ will be zero and some rows of $\boldsymbol{B}_v$ will be null. The null rows correspond to variables in $\boldsymbol{v}$ that do not depend on any other variables and are consequently exogenous. Let $\boldsymbol{v}_x$ be a $t \times 1$ vector formed from $\boldsymbol{v}$ by replacing all elements corresponding to non-null rows of $\boldsymbol{B}_v$ by zeros. Thus $\boldsymbol{v}_x$ consists of exogenous variables (including error terms) in $\boldsymbol{v}$ with endogenous variables in $\boldsymbol{v}$ replaced by zeros.

The comprehensive RAM data model (Browne and Mels 1990) is

$$\boldsymbol{v} = \boldsymbol{B}_v \boldsymbol{v} + \boldsymbol{v}_x \tag{4.74}$$

where $Cov(\boldsymbol{v}_x, \boldsymbol{v}_x') = \boldsymbol{\Phi}_v$. Since some elements of $\boldsymbol{v}_x$ are constants, some rows and corresponding columns of $\boldsymbol{\Phi}_v$ will be null. In the compact RAM, error terms are excluded from $\boldsymbol{v}$ and replace some of the null elements of $\boldsymbol{v}_x$.

The RAM data model in (4.74) generates the following structure for the covariance matrix of all variables, manifest and latent, in the system:

$$\boldsymbol{\Upsilon} = Cov(\boldsymbol{v}, \boldsymbol{v}') = (\boldsymbol{I} - \boldsymbol{B}_v)^{-1}\boldsymbol{\Phi}_v(\boldsymbol{I} - \boldsymbol{B}_v')^{-1} \tag{4.75}$$

Non-null elements of $\boldsymbol{\Phi}_v$ are variances and covariances of exogenous variables in $\boldsymbol{v}$ and will all be present in the same positions in $\boldsymbol{\Upsilon}$. These may be referred to inherent variances and covariances as they are inherent in the definition of the exogenous variables. Zero elements of $\boldsymbol{\Phi}_v$ are replaced in $\boldsymbol{\Upsilon}$ by variances and covariances of endogenous variables as implied by the data model (4.74). These may be referred to as implied variances and covariances.

Variances and covariances of manifest variables are included in $\boldsymbol{\Upsilon}$. These are extracted from $\boldsymbol{\Upsilon}$ in the following manner to define the covariance structure for manifest variables. Let p be the number of manifest variables in $\boldsymbol{v}$. Suppose that $\boldsymbol{y}$ is a $p \times t$ selection matrix, with rows selected from the $t \times t$ identity matrix, $\boldsymbol{I}_t$, such that the elements of

$$\boldsymbol{v}_m = \boldsymbol{J}\boldsymbol{v} \tag{4.76}$$

are all manifest variables. For ease of exposition, it will be assumed here that variables in $\boldsymbol{v}$ are ordered so that the first p elements are manifest variables. Then $\boldsymbol{J}$ consists of the first p rows of $\boldsymbol{I}_t$. The covariances structure for the manifest variables may then be expressed as

$$\boldsymbol{\Sigma} = Cov(\boldsymbol{v}_m , \boldsymbol{v}_m') = \boldsymbol{J}\boldsymbol{\Upsilon}\boldsymbol{J}' = \boldsymbol{J}(\boldsymbol{I} - \boldsymbol{B}_v)^{-1}\boldsymbol{\Phi}_v(\boldsymbol{I} - \boldsymbol{B}_v')^{-1}\boldsymbol{J}' \tag{4.77}$$

A similar covariance structure holds for the compact RAM.

Since $\boldsymbol{J}$ is a fixed matrix, there are only two matrices with free elements, $\boldsymbol{B}_v$ and $\boldsymbol{\Phi}_v$, in the RAM covariance structure (4.77) as opposed to the eight matrices with free elements in the LISREL covariance structure (4.71)–(4.73). The two models are, however, equivalent. Clearly any submodel in RAM may be expressed as a LISREL model by defining $\boldsymbol{\Gamma} = \boldsymbol{0}, \boldsymbol{\Phi}_x = \boldsymbol{0}, \boldsymbol{\Lambda}_y = \boldsymbol{J}, \boldsymbol{B}_y = \boldsymbol{B}_v, \boldsymbol{\Psi} = \boldsymbol{\Phi}_v$ and $\boldsymbol{\Theta}_{u_y} = \boldsymbol{0}$ in (4.71) and omitting (4.72) and (4.73). Conversely, LISREL is a special case of RAM if $\boldsymbol{v}$ and $\boldsymbol{v}_x$ are defined as

$$\boldsymbol{v} = \begin{bmatrix} \boldsymbol{y} \\ \boldsymbol{x} \\ \boldsymbol{z}_y \\ \boldsymbol{z}_x \\ \boldsymbol{e} \\ \boldsymbol{u}_y \\ \boldsymbol{u}_x \end{bmatrix}, \quad \boldsymbol{v}_x = \begin{bmatrix} \boldsymbol{0} \\ \boldsymbol{0} \\ \boldsymbol{0} \\ \boldsymbol{z}_x \\ \boldsymbol{e} \\ \boldsymbol{u}_y \\ \boldsymbol{u}_x \end{bmatrix} \tag{4.78}$$

so that $\boldsymbol{B}_v$ and $\boldsymbol{\Phi}_v$ in (4.77) become

$$\boldsymbol{B}_v = \begin{bmatrix} \boldsymbol{0} & \boldsymbol{0} & \boldsymbol{\Lambda}_y & \boldsymbol{0} & \boldsymbol{0} & \boldsymbol{I} & \boldsymbol{0} \\ \boldsymbol{0} & \boldsymbol{0} & \boldsymbol{0} & \boldsymbol{\Lambda}_x & \boldsymbol{0} & \boldsymbol{0} & \boldsymbol{I} \\ \boldsymbol{0} & \boldsymbol{0} & \boldsymbol{B}_y & \boldsymbol{\Gamma} & \boldsymbol{I} & \boldsymbol{0} & \boldsymbol{0} \\ \boldsymbol{0} & & & \cdots & & & \boldsymbol{0} \\ \vdots & & & & & & \vdots \\ \boldsymbol{0} & & & \cdots & & & \boldsymbol{0} \end{bmatrix} \tag{4.79}$$

$$\boldsymbol{\Phi}_v = \begin{bmatrix} 0 & & & & & & \\ 0 & 0 & & & & & \\ 0 & 0 & 0 & & & & \\ 0 & 0 & 0 & \boldsymbol{\Phi}_x & & & \\ 0 & 0 & 0 & 0 & \boldsymbol{\Psi} & & \\ 0 & 0 & 0 & 0 & 0 & \boldsymbol{\Theta}_{u_y} & \\ 0 & 0 & 0 & 0 & 0 & 0 & \boldsymbol{\Theta}_{u_x} \end{bmatrix} \tag{4.80}$$

The eight LISREL matrices have been replaced by two large sparse RAM matrices. In implementing RAM in a computer program, sparse matrix algebraic methods are used to avoid storage of null elements of matrices and avoid multiplications by zero. The sparse nature of $\boldsymbol{B}_v$ and $\boldsymbol{\Phi}_v$ does not lead to computational inefficiency if appropriate steps are taken. It is not necessary to store the selection matrix $\boldsymbol{J}$ at all, since $\boldsymbol{J\Upsilon J}'$ is notation for the leading $p \times p$ submatrix of $\boldsymbol{\Upsilon}$, and $\boldsymbol{J}$ has been used here primarily to show relationships between RAM and LISREL. Some of the null elements in (4.79) and (4.80) may be allowed to be non-null to define models that would not be permissible in LISREL if it were used as originally intended.

Some structural equation models may not be identified in that different parameter values will yield the same value of $\boldsymbol{\Sigma}$. The investigation of the identification of a model can involve some complex algebra. Bekker, Merckens, and Wansbeek (1994) have recently provided a comprehensive treatment of identification, accompanied by a diskette with programs to carry out the necessary algebra by computer for the model of interest.

There is, however, a simple necessary condition for identification which should be taken into account when specifying a structural equation model. This is that the scale of every latent variable should be specified, either by fixing the variance of the latent variable or by fixing an appropriate path coefficient. It is usual to fix these parameters to be equal to one, although other nonzero values would be suitable. For simplicity of exposition it will be assumed throughout that the fixed values are one. In terms of the RAM, for every latent variable, $v_j, j > p$, either the jth diagonal element of $\boldsymbol{\Upsilon}$ should be fixed at one, or at least one element in the jth column of $\boldsymbol{B}_v$ should be fixed at one. Path coefficients are regression weights, and consequently depend on the scales of both the dependent variable and the explanatory variable. If path coefficients are to be compared, therefore, there are often advantages in standardizing path coefficients by fixing latent variable variances at one (cf. Bollen 1989, pp. 123–126). This avoids the difficulties in interpretation when latent variable variances are allowed to assume arbitrary values that depend on the arbitrary choice of a path coefficient to set equal to one.

In many applications of structural equation modeling, the scales of manifest variables are arbitrary. In situations like this it is advantageous to adopt the RAM for the correlation matrix so that [cf. equation (4.5)] the covariance structure becomes

$$\boldsymbol{\Sigma} = \boldsymbol{D}_{\sigma_d} \boldsymbol{J\Upsilon J}' \boldsymbol{D}'_{\sigma_d} \tag{4.81}$$

where

$$\mathrm{Diag}[\boldsymbol{P}] = \mathrm{Diag}[\boldsymbol{J\Upsilon J}'] = \boldsymbol{I} \tag{4.82}$$

Again (4.82) involves fixing diagonal elements of $\boldsymbol{\Upsilon}$ at one.

Fixing a diagonal element of $\boldsymbol{\Upsilon}$ at one is straightforward when it is an inherent variance and equal to a diagonal element of $\boldsymbol{\Phi}_v$. It will then be a parameter in the model. The problem is less straightforward if the diagonal element of $\boldsymbol{\Upsilon}$ is an implied variance as this then is a nonlinear function of elements and $\boldsymbol{B}_v$ and $\boldsymbol{\Phi}_v$. It can, however, be handled effectively using the procedure for constrained estimation described in Section 3, with a typical iteration given by (4.60). An alternative reparameterization approach (McDonald, Parker, and Ishizuka 1993) may be employed in situations where the model is recursive, that is $\boldsymbol{B}_v$ can be chosen to be lower triangular by means of a suitable ordering of elements of $\boldsymbol{v}$.

It is convenient to describe a structural equation model by means of a path diagram. A very well known early application of structural equation modeling by Jöreskog and Sörbom (1986) will be employed as an example, but modified slightly, in order to illustrate recent developments. A study of the stability of attitudes over time was conducted by Wheaton, Muthén, Alwin, and Summers (1977). Attitude scales measuring Anomia and Powerlessness were administered to a sample consisting of 932 persons in 1967 and in 1971. A Socioeconomic Index and the number of years of schooling were also recorded. Jöreskog and Sörbom (1986) provided a model for this study. The manifest variables Anomia (ANOMIA67, ANOMIA71) and Powerlessness (POWERLS67, POWERLS71), were regarded as indicators of the latent variable Alienation (ALNTN67, ALNTN71). The Socioeconomic Index (SEI) and number of years of schooling (EDUCTN) were regarded as indicators of the latent variable socioeconomic status (SES). The model represented in Figure 1 is an adaption of Jöreskog and Sörbom (1986, Fig. III.4). ALNTN67 depends only on SES while ALNTN71 depends on SES and ALNTN67.

Manifest variables are enclosed in rectangles or squares (e.g. ANOMIA67) and latent variables are enclosed in ellipses or circles (e.g. ALNTN67). Error terms (e.g. E1, Z1) are regarded as latent variables that emit single arrows. The total number of variables in the system is $t = 17$ of which $p = 6$ are manifest. Single-headed arrows represent path coefficients or regression weights while curved double-headed arrows represent variances if both arrowheads touch the same object and covariances if they touch different objects. There is a one to one correspondence between the path diagram and the RAM. Each object (rectangle or circle) represents an element of $\boldsymbol{v}$, or variable in the system. Manifest variables (squares) are from the first p elements of $\boldsymbol{v}$. Each single-headed arrow in the path diagram represents a non-null element of $\boldsymbol{B}_v$; if no number is assigned to the arrow (e.g. ANOMIA67 $\leftarrow$ ALNTN67) the element is a free parameter, and if a number is assigned to the arrow (e.g. 1.0 for ANOMIA67 $\leftarrow$ E1) the element of $\boldsymbol{B}_v$ is fixed at that value. The variable that emits the arrow represents the column of $\boldsymbol{B}_v$, and the variable that receives the arrow represents the row. Each double-headed arrow represents an element of the symmetric matrix $\boldsymbol{\Phi}_v$; if the two arrowheads touch the same object (e.g. E1 $\leftrightarrow$ E1) a diagonal element of $\boldsymbol{\Phi}_v$ is represented, and if the two arrowheads touch different objects (e.g. E1 $\leftrightarrow$ E3) the corresponding two equal off-diagonal elements of $\boldsymbol{\Phi}_v$ are represented. Again, numbers assigned to arrows (e.g. 1.0 for SES $\leftrightarrow$ SES) represent fixed elements of $\boldsymbol{\Phi}_v$.

As mentioned earlier, the scale of each latent variable has to be set for identification purposes. In terms of the path diagram this means that either a number must be assigned

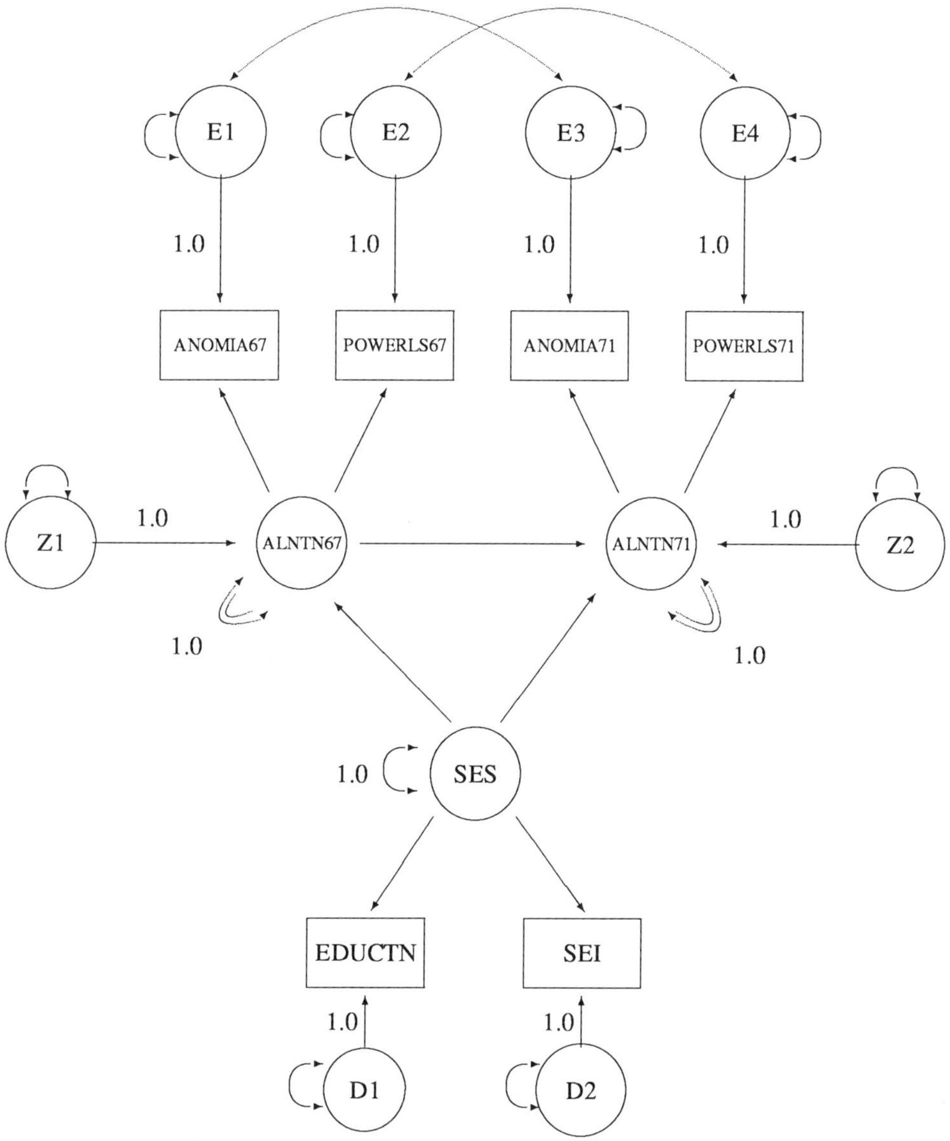

FIGURE 1. Path diagram for the Wheaton et al. (1977) study of stability of alienation.

to the two-headed arrow with both ends on the circle or a number must be assigned to a single-headed arrow emitted by the circle. A circle that only emits single-headed arrows and receives none (e.g. SES, E1) represents an exogenous variable. Constraining its variance to one implies fixing a diagonal element of $\boldsymbol{\Phi}_v$ to one. On the other hand a circle that receives at least one arrow (e.g. ALNTN67, ALNTN71) represents an endogenous variable and its variance is implied. Constraining an implied variance to one involves the imposition of nonlinear constraints on parameters. In the path diagram the two implied variances constrained to one (for ALNTN67 and ALNTN71) are represented by double-lined arrows and the single inherent variance constrained to one (SES) by a single arrow. The variances of all common factors (circles that emit more than one arrow) have been fixed for identification purposes. This facilitates the assessment of relative dependencies (cf. Bollen 1989, p. 124) of different factors on manifest variables. Following common practice, the path coefficient has been fixed to one for error terms (circles that emit a single arrow).

In the compact RAM, the distinction between an error and the associated dependent variable is blurred. An error is regarded as that part of the dependent variable that is not accounted for by the explanatory variables, so that error variances are regarded as partial variances and error covariances as partial covariances. Consequently, errors are omitted from the path diagram and error variance or covariance paths are assigned to the associated dependent variables (cf. McDonald 1985, sec. 4.2).

In order to allow for possibly arbitrary differences in the scales of the manifest variables (ANOMIA, POWERLS, EDUCTN, SEI) their variances will also be scaled to one by using the scaled form of the RAM given in (4.81). This does involve some loss of information as differences in variances between the two repeated applications of the Anomia scale could be of interest, as could differences between the two repeated applications of the Powerlessness scale. The sample correlation matrix is presented in Table 1.

Table 1. Stability of alienation: sample correlation matrix

	ANOMIA67	POWRLS67	ANOMIA71	POWRLS71	EDUCTN	SEI
ANOMIA67	1.00					
POWRLS67	0.66	1.00				
ANOMIA71	0.56	0.47	1.00			
POWRLS71	0.44	0.52	0.67	1.00		
EDUCTN	-0.36	-0.41	-0.35	-0.37	1.00	
SEI	-0.30	-0.29	-0.29	-0.28	0.54	1.00

SOURCE. Wheaton, Muthén, Alwin and Summers (1977).

It is permissible to fit the covariance structure (4.81) to a correlation matrix since all scale changes are absorbed by the nuisance parameters σ_d in the diagonal of $\boldsymbol{D}_{\sigma_d}$. Maximum Wishart Likelihood estimates are shown in Table 2, together with associated standard error estimates and 90% one-at-a-time confidence intervals.

In the case of error variances, the confidence intervals are asymmetric. If $\hat{\psi}_i$ is the variance estimate, a symmetric confidence interval is obtained around $\ln(\hat{\psi}_i^{-1} - 1)$ and trans-

Table 2. Maximum likelihood estimates of free parameters

Path			Point Estimate	Standard Error	90% Confidence Interval
ANOMIA67	$\leftarrow$	ALNTN67	0.77	0.025	(0.73;0.82)
POWRLS67	$\leftarrow$	ALNTN67	0.85	0.026	(0.81;0.89)
ANOMIA71	$\leftarrow$	ALNTN71	0.81	0.026	(0.76;0.85)
POWRLS71	$\leftarrow$	ALNTN61	0.83	0.027	(0.79;0.88)
SEI	$\leftarrow$	SES	0.64	0.030	(0.59;0.69)
EDUCTN	$\leftarrow$	SES	0.84	0.032	(0.79;0.89)
ALNTN67	$\leftarrow$	SES	-0.56	0.035	(-0.62;-0.51)
ALNTN71	$\leftarrow$	ALNTN67	0.57	0.041	(0.50;0.63)
ALNTN71	$\leftarrow$	SES	-0.21	0.045	(-0.28;-0.13)
E1	$\leftrightarrow$	E1	0.40	0.039	(0.34;0.47)
E1	$\leftrightarrow$	E3	0.13	0.026	(0.09;0.18)
E2	$\leftrightarrow$	E2	0.27	0.044	(0.21;0.36)
E2	$\leftrightarrow$	E4	0.04	0.027	(-0.01;0.08)
E3	$\leftrightarrow$	E3	0.35	0.042	(0.29;0.43)
E4	$\leftrightarrow$	E4	0.31	0.044	(0.24;0.39)
D1	$\leftrightarrow$	D1	0.29	0.054	(0.22;0.39)
D2	$\leftrightarrow$	D2	0.59	0.039	(0.53;0.66)
Z1	$\leftrightarrow$	Z1	0.68	0.039	(0.62;0.74)
Z2	$\leftrightarrow$	Z2	0.50	0.033	(0.45;0.56)

formed back (cf. Browne 1982, formula (1.6.41)). This prevents the confidence interval ever covering inadmissible values of the parameter (i.e. $\psi < 0$ or $\psi > 1$).

The comparison of confidence intervals makes it less tempting to draw attention to trivial differences (although this does not provide statistical tests with specified confidence coefficients). The influence of SES on ALNTN67 is negative and larger in magnitude than the influence of SES on ALNTN71. Also the influence of ALNTN67 on ALNTN71 is stronger than that of SES. The use of standardized path coefficients throughout makes the comparison of relative influences possible. It can be seen that EDUCTN is a somewhat stronger marker of SES than SEI in this investigation.

The value of the likelihood ratio test statistic is 4.74, corresponding to an exceedance probability of 0.315, assuming a chi-squared distribution with four degrees of freedom. Clearly the model cannot be rejected. This does not mean that the model is correct, particularly since there are only four degrees of freedom and the power (cf. Satorra and Saris 1985) of the test will be low. It is instructive to make use of a measure of goodness of fit of the model.

Many measures of goodness of fit are available (e.g. Bollen 1989, pp. 269–279). A recent collection of various authors' views on the assessment of fit may be found in Bollen and Long (1993). Here an approach, due to Steiger and Lind (1980) and treated in detail in Browne and Cudeck (1993), will be applied. Measures of goodness of fit of a model based on a sample are subject to variation. Consequently, it is helpful to consider a measure of goodness of fit of the model in the population and to make the use of a confidence interval on this measure to avoid losing sight of sampling variation.

Under assumptions of population drift (Section 2.1) and a correctly specified discrepancy function (Section 2.2) the asymptotic distribution of the test statistic, $n \times \hat{F}$, is noncentral chi-squared with a noncentrality parameter that may be expressed (Satorra and Saris 1985, p. 85; Steiger, Shapiro, and Browne 1985, theorem 1) as $\lambda = n \times F_0$, where F_0 is the minimum value of the discrepancy function when the model is fitted to the population covariance matrix Σ_0. A measure of lack of fit of the model relative to its degrees of freedom, d, proposed by Steiger and Lind (1980), is the root mean square error of approximation

$$\varepsilon_a = \sqrt{\frac{F_0}{d}} \tag{4.83}$$

A value of $\varepsilon_a \leq 0.05$ is regarded as indicating close fit of the model.

A confidence interval on ε_a may be obtained assuming a noncentral chi-squared distribution for $n \times \hat{F}$ (Steiger 1990; Browne and Cudeck 1993). In the present example, the 90% confidence interval is (0;0.01). Since the lower bound of this confidence interval is equal to zero, the null hypothesis, $H_0 : \varepsilon_a = 0$, that the model fits perfectly is not rejected at the 5% level, as was seen earlier. More important, however, is the fact that the upper limit of this confidence interval is less than 0.05. Since this would happen with a probability of at most 0.05 if the fit of the model in the population were not close, $\varepsilon \geq 0.05$, it can reasonably be assumed that the model fits closely in the population.

Close fit of a model does not necessarily imply that the model is a good one (Cliff, 1983). It should be borne in the mind that there may be other equivalent models that fit as closely. Any choice among equivalent models has to rely on the judgment of the investigator. Bekker, Merckens, and Wansbeck (1994) have provided means for investigating the equivalence of models. Some simple sufficient conditions for models to be equivalent have been given by Lee and Hershberger (1990), and these can be used to generate equivalent models. In an investigation of published studies in different substantive areas, MacCallum, Uchino, Wegener, and Fabrigar (1993) have found that equivalent models are often disregarded and that sometimes equivalent models that are as plausible as the original model can be found.

An example of a model that is equivalent to the model in Figure 1 is shown in Figure 2. The directions of the paths ALNTN67 $\leftarrow$ SES and ALNTN71 $\leftarrow$ SES have been reversed to SES $\leftarrow$ ALNTN67 and SES $\leftarrow$ ALNTN71 and appropriate changes to error terms have been made.

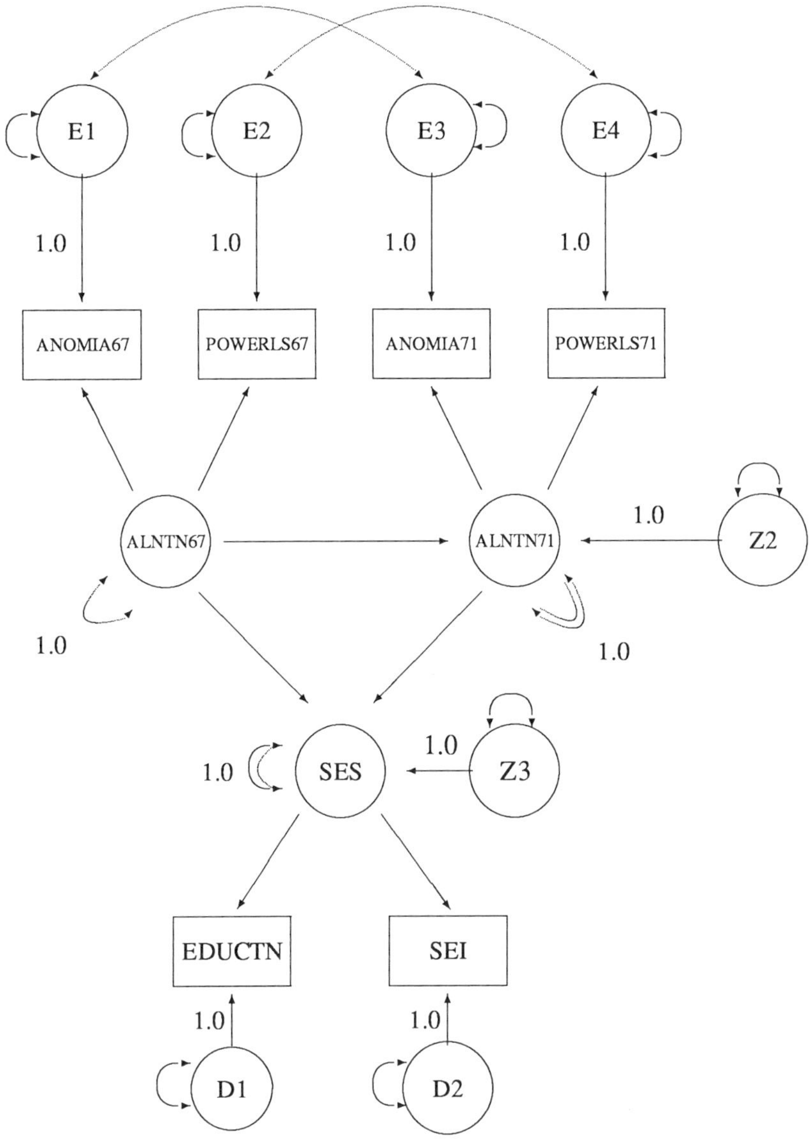

FIGURE 2. Path diagram of an equivalent model for the stability of alienation.

The minimum WML discrepancy function value for the equivalent model is 0.00509029 and is identical to all eight figures reported to that for the original model. Maximum likelihood estimates of new parameters or parameters that have changed in value are shown in Table 3.

The two models have quite different interpretations but have exactly the same goodness

Table 3. Equivalent model: changed parameter estimates

Path			Point Estimate	Standard Error	90% Confidence Interval
SES	$\leftarrow$	ALNTN67	-0.38	0.056	(-0.47 ; -0.29)
ALNTN71	$\leftarrow$	ALNTN67	0.68	0.026	(0.64 ; 0.73)
SES	$\leftarrow$	ALNTN71	-0.27	0.056	(-0.36 ; -0.17)
Z3	$\leftrightarrow$	Z3	0.65	0.039	(0.58 ; 0.71)
Z2	$\leftrightarrow$	Z2	0.53	0.035	(0.47 ; 0.59)

of fit. No statistical means for choosing between the two models is possible. On the other hand, the second model where a state of mind (alienation) affects socioeconomic status is clearly less plausible than the original model where socioeconomic status affects the state of mind. Sound judgment would favor the original model.

In some situations, structural equation models that affect the manifest variable mean vector are employed. The RAM can be adapted to allow for mean vectors with a linear structure (McArdle and McDonald 1984, sec. 4). An example using ordered categorical outcomes is given in Section 5.6.

A great deal of attention has been devoted to structural equation modeling for latent variables since the first models were proposed in the early 1970s. An extensive bibliography has been provided by Austin and Wolfe (1991).

4.3 Other Mean and Covariance Structures

Although structural equation models with latent variables have dominated applications, there are other models that are useful in some circumstances. Some of these will be reviewed briefly here. Parameter estimates and goodness of fit test statistics may be computed using methods described in Section 3 (see also Browne and Du Toit 1992).

Models for Repeated Measurements Over Time

Consider situations where repeated measurements are made at equally spaced intervals over time on each of a number of subjects or experimental units. This occurs in panel studies where the same measurement is made repeatedly at long time intervals, and in learning experiments where measurements are made repeatedly at short intervals during a learning process. Two types of model will be considered for situations of this type: latent curve models and time series models.

Latent curve models Latent curve models are intended for situations where the repeated measurements appear to follow smooth and fairly similar trends over time for each subject but there are variations in initial performance, final performance, and rate of change between subjects. They originated with adaptations of principal component methods (Rao 1958; Tucker 1958) but now have developed to models that yield testable structures for the mean vector and covariance matrix (McArdle 1988; Meredith and Tisak 1990).

The latent curve data model is

$$\boldsymbol{y} = \boldsymbol{\Lambda z} + \boldsymbol{u} \tag{4.84}$$

where the $p \times 1$ vector variate $\boldsymbol{y}$ represents repeated measurements made at equal intervals over time on a typical person, each column of the $p \times m$ parameter matrix $\boldsymbol{\Lambda}$ represents p sequential values of a basis curve, the elements of the $m \times 1$ latent vector variate $\boldsymbol{z}$ represents unobservable weights on the basis curve for the person, and the $p \times 1$ vector variate $\boldsymbol{u}$ represents unobservable measurement errors. Let $E(\boldsymbol{z}) = \boldsymbol{\zeta}$, $Cov(\boldsymbol{z}, \boldsymbol{z}') = \boldsymbol{\Phi}$, and $Cov(\boldsymbol{u}, \boldsymbol{u}') = \boldsymbol{\Psi}$ and assume that $Cov(\boldsymbol{z}, \boldsymbol{u}') = \boldsymbol{0}$. It then follows that the mean and covariance structures for $\boldsymbol{y}$ are

$$\boldsymbol{\mu} = E(\boldsymbol{y}) = \boldsymbol{\Lambda\zeta} \tag{4.85}$$

$$Cov(\boldsymbol{y}, \boldsymbol{y}') = \boldsymbol{\Lambda\Phi\Lambda}' + \boldsymbol{\Psi} \tag{4.86}$$

The model differs from the factor analysis model (Section 4.1) in that a structure (4.85) is generated for the mean vector, but the covariance structure (4.86) is of the same form as factor analysis covariance structure (4.62). There is, however, a substantial difference in the manner in which the elements of $\boldsymbol{\Lambda}$ are regarded. In latent curve models the elements of each column of $\boldsymbol{\Lambda}$ are regarded as function values following some smooth trend, and there is no question of a simple structure with scattered null elements as in factor analysis.

If elements of columns of $\boldsymbol{\Lambda}$ are regarded as values of some unspecified function and are estimated separately, and $\boldsymbol{\Psi}$ is assumed to have a simple linear pattern such as a diagonal matrix, then the latent curve model fits in the LISREL or RAM frameworks and available programs may be employed (McArdle 1978; Meredith and Tisak 1990). If more than one column of $\boldsymbol{\Lambda}$ consists of free elements, however, there is a rotation problem that has not been satisfactorily resolved. Practical applications of this approach have therefore been either limited to a $p \times 1$ $\boldsymbol{\Lambda}$ representing a single basis curve, or to a $p \times 2$ $\boldsymbol{\Lambda}$ with a column of unit elements corresponding to an intercept, and a second column with free elements (Meredith and Tisak 1990).

One way of specifiying functional forms for the basis curves (Browne and Du Toit 1991; Browne 1993) is to assume that the elements of $\boldsymbol{\mu}$ follow a specified curve, such as a Gompertz, logistic, or exponential curve, and to apply a first order Taylor expansion to obtain a $\boldsymbol{\Lambda}$ whose columns are partial derivatives of the mean curve. The number of columns, m, of $\boldsymbol{\Lambda}$ then is equal to the number of parameters in the mean curve. Both $\boldsymbol{\zeta}$ and $\boldsymbol{\Lambda}$ are functions of the same m parameters. It has been found (Browne 1993) in practical examples that the specification of $\boldsymbol{\Psi}$ as a covariance matrix for a stationary autoregressive process is desirable.

Fixed mean curves with time series deviations An alternative model for repeated measurements assumes there is a common fixed curve and that individual deviations from this curve follow a common autoregressive time series with moving average residuals (Du Toit 1979; Browne and Du Toit 1992). The data model is of the form

$$\boldsymbol{y} = \boldsymbol{\mu} + \boldsymbol{e}$$

where

$$\boldsymbol{\mu} = \boldsymbol{f}(\boldsymbol{\vartheta}) \tag{4.87}$$

is a $p \times 1$ vector with the tth element given by $\mu_t = f(t, \vartheta)$ where f is a trend function such as a logistic function with parameter vector $\boldsymbol{\vartheta}$. The deviation vector variate, $\boldsymbol{e}$, is assumed to have elements generated by an ARMA(a, b) process with homogeneous autoregressive weights and homogeneous moving average weights but possibly nonhomogeneous white noise variances:

$$e_t - \alpha_1 e_{t-1} - \cdots - \alpha_a e_{t-a} = u_t - \beta_1 u_{t-1} - \cdots - \beta_b u_{t-b} \tag{4.88}$$

where the u_t are independently distributed white noise variates with null means. The diagonal covariance matrix of $\boldsymbol{u} = (u_1, \ldots, u_p)'$ will be represented by $\boldsymbol{D}_\psi$. If the white noise variances are assumed to be homogeneous then $\boldsymbol{D}_\psi = \psi \boldsymbol{I}$. Alternatively, the diagonal elements of $\boldsymbol{D}_\psi$ may be assumed to follow a parametric trend function.

The model generates the structure (4.87) for $\boldsymbol{\mu} = E(\boldsymbol{y})$ and the covariance structure

$$Cov(\boldsymbol{y}, \boldsymbol{y}') = \boldsymbol{\Sigma} = \boldsymbol{T}_\alpha^{-1}(\boldsymbol{J}_s'\boldsymbol{\Phi}\boldsymbol{J}_s + \boldsymbol{T}_\beta \boldsymbol{D}_\psi \boldsymbol{T}_\beta')\boldsymbol{T}_\alpha^{-1'} \tag{4.89}$$

where $\boldsymbol{J}_s'$ is a $p \times s$ matrix that consists of the first $s = \max(a, b)$ columns of $\boldsymbol{I}_p$, $\boldsymbol{\Phi}$ is an $s \times s$ initial state covariance matrix, $\boldsymbol{T}_\alpha$ is a lower triangular Toeplitz matrix with a unit main diagonal and a non-null subdiagonals below it, and $\boldsymbol{T}_\beta$ is similarly defined with b non-null subdiagonals. For example, if $p = 5$, $a = 2$, and $b = 1$ then

$$\boldsymbol{T}_\alpha = \begin{pmatrix} 1 & 0 & 0 & 0 & 0 \\ -\alpha_1 & 1 & 0 & 0 & 0 \\ -\alpha_2 & -\alpha_1 & 1 & 0 & 0 \\ 0 & -\alpha_2 & -\alpha_1 & 1 & 0 \\ 0 & 0 & -\alpha_2 & -\alpha_1 & 1 \end{pmatrix}, \quad \boldsymbol{T}_\beta = \begin{pmatrix} 1 & 0 & 0 & 0 & 0 \\ -\beta_1 & 1 & 0 & 0 & 0 \\ 0 & -\beta_1 & 1 & 0 & 0 \\ 0 & 0 & -\beta_1 & 1 & 0 \\ 0 & 0 & 0 & -\beta_1 & 1 \end{pmatrix} \tag{4.90}$$

If the ARMA process prior to the first observation is assumed to be the same as after the first observation and if white noise variances are assumed to be homogeneous then $\boldsymbol{\Phi}$ is a function (Du Toit 1969) of the α_i, β_j, and ψ. If, however, these assumptions are not made, the distinct elements of $\boldsymbol{\Phi}$ are treated as additional parameters to be estimated and the estimation process is simplified. A practical example of the application of this model may be found in Browne and Du Toit (1991).

A related covariance structure with homogeneous white noise variances and a null initial state covariance matrix in (4.89) was suggested by Anderson (1975, equation (5.2)). Another related covariance structure is the quasi-Wiener Simplex structure with nonhomogeneous autoregressive weights, nonhomogeneous moving average weights, nonhomogeneous white noise variances, and a null initial state covariance matrix (Jöreskog 1970).

Direct Product Covariance Structures

Because of the use of the word "factor," with different meanings, in factor analysis and in the analysis of variance, confusion can arise. In generalizability theory (Cronbach, Gleser, Nanda, and Rajaratnam 1972) it has become common practice to use the term "facet," introduced by Guttman (1954), instead of the analysis of variance "factor," and "condition" of a facet instead of "level" of a factor. This terminology will be adopted here to avoid any confusion with the factor analysis models considered in Section 4.1.

Consider a situation in which measurements are constructed according to a crossed two-facet design in which each condition of one facet, the T-facet, is combined with each condition of the other facet, the M-facet. In some situations the T-facet refers to traits and the M-facet to methods for measuring these traits. In other situations, the T-facet refers to attributes measured, and the M-facet to occasions on which measurements are taken. If there are t conditions of the T-facet and m conditions of the M-facet, the vector variate, $\boldsymbol{y}$, representing observed variables, consists of $p = t \times m$ elements. A typical element of $\boldsymbol{y}$, representing a measurement taken under condition r of the T-facet and condition s of the M-facet will be represented by $y(T_r, M_s)$. It will be assumed that the elements of $\boldsymbol{y}$ are arranged with conditions of the T-facet nested within conditions of the M-facet so that

$$\boldsymbol{y} = (y(T_1, M_1), y(T_2, M_1), \ldots, y(T_t, M_1), \ldots y(T_1, M_m), \ldots, y(T_t, M_m))$$

It is helpful to extract information concerning interrelationships of conditions within the T-facet as well as interrelationships of conditions within the M-facet from the $tm \times tm$ covariance matrix $\boldsymbol{\Sigma} = Cov(\boldsymbol{y}, \boldsymbol{y}')$. Swain (1975b) suggested the covariance structure

$$\boldsymbol{\Sigma} = \boldsymbol{\Sigma}_M \otimes \boldsymbol{\Sigma}_T \tag{4.91}$$

where the $\otimes$ indicates a right direct (Kronecker) product, $\boldsymbol{\Sigma}_M$ is a $m \times m$ positive definite parameter matrix representing covariances among conditions of the M-facet, and $\boldsymbol{\Sigma}_T$ is a $t \times t$ positive definite parameter matrix representing covariances among conditions of the T-facet.

In Browne (1984b) the direct product model was extended to make it independent of the scale of the measurements and to allow for measurement error. The following "composite direct product" model was proposed:

$$\boldsymbol{\Sigma} = \boldsymbol{D}_\zeta(\boldsymbol{P}_M \otimes \boldsymbol{P}_T + \boldsymbol{D}_\psi)\boldsymbol{D}_\zeta \tag{4.92}$$

with the option of the further decompositions

$$\boldsymbol{D}_\zeta = \boldsymbol{D}_{\zeta_T} \otimes \boldsymbol{D}_{\zeta_M} \tag{4.93}$$

and

$$\boldsymbol{D}_\psi = \boldsymbol{D}_{\psi_T} \otimes \boldsymbol{D}_{\psi_M} \tag{4.94}$$

The diagonal elements of the diagonal matrix $\boldsymbol{D}_\zeta$ represent scale factors that may be decomposed into T and M components in (4.93) and diagonal elements of $\boldsymbol{D}_\zeta^2\boldsymbol{D}_\psi$ are error variances. In (4.92) P_M and P_T are positive definite matrices with unit diagonal elements representing M-facet true score intercorrelations and T-facet true score intercorrelations respectively. The direct product model (4.91) is a special case of the composite direct product model (4.92) obtained by imposing (4.93) and taking D_ψ to be a null matrix.

Extensions and further information on the model are given in Cudeck (1988) and a systematic treatment of special cases is contained in Bagozzi and Yi (1991).

Circumplex Structures

In some situations a circular ordering of variables is possible such that variables that are close in this ordering are more highly correlated than variables that are distant. Guttman (1954) referred to a correlation matrix satisfying a pattern of this sort as a "circumplex."

Anderson (1962) proposed a circular stochastic process model for the circumplex. If error of measurement is taken into account, the covariance structure is of the form

$$\Sigma = D_\zeta(P + D_\psi)D_\zeta \tag{4.95}$$

where D_ζ is a diagonal scaling matrix, $D_\zeta^2 D_\psi$ represents error variances, and P represents a true score correlation matrix with elements satisfying the correlation function

$$\rho_{ij} = \frac{\cosh(\lambda|\theta_i - \theta_j| - \pi)}{\cosh(\lambda\pi)}$$

where θ_i is a polar angle in radians locating the true score for the ith test on the circumference of the circle and λ is a parameter governing the correlation at a separation of π radians. This correlation function allows only for positive correlation coefficients.

An alternative correlation function, based on a Fourier series, that allows for negative correlations is given in Browne (1992). Practical examples are provided in Browne and Du Toit (1992) and Browne (1992).

5 Mean and Covariance Structures with Nonmetric Dependent Variables

In the previous sections we have analyzed only dependent variables that are metric where the distances between the outcomes in a variable are known. Typical examples are variables such as income and professional experience measured in months. In this section we extend the mean and covariance structures to include dependent variables that are metric and/or metrically classified and/or one-sided or double-sided censored and/or dichotomous and/or ordered categorical. For definitions and examples of such variables, the reader is referred to Sections 7 and 8 of Chapter 3.

The mean- and covariance-structure models previously considered are extended in the following way. First, instead of considering dependent and explanatory variables simultaneously in an unconditional mean and covariance structure model, we consider conditional mean and covariance structure models where we distinguish explicitly between dependent and explanatory variables. Second, instead of formulating a conditional mean and covariance structure model for the observed dependent variables directly, we formulate such a model for a vector of unobserved dependent variables, all of which are assumed to be metrically scaled. Third, each of these unobserved variables is connected to a corresponding observed variable through one of the threshold models discussed in Sections 7 and 8 of Chapter 3 under the assumption of multivariate normality of the error terms in the mean and covariance structure model. These threshold models generate the usual linear regression, tobit, and probit models of Chapter 3 for each dependent variable. In addition, the assumption of multivariate normality and these threshold models generate polyserial and polychoric covariances and correlations as estimates of the covariances and correlations between the error terms of the conditional mean and covariance structure.

The inclusion of threshold models in mean and covariance structures generates additional identification and estimation problems. The additional identification problems are discussed in the subsection on threshold models. The estimation procedure proposed here is a combination of marginal maximum likelihood estimation and minimum distance estimation. (Both estimation methods are discussed more generally in Section 3 of Chapter 3.) Model specification, identification problems, and estimation are illustrated by an analysis of two-wave panel data of achievement in and attitude toward high school mathematics from the Longitudinal Study of American Youth (LSAY) (Miller, Hoffer, Suchner, Brown, and Nelson 1992).

5.1 Unconditional and Conditional Mean and Covariance Structures

Let $\boldsymbol{z}_i = (\boldsymbol{y}_i', \boldsymbol{x}_i')'$ be a $(p + r) \times 1$ random vector. The $p \times 1$ vector $\boldsymbol{y}_i$ denotes the vector of metrically scaled dependent variables and the $r \times 1$ vector $\boldsymbol{x}_i$ denotes a vector of explanatory variables that may be metrically scaled or zero/one coded dummy variables. An unconditional mean and covariance structure model is given by

$$E(\boldsymbol{z}_i) = \boldsymbol{\gamma}_u(\tilde{\boldsymbol{\vartheta}}) \,, \quad V(\boldsymbol{z}_i) = \boldsymbol{\Sigma}_u(\tilde{\boldsymbol{\vartheta}}) \tag{4.96}$$

where $\boldsymbol{\gamma}_u \sim (p + r) \times 1$ is the unconditional mean and $\boldsymbol{\Sigma}_u \sim (p + r) \times (p + r)$ is the unconditional covariance matrix of $\boldsymbol{z}_i$. Both the mean vector and the covariance matrix are functions of a $q_u \times 1$ vector $\tilde{\boldsymbol{\vartheta}}$ of parameters that are to be estimated from a (random) sample of data points $\boldsymbol{z}_i$, $i = 1, \ldots, n$. The parameter vector depends on the specification of the mean and covariance structure as shown in Sections 1 to 4. In addition, it has been assumed that the parameterization in $\tilde{\boldsymbol{\vartheta}}$ is second order identified, that is the equalities $\boldsymbol{\gamma}_u(\tilde{\boldsymbol{\vartheta}}_1) = \boldsymbol{\gamma}_u(\tilde{\boldsymbol{\vartheta}}_2)$ and $\boldsymbol{\Sigma}_u(\tilde{\boldsymbol{\vartheta}}_1) = \boldsymbol{\Sigma}_u(\tilde{\boldsymbol{\vartheta}}_2)$ imply that $\tilde{\boldsymbol{\vartheta}}_1 = \tilde{\boldsymbol{\vartheta}}_2$. No assumptions about the form of the joint distribution of $\boldsymbol{y}_i$ and $\boldsymbol{x}_i$ need to be made.

In a conditional mean and covariance structure model, the dependent and the explanatory variables are separated and the mean and covariance structure is defined for $\boldsymbol{y}_i$ conditional on $\boldsymbol{x}_i$. The model is now given by

$$E(\boldsymbol{y}_i|\boldsymbol{x}_i) = \boldsymbol{\gamma}(\boldsymbol{\vartheta}) + \boldsymbol{\Pi}(\boldsymbol{\vartheta})\boldsymbol{x}_i \,, \quad V(\boldsymbol{y}_i|\boldsymbol{x}_i) = \boldsymbol{\Sigma}(\boldsymbol{\vartheta}) \tag{4.97}$$

where $\gamma \sim p \times 1$ is a vector of regression constants, $\boldsymbol{\Pi} \sim p \times r$ is a matrix of regression coefficients, and $\boldsymbol{\Sigma} \sim p \times p$ is the conditional covariance matrix, that is, the covariance matrix of errors in a multivariate regression. The regression coefficients γ, $\boldsymbol{\Pi}$, and the covariance matrix $\boldsymbol{\Sigma}$ are the reduced form parameters of the conditional mean and covariance structure. Again, $\boldsymbol{\vartheta}$ is the $q \times 1$ vector of parameters to be estimated from the data. To distinguish $\boldsymbol{\vartheta}$ from the reduced form parameters, $\boldsymbol{\vartheta}$ is called the vector of fundamental parameters. Second order identification of $\boldsymbol{\vartheta}$ is assumed as before and no additional distributional assumptions need to be made about the form of the conditional distribution of $\boldsymbol{y}_i$ given $\boldsymbol{x}_i$.

A brief illustration of a typical parameterization of a conditional model is provided by a simultaneous equation model in a latent variable vector $\boldsymbol{\eta}_i$ coupled with a factor analytic model for $\boldsymbol{y}_i$:

$$\boldsymbol{\eta}_i = \boldsymbol{B}\boldsymbol{\eta}_i + \boldsymbol{\Gamma}\boldsymbol{x}_i + \boldsymbol{\zeta}_i \tag{4.98}$$

where $\boldsymbol{B}$ is the matrix of regression coefficients for the endogenous variables, $\boldsymbol{\Gamma}$ is the matrix of regression coefficients for the explanatory variables, and $\boldsymbol{\zeta}_i$ is a vector of disturbances with expected value $\boldsymbol{0}$ and covariance matrix $\boldsymbol{\Psi}$. Additionally,

$$\boldsymbol{y}_i = \boldsymbol{\nu} + \boldsymbol{\Lambda}\boldsymbol{\eta}_i + \boldsymbol{\delta}_i \tag{4.99}$$

where $\boldsymbol{\nu}$ is the vector of regression constants for $\boldsymbol{y}_i$, $\boldsymbol{\Lambda}$ is the matrix of factor loadings, and $\boldsymbol{\delta}_i$ is the measurement error with expected value $\boldsymbol{0}$ and covariance matrix $\boldsymbol{\Theta}$. The parameter vector contains the free elements in $\boldsymbol{B}, \boldsymbol{\Gamma}, \boldsymbol{\Psi}, \boldsymbol{\nu}, \boldsymbol{\Lambda}$, and $\boldsymbol{\Theta}$. The reduced form parameters are then given by:

$$\gamma(\boldsymbol{\vartheta}) = \boldsymbol{\nu} \tag{4.100}$$

$$\boldsymbol{\Pi}(\boldsymbol{\vartheta}) = \boldsymbol{\Lambda}(\boldsymbol{I} - \boldsymbol{B})^{-1}\boldsymbol{\Gamma} \tag{4.101}$$

$$\boldsymbol{\Sigma}(\boldsymbol{\vartheta}) = \boldsymbol{\Lambda}(\boldsymbol{I} - \boldsymbol{B})^{-1}\boldsymbol{\Psi}(\boldsymbol{I} - \boldsymbol{B})'^{-1}\boldsymbol{\Lambda}' + \boldsymbol{\Theta} \tag{4.102}$$

Obviously, every unconditional mean and covariance structure can be written as a conditional model by declaring all variables as dependent variables and setting $\boldsymbol{\Pi} = \boldsymbol{0}$. On the other hand, the conditional mean and covariance structure can be embedded in the unconditional model. Let $\boldsymbol{\mu}_x$ and $\boldsymbol{\Omega}_x$ denote the expected value and the covariance matrix of the vector $\boldsymbol{x}_i$ of explanatory variables. Then the unconditional expected value and covariance matrix of the joint vector $\boldsymbol{z}_i$ under the formulation of the conditional model is given by:

$$\gamma_u = \begin{pmatrix} \gamma + \boldsymbol{\Pi}\boldsymbol{\mu}_x \\ \boldsymbol{\mu}_x \end{pmatrix}, \quad \boldsymbol{\Sigma}_u = \begin{pmatrix} \boldsymbol{\Pi}\boldsymbol{\Omega}_x\boldsymbol{\Pi}' + \Sigma & \boldsymbol{\Pi}\boldsymbol{\Omega}_x \\ \boldsymbol{\Omega}_x\boldsymbol{\Pi}' & \boldsymbol{\Omega}_x \end{pmatrix} \tag{4.103}$$

The parameter vector $\boldsymbol{\vartheta}_u$ of the unconditional model contains as components the regression parameters of the conditional model as well as the expected value and the covariance matrix of the explanatory variables. Hence, a multivariate regression model can be parameterized as an unconditional mean and covariance structure. Estimation of the reduced form parameters $\boldsymbol{\gamma}, \boldsymbol{\Pi}, \boldsymbol{\Sigma}$ may be performed with maximum likelihood (ML) estimation for the parameters of the conditional or the unconditional model. The estimator of the asymptotic covariance matrix may have to be adjusted (cf. Section 1.3) if the assumption of conditional multivariate normality of $\boldsymbol{y}_i$ given $\boldsymbol{x}_i$ does not hold. Note, however, that $\boldsymbol{\mu}_x$ and $\boldsymbol{\Omega}_x$ are not explicitly estimated in the conditional formulation.

This equivalence of estimation of the reduced form parameters does not hold if the assumption of metrically scaled dependent variables is given up and threshold models for the inclusion of nonmetric dependent variables are included in the model. Before this problem can be discussed further, we introduce the conditional mean and covariance structure for unobserved dependent variables collected in the $p \times 1$ vector $\boldsymbol{y}_i^\star$:

$$\boldsymbol{y}_i^\star = \boldsymbol{\gamma}(\boldsymbol{\vartheta}) + \boldsymbol{\Pi}(\boldsymbol{\vartheta})\boldsymbol{x}_i + \boldsymbol{\epsilon}_i, \qquad \boldsymbol{\epsilon}_i \sim \mathcal{N}(\boldsymbol{0}, \boldsymbol{\Sigma}(\boldsymbol{\vartheta})) \tag{4.104}$$

As opposed to the previous case, with observed vector $\boldsymbol{y}_i$ we have now made the assumption that $\boldsymbol{y}_i^\star$ is not observed and that the error term $\boldsymbol{\epsilon}_i$ follows a p-dimensional multivariate normal distribution. Therefore, the conditional distribution of $\boldsymbol{y}_i^\star$ given $\boldsymbol{x}_i$ is multivariate normal with expected value $\boldsymbol{\gamma}+\boldsymbol{\Pi}\boldsymbol{x}_i$ and covariance matrix $\boldsymbol{\Sigma}$. $\boldsymbol{\vartheta}$ is again the q-dimensional vector of fundamental parameters. No assumptions are made about the distribution of $\boldsymbol{x}_i$.

The assumption of conditional multivariate normality implies that each individual dependent variable $y_{ij}^\star$ is conditionally univariate normal with

$$y_{ij}^\star = \gamma_j(\boldsymbol{\vartheta}) + \boldsymbol{\Pi}_{j.}(\boldsymbol{\vartheta})\boldsymbol{x}_i + \epsilon_{ij}, \qquad \epsilon_{ij} \sim \mathcal{N}(0, \sigma_{jj}(\boldsymbol{\vartheta})) \tag{4.105}$$

where γ_j is the regression constant for the jth variable, $\boldsymbol{\Pi}_{j.}$ is the jth row of $\boldsymbol{\Pi}$, and σ_{jj} is the error variance for the jth dependent variable. Additionally, the assumption of conditional multivariate normality implies bivariate conditional normality:

$$\begin{pmatrix} \epsilon_{ij} \\ \epsilon_{it} \end{pmatrix} = \mathcal{N}\left(\begin{bmatrix} 0 \\ 0 \end{bmatrix}, \begin{bmatrix} \sigma_{jj} & \sigma_{jt} \\ \sigma_{jt} & \sigma_{tt} \end{bmatrix} \right) \tag{4.106}$$

The conditional variances and the covariance for the jth and tth variable are denoted by $\sigma_{jj}, \sigma_{jt}, \sigma_{tt}$. Since the variables $y_{ij}^\star$ are not observed, the scale of $y_{ij}^\star$ and therefore the variance are arbitrary. Hence, the covariances cannot be estimated directly.

5.2 Inclusion of Threshold Models

Nonmetric variables are now included by mapping the unobserved variable $y_{ij}^\star$ onto the observed variable y_{ij} through one of the following observation rules using a threshold model, as in Sections 7 and 8 of Chapter 3. For convenience, the case index $i = 1, \ldots, n$ is omitted. Each observed dependent variable y_j is a function $y_j = c_j(y_j^\star, \boldsymbol{\tau}_j)$ of the corresponding unobserved variable $y_j^\star$ and a vector $\boldsymbol{\tau}_j$ of known or unknown thresholds. If the thresholds are known, they are denoted by t instead of τ.

- *Metrically* scaled dependent variable y_j:

$$y_j = y_j^\star \tag{4.107}$$

  that is, $y_j^\star$ is observed directly.

- *Metrically classified* dependent observed variable y_j with class boundaries $t_{j,1} < t_{j,2} < \ldots < t_{j,K_j}$ known a priori, and categories $y_j = 1, \ldots, K_j + 1$ (Stewart 1983):

$$y_j = k \iff y_j^\star \in [t_{j,k-1}, t_{j,k}) \tag{4.108}$$

with $[t_{j,0}, t_{j,1}) = (-\infty, t_{j,1})$ and $t_{j,K_j+1} = +\infty$.

Note that the assumption of conditional univariate normality implied by the conditional multivariate normality of ϵ is crucial not only for finding an estimation procedure but also for formulating the model. Once the distributional assumption is made, the probability that $y_j = k$ can be computed from this assumption and the model and estimation procedure of Subsection 8.2 of Chapter 3 can be used to estimate the reduced form parameters γ_j, $\boldsymbol{\Pi}_{j\cdot}$, and σ_{jj} for the jth component of $\boldsymbol{y}^\star$.

- *One-sided censored* dependent variable y_j with a priori known threshold value $t_{j,1}$:

$$y_j = \begin{cases} y_j^\star & \text{if } y_j^\star > t_{j,1} \\ t_{j,1} & \text{if } y_j^\star \leq t_{j,1} \end{cases} \tag{4.109}$$

Again, the assumption of conditional normality is crucial for the model formulation to define the probability that $y_j^\star$ does not cross the known threshold $t_{j,1}$. This can be called a tobit relation (Tobin 1958). The reduced form parameters for the jth component of $\boldsymbol{y}^\star$ can be estimated by using the model formulation and ML estimation procedure for the tobit model, as discussed in Subsection 7.4 of Chapter 3.

- *Double-sided censored* dependent variable y_j with two threshold values $t_{j,1} < t_{j,2}$ known a priori:

$$y_j = \begin{cases} t_{j,1} & \text{if } y_j^\star \leq t_{j,1} \\ y_j^\star & \text{if } t_{j,1} < y_j^\star < t_{j,2} \\ t_{j,2} & \text{if } y_j^\star \geq t_{j,2} \end{cases} \tag{4.110}$$

This relation is called a Two-Limit Probit Relation; see Rosett and Nelson (1975) and also Section 8.2 of Chapter 3.

- *Ordered categorical* dependent variable y_j with unknown threshold values $\tau_{j,1} < \tau_{j,2} < \ldots < \tau_{j,K_j}$, which must be estimated, in contrast to metrically classified variables, and with ordered categories $y_j = 1, \ldots, K_j + 1$:

$$y_j = k \iff y_j^\star \in [\tau_{j,k-1}, \tau_{j,k}) \tag{4.111}$$

with $[\tau_{j,0}, \tau_{j,1}) = (-\infty, \tau_{j,1})$ and $\tau_{j,K_j+1} = +\infty$. This is called an Ordinal Probit Relation (McKelvey and Zavoina 1975).

The distributional assumption of conditional normality specifies the probit model by defining the probability that $y_j = k$. The reduced form parameters γ_j and $\boldsymbol{\Pi}_{j\cdot}$, may be estimated using ML estimation for the ordered probit model. The parameters γ_j, $\boldsymbol{\Pi}_{j\cdot}$, and σ_{jj} of the reduced form for an ordered categorical variable y_j cannot be identified without further restrictions because the threshold values τ_j are unknown and the variance of an ordered categorical variable is not defined, unless additional information is provided, for instance, by panel data. To clarify this issue, we consider a univariate ordered categorical variable y_{i1} with outcomes $k = 1, \ldots, K + 1$, and explanatory variables $\boldsymbol{x}_{i1}$

$$y_{i1}^{\star} = \gamma_1 + \boldsymbol{\Pi}_{1\cdot}\boldsymbol{x}_{i1} + \epsilon_{i1}, \quad \epsilon_{i1} \sim \mathcal{N}(0, \sigma_{11}) \tag{4.112}$$

with the observation rule $y_{i1} = k \iff \tau_{1,k-1} < y_{i1}^{\star} \leq \tau_{k,1}, \quad k = 1, \ldots, K + 1$. The probability that $y_{i1} = k$ is therefore given by

$$\Pr(y_{i1} = k) = \Pr(\tau_{1,k-1} < y_{i1}^{\star} \leq \tau_{1,k}) = \Pr(y_{i1}^{\star} \leq \tau_{1,k}) - \Pr(y_{i1}^{\star} \leq \tau_{1,k-1}) \tag{4.113}$$

Considering only $\Pr(y_{i1}^{\star} \leq \tau_{1,k})$ yields, by the normality assumption:

$$\Pr(y_{i1}^{\star} \leq \tau_{1,k}) = \Pr(\epsilon_{i1} \leq \tau_{1,k} - \gamma_1 - \boldsymbol{\Pi}_{1\cdot}\boldsymbol{x}_{i1}) = \Phi\left(\frac{\tau_{1,k} - \gamma_1 - \boldsymbol{\Pi}_{1\cdot}\boldsymbol{x}_{i1}}{\sqrt{\sigma_{11}}}\right) \tag{4.114}$$

This probability does not change if a scalar c is added to $\tau_{1,k}$ and γ_1 and/or if the numerator and the denominator are both multiplied by a scalar d. Therefore, only the difference $\tau_{1,k} - \gamma_1$ is identified and $\tau_{1,k}$, γ_1, and $\boldsymbol{\Pi}_{1\cdot}$ are identified only up to a scalar. As arbitrary identification rules, we set $\tau_{1,1}$ to 0 and σ_{11} to one.

If, in addition to y_{i1}, the same variable is observed a second time as y_{i2} and the linear model for $y_{i2}^{\star}$ is given by

$$y_{i2}^{\star} = \gamma_2 + \boldsymbol{\Pi}_{2\cdot}\boldsymbol{x}_{i2} + \epsilon_{i2}, \qquad \epsilon_{i2} \sim \mathcal{N}(0, \sigma_{22}) \tag{4.115}$$

then the probability $\Pr(y_{i2}^{\star} \leq \tau_{2,k})$ may be written as:

$$\Pr(y_{i2}^{\star} \leq \tau_{2,k}) = \Phi\left(\frac{\tau_{2,k} - \gamma_2 - \boldsymbol{\Pi}_{2\cdot}\boldsymbol{x}_{i2}}{\sqrt{\sigma_{22}}}\right) \tag{4.116}$$

The same identification problems occur in (4.114) and (4.116). Even, if $\tau_{2,1}$ is set to zero, equality constraints like $\tau_{1,k} = \tau_{2,k}$, $k > 2$, $\gamma_1 = \gamma_2$ and $\boldsymbol{\Pi}_{1\cdot} = \boldsymbol{\Pi}_{2\cdot}$ cannot be tested if σ_{22} is unknown. Setting σ_{22} equal to $\sigma_{11} = 1$ may not be useful since the errors may have different variances over time. Then, only proportionality restrictions such as $\tau_{2,k} = \tau_{1,k}/\sqrt{\sigma_{22}}$, $\gamma_2 = \gamma_1/\sqrt{\sigma_{22}}$ and $\boldsymbol{\Pi}_{2\cdot} = \boldsymbol{\Pi}_{1\cdot}/\sqrt{\sigma_{22}}$ can be formulated and tested (cf. Sobel and Arminger 1992). On the other hand, the proportionality constant $\alpha_2 = 1/\sqrt{\sigma_{22}}$ can be estimated from the ordinal probit regression for y_{i2} if it is assumed that $\tau_{2,k} = \tau_{1,k}$ and/or $\gamma_2 = \gamma_1$ and/or $\boldsymbol{\Pi}_{2\cdot} = \boldsymbol{\Pi}_{1\cdot}$. In this case one finds:

$$\begin{aligned}\Pr(y_{i2}^\star \le \tau_{1,k}) &= \Phi\left(\frac{\tau_{1,k} - \gamma_1 - \boldsymbol{\Pi}_{1\cdot}\boldsymbol{x}_{i2}}{\sqrt{\sigma_{22}}}\right) \\ &= \Phi\left(\alpha_2\tau_{1,k} - \alpha_2\gamma_1 - \alpha_2\Pi_{1\cdot}\boldsymbol{x}_{i2}\right) \end{aligned} \tag{4.117}$$

If more than two panel waves are observed, a proportionality coefficient α_t for each wave $t > 1$ will be used. The assumption that σ_{11} and σ_{22} are equal and can therefore be set to 1 is meaningful if covariance stationarity is assumed. This assumption will therefore be important for time series of ordered categorical data. However, in panel data, the variances σ_{tt} may depend on the time point t because σ_{tt} here describes the conditional variance of the population at time t.

If the regression constant γ_j is set to zero, then $\tau_{j,1}$ is a free parameter. The relationship between the threshold parameters $\tau_{j,1}, \tau_{j,2}, \ldots, \tau_{j,K}$, when a regression constant γ_j is included, and the threshold parameters $\tau_{j,1}^\star, \tau_{j,2}^\star, \ldots, \tau_{j,K}^\star$, when the model does not include a regression constant, is:

$$\tau_{j,1}^\star = -\gamma_j \quad , \quad \tau_{j,2}^\star = \tau_{j,2} - \gamma_j \quad , \ldots , \quad \tau_{j,K}^\star = \tau_{j,K} - \gamma_j \tag{4.118}$$

To conclude this subsection, we note that the reduced form parameters γ_j, $\boldsymbol{\Pi}_{j\cdot}$, and σ_{jj} can be estimated by using univariate OLS, tobit or probit regression individually for each component of $\boldsymbol{y}^\star$. However, we have not yet formulated a model for the association between two nonmetric observed variables y_j and y_t.

5.3 Conditional Polyserial and Polychoric Covariance and Correlation Coefficients

The assumption that $y_{ij}^\star$ and $y_{it}^\star$ given $\boldsymbol{x}_i$ are bivariate normal with covariance σ_{jt} allows us to compute the probability that the observed variables y_{ij} and y_{it} take on the values k and l respectively given $\boldsymbol{x}_i$. If both variables are ordered categorical, this probability is given by:

$$P(y_{ij} = k, y_{it} = l|\boldsymbol{x}_i) = \int_{\tau_{j,(k-1)}}^{\tau_{j,(k)}} \int_{\tau_{t,(l-1)}}^{\tau_{t,(l)}} \varphi(y_j^\star, y_t^\star|\mu_{ij}, \sigma_{jj}, \mu_{it}, \sigma_{tt}, \sigma_{jt}) dy_t^\star dy_j^\star \tag{4.119}$$

in which $\mu_{ij} = \gamma_j + \boldsymbol{\Pi}_{j\cdot}\boldsymbol{x}_i$, $\mu_{it} = \gamma_t + \boldsymbol{\Pi}_{t\cdot}\boldsymbol{x}_i$ and $\varphi(y_t^\star, y_j^\star|\mu_j, \sigma_{jj}, \mu_t, \sigma_{tt}, \sigma_{tj})$ is the bivariate normal density function. Note that when y_{ij} and y_{it} are ordinal, $\hat{\sigma}_{tt} = \hat{\sigma}_{jj} = 1$. Hence, σ_{tj} is a correlation coefficient.

If both observed variables are dichotomous and no explanatory variables are available, the thresholds or respectively the regression constants and the correlation coefficient σ_{jt} may be estimated from the relative frequencies of the 2×2 table formed by y_j and y_t. The correlation coefficient σ_{jt} is then called the tetrachoric correlation coefficient (Pearson 1900). The tetrachoric correlation coefficient has been extended to ordered categorical variables (cf. Olsson 1979) and is then called the polychoric correlation coefficient. Since we have conditioned y_{ij} and y_{it} on $\boldsymbol{x}_i$, the correlation coefficient in the conditional model is called the conditional polychoric correlation coefficient.

If the observed variable y_{ij} is metrically scaled and the variable y_{it} is ordered categorical, the covariance σ_{jt} is called the conditional polyserial covariance. The corresponding correlation coefficient is the conditional polyserial correlation (Olsson, Drasgow, and Dorans 1982). Note however that the correlation coefficient and the covariance are not equal in general, because the variance σ_{jj} of the metrically scaled variable y_{ij} may be different from one. If there are no explanatory variables and the ordered categorical variable is dichotomous, the conditional polyserial correlation coefficient becomes the familiar biserial correlation.

In the same way, correlations and covariances between any observed variables with observation rules of Section 5.2 may be formulated. The assumption of conditional multivariate normality of $\boldsymbol{y}^{\star}$ given $\boldsymbol{x}_i$ is used throughout.

After the introduction of threshold models and polychoric and polyserial covariances and correlations we return to the distinction between unconditional and conditional mean and covariance structures. Until now we have formulated the model only conditionally making the assumption of conditional multivariate normality without distributional assumptions about $\boldsymbol{x}_i$. However, if we use the unconditional formulation, two disadvantages occur. The first is that the use of threshold models and polyserial and polychoric covariances is then based on the assumption that $\boldsymbol{y}_i^{\star}$ and $\boldsymbol{x}_i$ are jointly multivariate normal, which is doubtful and often implausible, for example, when $\boldsymbol{x}_i$ contains dummy variables. Also, joint multivariate normality is a much more restrictive assumption than conditional multivariate normality. The second disadvantage is that in addition to γ, $\boldsymbol{\Pi}$, and $\boldsymbol{\Sigma}$, the parameters in $\boldsymbol{\mu}_x$ and $\boldsymbol{\Omega}_x$ must be estimated. This disadvantage usually causes problems in estimating the asymptotic covariance matrix of polyserial and polychoric covariances and correlations when many explanatory variables are used.

5.4 Estimation

The parameters to be estimated in a conditional mean and covariance structure model for a vector of metrically and/or nonmetrically scaled variables consist of the unknown thresholds and the parameterization for the reduced form parameters γ, $\boldsymbol{\Pi}$, and $\boldsymbol{\Sigma}$. The free parameters to be estimated are collected in a $q \times 1$ vector $\boldsymbol{\vartheta}$. In the model of (4.98) and (4.99), $\boldsymbol{\vartheta}$ contains the free parameters in $(\boldsymbol{\tau}_1, \ldots, \boldsymbol{\tau}_p, \boldsymbol{B}, \boldsymbol{\Gamma}, \boldsymbol{\Psi}, \boldsymbol{\nu}, \boldsymbol{\Lambda}$, and $\boldsymbol{\Theta})$. We now describe one general approach to the estimation of these parameters which is a modified maximum likelihood approach.

Likelihood-Function

The likelihood of $\boldsymbol{\vartheta}$ based on the assumption of an iid sample $(\boldsymbol{y}_i, \boldsymbol{x}_i)$, $i = 1, \ldots, N$ is:

$$L(\boldsymbol{\vartheta}) = \prod_{i=1}^{N} P(\boldsymbol{y}_i | \boldsymbol{x}_i) \qquad (4.120)$$

The individual probabilities are defined by the assumption of conditional multivariate normality

$$P(\boldsymbol{y}_i|\boldsymbol{x}_i) = \int_{c_1^-(y_{i1}|\boldsymbol{\tau}_1)} \cdots \int_{c_p^-(y_{ip}|\boldsymbol{\tau}_p)} \varphi(\boldsymbol{y}^\star|\gamma(\boldsymbol{\vartheta}) + \boldsymbol{\Pi}(\boldsymbol{\vartheta})\boldsymbol{x}_i, \boldsymbol{\Sigma}(\boldsymbol{\vartheta}))d\boldsymbol{y}^\star \tag{4.121}$$

where $\varphi(\boldsymbol{y}^\star|\boldsymbol{\mu}, \boldsymbol{\Sigma})$ denotes again the multivariate normal density with expected value $\boldsymbol{\mu}$ and covariance matrix $\boldsymbol{\Sigma}$. The domain of integration for each component $y_{ij}^\star$ is the set of all $y_{ij}^\star$ which is mapped onto the observed value of $y_{ij} = c_j(y_{ij}^\star|\boldsymbol{\tau}_j)$ depending on the threshold vector $\boldsymbol{\tau}_j$. This set is denoted by $c_j^-(y_{ij}|\boldsymbol{\tau}_j)$.

To estimate the parameter vector $\boldsymbol{\vartheta}$ with full maximum-likelihood on the basis of this density function, it is necessary to compute p-dimensional normal distribution integrals; for $p > 3$, this is very cumbersome. However, if we consider just one dependent variable y_{ij} given $\boldsymbol{x}_i$ the univariate marginal density $P(y_{ij}|\boldsymbol{x}_i)$, depending on the measurement level of the ith case, is either a univariate normal distribution density or a marginal probability determined by integration of a univariate normal density. With two dependent variables y_j and y_t, the common density $P(y_{ij}, y_{it}|\boldsymbol{x}_i)$ is either a bivariate normal distribution density or is determined by integration over a bivariate normal density. Since the normal distribution is characterized completely by the means and variances of each variable $y_j^\star$ and the covariances of each pair of variables $y_j^\star$ and $y_t^\star$ for given $\boldsymbol{x}$, Muthén (1984) suggests an estimation strategy based solely on the computed densities and probabilities from univariate and bivariate normal distribution densities. Reduced form parameters $\boldsymbol{\tau}$, γ, $\boldsymbol{\Pi}$, and $\boldsymbol{\Sigma}$ can thus be estimated consistently. Estimation of the fundamental parameters $\boldsymbol{\vartheta}$ is then based on the estimated reduced form parameters. Proofs of consistency and asymptotic normality of parameter estimates obtained by this procedure are found in Küsters (1987, 1990).

Alternative estimation strategies employ Gauss-Hermite integration to compute the individual probabilities (Bock and Gibbon 1994) or switch to a Bayesian framework and describe the a posteriori distributions of the parameters using the Gibbs sampler (Muthén and Arminger 1994).

Marginal ML Estimation of the Reduced Form Parameters

The parameters estimated in the first stage are the thresholds denoted by the vector $\boldsymbol{\tau}_j$, the regression constant γ_j, the regression coefficients $\boldsymbol{\Pi}_{j\cdot}$ and the variance σ_{jj} for each component y_j, $j = 1, \ldots, p$. The marginal ML estimation procedure is based on conditional univariate normality. The computations are performed using ordinary univariate regression, tobit and ordinal probit regression (see Sections 7 and 8 of Chapter 3).

In the second stage we estimate the polychoric and polyserial correlations and covariances of the error terms in the reduced form equations. Note that in this stage the covariances are estimated without parametric restrictions. Since the errors are assumed to be normally distributed, and strongly consistent estimators of the reduced form coefficients have already been obtained in the first stage, the estimation problem reduces to maximizing the loglikelihood function

$$l_{jt}(\sigma_{jt}) = \sum_{i=1}^{n} \ln P(y_{ij}, y_{it}|\boldsymbol{x}_i, \hat{\boldsymbol{\tau}}_j, \hat{\gamma}_j, \hat{\boldsymbol{\Pi}}_{j\cdot}, \hat{\sigma}_{jj}, \hat{\boldsymbol{\tau}}_t, \hat{\gamma}_t, \hat{\boldsymbol{\Pi}}_{t\cdot}, \hat{\sigma}_{tt}, \sigma_{jt}) \tag{4.122}$$

in which $P(y_{ij}, y_{it}|\boldsymbol{x}_i, \hat{\boldsymbol{\tau}}_j, \hat{\gamma}_j, \hat{\boldsymbol{\Pi}}_{j\cdot}, \hat{\sigma}_{jj}, \hat{\boldsymbol{\tau}}_t, \hat{\gamma}_t, \hat{\boldsymbol{\Pi}}_{t\cdot}, \hat{\sigma}_{tt}, \sigma_{jt})$ is the bivariate probability of y_{ij} and y_{it} given $\boldsymbol{x}_i$ and the reduced form coefficients. A typical example of this bivariate probability is the case when y_{ij} and y_{it} are both ordered categorical. Then the probability that $y_{ij} = k$ and $y_{it} = l$ is given by:

$$P(y_{ij} = k, y_{it} = l|\boldsymbol{x}_i) = \int_{\hat{\tau}_{j,(k-1)}}^{\hat{\tau}_{j,(k)}} \int_{\hat{\tau}_{t,(l-1)}}^{\hat{\tau}_{t,(l)}} \varphi(y_j^\star, y_t^\star|\hat{\mu}_{ij}, \hat{\sigma}_{jj}, \hat{\mu}_{it}, \hat{\sigma}_{tt}, \sigma_{jt})dy_t^\star dy_j^\star \quad (4.123)$$

in which $\hat{\mu}_{ij} = \hat{\gamma}_j + \hat{\boldsymbol{\Pi}}_{j\cdot}\boldsymbol{x}_i, \hat{\mu}_{it} = \hat{\gamma}_t + \hat{\boldsymbol{\Pi}}_{t\cdot}\boldsymbol{x}_i$ and $\varphi(y_j^\star, y_t^\star|\mu_j, \sigma_{jj}, \mu_t, \sigma_{tt}, \sigma_{jt})$ is the bivariate normal density function. Note that for ordered categorical variables $\hat{\sigma}_{jj} = \hat{\sigma}_{tt} = 1$ because of the chosen identification restriction. The loglikelihood function $l_{jt}(\sigma_{jt})$ has to be modified accordingly if variables with other measurement levels are used.

The estimation of the polychoric and polyserial correlations is numerically difficult if the true correlation is near the boundary values -1 or $+1$. To increase numerical stability, the correlations should be transformed using Fisher's z-transformation with:

$$z = \ln \frac{1+\rho}{1-\rho} \quad (4.124)$$

The estimated thresholds $\hat{\boldsymbol{\tau}}_j$, the reduced form coefficients $\hat{\gamma}_j$ and $\hat{\boldsymbol{\Pi}}_{j\cdot}$, the variances $\hat{\sigma}_{jj}$, and the covariances $\hat{\sigma}_{jt}$ from all components are then collected in a $\tilde{p} \times 1$ vector $\hat{\boldsymbol{\kappa}}_n$ which depends on the sample size n. For the final estimation stage, a strongly consistent estimate of the asymptotic covariance matrix $\boldsymbol{W}$ of $\hat{\boldsymbol{\kappa}}_n$ is computed. This estimate is denoted by $\hat{\boldsymbol{W}}_n$. The asymptotic covariance matrix $\boldsymbol{W}$ is difficult to derive since the estimates of $\hat{\sigma}_{jt}$ of the second stage depend on the estimated coefficients $\hat{\boldsymbol{\tau}}_f, \hat{\gamma}_f, \hat{\boldsymbol{\Pi}}_{f\cdot}, \hat{\sigma}_{ff}, f = j, t$ of the first stage. The various elements of the asymptotic covariance matrix $\boldsymbol{W}$ and a consistent estimator $\boldsymbol{W}_n$ are given in Küsters (1987,1990) and Muthén and Satorra (1994). If all dependent variables are metrically scaled and an unconditional covariance structure is used, the matrix $\boldsymbol{W}_n$ is the weight matrix in the ADF method described in Section 1.3.

Minimum Discrepancy Estimation of the Fundamental Parameters

In the third stage the vector $\boldsymbol{\kappa}$ of thresholds, the reduced form regression coefficients and the reduced form covariance matrix is written as a function of the structural parameters of interest, collected in the parameter vector $\boldsymbol{\vartheta}$. The parameter vector $\boldsymbol{\vartheta}$ is then estimated by minimizing the discrepancy function (cf. Section 1.3)

$$Q_n(\boldsymbol{\vartheta}) = (\hat{\boldsymbol{\kappa}}_n - \boldsymbol{\kappa}(\boldsymbol{\vartheta}))'\hat{\boldsymbol{W}}_n^{-1}(\hat{\boldsymbol{\kappa}}_n - \boldsymbol{\kappa}(\boldsymbol{\vartheta})) \quad (4.125)$$

based on the asymptotic normality of the estimators of the reduced form coefficients. The vector $\hat{\boldsymbol{\kappa}}_n$ is asymptotically normal with expected value $\boldsymbol{\kappa}(\boldsymbol{\vartheta})$ and covariance matrix $\boldsymbol{W}$ estimated by $\hat{\boldsymbol{W}}_n$. The general parameter restrictions discussed in Section 3 of Chapter 3 and in Section 3 of this chapter may be used, allowing a flexible parameterization of thresholds and the conditional mean and covariance structure.

The quadratic form $Q_n(\boldsymbol{\vartheta})$ is centrally chi-square distributed with $\tilde{p} - q$ degrees of freedom if the model is specified correctly and the sample size is sufficiently large. The number

$\tilde{p}$ indicates the number of elements in $\hat{\boldsymbol{\kappa}}_n$ while q is the number of elements in $\boldsymbol{\vartheta}$. Therefore Q_n can be used as a goodness-of-fit measure for the fit of the chosen parameterization to the reduced form parameters. Statistics that are derived from this chi-square statistic such as the RMSEA discussed in Section 4.2 can be computed as well. The goodness-of-fit for individual components should be computed using the coefficients of determination in Sections 7 and 8 of Chapter 3.

Since $\hat{\boldsymbol{W}}_n$ is a strongly consistent estimate of $\boldsymbol{W}$, the minimum distance estimator $\hat{\boldsymbol{\vartheta}}$ of $\boldsymbol{\vartheta}$ is asymptotically normal with

$$\hat{\boldsymbol{\vartheta}} \overset{A}{\sim} \mathcal{N}\left(\mathbf{0}, \left[\frac{\partial \boldsymbol{\kappa}}{\partial \boldsymbol{\vartheta}} \hat{\boldsymbol{W}}_n^{-1} \left(\frac{\partial \boldsymbol{\kappa}}{\partial \boldsymbol{\vartheta}}\right)'\right]^{-1}\right) \qquad (4.126)$$

The first derivatives of $\boldsymbol{\kappa}$ with respect to $\boldsymbol{\vartheta}$ are evaluated at $\hat{\boldsymbol{\vartheta}}$. The asymptotic normality of $\hat{\boldsymbol{\vartheta}}$ is the basis for computing likelihood ratio, Lagrange multiplier, and Wald statistics to test specific hypotheses about $\boldsymbol{\vartheta}$.

Variance Standardization

The covariance structure for $\boldsymbol{\Sigma}(\boldsymbol{\vartheta}) \sim p \times p$ in the conditional model is often specified as a multiplicative model of the form:

$$\boldsymbol{\Sigma}(\boldsymbol{\vartheta}) = \boldsymbol{A}\boldsymbol{\Omega}(\boldsymbol{\vartheta})\boldsymbol{A}' \qquad (4.127)$$

Here, $\boldsymbol{\Omega}(\boldsymbol{\vartheta})$ denotes the covariance matrix of the error terms of the structural form and $\boldsymbol{A}$ is a transformation matrix that maps $\boldsymbol{\Omega}(\boldsymbol{\vartheta})$ onto the reduced form covariance matrix $\boldsymbol{\Sigma}(\boldsymbol{\vartheta})$. A typical example is the simultaneous equation model $\boldsymbol{y}^\star = \boldsymbol{B}\boldsymbol{y}^\star + \boldsymbol{\Gamma}\boldsymbol{x} + \boldsymbol{\epsilon}$ with $V(\boldsymbol{\epsilon}) = \boldsymbol{\Omega}(\boldsymbol{\vartheta})$ where $\boldsymbol{\Sigma}(\boldsymbol{\vartheta}) = (\boldsymbol{I} - \boldsymbol{B}(\boldsymbol{\vartheta}))^{-1}\boldsymbol{\Omega}(\boldsymbol{\vartheta})(\boldsymbol{I} - \boldsymbol{B}(\boldsymbol{\vartheta}))^{-1'}$. Here, $\boldsymbol{A} = (\boldsymbol{I} - \boldsymbol{B}(\boldsymbol{\vartheta}))^{-1}$. If all observed dependent variables are ordered categorical, then the values of $\boldsymbol{\vartheta}$ must be chosen in such a way that $\text{diag}(\boldsymbol{\Sigma}(\boldsymbol{\vartheta})) = \boldsymbol{e}_p$, that is a $p \times 1$ vector of ones. Hence, we have:

$$\text{diag}(\boldsymbol{\Sigma}(\boldsymbol{\vartheta})) = \text{diag}(\boldsymbol{A}\boldsymbol{\Omega}(\boldsymbol{\vartheta})\boldsymbol{A}') = \boldsymbol{e}_p \qquad (4.128)$$

If the ith row of $\boldsymbol{A}$ is denoted by $\boldsymbol{A}_i$, we obtain

$$\text{diag}(\boldsymbol{A}\boldsymbol{\Omega}(\boldsymbol{\vartheta})\boldsymbol{A}') = \text{diag}\begin{bmatrix} \boldsymbol{A}_1\boldsymbol{\Omega}\boldsymbol{A}_1' & \cdots & \cdots \\ \cdots & \ddots & \cdots \\ \cdots & \cdots & \boldsymbol{A}_p\boldsymbol{\Omega}\boldsymbol{A}_p' \end{bmatrix} = \begin{bmatrix} \boldsymbol{A}_1\boldsymbol{\Omega}\boldsymbol{A}_1' \\ \vdots \\ \boldsymbol{A}_p\boldsymbol{\Omega}\boldsymbol{A}_p' \end{bmatrix} = \boldsymbol{e}_p \qquad (4.129)$$

Therefore $\boldsymbol{A}_i\boldsymbol{\Omega}\boldsymbol{A}_i' = 1$ must apply to all $i = 1, \cdots, p$. We now apply standard rules for the trace (tr) of matrices and for vectorized (vec) matrices. The trace of a matrix may be written as $\text{tr}(\boldsymbol{A} \cdot \boldsymbol{B}) = \text{tr}(\boldsymbol{B} \cdot \boldsymbol{A}) = (vec(\boldsymbol{A}'))' \cdot vec(\boldsymbol{B})$ where $vec(\boldsymbol{B})$ stacks the columns of the matrix $\boldsymbol{B}$ above each other.

$$\boldsymbol{A}_i\boldsymbol{\Omega}\boldsymbol{A}_i' = \text{tr}(\boldsymbol{A}_i\boldsymbol{\Omega}\boldsymbol{A}_i') = \text{tr}(\boldsymbol{A}_i'\boldsymbol{A}_i\boldsymbol{\Omega}) = (vec(\boldsymbol{A}_i'\boldsymbol{A}_i))' \cdot vec(\boldsymbol{\Omega}) = 1 \qquad (4.130)$$

Combining the equations for all $i = 1, \cdots, p$ we obtain:

$$\begin{bmatrix} (vec(\boldsymbol{A}_1'\boldsymbol{A}_1))' \\ \vdots \\ (vec(\boldsymbol{A}_p'\boldsymbol{A}_p))' \end{bmatrix} \cdot vec(\boldsymbol{\Omega}) = \tilde{\boldsymbol{A}} \cdot vec(\boldsymbol{\Omega}) = \boldsymbol{e}_p \tag{4.131}$$

Here, $\tilde{\boldsymbol{A}}$ represents a $p \times p^2$ matrix.

$\boldsymbol{\Omega}$ contains elements which have to be restricted due to variance normalization as well as elements which do not have to be restricted to enforce the variance normalization. To compute the restricted parameters, $vec(\boldsymbol{\Omega})$ is separated into a fixed part $\boldsymbol{\Omega}_R$ to be restricted and a free part $\boldsymbol{\Omega}_F$. The separation must be performed so that

$$vec(\boldsymbol{\Omega}) = vec(\boldsymbol{\Omega}_F) + vec(\boldsymbol{\Omega}_R) \tag{4.132}$$

Here, $vec(\boldsymbol{\Omega}_F)$ is a $p^2 \times 1$ vector with zeros in the places for the restricted parameters and $vec(\boldsymbol{\Omega}_R)$ represents a $p^2 \times 1$ vector with zeros in the places for the free parameters.

In general, only the diagonal elements of $\boldsymbol{\Omega}$ must be restricted to enforce the variance normalizations, so that an *expansion matrix* $\boldsymbol{E}$ ($p^2 \times p$) with the property

$$vec(\boldsymbol{\Omega}_R) = \boldsymbol{E} \cdot \text{diag}(\boldsymbol{\Omega}) \tag{4.133}$$

can be formed. Then

$$vec(\boldsymbol{\Omega}) = vec(\boldsymbol{\Omega}_F) + \boldsymbol{E} \cdot \text{diag}(\boldsymbol{\Omega}) \tag{4.134}$$

Substitution yields:

$$\tilde{\boldsymbol{A}} \cdot (vec(\boldsymbol{\Omega}_F) + \boldsymbol{E} \cdot \text{diag}(\boldsymbol{\Omega})) = \boldsymbol{e}_p \tag{4.135}$$

Since only the elements in diag($\boldsymbol{\Omega}$) must be restricted, this equation system must be solved for diag($\boldsymbol{\Omega}$). This yields

$$\text{diag}(\boldsymbol{\Omega}) = \left(\tilde{\boldsymbol{A}}\boldsymbol{E}\right)^{-1} \left(\boldsymbol{e}_p - \tilde{\boldsymbol{A}} \cdot vec(\boldsymbol{\Omega}_F)\right) \tag{4.136}$$

as the general formula for computing the elements to be restricted to enforce variance normalization. In this way, the variance normalizations for ordered categorical variables can always be imposed.

5.5 Multigroup Analysis

The model specification and estimation for mean and covariance structures with nonmetric dependent variables may be carried over to the analysis of G groups with samples $(\boldsymbol{y}_i^{(g)}, \boldsymbol{x}_i^{(g)})$, $i = 1, \ldots, N^{(g)}$ where $N^{(g)}$ cases are observed in each group g. For each of the groups, a separate model is formulated. For the reduced form parameters we get G group-specific structures of the following form:

$$\gamma^{(g)} = \gamma^{(g)}(\boldsymbol{\vartheta}) \tag{4.137}$$

$$\boldsymbol{\Pi}^{(g)} = \boldsymbol{\Pi}^{(g)}(\boldsymbol{\vartheta}) \tag{4.138}$$

$$\boldsymbol{\Sigma}^{(g)} = \boldsymbol{\Sigma}^{(g)}(\boldsymbol{\vartheta}) \tag{4.139}$$

For each group, the reduced form parameters as well as the asymptotic covariance $\boldsymbol{W}^{(g)}$ are estimated separately. To estimate the simultaneous parameter vector $\boldsymbol{\vartheta}$ for all groups, the quadratic function is generalized to the simultaneous analysis of several groups:

$$Q(\boldsymbol{\vartheta}) = \sum_{g=1}^{G} (\hat{\boldsymbol{\kappa}}_n^{(g)} - \boldsymbol{\kappa}^{(g)}(\boldsymbol{\vartheta}))' \hat{\boldsymbol{W}}_n^{(g)^{-1}} (\hat{\boldsymbol{\kappa}}_n^{(g)} - \boldsymbol{\kappa}^{(g)}(\boldsymbol{\vartheta})) \tag{4.140}$$

Here the parameters of *all groups* to be estimated are collected in the vector of structural parameters $\boldsymbol{\vartheta}$. The relations of the structural parameters between the groups (e.g. identity relations) can be taken into account by using the restrictions described in Section 3 of Chapter 3. Thus the structural parameters of several populations can be estimated simultaneously by minimizing the modified quadratic form. The factor analysis of several populations (Jöreskog 1971) is a typical example of multigroup analysis. The analysis of models with missing or incomplete data is another application of the multiple group option. In such cases data from several populations are combined without the same set of variables being available for all populations. Such problems can be analyzed by means of the group-specific model without reducing the data sets to those variables which are available for all groups. Multiple group estimation with incomplete data is discussed by Allison (1987), Arminger and Sobel (1990), and Muthén, Kaplan, and Hollis (1987).

5.6 Example: Achievement in and Attitude toward High School Mathematics

To illustrate the models, the identification problems, and the estimation method discussed above, data from the Longitudinal Study of American Youth (LSAY) (Miller et al. 1992) are used. A more detailed analysis using seven waves is found in Arminger (1994). The dependent variables of interests are math score (M) and attitude toward math (A). The math score is observed directly, the attitude is measured by three items, I1, I2, and I3. These items are five-point Likert scales: I1, "How much do you like your math class?"; I2, "How clear is your math class teacher explaining the material?"; and I3, "How difficult or easy is the math course for you?". The lower categories indicate liking and the higher categories indicate disliking. The math scores were measured in fall 1987 and fall 1988 and are denoted by M-T1 and M-T2. The attitude items were collected in fall 1987 and spring 1988. These variables are denoted by I1-T1, I2-T1, I3-T1, and I1-T2, I2-T2, I3-T2. The sample size of the data set for the analysis is 1519. As background variable, the parents' socioeconomic status (SES) is used. Since the only explanatory variable SES is approximately normally distributed we assume that the variables for SES, M-T1, and M-T2 and the underlying variables for I1-T1, I2-T1, I3-T1, I1-T2, I2-T2, and I3-T2 are jointly multinormally distributed and use therefore the unconditional formulation in (4.103) for the model specification of the mean and covariance structure.

We are interested in estimating a model that describes the relationship of math score and attitude toward math over the time points given here. Specifically we look for the regression of math score on attitude and the regression of attitude on math score controlling for the level in both variables achieved at the first time point and for the socioeconomic background. Before we write a model for the unconditional mean and covariance structure, we report the reduced form parameters estimated with the program MECOSA (Schepers and Arminger 1992).

Estimation of Reduced Form Parameters

The threshold parameters are given in Table 4. Since the regression constants are estimated, the first threshold for each ordered categorical variable is set to zero.

Table 4. Reduced form thresholds

Variables	τ_1	τ_2	τ_3	τ_4
I1-T1	0.000	0.676	1.291	1.547
I2-T1	0.000	0.693	1.172	1.405
I3-T1	0.000	0.808	1.436	1.786
I1-T2	0.000	0.803	1.357	1.659
I2-T2	0.000	0.680	1.113	1.402
I3-T2	0.000	0.738	1.362	1.710

Since we are analyzing the unconditional form of a mean and covariance structure, the regression constants given in Table 5 can be interpreted as unconditional means.

The math score is higher at the second time point than at the first time point. Comparing the means for the variables underlying I1-T2, I2-T2, and I3-T2 with the means for I1-T1, I2-T1, and I3-T1 shows that the students have a slightly better attitude toward math at the second time point than at the first time point.

Table 5. Reduced form regression constants

Variables	Parameters
SES	0.080
M-T1	5.331
M-T2	5.590
I1-T1	0.108
I2-T1	-0.351
I3-T1	0.356
I1-T2	0.332
I2-T2	-0.090
I3-T2	0.498

Finally, the reduced form covariances are given in Table 6.

Table 6. Reduced form covariances

	SES	M-T1	M-T2	I1-T1	I2-T1	I3-T1	I1-T2	I2-T2	I3-T2
SES	0.548	0.196	0.189	-0.019	0.039	0.019	-0.004	0.020	0.008
M-T1	0.196	0.861	0.737	-0.106	-0.008	-0.036	-0.109	-0.098	-0.094
M-T2	0.189	0.737	0.914	-0.099	0.010	-0.034	-0.113	-0.104	-0.088
I1-T1	-0.019	-0.106	-0.099	1.000	0.539	0.583	0.525	0.277	0.362
I2-T1	0.039	-0.008	0.010	0.539	1.000	0.401	0.366	0.467	0.262
I3-T1	0.019	-0.036	-0.034	0.583	0.401	1.000	0.407	0.219	0.463
I1-T2	-0.004	-0.109	-0.113	0.525	0.366	0.407	1.000	0.537	0.518
I2-T2	0.020	-0.098	-0.104	0.277	0.467	0.219	0.537	1.000	0.400
I3-T2	0.008	-0.094	-0.088	0.362	0.262	0.463	0.518	0.400	1.000

Note that the variables SES, M-T1, and M-T2 are metrically scaled. Therefore, the variances can be estimated (first three diagonal entries). The variables I1-T1 to I3-T2 are ordered categorical. Therefore, the variances are set to one. The covariances between the first three and the other variables are therefore polyserial covariances computed by multiplying the polyserial correlation coefficients with the standard deviations of SES, M-T1, and M-T2. Within the variables I1-T1 to I3-T2, only polychoric correlations can be computed since the variances have been set to one.

Specification of the Structural Equation Model

We first describe a structural equation model for the variable η_{i1} for SES, η_{i2} and η_{i3} for math scores M-T1 and M-T2, and the variables η_{i4} and η_{i5} for the unobserved attitude variables A-T1 and A-T2. The model structure is depicted graphically in Figure 3.

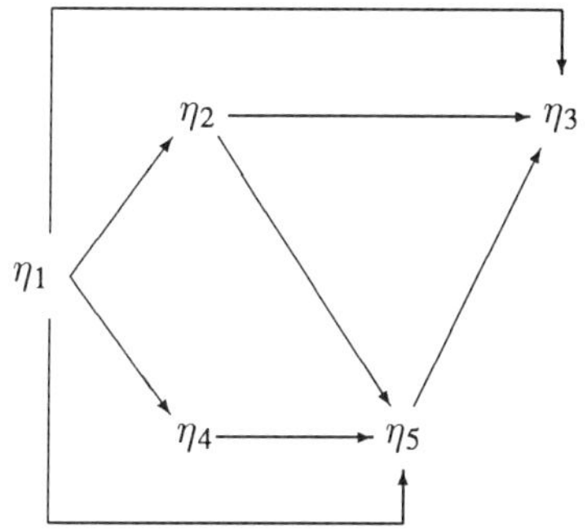

FIGURE 3. Autoregressive structure of math scores and attitude toward math.

It is assumed that the regression coefficients of η_2 - η_5 on η_1 are nonzero, that is SES is a background variable for both math score and attitude at both time points. For the math score, we assume a first order autoregressive process and therefore let η_3 depend on η_2. A similar assumption is made for the attitude variable where η_5 is conditioned on η_4. We also condition η_3 on η_5 and η_5 on η_2 to analyze the mutual dependence of math score and attitude toward math. The structural equation model is therefore given by

$$\boldsymbol{\eta}_i = \boldsymbol{B}\boldsymbol{\eta}_i + \boldsymbol{\zeta}_i, \quad \text{with } \boldsymbol{\zeta}_i \sim \mathcal{N}(\mathbf{0}, \boldsymbol{\Psi}) \tag{4.141}$$

Since the math scores and the attitudes are conditioned on SES, ζ_{i1} is assumed to be uncorrelated with ζ_{i2} - ζ_{i5}. Since we have formulated an autoregressive model for the math scores and the attitude toward math we also assume that there is no additional serial correlation in the error terms. Therefore, the matrix $\boldsymbol{\Psi}$ is diagonal. Under these assumptions, $\boldsymbol{B}$ and $\boldsymbol{\Psi}$ may be written as:

$$\boldsymbol{B} = \begin{pmatrix} 0 & 0 & 0 & 0 & 0 \\ \beta_{21} & 0 & 0 & 0 & 0 \\ \beta_{31} & \beta_{32} & 0 & 0 & \beta_{35} \\ \beta_{41} & 0 & 0 & 0 & 0 \\ \beta_{51} & \beta_{52} & 0 & \beta_{54} & 0 \end{pmatrix}, \quad \boldsymbol{\Psi} = \begin{pmatrix} \psi_{11} & 0 & 0 & 0 & 0 \\ 0 & \psi_{22} & 0 & 0 & 0 \\ 0 & 0 & \psi_{33} & 0 & 0 \\ 0 & 0 & 0 & \psi_{44} & 0 \\ 0 & 0 & 0 & 0 & \psi_{55} \end{pmatrix}$$

Specification of the Measurement Model

The measurement model for the underlying variables $y^{\star}_{ij}$, $j = 1, \ldots, 9$ is specified as

$$\boldsymbol{y}^{\star}_i = \boldsymbol{\nu} + \boldsymbol{\Lambda}\boldsymbol{\eta}_i + \epsilon_i, \quad \text{with } \epsilon_i \sim \mathcal{N}(\mathbf{0}, \boldsymbol{\Theta}) \tag{4.142}$$

where $\boldsymbol{\eta}_i$ is again the 5×1 vector of SES, M-T1, M-T2, A-T1, and A-T2. The vector $\boldsymbol{\nu}$ of regression constants is specified as $\boldsymbol{\nu} = (\nu_1, \ldots, \nu_9)'$ without further constraints.

The 9×5 matrix $\boldsymbol{\Lambda}$ of factor loadings is specified as:

$$\boldsymbol{\Lambda} = \begin{pmatrix} 1 & 0 & 0 & 0 & 0 \\ 0 & 1 & 0 & 0 & 0 \\ 0 & 0 & 1 & 0 & 0 \\ 0 & 0 & 0 & 1 & 0 \\ 0 & 0 & 0 & \lambda_{54} & 0 \\ 0 & 0 & 0 & \lambda_{64} & 0 \\ 0 & 0 & 0 & 0 & \alpha_1 \\ 0 & 0 & 0 & 0 & \alpha_2\lambda_{54} \\ 0 & 0 & 0 & 0 & \alpha_3\lambda_{64} \end{pmatrix} \tag{4.143}$$

The variables SES, M-T1, and M-T2 are observed directly. The corresponding factor loadings $\lambda_{11}, \lambda_{22}, \lambda_{33}$ are therefore set to one. The item I1-T1 is chosen as reference item and λ_{44} is therefore also set to one. The factor loadings λ_{54} and λ_{64} for items I2-T1 and I3-T1 vary freely. The items I1-T2, I2-T2, and I3-T2 are assumed to have the same connotation as I1-T1, I2-T1, and I3-T1 and should therefore have the same factor loadings. However, since all items are ordered categorical, the variances in the reduced form have been set to one for identification. Hence, the factor loadings are only identified up to scale (cf. Subsection 5.2 for the reduced form regression coefficients). If it is assumed that the reduced form error variances of the items I1, I2, and I3 may differ over time, the factor loadings for the second time point must be multiplied by the proportionality constants α_1, α_2, and α_3 (as above).

The covariance matrix $\boldsymbol{\Theta}$ of measurement errors $\boldsymbol{\epsilon}_i$ is specified as:

$$\boldsymbol{\Theta} = \begin{pmatrix} 0 & 0 & 0 & 0 & 0 & 0 & 0 & 0 & 0 \\ 0 & 0 & 0 & 0 & 0 & 0 & 0 & 0 & 0 \\ 0 & 0 & 0 & 0 & 0 & 0 & 0 & 0 & 0 \\ 0 & 0 & 0 & \vartheta_{44} & 0 & 0 & \vartheta_{47} & 0 & 0 \\ 0 & 0 & 0 & 0 & \vartheta_{55} & 0 & 0 & \vartheta_{58} & 0 \\ 0 & 0 & 0 & 0 & 0 & \vartheta_{66} & 0 & 0 & \vartheta_{69} \\ 0 & 0 & 0 & \vartheta_{74} & 0 & 0 & \vartheta_{77} & 0 & 0 \\ 0 & 0 & 0 & 0 & \vartheta_{85} & 0 & 0 & \vartheta_{88} & 0 \\ 0 & 0 & 0 & 0 & 0 & \vartheta_{96} & 0 & 0 & \vartheta_{99} \end{pmatrix} \tag{4.144}$$

The parameters ϑ_{47}, ϑ_{58}, and ϑ_{69} are the covariances of the measurement errors for the items I1, I2, and I3 over time. The error variances for SES, M-T1, and M-T2 are zero because these variables have been observed directly. The error variances $\vartheta_{44}, \vartheta_{55}, \vartheta_{66}, \vartheta_{77}, \vartheta_{88}$, and ϑ_{99} are not free parameters because the variances of the reduced form covariance matrix are set to one for ordered categorical variables. The reduced form of the model is given by:

$$E(\boldsymbol{y}^\star) = \gamma(\boldsymbol{\vartheta}) = \boldsymbol{\nu}\ , \quad V(\boldsymbol{y}^\star) = \boldsymbol{\Sigma}(\boldsymbol{\vartheta}) = \boldsymbol{\Lambda}(\boldsymbol{I} - \boldsymbol{B})^{-1}\boldsymbol{\Psi}(\boldsymbol{I} - \boldsymbol{B})'^{-1}\boldsymbol{\Lambda}' + \boldsymbol{\Theta} \tag{4.145}$$

Therefore, the variance standardization algorithm of Subsection 5.4 has to be employed.

Specification of the Threshold Model

The ordered categorical variables y_{i4} (I1-T1)–y_{i9} (I3-T2) are connected to the unobserved variables $y^*_{i4} - y^*_{i9}$ through threshold models with three unknown thresholds $\tau_{j,2}$, $\tau_{j,3}$, and $\tau_{j,4}$, $j = 4, \ldots, 9$. The first threshold is restricted to zero for identification. Since the same items I1, I2, and I3 are measured at two time points, the thresholds for I1-T1 and I1-T2, I2-T1, and I2-T2, and I3-T1 and I3-T2 are set equal. However, to take account of the fact that the reduced form error variances may vary over time, the thresholds for I1-T2, I2-T2, and I3-T2 have to be multiplied by the proportionality coefficients α_1, α_2, and α_3, like the factor loadings.

Estimation Results

The estimated reduced form coefficients $\hat{\boldsymbol{\tau}}_j, j = 4, \ldots, 9$ (thresholds), $\hat{\gamma}$ and $\hat{\Sigma}$ are collected in a vector $\hat{\boldsymbol{\kappa}}$ with 66 elements. The asymptotic covariance matrix of $\hat{\boldsymbol{\kappa}}$ has therefore $66 \cdot 67/2 = 2211$ elements. The elements of the vector $\boldsymbol{\vartheta}$ of fundamental parameters are the thresholds $\boldsymbol{\tau}_4$, $\boldsymbol{\tau}_5$, and $\boldsymbol{\tau}_6$ for the items I1-T1, I2-T1, and I3-T1, the proportionality coefficients $\alpha_1, \alpha_2, \alpha_3$, and the free elements of $\boldsymbol{B}, \boldsymbol{\Psi}, \boldsymbol{\nu}, \boldsymbol{\Lambda}$, and $\boldsymbol{\Theta}$. The number of elements in $\boldsymbol{\vartheta}$ is 39, yielding $66 - 39 = 27$ degrees of freedom. The parameter estimates of $\boldsymbol{B}$ and $\boldsymbol{\Psi}$ for the structural model are given in Table 7. The chi-square statistic is 72.713 indicating a lack of fit at the 0.05 test level. However, the RMSEA index is 0.033,

Table 7. Regression coefficients, error variances, and coefficients of determination in the structural equation model

Variables	SES	M-T1	M-T2	A-T1	A-T2
SES	0.000	0.000	0.000	0.000	0.000
M-T1	0.365	0.000	0.000	0.000	0.000
	(11.552)[a]				
M-T2	0.048	0.839	0.000	0.000	-0.033
	(2.362)	(43.528)			(-1.747)
A-T1	-0.007	0.000	0.000	0.000	0.000
	(-0.191)				
A-T2	0.035	-0.098	0.000	0.629	0.000
	(1.205)	(-4.182)		(23.809)	
Variances	0.541	0.795	0.282	0.754	0.362
R^2	0.000	0.083	0.691	0.000	0.458

[a] z-values are given in parentheses.

indicating a close fit (cf. Section 4.2).

The estimated regression coefficients reveal strong autoregression for each of the dependent variables. The relationship between math score and attitude, as captured in the lagged regression coefficients of M-T1 to A-T2 and A-T2 to M-T2, is small. Therefore, no dependence of math score and attitude on each other over time is indicated by the model. The regression coefficient for SES is strong only at the first time point for the math score. Given M-T1, all other regression coefficients for SES are quite small. The coefficient of determination is zero for SES because this variable has only been used as an explanatory variable. The coefficients of determination for M-T1 and A-T1 are low indicating that SES may not be a very good predictor of math score and attitude. The coefficients of determination for M-T2 and A-T2 are quite high, due to the high autoregression coefficients.

The regression constants $\boldsymbol{\nu}$ of the measurement model are given in Table 8. These constants can be interpreted in the same way as the reduced form regression constants described previously.

Table 8. Regression constants of measurement model

Variables	Parameter estimates	z - values
SES	0.075	(3.987)
M-T1	5.319	(222.904)
M-T2	5.584	(221.315)
I1-T1	0.111	(3.533)
I2-T1	-0.353	(-11.083)
I3-T1	0.345	(10.785)
I1-T2	0.312	(9.735)
I2-T2	-0.088	(-2.765)
I3-T2	0.490	(14.831)

Before we turn to the matrix of factor loadings, we give the estimates of the proportionality coefficients in Table 9. Since we are interested in whether or not these coefficients are significantly different from one we give the the z-values for the test $H_0 : \alpha_j = 1, j = 1, 2, 3$ against $H_1 : \alpha_j \neq 1$. The null hypotheses cannot be rejected at the 0.05 test level indicating that these parameters should be restricted in a more refined model.

Table 9. Proportionality coefficients

Parameters	estimates	z - values
α_1	1.049	(1.372)
α_2	1.013	(0.322)
α_3	0.956	(1.439)

With these proportionality coefficients, the estimate of the matrix Λ of factor loadings is given in Table 10.

Table 10. Estimates of factor loadings

Variables	SES	M-T1	M-T2	A-T1	A-T2
SES	1.000	0.000	0.000	0.000	0.000
M-T1	0.000	1.000	0.000	0.000	0.000
M-T2	0.000	0.000	1.000	0.000	0.000
I1-T1	0.000	0.000	0.000	1.000	0.000
I2-T1	0.000	0.000	0.000	0.753	0.000
				(21.258)[a]	
I3-T1	0.000	0.000	0.000	0.772	0.000
				(20.747)	
I1-T2	0.000	0.000	0.000	0.000	1.049
I2-T2	0.000	0.000	0.000	0.000	0.763
I3-T2	0.000	0.000	0.000	0.000	0.738

[a]Note that only λ_{54} and λ_{64} are free parameters. Therefore, z-values (in parentheses) are shown only for these parameters.

The estimated covariance matrix of measurement errors is shown in Table 11.

Table 11. Covariance matrix of measurement errors

	SES	M-T1	M-T2	I1-T1	I2-T1	I3-T1	I1-T2	I2-T2	I3-T2
SES	0.000	0.000	0.000	0.000	0.000	0.000	0.000	0.000	0.000
M-T1	0.000	0.000	0.000	0.000	0.000	0.000	0.000	0.000	0.000
M-T2	0.000	0.000	0.000	0.000	0.000	0.000	0.000	0.000	0.000
I1-T1	0.000	0.000	0.000	0.246	0.000	0.000	0.027	0.000	0.000
I2-T1	0.000	0.000	0.000	0.000	0.573	0.000	0.000	0.232	0.000
I3-T1	0.000	0.000	0.000	0.000	0.000	0.550	0.000	0.000	0.180
I1-T2	0.000	0.000	0.000	0.027	0.000	0.000	0.264	0.000	0.000
I2-T2	0.000	0.000	0.000	0.000	0.232	0.000	0.000	0.611	0.000
I3-T2	0.000	0.000	0.000	0.000	0.000	0.180	0.000	0.000	0.636

The covariance between I1-T1 and I1-T2 is not significant at the 0.05 test level (z = 1.112). The other covariances, between I2-T1 and I2-T2 ($z = 9.623$) and I3-T1 and I3-T2, ($z = 8.817$) are significant indicating a significant correlation of the measurement errors over time.

The estimated thresholds are given in Table 12. The threshold values for I1-T2, I2-T2, and I3-T2 are obtained by multiplying the thresholds of I1-T1 with $\hat{\alpha}_1$, I2-T1 with $\hat{\alpha}_2$, and I3-T1 with $\hat{\alpha}_3$.

Table 12. Estimates of threshold values

Variables	τ_1	τ_2	τ_3	τ_4
I1-T1	0.000	0.708	1.274	1.543
I2-T1	0.000	0.668	1.106	1.361
I3-T1	0.000	0.767	1.396	1.747
I1-T2	0.000	0.743	1.337	1.619
I2-T2	0.000	0.677	1.121	1.379
I3-T2	0.000	0.736	0.334	1.669

Finally, we give the coefficients of determination for the items I1-T1 to I3-T2 in Table 13. These coefficients are computed with the formula of McKelvey and Zavoina (1975) discussed in Section 7.3 of Chapter 3.

Table 13. Coefficients of determination

Items	I1-T1	I2-T1	I3-T1	I1-T2	I2-T2	I3-T2
R^2	0.754	0.428	0.450	0.736	0.389	0.364

The coefficients of determination show that the items I1-T1 and I1-T2 best capture the attitude variable while the measurement error is much greater in the other items.

6 Software

A fair number of programs for computing estimates of the parameters of mean and covariance structures are available. We first describe briefly the programs that we have some experience in using and then give references for other computer programs.

EQS 4.0 is the most recent version of the EQS program (Bentler 1989, 1994). The program provides the computation of Pearson, polyserial, and polychoric correlations and the corresponding covariances from raw data for the unconditional model. Complicated hierarchical models may be formulated for the expected covariance matrix. The estimation procedures discussed in Section 1.3 may be used. The program provides a great number of test statistics and other diagnostic tools. The program handles multiple groups.

LISCOMP (Muthén 1988) is a program allowing structural equation modeling with a mixture of dichotomous, ordered categorical, censored, and normal or non-normal continuous dependent variables. Threshold, intercept, slope, and (residual) covariance and correlation structures in multiple groups can be handled, allowing conditioning on exogenous observed variables. A scaling matrix allows for comparison of latent response variable variances for categorical variables in multiple-group and longitudinal analysis. For the general case, weighted least-squares estimation is used. For bivariate categorical responses, full-information maximum-likelihood-estimation is also possible. Normality testing is available for polychoric correlations. Monte Carlo features are included.

LISREL 8 (Jöreskog and Sörbom 1993) is the most recent version of the LISREL program. The LISREL preprocessor PRELIS 2 allows one to compute polyserial and polychoric correlations and covariances for the conditional and the unconditional model. The covariance matrix obtained can then be analyzed within the LISREL 8 program, using the LISREL model discussed in Section 4.2. It provides the estimation procedures discussed in Section 1.3. Individual parameters can be restricted in a very flexible way allowing not only equality, but also more complicated restrictions. The program handles multiple groups.

MECOSA (Schepers and Arminger 1992) is a program written in GAUSS to analyze complex mean and covariance structures with metric and/or nonmetric dependent variables. The conditional mean and covariance matrices of dependent variables can be structured as arbitrary differentiable functions of the parameters of interest. Therefore, structures with nonlinear restrictions on the parameters such as nonlinear simultaneous probit models or models which are formulated in terms of the inverse of a covariance matrix can be estimated. The variable types include metric, one- and double-sided censored, classified, dichotomous, and ordered categorical variables. The program can deal with multiple groups. For the computation of the conditional polyserial and polychoric covariances and correlations, Fisher's z-transformation is used.

RAMONA (Browne, Mels and Coward 1994) implements McArdle and McDonald's (1984) Reticular Action Model, given in (4.77), and has the facility of handling a correlation matrix though use of (4.82) in conjunction with the nonlinear equality constraints (4.81). It can also constrain endogenous latent variable variances to be equal to one and consequently is able to provide standard errors for standardized path coefficients and associated confidence intervals. The specific treatment of correlation structures in RAMONA enables the program to avoid common mistakes associated with the analysis of correlation matrices (Cudeck 1989). RAMONA was employed for carrying out the analyses reported in Section 4.2. The model is conveyed to the program using instructions, coded directly from the path diagram, of the form ANOMIA67 $\leftarrow$ ALNTN67 (see Table 2). There is no use of algebraic notation. Similar notation is used for output, as may be seen in Table 2. One-at-a-time confidence intervals are provided for all parameters in the model as well as for population values of the fit measures employed.

Some additional computer programs are AMOS by Arbuckle (1994), CALIS by Hartmann (1992), COSAN by Fraser and McDonald (1988), LINCS by Schoenberg and Arminger (1988), Mx by Neale (1991), and SEPATH by Steiger (1994).

REFERENCES

Aitchison, J., and Silvey, S. D. (1960), "Maximum Likelihood Estimation Procedures and Associated Tests of Significance," *Journal of the Royal Statistical Society*, Series B, 22, 154–171.

Albert, A. A. (1944), "The Minimum Rank of a Correlation Matrix," *Proceedings of the National Academy of Science*, 30, 144–146.

Allison, P. D. (1987), "Estimation of Linear Models with Incomplete Data," *Sociological Methodology 1987*, ed. C. C. Clogg, Washington, DC: American Sociological Association, 71–103.

Amemiya, Y., and Anderson, T. W. (1990), "Asymptotic Chi-Square Test for a Large Family of Factor Analysis Models," *Annals of Statistics*, 18, 1453–1463.

Anderson, T. W. (1960), "Some stochastic process models for intelligence test scores," in: K. J. Arrow, S. Karlin, and P. Suppes (eds.), *Mathematical Methods in the Social Sciences*, pp. 205–220. Stanford: Stanford University Press.

Anderson, T. W., and Amemiya, Y. (1988), "The Asymptotic Distribution of Estimators in Factor Analysis under General Conditions," *Annals of Statistics*, 16, 759–771.

Anderson, T. W., and Rubin, H. (1956), "Statistical Inference in Factor Analysis," in J. Neyman (ed.), *Proceedings of the Third Berkeley Symposium on Mathematical Statistics and Probability*, Vol. V, Berkeley, University of California Press, 111–150.

Arbuckle, J. (1994), *AMOS 3.5: Analysis of Moment Structures*, Chicago IL: Smallwaters Corporation.

Archer, C. O., and Jennrich, R. I. (1973), "Standard Errors for Orthogonally Rotated Factor Loadings," *Psychometrika*, 38, 581–592.

Arminger, G. (1994), "Dynamic Factor Models for the Analysis of Ordered Categorical Panel Data," unpublished manuscript, Department of Economics, University of Wuppertal.

Arminger, G., and Schoenberg, R. J. (1989), "Pseudo maximum likelihood estimation and a test for misspecification in mean- and covariance-structure models," *Psychometrika*, 54, 409–425.

Arminger, G., and Sobel, M. (1990), "Pseudo maximum likelihood estimation of mean and covariance structures with missing data," *Journal of the American Statistical Association*, 85, 195–203.

Austin, J., and Wolfle, L. M. (1991), "Annotated bibliography of structural equation modeling: technical work," *British Journal of Mathematical and Statistical Psychology*, 44, 93–152.

Bagozzi, R. P., and Yi, Y. (1992), "Testing hypotheses about methods, traits and communalities in the direct product model," *Applied Psychological Measurement*, 16, 373–380.

Bekker, P. A., Merckens, A., and Wansbeck, T. (1994), *Identification, Equivalent Models and Computer Algebra*, Boston: Academic Press.

Bentler, P. M. (1989), *Theory and Implementation of EQS*, Los Angeles: BMDP Statistical Software Inc.

——— (1994), *EQS 4.0*, BMDP Statistical Software, Los Angeles.

Bentler, P. M., and Dijkstra, T. (1985), "Efficient Estimation via Linearization in Structural Models," in: P. Krishnaiah (ed.), *Multivariate Analysis VI*, Amsterdam: Elsevier, 9–42.

Bentler, P. M., and Weeks, D. G. (1980), "Linear Structural Equations with Latent Variables," *Psychometrika*, 45, 289–308.

Bock, R. D., and Gibbons, R. D. (1994), "High-dimensional Multivariate Probit Analysis," unpublished manuscript, Department of Psychology, University of Chicago.

Bollen, K. A. (1989), *Structural Equations with Latent Variables*, New York: Wiley.

Bollen, K. A., and Long, J. S. (eds.) (1993), *Testing Structural Equation Models*, Newbury Park: Sage.

Browne, M. W. (1974), "Generalised least squares estimators in the analysis of covariance structures," *South African Statistical Journal*, 8, 1–24. [Reprinted in D.J. Aigner and A.S. Goldberger (eds.), *Latent Variables in Socio-Economic Models*, pp. 205–226. Amsterdam: North Holland, 1977.]

——— (1982), "Covariance structures," in D.M. Hawkins (ed.), *Topics in Applied Multivariate Analysis*, pp. 72–141, Cambridge: Cambridge University Press.

——— (1984a), "Asymptotically distribution free methods in the analysis of covariance structures," *British Journal of Mathematical and Statistical Psychology*, 37, 62–83.

——— (1984b), "The decomposition of multitrait-multimethod matrices," *British Journal of Mathematical and Statistical Psychology*, 37, 1–21.

——— (1990), "Asymptotic robustness of normal theory methods for the analysis of latent curves," in W. Fuller and P. Brown (eds.) *Statistical Analysis of Measurement Error Models and Applications*, pp. 211–225, American Mathematical Society: Contemporary Mathematics Series, 112.

——— (1993), "Structured Latent Curve Models," in C. M. Cuadras and C. R. Rao (eds.), *Multivariate Analysis: Future Directions 2*, pp. 171–198, Amsterdam: North Holland.

Browne, M. W., and Cudeck, R. (1993), "Alternative ways of assessing model fit," in: Bollen, K. A., and Long, J. S. (eds.), *Testing Structural Equation Models*, Newbury Park: Sage, 136–162.

Browne, M. W., and Du Toit, S. H. C. (1991), "Models for learning data," in L. M. Collins and J. Horn (eds.), *Best Methods for the Analysis of Change*, pp. 47–68, Washington: American Psychological Association.

——— (1992), "Automated fitting of nonstandard models," *Multivariate Behavioral Research*, 27, 269–300.

Browne, M. W., and Mels, G. (1990), *RAMONA PC User's Guide*. Report: Department of Statistics, University of South Africa.

Browne, M. W., Mels, G., and Coward, M. (1994), "Path Analysis: Ramona," *SYSTAT for DOS: Advanced Applications, Version 6 Edition*, Evanston, IL: Systat Inc., 163–224 (in press).

Browne, M. W., and Shapiro, A. (1988), "Robustness of normal theory methods in the analysis of linear latent variate models," *British Journal of Mathematical and Statistical Psychology*, 41, 193–208.

Cliff, N. (1983), "Some cautions concerning the application of causal modeling methods," *Multivariate Behavioral Research*, 18, 115–126.

Cronbach, L. J., Gleser, G. C., Nanda, H., and Rajaratman, N. (1972), *The Dependability of Behavioral Measurements: Theory of Generalizability for Scores and Profiles*, New York: Wiley.

Cudeck, R. (1988), "Multiplicative models and MTMM matrices," *Journal of Educational Statistics*, 13, 131–147.

———(1989), "Analysis of correlation matrices using covariance structure models," *Psychological Bulletin*, 105, 317–327.

———(1991), "Noniterative factor analysis estimators with algorithms for subset and instrumental variable selection," *Journal of Educational Statistics*, 16, 35–52.

Cudeck, R., and O'Dell, L. L. (1994), "Applications of standard error estimates in unrestricted factor analysis: Significance tests for factor loadings and correlations," *Psychological Bulletin*, 115, 475–487.

Dijkstra, T. K. (1983), "Some comments on maximum likelihood and partial least squares methods," *Journal of Econometrics*, 22, 67–90.

———(1992), "On statistical inference with parameter estimates on the boundary of the parameter space," *British Journal of Mathematical and Statistical Psychology*, 45, 289–309.

Du Toit, S. H. C. (1979), *The Analysis of Growth Curves*, Ph. D. dissertation, Department of Statistics, University of South Africa.

Fraser, C., and McDonald, R. P. (1988), "Covariance Structure Analysis," *Multivariate Behavioral Research*, 23, 263–265.

GAUSS, Version 3.1 (1993), *Systems and Graphics Manual*, Aptech Systems, Kent, Washington.

Gill, P. E., Murray, W., and Wright, M. H. (1981), *Practical Optimisation*, London: Academic Press.

Guttman, L. (1954), "A new approach to factor analysis: the radex," in P. F. Lazarsfeld (ed.), *Mathematical Thinking in the Social Sciences*, pp. 258–348, Glencoe, IL: The Free Press.

Hartmann, W. M. (1992), *The CALIS procedure: Extended User's Guide,* Cary, NC: The SAS Institute.

Ihara, M., and Kano, Y. (1986), "A new estimator of the uniqueness in factor analysis," *Psychometrika*, 51, 563–566.

Jennrich, R. I. (1970), "An asymptotic chi-square test for the equality of two correlation matrices," *Journal of the American Statistical Association*, 65, 904–912.

Jennrich, R. I., and Clarkson, D.B. (1980), "A feasible method for standard errors of estimate in maximum likelihood factor analysis," *Psychometrika*, 45, 237–247.

Jennrich, R. I., and Robinson, S. M. (1969), "A Newton-Raphson algorithm for maximum likelihood factor analysis," *Psychometrika*, 34, 111–123.

Jennrich, R. I., and Sampson, P. F. (1966), "Rotation for simple loadings," *Psychometrika*, 31, 313–323.

———(1968), "Application of stepwise regression to nonlinear estimation," *Technometrics*, 10, 63–72.

Jöreskog, K. G. (1967), "Some contributions to maximum likelihood factor analysis," *Psychometrika*, 32, 443–482.

———(1969), "A general approach to confirmatory maximum likelihood factor analysis," *Psychometrika*, 34, 183–202.

———(1970), "Estimation and testing of simplex models," *British Journal of Mathematical and Statistical Psychology*, 23, 121–145.

———(1973), "A general method for estimating a linear structural equation system," in: A. S. Goldberger and O. D. Duncan (eds.), *Structural Equation Models in the Social Sciences*, New York: Academic Press, pp. 85–112.

Jöreskog, K. G., and Sörbom, D. (1986), *LISREL VI: Analysis of Linear Structural Relationships by Maximum Likelihood, Instrumental Variables and Least Squares Methods, Users Guide*, Mooresville, IN: Scientific Software.

———(1993), *LISREL 8: Structural Equation Modeling with the SIMPLIS Command Language*, Hillsdale, NJ: Erlbaum.

Kaiser, H. F. (1958), "The Varimax Criterion for Oblique Rotation in Factor Analysis," *Psychometrika*, 23, 187–200.

Kano, Y. (1986), "Conditions on Consistency of Estimators in Covariance Structure Model," *Journal of the Japanese Statistical Society*, 13, 137–144.

———(1990), "Non-iterative Estimation and the Choice of the Number of Factors in Exploratory Factor Analysis," *Psychometrika*, 55, 277-291.

Keesling, J. W. (1972), "Maximum Likelihood Approaches to Causal Analysis," unpublished Ph.D. dissertation. Department of Education, University of Chicago.

Kiiveri, H. T. (1987), "An Incomplete Data Approach to the Analysis of Covariance Structures," *Psychometrika*, 52, 539–554.

Küsters, U. (1987), *Hierarchische Mittelwert- und Kovarianzstrukturmodelle mit nichtmetrischen endogenen Variablen*. Heidelberg: Physica Verlag.

———(1990), "A note on sequential ML estimates and their asymptotic covariance matrices," *Statistical Papers 31*, 131–145.

Lawley, D. N. (1940), "The Estimation of Factor Loadings by the Method of Maximum Likelihood," *Proceedings of the Royal Society of Edinburgh*, 60, 64–82.

Lee, S., and Hershberger, S. (1990), "A simple rule for generating equivalent models in covariance structure modeling," *Multivariate Behavioral Research*, 25, 313–334.

Lee, S. Y., and Jennrich, R. I. (1979), "A Study of Algorithms for Covariance Structure Analysis with Specific Comparisons Using Factor Analysis," *Psychometrika*, 44, 99–113.

Lord, F. M., and Novick, M. R. (1968), *Statistical Theories of Mental Test Scores*. Reading, Massachusetts: Addison Wesley.

MacCallum, R. C., Wegener, D. T., Uchino, B. N., and Fabrigar, L. R. (1993), "The problem of equivalent models in applications of covariance structure analysis," *Psychological Bulletin*, 114, 185–199.

Magnus, J. R., and Neudecker, H. (1988) *Matrix Differential Calculus with Applications in Statistics and Econometrics*. New York: Wiley.

Mardia, K.V. (1970), "Measures of multivariate skewness and kurtosis with applications," *Biometrika*, 57, 519–530.

McArdle, J. J. (1970), "The development of general multivariate software," in: J. J. Hirschbuhl (ed.), *Proceedings of the Association for the Development of Computer-based Instructional Systems*, Akron, OH: University of Akron Press.

McArdle, J. J., and McDonald, R. P. (1984), "Some algebraic properties of the Reticular Action Model for moment structures," *Journal of Mathematical and Statistical Psychology*, 37, 234–251.

McDonald, R. P. (1975), *Factor Analysis and Related Methods*, Hillsdale, NJ: Erlbaum.

———(1980), "A simple comprehensive model for the analysis of covariance structures: some remarks on applications," *British Journal of Mathematical and Statistical Psychology*, 33, 161–183.

McDonald, R. P., Parker, P. M., and Ishizuka, T. (1993), "A scale invariant treatment for recursive path models," *Psychometrika*, 58, 431–443.

McKelvey, R. D., and Zavoina, W. (1975), "A Statistical Model for the Analysis of Ordinal Level Dependent Variables," *Journal of Mathematical Sociology 4*, 103–120.

Miller, J. D., Hoffer, T., Suchner, R. W., Brown, K. G., and Nelson, C. (1992), *LSAY Codebook: Student, Parent, and Teacher Data for Cohort Two for Longitudinal Years One through Four (1987 - 1991)*, Vol. 2, Northern Illinois University, De Kalb, 60115 - 2854.

Muthén, B. O. (1984), "A General Structural Equation Model with Dichotomous, Ordered Categorical, and Continuous Latent Variable Indicators," *Psychometrika 49*, 115–132.

———(1988), *LISCOMP: Analysis of Linear Structural Equations with a Comprehensive Measurement Model*, Mooresville, IN: Scientific Software.

Muthén, B. O., and Arminger, G. (1994), "Bayesian Latent Variable Regression for Binary and Continuous Response Variables Using the Gibbs Sampler," unpublished manuscript, Graduate School of Education, University of California, L.A.

Muthén, B. O., and Kaplan, D. (1992), "A comparison of some methodologies for the factor analysis of non-normal Likert variables: A note on the size of the model," *British Journal of Mathematical and Statistical Psychology*, 45, 19–30.

Muthén, B. O., and Satorra, A. (1994), "Technical Aspects of Muthén's LISCOMP Approach to Estimation of Latent Variable Relations with a Comprehensive Measurement Model," under review in *Psychometrika.*

Neale, M. C. (1991), *Mx: Statistical Modeling*, Department of Human Genetics, Box 3 MCV, Richmond, VA 23298.

Olsson, U. (1979), "Maximum Likelihood Estimation of the Polychoric Correlation Coefficient," *Psychometrika*, 44, 443–460.

Olsson, U., Drasgow F., and Dorans, N. J. (1982), "The Polyserial Correlation Coefficient," *Psychometrika*, 47, 337–347.

Pearson, K. (1900), "Mathematical Contributions to the Theory of Evolution in the Inheritance of Characters Not Capable of Exact Quantitative Measurement," *VIII. Philosophical Transactions of the Royal Society A*, 195, 1075–1090.

Rao, C. R. (1958), "Some statistical methods for comparison of growth curves," *Biometrika*, 14, 1–17.

Rosett, R. N., and Nelson, F. D. (1975), "Estimation of the two-limit probit regression model," *Econometrica*, 43, 141–146.

Satorra, A. (1989), "Alternative test criteria in covariance structure analysis: a unified approach," *Psychometrika*, 54, 131–151.

——(1993), "Asymptotic robust inferences in multi–sample analysis of augmented–moment structures," in C. M. Cuadras and C. R. Rao (eds.), *Multivariate Analysis: Future Directions 2*, 211–229, Amsterdam: North Holland.

Satorra, A., and Saris, W. E. (1985), "The power of the likelihood ratio test in covariance structure analysis," *Psychometrika*, 50, 83–90.

Shapiro, A. (1983), "Asymptotic distribution theory in the analysis of covariance structures," *South African Statistical Journal*, 17, 33–81.

——(1984), "A Note on the Consistency of Estimators in the Analysis of Moment Structures," *British Journal of Mathematical and Statistical Psychology*, 37, 84–88.

——(1985a), "Asymptotic Equivalence of Minimum discrepancy Estimators to GLS Estimators," *South African Statistical Journal*, 19, 73–81.

——(1985b), "Asymptotic distribution of test statistics in the analysis of moment structures under inequality constraints," *Biometrika*, 72, 133–184.

——(1987), "Robustness Properties of the MDF Analysis of Moment Structures," *South African Statistical Journal*, 21, 39–62.

Shapiro, A., and Browne, M. W. (1989), "On the Asymptotic Bias of Parameters under Parameter Drift," *Statistics and Probability Letters*, 7, 221–224.

——(1990), "On the Treatment of Correlation Structures as Covariance Structures," *Linear Algebra and its Applications*, 127, 567–587.

Schepers, A., and Arminger, G. (1992), *MECOSA: A Program for the Analysis of General Mean- and Covariance Structures with Non-Metric Variables, User Guide*, Frauenfeld: SLI-AG, Zürcher Str. 300, CH-8500 Frauenfeld, Switzerland.

Schoenberg, R., and Arminger, G. (1988), *LINCS2: A Program for Linear Covariance Structure Analysis*, Kensington, MD: RJS Software.

Sobel, M., and Arminger, G. (1992), "Modeling Household Fertility Decisions: A Nonlinear Simultaneous Probit Model," *Journal of the American Statistical Association*, 87, 38–47.

Steiger, J. H. (1990), "Structural model evaluation and modification: an interval estimation approach," *Multivariate Behavioral Research*, 25, 173–180.

——(1994), *Structural Equation Modeling: Technical Documentation*, StatSoft: STATISTICA (in press).

Steiger, J. H., and Lind, J. C. (1980), "Statistically based tests for the number of common factors," paper presented at the annual meeting of the Psychometric Society, Iowa City, Iowa.

Steiger, J. H., Shapiro, A., and Browne, M. W. (1985), "On the multivariate asymptotic distribution of sequential chi–square test statistics," *Psychometrika*, 50, 253–264.

Stewart, M. B. (1983), "On Least Squares Estimation When the Dependent Variable Is Grouped," *Review of Economic Studies*, 50, 737–753.

Swain, A. J. (1975a), "A Class of Factor Analysis Estimation Procedures with Common Asymptotic Properties," *Psychometrika*, 40, 315–335.

——(1975b), *Analysis of Parametric Structures for Variance Matrices*, unpublished Ph.D. dissertation, University of Adelaide.

Tobin, J. (1958), "Estimation of Relationships for Limited Dependent Variables," *Econometrica*, 26, 24–36.

Tucker, L. R. (1958), "Determination of parameters of a functional relation by factor analysis," *Psychometrika*, 23, 19–23.

Wald, H. (1943), "Tests of Statistical Hypotheses Concerning Several Parameters When the Number of Observations Is Large," *Transactions of the American Mathematical Society*, 54, 426–482.

Wiley, D. E. (1973), "The Identification Problem for Structural Equation Models with Unmeasured Variables," in A. S. Goldberger and O. D. Duncan (eds.), *Structural Equation Models in the Social Sciences*, New York: Academic Press, 69–83.

Chapter 5

The Analysis of Contingency Tables

MICHAEL E. SOBEL

1 Introduction

Social and behavioral scientists routinely use statistical models to make inferences about the distribution of one or more dependent variables (which may be observed and/or unobserved), conditional on a set of independent variables. Often, this conditional distribution is assumed to be absolutely continuous, and in many instances, normal.

The models discussed in this chapter differ from the characterization above in two important respects. First, they may be viewed as models for the univariate or joint distribution of a set of random variables, and therefore it is not necessary to distinguish between dependent and independent variables. If desired, this distinction can be brought into the analysis by using the relationship between the joint and conditional distributions. Second, the distribution of the random variables in a log-linear model is discrete. In particular, the data are usually assumed to follow either the Poisson, multinomial, or independent (product) multinomial distributions.

Categorical data arise in numerous ways. Most notable is the case where measurements on one or more variables are taken on a nominal scale (scales). Here, log-linear models can be used to test hypotheses about the distribution of the variable(s). For example, in the univariate case one might ask whether the observed counts are consistent with the hypothesis of equiprobability, and in the bivariate case, whether the counts in the two-way cross-classification table are consistent with the hypothesis of statistical independence. Both questions may be addressed without using log-linear models, but when the null hypothesis is rejected, the log-linear model can be used to address the types of departures from the null. The same models may also be used when one or more variables are measured on

MICHAEL E. SOBEL • Department of Sociology, University of Arizona, Tucson, Arizona 85721, USA. •
For helpful comments and discussion, I am grateful to Gerhard Arminger, Clifford C. Clogg, David Draper, Scott Eliason, and J. Scott Long.

Handbook of Statistical Modeling for the Social and Behavioral Sciences, edited by Gerhard Arminger, Clifford C. Clogg, and Michael E. Sobel. Plenum Press, New York, 1995.

an ordinal scale. Here, models for ordinal data can be used to test the types of hypotheses above and in addition, to estimate the distances between categories of the ordered variable(s). Similar models may be used when one or more of the variables is measured on a partially ordered scale, or when interval or ratio data have been discretized. Although the practice of discretizing interval or ratio data can lead to a loss of information, in many instances investigators only have access to data in this form. Finally, log-linear models are also used to analyze counts; here the data are measured on an absolute scale.

Compared to regression methods, the log-linear model is relatively new, although its origins can be traced to the work of Pearson and Yule in the early twentieth century. Pearson (1900) derived the χ^2 test for the null hypothesis that the data are drawn from a multinomial distribution with known probabilities, and he also proposed a χ^2 test for the null hypothesis that two discrete random variables are independent. The latter test, based on the wrong degrees of freedom, was subsequently corrected by Fisher (1922), and the corrected version is the familiar Pearson χ^2 test for independence that is taught in basic statistics courses. In the same year, Yule's Q (Yule 1900) was proposed as a measure of the association between two dichotomous variables. This measure is a function of the cross product (or odds) ratio, which is the basic measure of association used in log-linear models. Despite his contributions to the statistical theory underlying log-linear models, Pearson (Pearson and Heron 1913) staunchly opposed Yule's proposals. Whereas Yule argued that the observed variables in a cross-classification were the variables of interest, Pearson argued that the observed variables were merely discretized versions of two continuous random variables with a bivariate normal distribution. This was the basis for the tetrachoric correlation coefficient (Pearson 1904). Thus, Pearson might be viewed as the forefather of multivariate probit models, while Yule might be viewed as the forefather of the log-linear model. Several important subsequent developments include Bartlett (1935), who examined the two by two by two table, using conditional odds ratios to define the concept of three-factor interaction, Roy and Kastenbaum (1956), who extended Bartlett's work, and Edwards (1963). A key breakthrough is Birch (1963), who expressed the log-linear model in its current ANOVA-like form, and developed the basic asymptotic theory under Poisson, multinomial and product multinomial sampling.

For the past 30 years, categorical data analysis has been an exceptionally active area of statistical research. At this point, an inclusive survey of the log-linear model alone and its connections with other areas of statistical research would require a lengthy book, as opposed to a short chapter. Fortunately, a number of book-length treatments and specialized survey articles exist that (taken together) provide a comprehensive treatment. Some of the booklength treatments include: (1) Agresti (1984), a text focusing on methods for ordinal discrete data; (2) Agresti (1990), a graduate text that covers the basic log-linear model and the theory underlying it, logistic regression and generalized linear models, as well as applications to survival analysis and longitudinal analysis; (3) Anderson (1990), a graduate text emphasizing inference in exponential families, with a chapter on latent structure analysis; (4) Bishop, Fienberg, and Holland (1975), a classic text offering a comprehensive introduction to the log-linear model and to the literature on discrete data prior to 1975; (5) Christensen (1990), a graduate text requiring less mathematical sophistication than Agresti (1990), but more than Fienberg (1980), with material on graphical models; (6) Clogg and

Shihadeh (1994), an introductory level treatment of methods for ordinal data; (7) Fienberg (1980), an excellent introductory level treatment, though now somewhat dated; (8) Goodman (1984), a collection of previously published articles by Goodman and Clogg on models for ordinal discrete data; (9) Haberman (1974a), a very mathematical treatment; (10) Haberman (1978, 1979), a two-volume set divided into introductory and more advanced topics, including material on Haberman's latent class model; (11) Plackett (1981), a rather theoretical introduction to the subject.

Here, space limitations preclude consideration of a number of relevant topics in the literature, including log-linear models for rates (Haberman 1978), survival analysis (Allison 1982; Laird and Oliver 1981), correspondence analysis (Benzécri 1969; Benzécri and Gopalan 1991; Goodman 1991), missing data (Little and Schenker, Chapter 2), latent trait and latent class models (Clogg, Chapter 6), network analysis (Wasserman and Faust 1994), marginal models and longitudinal categorical data (Liang, Zeger, and Qaquish 1992; Fitzmaurice, Laird, and Rotnitsky 1993; Gilula and Haberman 1994), and the usual approach to causal modeling in the literature. In addition, attention is confined to models where all the variables are discrete, thereby excluding logistic regression and probit models (Agresti 1990; Amemiya 1981; Arminger, Chapter 3; Cox and Snell 1989; Maddala 1983; McCullagh and Nelder 1989). A number of other important topics are mentioned only in passing; when this is the case, citations to the literature are given.

The paper proceeds as follows. Section 2 uses several examples to introduce the reader to the log-linear model. Section 3 discusses the use of the odds ratio as a measure of association in two-way and three-way tables. Section 4 introduces models for the two-way table, Section 5 extends the discussion to the three-way table, and Section 6 takes up the case of higher-way tables. Section 7 discusses estimation theory for the models. Section 8 discusses residuals and model selection procedures, and Section 9 discusses computer programs that can be used to fit the models considered herein.

2 Introductory Examples

2.1 Some Models for Univariate Distributions

In this section, the general log-linear model is introduced. Maximum-likelihood procedures are used to estimate the model parameters, and goodness-of-fit tests are also introduced. Throughout, the data are assumed to be drawn from a multinomial distribution. In Section 7, these procedures will be justified and the distributional assumptions extended to Poisson sampling and independent multinomial sampling.

Let $M(n, \boldsymbol{\pi})$ denote a multinomial distribution with T categories, where n denotes the sample size, $\boldsymbol{\pi}' = (\pi_1, \ldots, \pi_T)$, $0 < \pi_t < 1, t = 1, \ldots, T$, and $\sum_{t=1}^{T} \pi_t = 1$. The probability distribution is

$$p(\boldsymbol{n}; \boldsymbol{\pi}) = \frac{n!}{\prod_{t=1}^{T} n_t!} \prod_{t=1}^{T} \pi_t^{n_t}, \tag{5.1}$$

where $\boldsymbol{n} = (n_1, \ldots, n_T)'$, and n_t is (for now) a positive integer such that $\sum_{t=1}^{T} n_t = n$. The log-likelihood function corresponding to (5.1) is

$$\begin{aligned} \ell(\boldsymbol{\pi};\boldsymbol{n}) &= \log n! - \sum_{t=1}^{T} \log n_t! + \sum_{t=1}^{T} n_t \log \pi_t \\ &= \log n! - \sum_{t=1}^{T} \log n_t! - n\log n + \sum_{t=1}^{T} n_t \log m_t, \end{aligned} \tag{5.2}$$

where $m_t = n\pi_t$ is the expected count in the t-th category. Note that $\sum_{t=1}^{T} m_t = n$.

A log-linear model for m_t is linear in the parameters on the logarithmic scale:

$$\log m_t = \boldsymbol{x}_t'\boldsymbol{\beta}, \tag{5.3}$$

where $\boldsymbol{x}_t = (1, \boldsymbol{z}_t')'$ is a $p \times 1$ vector of real numbers, and $\boldsymbol{\beta} = (\beta_0, \boldsymbol{\beta}_1')'$ is a $p \times 1$ parameter vector to be estimated. The model may also be written as

$$\log \boldsymbol{m} = \boldsymbol{X}\boldsymbol{\beta}, \tag{5.4}$$

where $\boldsymbol{m} = (m_1, \ldots, m_T)'$, and $\boldsymbol{X} = (\boldsymbol{x}_1, \ldots, \boldsymbol{x}_T)'$ is the model matrix. In this chapter, $\boldsymbol{X}$ is assumed to have full column rank p.

Substitution of (5.3) into (5.2) yields the log-likelihood expressed as a function of $\boldsymbol{\beta}$:

$$\ell(\boldsymbol{\beta};\boldsymbol{n}) = \log n! - \sum_{t=1}^{T} \log n_t! - n\log n + \sum_{t=1}^{T} n_t \boldsymbol{x}_t'\boldsymbol{\beta}. \tag{5.5}$$

Note that only the last term of (5.5) involves the parameters. This is the kernel of the log-likelihood.

The maximum likelihood estimator (MLE) of $\boldsymbol{\beta}$ is obtained by maximizing (5.5) or equivalently, the kernel of (5.5) with respect to $\boldsymbol{\beta}$ under the constraint $\sum_{t=1}^{T} m_t = n$. To impose this constraint in the maximization problem, the method of Lagrange multipliers can be used. Alternatively, $\exp(\beta_0 + \boldsymbol{z}_t'\boldsymbol{\beta}_1)$ can be substituted for m_t in the constraint equation, whence

$$\beta_0 = \log n - \log \sum_{t=1}^{T} \exp(\boldsymbol{z}_t'\boldsymbol{\beta}_1). \tag{5.6}$$

Substitution of (5.6) into (5.5) yields

$$\ell^*(\boldsymbol{\beta}_1;\boldsymbol{n}) = \log n! - \sum_{t=1}^{T} \log n_t! + \sum_{t=1}^{T} n_t(\boldsymbol{z}_t'\boldsymbol{\beta}_1 - \log \sum_{t=1}^{T} \exp(\boldsymbol{z}_t'\boldsymbol{\beta}_1)), \tag{5.7}$$

and maximization of (5.7) with respect to $\boldsymbol{\beta}_1$ yields the MLE $\hat{\boldsymbol{\beta}}_1$ of $\boldsymbol{\beta}_1$. The MLE of β_0 is then $\hat{\beta}_0 = \log n - \log \sum_{t=1}^{T} \exp(\boldsymbol{z}_t'\hat{\boldsymbol{\beta}}_1)$. Note that the parameter β_0 is a function of the other parameters and sample size. Since the sample size is fixed under multinomial sampling, this parameter is redundant, and its inclusion is necessary only to insure that the constraints are satisfied. It follows that the MLE of $\log m_t$ is $\boldsymbol{x}_t'\hat{\boldsymbol{\beta}}$ and the MLE $\hat{m}_t$ of m_t is $\exp(\boldsymbol{x}_t'\hat{\boldsymbol{\beta}})$.

It is also worth noting that the log-linear model above is equivalent to the multinomial response model

$$\pi_t = \frac{\exp(z_t'\beta_1)}{\sum_{t=1}^T \exp(z_t'\beta_1)}. \tag{5.8}$$

To see this, substitute (5.6) into (5.3) and rearrange.

Testing the fit of the model requires comparison of $\hat{m}_t$ with n_t, $t = 1, \ldots, T$. A number of statistics can be used for this purpose (Read and Cressie 1988). Here, three statistics are used. The first two are the familiar Pearson χ^2 and the likelihood ratio χ^2. These are, respectively:

$$X^2 = \sum_{t=1}^T \frac{(n_t - \hat{m}_t)^2}{\hat{m}_t}, \tag{5.9}$$

$$L^2 = 2\sum_{t=1}^T n_t \log(\frac{n_t}{\hat{m}_t}). \tag{5.10}$$

For future reference, note that the value of β that maximizes (5.5) is identical to the value that minimizes $2\sum_{t=1}^T n_t \log(\frac{n_t}{m_t})$.

The third statistic is the power divergence statistic

$$D^2(\lambda) = \frac{2}{\lambda(\lambda+1)} \sum_{t=1}^T n_t((\frac{n_t}{\hat{m}_t})^\lambda - 1). \tag{5.11}$$

Read and Cressie (1988) discuss this statistic and suggest using the value $\lambda = 2/3$, a compromise between X^2 and L^2; hereafter, this value is utilized and the statistic $D^2(2/3)$ is written as D^2.

Under the null hypothesis that the model holds, all three statistics are asymptotically distributed (as $n \to \infty$, where T is fixed) χ^2 with $(T-1)-(p-1) = T-p$ degrees of freedom (df). Note that $p-1$ is the number of functionally independent parameters estimated under the model. Note also that as p is the number of linearly independent parameters estimated under the model, the df may be calculated as $T - p$ = number of cells − number of linearly independent parameters estimated.

Example One: Liberalism/Conservatism

The 1977 General Social Survey (GSS), conducted by the National Opinion Research Center (Davis 1980), asked respondents to place themselves on a seven-point scale ranging from extremely liberal to extremely conservative. Excluding the 77 respondents who either did not answer the question at all or who did not know where to place themselves on the scale, the response categories and sample counts (in parentheses) for this discrete random variable (Y) are: extremely liberal (37), liberal (169), slightly liberal (214), moderate, middle of the road (564), slightly conservative (251), conservative (179), extremely conservative (39). See also Clogg and Shockey (1988), who analyzed responses to the same item using the 1982 GSS.

A "saturated" log-linear model for the data is given by (5.4) with $\boldsymbol{\beta} = (\beta_0, \beta_1, \beta_2, \beta_3, \beta_4, \beta_5, \beta_6)'$, and model matrix

$$\begin{pmatrix} 1 & 1 & 0 & 0 & 0 & 0 & 0 \\ 1 & 0 & 1 & 0 & 0 & 0 & 0 \\ 1 & 0 & 0 & 1 & 0 & 0 & 0 \\ 1 & 0 & 0 & 0 & 1 & 0 & 0 \\ 1 & 0 & 0 & 0 & 0 & 1 & 0 \\ 1 & 0 & 0 & 0 & 0 & 0 & 1 \\ 1 & -1 & -1 & -1 & -1 & -1 & -1 \end{pmatrix}$$

Alternatively, the model may be written in the usual ANOVA fashion:

$$\log m_t = \mu + \alpha_t^Y, \tag{5.12}$$

$t = 1, \ldots, 7$. In this form, the model is overparameterized. To identify the model, it is usual to set α_t^Y to 0 for some t, or to impose the constraint $\sum_{t=1}^{7} \alpha_t^Y = 0$. For the previous model matrix, the relationship between $\boldsymbol{\beta}$ and the parameters of (5.12) is

$$\mu = \beta_0;\ \alpha_t^Y = \beta_t, t = 1, \ldots, 6;\ \alpha_7^Y = -\sum_{t=1}^{6} \beta_t.$$

Under the saturated model, six independent parameters $(\alpha_1^Y, \ldots, \alpha_6^Y)$ are estimated. The df are therefore $(7-1) - 6 = 0$. It is not hard to see that under the saturated model, $n_t = \hat{m}_t$, that is, the sample counts are fitted perfectly.

Typically, a substantive investigator is not interested in the saturated model but in various more restricted alternatives. These may often be obtained by imposing additional structure on the model parameters. For example, if $\alpha_t^Y = 0$ for every t, the model

$$\log m_t = \mu \tag{5.13}$$

corresponding to the uniform multinomial distribution is obtained. Under (5.13), all response categories are equally probable, $\mu = \hat{\mu} = \log(n/7) = \log(1453/7)$, and the null hypothesis of a uniform multinomial can be tested using either X^2, L^2, or D^2 on 6 df. The values of these statistics are, respectively, 855.42, 908.81, and 868.57, indicating that (5.13) is untenable.

More interesting hypotheses can be tested if the ordinal nature of the response variable is taken into account. For example, Haberman (1978) and Clogg and Shockey (1988) consider log-linear models for univariate ordered responses in which the probability distribution is symmetric about the "middle" category(ies). For these data, T is odd and the hypothesis of symmetry is: $\pi_1 = \pi_7$, $\pi_2 = \pi_6$, $\pi_3 = \pi_5$. The symmetry model can be estimated by imposing suitable restrictions on the parameters of (5.12) , as in Haberman (1978) or Clogg and Shockey (1988). However, other parameterizations may be more informative; for example, consider:

$$\log m_t = \beta_0 + z_{t1}\beta_1 + z_{t2}\beta_2 + z_{t3}\beta_3, \tag{5.14}$$

where $z_{t1} = 0$ if $t = 4$, 1 otherwise, $z_{t2} = 0$ if $t = 3, 4, 5$, and 1 otherwise, $z_{t3} = 1$ if $t = 1, 7$, and 0 otherwise. Under (5.14), β_1 is the logarithm of the odds (also called the log odds or logit) on responding in the third (or fifth) category, relative to the fourth category, β_2 is the log odds on responding in the second (or sixth) category, relative to the third (fifth) category, and β_3 is the log odds on responding in the first (seventh) category, relative to the second (sixth) category. Note how this parameterization exploits the ordinal nature of the response by using the logits between adjacent categories. Note also that β_1, β_2, and β_3 convey valuable information about the shape of the probability distribution. For example, if $\beta_1 < 0$, $\beta_2 < 0$, $\beta_3 < 0$, the "middle" category is modal, and the extreme categories are least probable. If $\beta_1 > 0$, $\beta_2 > 0$, $\beta_3 > 0$, the extreme categories are modal, and the middle category is least probable. This latter pattern might occur, for example, in studying attitude polarization.

When symmetry holds, a number of restricted versions may also be of interest. For example, a quadratic log-linear model of symmetry is (Clogg and Shockey 1988; Haberman 1978)

$$\log m_t = \beta_0 + z_{t1}\beta_1, \tag{5.15}$$

where $z_{t1} = (t-4)^2$. One might wish to think of the variable z_{t1} as a measure of the squared distance between the t-th category and the middle category. Such an interpretation is tantamount to the assumption that the categories are equally spaced, an assumption for which little justification is available, given the level of measurement corresponding to the response variable in this application. Alternatively, as suggested by the previous material, one might want to entertain the hypothesis $\beta_1 = \beta_2 = \beta_3$ in equation (5.14). This is equivalent to the hypothesis that the adjacent category logits are equal, that is,

$$\log \frac{m_7}{m_6} = \log \frac{m_6}{m_5} = \log \frac{m_5}{m_4} = \log \frac{m_3}{m_4} = \log \frac{m_2}{m_3} = \log \frac{m_1}{m_2} = \beta. \tag{5.16}$$

This model corresponding to (5.16) might be viewed as a univariate analog of the uniform association model (Duncan 1979; Goodman 1979) for the two-way table. Thus, I shall refer to this model as a uniform logit model with symmetry. The model can also be obtained from (5.15) by letting $z_{t1} = \mid t - 4 \mid$, showing that it is a univariate analog of the fixed distance model for the two-way table (Goodman 1972; Haberman 1974a).

The symmetry models described above were fit to the GSS data. Computations were performed using release 3.77 of the computer program GLIM (Numerical Algorithms Group 1986). The model of (5.15) failed to fit ($X^2 = 111.10$, $L^2 = 113.44$, $D^2 = 111.73$, df = 5), as did the uniform logit model with symmetry ($X^2 = 63.72$, $L^2 = 62.75$, $D^2 = 63.32$, df = 5). However, the unrestricted version of symmetry fits the data well ($X^2 = 3.28$, $L^2 = 3.29$, $D^2 = 3.28$, df = 3). The ML estimates of β_1, β_2, and β_3 are (asymptotic standard errors in parentheses), respectively: $-.886$ (.063), $-.290$ (.071) and -1.522 (.127). The estimates indicate that the middle category is modal, followed by the categories adjacent to the middle, the conservative and liberal categories, and finally the extreme categories.

An alternative parameterization, motivated by Goodman's log-bilinear models for two-way tables (Goodman 1979, 1981, 1985, 1986), permits the assignment of scores to the categories of the ordinal variable. Rather than making an equal-spacing assumption, as above, assume a log-nonlinear model of symmetry:

$$\log m_t = \mu + v_{t1}\beta_1, \tag{5.17}$$

where $v_{41} = 0$ and $v_{11} = v_{71}, v_{21} = v_{61}, v_{31} = v_{51}$ are unknown distances (symmetric about the middle score, which, without loss of generality, is arbitrarily taken to be 0) to be estimated from the data. Because $v_{t1}\beta_1 = (v_{t1}\gamma)(\beta_1/\gamma)$ for any value of γ other than 0, the parameters of (5.17) are not identified. Setting $v_{11} = v_{71} = 1$, that is, making the distance between the middle and extreme categories 1, allows the distances v_{21} and v_{31} to be determined. Because (5.17) and (5.14) are equivalent, parameter estimates from (5.14) can be used to solve for estimates of (5.17). Thus, computer programs for fitting log-linear models can be used to fit this nonlinear model. (In more general cases, special programs should be used to estimate models that are nonlinear in the parameters). Finally, for future reference note that the scores used here do not incorporate sample information, as opposed to scores obtained by placing restrictions on the sample moments.

Estimates (standard errors in parentheses obtained by the delta method) for the parameters of the log-nonlinear model of symmetry are $\hat{v}_{31} = .329(.025)$, $\hat{v}_{21} = .436(.029)$, and $\hat{\beta}_1 = -2.698(.123)$. The unequal spacing of the distances reveals why both the uniform logit model of symmetry and the model of (5.15) fail to hold for these data. Relatively speaking, the slightly liberal (slightly conservative) categories and liberal (conservative) categories are not far from one another, and both sets of categories are far from the respective extreme categories.

The example above introduces the basic theory underlying the log-linear model and provides a rationale for modeling the data in a simple context. Before attention is turned to more complex models and issues concerning the measurement of association, it is worth noting that the models considered above are easily extended to the case where symmetry fails to hold and there are one or more natural middle categories (Clogg and Shockey 1988; Haberman 1978). Covariates can also be incorporated (Sobel 1994). In the uniform logit model with symmetry and the log-nonlinear model of symmetry proposed above, the scores are symmetric about the score(s) for the middle category (categories). In the case where symmetry fails to hold, these models can be extended in a number of ways while retaining the symmetric scoring. In the case of an ordered response where one or more natural middle categories cannot be identified (implying that symmetry would not be of interest), these two models are also readily generalized. For example, a uniform logit model for an ordered response can be constructed as

$$\log \frac{m_t}{m_{t-1}} = \lambda_1, \tag{5.18}$$

$t = 2, \ldots, T$, and a nonlinear logit model for an ordered response as

$$\log \frac{m_t}{m_{t-1}} = (v_{t1} - v_{t-1,1})\lambda_1^*, \tag{5.19}$$

where the scores are subject to two restrictions, for example, $\sum_{t=1}^{T} v_{t1} = 0$ and $\sum_{t=1}^{T} v_{t1}^2 = 1$. Note, however, that if the $\hat{v}_{t1}, t = 1, \ldots, T$ are not consistent with the restriction $v_{11} \leq$

$v_{21} \leq \ldots \leq v_{T1}$, one would not want to use the model to assign scores to the ordered response variable. In the case where one suspects an ordering is present but does not know the ordering, the nonlinear logit model might be used to establish such an ordering.

The model described by (5.18) is very restrictive, implying that either the first or the T-th category is modal. Similarly, in the model of equation (5.19), either the first or T-th category is modal or the scores do not satisfy the inequality restrictions.

A less restrictive approach in the ordered case would be to model the cumulative probabilities (expected frequencies) by means of the threshold model:

$$\Pr(Y \leq t) = \frac{\exp(v_{t1})}{1 + \exp(v_{t1})}, \tag{5.20}$$

where v_{t1}, $t = 1, \ldots, T - 1$ is an unknown score such that $Y \leq t$ if and only if a logistic random variable $V \leq v_{t1}$, $t = 1, \ldots, T - 1$. In turn, the threshold model corresponds to the cumulative logit model:

$$\log \frac{\Pr(Y \leq t)}{\Pr(Y > t)} = v_{t1} \tag{5.21}$$

where $t = 1, \ldots, T - 1$. For further material on the cumulative logit model, see McCullagh and Nelder (1989) or Agresti (1990).

2.2 Measuring Association in the Two-by-Two Table: The Odds Ratio

Although many other interesting models for univariate responses can be formulated, investigators are typically interested in studying the association among two or more random variables. The first question is how this association should be defined.

Association between two random variables can be measured in various ways. Users of regression models are most familiar with the correlation coefficient, which is a natural measure of association for the bivariate normal distribution (Anderson 1984). For metric random variables that are not bivariate normal, the correlation coefficient between two random variables Y and Z (ρ_{YZ}) still retains some desirable properties, including (a) symmetry ($\rho_{YZ} = \rho_{ZY}$), (b) $\rho_{YZ} = 0$ if Y and Z are independent (hereafter written as $Y \| Z$), (c) invariance (up to a change in sign) under linear transformations of Y and/or Z, and (d) the correlation inequality $-1 \leq \rho_{YZ} \leq 1$, with equality holding if and only if Z is a linear transformation of Y.

For discrete data, however, matters are different. First, bivariate normal theory cannot be used to justify the correlation coefficient. Second, a single correlation coefficient is no longer a desirable measure of association in general. For example, if a discrete variable Y has two categories and a discrete variable Z has three categories, the value of the correlation coefficient will depend upon the category scores assigned to the two variables (except under independence), and insofar as such assignment is arbitrary, as with nominal and ordinal data, the magnitude of a correlation coefficient will not indicate the amount of association present.

The basic measure of association used in log-linear models is the odds ratio (or cross-product ratio). These are defined with respect to two-by-two subtables. For an $I \times J$

contingency table, there are $(I-1)\times(J-1)$ nonredundant cross-product ratios (see Section 3.1).

To motivate the odds ratio, let Y and Z denote two binary random variables. The joint distribution of Y and Z can be displayed in the form of a two-by-two contingency table:

$$\begin{bmatrix} \pi_{11}^{YZ} & \pi_{12}^{YZ} \\ \pi_{21}^{YZ} & \pi_{22}^{YZ} \end{bmatrix}$$

where $\pi_{ij}^{YZ} > 0$ denotes the probability that the row variable Y is at level i and the column variable Z is at level j, $i = 1, 2; j = 1, 2$. The odds on level 1 of Y, as opposed to level 2, given $Z = j$, is defined as:

$$\Omega_{(1:2)j}^{YZ} = \frac{\pi_{1|j}^{Y|Z}}{\pi_{2|j}^{Y|Z}} = \frac{(\pi_{1j}^{YZ}/\pi_j^Z)}{(\pi_{2j}^{YZ}/\pi_j^Z)} = \frac{\pi_{1j}^{YZ}}{\pi_{2j}^{YZ}}, \tag{5.22}$$

where π_j^Z, $j = 1, 2$, denotes the marginal probability of Z. In epidemiological contexts, this measure is also sometimes called the relative risk. The log odds or logit is obtained, as in Section 2.1, by taking the natural logarithm of the odds. A question that naturally arises is whether the conditional distributions are the same in each column. If so, the two logits are identical, and the ratio

$$\theta_{(12)(12)}^{YZ} = \frac{\Omega_{(1:2)1}^{YZ}}{\Omega_{(1:2)2}^{YZ}}, \tag{5.23}$$

also known as the odds ratio or cross-product ratio, takes on the value 1. Values of the odds ratio greater (less) than 1 correspond to the case $\pi_{1|1}^{Y|Z} > \pi_{1|2}^{Y|Z}$ $(\pi_{1|1}^{Y|Z} < \pi_{1|2}^{Y|Z})$.

The odds ratio for the two-by-two table has a number of properties analogous to those of the correlation coefficient that are easily derived from (5.22) and (5.23). First, it is symmetric, that is, $\theta_{(12)(12)}^{YZ} = \theta_{(12)(12)}^{ZY}$. Second, it takes the value 1 (0 on the logarithmic scale) if and only if $Y \perp\!\!\!\perp Z$. Third, if the rows (or the columns) of the table are interchanged, the reciprocal of the original odds ratio is obtained. On the logarithmic scale, this is the negative of the original logarithm. This property is analogous to (c) above for the correlation coefficient. Property (d) does not hold for the odds ratio, which lies in the interval $(0, \infty)$ (or $[0, \infty]$) if probabilities of 0 are allowed), but it is easy to renormalize the odds ratio (or its logarithm) so that (d) holds, if desired. Yule's Q, for example, is one such normalization.

The odds ratio has an additional property that the correlation coefficient does not have. Specifically, if the entries in row i are multiplied by positive scalars r_i and the entries in column j by positive scalars c_j, and the π_{ij}^{YZ} of the original table are replaced with the new entries $r_i c_j \pi_{ij}^{YZ}$, the new table (which can be renormalized so that the entries sum to 1) has the same odds ratio as the original table, even when the marginal distributions of the original table do not match the marginal distributions of the new table. Thus, the odds ratio is often said to be a measure of association that is "independent" of the marginal distributions. Yule (1912) argued that measures of association with this property are desirable.

To understand this additional property better, note that equations (5.22) and (5.23) imply that two tables with the same conditional distribution $Y \mid Z$ or $Z \mid Y$ have identical values of the odds ratio (the converse is not true). Thus, if one wants to say that tables with the same conditional distribution feature the same amount of association, the correlation coefficient is not an acceptable measure of association (because its value depends on the marginal distributions), but the odds ratio is. This is essentially the argument given by Edwards (1963).

Choosing between the correlation coefficient and the odds ratio as the basic measure of association in the two-by-two table hinges primarily on the desirability of this additional property (and secondarily on the correlation inequality when 0 probabilities are permitted). To see this, note that as both measures are symmetric, a choice cannot be made on these grounds. Further, for the two-by-two table, the correlation coefficient is invariant with respect to monotone transformations of Y and Z, and takes the value 0 if and only if Y and Z are independent.

Use of the odds ratio as the fundamental measure of association for discrete data has additional implications. From equations (5.22) and (5.23) it is evident that the odds ratio can be written in terms of the joint distribution of Y and Z, the conditional distribution $Y \mid Z$, or the conditional distribution $Z \mid Y$. Thus, $\theta^{YZ}_{(12)(12)}$ can be estimated by sampling from the joint distribution of Y and Z, or by taking independent samples from $Y \mid Z = 1$ and $Y \mid Z = 2$, or by taking independent samples from $Z \mid Y = 1$ and $Z \mid Y = 2$. Thus, the same characteristic can be estimated under random sampling from the joint distribution, or stratified random sampling from the conditional distribution of $Y \mid Z$ ($Z \mid Y$), irrespective of the stratum weights, which might not be known (Cornfield 1956). This relationship is exploited in epidemiological studies that are retrospective (sometimes called case-control studies), as opposed to prospective. But see also Goddard (1991) for some problems stemming from the uncritical usage of this relationship.

Further insight is obtained by considering the sample estimate of $\theta^{YZ}_{(12)(12)}$,

$$\hat{\theta}^{YZ}_{(12)(12)} = \frac{n_{11}n_{22}}{n_{12}n_{21}}, \tag{5.24}$$

where $n_{ij}, i = 1,2; j = 1,2$ is the sample count in the (ij)th cell. (If n_{12} and or n_{21} is 0, $\hat{\theta}^{YZ}_{(12)(12)} = \infty$.) For the multinomial case, n_{ij} is the MLE (when there are no restrictions) of the expected count $m_{ij} = n\pi^{YZ}_{ij}$, and thus $\hat{\theta}^{YZ}_{(12)(12)}$ is the MLE of $\theta^{YZ}_{(12)(12)}$. In the case where the row totals are fixed by the sampling design to n^{Y}_{i}, $i = 1,2$, and independent samples from the row multinomials (in this case binomial) with probabilities $\pi^{Z|Y}_{j|i}$ are taken, n_{ij} is the MLE of the expected count $m_{ij} = n^{Y}_{i}\pi^{Z|Y}_{j|i}$, and as before, $\hat{\theta}^{YZ}_{(12)(12)}$ is the MLE of $\theta^{YZ}_{(12)(12)}$. In the literature, this sampling scheme is referred to as product (or independent) multinomial sampling. Similar remarks apply when the column totals are fixed by the sampling design. The independent multinomial sampling scheme also arises when the row (column) variable is not a random variable (for example, time) and independent multinomial samples are taken at different levels of this variable.

Example 2: Wife's Desire and Fertility

To illustrate the connection between the odds ratio and the log-linear model in the two-by-two table, I reconsider data from the Princeton fertility study (Bumpass and Westoff

1970) on the relationship between desires for a third child and subsequent fertility, using a sample of families where the second child was born in September 1956. Table 1 displays the relationship between subsequent fertility (Y), wife's reported desire for another child (Z) and husband's reported desire for another child (W). The measures Z and W were collected in 1957, and the measure Y between 1963 and 1967. In the analysis below, the table is collapsed across the variable W, and attention focuses exclusively on the bivariate relationship between Y and Z, that is, no attempt is made to draw inferences about the trivariate relationship from the bivariate relationship. To justify this analysis, note that if the trivariate distribution is multinomial, the bivariate distribution is also multinomial. In Section 5, the trivariate relationship is also considered and conditions under which inferences about certain aspect of the trivariate relationship can be obtained from the collapsed table are discussed.

Table 1. Third Child by Husband's and Wife's Desire

		W = Husband's Desire	
Y = Third Child	Z = Wife's Desire	No	Yes
No	No	80	24
	Yes	27	37
	No	49	37
Yes	Yes	39	338

The null hypothesis that wife's desire for a third child and subsequent fertility are independent can be tested using either X^2, L^2, or D^2 on 1 df. Alternatively, one can use the sample odds ratio of (5.24) to assess the amount of association in the table. This yields $\hat{\theta}^{YZ}_{(12)(12)} = (104 \times 377)/(64 \times 86) = 7.124$ ($\log \hat{\theta}^{YZ}_{(12)(12)} = 1.963$), suggesting a "strong" association between wife's desire and subsequent fertility. Under multinomial sampling (or indpendent multinomial sampling) the estimate above is ML, and application of the delta method yields an asymptotic standard error of .199 for $\log \hat{\theta}^{YZ}_{(12)(12)}$ (Woolf 1955). This yields the 95% confidence interval $1.963 \pm (1.96)(.199) = (1.570, 2.353)$; if desired, a 95% confidence interval for $\theta^{YZ}_{(12)(12)}$ can be obtained by exponentiating the lower and upper limits of the previous interval. Clearly, the null hypothesis of independence is not tenable. Loglinear models may also be used to obtain the inference above. Under multinomial sampling the hypothesis of independence is: $m_{ij} = m_i^Y m_j^Z / n$, whence

$$\log m_{ij} = -\log n + \log m_i^Y + \log m_j^Z. \tag{5.25}$$

Under this hypothesis, because the marginal totals add to n, m_1^Y and m_1^Z determine all the m_{ij}^{YZ}. Thus, there are two functionally independent parameters to estimate and hence the df are (3 − 2) = 1. A reparameterization of (5.25) yields the model of independence in its usual form:

$$\log m_{ij} = \mu + \alpha_i^Y + \alpha_j^Z, \tag{5.26}$$

where $\alpha_1^Y + \alpha_2^Y = 0$, $\alpha_1^Z + \alpha_2^Z = 0$.

Equivalently, (5.26) may be written as

$$\log m_t = \beta_0 + \beta_1 z_{t1} + \beta_2 z_{t2}, \tag{5.27}$$

where $z_{t1} = 1$ if $Y = 1$ (no third child), -1 otherwise, $z_{t2} = 1$ if $Z = 1$ (wife does not desire another child), -1 otherwise.

In the literature, it is usual to write the saturated model for the two by two table in the form

$$\log m_{ij} = \mu + \alpha_i^Y + \alpha_j^Z + \alpha_{ij}^{YZ}, \tag{5.28}$$

with

$$\sum_{i=1}^{2} \alpha_i^Y = \sum_{j=1}^{2} \alpha_j^Z = \sum_{i=1}^{2} \alpha_{ij}^{YZ} = \sum_{j=1}^{2} \alpha_{ij}^{YZ} = 0. \tag{5.29}$$

Alternatively, the saturated model may be written as

$$\log m_t = \beta_0 + \beta_1 z_{t1} + \beta_2 z_{t2} + \beta_3 z_{t3}, \tag{5.30}$$

where z_{t1} and z_{t2} are defined as in (5.27), and $z_{t3} = z_{t1} \times z_{t2}$. It is easy to see that $4\beta_3 = \log \theta_{(12)(12)}^{YZ}$. Thus, the coefficient for the interaction term measures the association in the table and the null hypothesis of independence is equivalent to the hypothesis $\beta_3 = 0$.

The relationship between the parameters of (5.28) and (5.29) and the β parameters of (5.30) is:

$$\mu = \beta_0;\ \alpha_1^Y = \beta_1;\ \alpha_2^Y = -\beta_1;\ \alpha_1^Z = \beta_2;\ \alpha_2^Z = -\beta_2;$$
$$\alpha_{11}^{YZ} = \alpha_{22}^{YZ} = \beta_3;\ \alpha_{12}^{YZ} = \alpha_{21}^{YZ} = -\beta_3.$$

The results obtained by using ML to estimate the saturated log-linear model are identical to those already obtained.

Before taking up the more general context of the $I \times J$ contingency table, it is worth considering one more model for these data. This is the model of symmetry for the two-by-two table, which specifies $m_{12} = m_{21}$, or equivalently, $\pi_{12} = \pi_{21}$. Under this model, the probability that the wife desires another child and does not experience subsequent fertility is equal to the probability that the wife does not desire another child and experiences subsequent fertility. Note that under this model, the marginal distributions of Y and Z must be identical. It is easy to see that the model can be obtained from (5.28) by imposing the restriction $\alpha_i^Y = \alpha_i^Z$ or from (5.3) by imposing the restriction $\beta_1 = \beta_2$. Hence the df are 1. For these data, the null hypothesis of symmetry is not rejected at the .05 level of significance ($X^2 = 3.23$, $L^2 = 3.24$, $D^2 = 3.23$). The χ^2 test for symmetry in the two-by-two table is often called McNemar's test, after McNemar (1947).

3 Odds Ratios for Two- and Three-Way Tables

3.1 Odds Ratios for Two-Way Tables

Let Y denote a discrete random variable with I categories, $i = 1, \ldots, I$, Z a discrete random variable with J categories, $j = 1, \ldots, J$, and $\{\pi_{ij}^{YZ}\}$ the two way cross-classification with (ij)th entry $\pi_{ij}^{YZ} = \Pr(Y = i, Z = j) > 0$. As in Section 2, association in the two-way table can be measured in various ways, and a number of summary measures have been proposed (Agresti 1984; Goodman and Kruskal 1979). Altham's (1970a) generalization of Edward's (1963) argument to the $I \times J$ cross-classification table justifies a different approach and leads to use of the odds ratios:

$$\theta_{(r_1,r_2)(c_1,c_2)}^{YZ} = \frac{\pi_{r_1c_1}^{YZ}\pi_{r_2c_2}^{YZ}}{\pi_{r_1c_2}^{YZ}\pi_{r_2c_1}^{YZ}}, \tag{5.31}$$

where r_1 and r_2 (c_1 and c_2) are, respectively, two row (column) indices. Above and beyond the properties previously discussed, the complete set of odds ratios has the following desirable properties: (e) the marginal distributions of the table and the nonredundant odds ratios uniquely determine the joint distribution (Sinkhorn 1967); (f) any set of $(I-1) \times (J-1)$ nonredundant odds ratios (a basic set) determines all other odds ratios that can be formed; (g) $Y \| Z$ if and only if all odds ratios are 1.

Properties (f) and (g) are easy to show and frequently used. To obtain (f) consider the basic set of $(I-1) \times (J-1)$ local odds ratios (used subseqently in connection with models for ordered discrete variables):

$$\theta_{ij}^{YZ} = \frac{\pi_{ij}^{YZ}\pi_{i+1,j+1}^{YZ}}{\pi_{i,j+1}^{YZ}\pi_{i+1,j}^{YZ}}, \tag{5.32}$$

$i = 1, \ldots, I-1; j = 1, \ldots, J-1$. Without loss of generality, in (5.31) let $r_1 < r_2$, $c_1 < c_2$ and note that

$$\theta_{(r_1,r_2)(c_1,c_2)}^{YZ} = \prod_{i=r_1}^{r_2-1}\prod_{j=c_1}^{c_2-1}\theta_{ij}^{YZ}.$$

To see that $Y \| Z$ implies odds ratios with value 1, substitute $\pi_{ij}^{YZ} = \pi_i^Y\pi_j^Z$, $i = 1, \ldots, I; j = 1, \ldots, J$ into (5.31). To obtain the converse, note that

$$\theta_{(r_1,r_2)(c_1,c_2)}^{YZ} = 1 \Rightarrow \pi_{r_1c_1}^{YZ}\pi_{r_2c_2}^{YZ} = \pi_{r_1c_2}^{YZ}\pi_{r_2c_1}^{YZ} \Rightarrow \pi_{r_1}^{Y}\pi_{c_1}^{Z} =$$
$$\sum_{c_2=1}^{J}\sum_{r_2=1}^{I}\pi_{r_1,c_2}^{YZ}\pi_{r_2,c_1}^{YZ} = \sum_{c_2=1}^{J}\sum_{r_2=1}^{I}\pi_{r_1,c_1}^{YZ}\pi_{r_2,c_2}^{YZ} = \pi_{r_1,c_1}^{YZ}.$$

Finally, note (as in Section 2.2) that the odds ratios can be written in terms of the conditional distribution $Y \mid Z$ ($Z \mid Y$). Thus, as before, these quantities can be estimated by sampling from the joint distribution of Y and Z, the conditional distribution $Y \mid Z$ or the conditional distribution $Z \mid Y$.

3.2 Odds Ratios for Three-Way Tables

Let Y denote the row variable with I categories, $i = 1, \ldots, I$, let Z and W denote, respectively, the column and layer variables with J and K categories, $j = 1, \ldots, J$; $k = 1, \ldots, K$, and let $\{\pi_{ijk}^{YZW}\}$ denote the three-way cross-classification with (ijk)th entry $\pi_{ijk}^{YZW} = \Pr(Y = i, Z = j, W = k) > 0$. Altham's (1970b) results for the three-way table leads naturally to consideration of conditional odds ratios. These measure the association between two variables at a fixed level of the third variable. For example, the conditional association between Y and Z is measured by $K(I-1)(J-1)$ nonredundant conditional odds ratios:

$$\theta_{(r_1,r_2)(c_1,c_2)(k)}^{YZ|W} = \frac{\pi_{r_1 c_1 k}^{YZW} \pi_{r_2 c_2 k}^{YZW}}{\pi_{r_1 c_2 k}^{YZW} \pi_{r_2 c_1 k}^{YZW}} = \frac{\pi_{r_1,c_1|k}^{YZ|W} \pi_{r_2,c_2|k}^{YZ|W}}{\pi_{r_1,c_2|k}^{YZ|W} \pi_{r_2,c_1|k}^{YZ|W}}. \tag{5.33}$$

Similarly, the conditional association between Y and W is measured by $J(I-1)(K-1)$ nonredundant conditional odds ratios

$$\theta_{(r_1,r_2)(\ell_1,\ell_2)(j)}^{YW|Z}, \tag{5.34}$$

and the conditional association between Z and W is measured by forming a nonredundant set of $I(J-1)(K-1)$ conditional odds ratios.

In equation (5.33), there are three cases of interest.

1. The conditional association between Y and Z depends on W, that is, for some combination of rows and columns,

$$\theta_{(r_1,r_2)(c_1,c_2)(k)}^{YZ|W} \neq \theta_{(r_1,r_2)(c_1,c_2)(k')}^{YZ|W}. \tag{5.35}$$

   For this case, using the definition of a conditional odds ratio, it follows that the conditional association between Y and W (Z and W) depends on Z (Y).

2. For all combinations of rows and columns,

$$\theta_{(r_1,r_2)(c_1,c_2)(k)}^{YZ|W} = \theta_{(r_1,r_2)(c_1,c_2)(k')}^{YZ|W}, \; k = 1, \ldots, K, \tag{5.36}$$

   and for some combination of rows and columns, the conditional odds ratios has a value other than 1. From (5.36) and case 1, it follows that the conditional association between Y and W (Z and W) is constant across levels of Z (Y).

3. Here, case 2 holds, and in addition, all conditional odds ratios in (5.36) are 1. This case applies if and only if Y and Z are conditionally independent, given W (hereafter denoted $Y \underline{\|} Z \mid W$). Note that $Y \underline{\|} Z \mid W$ does not imply $Y \underline{\|} W \mid Z$ or $Z \underline{\|} W \mid Y$, so it is not generally true for this case that the odds ratios in (5.34) have value 1. Several additional cases are also of interest.

4. Case 3 holds, and in addition, all the conditional odds ratios in one of the other sets are 1. Thus, $Y \underline{||} Z \mid W$ and, for example, $Y \underline{||} W \mid Z$. This is is equivalent to $Y \underline{||} (Z, W)$.

5. Case 3 holds, and in addition, all the conditional odds ratios in (5.34) are 1. From case 4, $Y \underline{||} (Z, W)$ and $W \underline{||} (Y, Z)$, whence Y, Z, and W are independent.

   Finally, note that (as before) the odds ratios can be written in terms of conditional distributions. Thus, for example, (5.31) can be computed from the trivariate distribution, the distribution $(Y, Z) \mid W$, the distribution $Y \mid (W, Z)$, and so on.

4 Models for the Two-Way Table

In this section, interest focuses primarily on using the log-linear model to make inferences about the structure of association in the $I \times J$ table. As before, ML estimation is used for the examples, and essentially the same results are obtained under multinomial sampling, independent multinomial sampling, and sampling from IJ independent Poisson distributions, as shown in Section 7. Intuitively, the connection between the multinomial and product multinomial case results from the fact that the probability distribution of the latter can be written as a product of multinomials, and the likelihood can be maximized subject to the constraint that the expected counts add to the sample size within each multinomial. The connection to the Poisson case results from the well-known facts (see Bishop et al. 1975 or Agresti 1990) that if IJ independent Poisson random variables are sampled, conditioning on the random sample size n yields the multinomial distribution, while conditioning on the random row or column totals yields the product multinomial.

4.1 Basic Models

To begin, consider the case of independence in the $I \times J$ table. Under multinomial sampling, the MLE of m_{ij} is $np_i^Y p_j^Z$ and substitution of $\hat{m}_{ij}$ into (5.9), (5.10), or (5.11) yields a test for independence on $(I-1) \times (J-1)$ df. The df may be obtained, following the argument in Section 2.2.1 [see (5.25)] by noting that because m_i^Y and m_j^Z must add to n, there are $(I-1)$ nonredundant m_i^Y and $(J-1)$ nonredundant m_j^Z. Thus, the df are $(IJ-1)-(I-1+J-1)=(I-1)(J-1)$. For the product multinomial case, with fixed column totals, the row distributions in each column are identical if $m_{ij} = n_j^Z \pi_i^Y = n(n_j^Z/n)\pi_i^Y$. For this case, the df are $(I-1) \times J - (I-1) = (I-1) \times (J-1)$, as in the multinomial case. In addition, because $p_j^Z = n_j^Z/n$, the same test for independence is obtained under both sampling schemes.

As in Section 2.2.1, the log-linear model may also be used to test the hypothesis of independence and to model departures from independence. Following the argument there, the log-linear model of independence can be written as in (5.26), with

$$\sum_{i=1}^{I} \alpha_i^Y = 0; \sum_{j=1}^{J} \alpha_j^Z = 0.$$

The saturated log-linear model for the $I \times J$ table is then given by (5.28), with

$$\sum_{i=1}^{I} \alpha_i^Y = 0; \sum_{j=1}^{J} \alpha_j^Z = 0; \sum_{i=1}^{I} \alpha_{ij}^{YZ} = 0, j = 1, \ldots, J; \sum_{j=1}^{J} \alpha_{ij}^{YZ} = 0, i = 1, \ldots, I.$$

The $(I-1)(J-1)$ linearly independent interaction parameters (α_{ij}^{YZ}) measure the association in the table. To see this, note that

$$\alpha_{r_1c_1}^{YZ} + \alpha_{r_2c_2}^{YZ} - \alpha_{r_1c_2}^{YZ} - \alpha_{r_2c_1}^{YZ} = \log \theta_{(r_1,r_2)(c_1,c_2)}^{YZ},$$

and all interactions are 0 if and only if all the logged odds ratios are 0.

The log-linear model for the two-way table can also be used to model various types of equiprobability hypotheses. In Section 4.2, an equiprobability hypothesis (symmetry) is discussed. Here Y and Z are associated. An equiprobability hypotheses in which Y and Z are not associated is given by the model

$$\log m_{ij} = \mu + \alpha_j^Z, \tag{5.37}$$

which corresponds to independence of Y and Z with $\Pr(Y = i) = 1/I$. It is important to note that the model of (5.37) is not appropriate under product multinomial sampling, with fixed row totals. This may be seen from the likelihood function for product multinomial sampling, in which the row totals are fixed. For this case, the terms α_i^Y should always be included. Similar remarks apply to the α_j^Z when the column totals are fixed by the sampling plan.

When Y is a dependent variable, and Z an independent variable, interest centers on the conditional distribution $Y \mid Z$. Similarly, when Z is dependent and Y independent, the conditional distribution $Z \mid Y$ is of interest. For the case where Y is a dichotomous dependent variable, the distribution of Y at level j of Z is binomial, with canonical parameter given by the logit $\log(\pi_{1|j}^{Y|Z}/\pi_{2|j}^{Y|Z})$. This leads naturally to the logit model:

$$\log \frac{\pi_{1|j}^{Y|Z}}{\pi_{2|j}^{Y|Z}} = \lambda + \lambda_j^Z, \tag{5.38}$$

where the λ_j^Z parameters are subject to one linear constraint, for example, $\sum_{j=1}^{J} \lambda_j^Z = 0$. Other constraints could also be placed on the λ_j^Z, as in the subsequent example. The logit model is a special case of the log-linear model and may be obtained from (5.28) by using

$$\log \frac{\pi_{1|j}^{Y|Z}}{\pi_{2|j}^{Y|Z}} = \log m_{1j} - \log m_{2j} = 2\alpha_1^Y + 2\alpha_{1j}^{YZ}.$$

Thus, in the logit model Y does not depend on Z if and only if $\lambda_j^Z = 0, j = 1, \ldots, J$, that is, if and only if Y and Z are independent.

The logit model above is easily extended to the $I \times J$ table. For this case, the distribution of Y at each level j of Z is multinomial (product multinomial across levels of Z), and a natural model to consider is:

$$\log \frac{\pi_{i|j}^{Y|Z}}{\pi_{i'|j}^{Y|Z}} = \lambda_i^Y + \lambda_{ij}^{YZ}, \tag{5.39}$$

where the λ_{ij}^{YZ} are subject to $(I-1)$ linear restrictions, for example, $\sum_{j=1}^{J}\lambda_{ij}^{YZ}=0$, i' is a fixed reference point, and i takes on all values other than i'. As above, the logit model is a special case of the log-linear model and the parameters of the latter are related to the parameters of the log-linear model:

$$\lambda_i^Y = \alpha_i^Y - \alpha_{i'}^Y, \lambda_{ij}^{YZ} = \alpha_{ij}^{YZ} - \alpha_{i'j}^{YZ}.$$

The model of (5.39) is called the baseline category logit model (Agresti 1990). In some instances, there will be a natural choice for the baseline category i'. When this is the case, utilization of this category as baseline will simplify the substantive interpretation of the parameters estimated. However, it does not really matter which category is chosen as baseline, for if some other category (say i^*) is chosen as baseline, the parameters of this new baseline category logit model are a linear transformation of the parameters of the baseline category logit model, with category i' as baseline. Further, the χ^2 tests and df are identical in both cases. The baseline category logit model described above treats Y as nominal; for ordered dependent variables, other types of logit models will be considered subsequently.

Example 3: Causal Analysis, the Logit Model, and Imitative Suicide

To illustrate log-linear and logit models for the two-way table, and to tie these models to the subject of causal inference as well, I reconsider data analyzed by Bollen and Phillips (1982). These authors collected data on the number of suicides in the United States during the week before and the week after a nationally publicized suicide story, for seven different periods. They hypothesized that more suicides should take place the week after. The data are reported in Table 2. Since the ordering of the periods is not exploited in the following analysis, the original text can be consulted for the actual dates.

Table 2. Number of Pre and Post Story Suicides

	Z = Story Period						
Y = Week	1	2	3	4	5	6	7
Before = 1	444	435	514	462	572	501	575
After = 2	554	528	487	482	593	553	550

Using a one-tailed t-test on the differences $(n_{2j} - n_{1j})$, the authors rejected the null hypothesis of no story effect ($t = 1.725$, $df = 6$, $p = .068$). Although the use of this test can be criticized, the intent here is rather to show how more refined inferences can be made using the log-linear model.

To justify the subsequent analysis, suppose that in the week before the jth suicide story is publicized, each member r_j of the population, $r_j = 1, \ldots, N_j$, has propensity $\pi_{1jr}^{Y_Z}$ to commit suicide during the week, and let $\pi_{1j}^{Y_Z}$ denote the average of the individual propensities. (The notation Y_Z is used here to indicate that Z is not regarded as a random variable in the analysis.) The propensity parameter $\pi_{1jr}^{Y_Z}$ may be thought of as the parameter of a Bernoulli random variable Y_{1jr}. Similarly, define the set of random variables Y_{2jr}, the propensities $\pi_{2jr}^{Y_Z}$, and the average propensity $\pi_{2j}^{Y_Z}$ referring to the week after. Finally, suppose that the random variables defined above are independent (across weeks and persons). Under the assumptions above, with large N_j and small values of the average propensities, the counts n_{ij} in Table 2 are approximately independent Poisson (m_{ij}) random variables (Feller 1968).

In the literature on causal inference (see Chapter 1), the unit causal effect (under the setup here) of the j-th suicide story on the r_j-th member of the population would usually be defined as $y_{2jr} - y_{1jr}$, where the lower-case notation is used to denote the realization of the respective random variables. Clearly, this effect cannot be observed for persons committing suicide during the period before the suicide story. Alternatively, one might prefer to define a different quantity, which I shall call the unit causal propensity, as: $\pi_{2jr}^{Y_Z} - \pi_{1jr}^{Y_Z}$. This is not observable, either. However, under the assumptions above, inferences about the total effect of the j-th suicide story, defined as some function of m_{2j} and m_{1j}, $j = 1, \ldots, J$, can be made by using the Poisson log-linear model. Equivalently, because the distribution $n_{2j} \mid (n_{1j} + n_{2j})$ is binomial with parameter $\pi_j^{*Y_Z} = m_{2j}/(m_{1j} + m_{2j})$, the logit model can be used to obtain such inferences. Note also that $\pi_j^{*Y_Z} \approx \pi_{2j}^{Y_Z}/(\pi_{1j}^{Y_Z} + \pi_{2j}^{Y_Z})$, as the population at risk during the two-week interval is roughly constant.

Four hypotheses about the effects of stories on suicide are considered. The most restrictive (presented for illustrative purposes only) is $m_{ij} = m$ for all i and j, corresponding to the model of (5.13). This hypothesis is not tenable ($X^2 = 64.68$, $L^2 = 65.28$, $D^2 = 64.86$, df = 13). Of greater interest is the hypothesis of no story effect, $m_{1j} = m_{2j}$ for all j. The appropriate model is given by (5.37), where the seven α_j^Z parameters are subject to one linear constraint, for example, $\sum_{j=1}^{7} \alpha_j^Z = 0$ or $\alpha_1^Z = 0$. The fit of this model is not satisfactory either ($X^2 = 25.76$, $L^2 = 25.80$, $D^2 = 25.77$, df = 7), leading to the conclusion that there is a story effect, at least for some period. Next is the hypothesis of a constant effect across periods. The corresponding model is given by (5.26) (the model of independence in the multinomial context). The constant effect hypothesis is also rejected ($X^2 = 17.57$, $L^2 = 17.58$, $D^2 = 17.57$, df = 6) at the .05 level.

Finally, the model with variable effects is equivalent to (5.28), the equation for the saturated log-linear model. Because the effects are not constant, I reparameterized the saturated model as follows:

$$\log m_{ij} = \mu + \alpha_j^Z + \alpha_{ij}^{Y_Z}, \tag{5.40}$$

where $\alpha_1^Z = 0$, $\alpha_{1j}^{Y_Z} = 0$, $j = 1, \ldots, J$. Under this parameterization, $\alpha_{2j}^{Y_Z}$ is the total effect of the suicide story in the j-th period, when this effect is defined as $\log(m_{2j}/m_{1j})$. This quantity is the log of the ratio of two Poisson means, and if inferences about the ratio are desired, one might prefer to define the effect as $\exp(\alpha_{2j}^{Y_Z})$. (See Cox and Hinkley 1974 for related material.) Parameter estimates (asymptotic standard errors in parentheses) of the

α_{2j}^{Yz} are: $\hat{\alpha}_{21}^{Yz} = .221\ (.064)$, $\hat{\alpha}_{22}^{Yz} = .194\ (.065)$, $\hat{\alpha}_{23}^{Yz} = -.054\ (.063)$ $\hat{\alpha}_{24}^{Yz} = .042\ (.065)$, $\hat{\alpha}_{25}^{Yz} = .036\ (.059)$, $\hat{\alpha}_{26}^{Yz} = .099\ (.062)$, $\hat{\alpha}_{27}^{Yz} = -.044\ (.060)$. Evidently, there are story effects in the first two periods only, and these are in the hypothesized direction. Taken in toto, the findings here cast some doubt on the original hypothesis, and suggest the need for further investigation. One possibility is that the first two periods (or the data from these periods) are aberrant in some fashion. Another is that behavior is imitated under some conditions, but not others. If this is the case, future work might attempt to specify conditions under which imitative behavior does and does not occur, leading to refinement of the original hypothesis.

The connection between the log-linear model and the binomial logit model in this example is straightforward (Clogg and Eliason 1987). Here the logit model is

$$\log \frac{\pi_j^{*Yz}}{1 - \pi_j^{*Yz}} = \lambda + \lambda_j^Z, \tag{5.41}$$

and equating $\log(m_{2j}/m_{1j})$ with $\log(\pi_j^{*Yz}/(1 - \pi_j^{*Yz}))$ gives $\lambda = 0$, $\lambda_j^Z = \alpha_{2j}^{Yz}$.

4.2 Models for Square Tables

Tables with a one-to-one correspondence between the row and column variable arise in numerous contexts. For example, when two judges rate a sample of subjects on a discrete variable, it is of interest to cross-classify the responses to study the agreement between judges (Fleiss 1981; Sobel 1988; Uebersax 1991, 1992). Similarly, in studies where control and treatment units are matched and a discrete outcome is measured, a square table also arises (Cox 1958; McCullagh 1982). In survey research, it is common to compare two or more forms of an identically coded discrete item (for example, Reiser, Wallace, and Schuessler 1986). In studies of occupational mobility, the cross-classification of son's occupation by father's occupation, known as the mobility table, is of interest (Sobel, Hout, and Duncan 1985), and in studies of married couples, religious homogamy might be of interest (Johnson 1980). In panel studies, respondents may be compared at two or more points in time on a discrete variable (Agresti 1990; Bishop et al. 1975; Clogg, Eliason, and Grego 1990; Hagenaars 1990).

In each of the cases above, $I = J$, and as in example 2 of Section 2.2.1, a number of additional questions about the relationship between Y and Z arise naturally. One type of question concerns solely the marginal distributions of the rows and columns. Of particular interest is the hypothesis that the marginal distributions are identical, in which case marginal homogeneity (MH) is said to hold. Stuart (1955) first gave a test for MH, and Madansky (1963) discussed ML estimation of this model (which is not a log-linear model). For further material, see Firth (1989), who also shows how to use the computer program GLIM to fit MH.

A second type of question focuses on the structure of association in the table. For example, in square tables, independence (IN) often fails to hold because the counts cluster on the diagonal. Apart from this clustering, there may be no further relationship. This leads to the model of quasi-independence (QI) (Goodman 1968):

$$\log m_{ij} = \mu + \alpha_i^Y + \alpha_j^Z + \delta_{ij}\alpha_{ij}^{YZ}, \tag{5.42}$$

where $\delta_{ij} = 1$ if $i = j$, 0 otherwise. Under this model $(I - 1)$ linearly independent α_i^Y, $(I - 1)$ linearly independent α_j^Z, and I linearly independent α_{ij}^{YZ} are estimated. The df are thus $I^2 - 1 - [2(I - 1) + I] = I^2 - 3I + 1$.

Third, one might be interested in both the marginal distributions and the structure of association in the table. This leads to consideration of other models. In Section 2.2, the model of symmetry for the two-way table was considered. Under this model, MH holds. More generally, for the $I \times J$ table, the model of symmetry specifies $m_{ij} = m_{ji}$ for all i and j. Here symmetry implies MH, and it also implies

$$\theta_{(r_1,r_2)(c_1,c_2)}^{YZ} = \theta_{(c_1,c_2)(r_1,r_2)}^{YZ}, \tag{5.43}$$

which means that the pattern of association is restricted under this model. More restrictive patterns of association may also be considered.

The most restrictive log-linear model of interest that preserves marginal homogeneity and independence is the model called "symmetry + independence" (SI) by Goodman (1985). This model was previously discussed by Hope (1982) and Sobel (1983), and it can be obtained from (5.26) by imposing the restriction $\alpha_i^Y = \alpha_j^Z$, if $i = j$. The df are thus $(I^2 - 1) - (I - 1) = I(I - 1)$. SI can fail to hold either because symmetry fails to hold and/or independence fails to hold. If symmetry holds and independence fails only because the counts cluster on the diagonal, the model of symmetry + quasi-independence (SQI) holds. Under SQI, I more parameters are estimated than under SI, and thus the df are $I(I-2)$. SQI is a special case of the more general model of symmetry (S), provided $I > 3$. Symmetry can be be obtained from (5.28) by imposing the restrictions $\alpha_i^Y = \alpha_j^Z$ if $i = j$, $\alpha_{ij}^{YZ} = \alpha_{ji}^{YZ}$. The df may be obtained by counting the number of functionally independent parameters estimated under the model, and subtracting from $I^2 - 1$, as in Section 2.1. Since $I(I+1)/2\ m_{ij}$ are estimated, and these are subject to the constraint that $\sum_{i=1}^{I}\sum_{j=1}^{I} m_{ij} = n$, the df are $(I^2 - 1) - [(I(I+1)/2) - 1] = I(I - 1)/2$.

Symmetry may also fail to hold. For example, in studies of occupational mobility, the distribution of the respondents' current job is typically different than the distribution of respondents' social origins (indexed by father's occupation). In addition, the counts cluster on the diagonal (occupational inheritance). Thus, QI is of interest. Of greater interest is quasi-symmetry (QS) (Sobel et al. 1985). QS was proposed by Caussinus (1965), and it can be written as

$$\log m_{ij} = \mu + \alpha_i^Y + \alpha_j^Z + \alpha_{ij}^{YZ}, \tag{5.44}$$

with $\alpha_{ij}^{YZ} = \alpha_{ji}^{YZ}$. Comparison of S and QS indicates that under QS, $I - 1$ additional linearly independent parameters are estimated. The df are therefore $I(I - 1)/2 - (I - 1) = (I-1)(I-2)/2$. The comparison also reveals that if S holds, QS and marginal homogeneity hold. The converse is also true (see Bishop et al. 1975 for a proof). Thus, when QS holds, the null hypothesis of marginal homogeneity can be tested by comparing the fit of S with QS by means of the conditional likelihood ratio test $(L^2(S) - L^2(QS))$. Under the null hypothesis, this statistic follows a χ^2 distribution with $(I - 1)$ df in large samples. Tests could also be based on the difference between the Pearson χ^2 statistics or the differences in D^2 (tests using these differences can give negative values), and Wald tests could also be used. QS can also be motivated in other ways (McCullagh 1982), and the model permits a number of other descriptions. For example, Goodman (1985) refers to QS as symmetric association, by virtue of the fact that the local odds ratios $\{\theta_{ij}\}$ satisfy the relationship $\theta_{ij} = \theta_{ji}$. (Of course, (5.43) holds under QS, and the local odds ratios are just a particular set.)

Finally, for the case when QS fails to hold, a number of unsaturated models have been proposed in the literature. Most of these models assume the row and column variables are ordinal. For further details, see Goodman (1972; 1984), Yamaguchi (1990), and Section 4.3. Figure 1 uses an implication graph to display the relationships between the special cases of QS discussed. The directed arrow indicates that one model implies another. Because implication is transitive, only the first implication in a chain is depicted. In this graph, $I \geq 4$, as several of the models would be identical otherwise.

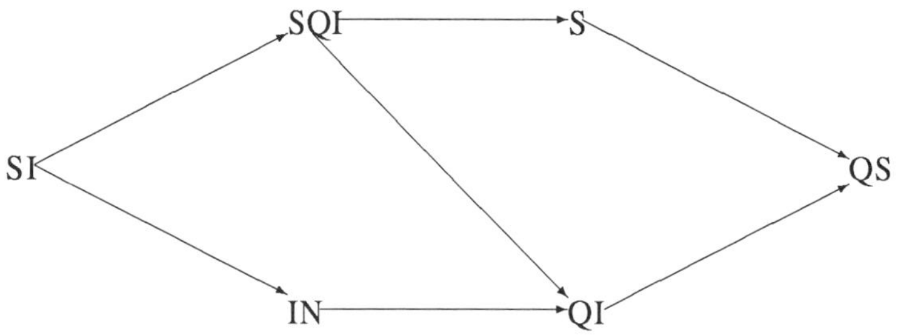

Figure 1. Implication Graph for Some Models for the Square Table

Example 4: Occupational Mobility in Brazil

To illustrate the models of Section 4.2, I reconsider the 1956 intergenerational occupational mobility table for São Paulo, Brazil (Hutchinson 1958), subsequently analyzed by Sobel et al. (1985). The data, presented in Table 3, are taken from the original source. Although Sobel et al. (1985) analyzed the correct table, the Brazilian table they present contains a typographical error in the (34)th cell. The occupational categories are: (1) professional and high administrative, (2) managerial and executive, (3) intermediate nonmanual, (4) lower nonmanual, (5) skilled manual, and (6) semiskilled and unskilled manual. Note that the occupations are ordered by status, and thus the table is doubly ordered. In Section 4.3.1, this ordering is explicitly taken into account, but here the analysis itself treats the row and column variables as nominal.

The models described in Section 4.2 were fit to these data. Table 4 presents the goodness-of-fit statistics and df for each model. Inspection of the table reveals that none of the models fit the data, at the .05 level, with the exception of QS. Because QS fits the data, the conditional likelihood ratio test could be used to test for marginal homogeneity. The null hypothesis would be rejected $((L^2(S) - L^2(QS)) = 223.14 - 14.16 = 208.98$, df = 5).

Table 3. Intergenerational Mobility Table, São Paulo, Brazil, 1956

	Z = Son's Occupation					
Y = Father's Occupation	1	2	3	4	5	6
1	33	12	10	3	0	0
2	14	25	16	3	2	0
3	13	16	68	21	16	1
4	6	30	39	74	61	7
5	5	16	26	45	132	24
6	1	9	29	41	142	116

Additional insight can be obtained (when QS holds) by reparameterizing (5.44) as

$$\log m_{ij} = \eta_i + \eta_j + \alpha_j^Z + \alpha_{ij}^{YZ}, \tag{5.45}$$

where $\alpha_1^Z = 0$, $\alpha_{ij}^{YZ} = \alpha_{ji}^{YZ}$, and $\alpha_{ij}^{YZ} = 0$ if $i = j$, $i = 1, \ldots, 6$. Under this parameterization, due to Sobel et al. (1985), symmetry holds if and only if $\alpha_j^Z = 0$, $j = 1, \ldots, J$, and under QS $\log(m_{ij}/m_{ji}) = \alpha_j^Z - \alpha_i^Z$. Thus, any asymmetry in the table due to the different marginal distributions is captured by the α_j^Z parameters and these parameters account for any and all asymmetries. This interpretation does not hold when QS fails to hold. Under QS, note also that differences between the α_j^Z are invariant with respect to the manner in which the α_j^Z are identified. Second, recall that a prevalent feature of the mobility table and many other square tables is the tendency for the counts to cluster on the diagonal. The usual odds ratios and interaction parameters used in log-linear modeling do not shed light on this tendency. However, by taking $r_1 = c_1$, $r_2 = c_2$ in equation(5.31), one obtains diagonal odds ratios of the form $(m_{ii}m_{jj}/m_{ij}m_{ji})$. These odds ratios measure association with respect to the tendency for the counts to cluster on the diagonal (Sobel et al. 1985), and the set $\Theta_D = \{\theta_{(ij)(ij)}^{YZ}, i < j\}$ of diagonal odds ratios suffices to characterize the association in the table when QS holds (Sobel 1988). Under the parameterization used in (5.45), it is clear that $-2\alpha_{ij}^{YZ} = \log \theta_{(ij)(ij)}^{YZ}$; thus the α_{ij}^{YZ} measure the association in the table inversely to the diagonal odds ratios. These diagonal odds ratios are also useful for studying agreement (Darroch and McCloud 1986; Sobel 1988).

Table 4. Goodness of Fit and *df* for the Brazilian Data

Model	X^2	L^2	D^2	*df*
SI	835.64	766.32	772.08	30
SQI	440.17	421.18	420.54	24
S	193.50	223.14	199.22	15
IN	765.92	645.20	689.07	25
QI	257.85	231.07	238.52	19
QS	12. 25	14.16	12.64	10

Table 5 reports ML estimates (standard errors in parentheses) of the α_j^Z and α_{ij}^{YZ} parameters for the Brazilian data, under QS. Recalling that differences between the α_j^Z parameters reflect the impact of marginal heterogeneity on asymmetry, the estimates in the first column reveal the prevalence of upward social mobility in Brazil. The association between father's occupation and son's occupation is "strong" and the strength of the association is directly related to the distance between fathers and son's occupation, as evidenced, for example, by comparing quantities like $\hat{\alpha}_{16}^{YZ}$ = -5.719 and $\hat{\alpha}_{45}^{YZ}$ = -.628. (Users who are accustomed to thinking in terms of the log odds ratios might prefer to use the relationship $-2\alpha_{ij}^{YZ} = \log \theta_{(ij)(ij)}^{YZ}$ to draw the conclusions above from examining $-2\hat{\alpha}_{16}^{YZ} = 11.438$ and $-2\hat{\alpha}_{45}^{YZ} = 1.256$.) For further interpretation of these data, see Sobel et al. (1985) and Section 4.3.1.

4.3 Models for Ordinal Variables

The log-linear models in Sections 4.1 and 4.2 treat the variables as nominal. If one or both variables are ordinal, more parsimonious and enlightening models can be used. In Section 2, ordering in the univariate context was taken into account by imposing parametric structure on the adjacent category logits, and in Section 4.1, the logit model was extended to the two-way table [see equation (5.39)] with the row variable Y dependent on the column variable Z.

When a dichotomous row variable Y is treated as dependent, and the column variable Z is ordered, with known scores v_j, $j = 1, \ldots, J$, it would be natural (as in logistic regression) to regress the logit on these scores, replacing the binomial logit model of (5.38) by

$$\log \frac{\pi_{1|j}^{Y|Z}}{\pi_{2|j}^{Y|Z}} = \lambda + \gamma v_j. \tag{5.46}$$

Table 5. Parameter Estimates for the Brazilian Data, Under QS

	Parameter					
Column	α_j^Z	α_{1j}^{YZ}	α_{2j}^{YZ}	α_{3j}^{YZ}	α_{4j}^{YZ}	α_{5j}^{YZ}
2	.048	-.793				
	(.293)	(.237)				
3	-.487	-1.445	-.982			
	(.293)	(.237)	(.214)			
4	-1.422	-2.630	-1.207	-.966		
	(.314)	(.362)	(.224)	(.160)		
5	-1.221	-3.449	-2.043	-1.573	-.628	
	(.320)	(.466)	(.270)	(.175)	(.122)	
6	-3.100	-5.719	-3.405	-2.463	-1.668	-.788
	(.358)	(1.017)	(.379)	(.228)	(.179)	(.123)

Without loss of generality, the column scores are taken (after rearrangement of columns, if necessary) so that $v_1 \leq v_2 \leq \cdots \leq v_J$.

Analogously, when Y is a polytomous variable treated as dependent, a natural extension of the baseline category logit model of (5.39) is the model

$$\log \frac{\pi_{i|j}^{Y|Z}}{\pi_{i'|j}^{Y|Z}} = \lambda_i^Y + \gamma_i^Y v_j, \tag{5.47}$$

where $i = 1, \ldots, i' - 1, i' + 1, \ldots, I$. In this model, (holding i fixed) the J logits depend on the column scores, as in the dichotomous case. In addition, this dependence is now allowed to vary across rows. A natural question to ask is whether this dependence is invariant across rows. To address this question, the model (5.47) can be reexpressed as

$$\log \frac{\pi_{i|j}^{Y|Z}}{\pi_{i'|j}^{Y|Z}} = \lambda_i^Y + \gamma\gamma_i^{*Y} v_j, \tag{5.48}$$

and the hypothesis that all the γ_i^{*Y} are 1 can be tested. When the null hypothesis holds, for $i = 1, \ldots, i' - 1, i' + 1, \ldots, I$, the set of J baseline category logits are parallel, when plotted against the column scores. When, in addition, the λ_i^Y are identical, the $I - 1$ parallel lines reduce to a single line.

Further interpretation of the γ_i^{*Y} parameters of (5.48) is facilitated by comparing the logits for adjacent columns j and $j + 1$ under the model. This leads [see (5.22) and (5.23)] to consideration of the log odds ratios $\log \theta^{YZ}_{(i'i)(j,j+1)} = \gamma\gamma_i^{*Y}(v_{j+1} - v_j)$. These odds ratios, which incorporate the information on the ordering of the column variable, may be called partially (in the columns) local, in contrast to the local odds ratios of (5.32). Whereas the latter set of odds ratios is a natural set to consider when both the row and column variables are ordered, the partially local odds ratios are a natural set of odds ratios to consider when only one of the variables is ordered.

From the expression for $\theta^{YZ}_{(i'i)(j,j+1)}$, it follows that

$$\frac{\pi^{Z|Y}_{j+1|i}}{\pi^{Z|Y}_{j+1|i'}} = \left(\frac{\pi^{Z|Y}_{j|i}}{\pi^{Z|Y}_{j|i'}}\right) \exp(\gamma\gamma_i^{*Y}(v_{j+1} - v_j)), \tag{5.49}$$

and from (5.49), it follows that when $\gamma\gamma_i^{*Y} > 0$ (< 0), the random variable $(Z \mid Y = i)$ is stochastically larger (smaller) than the random variable $(Z \mid Y = i')$. To see this, note that $\gamma\gamma_i^{*Y} > 0$ implies

$$\frac{\pi^{Z|Y}_{j+1|i}}{\pi^{Z|Y}_{j+1|i'}} \geq \frac{\pi^{Z|Y}_{j|i}}{\pi^{Z|Y}_{j|i'}}, j = 1, \ldots, J-1, \Rightarrow \frac{\pi^{Z|Y}_{J|i}}{\pi^{Z|Y}_{J|i'}} \geq \frac{\pi^{Z|Y}_{J-1|i}}{\pi^{Z|Y}_{J-1|i'}} \geq \cdots \geq \frac{\pi^{Z|Y}_{1|i}}{\pi^{Z|Y}_{1|i'}}.$$

Next, if $\pi^{Z|Y}_{1|i} > \pi^{Z|Y}_{1|i'}$ either $\sum_{j=1}^{J} \pi^{Z|Y}_{j|i} > 1$ or $\sum_{j=1}^{J} \pi^{Z|Y}_{j|i'} < 1$. Thus, $\pi^{Z|Y}_{1|i} \leq \pi^{Z|Y}_{1|i'}$, and if equality holds, $\pi^{Z|Y}_{j|i} = \pi^{Z|Y}_{j|i'}$ $j = 1, \ldots, J$.

It suffices now to consider the case where $\pi^{Z|Y}_{1|i} < \pi^{Z|Y}_{1|i'}$. Let

$$j^* = \max\{j : \frac{\pi^{Z|Y}_{j|i}}{\pi^{Z|Y}_{j|i'}} \leq 1\}.$$

Clearly, $j^* < J$, as otherwise either

$$\sum_{j=1}^{J} \pi^{Z|Y}_{j|i} < 1 \quad \text{or} \quad \sum_{j=1}^{J} \pi^{Z|Y}_{j|i'} > 1.$$

For $j \leq j^*$, clearly $\Pr(Z \leq j \mid Y = i) < \Pr(Z \leq j \mid Y = i')$. For $j > j^*$, the same result holds. To see this, note that

$$\Pr(Z \leq j \mid Y = i) > \Pr(Z \leq j \mid Y = i') \Rightarrow \sum_{j=1}^{J} \pi^{Z|Y}_{j|i} > 1 \quad \text{or} \quad \sum_{j=1}^{J} \pi^{Z|Y}_{j|i'} < 1.$$

Thus, $\gamma\gamma_i^{*Y}$ measures the extent to which the independent variable is stochastically larger (smaller) in the ith, relative to the i'th row, and in the case where all the $\gamma_i^{*Y} = 1$, the independent variable is stochastically larger (smaller) in each row, relative to the i'th, to the same extent. In this case, the distribution of the independent variable is identical within rows $i = 1, \ldots, i'-1, i'+1, \ldots, I$.

Having treated the case of an ordered independent variable and a polytomous dependent variable, I now take up the case where the independent variable is polytomous and the dependent variable is ordered. The treatment of this case is less transparent. However, by reexpressing (5.49) as

$$\frac{\pi^{Z|Y}_{j+1|i}}{\pi^{Z|Y}_{j|i}} = \left(\frac{\pi^{Z|Y}_{j+1|i'}}{\pi^{Z|Y}_{j|i'}}\right) \exp(\gamma\gamma^{*Y}_i (v_{j+1} - v_j)), \tag{5.50}$$

it is apparent that the model (5.48) leads also to a description of the distribution $Z \mid Y$. Taking the natural logarithm of (5.50) leads straightforwardly to an adjacent category logit model:

$$\log \frac{\pi^{Z|Y}_{j+1|i}}{\pi^{Z|Y}_{j|i}} = \lambda^{*Z}_j + \gamma\gamma^{*Y}_i (v_{j+1} - v_j), \tag{5.51}$$

where

$$\lambda^{*Z}_j = \log \frac{\pi^{Z|Y}_{j+1|i'}}{\pi^{Z|Y}_{j|i'}}. \tag{5.52}$$

Thus the previous treatment also suggests how ordered dependent variables can be handled. In either the case of a polytomous dependent variable and an ordered independent variable or the case of a polytomous independent variable and an ordered dependent variable, the polytomous variable might itself be ordered, and thus the results above also suggest how the case where both variables are ordered should be handled. Discussion of this point is momentarily deferred.

The foregoing models for ordered variables can be motivated in other (less direct) ways as well (for example, see Agresti 1990; Goodman 1984). For example, the baseline category logit model of (5.48) may first be reexpressed as a model for the expected counts:

$$\log m_{ij} = \mu + \alpha^Y_i + \alpha^Z_j + \beta\beta^Y_i v_j, \tag{5.53}$$

with $\beta(\beta^Y_i - \beta^Y_{i'}) = \gamma\gamma^{*Y}_i$. Equation (5.53) can then be used to be used to obtain an adjacent category logit model that is a reparameterization of (5.51). Alternatively, starting with the model of (5.53), logit models for the distributions $Y \mid Z$ and $Z \mid Y$ may be obtained. (Note also that (5.53) may be viewed as a generalization of some of the ordered models of Section 2.)

In (5.53), the scores on the ordinal column variable are treated as known. In this case, with the usual constraints on the α^Y_i and α^Z_j, only $I - 2$ of the β^Y_i parameters are linearly independent, and thus the df are $(IJ-1)-(I-1+J-1+1+I-1) = (I-1)(J-2)$. This model, proposed by Simon (1974) and Haberman (1974b), has been called the row effects model. (Given the way the model is motivated here, the nomenclature might not seem natural, but I shall retain it nevertheless.) Similarly, the special case where all the β^Y_i are 1 might be called the uniform row effects model. For this model, the df are $(I-1)(J-1)-1$. When the row effects model holds, the null hypothesis that the uniform row effects model holds can be tested by means of a conditional likelihood ratio test on $(I-2)$ df. Other tests could also be constructed. Further material on this model follows shortly.

In many instances, it is only known that the column variable is ordinal, but the column scores are unknown. If, in equation (5.53), the column scores are treated as parameters to be estimated, the resulting model is not a log-linear model. For this case, the location and scale of the v_j can be changed without altering the m_{ij}, and thus, only $(J-2)$ more linearly independent parameters are estimated under this model than under the row effects model. The df are thus $(I-2)(J-2)$. The v_j can be restricted in various ways to achieve identifiability, and the columns may be reordered (if necessary) so that $v_1 \leq v_2 \leq \ldots \leq v_J$. Because the model is not log-linear, some additional difficulties arise in estimation and testing of the model. In particular, if (5.53) holds with unkown column scores, the conditional likelihood ratio test for testing against independence does not have an asymptotic χ^2 distribution (Haberman 1981). Other differences are treated in Section 7.

Reversing the roles of Y and Z leads to analogous models. Thus, for example, the column effects model and the uniform column effects model are obtained, with row scores $u_1 \leq u_2 \leq \ldots \leq u_I$.

So far, the case where one variable is ordered and the other variable is unordered has been treated. When both variables are ordered and Y is dependent, it is natural to incorporate this additional information into the analysis by replacing the baseline category logits in (5.48) with the adjacent category logits $\log(\pi_{i+1|j}^{Y|Z}/\pi_{i|j}^{Y|Z})$. As before, this leads to the model of equation (5.53), and the column scores may be treated as known constants or as parameters to be estimated. In addition, consideration of an adjacent logit model for the distribution $Z \mid Y$ yields a model for the expected counts:

$$\log m_{ij} = \mu + \alpha_i^Y + \alpha_j^Z + \beta\beta_j^Z u_i, \tag{5.54}$$

where the row scores may be treated as known constants or as parameters to be estimated. The models of (5.53) and (5.54) are identical, and comparison indicates that in the case where the row and column scores are both unknown, the v_j of (5.53) and the β_j^Z of (5.54) may be taken to be identical. Similarly, the β_i^Y of (5.53) and the u_i of (5.54) may be taken to be identical. Thus, one might write either of these equivalent models as

$$\log m_{ij} = \mu + \alpha_i^Y + \alpha_j^Z + \beta u_i v_j, \tag{5.55}$$

where both u_i, $i = 1, \ldots, I$ and v_j, $j = 1, \ldots, J$ are unknown sets of scores to be estimated (Goodman 1979). Andersen (1980) also proposed this model. Following Goodman's (1985) terminology, I will call this the RC(1) model. The nomenclature stems from the fact that the RC(1) model is a special case of a model called the RC(M) model, which permits the assignment of M sets of row and column scores. Goodman (1981) noted that under the RC(1) model, when the row and column scores are ordered as previously described (which might require rearrangement of rows and/or columns) and $\beta > 0$, the row variable is stochastically higher in the jth column than in the j'th column when $j > j'$. Similar remarks apply to the column variable. Note also that in this case, the local odds ratios are at least 1. Yule (1906) referred to tables which exhibit this property (possibly after reordering of rows and/or columns) as isotropic. The results on stochastic ordering do not extend to the RC(M) models.

Because the location and scale of the v_j and the u_i in the RC(1) model are arbitrary and must be restricted in order to identify the model, it is important to note that both the parameter β and the scores depend on the scoring method used (Becker and Clogg 1989), although the estimated frequencies (and hence the value of the fit statistics) do not depend upon this choice. Two types of restrictions have been widely used. Goodman (1979) uses the restrictions

$$\sum_{i=1}^{I} u_i = 0, \sum_{i=1}^{I} u_i^2 = 1, \sum_{j=1}^{J} v_j = 0, \sum_{j=1}^{J} v_j^2 = 1.$$

Another popular choice (Goodman 1981) sets the sample mean of the row and column scores to 0, and the sample variance of the row and column scores to 1, using sample marginal distributions as weights:

$$\sum_{i=1}^{I} u_i p_i^Y = 0, \sum_{i=1}^{I} u_i^2 p_i^Y = 1, \sum_{j=1}^{J} v_j p_j^Z = 0, \sum_{j=1}^{J} v_j^2 p_j^Z = 1.$$

While this choice appears attractive, and was used by Goodman to develop the relationship between the RC(M) models and canonical correlation models, Becker and Clogg (1989) point out that such a choice reintroduces the marginal distributions into the analysis, that is, the interaction parameters are now confounded with the marginal distributions. Finally, two other remarks should also be made about this choice. First, the marginal distributions in question are not population distributions but sample distributions, and this introduces additional variability into the estimates of the scores that should be taken into account. Second, it is difficult to justify this choice for the product multinomial case. For example, with the marginal distribution of Y fixed by the sampling design, weights that use the row distribution in the data appear to have little appeal. If the distribution of Y in the population were known and used instead, this remark would not apply. For additional material on some of these matters, see Clogg and Rao (1991) and Goodman (1991).

The RC(1) model subsumes, in addition to the row effects model and the column effects model, a number of other useful log-linear models. The most general of these is the linear by linear interaction model. In this model, also given by (5.55), both the row scores and the column scores are known. The parameter β accounts for the association in the table (in the sense that independence holds if and only if $\beta = 0$), and the df are therefore $(I-1)(J-1)-1$.

When the row scores are known, and the column scores are unkown but equally spaced, ie., $v_{j+1} - v_j = v$, $j = 1, \ldots, J-1$, v can be taken to be 1 (as $\beta v_j = \beta v/(v/v_j)$), which yields the model

$$\log m_{ij} = \mu + \alpha_i^Y + \alpha_j^Z + (j - P)u_i\beta, \tag{5.56}$$

where P is an arbitrary real number. In most applications, P has been chosen to be either 0 or $(J+1)/2$. Under this model,

$$\log \frac{m_{i+1,j}}{m_{ij}} = \lambda_i^Y + (j - P)(u_{i+1} - u_i)\beta \tag{5.57}$$

and

$$\log \frac{m_{i,j+1}}{m_{ij}} = \lambda_j^Z + u_i\beta. \tag{5.58}$$

Whereas the adjacent category logits for the rows depend on both the row and column scores, the adjacent category logits for the columns depend only on the row scores.

Similar models are also obtained when the column scores are known and the row scores are unknown but equally spaced. In the case where both the row and column scores are equally spaced, the model

$$\log m_{ij} = \mu + \alpha_i^Y + \alpha_j^Z + (i - Q)(j - P)\beta \tag{5.59}$$

is obtained, where P and Q are arbitrary real numbers. This model (Duncan 1979; Goodman 1979) is called the uniform association (UA) model because all the local odds ratios of equation (5.32) have value $\exp(\beta)$.

The models for ordinal discrete variables considered thus far have been motivated by considering adjacent category logits. Equivalent models can be motivated by considering models for the local odds ratios (Goodman 1979). Thus, (5.54) is equivalent to a model for the local log odds ratio of the form

$$\log \theta_{ij}^{YZ} = \beta(u_{i+1} - u_i)(v_{j+1} - v_j) \tag{5.60}$$

where $i = 1, \ldots, I - 1; j = 1, \ldots, J - 1$, and all the models previously considered can be developed from (5.60). In this equation, it is natural to interpret β as a uniform association parameter for equally spaced scores, and to interpret the scores as adjustments for the lack of equal spacing among the categories of the row and/or column variable.

Many other models for ordered discrete variables may be considered. For example, the linear by linear interaction model can be generalized in another direction when the odds ratio has the form

$$\theta_{ij}^{YZ} = \beta^{*(u_{i+1}-u_i)(v_{j+1}-v_j)} \frac{\beta_{i+1}^{*Y(v_{j+1}-v_j)}}{\beta_i^{*Y}} \frac{\beta_{j+1}^{*Z(u_{i+1}-u_i)}}{\beta_j^{*Z}}, \tag{5.61}$$

where $i = 1, \ldots, I - 1; j = 1, \ldots, J - 1$, and the scores are taken as known. The corresponding log-linear model is:

$$\log m_{ij} = \mu + \alpha_i^Y + \alpha_j^Z + u_i v_j \beta + v_j \beta_i^Y + u_i \beta_j^Z, \tag{5.62}$$

where $\beta = \log \beta^*$, $\beta_i^Y = \log \beta_i^{*Y}$, $\beta_j^Z = \log \beta_j^{*Z}$, from which it is easy to see that the row effects model and the column effect models previously discussed are also special cases of this model [hereafter the R+C model, using the nomenclature of Goodman (1981)]. Under this model, in addition to the usual restrictions on the α_i^Y and α_j^Z parameters, two linear restrictions must be imposed on each of β_i^Y and β_j^Z, and thus the df are $(I - 2)(J - 2)$, as in the RC(1) model.

As indicated before, the RC(1) model can be generalized in various ways, and the RC(1) and R+C models can also be combined. The models above are also easily specialized to the case where the table is square (as below).

Information on ordering can also be exploited in different ways. Instead of building models for adjacent category logits, models for continuation ratio logits may be constructed. For the case where Y is an ordinal discrete variable, the continuation ratio logits are

$$\log \frac{\Pr(Y = i \mid Z = j)}{\Pr(Y \geq i \mid Z = j)}, i = 1, \ldots, I - 1; j = 1, \ldots, J.$$

For further discussion, see Fienberg (1980). A more common strategy is to use the cumulative logits:

$$\log \frac{\Pr(Y \leq i \mid Z = j)}{\Pr(Y > i \mid Z = j)}, i = 1, \ldots, I - 1; j = 1, \ldots, J.$$

Agresti (1990) considers a number of models for cumulative logits, including models for cumulative odds ratios. These models are not necessarily equivalent to log-linear models for the expected counts.

The cumulative logits are often used in conjunction with continuous explanatory variables as well. Treatment of this important case is beyond the scope of this chapter.

Example 4 Continued: Occupational Mobility in Brazil

To illustrate some of the models for ordered discrete variables, the Brazilian mobility table is reconsidered. Recall that QS fit the data well and that the occupations are ordered by status, from 1 (highest) to 6 (lowest). Under QS, counts on the diagonal are fitted exactly. The ordered models considered above do not allow for special treatment of the diagonal cells. By including special parameters for these cells (or by deleting these cells and applying the desired ordered model to the incomplete table), the ordered models previously considered are easily extended. Second, under QS, the α_{ij}^{YZ} are symmetric, that is, $\alpha_{ij}^{YZ} = \alpha_{ji}^{YZ}$. In (5.53), $\alpha_{ij}^{YZ} = \beta u_i v_j \neq \beta u_j v_i = \alpha_{ji}^{YZ}$, that is, when the row and column scores are not identical, the ordered models above do not imply QS. Similar remarks apply to the R+C model.

The foregoing suggests consideration of an ordered model with identical (homogeneous) row and column scores and diagonal parameters. Following Sobel et al. (1985), I consider the quasi-homogeneous R+C model with an equal-spacing assumption:

$$\log m_{ij} = \eta_i + \eta_j + \alpha_j^Z + \delta_{ij}\alpha_{ij}^{YZ} + (6-i)(6-j)\beta + (6-j)\beta_i + (6-i)\beta_j, \qquad (5.63)$$

with the η_i and α_j^Z defined as before, and $\beta_1 = \beta_6 = 0$.

The model of (5.63) fits the data well ($X^2 = 16.00$, $L^2 = 17.22$, $D^2 = 16.19$, df = 14), suggesting that the equal spacing assumption is acceptable. The parameter estimates of the β_i are, respectively (standard errors in parentheses): $\hat{\beta}_2 = -.198(.197)$, $\hat{\beta}_3 = -.108(.118)$, $\hat{\beta}_4 = -.108(.100)$, $\hat{\beta}_5 = -.194(.157)$, which suggests that the quasi-uniform association (QUA) model (Duncan 1979) will also fit these data. For this model, the df are 18, and the fit of the model is no worse, at the .05 level, than the fit obtained under the previous model (($X^2 = 17.56$, $L^2 = 19.56$, $D^2 = 18.02$). Under QUA, estimates of the α_j^Z are similar to those in Table 5, and thus are not reported. The estimate of β is .3682, with standard error .030. Unlike the uniform association model, the local odds ratios are not identical under QUA, and thus β does not admit the same interpretation under QUA and UA. However, the local odds ratios that do not involve diagonal cells have value $\exp(\beta)$ under QUA. The local odds ratios involving diagonal cells are easily computed under this model, and the previously given relationship between the local odds ratios and the odds ratios of (5.31), with $r_1 = c_1$, $r_2 = c_2$, can be used to obtain the diagonal odds ratios and/or interaction parameters considered in the previous analysis of the table, as in Sobel et al. (1985). Since these address the tendency for the counts to cluster on the diagonal, they are of interest in both the ordered and unordered case. For further details, see Sobel et al. (1985).

5 Models for the Three-Way Table

In this section, interest centers on using the log-linear model to assess the structure of association in the $I \times J \times K$ table. As before, the examples use maximum likelihood, and essentially the same results are obtained under multinomial sampling, product multinomial sampling, and Poisson sampling. After taking up the basic case, I consider models for tables with a one-to-one correspondence between two or more variables, and following this, models where one or more of the variables is ordinal.

5.1 Basic Models

For the three-way table, with row variable Y, column variable Z, and layer variable W, the saturated log-linear model is

$$\log m_{ijk} = \mu + \alpha_i^Y + \alpha_j^Z + \alpha_k^W + \alpha_{ij}^{YZ} + \alpha_{ik}^{YW} + \alpha_{jk}^{ZW} + \alpha_{ijk}^{YZW}. \tag{5.64}$$

Clearly, the model of (5.64) is overparameterized, and it is usual to identify the model by imposing the ANOVA-type restrictions:

$$\begin{aligned}\sum_{i=1}^{I} \alpha_i^Y &= \sum_{j=1}^{J} \alpha_j^Z = \sum_{k=1}^{K} \alpha_k^W = \sum_{i=1}^{I} \alpha_{ij}^{YZ} = \sum_{j=1}^{J} \alpha_{ij}^{YZ} = \sum_{i=1}^{I} \alpha_{ik}^{YW} = \sum_{k=1}^{K} \alpha_{ik}^{YW} \\ &= \sum_{j=1}^{J} \alpha_{jk}^{ZW} = \sum_{k=1}^{K} \alpha_{jk}^{ZW} = \sum_{i=1}^{I} \alpha_{ijk}^{YZW} = \sum_{j=1}^{J} \alpha_{ijk}^{YZW} = \sum_{k=1}^{K} \alpha_{ijk}^{YZW} = 0.\end{aligned}$$

Simple interpretations of the model parameters are obtained by connecting the model to the previous discussion of odds ratios for three-way tables. Throughout, attention is confined to hierarchical models, that is, models in which the presence of a higher-way term implies the presence of all lower-order relatives. For example, in a hierarchical model, inclusion of the term α_{ij}^{YZ} in a model implies inclusion of the lower-order terms α_i^Y and α_j^Z.

To begin, from (5.64),

$$\begin{aligned}\log \theta_{(r_1,r_2)(c_1,c_2)(k)}^{YZ|W} = {} & (\alpha_{r_1c_1}^{YZ} + \alpha_{r_2c_2}^{YZ} - \alpha_{r_1c_2}^{YZ} - \alpha_{r_2c_1}^{YZ}) + \\ & (\alpha_{r_1c_1k}^{YZW} + \alpha_{r_2c_2k}^{YZW} - \alpha_{r_1c_2k}^{YZW} - \alpha_{r_2c_1k}^{YZW}).\end{aligned} \tag{5.65}$$

From (5.65), it is evident that if the three-way interactions are 0, the conditional association between any two variables is constant across the third variable. The converse is also true. To see this, note that if the conditional association is constant across the layer variable,

$$(\alpha_{r_1c_1k}^{YZW} + \alpha_{r_2c_2k}^{YZW} - \alpha_{r_1c_2k}^{YZW} - \alpha_{r_2c_1k}^{YZW}) = (\alpha_{r_1c_1k'}^{YZW} + \alpha_{r_2c_2k'}^{YZW} - \alpha_{r_1c_2k'}^{YZW} - \alpha_{r_2c_1k'}^{YZW}),$$

$k' = 1, \ldots, K$. Summing over k' gives

$$(\alpha_{r_1 c_1 k}^{YZW} + \alpha_{r_2 c_2 k}^{YZW} - \alpha_{r_1 c_2 k}^{YZW} - \alpha_{r_2 c_1 k}^{YZW}) = 0,$$

and holding r_1 and c_1 fixed, summing over r_2 and c_2, $\alpha_{r_1 c_1 k}^{YZW} = 0$.

The foregoing shows that the saturated model corresponds to case 1 of Section 3.2, and the case where the three-way interactions vanish to case 2. The log-linear model corresponding to case 2 is called the partial association model. Because there are $(I-1)(J-1)(K-1)$ linearly independent three-way interactions in the saturated model, the df for the partial association model are also $(I-1)(J-1)(K-1)$.

From (5.65), it is also evident that when the three-way interactions are 0, and in addition, the two-way interactions between Y and Z vanish, case 3 of Section 3.2 is obtained. Thus, $Y \perp\!\!\!\perp Z \mid W$. The converse can be demonstrated by arguing along the lines above. For this log-linear model, the df are $(I-1)(J-1)(K-1) + (I-1)(J-1) = (I-1)(J-1)K$. Similarly, log-linear models with $Y \perp\!\!\!\perp W \mid Z$ (or $Z \perp\!\!\!\perp W \mid Y$) correspond to the case where the three-way interactions are 0 and the two-way interactions between Y and W (or Z and W) are 0. The df for these two models are, respectively, $(I-1)(K-1)J$ and $(J-1)(K-1)I$.

The foregoing suggests two ways to test the hypothesis $Y \perp\!\!\!\perp Z \mid W$. First, the log-linear model corresponding to this hypothesis could be compared to the saturated model, using any of the fit statistics in Section 1. Second, the log-linear model of conditional independence could be compared, by means of a conditional test, using any of the fit statistics in Section 1, with the partial association model. This test has df $(I-1)(J-1)$, in contrast to $(I-1)(J-1)K$, the df for the unconditional test. Unlike the unconditional test, however, the conditional test is valid only when the partial association model holds, but when this model holds, the conditional test is more powerful than the unconditional test. For further material on tests for conditional independence, see Agresti (1990).

From case 4 of Section 3.2. and the material in the preceeding paragraph, $Y \perp\!\!\!\perp (Z, W)$ if and only if the log-linear model

$$\log m_{ijk} = \mu + \alpha_i^Y + \alpha_j^Z + \alpha_k^W + \alpha_{jk}^{ZW} \tag{5.66}$$

holds. For this model, the df are $(I-1)(J-1)K + (I-1)(K-1)$. Models with $Z \perp\!\!\!\perp (Y, W)$ or $W \perp\!\!\!\perp (Y, Z)$ are constructed analogously.

Finally, when all the two-factor interactions are missing, it follows from case 5 of Section 3.2 and the material in the preceding paragraph that Y, Z, and W are mutually independent:

$$\Pr(Y = i, Z = j, W = k) = \Pr(Y = i)\Pr(Z = j)\Pr(W = k)$$

for all i, j, k.

In addition, it is sometimes of interest to consider log-linear models with one or more main effects absent, for example, the model

$$\log m_{ijk} = \mu + \alpha_i^Y + \alpha_j^Z + \alpha_{ij}^{YZ}. \tag{5.67}$$

Under this model, $m_{ijk} = m_{ijk'}$, $k' = 1, \ldots, K$, and $W \perp\!\!\!\perp (Y, Z)$; thus

$$\Pr(Y = i, Z = j, W = k) = \Pr(Y = i, Z = j)/K.$$

When a bivariate or univariate distribution is fixed by the sampling design, the appropriate distribution is the product multinomial, and as in Section 4, it is necessary to include the relevant parameters. For example, if the distribution of W is fixed by the sampling scheme, it is necessary to include all terms of the form α_k^W. For the product multinomial case, parameter interpretation is essentially the same, although this case sometimes suggests different hypotheses of interest, and researchers tend to use somewhat different language to describe the models above.

When one or more of the variables are viewed as dependent, it is natural to condition on the other variable(s). For example, under multinomial or Poisson sampling, with Y dependent, the distribution $Y \mid (Z, W)$ is product multinomial. As in Section 4, such considerations lead naturally to a baseline category logit model:

$$\log \frac{\pi_{i|jk}^{Y|ZW}}{\pi_{i'|jk}^{Y|ZW}} = \lambda_i^Y + \lambda_{ij}^{YZ} + \lambda_{ik}^{YW} + \lambda_{ijk}^{YZW}, \tag{5.68}$$

where i' is a fixed reference point, i takes on all values other than i', the λ_{ij}^{YZ} and the λ_{ik}^{YW} parameters are each subject to $(I-1)$ linear restrictions, and the λ_{ijk}^{YZW} are subject to $(I-1)(J+K-1)$ linear restrictions.

Example 5: Wife's Desire, Husband's Desire, and Future Fertility

To illustrate the models introduced, I consider the trivariate distribution of fertility (Y) by wife's desire (Z) by husband's desire (W), using the data in Table 1. Demographers have debated whether the inclusion of the desires of both partners predicts future fertility better than predictions based on the desires of one partner (usually the wife) only. To adress this particular question, one might look at the logit model:

$$\log \frac{\pi_{2|jk}^{Y|ZW}}{\pi_{1|jk}^{Y|ZW}} = \lambda + \lambda_j^Y + \lambda_k^W + \lambda_{jk}^{ZW}, \tag{5.69}$$

where, for each variable, a "yes" response corresponds to the second level of the variable, and a "no" response to the first level.

The null hypothesis that the husband's desire does not help to predict future fertility, net of wife's desire, might be assessed by testing whether λ_k^W and λ_{jk}^{ZW} vanish in (5.69). The corresponding log-linear model is

$$\log m_{ijk} = \mu + \alpha_i^Y + \alpha_j^Z + \alpha_k^W + \alpha_{jk}^{ZW} + \alpha_{ij}^{YZ}, \tag{5.70}$$

which states $Y \perp\!\!\!\perp W \mid Z$, or equivalently,

$$\Pr(Y = i \mid Z = j, W = k) = \Pr(Y = i \mid Z = j).$$

This null hypothesis is rejected, at the .01 level ($X^2 = 52.29$, $L^2 = 43.74$, $D^2 = 49.34$, df = 2).

If husbands are more influential than wives, it may be the case that the wife's desire, net of the husband's, does not help to predict future fertility. This suggests consideration of the model $Y \| Z \mid W$. This hypothesis is also rejected at the .01 level ($X^2 = 45.21$, $L^2 = 36.93$, $D^2 = 41.82$, df = 2).

Finally, the partial association model was also fit to these data. The null hypothesis would not be accepted at the .05 level, but it would be accepted at the .025 level ($X^2 = 4.64$, $L^2 = 4.54$, $D^2 = 4.60$, df = 1).

The results above suggest that inclusion of both partner's desires results in a better (from a statistical point of view) fit to the data. From a practical standpoint, it may be the case that models including only the wife's desire or the husband's desire predict future fertility nearly (subjectively) as well. To examine this issue further, the models above were compared with the baseline model $Y \| (Z, W)$ by using a pseudo R^2:

$$R^2 = \frac{L_0^2 - L_1^2}{L_0^2},$$

where L_0^2 is the value of L^2 for the baseline model, and L_1^2 is the value for the alternative. Note that one could also use X^2 or D^2 in the equation above. The R^2 statistic used here differs from the usual R^2 statistic in the literature (Haberman 1978), which uses the model of independence as a baseline. Sobel and Bohrnstedt (1985) discuss the choice of baseline models for fit statistics of this nature in covariance structure analysis and give reasons for employing baseline models other than independence. Their discussion carries over to this context as well.

Using $L_0^2 = 147.77$, comparison of the baseline with the model $Y \| W \mid Z$ yields $R^2 = .70$, indicating a relative reduction of L^2 of 70%. Similarly, comparison of the baseline with $Y \| Z \mid W$ gives $R^2 = .75$. These comparisons suggest that both husband's and wife's desire can be used to predict future fertility, that husband's desire is a slightly better predictor, and that further improvement can be made by using the desires of both partners. For ways of testing to see whether these relative reductions are significant, see Haberman (1982). For an alternative analysis (using an R^2 measure interpretable as a squared multiple correlation coefficient) that reaches the conclusion that the improvement from inclusion of both desire variables is smaller, see Sobel and Arminger (1992).

5.2 Collapsibility in Models for the Three-Way Table

In the previous analysis, the partial association model fit the data at the .025 level. If an investigator were to accept this model, a natural question arises. Since the partial association model is equivalent to the case where all sets of conditional odds ratios do not depend on the level of the third variable conditioned on, it is natural to ask whether some or all of the relationship between any two variables in the trivariate table can be estimated from the corresponding marginal (collapsed) table, that is, the table of counts obtained by summing over the third variable. This raises the issue called "collapsibility;" the analogous issue in regression is usually called "omitted variable bias."

An affirmative answer to the question above can be important for several reasons. First, if the three-way table is sparse but "collapsible" with respect to one or more two-way interactions, better estimates of the relationship between two variables can be obtained from the corresponding marginal table. In the case where the third variable is not of inherent substantive interest (for example, one is interested in the relationship between Y and Z and the variable W denotes different studies on the relationship between Y and Z), an investigator will want to know when it is possible to pool the data across studies. In other instances, the third variable may be unobserved, but an investigator may have information (say from other studies) about the trivariate relationship, and if "collapsibility" holds with respect to the two-way interaction, the two-way table can be used to make valid inferences about this interaction, even in the absence of the third variable. While there are good reasons to raise the issue of collapsibility (as above) an important caveat is also in order. I use regression analysis to illustrate the point. When the subject of omitted variable bias is considered, it is typically assumed that the full model is properly specified (often called true) and the reduced model is misspecified (often called false), and the fact that the parameters of the reduced model are not (except when special conditions hold) invariant with respect to inclusion of additional variables leads to the conclusion that these parameters are biased. In the social sciences, this argument is typically used to argue for the inclusion of additional "explanatory" variables. However, it is important to note that the reduced model may be properly specified, given the set of variables conditioned upon in that model. If interest focuses on the conditional expectation, given that set, then the parameters of the full model are not those of interest. In fact, it may well be the case that the reduced model is properly specified, given the set of variables conditioned upon there, while the full model can be misspecified, given the set of variables conditioned upon the full model. The upshot of the discussion is that investigators should first be clear about what it is they wish to estimate, and then ask whether or not the question of omitted variable bias is even relevant. Virtually identical remarks apply in the context here.

The "collapsibility" issue is somewhat complicated, and various different notions have been proposed. To keep matters simple, only conditions under which the parameters α_{12}^{*YZ} obtained from the table collapsed over W are identical to the parameters α_{12}^{YZ} of the three-way table are considered. When these sets of parameters are identical, I shall say that the three-way table is collapsible over W with respect to the two-factor interaction between Y and Z.

Although the focus here is limited, many of the issues in the literature can be illustrated with this case. However, it is also important to note that the issue of "collapsibility" can be examined in higher-way tables, and it is not necessary to restrict attention to only one set of parameters in the collapsed table, as here.

If $Y \perp\!\!\!\perp W \mid Z$ and/or $Z \perp\!\!\!\perp W \mid Y$, the three-way table is collapsible over W with respect to the two-factor interaction between Y and Z (Bishop et al. 1975). To see this, use equation (5.64), and note that

$$m_{ij} = \sum_{k=1}^{K} m_{ijk} = \exp(\mu + \alpha_i^Y + \alpha_j^Z + \alpha_k^W + \alpha_{ij}^{YZ})\eta_{ij},$$

$$\eta_{ij} = \sum_{k=1}^{K} \exp(\alpha_{ik}^{YW} + \alpha_{jk}^{ZW} + \alpha_{ijk}^{YZW}).$$

Next, note that when η_{ij} reduces to η_i or η_j, the desired result holds. Note also [see (5.65)] that this result implies that the estimates of conditional association under the full model and the estimates of association under the reduced model (called marginal association) are identical.

The condition above is sufficient, but not necessary. Whittemore (1978) gives examples where collapsibility over W with respect to the two-factor interaction between Y and Z holds and $\alpha_{ijk}^{YZW} \neq 0$. Because the three-factor interaction contains information about the relationship between Y and Z, she argues that it does not make sense to collapse over W in this case and she proposes the concept of "strict collapsibility," which combines collapsibility with vanishing three-way interaction. Necessary and sufficient conditions for strict collapsibility, as well as a test for strict collapsibility, are also given. Because strict collapsibility is not equivalent to a log-linear model, implementation of Whittemore's test requires the use of special software. Subsequently, Ducharme and Lepage (1986) defined a concept of strong collapsibility that is equivalent to the sufficient conditions above. Thus, strong collapsibility can be tested by comparing the appropriate restricted models for the three-way table with the saturated model.

Clogg, Petkova, and Shihadeh (1992) take a different approach, by exploiting the analogy between the generic notion of "collapsibility" and omitted variable bias in regression analysis. They argue, in line with Whittemore (1978), that it is not meaningful to speak of "collapsibility" in the presence of three-factor interaction, and they propose to test for what they call "collapsibility" (hereafter collapsibility$_1$) by comparing a single nested model (a null model) with a full model, in this case the partial association model. Provided the partial association model holds, these authors use a single test to compare the null model and the partial association model.

In the context where Y is regarded as a response, the restricted model used by Clogg et al. (1992) is a logit model equivalent to the log-linear model $Y \| W \mid Z$. The models $Z \| W \mid Y$ and $Y \| (Z, W)$, which would also be considered in the unconditional approach of Ducharme and Lepage (1986), are not considered in this approach. Therefore, for the logit model, strong collapsibility implies collapsibility$_1$, but collapsibility$_1$ does not imply strong collapsibility. This is an attractive feature of the Clogg et al. approach if one accepts the argument of Asmussen and Edwards (1983) (who take a somewhat different approach to the notion of "collapsibility") that when Y is a response, it is not meaningful to consider models such as $Z \| W \mid Y$.

However, for the multinomial log-linear model, the approach of Clogg et al. leads to use of

$$\log m_{ijk} = \mu + \alpha_i^Y + \alpha_j^Z + \alpha_{ij}^{YZ}$$

as the restricted model. Thus, collapsibility$_1$ holds only if $\pi_{ijk}^{YZW} = \pi_{ij}^{YZ}/K$. From this relationship, it is evident that collapsibility$_1$ is sufficient for strong collapsibility, hence strict collapsibility and collapsibility. Clearly, collapsibility$_1$ is not necessary for these other forms of "collapsibility." Thus, for log-linear models, collapsibility$_1$ is the strongest form examined. Note also that collapsibility$_1$ implies $\Pr(W = k) = (1/K)$, $k = 1, \ldots, K$, which suggests that collapsibility$_1$ is, in the context of the log-linear model, overly strong.

5.3 Models for Tables with a One-to-One Correspondence among Categories

In Section 4.2, contexts leading to square tables were discussed, and models for such tables were considered. Similar contexts also give rise to higher-way tables with a one-to-one correspondence between at least two of the variables in the analysis. Here, there are essentially two cases of interest: (1) comparable square tables are observed in two or more different settings, and the aim is to compare the bivariate relationship across settings, as, for example, in cross-national studies of occupational mobility; (2) a one-to-one correspondence between three or more variables exists, for example, three judges rate a sample of subjects on a discrete variable such as psychiatric diagnosis. A number of authors have proposed extensions of the models in Section 4.2 to the cases above (Bishop et al. 1975; Darroch 1986; Madansky 1963; McCullagh 1982; Sobel 1988).

In case 1, given the aim of the investigation and/or the sampling mechanism used, the distribution of settings is treated as fixed. Let W be the setting variable, which may be a compound variable formed from the cross-classification of two or more variables, and let Y and Z be two discrete variables with a one-to-one correspondence. The setting variable has K categories and Y and Z each have I categories. In Section 4.2, the most restrictive model considered was SI and the least restrictive QS. The natural extension of QS for this case is

$$\log m_{ijk} = \eta_{ik} + \eta_{jk} + \alpha_{jk}^{ZW} + \alpha_{ijk}^{YZW}, \tag{5.71}$$

where $\alpha_{ijk}^{YZW} = \alpha_{jik}^{YZW}$, and $\alpha_{ijk}^{YZW} = 0$, if $i = j$, which states that QS holds within each setting. The df for this model are $K(I-1)(I-2)/2$. Under this parameterization, the α_{jk}^{ZW} parameters are analogous to the α_j^Z parameters used in (5.45) and, within each setting, admit the same interpretation. Similar remarks apply to the α_{ijk}^{YZW} parameters of (5.71) and the α_{ij}^{YZ} of (5.45).

Analogously, SI is extended by requiring SI to hold in each setting. The df are therefore $KI(I-1)$, and SI may be obtained from (5.71) by setting the α parameters to 0. In like fashion, extensions of the other models depicted in Figure 1 are readily obtained.

The models above restrict the pattern of association and/or the marginal distributions within each setting, but do not (in general) feature any form of across-setting homogeneity. Using these models, an investigator could ascertain, for example, whether QS holds in each setting, but could not, were this the case, assess whether or not the association between Y and Z is identical across settings. To answer this latter question, models that constrain the pattern of assocation and/or the marginal distributions across settings must be considered, and a large number of models can be generated in this fashion. Restrictions on the association across settings are implemented by requiring, for any pair (k, k'), $\alpha_{ijk}^{YZW} = \alpha_{ijk'}^{YZW} = \alpha_{ij}^{YZ}$. Thus, any model with homogenous association is a special case of the partial association model. Similarly, restrictions on the marginal distributions are implemented by imposing the same type of equality restrictions on the α_{jk}^{ZW} and/or η_{ik} parameters. For further details and examples, see Sobel (1988).

For the second case, under multinomial sampling, with three variables, an I^3 table is observed, and a natural extension of QS is

$$\log m_{ijk} = \eta_i + \eta_j + \eta_k + \alpha_i^Y + \alpha_j^Z + \alpha_k^W + \alpha_{ijk}^{YZW}, \tag{5.72}$$

where

$$\alpha_{ijk}^{YZW} = \alpha_{ikj}^{YZW} = \alpha_{jik}^{YZW} = \alpha_{jki}^{YZW} = \alpha_{kij}^{YZW} = \alpha_{kji}^{YZW},$$

that is, the three-way interactions are "symmetric." To identify the model, $I+2$ restrictions are imposed on the set of α_i^Y, α_j^Z, and α_k^W, in addition to the restrictions $\alpha_{ijk}^{YZW} = 0$ if $i = j = k$. The df are $I^3 - ((2I-2) + \frac{(I+2)!}{3!(I-1)!})$. As in Section 4.2, under this parameterization, ratios of the main effect parameters are invariant with respect to the restrictions used to identify the model, and these ratios quantify the impact of marginal heterogeneity (between the three variables) on asymmetry of the m_{ijk}. Second,

$$\alpha_{ijk}^{YZW} = \frac{1}{3}\log\frac{m_{ijk}m_{kij}m_{jki}}{m_{iii}m_{jjj}m_{kkk}},$$

showing that the three-way interactions measure the total dependence in the table, with respect to the "consistent" cells.

Similarly, the natural extension of SI for this case is obtained by setting the α parameters in (5.72) to 0, and the df are thus $I^3 - I$. In like fashion, the other models depicted in Figure 1 are extended.

The models above restrict the pattern of association in the three-way table, and/or feature various forms of symmetry (exchangeability) between the three variables. For example, when $\alpha_i^Y = \alpha_i^Z = \alpha_i^W$, symmetry holds, that is, the three variables are exchangeable. If, however, $\alpha_i^Y = \alpha_i^Z$, Y and Z are exchangeable, but the set (Y, Z, W) is not exchangeable. When the three-way interaction $\alpha_{ijk}^{YZW} = \alpha_{ij}^{YZ} + \alpha_{ik}^{YW} + \alpha_{jk}^{ZW}$, analogues of these models that feature symmetric partial association are obtained. For further details on the models above and their properties, see Sobel (1988). Clogg (1982) also considered symmetric partial association models similar to the partial association models above.

Finally, note that the models just treated can be generalized for higher-way tables in an obvious fashion and combined with the models for the first case. An interesting generalization, due to Ten Have and Becker (1991), partitions the variables into sets and constructs log-linear models featuring exchangeability of all variables, or exchangeability of variables within sets.

5.4 Models for Tables With Ordered Variables

I now consider models for the three-way table with one or more variables ordinal, focusing attention on models that may be viewed as extensions of RC(1) and the special cases of RC(1) previously discussed. Parallelling the treatment in Section 5.3, the case where interest centers on comparing the association between Y and Z across K different settings is considered first.

Clogg (1982) extended the RC(1) model to the case of K settings as

$$\log m_{ijk} = \mu + \alpha_i^Y + \alpha_j^Z + \alpha_k^W + \alpha_{ik}^W + \alpha_{jk}^{ZW} + \beta_k u_{ik} v_{jk}. \tag{5.73}$$

Under (5.73), RC(1) holds in each setting, and $\beta_k u_{ik} v_{jk}$ is equal to $\alpha_{ij}^{YZ} + \alpha_{ijk}^{YZW}$ in (5.64). The model is appropriate for both multinomial and product multinomial sampling, with the marginal distribution of W fixed by the sampling design. As in Section 4.3, the scores u_{ik} and v_{jk} could be treated as known or unknown, or some of the scores could be known while others are unknown, and thus the model can be used, for example, when the row (column) variable is nominal, and the column (row) variable ordinal. Thus, (5.73) subsumes the case where a column (row) effects model holds in each setting, the case where linear-by-linear interaction holds in each setting, and the case where uniform association holds in each setting.

More parsimonious models may be obtained from (5.73) by imposing restrictions on the parameters. Of most interest is the hypothesis that the association between Y and Z is the same across settings. In this case $\beta_k u_{ik} v_{jk}$ does not depend on k, and thus, the three-way interaction is nil. When this partial association model holds, one might also wish to consider models with α_{ik}^{YW} and/or $\alpha_{jk}^{ZW} = 0$.

The model of (5.73) can also be extended to a model for the three-way table analogous to the RC(M) model for the two-way table (Becker and Clogg 1989). In addition, when the scores in (5.73) are known, a linear-by-linear association model is obtained, and as in Section 4.3, this suggests the alternative generalization [compare with (5.62)]

$$\log m_{ijk} = \mu + \alpha_i^Y + \alpha_j^Z + \alpha_k^W + \alpha_{ik}^W + \alpha_{jk}^{ZW} + \beta_k u_{ik} v_{jk} + \beta_{ik}^{YW} v_{jk} + \beta_{jk}^{ZW} u_{ik}, \quad (5.74)$$

also considered in Clogg (1982).

Finally, it should be noted, as in Section 4.3, that when there is a one-to-one correspondence between Y and Z, the counts will often cluster on the diagonal, and the models above should be modified to take this into account. This may be done by including, for each setting, parameters for the diagonal (essentially the strategy in the analysis of the Brazilian table), or alternatively, the other parameters of the model may be estimated by applying the model to the off-diagonal cells only. From the point of view here, the former approach is preferable; under the latter approach, the diagonal parameters would not be estimated, and one could not test the hypothesis that these parameters are homogenous across settings. Clearly, such a hypothesis must be tested before accepting the claim that the pattern of association in each of the K tables is identical. For illustrations using the models above, see Clogg (1982).

For the case when all three variables might be treated as ordinal, a natural extension of RC(1) to consider is:

$$\log m_{ijk} = \mu + \alpha_i^Y + \alpha_j^Z + \alpha_k^W + \alpha_{ij}^{YZ} + \alpha_{ik}^{YW} + \alpha_{jk}^{ZW} + \beta u_i v_j w_k. \quad (5.75)$$

Goodman (1979) proposed the use of this model using integer scoring as a generalization of the uniform association model. Here, integer scoring is not assumed. Under this model, the three-way interaction is more parsimoniously parameterized by means of the parameter β and the sets of scores u_i, v_j, w_k. As in Section 4.3, the scores could be treated as known or unknown, or some of the scores could be known while others are unknown, and the models may be applied when one, two, or all variables are ordinal. A number of desirable properties of the RC(1) model for the two-way table also carry over to this model as well, that is, it subsumes analogues to row effects and column effects (and now level) models, and to the linear-by-linear interaction (now linear by linear by linear) and uniform association models.

A drawback of this model is that when the partial association model holds, $\beta = 0$, and the model reduces to the standard model of partial association for the nominal table. Thus, any information on the ordinality of one or more of the variables is lost. This leads to consideration of models with special structure imposed on the two-way interactions as well, for example, the model

$$\log m_{ijk} = \mu + \alpha_i^Y + \alpha_j^Z + \alpha_k^W + \beta^{YZ} u_i v_j + \beta^{YW} u_i w_k + \beta^{ZW} v_j w_k + \beta^{YZW} u_i v_j w_k, \tag{5.76}$$

proposed by Agresti and Kezouh (1983).

In this model, which is a special case of the model in (5.75), the two-way interactions are now more parsimoniously modeled as $\beta^{YZ} u_i v_j$, and so on. As before, the sets of scores u_i, v_j, w_k may be known, unknown, or some may be known and others unknown. Thus, (5.76) also generalizes the RC(1) model and the attractive features of RC(1) carry over to this model, as previously discussed.

By imposing restrictions on the β parameters of (5.76) more parsimonious models are obtained, and provided the hierarchy principle is not violated, the resulting models also admit the interpretations imparted in Section 5.1. In the case where $\beta^{YZW} = 0$, a partial association model not equivalent to the nominal model is obtained, and if in addition, $\beta^{YW} = 0$, the resulting model implies the model $Y \| W \mid Z$ for nominal data previously considered. Note, however, that if one rejects the hypothesis $\beta_{YW} = 0$, this does not imply rejection of the model $Y \| W \mid Z$ for the nominal case. This is because the model imposes additional structure on the interactions, and it may be this additional structure that leads to rejection. Thus, an investigator might want to consider models that are intermediate to those of equations (5.75) and (5.76), for example, a partial association model of the form:

$$\log m_{ijk} = \mu + \alpha_i^Y + \alpha_j^Z + \alpha_k^W + \beta^{YZ} u_i v_j + \alpha_{ik}^{YW} + \beta^{ZW} v_j w_k. \tag{5.77}$$

When a one-to-one correspondence exists among two or more of the variables under consideration, the models above can be modified to take into account the tendency of the counts to cluster on the diagonal (two way and/or three way) and the scores may also be taken to be homogeneous. By imposing these types of restrictions, the models for ordinal variables considered here become special cases of the models for square tables considered in Section 5.2, and as in Section 4.3.1, this permits consideration of models that also take into account the ordinality of the variables. For illustrations using models of this nature, see Clogg (1982), Sobel (1988), and Becker (1990).

6 Higher-Way Tables

Log-linear models for higher-way tables are easily constructed by extending (5.64) to the case of four or more variables. However, in higher-way tables, there is an increase in interpretive difficulty. In addition, technical problems (for example, sparse tables) are more likely to be encountered, and the model selection process becomes more complicated. This section treat the first issue. The remaining difficulties are briefly discussed in Sections 7 and 8.

For higher-way tables, a shorthand notation is useful. For four variables Y, Z, W, and V, where V has categories $l = 1, \ldots, L$, let $(YZWV)$ denote the saturated model, (YZW, ZWV) the model with highest order terms α_{ijk}^{YZW} and α_{jkl}^{ZWV}, and so on. For hierarchical models, this notational device sets up a one-to-one correspondence between the model equation and the model denoted. Under the usual sampling schemes, the notation also sets up a one-to-one correspondence with the set of minimal sufficient statistics, for example, in the four-way table, (YZW, ZWV) indicates that the minimal sufficient statistics are the three-way totals

$$n_{ijk.} = \sum_{l=1}^{L} n_{ijkl}, n_{.jkl} = \sum_{i=1}^{I} n_{ijkl}.$$

In higher-way tables interpretation of models is facilitated by partitioning the variables into three subsets, treating the variables in each subset as a compound variable, and using the results of Section 5.1. To illustrate, in the five-way table, with variables Y, Z, W, V, U, and model $(YZVW, YZVU, UW)$ Y, Z, and V are grouped into the first subset, W and U are treated individually, and the model is the partial association model applied to the three subsets. Similarly, the model $(YZVW, YZVU)$ states that $U \perp\!\!\!\perp W \mid (Y, Z, V)$, and the model $(YZVW, U)$ states $U \perp\!\!\!\perp (YZVW)$.

Unfortunately, matters are not always this transparent. Consider instead the model (YZ, ZV, VU, UW) for the five-way table. Here, there are many ways to partition the variables into subsets, but no such arrangement yields a model with a unique interpretation. In this case, it is necessary to choose a partition, embed the model into a model with higher-way terms, and apply the interpretation that applies to the higher-way model. For example, variables Y and Z can be grouped into a subset, and variables U and W into a subset. Next, note that the model implies that the model (YZV, VUW) holds, and thus, that $(UW \perp\!\!\!\perp YZ) \mid V$.

By means of compounding variables and embedding models into larger models, the independence and conditional independence relations that apply in models with four or more variables can be obtained. However, the procedures described above are hit or miss, requiring the investigator to try out many different partitions and embeddings. In complicated settings, it would be easy to miss the best interpretation. Thus, it is important to formalize these procedures. The key to this formalization is to see which graphical models a particular log-linear model implies (Christenson 1990; Darroch, Lauritzen, and Speed 1980; Kiiveri and Speed 1982; Whittaker 1990).

Briefly, a graphical model is determined by the two-way interactions, and a model is graphical if it contains all the higher-way interactions that "generate" the two-way interactions. For example, if a model contains the two-way interactions between Y and Z, Z and V, Y and V, then, in order that the model be graphical, it must contain the three-way interaction between Y, Z, and V.

Graphical models admit simple interpretations in terms of independence and conditional independence, and these results can be read from the corresponding graph. For a given partition, with subsets 1, 2, and 3, Darroch et al. (1980) showed that the variables in subset 1 are conditionally independent of the variables in subset 3, given the variables in subset 2, if and only if every "chain" (defined in the usual graph-theoretic manner) between a variable in the first subset and the third has at least one vertex in the second subset, or more informally, if and only if every chain between a variable in the first and third subsets passes through one or more variables in the second subset.

7 Estimation Theory

So far, maximum likelihood has been used to estimate the models, assuming Poisson, multinomial, or product multinomial sampling. Some justification for these procedures is now given, followed by a more general discussion of estimation of models for discrete data. Numerical algorithms and computer programs for parameter estimation are considered in Section 9.

The intent here is to introduce the reader to some of the basic asymptotic results in the literature and the types of conditions needed to justify these, as opposed to providing rigorous proofs. After presenting the basic results for MLEs, ML estimation is placed within a broader framework that encompasses not only ML estimation, but a number of other estimation frameworks that have been used in conjunction with discrete data, including quasi-likelihood estimation, weighted least squares, and the pseudo-maximum likelihood estimator in Rao and Thomas (1988). Hopefully, the use of this framework will help to tie together a number of seemingly disparate threads in the literature.

I discuss only the conditions needed to obtain either BAN or CAN estimators, and do not justify the χ^2 tests used in the literature. However, by using standard arguments and Taylor expansions, these tests are readily derived once the asymptotic distribution of the estimator itself is obtained (for example, see Amemiya 1985). In addition, a number of other estimators are not discussed, for example, Bayesian estimators and kernel estimators. For further justification, as well as treatment of estimators and results not considered here, the reader should consult Agresti (1990), Birch (1963, 1964), Bishop et al. (1975), Cox (1984), Haberman (1974a), McCullagh and Nelder (1989), Rao (1973), and Read and Cressie (1988).

I begin by showing that, under certain conditions, the same MLEs and asymptotic standard errors for the multinomial and product multinomial models can be obtained under Poisson sampling. The results imply that separate treatments of the three cases is not required, and computer programs (for example, GLIM) for log-linear models for independent Poisson counts can be used to fit corresponding product multinomial or multinomial response models. Birch (1963) first showed the MLEs were the same under the three schemes, and Palmgren (1981) gave the results on the asymptotic standard errors. Both authors treated the case of product multinomial sampling, whereas I treat only the multinomial case; the extension from this case to the product multinomial case is straightforward. The presentation here differs from that of Palmgren, and the treatment is more direct. For an alternative approach that leads to a more unified treatment of the issues involved, see Lang (1994).

For sampling from T independent Poisson distributions, the log-likelihood, under the model

$$\log m_t = \delta + \boldsymbol{z}_t'\boldsymbol{\beta}_1, \tag{5.78}$$

is

$$\ell(\delta, \boldsymbol{\beta}_1; \boldsymbol{n}) = \sum_{t=1}^{T} n_t(\delta + \boldsymbol{z}_t'\boldsymbol{\beta}_1) - \sum_{t=1}^{T} \exp(\delta + \boldsymbol{z}_t'\boldsymbol{\beta}_1) - \sum_{t=1}^{T} \log n_t!. \tag{5.79}$$

Maximizing (5.79) first with respect to δ gives

$$\delta = \log n - \log \sum_{t=1}^{T} \exp(\boldsymbol{z}_t'\boldsymbol{\beta}_1). \tag{5.80}$$

[Compare with the expression for $\boldsymbol{\beta}_0$ in equation (5.6).] Substitution of (5.80) into (5.79) gives the concentrated log-likelihood, with kernel given by (5.7). This shows the same MLE of $\boldsymbol{\beta}_1$ is obtained under both schemes.

Second, under the Poisson scheme,

$$\frac{\partial \ell(\delta, \boldsymbol{\beta}_1; \boldsymbol{n})}{\partial(\delta, \boldsymbol{\beta}_1')'} = \boldsymbol{X}'(\boldsymbol{n} - \boldsymbol{m}), \tag{5.81}$$

where $\boldsymbol{X} = (\boldsymbol{1}, \boldsymbol{Z})$. Thus, $\sum_{t=1}^{T} \hat{m}_t = n$ and

$$-\frac{\partial^2 \ell(\delta, \boldsymbol{\beta}_1; \boldsymbol{n})}{\partial(\delta, \boldsymbol{\beta}_1')'\partial(\delta, \boldsymbol{\beta}_1')} = \boldsymbol{X}'\boldsymbol{D}(\boldsymbol{m})\boldsymbol{X}, \tag{5.82}$$

where $\boldsymbol{D}(\boldsymbol{m})$ is the $T \times T$ diagonal matrix with tt-th entry m_t. The covariance matrix of $(\delta, \boldsymbol{\beta}_1')'$ is thus estimated by $(\boldsymbol{X}'\boldsymbol{D}(\hat{\boldsymbol{m}})\boldsymbol{X})^{-1}$.

Partitioning (5.82) gives

$$-\frac{\partial^2 \ell(\delta, \boldsymbol{\beta}_1; \boldsymbol{n})}{\partial \delta^2} = \sum_{t=1}^{T} m_t, \tag{5.83}$$

$$-\frac{\partial^2 \ell(\delta, \boldsymbol{\beta}_1; \boldsymbol{n})}{\partial \delta \partial \boldsymbol{\beta}_1'} = \boldsymbol{m}'\boldsymbol{Z}, \tag{5.84}$$

$$-\frac{\partial^2 \ell(\delta, \boldsymbol{\beta}_1; \boldsymbol{n})}{\partial \boldsymbol{\beta}_1 \partial \boldsymbol{\beta}_1'} = \boldsymbol{Z}'\boldsymbol{D}(\boldsymbol{m})\boldsymbol{Z}. \tag{5.85}$$

Using standard results on partitioned inverses, the covariance matrix of $\hat{\boldsymbol{\beta}}_1$ is consequently estimated by

$$(\boldsymbol{Z}'\boldsymbol{D}(\hat{\boldsymbol{m}})\boldsymbol{Z} - n^{-1}((\boldsymbol{Z}'\hat{\boldsymbol{m}}\hat{\boldsymbol{m}}'\boldsymbol{Z})))^{-1} = n^{-1}(\boldsymbol{Z}'(\boldsymbol{D}(\hat{\boldsymbol{\pi}}) - \hat{\boldsymbol{\pi}}\hat{\boldsymbol{\pi}}')\boldsymbol{Z})^{-1}. \tag{5.86}$$

Differentiating (5.7) twice gives

$$\left(\frac{\partial^2 \ell^*(\boldsymbol{\beta}_1; \boldsymbol{n})}{\partial \boldsymbol{\beta}_1 \partial \boldsymbol{\beta}_1'} |_{\hat{\boldsymbol{\beta}}_1}\right) = \frac{n}{(\mathbf{1}'\boldsymbol{\gamma})}(\boldsymbol{Z}'\boldsymbol{D}(\boldsymbol{\gamma})\boldsymbol{Z}) - \frac{n}{(\mathbf{1}'\boldsymbol{\gamma})^2}(\boldsymbol{Z}'\boldsymbol{\gamma}\boldsymbol{\gamma}'\boldsymbol{Z}), \tag{5.87}$$

where $\mathbf{1}$ is the $T \times 1$ column vector with all entries one, $\gamma_t = \exp(\boldsymbol{z}_t'\boldsymbol{\beta}_1)$, $\boldsymbol{\gamma} = (\gamma_1, \ldots, \gamma_T)'$, and $\boldsymbol{D}(\boldsymbol{\gamma})$ is the $T \times T$ diagonal matrix with tt-th entry γ_t. At $\hat{\boldsymbol{\beta}}_1$, $\hat{m}_t = \exp(\hat{\beta}_0 + \boldsymbol{z}_t'\hat{\boldsymbol{\beta}}_1)$, and substitution of $\exp(\hat{\beta}_0)$ and $\exp(2\hat{\beta}_0)$ into the appropriate numerators and denominators of (5.87), evaluated at $\hat{\boldsymbol{\beta}}_1$ gives

$$\left(\frac{\partial^2 \ell^*(\boldsymbol{\beta}_1; \boldsymbol{n})}{\partial \boldsymbol{\beta}_1 \partial \boldsymbol{\beta}_1'} |_{\hat{\boldsymbol{\beta}}_1}\right)^{-1} = n^{-1}(\boldsymbol{Z}'(\boldsymbol{D}(\hat{\boldsymbol{\pi}}) - \hat{\boldsymbol{\pi}}\hat{\boldsymbol{\pi}}')\boldsymbol{Z})^{-1}, \tag{5.88}$$

the desired result.

Now, following Birch (1963), I proceed to show that, when $n_t > 0$ for all t, there is a global MLE. Inspection of (5.79), with m_t in place of $\exp(\delta + \boldsymbol{z}_t'\boldsymbol{\beta}_1)$, indicates (using $n_t > 0$) that the log-likelihood decreases without bound as $\log m_t$ increases or decreases without bound. This means that the log-likelihood is maximized in the interior of the parameter space. Second, because $\boldsymbol{X}$ has full column rank, and $m_t > 0$ at a maximum, the matrix of (5.82) is positive definite throughout the interior of the parameter space, whence (5.81) has a unique solution, corresponding to a maximum.

The results above require $n_t > 0$. For some cross-classifications, a given cell of the table must be empty, for example, if occupation is cross-classified by education, the number of doctors with less than a high school degree must be 0. In this case, $n_t = 0$ because $m_t = 0$, and the cell in question is said to be a "structural zero". Contingency tables with "structural zeroes" are said to be incomplete. In fitting models to incomplete tables, such cells make no contribution to the likelihood. If the model for the complete table is identified when such cells are ignored [see Bishop et al. (1975, chap. 5) on this issue], the previous results hold without modification. When $m_t > 0$ and $n_t = 0$, n_t is said to be a "sampling zero." In this case, the MLE may or may not exist. Clearly, the MLE will not exist when it is impossible to satisfy the first-order conditions (5.81) with $m_t > 0$. The problem of existence and uniqueness of MLEs has been addressed by Haberman (1974a), and the interested reader should consult this source for further details. Note that Haberman's results have already been used in the text, in the analysis of the Brazilian mobility table.

The MLE is also consistent and asymptotically normal. To see why this is so, consider the multinomial case, and let $\boldsymbol{\pi}_0$ and $\boldsymbol{\beta}_{10}$ denote the true values of $\boldsymbol{\pi}$ and $\boldsymbol{\beta}_1$, respectively. Here, the n observations can be treated as an independently and identically distributed (iid) sample from the $M(1, \boldsymbol{\pi})$ distribution, and by the multivariate central limit theorem for iid observations, the sample mean vector is a consistent, asymptotically normal estimator of $\boldsymbol{\pi}$, that is,

$$\sqrt{n}(\boldsymbol{p} - \boldsymbol{\pi}) \xrightarrow{\mathcal{D}} N(\mathbf{0}, \boldsymbol{D}(\boldsymbol{\pi}_0) - \boldsymbol{\pi}_0\boldsymbol{\pi}_0'), \tag{5.89}$$

where $\xrightarrow{\mathcal{D}}$ denotes convergence in distribution. Next, using a first-order expansion about $\hat{\beta}_1$,

$$\frac{\partial \ell^*(\boldsymbol{\beta}_1; \boldsymbol{n})}{\partial \boldsymbol{\beta}_1} |_{\boldsymbol{\beta}_{10}} = -\frac{\partial^2 \ell^*(\boldsymbol{\beta}_1; \boldsymbol{n})}{\partial \boldsymbol{\beta}_1 \partial \boldsymbol{\beta}_1'} |_{\boldsymbol{\beta}^*} (\hat{\boldsymbol{\beta}}_1 - \boldsymbol{\beta}_{10}), \tag{5.90}$$

where β^* denotes p (generally different) points on the line segment between $\hat{\beta}_1$ and β_{10}. Using (5.90),

$$\hat{\beta}_1 - \beta_{10} = -(\frac{\partial^2 \ell^*(\beta_1; n)}{\partial \beta_1 \partial \beta_1'} \mid_{\beta^*})^{-1} \frac{\partial \ell^*(\beta_1; n)}{\partial \beta_1} \mid_{\beta_{10}} =$$

$$((Z'(D(\pi(\beta_1)) - \pi(\beta_1)(\pi(\beta_1))')Z) \mid_{\beta^*})^{-1} Z'(\hat{p} - \pi_0). \quad (5.91)$$

Note that when the inverse does not exist, a generalized inverse may be used instead.

Consistency now follows from the boundedness of the first term on the right of (5.91) and the convergence in probability of $\hat{p}$ to π_0. Asymptotic normality of $\sqrt{n}(\hat{\beta}_1 - \beta_{10})$ follows from the facts that

$$\frac{1}{\sqrt{n}} \frac{\partial \ell^*(\beta_1; n)}{\partial \beta_1} \mid_{\beta_{10}}$$

is identical to the left-hand side of (5.89) and

$$Z'(D(\pi(\beta_1)) - \pi(\beta_1)\pi(\beta_1)') \mid_{\beta^*} Z \xrightarrow{\mathcal{P}} Z'(D(\pi_0) - \pi_0\pi_0')Z, \quad (5.92)$$

where $\xrightarrow{\mathcal{P}}$ denotes convergence in probability. The asymptotic covariance matrix of $\sqrt{n}(\hat{\beta}_1 - \beta_{10})$ is:

$$\text{plim}_{n\to\infty}(\frac{1}{n}\frac{\partial^2 \ell^*(\beta_1; n)}{\partial \beta_1 \partial \beta_1'} \mid_{\beta_{10}})^{-1} \lim_{n\to\infty} E(\frac{1}{\sqrt{n}} \frac{\partial \ell^*(\beta_1; n)}{\partial \beta_1} \mid_{\beta_{10}} \times$$

$$\frac{1}{\sqrt{n}} \frac{\partial \ell^*(\beta_1; n)}{\partial \beta_1'} \mid_{\beta_{10}}) \text{plim}_{n\to\infty}(\frac{1}{n}\frac{\partial^2 \ell^*(\beta_1; n)}{\partial \beta_1 \partial \beta_1'} \mid_{\beta_{10}})^{-1}], \quad (5.93)$$

which simplifies to

$$(Z'(D(\pi_0) - \pi_0\pi_0')Z)^{-1} \quad (5.94)$$

by virtue of (5.89), (5.91), and (5.92).

The foregoing results have been extended in a number of useful directions. First, the model for $\log m_t$ (or equivalently, $\log \pi_t$) need not be linear, for example, some of the ordinal models previously considered are not linear. In this case $\pi_t = f_t(\beta_1)$, where $\beta_1 \in \mathcal{B}$. Provided the model is strongly identified [given $\eta > 0$ and $d(\beta_1, \beta_{10}) > \eta$, there exists $\epsilon > 0$ such that $d(\boldsymbol{f}(\beta_1), \boldsymbol{f}(\beta_{10}) > \epsilon$, where d denotes Euclidean distance and $\boldsymbol{f} = (f_1, \ldots, f_T)')$], β_{10} lies in the interior of $\mathcal{B}$, π_t is continuously differentiable with respect to β_1 near β_{10}, and $(\partial\pi/\partial\beta_1') \mid_{\beta_{10}}$ has full column rank, the theory of local maximum likelihood estimation applies, that is, an MLE exists and is consistent and asymptotically normal. It should also be noted that this is true whether or not the f_t are twice continuously differentiable, that is, whether or not $\ell^*(\beta_1; n)$ is twice differentiable with respect to β_1.

Read and Cressie (1988) give even more general results for the sampling schemes above. They show, under the regularity conditions of Birch (1964), that the estimator defined by minimizing the discrepancy function

$$I^\lambda(n; m) = \frac{2}{\lambda(\lambda + 1)} \sum_{t=1}^{T} n_t((n_t/m_t)^\lambda - 1) \quad (5.95)$$

with respect to β_1, where λ is a known number in $(-\infty, \infty)$, is a BAN estimator. For the cases $\lambda = 0$ and $\lambda = -1$ the limit is used, and minimization of the discrepancy with $\lambda = 0$ corresponds to maximum likelihood estimation [see equation (5.10) and the material following it]. For the case $\lambda = 1$, the minimum χ^2 estimate (Neyman 1949) is obtained. These authors also show that the power divergence statistic of (5.11) is asymptotically χ^2 for all λ, and they recommmend, on the basis of both theoretical and Monte Carlo work, using the value $\lambda = 2/3$, especially for sparse tables.

The asymptotic results presented thus far are valid for the three sampling schemes previously discussed. For other cases, when such a scheme does not apply or is not known to apply, CAN estimators may be obtained, under suitable regularity conditions, using weaker assumptions, for example, assumptions about the first and second moments. The idea (see Amemiya 1985 for a nice treatment and precise conditions) is to replace a log-likelihood function $\ell(\boldsymbol{\vartheta}; \boldsymbol{n})$ by a twice continuously differentiable function $Q(\boldsymbol{\vartheta}; \boldsymbol{n})$ whose derivatives have properties that mimic various properties of the derivatives of the log-likelihood function, for example,

$$\hat{\boldsymbol{\vartheta}} - \boldsymbol{\vartheta}_0 = -\left(\frac{\partial^2 Q(\boldsymbol{\vartheta}; \boldsymbol{n})}{\partial \boldsymbol{\vartheta} \partial \boldsymbol{\vartheta}'} \Big|_{\boldsymbol{\vartheta}^*}\right)^{-1} \frac{\partial Q(\boldsymbol{\vartheta}; \boldsymbol{n})}{\partial \boldsymbol{\vartheta}} \Big|_{\boldsymbol{\vartheta}_0}, \tag{5.96}$$

where

$$\frac{1}{\sqrt{n}} \frac{\partial Q(\boldsymbol{\vartheta}; \boldsymbol{n})}{\partial \boldsymbol{\vartheta}} \Big|_{\boldsymbol{\vartheta}_0} \xrightarrow{\mathcal{D}} N(\mathbf{0}, \boldsymbol{B}(\boldsymbol{\vartheta}_0)), \tag{5.97}$$

$$\frac{1}{n} \frac{\partial^2 Q(\boldsymbol{\vartheta}; \boldsymbol{n})}{\partial \boldsymbol{\vartheta} \partial \boldsymbol{\vartheta}'} \Big|_{\boldsymbol{\vartheta}^*} \xrightarrow{\mathcal{P}} \boldsymbol{A}(\boldsymbol{\vartheta}_0) \tag{5.98}$$

for any $\boldsymbol{\vartheta}^*$ such that $\hat{\boldsymbol{\vartheta}}$ is consistent [$\boldsymbol{\vartheta}^*$ denotes p (the number of nonredundant parameters estimated) points on the line segment between $\hat{\boldsymbol{\vartheta}}$ and $\boldsymbol{\vartheta}_0$], and $\boldsymbol{A}(\boldsymbol{\vartheta}_0)$ is a negative definite matrix. Under these conditions,

$$\sqrt{n}(\hat{\boldsymbol{\vartheta}} - \boldsymbol{\vartheta}_0) \xrightarrow{\mathcal{D}} N(\mathbf{0}, (\boldsymbol{A}(\boldsymbol{\vartheta}_0))^{-1} \boldsymbol{B}(\boldsymbol{\vartheta}_0) (\boldsymbol{A}(\boldsymbol{\vartheta}_0))^{-1}. \tag{5.99}$$

For the MLE previously considered, the properties above clearly hold, and in addition $-\boldsymbol{A}(\boldsymbol{\vartheta}_0) = \boldsymbol{B}(\boldsymbol{\vartheta}_0)$, thereby explaining the simple form of the asymptotic covariance matrix in (5.94). Quasi-likelihood estimation (see McCullagh and Nelder 1989 for further details) is another special case. Here, $\boldsymbol{n}$ is assumed to have mean $\boldsymbol{m}$ and covariance matrix $\sigma^2 V(\boldsymbol{m})$, and $Q(\boldsymbol{\vartheta}; \boldsymbol{n})$ is taken to be a function with first derivative

$$\frac{\partial Q(\boldsymbol{\vartheta}; \boldsymbol{n})}{\partial \boldsymbol{\vartheta}} = \frac{\partial \boldsymbol{m}(\boldsymbol{\vartheta})}{\partial \boldsymbol{\vartheta}'} \boldsymbol{V}^{-1} (\boldsymbol{n} - \boldsymbol{m}(\boldsymbol{\vartheta}))/\sigma^2, \tag{5.100}$$

and

$$E\left(\frac{\partial^2 Q(\boldsymbol{\vartheta}; \boldsymbol{n})}{\partial \boldsymbol{\vartheta} \partial \boldsymbol{\vartheta}'} \Big|_{\boldsymbol{\vartheta}_0}\right) = \boldsymbol{A}(\boldsymbol{\vartheta}_0) = -\boldsymbol{B}(\boldsymbol{\vartheta}_0) = \frac{\partial \boldsymbol{m}(\boldsymbol{\vartheta})}{\partial \boldsymbol{\vartheta}'} \Big|_{\boldsymbol{\vartheta}_0} \boldsymbol{V}^{-1} \frac{\partial \boldsymbol{m}'(\boldsymbol{\vartheta})}{\partial \boldsymbol{\vartheta}} \Big|_{\boldsymbol{\vartheta}_0}. \tag{5.101}$$

Under suitable conditions, the normalized first derivative satisfies (5.97) and the normalized second derivative satisfies (5.98), and thus the quasi-likelihood estimator behaves like the ML estimator.

Other estimators with similar justification include the weighted least squares estimator (Grizzle, Starmer, and Koch 1969) and the pseudo-ML estimator described by Rao and Thomas (1988), which can be used to obtain CAN estimates in complex samples, under suitable conditions. The pseudo-ML estimator for the multinomial case is obtained by replacing the unweighted vector of sample proportions $\hat{\boldsymbol{p}}$ in the log-likelihood (due to stratification and clustering $\hat{\boldsymbol{p}}$ is not a consistent estimate of $\boldsymbol{\pi}$) by a consistent estimator $\tilde{\boldsymbol{p}}$. Solving the likelihood equations gives pseudo-ML estimates. The normalized first derivative obeys a central limit theorem with asymptotic covariance matrix

$$B(\beta_{10}) = \boldsymbol{Z}'\boldsymbol{V}\boldsymbol{Z}, \tag{5.102}$$

where $\boldsymbol{V}$ is the covariance matrix of $\tilde{\boldsymbol{p}}$ and the normalized second derivative converges in probability to

$$A(\beta_{10}) = -\boldsymbol{Z}'\boldsymbol{P}\boldsymbol{Z}, \tag{5.103}$$

where $\boldsymbol{P}$ is the multinomial covariance matrix for $\tilde{\boldsymbol{p}}$. In practice, $\boldsymbol{P}$ and $\boldsymbol{V}$ are unknown, and must be estimated. The former can be estimated as $(\boldsymbol{D}(\tilde{\boldsymbol{p}}) - \tilde{\boldsymbol{p}}\tilde{\boldsymbol{p}}')$. Rao and Thomas (1988) describe methods that can be used to obtain an estimate of $\boldsymbol{V}$.

8 Residual Analysis and Model-Selection Procedures

Thus far, goodness-of-fit statistics have been used to ascertain whether a given model fits the data, and also to informally select models. In this section, the use of residuals in model checking and model selection is briefly discussed, as is the general issue of model selection.

In regression analysis, the raw residual is, for any given observation, the difference between the value of the dependent variable (y_t) and its predicted value ($\hat{y}_t$). Typically, this residual is divided by s, where s^2 is an estimate of the error variance σ^2, and the standardized residuals $(y_t - \hat{y}_t)/s$ are compared to the standard normal distribution. However, the variance of the raw residual is not σ^2, but $\sigma^2(1 - \ell_{ii})$, where ℓ_{ii}, the leverage, is the $(ii)th$ element of the matrix $\boldsymbol{X}(\boldsymbol{X}'\boldsymbol{X})^{-1}\boldsymbol{X}'$, where X is the model matrix. Thus, comparing the standardized residuals to the standard normal is not entirely appropriate, and a more appropriate residual is the adjusted residual $(y_t - \hat{y}_t)/s\sqrt{1 - \ell_{ii}}$.

The analogue to the standardized residual in the log-linear model is the Pearson residual:

$$r_t = \frac{n_t - \hat{m}_t}{\sqrt{\hat{m}_t}}, \tag{5.104}$$

and its square is the contribution of the t-th cell to (5.9). Note that the denominator of r_t depends on t because the log-linear model is analogous to a heteroscedastic regression model. Asymptotically, the standardized residual has a mean of 0. However, like the standardized residual in regression, its variance is not 1, and comparison with the standard normal is not entirely appropriate. Thus, a more appropriate residual (Haberman 1973) for comparison with the standard normal is the adjusted residual $(n_t - \hat{m}_t)/\sqrt{\hat{m}_t(1 - \ell_{ii})}$, where ℓ_{ii} is the $(ii)th$ element of the matrix:

$$\boldsymbol{D}(\sqrt{\hat{\boldsymbol{m}}})\boldsymbol{X}(\boldsymbol{X}'\boldsymbol{D}(\hat{\boldsymbol{m}})\boldsymbol{X})^{-1}\boldsymbol{X}'\boldsymbol{D}(\sqrt{\hat{\boldsymbol{m}}}), \tag{5.105}$$

and $D(\sqrt{\hat{m}})$ is the $T \times T$ diagonal matrix with tt-th entry $\sqrt{\hat{m}_t}$.

Other residuals may also be considered, such as the deviance residual, the signed contribution of the t-th observation to (5.10). For further material on this subject and the relative performance of various residuals, see Pierce and Schafer (1986).

While residual analysis is useful for suggesting modifications to a given model, it is less useful for selecting models from a class of alternatives. For this problem, a number of strategies have been suggested. One strategy involves the use of procedures analogous to the stepwise methods sometimes used in conjunction with regression analysis. These procedures (Aitken 1979; Benedetti and Brown 1978; Goodman 1971; Wermuth 1976) are most useful in exploratory work with many variables (hence many possible models). However, there is no guarantee that use of these procedures will select a model that is either substantively meaningful or optimal by various statistical criteria. A second set of procedures involves comparison of models of interest with a baseline model. This leads to the use of R^2 type statistics, such as the pseudo R^2 statistic introduced in Section 5.1.1. A version of this statistic adjusted for degrees of freedom, analogous to the adjusted R^2 in regression analysis, can also be defined (Christensen 1990). As in regression analysis, the statistical rationale for the adjustment is weak. Other indices that have been used in conjunction with model selection include L^2/df, the ratio of the likelihood ratio chi-square under a model to its degrees of freedom (Haberman 1978) and the index of dissimilarity (Hauser, Koffel, Travis, and Dickinson 1975), which measures the percentage missclassification between the observed distribution and the distribution fitted under the model. Relative reductions in the index of dissimilarity have also been used to compare an unsaturated model of interest with a baseline model, usually a model of independence.

Most of the procedures above cannot be given strong justification by means of statistical theory. For an exception, see Haberman (1982). Alternative procedures for log-linear models based on a Bayesian approach have also been proposed (Raftery 1986, 1988; Spiegelhalter and Smith 1982). Here, two (or more) models are to be compared by means of their respective posterior probabilities. For comparing models M_0 and M_1, the Bayesian selects the model with higher posterior probability, for example, M_0 if

$$\frac{\Pr(M_0 \mid \boldsymbol{n})}{\Pr(M_1 \mid \boldsymbol{n})} > 1. \tag{5.106}$$

The quantity in (5.106) is called the posterior odds ratio, and it can be rewritten as

$$\frac{\Pr(M_0 \mid \boldsymbol{n})}{\Pr(M_1 \mid \boldsymbol{n})} = \frac{\Pr(\boldsymbol{n} \mid M_0)\Pr(M_0)}{\Pr(\boldsymbol{n} \mid M_1)\Pr(M_1)} = B_{01}\frac{\Pr(M_0)}{\Pr(M_1)}, \tag{5.107}$$

that is, as the product of the Bayes factor (B_{01}) with the prior odds ratio.

When both models have prior probability $1/2$, the Bayes factor is identical to the posterior odds ratio, and Bayes factors in excess of unity favor M_0.

Unlike the approaches to model selection previously discussed, the Bayesian approach derives from an explicit logic. In addition, this approach can be used to compare models that are not nested, whereas in the conditional likelihood ratio test, M_0 is nested under M_1.

For the case of a log-linear model M_0 parameter nested under M_1, Raftery (1986) showed, using the standard Jeffreys prior [the prior is proportional to $(\prod_{t=1}^{T} \pi_t)^{-1/2}$], that $-2 \log B_{01}$ is asymptotically equivalent in probability to the statistic

$$BIC = L^2 - (p_1 - p_0) \log n, \tag{5.108}$$

conditional on M_0, where $p_1 - p_0$ is the number of additional independent parameters estimated under M_1, as opposed to M_0. This is the model selection criterion derived by Schwarz (1978). Thus, values of BIC less than 0 favor selection of M_0, while values of BIC greater than 0 favor M_1. Note also that asymptotically equivalent approximations to L^2, such as X^2 and D^2, could be used as well.

From (5.108), it is evident that BIC penalizes models with more parameters. When model M_0 is nested under M_1, and obtained by imposing linear restrictions on the parameters of M_1, BIC is more likely to favor M_0 in large samples than the conditional likelihood ratio test. The same remark holds for the likelihood ratio test when M_1 is the saturated model. This makes BIC attractive to empirical researchers, who are often testing whether certain parameters of M_1 are 0, and typically prefer a more parsimonious and interpretable model. Unfortunately, these researchers often misuse this statistic to select simpler models, using it primarily when the usual measures of goodness-of-fit lead to rejection of the simpler model. Note also that for the case in point, as is evident from (5.107), assigning prior probability 1/2 to M_0 is equivalent to assigning positive probability to a subspace of the full parameter space that has Lebesgue measure 0 (Leamer 1983).

9 Software

This section describes several computer programs that can be used to fit the models discussed in this chapter. The listing is not comprehensive.

9.1 GLIM

GLIM is a program for fitting generalized linear models (Francis, Green, and Payne 1993). Both mainframe and PC versions are available. The program can be used to obtain MLEs for all the log-linear models described in this chapter as well as the model of marginal homogeneity. In addition, GLIM allows the user to input macros, and to write small programs for calculating many quantities that are not routinely available in many of the other programs. The models are fit by specifying the use of the Poisson error function and a model matrix. The model matrix may be given by the user or GLIM can be used to construct this matrix. GLIM uses the Fisher scoring algorithm (in this case, the algorithm reduces to the Newton- Raphson algorithm) to estimate the model parameters. For further information on the Newton-Raphson and/or Fisher scoring algorithms, see Kennedy and Gentle (1980) or McCullagh and Nelder (1989). GLIM routinely outputs L^2 and df for the model under consideration. Upon request, it outputs fitted values, parameter estimates and asymptotic standard errors, the asymptotic covariance matrix, Pearson residuals, and X^2. Adjusted residuals may also be obtained (Pierce and Shafer 1986) and D^2

may be obtained by using the calculate statement to (a) define the components of D^2 as: $d = (((yvariable/\%FV) * *(2/3)) - 1) * yvariable * 1.8$, where "yvariable"is replaced by the name of the dependent variable, and then (b) using the TABULATE statement to tabulate the d total. A number of authors (for example, Agresti and Kezouh 1983) have pointed out that the log-bilinear models considered in this chapter can also be fit in GLIM by exploiting the conditional linearity of the model. In general, I would not recommend this strategy, except in the case where the user is interested only in the parameter estimates. This is because at each step, the asymptotic covariance matrix that is given conditions on estimates that are treated incorrectly as known parameters.

9.2 BMDP

BMDP (Dixon 1992) has mainframe and PC versions of two programs that can be used to fit log-linear models. Program 4F is designed specifically for fitting log-linear models with all variables treated as nominal, and uses the iterative proportional fitting (IPF) algorithm. For a description and justification of this algorithm, see Bishop et al. (1975). The program can be used to ouptut L^2, fitted values, and Pearson residuals, parameter estimates and standard errors. In addition, stepwise procedures are available. Program 3R is a nonlinear regression program that allows the user to input a likelihood function. Thus, this program can be used to estimate log-linear models for ordinal variables, and log-nonlinear models. The program requires the user to input the log-likelihood and its first derivatives, and uses a Gauss-Newton algorithm. For fitting standard log-linear and log-bilinear models, other programs considered in this section are easier to use than 3R. However, many of these programs will not handle nonstandard models, for example, a log-linear model with linear restrictions that do not reduce to equality constraints or zero restrictions. The user can also input a different function, for example, the discrepancy function described by Read and Cressie (1988). In addition, the standard Newton-Raphson algorithm uses a step size of 1 at each iteration, whereas 3R will successively halve the step size at each iteration, if need be, to insure the likelihood does not decrease between iterations. This is useful when the standard Newton-Raphson algorithm fails to converge.

9.3 SAS

A procedure called CATMOD in SAS (SAS Institute, Inc. 1987) can be used to fit all the log-linear models discussed in this chapter, and it uses the Newton-Raphson algorithm. Like 3R, the step size is successively halved at each iteration, if need be. The program can be used in conjunction with ML estimation by specifying that ML is to be used, and the program can be used to ouptut L^2, fitted values, parameter estimates, and the covariance matrix of the parameter estimates. Both PC and mainframe versions are available. In addition, procedure GENMOD (SAS Institute Inc. 1993) can be used to fit generalized linear models. For further information on these models, see Section 9.1.

9.4 SPSS

SPSS (Norusis 1993; SPSS 1994) distributes mainframe and PC versions of two programs for fitting log-linear models, LOGLINEAR and HILOGLINEAR. The former can be used to obtain ML estimates of any of the log-linear models considered in this chapter. The program outputs L^2 and df and a variety of user defined optional output, such as fitted values, parameter estimates and standard errors, and Pearson and adjusted residuals. Program HILOGLINEAR is specifically for fitting hierarchical log-linear models with all variables treated as nominal, and uses the IPF algorithm. Stepwise selection procedures can also be used. However, unlike 4F, parameter estimates can be outputted in HILOGLINEAR only for saturated models. In addition, program GENLOG (SPSS, Inc. 1994) can be used to fit multinomial and Poisson log-linear models. The program uses the Newton-Raphson algorithm; output includes parameter estimates and standard errors, L^3, X^3, adjusted and deviance residuals. Currently, GENLOG runs under Windows; mainframe versions and versions for other operating systems are forthcoming.

9.5 GAUSS

GAUSS (Aptech Systems 1991a) is a matrix programming language for the PC. MARKOV (Aptech Systems 1991b) is a Gauss program written by J. Scott Long that contains a module for estimating log-linear models. The module uses the Newton-Raphson algorithm and can be used to fit any of the log-linear models described in this chapter. When all variables are treated as nominal, the program generates the appropriate model matrix. For the ordinal log-linear models, the user must supply the model matrix. The program can output parameter estimates and the asymptotic covariance matrix of the parameter estimates, df, L^2, X^2, D^2, BIC, and Pearson residuals.

9.6 CDAS

CDAS 3.50 (The Categorical Data Analysis System, 1990) is a PC program containing modules that permit estimation of any of the the models discussed in this chapter. For the ordinal log-linear models, the Newton-Raphson algorithm is used. For the log-nonlinear models, the unidimensional Newton method is used. The relevant progams output df, L^2, X^2, parameter estimates and standard errors, fitted values, Pearson residuals, and the index of dissimilarity. For log-linear models, adjusted residuals are also given. For the log-linear models, standard errors are obtained in the usual way (from the second derivates of the log-likelihood), but given the use of the unidimensional Newton method for the log-nonlinear models, standard errors for the estimates are obtained by jacknifing. CDAS 3.50 is maintained and distributed by Dr. Scott Eliason, Department of Sociology, University of Iowa, Iowa City, IA 52242.

9.7 S-Plus

S-Plus (Becker, Chambers, and Wilks 1988) is a programming language and interactive environment that runs under UNIX operating system or Windows. Generalized linear models can be fitted using the command "glm". For further information on generalized linear models, see Section 9.1. User supplied routines can also be written to fit log-linear models, using ML or some other objective function.

REFERENCES

Agresti, A. (1984), *Analysis of Ordinal Categorical Data*, New York: John Wiley.

——(1990), *Categorical Data Analysis*, New York: John Wiley.

Agresti, A., and Kezouh, A. (1983), "Association Models for Multidimensional Cross-Classifications of Ordinal Variables," *Communications in Statistics*, Ser. A, 12, 1261–1276.

Aitken, M. (1979), "A Simultaneous Test Procedure for Contingency Table Models," *Applied Statistics*, 28, 233–242.

Allison, P. D. (1982), "Discrete-Time Methods for the Analysis of Event Histories," in *Sociological Methodology, 1982*, ed. S. Leinhardt, San Francisco: Jossey-Bass, pp. 61–98.

Altham, P. M. E. (1970a), "The Measurement of Association of Rows and Columns for an $r \times s$ Contingency Table, " *Journal of the Royal Statistical Society*, Ser. B, 32, 63–73.

——(1970b), "The Measurement of Association in a Contingency Table: Three Extensions of the Cross-Ratios and Metric Methods," *Journal of the Royal Statistical Society*, Ser. B, 32, 395–407.

Amemiya, T. (1981), "Qualitative Response Models: A Survey," *Journal of Economic Literature*, 19, 1483–1536.

——(1985), *Advanced Econometrics*, Cambridge: Harvard University Press.

Andersen, E.B. (1980), *Discrete Statistical Models with Social Science Applications*, Amersterdam: North-Holland.

——(1990), *The Statistical Analysis of Categorical Data*, Berlin: Springer-Verlag.

Anderson, T.W. (1984), *An Introduction to Multivariate Statistical Analysis*, 2nd ed., New York: John Wiley.

Aptech Systems (1991a), *GAUSS: Programming Manual*, Kent, WA: Aptech Systems.

——(1991b), *MARKOV: A Statistical Environment for Gauss*, Aptech Systems: Kent, WA.

Asmussen, S., and Edwards, D. (1983), "Collapsibility and Response Variables in Contingency Tables," *Biometrika*, 70, 567–578.

Bartlett, M. S. (1935), "Contingency Table Interactions," *Journal of the Royal Statistical Society*, Suppl. 2, 248–252.

Becker, M. P. (1990), "Quasisymmetric Models for the Analysis of Square Contingency Tables,"*Journal of the Royal Statistical Society*, Ser. B, 52, 369–378.

Becker, M., and Clogg, C. C. (1989), "Analysis of Sets of Two-Way Contingency Tables Using Association Models," *Journal of the American Statistical Association*, 84, 142–151.

Becker, R. A., Chambers, J. M., and Wilks, R. A. (1988), *The New S Language*, Pacific Grove, CA: Wadsworth and Brooks/Cole.

Benedetti, J. K., and Brown, M. B. (1978), "Strategies for the Selection of Loglinear Models," *Biometrika*, 34, 680–686.

Benzécri, J. P. (1969), "Statistical Analysis as a Tool to Make Patterns Emerge From Data," in *Methodologies of Pattern Recognition*, ed. S. Watanabe, New York: Academic Press, pp. 35–73.

Benzécri, J. P., and Gopolan, T. K. (1991), *Correspondence Analysis Handbook*, New York: Marcel Dekker.

Birch, M. W. (1963), "Maximum Liklihood in Three-Way Contingency Tables," *Journal of the Royal Statistical Society*, Ser. B, 25, 220–233.

——— (1964), "A New Proof of the Pearson-Fisher Theorem," *Annals of Mathematical Statistics*, 35, 817–824.

Bishop, Y. V. V., Fienberg, S. E., and Holland, P. W. (1975), *Discrete Multivariate Analysis*, Cambridge, MA: MIT Press.

Bollen, K. A., and Phillips, D. P. (1982), "Imitative Suicides: A National Study of the Effects of Television News Stories," *American Sociological Review*, 47, 802–809.

Bumpass, L. L., and Westoff, C. F. (1970), *The Later Years of Childbearing*, Princeton, NJ: Princeton University Press.

Caussinus, H. (1965), "Contribution à l'analyse statistique des tableaux de correlation," *Annales de la Faculté des Sciences de l'Université de Toulouse*, 29, 77–182.

Christensen, R. (1990), *Log-Linear Models*, New York: Springer-Verlag.

Clogg, C. C. (1982), "Some Models for the Analysis of Association in Multiway Cross-Classifications Having Ordered Categories," *Journal of the American Statistical Society*, 77, 803–815.

Clogg, C. C., and Eliason, S. R. (1987), "Some Common Problems in Log-Linear Analysis," *Sociological Methods and Research*, 16, 8–44.

Clogg, C. C., Eliason, S. R., and Grego, J. M. (1990), "Models for the Analysis of Change in Discrete Variables," in *Statistical Methods in Longitudinal Research*, Vol. 2, ed. A. Von Eye, San Diego: Academic Press, pp. 409–441.

Clogg, C. C., Petkova, E., and E. S. Shihadeh (1992), "Statistical Methods for Analyzing Collapsibility in Regression Models," *Journal of Educational Statistics*, 17, 51–74.

Clogg, C. C., and Rao, C. R. (1991) Comment on "Measures, Models, and Graphical Displays in Cross-Classified Data," by L. A. Goodman, *Journal of the American Statistical Association*, 86, 1118–1120.

Clogg, C. C., and Shihadeh, E. S. (1994), *Statistical Models for Ordinal Variables*, Thousand Oaks, CA: Sage.

Clogg, C. C., and Shockey, J. W. (1988), "Multivariate Analysis of Discrete Data," in *Handbook of Multivariate Experimental Psychology*, eds. J. R. Nesselroade and R. B. Cattell, New York: Plenum Press, pp. 337–365.

Cornfield, J. (1956), "A Statistical Problem Arising From Retrospective Studies," in *Proceedings of the 3rd Berkeley Symposium*, ed. J. Neyman, Vol. 4, Berkeley: University of California Press, pp. 135–148.

Cox, C. (1984), "An Elementary Introduction to Maximum Likelihood Estimation for Multinomial Models: Birch's Theorem and the Delta Method," *American Statistician*, 38, 283–287.

Cox, D. R. (1958), "Two Further Applications of a Model for Binary Regression," *Biometrika*, 45, 562–565.

Cox, D. R., and Hinkley, D. V. (1974), *Theoretical Staistics*, London: Chapman and Hall.

Cox, D. R., and Snell, E. J. (1989), *Analysis of Binary Data*, 2nd ed., London: Chapman and Hall.

Darroch, J. N. (1986), "Quasi-Symmetry," in *Encyclopedia of Statistical Sciences*, Vol. 7, New York: John Wiley, pp. 469–473.

Darroch, J. N., Lauritzen, S. L., and Speed, T. P. (1980), "Markov Fields and Log-Linear Interaction Models for Contingency Tables," *Annals of Statistics*, 8, 522–539.

Darroch, J. N., and McCloud, P. I. (1986), "Category Distinguishability and Observer Agreement," *Australian Journal of Statistics*, 28, 371–388.

Davis, J. A. (1980), *Codebook for the 1980 General Social Survey*, Chicago: National Opinion Research Center.

Dixon, W. J. (1992), *BMDP Statistical Software Manual*, Vols. 1 and 2, Berkeley: University of California Press.

Ducharme, G. R., and Lepage, Y. (1986), "Testing Collapsibility in Contingency Tables," *Journal of the Royal Statistical Society*, Ser. B, 48, 197–205.

Duncan, O. D. (1979), "How Destination Depends on Origin in the Occupational Mobility Table," *American Journal of Sociology*, 84, 793–803.

Edwards, A. W. F. (1963), "The Measure of Association in a 2 × 2 Table," *Journal of the Royal Statistical Society*, Ser. A, 126, 109–114.

Eliason, S. (1990), *The Categorical Data Analysis System Version 3.50: User's Manual*, Department of Sociology, University of Iowa.

Feller, W. (1968), *An Introduction to Probability Theory and Its Applications*, Vol. 1, 3rd ed., New York: John Wiley.

Fienberg, S. E. (1980), *The Analysis of Cross-Classified Categorical Data*, 2nd ed., Cambridge: MIT.

Firth, D. (1989), "Marginal Homogeneity and the Superposition of Latin Squares," *Biometrika*, 76, 179–182.

Fisher, R. A. (1922), "On the Interpretation of Chi-Square From Contingency Tables and the Calculation of *P*," *Journal of the Royal Statistical Society*, 85, 87–94.

Fitzmaurice, G., Laird, N. M., and Rotnitsky, A. G. (1993), "Regression Models for Discrete Longitudinal Responses" (with discussion), *Statistical Science*, 8, 284–309.

Fleiss, J. L. (1981), *Statistical Methods for Rates and Proportions*, 2nd ed., New York: John Wiley.

Francis, B., Green, M., and Payne, C. (1993), *The GLIM System, Release 4 Manual*, Oxford: Clarendon Press.

Gilula, Z., and Haberman, S. (1994a), *Conditional Log-Linear Models for Analysing Categorial Panel Data*, Vol. 89, pp. 645–656.

Gilula, Z., and Haberman, S. (1994b), "Prediction Functions for Categorial Panel Data," forthcoming in *Journal of the American Statistical Association*.

Goddard, M. J. (1991), "Constructing Some Categorical Data Anomalies," *American Statistician*, 45, 129–134.

Goodman, L. A. (1968), "The Analysis of Cross-Classified Data: Independence, Quasi-Independence, and Interactions in Contingency Tables With or Without Missing Entries," *Journal of the American Statistical Association*, 63, 1091–1131.

——— (1971), "The Analysis of Multidimensional Contingency Tables: Stepwise Procedures and Direct Estimation Methods for Building Models for Multiple Classifications," *Technometrics*, 13, 33–61.

———(1972), "Some Multiplicative Models for the Analysis of Cross-Classified Data," in *Proceedings of the 6th Berkeley Symposium*, eds. L. Le Cam et al., Vol. 1, Berkeley: University of California Press, pp. 649–696.

———(1979), "Simple Models for the Analysis of Association in Cross- Classifications Having Ordered Categories," *Journal of the American Statistical Association*, 74, 537–552.

———(1981), "Association Models and Canonical Correlation in the Analysis of Cross-Classifications Having Ordered Categories," *Journal of the American Statistical Association*, 76, 320–334.

———(1984), *The Analysis of Cross-Classified Data Having Ordered Categories*, Cambridge: Harvard University Press.

———(1985), "The Analysis of Cross-Classified Data Having Ordered and/or Unordered Categories: Association Models, Correlation Models, and Asymmetry Models for Contingency Tables With or Without Missing Entries," *Annals of Statistics*, 13, 10–69.

———(1991), "Measures, Models, and Graphical Displays in the Analysis of Cross-Classified Data" (with discussion), *Journal of the American Statistical Association*, 86, 1085–1138.

Goodman, L. A., and Kruskal, W. H. (1979), *Measures of Association for Cross-Classifications*, New York: Springer-Verlag.

Grizzle, J. E., Starmer, C. F., and Koch, G. G. (1969), "Analysis of categorical data by linear models," *Biometrics*, 25, 489–504.

Haberman, S. J. (1973), "The Analysis of Residuals in Cross-Classification Tables," *Biometrics*, 29, 205–220.

———(1974a), *The Analysis of Frequency Data*, Chicago: The University of Chicago Press.

———(1974b), "Log-Linear Models for Frequency Tables with Ordered Classifications," *Biometrics*, 36, 589–600.

———(1978), *Analysis of Qualitative Data*, Vol. 1, New York: Academic Press.

———(1979), *Analysis of Qualitative Data*, Vol. 2, New York: Academic Press.

———(1981), "Tests for Independence in Two-Way Contingency Tables Based on Canonical Correlation and on Linear-by-Linear Interaction," *Annals of Statistics*, 9, 1178–1186.

———(1982), "The Analysis of Dispersion of Multinomial Responses," *Journal of the American Statistical Association*, 77, 568–580.

Hagenaars, J. (1990), *Categorical Longitudinal Data: Log-Linear Panel, Trend, and Cohort Analysis*, Newbury Park, CA: Sage.

Hauser, R. M., Koffel, J. N., Travis, H. P., and Dickinson, P. J. (1975), "Temporal Change in Occupational Mobility: Evidence for Men in the United States," *American Sociological Review* 40, 279–297.

Hope, K. (1982), "Vertical and Non-Vertical Class Mobility in Three Countries," *American Sociological Review*, 47, 99–113.

Hutchinson, B. (1958), "Structural and Exchange Mobility in the Assimilation of Immigrants to Brazil," *Population Studies*, 12, 111–120.

Johnson, R. A. (1980), *Religious Assortative Marriage in the United States*, New York: Academic.

Kennedy, W. J., Jr., and Gentle, J. E. (1980), *Statistical Computing*, New York: Marcel Dekker.

Kiiveri, H., and Speed, T. P. (1982), "Structural Analysis of Multivariate Data: A Review," in *Sociological Methodology, 1982*, ed. S. Leinhardt, San Francisco: Jossey-Bass, pp. 209–289.

Laird, N. M., and Oliver, D. (1981), "Covariance Analysis of Censored Survival Data Using Log-Linear Analysis Techniques," *Journal of the American Statistical Association*, 76, 231–240.

Lang, J. B. (1994), "An Alternative Approach to Showing the Equivalences Between Multinomial and Poisson Loglinear Models," *Technical Report* 227, Department of Statistics and Actuarial Science, University of Iowa.

Leamer, E. E. (1983), "Model Choice and Specification Analysis," in *Handbook of Econometrics*, Vol. 1, eds. Z. Griliches and M. D. Intriligator, Amsterdam: North Holland, pp. 285–330.

Liang, K. Y., Zeger, S. L., and Qaqish, B. (1992), "Multivariate Regression Analyses for Categorical Data (with discussion)," *Journal of the Royal Statistical Society*, Ser. B, 3–40.

Madansky, A. (1963), "Tests of Homogeneity for Correlated Samples," *Journal of the American Statistical Association*, 58, 97–119.

Maddala, G. S. (1983), *Limited-Dependent and Qualitative Variables in Econometrics*, Cambridge: Cambridge University Press.

McCullagh, P. (1982), "Some Applications of Quasisymmetry," *Biometrika*, 69, 303–308.

McCullagh, P., and Nelder, J. A. (1989), *Generalized Linear Models*, 2nd ed., London: Chapman and Hall.

McNemar, Q. (1947), "Note on the Sampling Error of the Difference Between Correlated Proportions or Percentages," *Psychometrika*, 12, 153–157.

Norusis, M. J. (1988), *SPSSX Advanced Statistics Guide*, 2nd edn, New York: McGraw-Hill.

——— (1993), *SPSS for Windows Advanced Statistics, Release 6.0*, Englewood Cliffs, NJ: Prentice-Hall.

Palmgren, J. (1981), "The Fisher Information Matrix for Log-Linear Models Arguing Conditionally in the Observed Explanatory Variables," *Biometrika*, 68, 563–566.

Pearson, K. (1900), "On a Criterion That a Given System of Deviations from the Probable in the Case of a Correlated System of Variables Is Such That It Can Be Reasonably Supposed to Have Arisen from Random Sampling," *Philosophical Magazine*, Ser. 5, 50, 157–175.

——— (1904), "Mathematical Contributions to the Theory of Evolution XIII: On the Theory of Contingency and Its Relation to Association and Normal Correlation," *Draper's Co. Research Memoirs, Biometric Series*, No. 1.

Pearson, K., and Heron, D. (1913), "On Theories of Association," *Biometrika*, 9, 159–315.

Pierce, D. A., and Schafer, D. W. (1986), "Residuals in Generalized Linear Models," *Journal of the American Statistical Association*, 81, 977–983.

Plackett, R. L. (1981), *The Analysis of Categorical Data*, 2nd ed., London: Griffin.

Raftery, A. E. (1986), "A Note on Bayes Factors for Log-Linear Contingency Table Models With Vague Prior Information," *Journal of the Royal Statistical Society*, Ser. B, 48, 249–250.

——— (1988) "Approximate Bayes Factors for Generalized Linear Models," Technical Report No. 121, Department of Statistics, University of Washington.

Rao, C. R. (1973), *Linear Statistical Inference and Its Applications*, 2nd edn, New York: Wiley.

Rao, J. N. K., and Thomas, D. R. (1988), "The Analysis of Cross-Classified Categorical Data from Complex Sample Surveys," in *Sociological Methodology, 1988*, ed. C. C. Clogg, Washington, DC.: American Sociological Association, pp. 213–269.

Read, T. R. C., and Cressie, N. A. C. (1988), *Goodness-of-Fit Statistics for Discrete Multivariate Data*, New York: Springer-Verlag.

Reiser, M., Wallace, M., and Schuessler, K. (1986), "Direction of Wording Effects in Dichotomous Social Life Feeling Items," in *Sociological Methodology,1986*, ed. N. B. Tuma, San Francisco: Jossey-Bass, pp. 1–25.

Roy, S. N., and Kastenbaum, M. A. (1956), "On the Hypothesis of No 'Interaction' in a Multiway Contingency Table," *Annals of Mathematical Statistics*, 27, 749–757.

SAS Institute, Inc. (1993), *SAS Technical Report* P-243, SAS/STAT Software: The GENMOD Procedure, Release 6.09, Cary, NC: SAS Institute Inc.

——— (1987), *SAS/STAT Guide for Personal Computers, Version 6.*, Cary, NC: SAS Institute Inc.

Schwarz, G. (1978), "Estimating the Dimension of a Model," *Annals of Statistics*, 6, 461–464.

Simon, G. (1974), "Alternative Analyses for the Singly-Ordered Contingency Table," *Journal of the American Statistical Association*, 69, 971–976.

Sinkhorn, R. (1967), "Diagonal Equivalence to Matrices with Prescribed Row and Column Sums," *American Mathematical Monthly*, 74, 402–405.

Sobel, M. E. (1983), "Structural Mobility, Exchange Mobility, and the Analysis of Occupational Mobility: A Conceptual Mismatch," *American Sociological Review* 48, 721–727.

——— (1988), "Some Models for the Multiway Contingency Table with a One-to-One Correspondence among Categories," in *Sociological Methodology, 1988*, ed. C. C. Clogg, Washington, DC.: American Sociological Association, pp. 165–192.

——— (1994), "Some Log-Linear and Log-Nonlinear Models for Ordinal Scales with Midpoints," unpublished manuscript.

Sobel, M. E., and Arminger, G. (1992), "Modeling Household Fertility Decisions: A Nonlinear Simultaneous Probit Model," *Journal of the American Statistical Association*, 87, 38–47.

Sobel, M. E., and Bohrnstedt, G. W. (1985), "Use of Null Models in Evaluating the Fit of Covariance Structure Models," in *Sociological Methodology, 1985*, ed. N. B. Tuma, San Francisco: Jossey-Bass, pp. 152–178.

Sobel, M. E., Hout, M., and Duncan, O. D. (1985), "Exchange, Structure and Symmetry in Occupational Mobility," *American Journal of Sociology*, 91, 359–372.

Spiegelhalter, D. J., and Smith, A. F. M. (1982), "Bayes Factors for Linear and Log-Linear Models With Vague Prior Information," *Journal of the Royal Statistical Society*, Ser. B, 44, 377–387.

SPSS, Inc. (1994), *SPSS 6.1 for Windows Update*, Englewood Cliffs, NJ: Prentice-Hall.

Stuart, A. (1955), "A Test for Homogeneity of the Marginal Distributions in a Two-Way Classification," *Biometrika*, 42, 412–416.

Ten Have, T. R., and Becker, M. P. (1991), "Multivariate Contingency Tables and the Analysis of Exchangeability," unpublished manuscript.

Uebersax, J. (1991), "Quantitative Methods for the Analysis of Observer Agreement: Toward A Unifying Model," Paper P-7686, The Rand Corporation.

——— (1992), "Modeling Approaches for the Analysis of Observer Agreement," *Investigative Radiology*, 27, 738–743.

Wasserman, S., and Faust, K. (1994), *Social Network Analysis: Methods and Applications*, New York: Cambridge University Press.

Wermuth, N. (1976), "Model Search Among Multiplicative Models," *Biometrics*, 32, 253–263.

Whittaker, J. (1990), *Graphical Models in Applied Multivariate Statistics*, New York: John Wiley.

Whittemore, A. S. (1978), "Collapsibility of Multidimensional Contingency Tables," *Journal of the Royal Statistical Society*, Ser. B, 40, 328–340.

Woolf, B. (1955), "On Estimating the Relation Between Blood Group and Disease," *Annals of Human Genetics (London)*, 19, 251–253.

Yamaguchi, K. (1990), "Some Models for the Analysis of Asymmetric Association in Square Contingency Tables with Ordered Categories," in *Sociological Methodology, 1990*, ed. C. C. Clogg, Oxford: Basil Blackwell, pp. 181–212.

Yule, G. U. (1900), "On the Association of Attributes in Statistics," *Philosophical Transactions of the Royal Society of London*, Ser. A, 194, 257–319.

——— (1906), "On a Property Which Holds Good for All Groupings of a Normal Distribution of Frequency for Two Variables, With Applications to the Study of Contingency Tables for the Inheritance of Unmeasured Qualities," *Proceedings of the Royal Society of London*, Ser. A, 77, 324–336.

——— (1912) "On the Methods of Measuring Association Between Two Attributes" (with discussion), *Journal of the Royal Statistical Society*, 75, 579–642.

Chapter 6

Latent Class Models

CLIFFORD C. CLOGG†

1 Introduction

This chapter on the latent class model has three purposes:

The latent class model (LCM) is introduced in a way that assumes little prior knowledge of the model. This introduction does, however, draw on other backgrounds, methodological or statistical, as do other chapters in this book. The goal is to show how the LCM arises naturally from the theory or the subject matter of social research, in many contexts at least. Many papers or books can serve as introductory treatments of LCMs as well as reviews of the literature: Andersen (1982, 1991), Bergan (1983), Goodman (1974b), Langeheine (1988), Langeheine and Rost (1988), Lazarsfeld and Henry (1968), McCutcheon (1987), Dillon and Goldstein (1984, chap. 10), and Schwartz (1986), among others. Because so many detailed introductions exist already, an abbreviated introduction should suffice here.

Recent developments since the late 1970s or so are covered. The field has progressed so rapidly that it is difficult to cover all important recent developments, so I will be selective. I reviewed the field nearly fifteen years ago (Clogg 1981—the paper was presented in 1978); this review will cover material developed since that time. The breakthroughs in the 1970s (Goodman 1974a, b; Haberman 1979) continue to be important. The work by Goodman and Haberman was primarily responsible for the tremendous growth in latent class analysis since the mid-1970s. But most of the survey papers listed above cover those innovations. A

CLIFFORD C. CLOGG • Department of Sociology and Department of Statistics, Pennsylvania State University, University Park, Pennsylvania 16802, USA. • Parts of this paper were given as an invited lecture at the International Conference on Social Science Methodology, 22-26 June 1992, in Trento, Italy. The author gratefully acknowledges support from the National Science Foundation (Grant No. SES-9011973), the National Institutes of Health (NICHD P-30 Grant to the Population Research Institute), and the Department of Economics, Bergische Universität Wuppertal, Germany, where the author was a Guest Professor in the summer of 1992. The author is indebted to G. Arminger, M. Croon, A. Formann, S. J. Haberman, J. Hagenaars, P. van der Heijden, T. Heinen, R. Langeheine, and M. Sobel for helpful comments.

† Deceased

Handbook of Statistical Modeling for the Social and Behavioral Sciences, edited by Gerhard Arminger, Clifford C. Clogg, and Michael E. Sobel. Plenum Press, New York, 1995.

cursory inspection of the list of references shows how much the field has grown since then. I refer to recent developments in the models, the methods, and the applications that seem important for next steps. I hasten to add that only methodological or statistical references have been included here. Many papers applying LCMs and giving methodological insights as well as substantive insights could have been added.

Finally, I cover new models that ought to be developed much further. My views on the importance of these models will be obvious, and I will try to give insights that might help guide the hard analytical work to come. I hasten to add that many new models will not be covered in detail. Work that should be emphasized in a longer treatment would include Croon's (1990) model for ordinal latent classes, Rost's (1990, 1991) blending of latent class models and Rasch models, and Kelderman and Macready's (1990) application of special latent class models for analysis of differential item functioning (or item bias analysis) (see also Kelderman 1989).

In short, I try to summarize the model as a general statistical model and give recent developments, but space constraints prevent full treatment of several important special models. As much as anything this paper is an outline of a research program for the future. Much more needs to be done to make latent class analysis useful as a general tool for multivariate analysis.

2 Computer Programs

Perhaps the best indicator of the growth in this methodological area since the mid-1970s is the availability of several computer programs, plus detailed manuals and examples, for the analysis of LCM's. Some of these are as follows:

MLLSA (Clogg 1977), based on Goodman(1974a, 1974b), plus enhancements (identifiability checks, multiple-group analysis) is described in McCutcheon (1987). For most models and most types of restricted models, the algorithm is essentially the same as the Expectation Maximization (EM) algorithm (Dempster, Laird, and Rubin 1977), although for some models an algorithm equivalent to the GEM (Little and Rubin 1987) is used when the calculations for the M step only approximate the maximization (see Mooijaart and van der Heijden 1992). MLLSA is now included as a module in the CDAS program (Eliason 1990).

LAT (Haberman 1979) and NEWTON (Haberman 1988) use a different formulation, described below, based on a log-linear decomposition. LAT is based on Fisher's method of scoring; NEWTON is a Newton-Raphson algorithm with adjustments for step-length from cycle to cycle. Model specification is carried out in terms of the design or model matrix for the relevant log-linear model as specified for the incompletely observed table, i.e., for the table cross–classifying the observed variables and the latent classes.

LCAG (Hagenaars 1990; Hagenaars and Luijkx 1988) includes most of the features available in MLLSA. Additional types of restrictions are possible, such as the capability of considering a model where the relationship among latent variables is developed as a log-linear model, which allows for the consideration of some special path models. LCAG has special provisions for handling missing data. MLLSA allows for multiple-group analysis as discussed below; LCAG allows for group comparisons by formulating special quasi-latent variables, however. The EM algorithm or an essentially equivalent algorithm is used, as in MLLSA.

Heinen (1993), in an important work covering latent class models as nonparametric or semiparametric analogues of common latent trait models, refers to other programs and provides comparisons among them.

Rost has several programs for analysis of LCM's, including LACORD (Rost 1988), which considers various models for partial-credit scored test items or ordinal indicators. See also Rost (1990, 1991).

PANMARK (Langeheine and van de Pol 1990; van de Pol and Langeheine 1990; van de Pol, Langeheine, and de Jong 1991), also based on the EM algorithm, estimates the same LCM models that are feasible in MLLSA (the overlap with LCAG is less clear). In addition, this program is suited for the analysis of discrete-time Markov chains where the Markovian property is posited to hold, in certain models, at the level of latent classes observed repeatedly in time.

While no general-purpose routine for LCMs has yet made its way into the main statistical or econometric software packages, this will undoubtedly occur soon. Moreover, it is possible to estimate and test LCMs using extensions of generalized linear models (so-called composite link functions) and the popular GLIM package (Arminger 1985; Palmgren and Ekholm 1985).

All of these programs use the principle of maximum likelihood estimation and employ modern algorithms for maximization. From 1950 (Lazarsfeld 1950) to the early 1970s no computational equipment existed whatsoever (except for special cases), with the effect that there was practically no substantive work using LCMs from the time of discovery until the late 1970s, when Goodman's contributions were recognized. The dramatic growth in applications or substantive work with LCMs since that time is no doubt due in large measure to the existence of good programs like those mentioned above.

3 Latent Class Models and Latent Structure Models

Latent class models comprise a subset of the general class of latent structure models. Besides LCMs the latter includes factor analysis models, covariance structure models, latent profile models, latent trait models (or models used in Item Response Theory), and others. Lazarsfeld and Henry (1968) attempted to organize these models in a common framework; see Bartholomew (1987) for a more recent attempt to do this. Both works are very successful in my estimation, although a fully general account of latent structure analysis, in modern terminology, has yet to be written. It is important to recognize that this chapter focuses on LCMs, not latent structure models or latent structure analysis as a whole. But the LCM is much more general than was previously supposed, so more of the latent variable terrain is covered than might be imagined at first glance.

Latent structure models can be organized in many ways. A simple way uses the following principles. First, what is the measurement scale of the observed or observable variables that will be treated as the multivariate observations (or indicators)? (Covariates or predictors may have different measurement scales, and their measurement properties are not important for defining types of latent structure models.) The possible scales are: continuous (and hence quantitative), restricted continuous (censored or truncated variables), categorical-dichotomous, categorical-nominal (multiple categories for each variable, but no ordering of categories), categorical-ordinal (ordinal levels such as a Likert scale), categorical-quantitative (levels are spaced with the spacings known in principle). There are other possibilities that mix the above types. A very common measurement scale in survey research, for example, is the partially ordered variable, a variable with most levels ordered but with one or more levels unordered. A Likert variable with five scale values (strongly agree, agree, neutral, disagree, strongly disagree) plus a Don't Know response (or possibly a No Answer response) is very common in survey research. In educational testing, the same variable type arises if partial credit scoring is used and an allowance is made for not answering the item. If the task were to predict just one such variable from a set of predictors, the different scales above lead to practically all of the main regression-type models in statistics (conventional linear regression, logit and probit models, tobit models, and so forth), including special models for handling sample selection (Know versus Don't Know can define a selection mechanism, for example). Latent variable models are just as much influenced by the measurement scale as are regression-type models. There are many different types of latent variable models for the same reason that there are many types of regression models depending on the form of the dependent variable.

Second, what distributional assumptions are made for the observed variables or observations on them? Note that the measurement scale for the observed variables is not usually an assumption as much as a fact. The type of measurement scale can be determined in principle once we know how the data were collected and organized, how the questions or items were presented to the subjects, and so forth. Models, of course, should be consistent with facts. But distributional assumptions are really assumptions, rarely facts like the scale of measurement. Some assumptions might be expressed in ways that mask what is really assumed. For example, summarizing the data first in terms of first and second moments and cross-moments, and then developing models for these moments, is closely related to the assumption of multivariate normality. Multinomial assumptions are usually implicit when categorical measurements are considered, but not always. It should be noted that the ordinary multinomial distribution makes no allowance for ordering or spacing of categories, but it is possible to make such allowances without assuming continuity, normality, or sufficiency of first and second moments.

Third, what is assumed about the measurement scale and the distribution of the latent variable(s)? We cannot observe, or directly measure, latent variables by definition, so both measurement scales and distributions are really assumptions as far as the latent variable is concerned. The simple facts of the matter are that latent structure (or latent variable) models up until very recently made three very simple assumptions about the latent variables: (1) the latent variable is continuous, with a distribution not specified explicitly (latent trait models, of which the Rasch model is a special case); (2) the latent variable is continuous and even normally distributed (factor analysis model or covariance structure models estimated with LISREL, LISCOMP, or MECOSA—with the latter the weaker assumption of conditional normality given predictors is made); (3) the latent variable is categorical-nominal (conventional latent class model for categorical observed variables, latent profile model for continuous observed variables) or perhaps categorical-ordinal (most scaling models fit here). The really exciting new developments in this area, to my way of thinking at least, pertain to LCMs with scaled latent classes (scaled multinomials), which can be regarded as nonparametric versions of the models with continuous latent variables (cf. Clogg 1988).

The careful reader might find as many as a dozen ways to categorize each of the three factors given above. The cross-classification of these would then lead to a large contingency table with each cell representing at least one type of latent variable model! There would be a few zero cells, but not many. It seems to me that a major task is to find linkages among the models that are practical. (There are linkages that are not practical; for example, saying that the continuous latent variable models are just latent class models with many, many latent classes is not helpful, although it is certainly true.) One way to do this is to assume that all latent variables are continuous, perhaps even normally distributed. This approach has been worked out very satisfactorily in the past decade. Another approach is to assume that latent variables are categorical, perhaps categorical and quantitative (or metric), which can be called a nonparametric or a semiparametric or simply a robust alternative. Hints at how this latter approach will work can be found in Becker and Clogg (1988) and Lindsay, Clogg, and Grego (1991). A thorough account is given in Heinen (1993) in the context of latent trait models.

4 Basic Concepts and Notation

We use Y to stand for an observed variable (not a covariate), and if we have J observed variables or items we use the notation, $Y_1, Y_2, \ldots, Y_J$. The term X is used for the latent variable. A common way to describe a density function for a random variable such as Y is $p_Y(y)$, where "Y" stands for the variable and "y" stands for the value that the variable takes on. It is also common to use Greek letters to stand for parameters to be estimated. In the latent class model, the parameters to be estimated are densities for discrete random variables, that is, probabilities. I suggest a notational style that will be easier to use than some others but which is still consistent with basic ideas concerning densities and parameters. (Compare the notational styles of Goodman (1974b), Formann (1992), Hagenaars (1990), and Haberman (1979), among others. The notation adopted here seems easier to use when considering LCM analogues to latent trait models, for example.) We use $\pi_X(t)$ and $\pi_{Y|X}(y)$ to stand for the basic or standard parameters in the LCM. These symbols will be consistent with conventional notation in statistics and we hope they will be easy to translate to other notational styles.

Of primary interest in the analysis of LCMs is the latent distribution, the distribution of the latent variable X. We suppose that X has T categories. The parameters describing this distribution are denoted as $\pi_X(t)$, for $t = 1, \ldots T$, and because these describe the marginal distribution of X, a discrete random variable, we must have $\sum_{t=1}^{T} \pi_X(t) = 1$, with $\pi_X(t) \geq 0$ for each t. Because of this, there will be at most $T - 1$ nonredundant parameters for X. In some cases with conventional LCMs, the categories of X will be ordered by imposing restrictions of some type, or an ordering will be imposed after examining parameter values. But usually the categories of X are unordered. Virtually all analyses of LCMs from Lazarsfeld (1950) through at least the mid-1980s characterized the latent variable X in this way. Because of this, LCMs have often been regarded as techniques for clustering data or building typologies, appropriate for the case where the levels of X are nominal (unordered).

To scale X we need to introduce *score* parameters. Let μ_t denote the scale value (metric) for all members of the t-th latent class. With nominal X (no score parameters), it suffices to include just one latent variable with T classes to cover all possibilities. For example, if two dichotomous latent variables exist, so that the latent distribution is viewed as a 2 x 2 table, we merely set $T = 4$. In this way, an LCM with $T > 2$ latent classes can always be regarded as a model with multiple latent variables. To scale X, on the other hand, means that we imagine X as a single (unidimensional) construct. We note that a scale is meaningless unless it is given a zero point and a unit point, or a location and scale. It will thus be possible to identify at most $T-2$ of the μ_t, which means that inference about metric, categorical latent variables begins when $T = 3$ latent classes are posited. (A dichotomous X can, of course, always be regarded as an extreme grouping of a quantitative variable, but the distance between the two levels is arbitrarily defined.)

We should alert ourselves to redundancies that can be viewed as identification problems. If for some t' we have $\pi_X(t') = 0$, then the t'-th latent class is void, and the model has $T-1$ and not T latent classes. For models with score parameters, we cannot distinguish classes t', t'' if $\mu_{t'} = \mu_{t''}$, and the model is once again equivalent to a model with $T-1$, not T, latent classes. (Latent classes t', t'' can be combined.) These difficulties can be relatively serious from a numerical standpoint, such as in determining the stability of estimation algorithms in cases where relatively small latent classes are modeled. It is not clear as yet, however, whether transforming the parameters, such as with a logistic transformation of the latent class proportions (cf. Formann 1985, 1992), brings much advantage.

With this change in the formulation allowing scores for the latent classes, the latent-class concept becomes very powerful indeed. We can regard X^* as the (true) continuous latent variable and X (with T classes and score values for the classes) as a discretized version of X^*. In principle, we ought to be able to approximate any X^* with a latent scaled-categorical variable provided T is large enough. But the value that T can take on is limited by what we observe—the number of Y variables and the number of categories used to measure each. Intuition leads one to expect that all of the available information about X^* or X reaches a saturation point with some (possibly large) value of T. But note that this approach does not assume normality of X^*; it is a nonparametric or semiparametric approach. Heinen (1993) gives serious consideration to this general approach.

The latent distribution is important, and usually we will prefer an approach that makes the parameters of the latent distribution explicit. That is, the latent class proportions $\pi_X(t)$ and the latent class scores μ_t ought to be regarded as fundamental parameters. The other parameters describe the relationship between X and the observed variables (the Ys). These parameters or functions of them have received the most attention in prior work. We use $\pi_{Y_j|X(t)}(y_j) = P(Y_j = y_j|X = t)$ with the understanding that the y_j denote the levels of j-th observed variable (e.g., $y_1 = 1, \ldots 5$, if the first item has five levels). Notice that these parameters must differ from the corresponding marginal distribution of Y_j, say, $p_{Y_j}(y_j)$, or else X cannot be identified from the data. (The case where the observed Ys are mutually independent is one special case; here X cannot be identified.) Again, we might wish to consider parameters that are functions of these conditional probabilities, and the two most common choices now are (a) log-cross-ratios (see Haberman 1979, 1988) and (b) logit-transformed probabilities (Formann 1992).

With these definitions of parameters (or probabilities or probability distributions) in mind, the LCM is defined by making two assumptions. First, latent classes are internally homogeneous (all individuals in a given latent class have the same probability distribution with respect to the Y variables), and second, the observed variables are conditionally independent given the level t of X. The first assumption is the same as used in the analysis of finite mixtures —the observed distribution is regarded as a mixture of T separate distributions with mixing weights $\pi_X(t)$; see Aitkin, Anderson, and Hinde (1981) and Titterington, Smith, and Makov (1985). Here, the idea is that the multinomial observed (the observable contingency table) is viewed as a weighted sum of a relatively small number of indirectly observed multinomials. Each of the latent classes corresponds to one of the indirectly observed multinomials. This assumption is always imposed. The most common way to modify this assumption is to consider models with additional latent classes, up to the limits imposed by identifiability or numerical stability.

The second assumption is the so-called axiom of local independence, which is used to define almost all latent structure models. It is possible to relax the latter assumption (cf. Clogg 1981a; Formann 1992) to some extent. One important new development in recent years is the formulation of models with some (carefully specified) local dependence. See Espeland and Handelman (1989) and Formann (1992). Sometimes models with some (carefully specified) local dependence might be appealing as substantive models; and such models can be useful to diagnose causes of lack of fit in models assuming local independence. And there are convincing substantive models where local dependence has to be allowed. For example, the mixed Markov model for repeated measures in a panel setting actually incorporate a special form of local dependence (van de Pol and Langeheine 1990). Also, Hagenaars (1988, 1990) gives models analogous to multiple-indicator, multiple-cause models where a very special form of local dependence is required by analogy to similar models for continuous variables derived from normal theory. The assumption of local independence, however, is implicit in almost all work with latent structure models regardless of the type, and we shall take this for granted below.

5 The Model Defined and Alternative Forms

Let $p_Y(y)$ denote the joint density of the observed variables, i.e., $Y = (Y_1, \ldots, Y_J)'$, a vector, and "y" stands for a cell in the table cross-classifying the J items. More generally, "y" just stands for a pattern of data observed for a unit, where the unit might be an individual or a group of individuals as with a cell of a contingency table. (When we use "Y" or "y" below, the representations stand for vectors, with the appropriate modifications. The usage should be clear from the context.) Let $\pi_{Y|X(t)}(y)$ denote the conditional density, i.e.,

$P(Y = y|X = t)$. By the axiom of local independence, we obtain

$$\pi_{Y|X(t)}(y) = \prod_{j=1}^{J} \pi_{Y_j|X(t)}(y_j). \tag{6.1}$$

The joint distribution of Y and X, call it $\pi_{Y,X}(y,t)$, is obtained as

$$\pi_{Y,X}(y,t) = \pi_X(t)\pi_{Y|X(t)}(y). \tag{6.2}$$

The LCM can now be written as

$$\pi_Y(y) = \sum_{t=1}^{T} \pi_{Y,X}(y,t). \tag{6.3}$$

This is the definition of the LCM used by Lazarsfeld and Goodman; only the notation has been changed slightly.

Suppose that just two categorical variables were available, say Y_1 and Y_2, having I and J levels each, so that the analysis pertains to an I x J contingency table. Here "y" refers to a cell, say cell $(i,j), i = 1,\ldots I, j = 1,\ldots, J$, and we can use the symbol π_{ij} to stand for $P_Y(y)$, where "y" is the set of responses for all indicators in the vector Y. (An example of this type will be considered shortly.) The T–class LCM in this case can be written as

$$\pi_{ij} = \sum_{t=1}^{T} \pi_X(t)\pi_{Y_1|X(t)}(i)\pi_{Y_2|X(t)}(j). \tag{6.4}$$

See, for example, Clogg (1981), Marsden (1985), Goodman (1987), de Leeuw and van der Heijden (1991), and van der Heijden, Mooijaart, and de Leeuw (1992). Here, cell (i,j) is the response pattern denoted by y in the more general expression. There are many other ways to write this simple model. For example, Clogg (1981) suggested writing $\pi_X(t)\pi_{Y_1|X(t)}(i)/\pi_{Y_1}(i) = \pi_{X|Y_1}(t)$, using the definition of conditional probabilities. The latter set of quantities can be regarded as the parameters of interest for the relationship between X and Y_1, and this parameterization plays a special role in latent budget analysis (see de Leeuw and van der Heijden 1992) as well as in the development of Markov–type models for the analysis of change.

5.1 Measuring Fit

We obtain expected frequencies as $F_{ij} = n\pi_{ij}$ in this case; estimates of these are compared to the observed frequencies, say f_{ij}, in the usual way using statistics such as: Pearson's goodness-of-fit statistic,

$$X^2 = \sum_y (f_y - \hat{F}_y)^2/\hat{F}_y$$

the likelihood-ratio statistic,

$$L^2 = 2\sum_y f_y \log(f_y/\hat{F}_y)$$

with the convention that $0\log(0) = 0$, or any other member of the Cressie-Read power-divergence family,

$$CR(\lambda) = 2[\lambda(\lambda+1)]^{-1}\sum_y f_y[(f_y/\hat{F}_y)^\lambda - 1].$$

[Pearson's statistic is obtained with $\lambda = 1$, and the likelihood-ratio statistic is obtained by letting λ approach 0. Cressie and Read (1984) suggest the value of the statistic with $\lambda = 2/3$.] These statistics possess null chi-squared distributions, asymptotically, although each statistic behaves differently with sparse data. (For sparse data, procedures analogous to those given in Clogg et al. 1991 can be adopted for analysis of LCMs.) Many popular indexes of fit are based on chi-squared statistics of one type or another.

Fit statistics not directly tied to the chi–squared distribution include the index of dissimilarity,

$$D = \sum_y |f_y - \hat{F}_y|/2n$$

The D quantity can be applied directly to the case with y denoting a response pattern but can be defined when y denotes an individual observation, although interpretation including inference is complicated for the latter case.

5.2 Alternative Forms of the Model

Here are some alternative ways to parameterize the model. It should be emphasized that these are reparameterizations of the same model, but they give different insights (e.g., with respect to how they shed light on the latent distribution), have very different computational implications, lead to different classes of restricted models, and, most important, lead to different generalizations. We shall call the parameterization of (6.1)-(6.3) the *classical* one, or the probability representation.

A straightforward representation familiar from log-linear analysis is based on (6.2). [Formula (6.3) is common to all representations.] The logarithm of the joint distribution involving X and Y is decomposed as

$$\log(\pi_{Y,X}(y,t)) = \lambda + \lambda_{X(t)} + \sum_{j=1}^{J} \lambda_{Y_j} + \sum_{j=1}^{J} \lambda_{Y_j X(t)}, \tag{6.5}$$

where notation for the levels of the Y variables has been suppressed for convenience. It is important to note that although this looks like a regular log-linear model with the Y_j's conditionally independent given the level t of X, this decomposition applies to the indirectly observed table. As with any log-linear model, regardless of the coding of effects, the lower-order terms do not refer to marginal distributions because there are higher-order relatives of those terms in the model. In short, the $\lambda_{X(t)}$ do not refer to the marginal distribution of X in general; these main-effect terms are not functions of $\pi_X(t)$ only. This formulation is due to Haberman (1977, 1979). The model is specified by giving the model matrix, and the programs built around this idea include LAT and NEWTON (Haberman 1988).

A slightly different formulation, due to Formann (see Formann 1985, 1992), takes the logit transformation of the parameters on the right-hand side of (6.2). For example, let $\phi_{X(t)} = \log(\pi_X(t)/\pi_X(T)), t = 1, \ldots, T-1$, and so forth. In this formulation, of course, the latent distribution is reproduced but on the logit scale. Similar transformations apply to the conditional probabilities defined above.

Other attractive frameworks include formulating a log-linear decomposition of the conditional distribution in (6.1), linked together with parameters for the marginal distribution of X. That is, retain the lambda parameters for everything except the X term in (6.5), for the conditional distribution. Finally, we might take advantage of the relationship,

$$\pi_{Y_j,X}(y_j, t) = \pi_X(t)\pi_{Y_j|X(t)}(y_j),$$

to formulate models that relate to the concepts of symmetry or quasisymmetry at the latent level (cf. Agresti 1992). A slight variation of this appears in latent budget analysis (van der Heijden, Mooijaart, and de Leeuw 1992). It appears that many other parameterizations are possible working from these different baselines. For some additional hints at reparameterization, see the GLIM reformulations given by Arminger (1985) and Palmgren and Ekholm (1985).

The various possibilities sketched above have very different implications for (a) multiple-group comparisons, (b) formulating restrictions and implementing restricted models in computational algorithms, (c) interpretation of results, and (d) actually doing latent class analysis with real data sets (the desirability of which should not be overlooked). To illustrate just the second point, note that imposing the restriction

$$\lambda_{Y_1(1),X(1)} = \lambda_{Y_2(2),X(2)}$$

is very different from imposing the restriction

$$\pi_{Y_1(1)|X(1)} = \pi_{Y_2(2)|X(2)} .$$

In most applications, the important quantities will be the parameters for the latent distribution and the parameters for the X–Y relationships. But in my experience, one of the most important practical uses of the LCM is prediction of membership in latent classes. So far we have looked at the marginal distribution of X and the conditional distribution of Y given X. We now consider the obverse question which can be answered from the distribution of X given Y. Applying the definition of conditional probability, we obtain

$$\pi_{X|Y}(t) = \pi_{Y,X}(y, t)/\pi_Y(y) . \tag{6.6}$$

Note that this defines a (discrete) distribution of X given $Y = y$, and this fact ought to be taken into account in generating predictions. For example, prediction rules and the method of accounting for measurement errors in predicting the discrete X will be different in general from the case where a continuous latent variable is predicted. Note also that the expression in (6.6) can be regarded as a posterior mean; for $Y = y$ we have the posterior distribution $\pi_{X|Y(y)}(t), t = 1, \ldots, T$. Some ideas on this appear in the section on prediction below. Note that the analogue to (6.6) for the two-way contingency table [see (6.4)] would be

$$\pi_{X|(Y_1=i,Y_2=j)}(t) = \pi_{Y_1,Y_2,X}(i,j,t)/\pi_{ij} .$$

In virtually all applications known to this author, the following steps are taken in predicting X from $Y = y$: (a) substitute sample estimates (MLEs) in (6.6), (b) predict for each y the value t of X for which the conditional probability estimate is maximized, (c) assign all cases with $Y = y$ to this modal value. These rules are sensible but they tend to exaggerate the degree of certainty in the prediction. See the section on prediction below.

Simple measures of the quality of the prediction such as the percent correctly allocated (accumulate correct or modal assignments across all cells y and express the result as a proportion of n) and the lambda measure of association between X and Y are easy to formulate (Clogg 1979, 1981a, b). To this author, the quality of the prediction provided by an LCM is just as important as goodness of fit. Prediction of the latent distribution, which can be very different from predicting the individual cell frequencies in the observed table, ought to play a more prominent role in latent structure analysis. When the goal is data reduction, perhaps as a first step in a more involved substantive analysis, goodness of prediction and goodness of fit work in opposite directions. To see this, note that the model of independence among the Ys is a one-class LCM and prediction is perfect for this model! Goodness of fit improves as we add more latent classes or remove restrictions but prediction will tend to worsen as more complexity is introduced. Assessing the quality of the prediction is very important, and it will be discussed from a new point of view later.

6 An Example: Latent Classes in the American Occupational Structure

6.1 Standard Latent Class Models for Two-Way Tables

Table 1 is a 5 x 5 occupational mobility table drawn from Knoke and Burke (1980, p. 67). This table was obtained by collapsing categories (and discarding the No Answer category—that is, censoring those not in the labor force) in the basic 18×18 intergenerational mobility table used in the famous study by Blau and Duncan (1967, p. 496). See Goodman and Clogg (1992) for details and analyses using a variety of models. Latent class models for two-way arrays of this type appear in Clogg (1981); also see Breiger (1981), Marsden (1985), and Hagenaars (1990, pp. 180–193).

Fit statistics for some elementary latent class models applied to these data appear in Table 2 (on p. 324). The models are H_0, the model of row-column independence or "perfect mobility" (equivalent to a one-class LCM, the natural baseline); H_1, the model of quasi-independence allowing effects for each of the cells on the main diagonal (this model can

Table 1. Occupational Mobility of American Men in 1963

Father's Occupational Status	Subject's Occupational Status 1	2	3	4	5
1	152	66	33	39	4
2	201	159	72	80	8
3	138	125	184	172	7
4	143	161	209	378	17
5	98	146	207	371	226

SOURCE: Blau and Duncan (1967, p. 496), as condensed in Knoke and Burke (1980, p. 67). The five occupational status categories pertain to (1) Professional and Managerial, (2) Clerical, Sales, and Proprietors, (3) Craftsmen, (4) Operatives and Laborers, (5) Farmers and Farm Laborers.

be viewed as a model with one "random" latent class and five "deterministic" latent classes for the cells on the main diagonal); H_2, the two-class LCM; and H_2', the two-class model with diagonal effects (this model can be viewed as $H_1 + H_2$). Let the variables now be denoted as O (origin level) and D (destination level), so the mobility table is the $O \times D$ cross-classification.

The quasi–independence model is usually expressed as

$$\pi_{ij} = \alpha_i \beta_j, i \neq j; \pi_{ij} = \alpha_i \beta_j \delta_i, i = j.$$

As shown in Clogg (1981), when $\delta_i > 1$ for each i (for each cell on the main diagonal), the quasi-independence model is a special deterministically restricted LCM with $T = I + 1$ latent classes. The LCM representation is

$$\pi_{ij} = \sum_{t=0}^{I} \pi_X(t) \pi_{A|X(t)}(i) \pi_{B|X(t)}(j),$$

with $\pi_{A|X(i)}(i) = \pi_{B|X(i)}(i) = 1, i = 1, \ldots, I$ (conditional probabilities set to zero for other combinations). In this LCM representation of the usual quasi–independence model, the 0-th latent class corresponds to a latent class with perfect mobility. The latent class proportions for $t = 1, \ldots, I$ can be related to the diagonal cell parameters (the δ_i). Model H_2' is obtained by adding another unrestricted latent class to this model. In Table 2 we also give the fit statistics for the RC association model, called H_{RC}; this model will be discussed later.

As we see, model H_2' fits the data very well; the index of dissimilarity between fitted and observed frequencies is .4%, indicating a dramatically good fit. [The frequencies represent estimates of population counts in ten thousands; chi-squared statistics will be influenced by the n assumed as the sample count, but the D value in (6.5) will not be so influenced.]

For models H_2 and H_2', simple counting to determine degrees of freedom leads to an error. There is an intrinsic problem of identification with the LCM applied to the two-way contingency table (and in some other types of tables as well). We can see this as follows. There are 25 cells in the table, but 24 are nonredundant since the frequencies sum to n (or the cell proportions sum to 1). For H_2, the number of parameters estimated appears to be 17 : 1 for the latent distribution, since $T = 2$, plus 16 for the conditional probabilities (4 nonredundant conditional probabilities for each of the four latent conditional distributions), giving a total of 17 parameters. Refer to the parameters on the right-hand side of (6.4); H_2 is this model with $T = 2$. This seems to indicate that there are 24 – 17 = 7 degrees of freedom (df), but actually the correct number is 9 df. There are two redundancies, as noted in Clogg (1981), proven in Goodman (1987), and studied in great detail in van der Heijden, Mooijaart, and de Leeuw (1992). These sources show how these dependencies arise and illustrate how the unidentified parameters can be restricted to form an identified model. Two restrictions must be imposed to identify parameters. Similar comments apply to model H_2'; ordinary counting gives an impression that there would be 2 df but there are actually 4 df. Note that an algorithm using full information from cycle to cycle, such as Fisher scoring or Newton-Raphson, would have great difficulty with a problem like this one where identifiability might not be known in advance. (Programs using these algorithms typically examine the rank of the estimated variance–covariance matrix at a given cycle and remove redundant parameter values when the rank is deficient. However, it can be difficult to distinguish between the case where a model is intrinsically under-identified regardless of the sample size and the case where a model is not estimable because the sample is not large enough, when this approach is used.) In many respects analyzing a two-way table ought to be the first order of business, so this simple example demonstrates the need for identification checks and careful selection of algorithms.

The parameter values for model H_2' (maximum likelihood estimates) are as follows. For the latent distribution, the model actually had $T = 7$ latent classes, two pertaining to random classes and five pertaining to deterministic classes. The deterministic classes can be described as latent stayer classes, one for each cell on the main diagonal. Label these classes $t = 3, \ldots, 7$, with the first two classes denoting the random classes. The parameter values were:

$$\hat{\pi}_X(t) : .33, .56, .02, .01, .02, .01, .06;$$

$$\hat{\pi}_{O|X=1}(i) : .21, .44, .23, .12, .00^*;$$

$$\hat{\pi}_{O|X=2}(i) : .00^*, .002, .16, .40, .44;$$

$$\hat{\pi}_{D|X=1}(j) : .40, .28, .14, .16, .02;$$

$$\hat{\pi}_{D|X=2}(j) : .12, .17, .25, .44, .02.$$

Entries followed by an asterisk (*) denote fixed restrictions; two restrictions were applied.

Table 2. Simple Latent Class Models Applied to the Data in Table 1

Models	df	X^2	L^2
H_0	16	875.03	830.98
H_1	11	269.06	255.14
H_2	9	239.85	262.92
H_2'	4	1.10	1.14
H_{RC}	4	1.81	1.89

NOTE: Calculations done with MLLSA computer program described, e.g., in McCutcheon (1987), using the algorithm presented in Goodman (1974b).

Notice how for the row variable (variable O) the conditional probabilities are restricted in a way that gives structure to the latent variable. We have used some information on the ordering of the categories to develop the restrictions. Level 1 is the highest and level 5 is the lowest of the occupational categories. So we have restricted parameter values with $\pi_{O|X(1)}(5) = .00$, implying that the first latent class cannot have recruited from the lowest occupational category, and vice versa. Similarly, the second latent class cannot have recruited from the highest occupational category, and vice versa (i.e., $\pi_{O|X(2)}(1) = .00$). A weak assumption on ordering of the row (origin) categories—really just the bottom and the top—allows the characterization of the first two latent classes as latent upper and latent lower.

6.2 Some Related Models

Functions of the parameters in the LCM, including functions of conditional probabilities like the following,

$$\pi_{X|O(i)}(t) = (\pi_{O|X}(i))(\pi_X(t))/p_O(i) \tag{6.7}$$

are used in the modification of LCMs for two-way tables called latent budget analysis (van der Heijden, Mooijaart, and de Leeuw 1992). Note that (6.7) is an application of Bayes Rule. Latent budget analysis is essentially a reparameterized version of standard latent class analysis for two-way tables. Different kinds of parameter restrictions become useful when the model is rewritten as above.

In recent years there has been renewed interest in methods that are closely related to the LCM. We comment now on three of these other methods. First is the method of graphical analysis (Whittaker 1990). This method essentially searches through multivariate data to find possible structures of conditional independence. If three categorical variables, A, B, and C, were considered, say, the method might entail fitting (or approximating the fit of) hierarchical log-linear models like $((AC), (BC))$, which is the hypothesis that A and B are conditionally independent given the levels of variable C. The LCM is a graphical model once we replace C by X. (At the latent level, the model can indeed be represented in this way.) In a sense, the LCM searches for a conditional independence structure, like graphical modeling, but does not restrict attention to observable conditioning variables. In fact, the LCM can be suggested as a tool for cases where graphical models might be used to describe relationships among categorical variables. We merely find a particular LCM that fits the data well and then find the observed variable or set of observed variables most closely associated with the latent variable X in the model. Variations of the procedures used for predicting latent classes from the observed distribution can be used to predict, say, the particular observed item (or set of items) most closely associated with X. A strategy similar to graphical modeling would be, for example, to first find an LCM that is satisfactory and then to find the observed variable or set of variables (e.g., A, B, C, AB, AC, etc.) most closely related to X. We note in passing that virtually the same comments apply if continuous variables are considered, in which case standard factor analysis models become tools for graphical analysis also.

As noted earlier, the independence model (H_0) can be regarded as a one-class LCM. Note that under this model, for the two-way table, $\pi_{ij} = p_{Y_1}(i)\, p_{Y_2}(j)$. Converting to row (or column) conditional distributions shows that the matrix or table has rank one. A correlation model consistent with a matrix or table of rank two is:

$$\pi_{ij} = p_{Y_1}(i)p_{Y_2}(j)(1 + \rho u_i v_j) \tag{6.8}$$

This model can be called a correlation model because ρ is the value of the correlation between the row and column variables, with scores u_i and v_j used to score the row and column levels, respectively. (In the population, if this model is true, the scores arise as eigenvectors in a transformed version of the table of probabilities.) The two-class LCM also represents a matrix or table of rank two. The parameters of the above correlation model, subject to usual restrictions of identifiability, can be translated into the parameters of the LCM (Goodman 1987) and vice versa (de Leeuw and van der Heijden 1991). A closely related model is the RC association model (Goodman 1984):

$$\pi_{ij} = \alpha_i \beta_j \exp(\phi \mu_i \nu_j)\,. \tag{6.9}$$

This model can be viewed as a rank two representation of certain other functions of the cell probabilities, but the main point is that the structure of the correlation model and the structure of the association model are similar. Note that if the data are close to independence, then $\alpha_i \approx p_{Y_1}(i)$, $\beta_j \approx p_{Y_2}(j)$, and by a first-order Taylor expansion ($\exp(h) \approx 1 + h$), the two models are essentially the same. The fit of the RC model including a parameter for each cell on the main diagonal is also given in Table 2, and we see that it fits almost as well as the preferred LCM. The degrees of freedom are the same; the number of parameters is the same. We often find close agreement between the two kinds of models in practice.

The models in (6.8) or (6.9) appear to be radically different from the LCM representation in (6.4). The conceptual basis *appears* to be radically different as well. For the correlation or association models, we usually imagine that the row variable is a categorization of a continuous variable (Y_1 is a grouped version of Y_1^*, say). The u_i for the correlation model or the μ_i for the association model represent scores at the latent level indicating, for example, how the grouping of the row variable might have occurred. Similar comments apply to the column variable. The LCM, in contrast, posits a categorical latent variable. But the models (LCM, correlation, association) are virtually indistinguishable for many sets of data! This evident duality between so-called discrete theories of measurement (LCMs) and so-called continuous theories is very important to recognize. It is perhaps one of the most exciting things about this area of methodology.

The correlation model in (6.8) is naturally extended to

$$\pi_{ij} = p_{Y_1}(i)p_{Y_2}(j)(1 + \sum_{m=1}^{M} \rho_m u_{im} v_{jm}),$$

where $M \leq \min(I, J) - 1$. The quantity M is often referred to as the dimension of the correlation structure. It corresponds to the number of eigenvalues (or correlations) that can be extracted from a slightly transformed version of the original table of probabilities. The association model in (6.9) is naturally extended to

$$\pi_{ij} = \alpha_i \beta_j \exp(\sum_{m=1}^{M} \phi_m \mu_{im} \nu_{jm}).$$

Connections between these models and their relationship to regular LCMs for two-way arrays are provided in Goodman (1984, 1987) and especially in Goodman (1991). The latter source gives many references to the area known as correspondence analysis. The correlation model is consistent with an LCM with $T = M + 1$ latent classes; both give a rank M representation of the data array. The association model above is analogous to the LCM with $M+1$ latent classes. In the correlation model, uniqueness is ensured by enforcing across-dimension orthogonality constraints on the eigenvectors. That is, the u_{im} are forced to be orthogonal to the $u_{im'}$, $m \neq m'$, and the v_{jm} are forced to be orthogonal to the $v_{jm'}$, $m \neq m'$. In addition the eigenvalues are defined uniquely given these orthogonality constraints as $\rho_1 > \rho_2 > \ldots > \rho_M$.

The T-class LCM (with $T = M + 1$) is quite intractable numerically for large T. We need to consider methods for estimating the general LCM with ordinal-type constraints ensuring that $\pi_X(1) > \pi_X(2) > \ldots > \pi_X(T)$. Ideally, we would like to determine the latent class proportions for a given T-class model where this constraint is imposed and in addition the largest possible values of these proportions except the last are used. (Note that this would uniquely define the mixing weights.) To my knowledge, this has not been done, although some developments in this direction are reported below in the subsection on latent class evaluation models.

7 Research Contexts Giving Rise to Latent Classes and Latent Class Models

In this section the goal is to give particular scientific or statistical contexts where the notion of latent classes is sensible. In many of these cases, the assumptions of the LCM given above are also sensible, even exactly correct in some instances. In social research, it has been customary to introduce LCMs as measurement error models (cf. Clogg and Sawyer 1981; Goodman 1974a; Hagenaars 1988, 1990; Schwartz 1987; Shockey 1988; and many others). It would take up too much space to cover all of the measurement-error methodology that has built up around the LCM, but we note that many of the references cited in this paper pertain to this general subject. We want to downplay this to some extent and emphasize other interpretations or possibilities, so a very selective subset of topics that might represent new directions for research is covered.

7.1 Medical Diagnosis

Suppose the goal is to analyze n independent blood samples. Each sample is either HIV+ or not, but the true classification is unknown. Presence or absence of HIV+ can be called X; and X is dichotomous, say with levels $t = 1, 2$. Next suppose that J different "fixed" tests (or even identical tests applied repeatedly) are applied *independently* to the blood samples. Call the tests $Y_1, \ldots, Y_J$ with $y_j = 1$ (for HIV+ test), $= 2$ (for not HIV+). The observed data form a 2^J contingency table. The LCM model described by (6.1)–(6.3) with $T = 2$ holds, i.e., this model is the correct model for the data. To see this, suppose that an observation with true value $X = 1$ is tested. The J items will have different probabilities of detecting presence of HIV+, but the detection rate for any particular test does not depend on the value of the test result for any other test. A similar argument applies to cases with true value $X = 2$. In this case, the LCM is a natural representation of the data. When diagnosis leads to presence/absence distinctions of this type, with independently applied tests, it is difficult to understand why any other model would be considered. Some inferences from the LCM might be complicated by the sampling scheme (how were the n blood samples obtained? what is the universe to which inferences generalize?), but this does not invalidate the model.

For this situation, the classical formulation of the LCM is appropriate. We would be interested in the latent distribution, $\pi_X(t)$; the so-called false-positive rates for the tests, $\pi_{Y_j|X=2}(1)$ (the probability that the j-th test is positive when in fact the case is negative); the similarly defined false-negative rates; and the prediction of latent classes (true HIV+ or not) from the results of several tests. Equality restrictions on the error rates can be used to advantage for all of these purposes. For example, we might consider imposing the constraint,

$$\pi_{Y_1|X(2)}(1) = \pi_{Y_2|X(2)}(2), \qquad (6.10)$$

which equates the false-positive rate for the tests (or items) 1 and 2. A nested comparison between the model with this constraint and the model without this constraint using the likelihood-ratio criterion can be used to test the constraint. A similar constraint can be applied to test false-negative rates. We might constrain the false-positive rate and the false-negative rate for a given test or item, for example,

$$\pi_{Y_1|X(1)}(2) = \pi_{Y_1|X(2)}(1). \tag{6.11}$$

Perhaps calibration can be done by applying tests independently to a group for which the true results are known. This might lead to restrictions of various kinds (e.g., on the latent distribution) and/or to multiple-group analysis. It might be known in advance that test 1, or Y_1, has a false-positive rate of .05, and imposing a *fixed* restriction on this quantity ought to be considered. It is possible that a model with $T = 3$ latent classes would be considered, for example if the two-class model did not fit the data. The extra latent class might refer to cases difficult to diagnose, or to "heterogeneity" not captured completely by the restriction to two latent classes. (Note that adding latent classes is a nonparametric way to model extra variation or heterogeneity.) In short, the LCM is natural for this situation, and the methodology of latent class analysis is directly relevant. For details, examples, and successful applications in similar settings, see Rindskopf and Rindskopf (1986); Uebersax and Grove (1990); Young (1983); and Young, Tanner, and Meltzer (1982). Many problems in medical diagnosis and related contexts give rise to the LCM whether this fact is recognized or not.

7.2 Measuring Model Fit with Latent Class Evaluation Models

Rudas, Clogg, and Lindsay (1994) consider a novel use of latent class techniques that gives a new index of model fit for quite general settings. Suppose that we are interested in a particular model H, such as row-column independence, for a given contingency table. We might try to fit various regular latent class models to the data, proceeding with standard latent class analysis. But if the goal is to assess model H, the latent class models that would normally be considered would not necessarily give us any direct evidence about the suitability of this model. We can utilize latent class ideas to arrive at a convenient index of model fit, and thus an answer to the question of how well H describes the data, in the following way.

Let $\pi = \pi_X(2)$ and $1 - \pi = \pi_X(1)$ define the latent class distribution for a two-class model. Now suppose that the following saturated two-class model is considered

$$P_Y(y) = (1 - \pi)\pi_{Y|X(1)}(y) + \pi P_{Y|X(2)}(y) \tag{6.12}$$

where model H is now applied to the first latent class (i.e., to $\pi_{Y|X(1)}(y)$) and where the conditional probability distribution for the second latent class (i.e., $P_{Y|X(2)}(y)$) is unrestricted, which is why we use slightly different notation. In this formulation, the unrestricted distribution for the second component can be viewed as a representation of unmodeled heterogeneity, i.e., variation not described by the model H. There is a vast literature on modeling "unobserved heterogeneity," and this approach can be viewed as a general alternative that is completely nonparametric. Note that the regular assumptions of latent class analysis, namely independence among the Ys within each latent class, are not used. (If H is the model of independence or mutual independence, then local independence would apply only in the first latent class.) The model is nevertheless a two-point mixture in spite of relaxed assumptions for the second latent class, and so falls in the general class of models considered in this chapter. Denote the above model as H_π, where π is the proportion in the second latent class, i.e., the latent class where the original model does not hold. Note that $H_{\pi=0} = H$ (all observations derive from the first latent class) and that $H_{\pi=1}$ is the unrestricted or saturated model (all observations derive from the second latent class). Because π is a latent class proportion, it naturally ranges over the interval $[0, 1]$, and so this quantity provides a convenient metric for interpretation.

In the above form, the model parameters are not identifiable, as earlier results on latent class analysis of the American occupational mobility table would indicate. We define the parameter π^* as the *minimum* value of π for which H_π is saturated. The minimum can be defined uniquely. Thus, the model in (6.12) gives a parameter, π^*, which gives the minimum possible fraction of the distribution that is "outside" the model H originally considered. The parameter π^* is a convenient index of the lack of fit of H; values close to zero provide evidence for H, and values close to one provide evidence against H.

The theory, computational algorithms, and comparisons to other indexes of fit or lack of fit are taken up in Rudas et al. (1994). It should be pointed out that computation is related to but different from computation in conventional LCMs; however, a variation of the EM algorithm can be used.

For the occupational mobility table in Table 1, the independence model (say, H) had $L^2 = 830.98$ on 16 *df*. The maximum likelihood estimate of π^* is .314, and a 95% lower limit calculated by approximate methods given in Rudas et al. is .282. In other words, 31% of the population (estimate) is outside the model of independence (or the model of perfect mobility), or 69% can be described by independence ("random" mobility structure). Either quantity can be viewed as a measure of the lack of fit of H, which is independence in this case, or as an index of "structure" in the table. Compare and contrast the present analysis with the conventional latent class analysis given earlier. It appears possible to utilize this framework to measure or index social structure in many other situations where indexes of segregation (index of dissimilarity, scaled versions of this popular index, Gini coefficients, and so on) are used. In many of these other cases, independence or homogeneity is the natural baseline or null model to which the "fit" of the data is compared using these indexes.

The model in (6.12) is a robust model in the sense that it is always true. It gives a new definition of *residuals*, because it places lack of fit in a single, absorbing latent class, here called the second latent class with minimum proportion π^*. The mixture idea is not here confined to contingency tables, as I hope the general notation makes clear. An interesting application is to consider the above formulation for the bivariate normal distribution with correlation ρ, to take one example. The case where $\rho = 0$ corresponds to independence, and we find that

$$\pi^* = 1 - [(1 - |\rho|)/(1 + |\rho|)]^{1/2}. \tag{6.13}$$

[Note the connection of logit(π^*) to Fisher's Z transformation.] Had the mobility table in Table 1 been viewed as a discrete realization of a bivariate normal distribution, for example, the value of $\pi^* = .314$ corresponds to a ρ value of .36. Note that we utilize the given two-class model without any regard whatsoever for philosophical notions about latent (categorical) variables. The mixture model is a robust method of parameterizing lack of fit, and it gives a convenient index that can be interpreted easily. The approach can be extended in many directions. Suppose we wish to index departures from independence in any setting where two or more distributions might be compared, such as occupational distributions for each sex (to measure so-called sex segregation). The above index can be readily used, and it avoids the pitfalls of using (a) residuals or other measures derived from the independence model, which usually does not fit the data, (b) measures of association that may or may not be appropriate for the problem at hand, or (c) models of nonindependence that require specifying parameters of interest. Of course, we are in favor of modeling data, so the strategy in (c) is always to be recommended, but it may be more convenient to index departures from independence (or some other model) by using the convenient mixture-based approach, given above. This use of latent class ideas is closely related to the semiparametric approach to be discussed later in connection with Rasch's model.

7.3 Rater Agreement

Suppose now that two or more raters (J raters in all) are assigned the task of rating n subjects. The subjects might be patients recovering from surgery, writing samples from students, and so forth. The raters are assumed to to be selected independently from some relevant universe of raters, and it is assumed that the ratings are done independently also (for example, rater 2 does not get to see the reports made by rater 1). There is also agreement that a certain number of rating categories (say, I) will be used by each rater; perhaps just $I = 2$ categories will be used, corresponding to pass or fail, acceptable or unacceptable, mastery or nonmastery (see Dayton and Macready 1980, and references cited there). The LCM model with $T = I$ is sensible, for reasons analogous to those given earlier. Taking account of the randomness in selecting raters is another matter, but this is an inferential issue, not something that alters the model as such.

If just two raters were used, the form of the data would be analogous to the two-way table on occupational mobility considered earlier. There is a nontrivial identification problem if just two raters are used, giving an observed two–way contingency table. Having more than two raters is useful to avoid identification problems in LCM analyses of inter-rater agreement. Conventional methodology for assessing reliability or consistency in a linear-model framework argues for this as well, so that random effects for inter-rater variability can be estimated. There are many ways to structure the LCM—different ways to impose restrictions—in this case; see Agresti (1988, 1990), Clogg (1979), Espeland and Handelman (1989) and, especially, Uebersax (1993) for details. Some, but not all, of the models covered in the latter source can be formulated in terms of LCMs.

Consider an example where J independent raters evaluate n subjects and must record $y = 0$ or 1 (e.g., fail or pass) for each subject. Each rater makes some error (probabilistic) in rating, but the raters act independently. A relatively simple model would have $T = 2$ latent classes. We would try to answer several questions, typically, with such data:

1. What proportion of subjects really passed, taking account of the measurement error?

In other words, what is the *latent distribution*?

2. Did the raters act in the same way (or were the raters *interchangeable*)?

3. Is there evidence for additional sources of systematic error structure?

4. How can we order or rank the raters? Can we order or rank (or scale) the subjects?

There are many other questions that could arise, especially if the possible responses were polytomous instead of dichotomous. Little work has been done on this more general problem except in the special case where the number of raters (or items) is equal to two or three. With polytomous responses, perhaps even ordered responses, the possibility of using scores for the Y categories, either fixed or random, or of using scores for latent classes (e.g., μ_t values) ought to be considered. Models given in Agresti (1988) could be extended to LCMs, for example.

The natural way to answer these questions with LCMs is as follows. A two-class latent structure is called for because the basic idea is that subjects either passed or failed. The axiom of local independence is appropriate because raters act independently (assuming no other complications). The answer to the first question is thus given by the $\pi_X(t)$, $t = 1, 2$ (or $t = 0, 1$). Note that $\pi_X(1)$, say, is a kind of average of the proportions $p_j(1) =$ Pr(rater j scores a randomly chosen subject at $y = 1$). Question 2 is formalized by imposing restrictions

$$\pi_{Y_j|X(t)}(0) = \pi_{Y_j'|X(t)}(0) ,$$

for different raters j and j', for each latent class t, and

$$\pi_{Y_j|X(t)}(1) = \pi_{Y_j'|X(t)}(1)$$

for different raters j and j', for each latent class t. These restrictions allow for different error rates for "fail" compared to "pass" but impose constant error rates across raters for each type of error. (Even more restrictive assumptions such as the same error rates for both types of errors could be imposed.) So the answer to the second question can be given in terms of the fit of this highly restricted model, perhaps via a likelihood-ratio comparison with one or more models without such stringent restrictions.

We have assumed that a two-class model is valid given the structure of the general rater agreement problem. Now suppose that the two-class model does not fit the data. What can be concluded? If $T > 2$ latent classes are required for an acceptable fit, the extra class or classes can be viewed as additional sources of heterogeneity that the original model does not capture. We might, for example, consider the model with restrictions above imposed for the first two latent classes, and then include a third latent class where no restrictions are applied. The additional latent class might correspond to cases where uncertain ratings arise, where raters have difficulty with the problem of rating particular individuals. This extra latent class is somewhat analogous to the latent "unscalable class" in scaling models (see below) [see Clogg and Goodman (1986), Clogg and Sawyer (1981), and Goodman (1975).] The additional heterogeneity would be modeled *nonparametrically* with one or more additional latent classes. So analysis and model respecification of this sort would answer question 3.

The fourth question above requires more work. First, if the highly restricted model defined above holds, then the raters do not differ (i.e., are exchangeable). In the context of the two-class model, failure of this set of restrictions implies systematic differences among raters. We can choose a variety of methods to measure or rank the raters or items in this case. Perhaps the most natural method is to use the odds-ratio or its logarithm for the association at the latent level between X (the true distribution) and a given rater, here called Y_j. We could thus use the quantities,

$$\theta_j = (\pi_{Y_j|X(1)}(1)\pi_{Y_j|X(2)}(2))/(\pi_{Y_j|X(1)}(2)\pi_{Y_j|X(2)}(1)), j = 1, ..., J,$$

or functions of these, such as Yule's Q. There are of course other ways to rank or scale the raters; note that under the baseline restrictions given above these θ_j values would all be the same. To rank the subjects without imposing parametric assumptions of any kind, the following strategy can be suggested. Take an individual with response pattern y, where y corresponds to a pattern like $(1, 2, 2, 2, 1, 1)$, say, and calculate

$$\phi_y = \log(\pi_{X|Y(y)}(1)/\pi_{X|Y(y)}(2)) . \tag{6.14}$$

That is, use the logit of the probability that $X = 1$. The probabilities themselves could also be used. Ranking subjects is more difficult if more than two latent classes are posited. It might be possible to gain something by assuming a *distribution* for the probabilities or for functions of them, but to our knowledge this has not been explored systematically for the general case.

7.4 Latent Class Models for Missing Categories

A very important use of LCMs arises in modeling contingency-table data where some of the respondents do not give standard or expected responses for one or more variables. See Little and Rubin (1987) for examples of contingency tables of this type, where log-linear models with supplemental margins are analyzed using conventional assumptions. (These assumptions are missing at random (MAR) or missing completely at random (MCAR).) Winship and Mare (1989) show how such cases can be modeled with LCMs, and Hagenaars (1990) shows how the analysis can proceed with a slightly modified version of the original model (using supplemental tables); for some related material, see Clogg (1984). The approach appears to have great utility in modeling selection and related factors in social research. For example, suppose that we examine degree of marital homogamy with a sample including married couples (only their choices of marriage partners are available) but also including unmarried persons (those whose choices have not been made). There are several ways to consider missing category codes in data that might result from such an analysis. The basic idea is to create a contingency table at the latent level, one corresponding to the analogue of our table cross-classifying Y with X, where now X or at least components of it are latent (or "missing"). A simple illustration is the ubiquitous Don't Know or Not Ascertained code in surveys. Suppose the observed variable Y is something like arrested versus not arrested. We want to know the distribution of Y for those whose arrest status was not ascertained. The table at the latent level incorporates this information. Winship and Mare give methods based on the log-linear formulation (many of their results could be obtained with restricted versions of the classical formulation) and use Haberman's (1988) NEWTON program for calculations.

8 Exploratory Latent Class Analysis and Clustering

The earlier example on the American occupational mobility (Table 1) illustrated identification problems in the context of a seemingly simple example where LCMs might be used. In many cases where LCMs might be utilized, however, we need not be so concerned with the identification problem, and at any rate some programs (such as MLLSA) contain checks for local identifiability that protect the user from problems of this type. The identifiability checks currently available depend on the algorithm used for estimation. If a method like scoring (LAT) or Newton-Raphson (NEWTON) is used, identification problems arise as rank problems in inverting the Hessian matrix (or observed or expected information matrix) at successive cycles in the procedure. It is possible that ill-conditioned problems have different numerical outcomes for these two methods, and a problem is that one does not know whether the apparent divergence in the algorithm is due to rank problems for the particular dataset involved or to *intrinsic* underidentification. An example of intrinsic underidentification is the two-class LCM applied to the two-way contingency table: this model is always underidentified (by two constraints). If an EM algorithm or related algorithm is used, typically the information matrix or Hessian is not calculated. Note that if the model is underidentified, the information matrix or the Hessian will not be of full rank and so cannot be inverted in the usual way. The principle attraction of the EM algorithm is that matrices of second derivatives are not required. The EM algorithm will give a solution—both parameter values and expected frequencies or probabilities—whether the model is identified or not. It is useful now to recognize that the algorithms find parameter values that maximize the log-likelihood, and LCMs and other mixture models often have log-likelihoods that are "flat" in one or more directions, which is one way to view the identification problem in samples.

Goodman (1974a, b) recommended using a Jacobian-like matrix to examine the shape of the maximized log-likelihood. By results on the calculus of many variables, a function is unique (the log-likelihood has a unique maximum) when the Jacobian of the transformation has full rank. This matrix can be defined as follows. Let

$$\boldsymbol{P} = p_Y(y), \quad y = 1, \ldots, d \,,$$

where here only we adopt the convention that y corresponds to one of the d possible response patterns. $\boldsymbol{P}$ is a column vector, and since the elements of this vector sum to unity, we can exclude the last value, giving a vector of length $d - 1$. (The contingency table is of size d.) Next define the vector of *nonredundant* parameter values, excluding parameters defined as complements of other parameters, for example, as

$$\boldsymbol{\Pi} = [\pi_X(1), \ldots, \pi_X(T-1), \pi_{Y_1}(y), \ldots, \pi_{Y_J}(y)]' \quad .$$

If it is thought that there are q nonredundant elements in $\boldsymbol{\Pi}$, then this column vector has length q. That is, we wish to check whether the number of degrees of freedom can be obtained by simple counting [$df = (d - 1) - q$]. Next form the matrix

$$M = \frac{\partial \boldsymbol{P}}{\partial \boldsymbol{\Pi}}$$

a matrix of order $(d - 1) \times q$. This matrix will be of full column rank ($= q$) if the parameters are identifiable. In practice, we evaluate the matrix M and its rank at the solution obtained. This is done in MLLSA and in some other programs. Tolerance for evaluating the rank must be handled carefully. With checks like the above and other signals about convergence rates, boundary values, and so on, we can diagnose the identifiability of the model easily given fast modern computation. Note that each of these procedures is equivalent to examining the curvature of the log-likelihood function at the maximum or in the approach to the maximum.

In usual cases we merely proceed by finding the "best" LCM for a given set of data, and if we have no prior information we should begin analysis with unrestricted models, that is, models where no restrictions, such as equality restrictions, are imposed on the model parameters. In such cases, the problem will be to find the "best" model using criteria of goodness-of-fit or related criteria. That is, we try to find the simplest model that fits the data to an acceptable degree, or the smallest number T of latent classes that is consistent with the data.

A recent study by Hogan, Eggebeen, and Clogg (1993) provides a realistic example that illustrates most of the issues involved. Four items pertaining to *giving* support (monetary or material, care, work around the household, companionship—call these Y_1, Y_2, Y_3, Y_4) and four analogous items pertaining to *receiving* support (of the same kinds—call these Y_5–Y_8) were asked of respondents in the National Survey of Families and Households. The sample selected for analysis pertained to $n = 4711$ of those surveyed who had at least one surviving parent, so the giving or exchanging pertains to giving to parents or receiving from parents. The number of items is thus $J = 8$, and the items were scored as $y = 0$ (exchange absent) or $y = 1$ (exchange present); this gives a 2^8 contingency table with 256 cells. Unrestricted LCMs with $T = 1, 2, 3, 4, 5$ were considered with the MLLSA program, and their fit statistics are given in Table 3.

Fit statistics given in this table are just a few of many that could have been selected, but we think these are among the most promising ones that can be recommended for general use. The likelihood-ratio statistic (L^2) seems to indicate that either the model with four latent classes or the model with five latent classes is satisfactory. The P value for the former is .068, and for the latter it is essentially .5. [Dividing L^2 by the corresponding df, giving an F statistic (Haberman 1979), leads to reasonable values for both of the latter two models.] D values indicate that not much is gained in fit by increasing the number of latent classes from four to five (4.8% in the former versus 4.2% in the latter). The BIC values, on the other hand, give a clear preference for the model with four latent classes. That is, in terms of the penalty imposed by this popular fit index, adding additional parameters does lead to enough of a decrease in the chi-squared value. In this case, most of the evidence thus points to the four-class model as an acceptable summary of the data. This conclusion was corroborated by the prediction indexes in Clogg and Sawyer (1981); the percent of subjects

Table 3. Fit Statistics for Unrestricted LCMs Applied to Data on Intergenerational Exchanges

$T =$	df	L^2	D	BIC
1	247	4145.5	.366	2056.5
2	238	728.4	.115	−1284.5
3	231	538.8	.095	−1414.9
4	222	253.5	.048	−1624.1
5	213	213.6	.042	−1587.9

NOTE: D is the index of dissimilarity between observed and estimated expected frequencies; BIC is the Bayesian Information Criterion; the model with $T = 1$ is the model of mutual independence.

correctly allocated in the latent classes was 88.2% for the four-class model and 82.9% for the five-class model, while the lambda measure of prediction was .762 for the former model, .655 for the latter model. Thus, predictability is more certain with the four–class model. In addition, the model with four latent classes appeared to be more interpretable than the model with five latent classes. Considerations of fit, parsimony, predictability, and interpretability all favor the four-class model in this particular case. See Hogan, Eggebeen, and Clogg (1993) for further details.

This example illustrates some potential problems with so-called boundary solutions that require comment. Proceeding from a model with T latent classes to one with $T + 1$ latent classes formally adds $J + 1$ parameters, in the case where each Y_j is dichotomous. (That is, we add one latent class proportion, $\pi_X(T + 1)$, say, and J conditional probabilities, $\pi_{Y_j|X(T+1)}(1)$, for $j = 1, \ldots, J$.) This is reflected, for example, in the *df* values for the independence model (247) compared to the two-class model ($df = 247 - (8 + 1) = 238$). Two boundary solutions were obtained for the three-class model, however, which accounts for the discrepancy in the comparison of that model's df value with the df value of the two-class model. The same boundary values were obtained for the other models. As in Goodman (1974a, b), we adopt the convention of adjusting the degrees of freedom for the number of boundary values actually estimated. To estimate these models, we start with different sets of initial values and check whether they all lead to the same solution (with boundary solutions in the more complicated models). This checking should always be done, and with fast computing now available it is easy to do this. To save space, many of the details of the analysis, such as parameter values and predictions of latent class membership from the observed y patterns, will not be given. However, we illustrate one potential pitfall in drawing inferences from comparisons of likelihood ratio statistics.

Suppose one wished to test the four-class model against the five-class model. The likelihood-ratio comparison looks straightforward because the two models are indeed nested (the four-class model is a special case of the five-class model). The ordinary comparison would lead to a difference of 39.9 ($= 253.5 - 213.6$), with an apparent df of 9($= 222 - 213$). But this comparison, while certainly a likelihood-ratio comparison formed in the usual way, does not lead to the standard chi-squared distribution under the null. The problem is that the "full" model is obtained from the "reduced" model in this case by setting one parameter at the boundary value. That is, the four-class model is obtained from the five-class model by constraining one latent class proportion in the five–class model to equal zero. Moreover, the parameters for the conditional probabilities in the void class are not properly defined in the "reduced" model. That is, these conditional probabilities could take on any value, not just zero, in the reduced model. Likelihood-ratio comparisons of nested models simply do not lead to *chi-squared* statistics in cases where the reduced model is obtained from the full model by placing parameters at their boundary values or when parameters are not defined appropriately. Solving this problem has proved difficult, and this is one reason why we are forced to rely on fit indexes like BIC and others to make inferences about models with different numbers of latent classes. The difference in likelihood-ratio statistics surely appears to be significant (the statistic probably behaves like a chi-squared statistic on somewhat fewer df), but we should not rely on the usual comparisons to make precise statements. On the other hand, likelihood–ratio comparisons involving models with the same number of latent classes are valid, asymptotically, in the analysis of LCMs, except in the special case where boundary solutions arise in one or more of the models being compared. The four latent classes were judged to correspond to "low exchangers" (latent class one, with estimated proportion .504), "advice givers" (latent class two, with proportion .155), "high exchangers" (latent class three, with proportion .126), and "receivers" (latent class four, with proportion .215). Note that the labeling of latent classes derives from the patterns observed in the conditional probabilities (not reported), and it is only claimed that the latent classes are nominal (not ordinal); the latent classes correspond to "clusters" in the data. Over half of the population, according to this definition in terms of LCMs, does not engage in intergenerational exchange as measured by the eight items involved. As in most applications of LCMs in social research, the model and the analysis that it provides can be used to advantage in creating a typology, or classification, for use as an explanatory variable or as a predictor in other analyses. Indeed, LCMs were used for this purpose in the source. It is natural therefore to turn to the problem of prediction.

9 Predicting Membership in Latent Classes

A very important practical consideration is predicting latent class membership using a given LCM (whether a restricted or an unrestricted version of the model) from the observed response patterns. For grouped data, a response y corresponds to a cell in the relevant contingency table; for ungrouped data, y is merely the response pattern for an individual. We now wish to distinguish sharply between the relevant conditional probability (or distribution) given in (6.6), namely, $\pi_{X|Y(y)}(t)$, for $t = 1, \ldots, T$ (and given y), and the sample estimate of this quantity or of this conditional distribution. We first review the standard method of prediction that is used, to the best of our knowledge, in all software routines that generate predictions from LCMs.

For each y, find the value of t, say t^*, that corresponds to

$$t_y^* = \max_t(\pi_{X|Y(y)}(t)) \,. \tag{6.15}$$

Next, assign the individual (or all respondents) with pattern y to t_y^*, using a modal prediction rule. This, or any sensible prediction rule, produces a classification of subjects into T groups corresponding to the latent classes, and reduces the data dramatically. (In the example in the previous section, the original cross-classification having $d = 256$ cells was reduced to one having $T = 4$ cells, for an effective reduction in the dimension of the data of 98%.) The quality of the prediction can be assessed in a variety of ways. For example, we can compare the predictions to the errors incurred; the error rate for pattern y is

$$\epsilon_y = 1 - \pi_{X|Y=y}(t_y^*) \tag{6.16}$$

under this rule of modal assignment. The cumulated number or expected number of errors, $\sum_y \epsilon_y F_y$, is a convenient index of predictability; this can be expressed in percentage terms by dividing by n, the sample size. If the model is true, it seems reasonable to use the expected frequency F_y in the above; however, we might use the observed frequency instead. (If we were dealing with ungrouped data, the observed frequencies would be zeroes and ones, which might argue for using the expected frequencies.) The modal assignment rule is consistent with the Goodman-Kruskal lambda measure of association, with the latent class distribution $\pi_X(t)$ representing the marginal distribution that would be used to predict X if Y were not known (hypothetically at least), and the error rates of (6.16) define the conditional prediction errors. Note that if we used proportional prediction rather than modal prediction, we would be led to the Goodman-Kruskal tau (asymmetric version) to measure prediction errors.

These deterministic rules assume that the parameters are known, whereas in fact they are estimated. A simple expedient is to replace the population parameters by their maximum likelihood esimates and then to proceed as in (6.15) or (6.16). Doing this creates a major difficulty in that all of the uncertainty in making the prediction of X is not taken into account. The above procedures using sample analogues of the prediction rules that would be used if the parameters were known can be found in Lazarsfeld and Henry (1968), Goodman (1974b), and Clogg (1979) or Clogg and Sawyer (1981). See the latter source for the indexes described here. These methods do not take account of the uncertainty in prediction owing to two primary facts. First, the model does not usually give zero error rates for prediction, at least for most cells or y patterns. So the predicted X should be regarded as the true X contaminated with measurement error; the error rates given above define the structure of the measurement error. Second, sample estimates are used in practical work, and there is uncertainty because of sampling error. We suggest an approach based on the multiple-imputation methodology associated with Rubin (1987).

A halfway implementation of the multiple-imputation method would be to generate M predictions for each subject (not just for each response pattern). Array the relevant conditional probabilities in cumulative form

$$(\pi_{X|Y}(1), \pi_{X|Y}(1) + \pi_{X|Y}(2), \ldots, 1) .$$

Call this these points $G_y(1), \ldots, G_y(T)$, with $G_y(T) = 1$. Then draw a random number from the uniform distribution on [0, 1], say u. If $u \leq G_y(1)$, assign the subject to latent class 1; if $G_y(1) < u \leq G_y(2)$ assign the subject to latent class 2; and so forth. Repeat the process M times for each subject, drawing a new random number independently each time; $M = 5$ ought to suffice based on use of this method elsewhere (see Rubin 1987). In effect, M different or potentially different predictions, say X_m for $m = 1, \ldots, M$, are used to represent the uncertainty in predicting X. If the predicted latent classification is used either as a predictor or as a dependent variable in subsequent analyses, then the multiple imputations are used in the standard way. That is, do the analysis with each of the M predictions of X, and assess both between and within variability for standard errors or tests of significance. For somewhat related results, see Clogg et al. (1991). This approach takes account of the uncertainty in assigning respondents to latent classes of the first kind mentioned above. A full multiple-imputation procedure taking account of sampling variability as well would specify a distribution for the cumulative probabilities. For example, we might take the logistic transform of the relevant conditional probabilities, assume normality of those quantities, and then draw from a normal posterior density to generate the M imputations required. This approach is satisfactory in principle; however, the sampling distribution may be far from normal even after employing the logistic transform, particularly if T is relatively large. An additional complication is that existing programs do not give the variance–covariance matrix associated with $\hat{\pi}_{X|Y}(t)$, although this could be derived from code produced in LAT or NEWTON. A simple expedient worth trying is to assume that the logit–transformed probability estimates behave as ordinary logits. In ordinary situations, $\log(\hat{p}/(1 - \hat{p}))$ is approximately normal with variance proportional to $(p(1 - p))^{-1}$. This approach would ignore the smoothing provided by the model and would give estimates of variance that are too large. More research is required to take into account the uncertainty in prediction of latent class membership. This is a pressing matter because in practical work the predicted X is often the data reduction sought for subsequent analyses. It is important to reflect the uncertainty in the prediction in modeling routines applied to predicted latent classes.

We note that the above description of prediction applies to any type of LCM, and often one of the goals of formulating restricted models is to structure the latent class prediction. A simple example based on the two-class quasi-latent structure applied to the American occupational mobility table illustrates the conventional way of predicting. The assignment into latent classes along with error rates for this model applied to the data in Table 1 appears in Table 4. Notes to the table describe the numbering of the latent classes. The assignment gives added meaning to the latent class variable; for example, latent class one is obviously an upper latent class and latent class two is obviously a lower latent class, because the two latent classes cover the upper and lower parts of the table, respectively. The model (with this prediction rule) correctly allocates 85.5% of the subjects into latent classes, and the lambda measure of prediction of X from the observed y values (i.e., from the cells) is .674. These indexes indicate that the prediction is satisfactory, but of course the magnitudes have to be interpreted in a comparative way, with other data sets or other models in mind for comparison. See Clogg (1981) for more details and see van der Heijden, Mooijaart, and de Leeuw (1992) for related results.

Table 4. Occupational Mobility of American Men in 1963
Predicted Latent Classes from Model H_2' of Table 2

	Subject's Occupational Status				
Father's Occupational Status	1	2	3	4	5
1	1	1	1	1	1
$\hat{\epsilon}$	.34	.00	.00	.00	.00
2	1	1	1	1	1
$\hat{\epsilon}$	.00	.13	.01	.02	.01
3	1	1	5	2	2
$\hat{\epsilon}$	.25	.41	.60	.24	.42
4	2	2	2	2	2
$\hat{\epsilon}$	.37	.22	.09	.12	.13
5	2	2	2	2	7
$\hat{\epsilon}$	.00	.00	.00	.00	.07

NOTE: Latent classes 1 and 2 correspond to the latent upper and latent lower classes, respectively; in these classes mobility is random. Latent classes 3–7 correspond to latent stayers in each of the respective diagonal cells. For example, the (5, 5) cell corresponding to Farm origin and Farm destination is predicted to consist of latent stayers in this category; and only 7% of subjects in this cell derive from the other latent classes. Refer to Table 1 and notes there.

10 Latent Class Models in Multiple Groups: Categorical Covariates in Latent Class Analysis

We now turn to the important problem of comparing latent structures via LCMs in different groups, assumed to be independently sampled. The groups might correspond to gender groups, to different time periods, to different countries, and so on. A simple method is to merely estimate separate models for each group and then make informal comparisons based on the parameter values, latent class predictions, or other aspects of the models. An altogether different approach called *simultaneous latent structure analysis* was put forth in Clogg and Goodman (1984, 1985, 1986), and several of the programs mentioned in the second section of this paper perform analyses of this general kind, most using the Clogg–Goodman formulation.

We suppose that the same indicators $Y_1, \ldots, Y_J$ are available in each of G groups, with g denoting a given group, $g = 1, \ldots, G$. The groups might correspond to subpopulations (nations, males and females, and so on). Or they might correspond to time periods, assuming that we have independent samples at each time period. Sticking to our convention of labelling parameters, there are now two key sets of parameters for comparing latent structures across groups. The first set pertains to the latent distribution, namely

$$\pi_{X|G(g)}(t) = Pr(X = t|G = g). \tag{6.17}$$

The second set pertains to the conditional probabilities, now conditioned as well on membership in a given group, namely

$$\pi_{Y_j|X(t),G(g)}(y_j) = Pr(Y_j = y_j|X = t, G = g). \tag{6.18}$$

In words, the first set gives the latent distribution (of X) in the g-th group, and the second set measures the association between X and the indicators group by group. Applying the LCM separately to each group merely estimates the parameters given in (6.17) and (6.18). Note that to be valid for across-group comparisons, we would normally require the same number of latent classes in each group, the same Y indicators (comparability of measures might be difficult to enforce in practice), and, if restricted LCMs are used, the same general kind of restrictions ought to be enforced in each group, or otherwise the meaning of the latent variable might be different across groups. (To see the latter point, try to imagine how one might compare a two-class structure in one group with a four-class structure in another group; this is possible but not recommended in general.)

In fact, the parameter definition given above for introducing a group variable (or covariate) leads to a model built along the lines used in (6.1)–(6.3), which defined the ordinary LCM for a single group. The relevant quantities to be modeled here are the *conditional* probabilities,

$$\pi_{Y|G(g)}(y) = Pr(Y_1 = y_1, \ldots, Y_J = y_J|G = g) .$$

Recall that y stands for an entire response pattern and Y for the entire set of J indicators. The simultaneous or multiple-group LCM can now be written as

$$\pi_{Y|G(g)}(y) = \sum_{t=1}^{T} \pi_{X|G=g}(t) \prod_{j=1}^{J} \pi_{Y_j|X=t,G=g}(y_j). \tag{6.19}$$

Clogg and Goodman show how this model can be written as a single-group latent structure by constructing a new latent variable $W = X \times G$. W is now a partially latent variable with $T \times G$ latent classes. (This method of crossing a latent variable with an observed variable has many uses.) The simultaneous latent structure model can be written as a deterministically restricted single-group LCM as

$$\pi_{Y,G}(y,g) = \sum_{s=1}^{S} \left\{ \pi_W(s) \left[\prod_{j=1}^{J} \pi_{Y_j|W=s}(y_j) \right] \pi_{G|W=s}(g) \right\}, \tag{6.20}$$

where $S = TG$ and where the other quantities are defined as in the single-group model. To complete the translation, the additional set of conditional probabilities (variable G is treated ostensibly as another indicator above) are restricted as

$$\pi_{G=g|W=s}(1) = 1; s = 1, \ldots, T;$$

$$\pi_{G=g|W=s}(2) = 1; s = T+1, \ldots, 2T;$$

$$\vdots$$

$$\pi_{G=g|W=s}(G) = 1; s = (G-1)T+1, \ldots, S(=TG).$$

With these restrictions imposed, which essentially define the first T classes of W as the latent classes for the first group, the next T classes of W as the latent classes for the second group, and so on, only a rescaling of the latent class proportions is required to retrieve the parameters of the simultaneous LCM. This means that restricted multiple–group LCMs can be developed in the same way, and using the same algorithms (taking account of the rescaling of the latent distribution) that would be used for the single–group case. See, for example, Clogg and Goodman (1984) for details.

Simultaneous LCMs provide new opportunities for comparisons among groups, and there is still much methodological work to be done. The types of questions that can be asked can be illustrated with examples of the types of across–group constraints that can be considered. Suppose we wish to test whether item Y_1 is an equally reliable indicator of X in groups one and two (such as males and females). Then we impose the constraint

$$\pi_{Y_1|X,G(1)}(y_1) = \pi_{Y_1|X,G(2)}(y_1)\ ,$$

for some or all values of $X = t$ and/or some or all values of y_1. Comparisons of such constrained models with the corresponding models without those contraints test the differential performance of the item in different groups. Variations of this idea could be used, for example, to test for so-called differentialitem functioning in educational testing (see Hambleton, Swaminathan, and Rogers, 1991). Next, suppose we want to test whether the latent distribution is the same or different between groups one and two, say. Then we impose the constraint

$$\pi_{X|G(1)}(t) = \pi_{X|G(2)}(t), \quad t = 1, \ldots, T, \tag{6.21}$$

or possibly just subsets of this general condition. It is an open question whether one should consider such homogeneity models if there is not also homogeneity in the conditional probabilities $\pi_{Y_j|X}(y_j)$. That is, do comparisons of latent structure make sense if we are not assured that the items measure X "in the same way" in the groups compared? This is a complicated issue, and analogous problems arise in multiple-group versions of covariance structure or factor analysis models. It is possible to estimate models with such constraints and to perform routine statistical tests based on likelihood-ratio comparisons. A very important point is that comparisons of latent structure, such as comparisons of latent distributions, are nonparametric; no assumptions on distributions or on the sufficiency of means and variances are required.

We note in passing that the formulation above depends heavily on the classical definition of parameters in LCMs. If Formann's logit-transformed version of the model parameters is used, the strategy can be employed with little modification. However, if Haberman's log-linear representation is used, with parameters like λ^{XG} denoting the group-by-X interaction, then it is much more difficult to derive constrained models or to perform tests of the above sort. The reason is that the X–G interaction in a log-linear representation is not a direct function of the $\pi_{X|G}(t)$ quantities. In general, when a log-linear model contains higher-order relatives of particular terms included in lower-order terms, the parameters for the lower-order terms do not refer any longer to *marginal* relationships, such as differences in marginal distributions in X across levels of G, and this makes the log-linear representation difficult to use for analysis of multiple-group LCMs. See Haberman (1979) and compare results there with Clogg and Goodman (1984, 1985, 1986), and see also Hagenaars (1990, 1993), and van der Heijden et al. (1992). The latter sources provide general strategies for including categorical covariates in path models based on the latent class model.

The approach just outlined is important from a methodological point of view because it shows how to formulate LCMs with qualitative or categorical covariates. We would always condition on the levels of variables regarded as covariates, which has been done above implicitly, and this method gives a general approach for analyzing qualitative covariates. At the same time, however, we do not have technology for dealing with many different qualitative covariates. Further work on the technology, including procedures for handling continuous covariates, ought to be given high priority. By analogy to the above, we ought to be able to handle a continuous covariate Z by allowing the logistic-transformed distribution for X or the logistic-transformed conditional distibutions of $Y_j|X$ to depend linearly on the levels of Z, for example. A simple case with dichotomous X would model the logits for the X distribution as $\log(\pi_X(1)/\pi_X(2)) = \alpha + \beta Z$. See Dayton and Macready (1988) for one attractive method of handling continuous covariates; in my opinion, the general LCM is quite intractable numerically when formulated for continuous covariates, and I think that the most progress will be made by settling on various restricted models where the relationship between indicators and X is constrained to some degree. It is also important to note that problems in numerical stability and assessing fit with continuous covariates have to be reckoned with.

11 Scaling, Measurement, and Scaling Models as Latent Class Models

In this section we briefly review how LCMs can or might be used to measure or scale. My philosophy of how this ought to be done appears in Clogg (1988) and in Lindsay et al. (1991). To fix concepts, we assume that scaling or measurement pertains to those situations where X is at least an ordinal-level classification. Ideally, of course, measurement refers to cases where X pertains to a quantitative classification.

11.1 Ordinal X

All examples treated earlier pertained to a nominal-level X or perhaps to an X that had weak ordering properties. For dichotomous Y_j's and $T > 2$, it will often be possible to order the latent classes a posteriori by using the pattern of conditional probabilities as a guide. Perhaps the simplest of methods is to suppose that y codes correspond to $y = 0$ for "low" and $y = 1$ for "high". Examine the prediction probabilities, $\pi_{Y_j|X(t)}(1)$ *as a set*, i.e., for $j = 1, \ldots, J$ and $t = 1, \ldots, T$. Pick a weight function, say $w_j, j = 1, \ldots, J$. The weights should reflect "importance of the item." Next calculate the sums,

$$Q_t = \sum_{j=1}^{T} w_j \pi_{Y_j|X(t)}(1), t = 1, \ldots, T . \tag{6.22}$$

Finally, rank order the latent classes according to the ordering of the sums. One could use instead the conditional probabilities, $\pi_{X|Y_j}(t)$, or functions of either set of quantities, such as logits. We could use weights such as $w_j = 1$ or some other set reflecting prior judgments on the importance of individual items.

For ordinal indicators, Clogg (1979) proposed the following model with ordinal X. Suppose each Y_j has just three levels for simplicity, so that the data comprise a 3^J contingency table, and suppose further that the levels of each Y_j correspond to "low", "medium", and "high", respectively. Now suppose that a T-class LCM is posited. The restrictions that define this model with ordinal latent classes are simply

$$\pi_{Y_j|X(3)}(1) = 0; \pi_{Y_j|X(1)}(3) = 0.$$

That is, the first latent class cannot contain individuals who respond at level $y = 3$ (high), and a response of $y = 2$ is regarded as a measurement error for members of this latent class. The third latent class cannot contain individuals who respond at level $y = 1$ (low); a response of $y = 2$ is regarded as a measurement error for members of this latent class.

Finally, in the second (middle) latent class, response errors in either direction are permitted. It will be appreciated that the first method is somewhat arbitrary: in a sample the ordering of latent classes will depend on chance fluctuations, and the ordering is not structured into the model a priori. The second method has to be generalized carefully when more than three possible (ordered) responses are available. There are many other ways to generate orderings of latent classes, but in my experience there is little practical payoff in doing so, at least with unrestricted LCMs. Researchers can in most instances determine orderings that are relevant by inspection of the parameter values.

An important insight from LCMs with ordinal X levels is as follows. The latent class proportions can be ordered, from low to high according to the ordering of X values, and the result can be used to fit an arbitrary distribution to the latent distribution estimated nonparametrically. Without loss of generality, suppose that the ordering corresponds to $t = 1, \ldots, T$. Then form the cumulative distribution function points as

$$\boldsymbol{\Pi}_X(1) = \pi_X(1);\ \boldsymbol{\Pi}_X(2) = \pi_X(1) + \pi_X(2);\ \ldots;\ \boldsymbol{\Pi}_X(T) = 1\ .$$

Sample estimates can be used and then any arbitrary distribution can be "fitted" to the latent distribution [for examples, see Clogg (1988)]. While this approach is not technically fancy, it brings the nonparametric spirit of LCMs together with a scaling idea, where the distributional assumption is imposed only after the modeling process is complete. The threshold values assuming a normal distribution with mean zero and variance unity, for example, are obtained as

$$t_1^* = \Phi^{-1}(\Pi_X(1));\ t_2^* = \Phi^{-1}(\Pi_X(2));\ \ldots\ .$$

Mean values within categories can be calculated also; these category scores or mean values can be used along with the latent class predictions in subsequent analyses.

Croon's model for ordinal latent classes (Croon 1990) can be viewed as a generalization of the above model. Croon's model appears to be the most general of those now available, in the sense that other assumptions that locate the latent classes on a metric are not assumed in developing the model or the inferences from it. There appear to be many instances where an ordinal X (but not a scaled X) ought to be assumed. For example, when the data pertain to preference *orderings*, it is natural to regard X as the latent preference ordering. Croon and Luijkx (1992), Boeckenholt and Boeckenholt (1991), and Boeckenholt (1992) develop special LCMs for this setting where only ordinal X is assumed.

11.2 Classical Scaling Models

An important class of restricted LCMs is the class of so-called scaling models. Lazarsfeld and Henry (1968) surveyed some of these, including the "latent distance" model; Goodman (1975) examined some of these models and proposed a new model [see also Clogg and Sawyer (1981)]. Formann (1988) considers an important class of models for "non-monotone" items, where conventional scaling ideas might not apply. Many of the standard models along with Goodman's modification of Guttman's model are considered in the multiple-group context as simultaneous, restricted LCMs in Clogg and Goodman (1986). Dayton and Macready (1980) show how Goodman's perspective can be combined with that of classical scaling models; they refer to a wide literature to which they have been the main contributors on "intrusion–omission" error models for the analysis of learning. Evidently many models of learning, where the Y_j's refer to "tasks" and responses refer to levels of "mastery", can be formulated in ways analogous to scaling models.

Classical scaling models appear to derive from Guttman's deterministic model of scaling. Guttman's model (or method), as well as the scaling models that derive from it, require (or search for) an *item ordering*. For dichotomous Y_j's, inspection of item marginals is often satisfactory, at least as an initial ordering. Suppose that there are $J = 4$ items with responses $y_j = 1, 2$, where 1 denotes "easy" and 2 denotes "hard". (If the items correspond to a test, then the response $y_j = 2$ corresponds to correct.) *If the items are ordered correctly* in terms of difficulty, then Guttman's model of perfect scaling would predict that only the following response patterns, called scale-type response patterns, would occur:

$$\begin{aligned} S_0 &= (1,1,1,1), S_1 = (2,1,1,1), S_2 = (2,2,1,1), \\ S_3 &= (2,2,2,1), S_4 = (2,2,2,2). \end{aligned} \tag{6.23}$$

That is, the first item is the easiest, the last item is the hardest (and is only completed correctly by those who correctly respond to the other items), and so forth. Response errors of one kind or another are assumed to produce departures from the expected Guttman pattern. Note that there are $J + 1$ scale-type response patterns with J dichotomous items; the response pattern S_t is such that t "correct" responses, in the "correct" order, are obtained. There are two main ways to model data where the Guttman pattern is expected but where response errors are modeled in a probabilistic way. The first may be called the response-error formulation.

Under the response-error formulation, suppose that there are $T = J + 1$ latent classes. Each latent class corresponds to one and only one scale type; label the latent classes as $t = 0, \ldots, T$, with $T = J$ with the redefinition of levels. The simplest scaling model (Proctor 1970) is defined with the following constraints applied to a T $(= J + 1)$ class LCM:

$$\pi_{Y_1|X(0)}(1) = \pi_{Y_2|X(0)}(1) = \pi_{Y_3|X(0)}(1) = \pi_{Y_4|X(0)}(1) =$$

$$\pi_{Y_1|X(1)}(2) = \pi_{Y_2|X(1)}(1) = \pi_{Y_3|X(1)}(1) = \pi_{Y_4|X(1)}(1) =$$

$$\pi_{Y_1|X(2)}(2) = \pi_{Y_2|X(2)}(2) = \pi_{Y_3|X(2)}(1) = \pi_{Y_4|X(2)}(1) =$$

$$\pi_{Y_1|X(3)}(2) = \pi_{Y_2|X(3)}(2) = \pi_{Y_3|X(3)}(2) = \pi_{Y_4|X(3)}(1) =$$

$$\pi_{Y_1|X(4)}(2) = \pi_{Y_2|X(4)}(2) = \pi_{Y_3|X(4)}(2) = \pi_{Y_4|X(4)}(2) .$$

Read these equations as follows: the first row pertains to the expected response pattern for persons who are in the first latent scale type, the second row pertains to the expected response pattern for persons who are in the second latent scale type, and so on. Because these conditional probabilities are equated, call their common value $1 - \alpha$, where α is the item error rate, assumed constant across latent types of individuals. Guttman's deterministic model says that $\alpha = 0$. Simple counting that takes account of the complements of probabilities reveals that there is just one parameter for the whole set of conditional probabilities. (Note that we cannot use likelihood-ratio tests or equivalent procedures such as $z = \hat{\alpha}/s(\hat{\alpha})$ to test whether $\alpha = 0$, because such a test pertains to a boundary value of the parameter.) The number of parameters is thus $J + 1$, i.e., J parameters for the latent distribution and one parameter total for the $\pi_{Y|X}(y)$.

Define α_{tj} as the error rate for the t-th latent class (or scale type) and the j-th item. We can define a variety of scaling models by picking restrictions on these that are in between the case of complete equality (Proctor's model) and the completely unrestricted $(J + 1)$–class LCM. For details, see Clogg and Goodman (1986) and references cited there.

Goodman (1975) considered a different formulation including an "unscalable class." This model posits $T = J + 1$ latent classes corresponding to the scale types in (6.23), but also one additional latent class, say class $T + 1$. In the extra latent class, no restrictions are imposed, so in this class responses are random, given the local independence assumption. For the four-item example, the conditional probabilities associated with the latent classes 0–4 in Goodman's model are restricted deterministically as follows:

$$\pi_{Y_1|X(0)}(1) = \pi_{Y_2|X(0)}(1) = \pi_{Y_3|X(0)}(1) = \pi_{Y_4|X(0)}(1) =$$

$$\pi_{Y_1|X(1)}(2) = \pi_{Y_2|X(1)}(1) = \pi_{Y_3|X(1)}(1) = \pi_{Y_4|X(1)}(1) =$$

$$\pi_{Y_1|X(2)}(2) = \pi_{Y_2|X(2)}(2) = \pi_{Y_3|X(2)}(1) = \pi_{Y_4|X(2)}(1) =$$

$$\pi_{Y_1|X(3)}(2) = \pi_{Y_2|X(3)}(2) = \pi_{Y_3|X(3)}(2) = \pi_{Y_4|X(3)}(1) =$$

$$\pi_{Y_1|X(4)}(2) = \pi_{Y_2|X(4)}(2) = \pi_{Y_3|X(4)}(2) = \pi_{Y_4|X(4)}(2) = 1.$$

With this model, there are $J + 1$ nonredundant latent class proportions and J nonredundant conditional probabilities (corresponding to the last latent class of unscalables), for a total of $2J + 1$ parameters. An unscalable class can be added to Proctor's model (Dayton and Macready 1980), or, for that matter, to any of the classical scaling models. But note that adding an unscalable latent class to take account of heterogeneity not modeled by the formulation of latent scale types and response errors creates a latent variable that is only partially ordered. In contrast, the classical scaling models based on modeling the expected scale-type response patterns in (6.23) produce an ordered X classification, assuming that the item ordering used to define the scale types is valid.

We illustrate the above models, and some related models to be discussed shortly, with the famous Stouffer–Toby data used in virtually all of the sources cited in this section. These data pertain to $J = 4$ dichotomous items, each eliciting either universalistic or particularistic responses in a given situation involving role conflict. Using the item ordering common to these sources gives the vector of frequencies, $(42, 1, 6, 2, 6, 1, 7, 2, 23, 4, 24, 9, 25, 6, 38, 20)$, $(n = 216)$. These frequencies correspond to cells $(1, 1, 1, 1), (1, 1, 1, 2), \ldots, (2, 2, 2, 2)$. Table 5 gives fit statistics for the models that we will discuss.

Table 5. Fit Statistics for Some Scaling Models Applied to the Stouffer–Toby Data

Model	Model Name	T	df	L^2
H_1	Independence	1	11	81.08
H_2	2–Class LCM	2	6	2.72
H_3	3–Class LCM	3	2	.42
M_1	Proctor	5	10	27.16
M_2	Goodman	6	6	.99
M_3	Proctor–Goodman	6	5	.94
M_4	Rasch	2	9	2.91
M_5	Rasch	3	8	1.09

NOTE: See text for description of models; T denotes the number of latent classes in the model. Model M_4 is the Rasch-type model with two latent classes of individuals while M_5 is the regular Rasch model estimated either by conditional maximum likelihood or as a three-class Rasch-type model.

It is natural to begin with the model of independence, equivalent to an LCM with one latent class, model H_1 in Table 5. If this model holds, then the items do not "measure." A clear rejection of this model here implies that the items might be used to measure some underlying trait or characteristic, degree of universalism in this instance. The conventional two-class model (H_2) could be used for scaling, because a dichotomy can always refer to "high" versus "low" (and scaling could be accomplished using some of the procedures given earlier). This model fits the data well, so no more than two latent classes are required. The three-class model fits the data almost perfectly (L^2 = .42 on 2 df). Proctor's model with five latent classes, each having an expected response pattern corresponding to one of the scale types in (6.23), does not fit well, implying that a single error-rate parameter constant across items and across latent types does not describe the data. (The error-rate parameter value is $\hat{\alpha}$ = .15.) In contrast, Goodman's scaling model fits the data extremely well; this model is formally equivalent to the model of quasi-independence "blanking" out the scale-type response patterns in (6.23). [Under this model, the proportion of respondents intrinsically unscalable is $\hat{\pi}_X(5)$ = .68, or 68% of the respondents are not scalable—a very large proportion. Such a high value for this index of scalability implies that the items do not conform very well to Guttman's model.] Finally, the Proctor–Goodman model proposed by Dayton and Macready (1980) also fits the data well. Under this model, the error-rate parameter is $\hat{\alpha}$ = .01 and the proportion intrinsically unscalable is $\hat{\pi}_X(5)$ = .67, results that are virtually indistinguishable from Goodman's model. Model M_5 is the Rasch model (see Hambleton et al. 1991), one of the popular models used in educational testing. The Rasch model, and the number of "latent classes" that can be identified with this model, will be taken up in the next subsection. For now note that we consider in Table 5 one Rasch-type model with just two (rather than three) latent classes, model M_4.

There are many other scaling models based on modeling response error processes formulated in terms of expected scale types, as in (6.23). Although classical scaling models of this sort are not used very much in testing, they have some uses in social research (such as measuring social distance, normative orderings of life events, etc.). Moreover, they give insights about how to use restricted LCMs for other types of problems. We note, for example, that Goodman's unscalable class corresponds in some respects to a "guessing" parameter in item-response theory (IRT) models.

Classical scaling models are difficult to generalize to cases with nondichotomous items. They depend too much on the existence of a prior ordering of items. Item difficulties are not usually known in advance and, indeed, one of the goals of scaling models is to *estimate* the item difficulties or orderings. Finally, the models are consistent with only an ordinal X, not with a quantitative X that would be demanded for scaling, and a scaled X value is difficult to justify unless added assumptions, like a distributional assumption, are imposed. These models do not, to my knowledge, justify creating a summated scale from the items, in spite of the fact that these models were conceived as statistical counterparts of Guttman's deterministic scaling model, which might justify a summated scale. On the positive side, these or similar models might be useful in survey research, as opposed to educational testing, where special structures of item-response patterns specified by social theory might be represented well by classical scaling models or their LCM counterparts. We now turn to the topic of scaling models based on the popular Rasch model and show how these models can be reformulated or modified as LCMs.

11.3 The Rasch Model and Related Models

To illustrate connections between Rasch models and LCMs, in this subsection we use some different notation. We let i denote the individual; j still refers to the item. Suppose now that each individual sampled has an "ability" or "propensity" called μ_i, for $i = 1, \ldots, n$. Of course, this person parameter has meaning only in the context of items to which the individual responds, the Y_j, $j = 1, \ldots, J$. We suppose that the items are dichotomous (i.e., $y_j = 0, 1$, corresponding to "fail or pass" or to "agree or do not agree"). Here only we collect the responses in a matrix, Y_{ij}, of order $n \times J$, with observed value y_{ij}. Now let p_{ij} denote the probability that person i gives response $y_j = 1$.

The Rasch model (see Clogg 1988, Formann 1985, and Lindsay et al. 1991) says that

$$p_{ij} = (\exp[y_{ij}(\mu_i - \theta_j)])/(1 + \exp(\mu_i - \theta_j)), \tag{6.24}$$

where the θ_j are the so-called item difficulties which order the items. (Note that in contrast to classical scaling models, a prior ordering of items is not used; the items are ordered and even scaled by the model.) This model has many attractive properties; see Hambleton et al. (1991) and references cited there. Rasch models form a natural benchmark for studying change as well as ability in a single cross-section, and LCMs can be formulated for analysis of change as well (see Formann 1993). The model posits a continuous latent variable with μ_i denoting the value of this variable for the i-th individual. There are both random-effects and fixed-effects versions of this model.

The likelihood for the i-th individual, assuming independence of item responses (the axiom of local independence is used because items are conditionally independent given the value of μ), is

$$L_i \propto \exp[y_{i+}\mu_i - \sum_{j=1}^{J} y_{ij}\theta_j] \tag{6.25}$$

Note that $y_{i+} = y_{i1} + \ldots + y_{iJ}$ for the i-th individual, and with zero–one scoring this quantity refers to the number right or the number of times the subject agrees. (The scores are sufficient statistics for the μ_i parameters.) The logarithm of this expression is a log-linear model, say

$$\log(P_{Y_i}(y_i)) = \lambda + s_y\mu_i + \sum_{j=1}^{J} z_{y_j}\theta_j \ , \tag{6.26}$$

with Y_i denoting the entire response pattern (a vector) for the i-th individual, and s_y denoting the score (the sum y_{i+}, and z_{y_j} an indicator variable, the same as y_{ij}). This fact has led to much interesting work on log-linear models and Rasch modeling [see, for example, Cressie and Holland (1983) and Tjur (1982)]. Because the score statistics take on only values in the set, $(0, 1, \ldots, J)$, conditioning on the sufficient scores as in conditional maximum likelihood is essentially the same as fitting the above log-linear model with the term, $s_y\mu_i$, replaced by a set of $J + 1$ effects for the possible score values. This fact is one of the main results in Tjur (1982).

We now consider a mixture approach to the above model, that is, an LCM of the Rasch-type. Now suppose that a discrete latent variable X having T classes drives the response process. This variable is described by latent class proportions as throughout this paper $(\pi_X(t), t = 1, \ldots, T)$ and category scores, say, $\mu_t, t = 1, \ldots, T$. In the notation used throughout this paper, the latent class model corresponding to (6.24) is

$$\pi_{Y_j|X(t)}(y_j) = (\exp[y_j(\mu_t - \theta_j)])/(1 + \exp(\mu_t - \theta_j)) \ . \tag{6.27}$$

The mixture (or LCM) likelihood, again based on the assumption of local independence, is

$$L_i^* = \sum_{t=1}^{T}[\pi_X(t)\prod_{j=1}^{J}\pi_{Y_j|X(t)}(y_j)] \ . \tag{6.28}$$

After some algebra, the term in brackets can be written as a regular log-linear model for the incomplete table cross-classifying Ys and X, giving a design matrix for use in a program like LAT or NEWTON. The above model with $T = 2$ was denoted as M_4 in Table 5, and it fits the data very well also. Although it is not directly apparent, this model places special restrictions on the conventional two-class LCM, and the restrictions are consistent with the data. (The likelihood-ratio comparison of M_4 and H_2 is relevant, giving a difference of $2.91 - 2.72 = .19$ on 3 df.) The parameter values for this model are

$$\hat{\pi}_X(1) = .718; \hat{\pi}_X(2) = .282$$

$$\hat{\mu}_1 = -1.941; \hat{\mu}_2 = 1.263$$

$$\hat{\theta}_1 = 2.866; \hat{\theta}_2 = 1.246; \hat{\theta}_3 = 1.323; \hat{\theta}_4 = .000$$

where the last item difficulty is restricted to zero for identification. Note that the item orderings are given by the θ values; the second and third items have approximately the same difficulty, although they suggest an item ordering different from the one used to develop the scaling models in the previous subsection.

A restricted LCM of the Rasch type is attractive in this case. Following Lindsay et al. (1991), we can determine that a restricted LCM of this type will be *equivalent* to Rasch's model for inferential purposes when certain regularity conditions are satisfied. In fact, whenever the number of latent classes in the above model satisfy the relationship, $T \geq (J+1)/2$, all of the available information in the data is captured by the LCM. When equality obtains, $T = (J+1)/2$, we say that the LCM "saturates" the relationship, in the sense that this value gives an exact reproduction of the continuous version of the model having nuisance parameters. For J odd (i.e., $3, 5, 7, 9, \ldots$), the value of T that saturates gives an exactly identified LCM (i.e., $2, 3, 4, 5, \ldots$). For an even number of items, parameters have to be restricted because the boundary value is then a fraction. That is, for J even we get $1.5, 2.5, 3.5$, and so on. The Rasch-type LCM with $T = 3$ latent classes is called M_5 in Table 5, and it gives the same fitted frequencies and indexes of fit, as well as the same item parameters, as does any procedure giving conditional maximum likelihood estimates for Rasch's model. The item difficulties under this latter model, equivalent to those of the conventional LCM analysis of Rasch's model, were practically the same as before.

There are several features of the above analysis that deserve comment. First, it seems natural for scaling to posit scaled latent classes, as with the μ_t values. Second, the problem of predicting latent classes can now be supplemented by predicting scores, or posterior means, for the respondents. Details are taken up in Lindsay et al. (1991), and see de Leeuw and Verhulst (1986) and Follman (1988). Third, for the Rasch model, which began with an hypothesis of a continuous latent trait, *statistical inference* with a very small number of latent classes [$T = (J+1)/2$ in this case] is identical to what we would have with a proper statistical analysis of the model assuming a continuous latent trait, except in some anomalous cases. The mathematical model is indeed different, and there can be no doubt that a model that posits a continuous latent variable is very different conceptually from a model that posits a (scaled) categorical latent variable. However, we cannot distinguish between the two very different formulations with the data! So we have another instance where a prior formulation of continuous latent variables cannot be distinguished from a prior formulation of a discrete latent variable. There are many other relationships of this kind in the latent variable area. For example, Bartholomew (1987, sec. 2.4) shows that a latent profile model for continuous measurements is indistinguishable, statistically, from a factor analysis model; the latent profile model is a latent class model for continuous measurements.

We hasten to add that the above formulation gives direct information about the latent distribution; we do not have to retrieve this information from various marginalization or conditioning arguments. An interesting byproduct of this formulation is that we can predict mean scores for all respondents, including those with extreme response patterns, $(1,1,1,1)$ and $(2,2,2,2)$ in this case. The approach can be generalized to deal with partial-credit scoring of test items, or for ordinal Y's. See Muraki (1992) and references cited there; this source covers models analogous to partial-credit two-parameter logistic models, the latter of which is discussed briefly next.

11.4 Extending Latent Class Models to Other Scaling Contexts

The so-called two-parameter logistic model for item analysis (Hambleton et al. 1991, p. 15) is

$$p_{ij} = (\exp[y_{ij} D\alpha_j(\mu_i - \theta_j)])/(1 + \exp[D\alpha_j(\mu_i - \theta_j]) . \tag{6.29}$$

This model is actually used more frequently than the Rasch model. In this expression, D is a constant ($= 1.7$) chosen to make the logistic function close to the Gaussian cumulative distribution function, and the α_j are so-called item discrimination parameters. Note that including the latter set of parameters can be viewed as a special type of interaction. This model is nonlinear in the parameters, even after using the logit transformation. It is somewhat analogous to log-multiplicative association models given in Goodman (1984); also see Clogg (1982). To my knowledge, a representation in terms of a finite mixture or restricted LCM has not been put forth. It seems likely, however, that this model can be represented with fewer latent classes than was the case for Rasch's model. That is, writing the above model as a conditional probability for the t–th latent class, with parameters μ_t replacing the μ_i, ought to lead to an equivalent *statistical* model with $T < (J+1)/2$. This seems to follow because while Rasch's model is additive on the logit scale (or log-linear as above), this model incorporates parameters that correspond somewhat to item–X interactions, and adding such interactions will decrease the number of latent classes that can be identified.

The three-parameter logistic model, which is used extensively in educational testing in the United States at least, adds to the above model a special guessing parameter. This model can be rewritten directly as an LCM also, and we note that the "latent guessing" class is somewhat analogous to Goodman's unscalable class covered above. This model, for the jth item, is

$$p_{ij} = \gamma_j + (1 - \gamma_j) \times (\exp[y_{ij} D\alpha_j(\mu_i - \theta_j)])/(1 + \exp[D\alpha_j(\mu_i - \theta_j)]) .$$

The γ_j values correspond to the distribution for the "unscalable" or "guessers" latent class. Indeed, in this form the model is a finite mixture model, with the second latent class described by the two-parameter logistic. Note the connection to Goodman's scaling model discussed earlier.

There are many connections between scaling models of the IRT variety and conventional models for factor analysis or covariance structure analysis. Clogg (1988) introduced a class of latent association models, analogous to the conditional association models in Clogg (1982), that in principle allows for a completely nonparametric approach to the statistical analysis. See also Rost (1990, 1991). Heinen (1993) considers LCM analogues to most common IRT models giving an approach that can no doubt be pursued much further. With these developments in mind, we can formulate special LCMs that represent models that might be considered with modifications of covariance structure analysis and estimated with modern programs like MECOSA, LISCOMP, EQS, and LISREL (cf. Chapter 4).

Suppose the Y_j's are dichotomous and that $T > 2$ latent classes are posited. An attractive model is

$$\log(P_{Y,X}(y,t)) = \lambda + \sum_{j=1}^{J} \lambda_{Y_j}(y_j) + \lambda_X(t) + \sum_{j=1}^{J} \phi_j \mu_t y_j \ . \tag{6.30}$$

Here, the scaled latent classes have scores to be estimated (the μ_t), much as with the LCM version of Rasch's model, and the item-by-X interactions are specified as linear-by-linear interactions. The parameters ϕ_j measure the associations of interest. If these are equated to each other, the Rasch-type model given above is obtained, which is another way to view the restrictions inherent in the Rasch model. If the scores for latent classes were fixed, then the above is a regular log-linear model at the latent level and can be estimated with LAT or NEWTON. Slightly different procedures are required for the general case because the model is not log-linear at the latent level, but is rather log-multiplicative. For polytomous, ordered Y variables, the model can be generalized in several ways. The y_j values in the interaction terms are replaced by score values, either fixed or random, denoting the scaled level of a particular observed variable. For Likert-type items with say five levels, we can use integer scoring so that each y_j takes on the values $0, 1, 2, 3, 4$. The model with these fixed scores is a generalization of the Andrich model (see Andrich 1979). If scores for levels of the observed variables are not available, models can be formulated with unknown scores, but to my knowledge such models have not been studied rigorously thus far. All models of this general type are analogous to covariance structure models with one latent factor, and they ought to be considered seriously in my judgment.

12 Conclusion

I hope that this survey shows that the LCM is a powerful model and that proper statistical analysis with both unrestricted and restricted models forms a flexible methodology that can be used in many areas, not only the areas in which LCMs were originally developed. There has been great progress in this area over the past two decades, and if current work is any indication, there is much in store for the future. I believe that further work on formulating models of interest for particular problems will be fruitful; these models will almost certainly take on the form of restricted LCMs. The unrestricted LCM has proven intractable for generalizations required for practical work, such as with including both continuous and categorical covariates. Another very fruitful avenue of research only touched on in the above review is formulating LCMs for dynamic processes. The work by Langeheine and van de Pol (1990) on latent Markov models represents this new development very well, and much more can be done. See Collins and Wugalter (1993) and Hagenaars (1993) for developments in the area of developmental or longitudinal modeling. A serious appraisal of these methods and models would require a chapter at least as long as the present one. To assess change, measurement error, and across-time reliability in categorica measurements (discrete states) appears to be well suited for the LCM setting. From a practical standpoint, computational equipment will have to be continually modified, and I believe that it is time to explore new algorithms and new software development plans for handling individual-level data, such as would be required for serious multivariate analysis of data of high dimension, including panel data. If past developments are a guide, we shall see a great many new developments in this area in the near future.

REFERENCES

Agresti, A. (1988), "A Model for Agreement Between Ratings on an Ordinal Scale," *Biometrics*, 44, 539–548.

——(1990), *Categorical Data Analysis*, New York: Wiley.

——(1992), "Computing Conditional Maximum Likelihood Estimates for Generalized Rasch Models Using Simple Loglinear Models with Diagonals Parameters," *Scandinavian Journal of Statistics*, 19, forthcoming.

Aitkin, M., Anderson, D., and Hinde, J. (1981), "Statistical Modelling of Data on Teaching Styles," *Journal of the Royal Statistical Society*, Ser. A, 144, 419–461.

Andersen, E. B. (1982), "Latent Structure Analysis: A Survey," *Scandinavian Journal of Statistics*, 9, 1–12.

——(1990), *The Statistical Analysis of Categorical Data* Rev. and enl. ed., Berlin: Springer-Verlag.

Andrich, D. (1979), "A Model for Contingency Tables Having an Ordered Response Classification," *Biometrics*, 35, 403–415.

Arminger, G. (1985), "Analysis of Qualitative Individual Data and of Latent Class Models With Generalized Linear Models," pp. 51-79 in P. Nijkamp, H. Leitner, and N. Wrigley, eds., *Measuring the Unmeasurable*, Dordrecht: Nijhoff Publishers.

Bartholomew, D. J. (1987), *Latent Variable Models and Factor Analysis*, London: Charles Griffin.

Becker, M. P., and Clogg, C. C. (1988), "A Note on Approximating Correlations from Odds Ratios," *Sociological Methods and Research*, 16, 407–24.

Bergan, J. R. (1983), "Latent-Class Models in Educational Research," pp. 305-360 in E. W. Gordon, ed., *Review of Research on Education*, Washington, DC.: American Educational Research Association.

Blau, P. M., and Duncan, O. D. (1967), *The American Occupational Structure*, New York: Wiley.

Boeckenholt, U. (1992), "Applications of Thurstonian Models to Ranking Data," pp. 157–172 in M. A. Fligner, and J. S. Verducci, eds., *Probability Models and Statistical Analysis for Ranking Data*, Berlin: Springer-Verlag.

Boeckenholt, U. and Boeckenholt, I. (1991), "Constrained Latent Class Analysis: Simultaneous Classification and Scaling of Discrete Choice Data," *Psychometrika*, 56, 699–716.

Breiger, R. L. (1981), "The Social Class Structure of Occupational Mobility," *American Journal of Sociology*, 87, 578–611.

Bye, B. V., and Schechter, E. S. (1986), "A Latent Markov Model Approach to the Estimation of Response Error in Multiwave Panel Data," *Journal of the American Statistical Association*, 81, 375–380.

Clogg, C. C. (1977), "Unrestricted and Restricted Maximum Likelihood Latent Structure Analysis: A Manual for Users," Working Paper 1977–09, Population Issues Research Center, Pennsylvania State University (Manual for MLLSA program)

——(1979), "Some Latent Structure Models for the Analysis of Likert-Type Data," *Social Science Research*, 8, 287–301.

———(1981a), "Latent Structure Models of Mobility," *American Journal of Sociology*, 86, 836–868.

———(1981b), "A Comparison of Alternative Models for Analyzing the Scalability of Response Patterns," pp. 240–280 in S. Leinhardt, ed., *Sociological Methodology 1981*, San Francisco: Jossey-Bass.

———(1981c), "New Developments in Latent Structure Analysis," pp. 214–246 in D. M. Jackson and E. F. Borgatta, eds., *Factor Analysis and Measurement in Sociological Research*, Beverly Hills, CA: Sage.

———(1982), "Some Models for the Analysis of Association in Multi-Way Cross-Classifications Having Ordered Categories," *Journal of the American Statistical Association*, 77, 803–815.

———(1984), "Some Statistical Models for Analyzing Why Surveys Disagree," pp. 319–366 in C. F. Turner and E. Martin, eds., *Surveying Subjective Phenomena*, Vol. 2, New York: Russell Sage Foundation.

———(1988), "Latent Class Models for Measuring," pp. 173–205 in R. Langeheine and J. Rost, eds., *Latent Trait and Latent Class Models*, New York: Plenum.

———(1992), "The Impact of Sociological Methodology on Statistical Methodology" (with discussion), *Statistical Science*, 7, 183–196.

Clogg, C. C., and Dajani, A. (1991), "Sources of Uncertainty in Modeling Social Statistics: An Inventory," *Journal of Official Statistics*, 7, 7-24.

Clogg, C. C., and Goodman, L. A. (1984), "Latent Structure Analysis of a Set of Multidimensional Contingency Tables," *Journal of the American Statistical Association*, 79, 762–71.

———(1985), "Simultaneous Latent Structure Analysis in Several Groups," pp. 81–110 in N. B. Tuma, ed., *Sociological Methodology 1985*, San Francisco: Jossey-Bass.

———(1986), "On Scaling Models Applied to Data from Several Groups," *Psychometrika*, 51, 123–35.

Clogg, C. C., Rubin, D. B., Schenker, N., Schultz, B., and Weidman, L. (1991), "Multiple Imputation of Industry and Occupation Codes in Census Public-Use Samples Using Bayesian Logistic Regression," *Journal of the American Statistical Association*, 86, 68–78.

Collins, L. M., and Wugalter, S. E. (1993), "Latent Class Models for Stage-Sequential Dynamic Latent Variables," *Multivariate Behavioral Research*, forthcoming.

Cressie, N., and Read, T. R. C. (1984), "Multinomial Goodness-of-Fit Tests," *Journal of the Royal Statistical Society*, Ser. B, 46, 440–464.

Cressie, N., and Holland, P. W. (1983), "Characterizing the Manifest Probabilities of Latent Trait Models," *Psychometrika*, 48, 129–141.

Croon, M. (1990), "Latent Class Analysis With Ordered Latent Classes," *British Journal of Mathematical and Statistical Psychology*, 43, 171–192.

Croon, M. and Luijkx, R. (1992), "Latent Structure Models for Ranking Data," pp. 53-74 in M. A. Fligner, and J. S. Verducci, eds., *Probability Models and Statistical Analyses for Ranking Data*, Berlin: Springer-Verlag.

Dayton, C. M., and Macready, G. D. (1980), "A Scaling Model With Response Errors and Intrinsically Unscalable Respondents," *Psychometrika*, 45, 343–356.

Dayton, C. M., and Macready, G. B. (1988), "Concomitant Variable Latent Class Models," *Journal of the American Statistical Association*, 83, 173–178.

de Leeuw, J., and Verhelst, N. (1986), "Maximum Likelihood Estimation in Generalized Rasch Models," *Journal of Educational Statistics*, 11, 183–196.

de Leeuw, J., and van der Heijden, P. G. M. (1991), "Reduced Rank Models for Contingency Tables," *Biometrika*, 78, 229–232.

Dempster, A. P., Laird, N. M., and Rubin, D. B. (1977), "Maximum Likelihood Estimation from Incomplete Data via the EM Algorithm," *Journal of the Royal Statistical Society*, Ser. B, 39, 1–38.

Dillon, W. R., and Goldstein, M. (1984), *Multivariate Analysis: Theory and Applications*, New York: Wiley.

Eliason, S. R. (1990), *The Categorical Data Analysis System. Version 3.50 User's Manual*, University of Iowa, Department of Sociology.

Espeland, M. A., and Handelman, S. A. (1989), "Using Latent Class Models to Characterize and Assess Relative Error in Discrete Measurements," *Biometrics*, 45, 587–599.

Follman, D. A. (1988), "Consistent Estimation in the Rasch Model Based on Nonparametric Margins," *Psychometrika*, 53, 553–562.

Formann, A. K. (1985), "Constrained Latent Class Models: Theory and Applications," *British Journal of Mathematical and Statistical Psychology*, 38, 87–111.

——— (1988), "Latent Class Models for Nonmonotone Dichotomous Items," *Psychometrika*, 56, 45–62.

——— (1992), "Linear Logistic Latent Class Analysis for Polytomous Data," *Journal of the American Statistical Association*, 87, 476–486.

——— (1993), "Measuring Change Using Latent Class Analysis," in A. von Eye, and C. C. Clogg, eds., *Analysis of Latent Variables in Developmental Research*, Newbury Park, CA: Sage.

Goodman, L. A. (1974a), "The Analysis of Systems of Qualitative Variables When Some of the Variables Are Unobservable. Part I—A Modified Latent Structure Approach," *American Journal of Sociology*, 79, 1179–1259.

——— (1974b), "Exploratory Latent Structure Analysis Using Both Identifiable and Unidentifiable Models," *Biometrika*, 61, 215–231.

——— (1975), "A New Model for Scaling Response Patterns: An Application of the Quasi-Independence Concept," *Journal of the American Statistical Association*, 70, 755–768.

——— (1984), *The Analysis of Cross-Classifications Having Ordered Categories*, Cambridge, Mass.: Harvard University Press.

——— (1987), "New Methods for Analyzing the Intrinsic Character of Qualitative Variables Using Cross-Classified Data," *American Journal of Sociology*, 93, 529–583.

Goodman, L. A., and Clogg, C. C. (1992), "New Methods for the Analysis of Occupational Mobility Tables and Other Kinds of Cross-Classifications," *Contemporary Sociology*, 21, 609–622.

Haberman, S. J. (1977), "Product Models for Frequency Tables Involving Indirect Observation," *Annals of Statistics*, 5, 1124–1147.

———(1979), *Analysis of Qualitative Data.* Vol. 2. New Developments, New York: Academic Press.

———(1988), "A Stabilized Newton-Raphson Algorithm for Loglinear Models for Frequency Tables Derived by Indirect Observation," pp. 193–212 in C. C. Clogg, ed., *Sociological Methodology 1988*, Washington, D.C.: American Sociological Association.

Hagenaars, J. A. (1986), "Symmetry, Quasi-Symmetry, and Marginal Homogeneity on the Latent Level," *Social Science Research*, 15, 241–255.

———(1988), "Latent Structure Models With Direct Effects Between Indicators," *Sociological Methods and Research*, 16, 379–405.

———(1990), *Categorical Longitudinal Data: Log-Linear Panel, Trend, and Cohort Analysis*, Newbury Park, CA: Sage.

Hagenaars, J. A., and Luijkx, R. (1987), "Manual for LCAG," Working Paper no. 17, Tilburg University, Department of Sociology.

Hambleton, R. K., Swaminathan, H., and Rogers, H. J. (1991), *Fundamentals of Item Response Theory*, Newbury Park, CA: Sage.

Heinen, T. (1993), *Discrete Latent Variable Models*, University of Tilburg Press.

Henry, N. W. (1983), "Latent Structure Analysis," pp. 497–504 in S. Kotz, and N. Johnson, eds., *Encyclopedia of Statistical Sciences*, Vol. 4, New York: Wiley.

Hogan, D. P., Eggebeen, D. J., and Clogg, C. C. (1993), "The Structure of Intergenerational Exchanges in American Families," *American Journal of Sociology*, 99, 1428–1458.

Kelderman, H. (1989), "Item Bias Detection Using Loglinear Item Response Theory," *Psychometrika*, 54, 681–697.

Kelderman, H., and Macready, G. (1990), "The Use of Loglinear Models for Assessing Item Bias Across Manifest and Latent Examinee Groups," *Journal of Educational Measurement*, 27, 307–327.

Knoke, D., and Burke, P. J. (1980), *Log-Linear Models*, Beverly Hills, CA: Sage.

Langeheine, R. (1988), "New Developments in Latent Class Theory," pp. 77–108 in R. Langeheine, and J. Rost, eds., *Latent Trait and Latent Class Models*, New York: Plenum.

Langeheine, R., and Pol, F. van de (1990), "A Unifying Framework for Markov Modeling in Discrete Space and Discrete Time," *Sociological Methods and Research*, 18, 416–441.

Langeheine, R., and Rost, J., eds. (1988), *Latent Trait and Latent Class Models*, New York: Plenum.

Lazarsfeld, P. F. (1950), "The Logical and Mathematical Foundation of Latent Structure Analysis," pp. 362–412 in E. A. Suchman, P. F. Lazarsfeld, S. A. Starr, and J. A. Clausen, eds., *Studies in Social Psychology in World War II, Vol. 4. Measurement and Prediction*, Princeton, NJ: Princeton University Press.

Lazarsfeld, P. F., and Henry, N. W. (1968), *Latent Structure Analysis*, Boston: Houghton Mifflin.

Lindsay, B., Clogg, C. C., and Grego, J. M. (1991), "Semi-parametric Estimation in the Rasch Model and Related Exponential Response Models, Including a Simple Latent Class Model for Item Analysis," *Journal of the American Statistical Association*, 86, 96–197.

Little, R. J. A., and Rubin, D. B. (1987), *Statistical Analysis With Missing Data*, New York: Wiley.

Macready, G. B., and Dayton, C. M. (1977), "The Use of Probabilistic Models in the Assessment of Mastery," *Journal of Educational Statistics*, 2, 99–120.

Marsden, P. V. (1985), "Latent Structure Models for Relationally Defined Social Classes," *American Journal of Sociology*, 90, 1002–1021.

McCutcheon, A. L. (1987), *Latent Class Analysis*, Beverly Hills, CA: Sage.

Mislevy, R. J., and Verhelst, N. (1990), "Modeling Item Responses When Different Subjects Employ Different Solution Strategies," *Psychometrika*, 55, 195–215.

Mooijaart, A., and van der Heijden, P. G. M. (1992), "The EM Algorithm for Latent Class Analysis With Equality Constraints," *Psychometrika*, 57, 261–270.

Muraki, E. (1992), "A Generalized Partial Credit Model: Application of an EM Algorithm," *Applied Psychological Measurement*, 16, 159–176.

Palmgren, J., and Ekholm, A. (1985), "GLIM for Latent Class Analysis," pp. 128–136 in R. Gilchrist, B. Francis, and J. Whittaker, eds., *Generalized Linear Models: Proceedings of the GLIM 85 Conference*, Berlin: Springer-Verlag.

Proctor, C. H. (1970), "A Probabilistic Formulation and Statistical Analysis of Guttman Scaling," *Psychometrika*, 35, 73-78.

Rindskopf, D., and Rindskopf, W. (1986), "The Value of Latent Class Analysis in Medical Diagnosis," *Statistics in Medicine*, 5, 21–27.

Rost, J. (1988), "Rating Scale Analysis with Latent Class Models," *Psychometrika*, 53, 327–348.

——— (1988), *LACORD—Latent Class Analysis for Ordinal Variables*, Kiel, Germany: Institute for Science Education.

——— (1990), "Rasch Models in Latent Classes: An Integration of Two Approaches to Item Analysis," *Applied Psychological Measurement*, 14, 271–282.

——— (1991), "A Logistic Mixture Distribution Model for Polychotomous Item Responses," *British Journal of Mathematical and Statistical Psychology*, 44, 75–92.

Rubin, D. B. (1987), *Multiple Imputation for Nonresponse in Surveys*, New York: Wiley.

Rudas, T., Clogg, C. C., and Lindsay, B. L. (1994), "A New Index of Fit Based on Mixture Methods for the Analysis of Contingency Tables," *Journal of the Royal Statistical Society*, Ser. B, forthcoming.

Schwartz, J. E. (1985), "The Neglected Problem of Measurement Error in Categorical Data," *Sociological Methods and Research*, 13, 435-66.

——— (1986), "A General Reliability Model for Categorical Data, Applied to Guttman Scales and Current Status Data," pp. 79–119 in N. B. Tuma, ed., *Sociological Methodology 1986*, San Francisco: Jossey-Bass.

Shockey, J. W. (1988), "Adjusting for Response Error in Panel Surveys: A Latent Class Approach," *Sociological Methods and Research*, 17, 65–92.

Tanner, M. A. (1991), *Tools for Statistical Inference: Observed Data and Data Augmentation Methods*, New York: Springer-Verlag.

——— (1985), "Modeling Agreement Among Raters," *Journal of the American Statistical Association*, 80, 175–180.

Titterington, D. M., Smith, A. F. M., and Makov, U. E. (1985), *Statistical Analysis of Finite Mixture Distributions*, New York: Wiley.

Tjur, T. (1982), "A Connection Between Rasch's Item Analysis Model and a Multiplicative Poisson Model," *Scandinavian Journal of Statistics*, 9, 23–30.

Tuch, S. A. (1983), "Analyzing Recent Trends in Prejudice Toward Blacks: Insights from Latent Class Models," *American Journal of Sociology*, 87, 130–142.

Uebersax, J. S. (1993), "Statistical Modeling of Expert Ratings on Medical Treatment Appropriateness," *Journal of the American Statistical Association*, 88, 421–427.

Uebersax, J. S., and Grove, W. M. (1990), "Latent Class Analysis of Diagnostic Agreement," *Statistics in Medicine*, 9, 559–572.

van de Pol, F. J. R., and Langeheine, R. (1990), "Mixed Markov Latent Class Models," pp. 213–247 in C. C. Clogg, ed., *Sociological Methodology 1990*, Oxford: Basil Blackwell.

van de Pol, F. J. R., Langeheine, R., and Jong, W. de (1991), "PANMARK User Manual: Panel Analysis Using Markov Chains," Version 2.2, Netherlands Central Bureau of Statistics, Voorburg.

van der Heijden, P. G. M., Mooijaart, A., and de Leeuw, J. (1992), "Constrained Latent Budget Analysis," Pp. 279–320 in P. Marsden, ed., *Sociological Methodology 1992*, Oxford: Blackwell.

Whittaker, J. (1990), *Graphical Models in Applied Multivariate Analysis*, New York : Wiley.

Winship, C. M., and Mare, R. D. (1989), "Loglinear Models with Missing Data: A Latent Class Approach," pp. 331–368 in C. C. Clogg, ed., *Sociological Methodology 1989*, Oxford: Blackwell.

Young, M. A. (1983), "Evaluating Diagnostic Criteria: A Latent Class Paradigm," *Journal of Psychiatric Research*, 17, 285–296.

Young, M. A., Tanner, M. A., and Meltzer, H. Y. (1982), "Operational Definitions of Schizophrenia: What Do They Identify?" *Journal of Nervous and Mental Disease*, 170, 443–447.

Chapter 7

Panel Analysis for Metric Data

CHENG HSIAO

1 Introduction

A cross-sectional data set refers to observations on a number of individuals at a given time. A time-series data set refers to observations made over time on a given unit. A panel (or longitudinal or temporal cross-sectional) data set follows a number of individuals over time. In recent years empirical studies that use panel data have become common. This is partly because the cost of developing panel or longitudinal data sets is no longer prohibitive. In some cases, computerized matching of existing administrative records can produce inexpensive longitudinal information, such as the Social Security Administration's Continuous Work History Sample (CWHS). In other cases, valuable longitudinal data bases can be generated by computerized matching of existing administrative and survey data, such as the University of Michigan's Panel Study of Income Dynamics (PSID) and the U.S. Current Population Survey. Even in cases where the desired longitudinal information can be collected only by initiating new surveys, such as the series of negative income tax experiments in the United States and Canada, the advance of computerized data management systems has made longitudinal data development cost-effective in the last 20 years (Ashenfelter and Solon 1982).

More importantly, panel data offer researchers many more possibilities than purely cross-sectional or time-series data. Like cross-sectional data, panel data describe each of a number of individuals. Like time-series data, they describe changes through time. By blending characteristics of both cross-sectional and time-series data, panel data can be used to (e.g. Hsiao 1985, 1986):

CHENG HSIAO • Department of Economics, University of Southern California, Los Angeles, California 90089–0253, USA. • This work was supported in part by NSF Grants SES 88-21205, 91-22481 and Irvine Faculty Research Fellowship. The author wishes to thank Michael Sobel and Gerhard Arminger for helpful comments and suggestions.

Handbook of Statistical Modeling for the Social and Behavioral Sciences, edited by Gerhard Arminger, Clifford C. Clogg, and Michael E. Sobel. Plenum Press, New York, 1995.

1. *Improve accuracy of parameter estimates.* Most estimators converge to the true parameters at the speed of the square root of the degrees of freedom. Panel data, by offering many more observations than purely cross-sectional or time-series data, can give more accurate parameter estimates.

2. *Lessen the problem of multicollinearity.* Many variables tend to move collinearly over time. The shortage of degrees of freedom and severe multicollinearity problems found in time-series data often frustrate investigators who wish to determine the individual influences of each explanatory variable. Panel data may lessen this problem by appealing to interindividual differences.

3. *Provide possibilities for reducing estimation bias.* A standard assumption in statistical specification is that the random-error term representing the effect of omitted variables is orthogonal to the included explanatory variables. Otherwise, the estimates are subject to omitted-variable bias when these correlations are not explicitly allowed for. Panel data provide means to eliminate or reduce the omitted-variable bias through various data transformations when the correlations between included explanatory variables and the random-error terms follow certain specific patterns (e.g. Chamberlain 1978; Hsiao 1985, 1986).

4. *Allow the specification of more complicated behavioral hypotheses.* Panel data, by blending the characteristics of interindividual differences across cross-sectional units and the intra-individual dynamics over time, often allow investigators to specify more complex and realistic models than that could have been done by using time-series or cross-sectional data alone (e.g. Appelbe, Dineen, Solvason, and Hsiao 1990; Griliches 1979; Heckman 1978, 1981a,b).

5. *Obtaining more accurate prediction of individual outcomes.* When people's behavior is similar in nature, namely they satisfy De Finette's (1964) exchangeability criterion, we can obtain more accurate prediction of individual outcome by "shrinking" the individual predictor toward the grand average of all the individual predictors (e.g. Hsiao, Mountain, Tsui, and Chan 1989; James-Stein 1961; Min and Zellner 1992).

However, new data sources also raise new issues. A standard assumption for statistical analysis is that data are generated from controlled experiments in which the outcomes are random variables with a probability distribution that is a smooth function of a small number of variables describing the conditions of the experiment. This may not be a bad approximation to macro relations, because the possibility exists that in aggregating individual units, the impacts of numerous factors not explicitly allowed for may average out and behave like a random disturbance term. In panel-data analysis, the emphasis is often on the individual unit. In explaining human behavior, the list of relevant factors may be extended ad infinitum. The effect of these factors that have not been explicitly allowed for may not necessarily behave as independently and identically distributed (iid) error terms across individual units and over time. In fact, one of the crucial issues in panel data analysis is how best to capture the interindividual differences and intra-individual dynamics that are not represented by the included explanatory variables.

To get an idea of the importance of correctly allowing for the unobserved heterogeneity across individual observations in panel data analysis, consider a simple example of regression analysis in which y is a linear function of z and other variables. Conditioning on z, if individual observations may be viewed as random draws from a common population, then we have

$$y_{it} = \gamma z_{it} + v_{it}, \quad i = 1, \ldots, N, t = 1, \ldots, T, \tag{7.1}$$

where the error term v_{it} represents the effects of all other variables. If individuals respond differently, given z, then we have

$$y_{it} = \gamma_i z_{it} + v_{it}, \quad i = 1, \ldots, N, t = 1, \ldots, T. \tag{7.2}$$

Thus, if (7.2) is the correct formulation, fitting (7.1) to the panel data will yield nonsense results. If (7.1) is the correct formulation, fitting (7.2) to the panel data, in general, will provide very little information about γ, because most panel data contain a large number of individuals and only a small number of time-series observations. When sample size increases in the sense of increasing N, (7.2) introduces the familiar incidental parameters problem (Neyman and Scott 1948), namely the number of parameters increases with the number of observations.

Furthermore, suppose that the error term or the effects of all omitted variables can be decomposed into two components: an individual time-invariant component, α_i, representing the effect of an individual's socioeconomic background, and a component u_{it} that represents the effects of all other variables and varies across i and over t, then we have

$$v_{it} = \alpha_i + u_{it}. \tag{7.3}$$

Next, there is the issue of whether the α_i should be treated as fixed constants (fixed effects) or random variables (random effects). If α_i are fixed constants, treating them as random leads to the misspecification of the intercept term. Thus, the estimate of (7.1) or (7.2) is subject to omitted variable bias. On the other hand, if α_i are random, treating them as fixed constants introduces N additional parameters, and thus reduces the efficiency of estimating (7.1) or (7.2).

Moreover, when α_i are treated as random, the composite error terms v_{it} can no longer be treated as independently, identically distributed. Suppose that α_i and u_{it} are independently distributed across i and $E(\alpha_i u_{it}) = 0$, then the covariance of v_{it} and v_{js} is of the form

$$E(v_{it}v_{js}) = \begin{cases} 0 & , \quad \text{if } i \neq j \\ E(\alpha_i^2) + E(u_{it}u_{is}) & , \quad \text{if } i = j. \end{cases} \tag{7.4}$$

In other words, the variance-covariance matrix of the error terms, v_{it} is no longer proportional to an identity matrix. Therefore, the least-squares estimator will not be as efficient as the generalized least-squares estimator.

In addition to the issue of the efficiency of the estimators, there is also the possibility that the omitted individual effects may be correlated with the included explanatory variables. For instance, consider the simple case of model (7.1) and (7.3). Suppose that the sampling distributions of the data are generated as in Figures 1 and 2, in which the broken circles represent the point scatter for an individual over time, and the broken straight lines represent the individual mean relations between y and z given α_i, that is, $E(y_{it}|z_{it}, \alpha_i) = \gamma z_{it} + \alpha_i$. Solid lines represent the regression of (7.1) using all NT observations when individual differences in α_i are ignored. As one can see from these figures, the pooled slope γ^* is different from γ. The direction of the pooled slope γ^* depends on γ and the correlation between α_i and z_{it}. For instance, in Figure 1, γ^* overestimates γ and in Figure 2, γ^* underestimates γ when α_i and z_{it} are positively correlated. In Figure 3, γ^* underestimates γ, when α_i and z_{it} are negatively correlated. In fact, a variety of circumstances may result depending on the scattering of the NT sample observations.

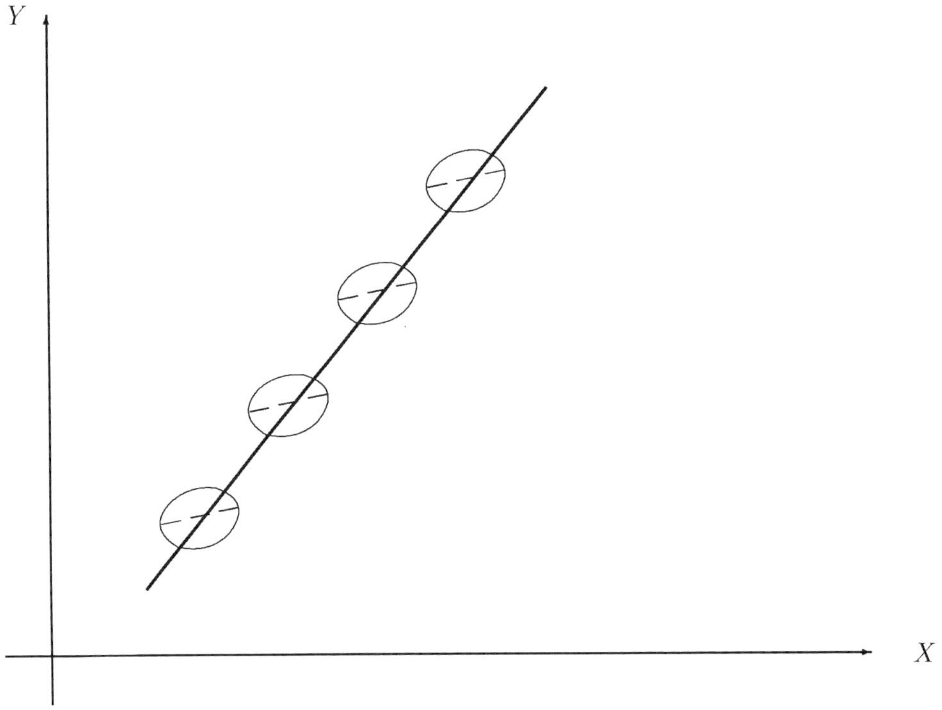

FIGURE 1

Because ignoring the unobserved heterogeneity in individual observations may directly affect the consistency and efficiency of the estimates, a number of models have been proposed to capture the unobserved heterogeneity. These include:

(i) A common model for all NT observations.

(ii) Different models for different individuals (and/or over time) [e.g. (7.3)].

(iii) Variable intercept models. Namely, there exist individual specific but time-invariant

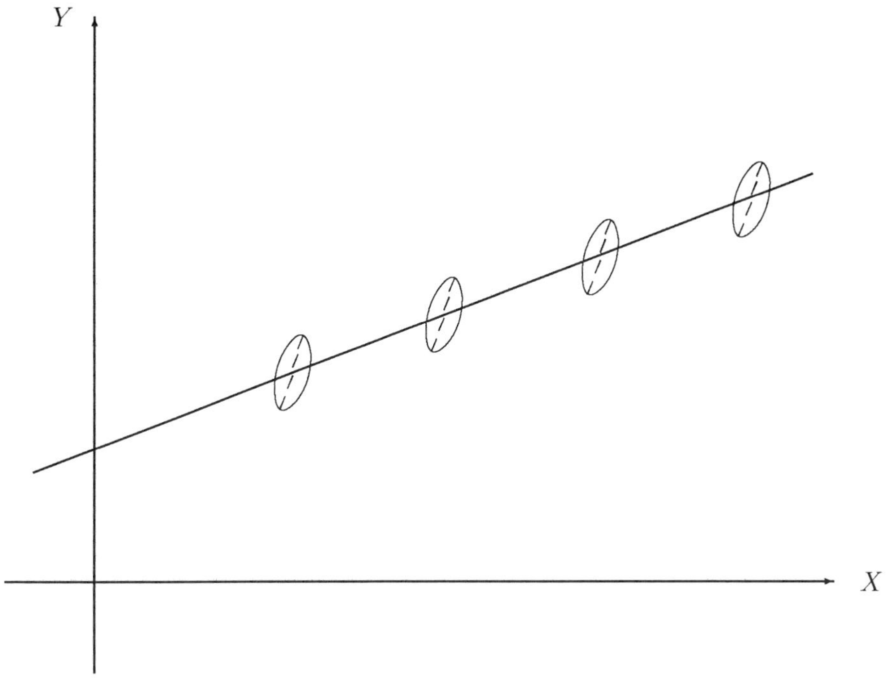

FIGURE 2

effects (e.g. α_i) and/or period specific and individual invariant effects, but the slope coefficients are the same for all individuals at all times [e.g. $\gamma_i = \gamma_j \ \forall\, i, j$ in (7.2)].

(iv) Error components model. The individual specific effects (α_i) and/or period-specific effects are treated as random variables.

(v) Random coefficients models. The coefficients of the model are different for different i [e.g. γ_i in (7.2)] and/or for different t. However, the differences are treated as different realizations of random variables that have a common probability distribution.

(vi) Mixed fixed and random coefficients models. That is, some of the coefficients are fixed and some of the coefficients are random variables with common means and a constant variance-covariance matrix.

For ease of exposition, in this paper we shall assume that there may be unobserved heterogeneity across i but not over t. We shall also only focus on the issue of modeling heterogeneity in a linear regression framework. The treatment of unobserved heterogeneity across i or over t is symmetric. All the results remain unchanged if there is unobserved heterogeneity over t but not across i with appropriate permutation of the data. Nor are the essential conclusions changed when both interindividual and intertemporal heterogeneity exist. For a more detailed discussion, see Hsiao (1986).

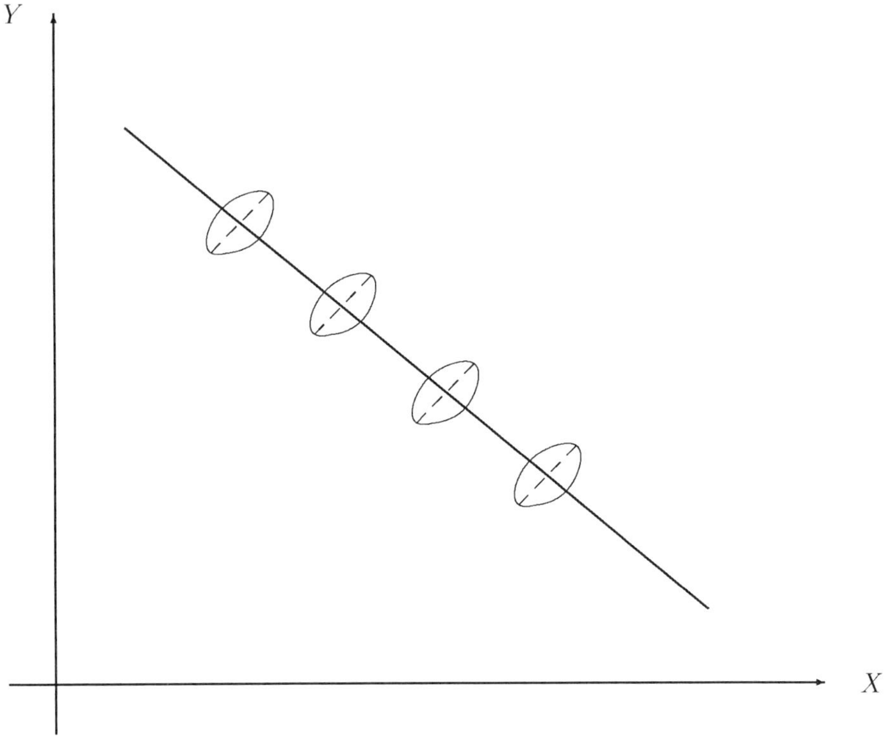

FIGURE 3

We begin with the assumption that the source of the individual heterogeneity is known. That is, we know the appropriate model, so the focus is on estimation. Then we shall discuss statistical procedures for discriminating among different formulations. In Section 2 we develop a general model that incorporates the cases above, and we derive Bayes estimates of the parameters because many different formulations can be viewed as a Bayesian linear regression model with different prior information on the parameters. We then provide detailed estimation methods for various special cases. In Section 3 we discuss two polar extreme cases — a common model for all observations and a different model for each individual, that is, choices (i) and (ii). Sections 4, 5, 6, and 7 discuss variable intercept [choice (iii)], error components [choice (iv)], random coefficients [choice (v)], and mixed fixed and random coefficients models [choice (vi)], respectively. In Section 8 we discuss issues of random versus fixed effects formulations. Conclusions are in Section 9.

2 A General Framework

2.1 The Basic Model

Suppose there are observations of $1 + k_1 + k_2$ variables $(y_{it}, \boldsymbol{x}'_{it}, \boldsymbol{z}'_{it})$ on N cross-sectional units over T time periods, where $i = 1, 2, \ldots, N$ and $t = 1, \ldots, T$. Let

$$\underset{NT\times 1}{\boldsymbol{y}} = \begin{pmatrix} \boldsymbol{y}_1 \\ \vdots \\ \boldsymbol{y}_N \end{pmatrix},$$

$$\underset{NT\times Nk_1}{\boldsymbol{X}} = \begin{pmatrix} \boldsymbol{X}_1 & \boldsymbol{0} & \ldots & \boldsymbol{0} \\ \boldsymbol{0} & \boldsymbol{X}_2 & & \vdots \\ \vdots & & \ddots & \\ \boldsymbol{0} & \ldots & & \boldsymbol{X}_N \end{pmatrix},$$

and

$$\underset{NT\times Nk_2}{\boldsymbol{Z}} = \begin{pmatrix} \boldsymbol{Z}_1 & \boldsymbol{0} & \ldots & \boldsymbol{0} \\ \boldsymbol{0} & \boldsymbol{Z}_2 & & \vdots \\ \vdots & & \ddots & \\ \boldsymbol{0} & \ldots & & \boldsymbol{Z}_N \end{pmatrix},$$

where

$$\underset{1\times T}{\boldsymbol{y}'_i} = (y_{i1}, \ldots, y_{iT}),$$

$$\underset{T\times k_1}{\boldsymbol{X}_i} = \begin{pmatrix} x_{1i1} & \ldots & x_{k_1i1} \\ \vdots & & \vdots \\ x_{1iT} & \ldots & x_{kiT} \end{pmatrix},$$

and

$$\underset{T\times k_2}{\boldsymbol{Z}_i} = \begin{pmatrix} z_{1i1} & \cdots & z_{k_2i1} \\ \vdots & & \vdots \\ z_{1iT} & \cdots & z_{k_2iT} \end{pmatrix}, \quad i = 1, \ldots, N.$$

We assume that

$$\boldsymbol{y} = \boldsymbol{X}\boldsymbol{\alpha} + \boldsymbol{Z}\boldsymbol{\gamma} + \boldsymbol{u}, \tag{7.5}$$

where $\boldsymbol{\alpha}$ and $\boldsymbol{\gamma}$ are $Nk_1 \times 1$ and $Nk_2 \times 1$ vectors, respectively, and the error term $\boldsymbol{u}$ is multivariate normally distributed with mean $\boldsymbol{0}$ and covariance matrix $\boldsymbol{C}_1$,

$$\underset{NT\times 1}{\boldsymbol{u}} = \begin{pmatrix} \boldsymbol{u}_1 \\ \vdots \\ \boldsymbol{u}_N \end{pmatrix} \sim \mathcal{N}(\boldsymbol{0}, \boldsymbol{C}_1), \tag{7.6}$$

and $\boldsymbol{u}_i' = (u_{i1}, \ldots, u_{iT})$. We let the $NT \times NT$ covariance matrix $\boldsymbol{C}_1$ be unrestricted to allow for correlations across cross-sectional units and over time. We assume that the variables $\boldsymbol{X}$ have random coefficients and let the $Nk_1 \times 1$ vector $\boldsymbol{\alpha}$ satisfy

$$\boldsymbol{\alpha} = \begin{pmatrix} \boldsymbol{\alpha}_1 \\ \vdots \\ \boldsymbol{\alpha}_N \end{pmatrix} = \boldsymbol{A}_1 \bar{\boldsymbol{\alpha}} + \boldsymbol{\epsilon} \tag{7.7}$$

where $\boldsymbol{\alpha}_i$ is a $k_1 \times 1$ vector, $i = 1, 2, \ldots, N$, $\boldsymbol{A}_1$ is an $Nk_1 \times m$ matrix with known elements, $\bar{\boldsymbol{\alpha}}$ is an $m \times 1$ vector of constants, and

$$\boldsymbol{\epsilon} \sim \mathcal{N}(\boldsymbol{0}, \boldsymbol{C}_2). \tag{7.8}$$

The variance-covariance matrix $\boldsymbol{C}_2$ is assumed to be nonsingular. The variables $\boldsymbol{Z}$ have fixed coefficients (effects). The $Nk_2 \times 1$ vector $\boldsymbol{\gamma}$ is assumed to satisfy

$$\boldsymbol{\gamma} = \begin{pmatrix} \boldsymbol{\gamma}_1 \\ \vdots \\ \boldsymbol{\gamma}_N \end{pmatrix} = \boldsymbol{A}_2 \bar{\boldsymbol{\gamma}} \tag{7.9}$$

where each $\boldsymbol{\gamma}_i$ is $k_2 \times 1$, $\boldsymbol{A}_2$ is an $Nk_2 \times n$ matrix with known elements and $\bar{\boldsymbol{\gamma}}$ is an $n \times 1$ vector of constants. Because $\boldsymbol{A}_2$ is known, (7.5) is formally identical to

$$\boldsymbol{y} = \boldsymbol{X\alpha} + \tilde{\boldsymbol{Z}}\bar{\boldsymbol{\gamma}} + \boldsymbol{u}, \tag{7.10}$$

where $\tilde{\boldsymbol{Z}} = \boldsymbol{Z A}_2$. However, in the form (7.10) we allow for various possible fixed parameters configurations. For instance, if we wish to allow $\boldsymbol{\gamma}_i$ to be different across cross-sectional units, we can let $\boldsymbol{A}_2 = \boldsymbol{I}_N \otimes \boldsymbol{I}_{k_2}$, where $\boldsymbol{I}_p$ denotes a $p \times p$ identity matrix. On the other hand, if we wish to constrain $\boldsymbol{\gamma}_i = \boldsymbol{\gamma}_j$, for all i and j, we can let $\boldsymbol{A}_2 = \boldsymbol{e}_N \otimes \boldsymbol{I}_{k_2}$, where $\boldsymbol{e}_N$ is an $N \times 1$ vector with each element equal to one.

We shall see later that many of the widely used models for pooling cross-section and time-series data are in fact special cases of (7.5) – (7.9).

2.2 A Bayes Solution

If we take the formulation (7.7), (7.8), and (7.9) to mean that there is prior information on the parameters $\boldsymbol{\alpha}$ (randomly distributed around the mean $\boldsymbol{A}_1\bar{\boldsymbol{\alpha}}$ with covariance matrix $\boldsymbol{C}_2$) and no information on the parameters $\bar{\boldsymbol{\gamma}}$, we can derive estimates of (7.10) by following a Bayes procedure (for an introduction to Bayes inference, see Judge, Griffiths, Hill, Lütkepohl, and Lee 1985). We shall first derive the results by assuming that $\boldsymbol{X}'\tilde{\boldsymbol{Z}} = \boldsymbol{0}$ and $\boldsymbol{C}_1 = \sigma^2 \boldsymbol{I}$, then derive the general results. This is partly for pedagogical reasons, and partly because some of the widely used panel data models are derivable from (7.10) by either assuming $\boldsymbol{X} = \boldsymbol{0}$ or $\tilde{\boldsymbol{Z}} = \boldsymbol{0}$. The orthogonality condition between $\boldsymbol{X}$ and $\tilde{\boldsymbol{Z}}$ ensures that the Bayes point estimates of $\boldsymbol{\alpha}$, $\bar{\boldsymbol{\gamma}}$ and $\bar{\boldsymbol{\alpha}}$ are not affected when either $\boldsymbol{X} = \boldsymbol{0}$ or $\tilde{\boldsymbol{Z}} = \boldsymbol{0}$. We assume that

A1. The prior distribution of $\boldsymbol{\alpha}$ and $\bar{\boldsymbol{\gamma}}$ are independent, that is

$$p(\boldsymbol{\alpha}, \bar{\boldsymbol{\gamma}}) = p(\boldsymbol{\alpha}) \cdot p(\bar{\boldsymbol{\gamma}}) \tag{7.11}$$

A2. $$p(\boldsymbol{\alpha}) \sim \mathcal{N}(\boldsymbol{A}_1\bar{\boldsymbol{\alpha}}, \boldsymbol{C}_2) \tag{7.12}$$

A3. There is no information about $\bar{\boldsymbol{\alpha}}$ and $\bar{\gamma}$, that is, $p(\bar{\boldsymbol{\alpha}})$ and $p(\bar{\gamma})$ are independent and

$$p(\bar{\boldsymbol{\alpha}}) \propto \text{constant}, \tag{7.13}$$
$$p(\bar{\gamma}) \propto \text{constant}, \tag{7.14}$$

THEOREM 1: Suppose that given $\boldsymbol{\alpha}$ and $\bar{\gamma}$,

$$\boldsymbol{y} \sim \mathcal{N}(\boldsymbol{X}\boldsymbol{\alpha} + \tilde{\boldsymbol{Z}}\bar{\gamma}, \sigma^2 \boldsymbol{I}_{NT}) \tag{7.15}$$

Under A.1 - A.3, if $\boldsymbol{X}'\tilde{\boldsymbol{Z}} = \boldsymbol{0}$, then:

(i) The posterior distribution of $\boldsymbol{\alpha}$ given $\bar{\boldsymbol{\alpha}}$ and $\boldsymbol{y}$ is $\mathcal{N}(\boldsymbol{\alpha}^*, \boldsymbol{D}_1)$, where

$$\boldsymbol{\alpha}^* = \left(\frac{1}{\sigma^2}\boldsymbol{X}'\boldsymbol{X} + \boldsymbol{C}_2^{-1}\right)^{-1}\left(\frac{1}{\sigma^2}\boldsymbol{X}'\boldsymbol{y} + \boldsymbol{C}_2^{-1}\boldsymbol{A}_1\bar{\boldsymbol{\alpha}}\right), \tag{7.16}$$
$$\boldsymbol{D}_1 = \left(\frac{1}{\sigma^2}\boldsymbol{X}'\boldsymbol{X} + \boldsymbol{C}_2^{-1}\right)^{-1}. \tag{7.17}$$

(ii) The posterior distribution of $\bar{\boldsymbol{\alpha}}$ given $\boldsymbol{y}$ is $\mathcal{N}(\bar{\boldsymbol{\alpha}}^*, \boldsymbol{D}_2)$, where

$$\bar{\boldsymbol{\alpha}}^* = \boldsymbol{D}_2\boldsymbol{A}_1'\boldsymbol{C}_2^{-1}\left(\frac{1}{\sigma^2}\boldsymbol{X}'\boldsymbol{X} + \boldsymbol{C}_2^{-1}\right)^{-1}\left(\frac{1}{\sigma^2}\boldsymbol{X}'\boldsymbol{y}\right), \tag{7.18}$$
$$\boldsymbol{D}_2 = \left\{\boldsymbol{A}_1'\left[\boldsymbol{C}_2 + \sigma^2(\boldsymbol{X}'\boldsymbol{X})^{-1}\right]^{-1}\boldsymbol{A}_1\right\}^{-1}. \tag{7.19}$$

(iii) The posterior distribution of $\bar{\gamma}$ given $\boldsymbol{y}$ is $\mathcal{N}(\bar{\gamma}^*, \boldsymbol{D}_3)$, where

$$\bar{\gamma}^* = \left(\tilde{\boldsymbol{Z}}'\tilde{\boldsymbol{Z}}\right)^{-1}\tilde{\boldsymbol{Z}}'\boldsymbol{y}, \tag{7.20}$$
$$\boldsymbol{D}_3 = \sigma^2\left(\tilde{\boldsymbol{Z}}'\tilde{\boldsymbol{Z}}\right)^{-1}. \tag{7.21}$$

(iv) The (unconditional) posterior distribution of $\boldsymbol{\alpha}$ given $\boldsymbol{y}$ is $\mathcal{N}(\tilde{\boldsymbol{\alpha}}, \tilde{\boldsymbol{D}}_1)$, where

$$\tilde{\boldsymbol{\alpha}} = \left[\frac{1}{\sigma^2}\boldsymbol{X}'\boldsymbol{X} + \boldsymbol{C}_2^{-1} - \boldsymbol{C}_2^{-1}\boldsymbol{A}_1\left(\boldsymbol{A}_1'\boldsymbol{C}_2^{-1}\boldsymbol{A}_1\right)^{-1}\boldsymbol{A}_1'\boldsymbol{C}_2^{-1}\right]^{-1}\left(\frac{1}{\sigma^2}\boldsymbol{X}'\boldsymbol{y}\right) \tag{7.22}$$
$$\tilde{\boldsymbol{D}}_1 = \left[\frac{1}{\sigma^2}\boldsymbol{X}'\boldsymbol{X} + \boldsymbol{C}_2^{-1} - \boldsymbol{C}_2^{-1}\boldsymbol{A}_1\left(\boldsymbol{A}_1'\boldsymbol{C}_2^{-1}\boldsymbol{A}_1\right)^{-1}\boldsymbol{A}_1'\boldsymbol{C}_2^{-1}\right]^{-1}. \tag{7.23}$$

PROOF: By Bayes Theorem,

$$\begin{aligned} p(\boldsymbol{\alpha}, \bar{\gamma}, \bar{\boldsymbol{\alpha}}|\boldsymbol{y}) &\propto p(\boldsymbol{y}|\boldsymbol{\alpha}, \bar{\gamma}, \bar{\gamma}) \cdot p(\boldsymbol{\alpha}, \bar{\gamma}, \bar{\boldsymbol{\alpha}}) \\ &= p(\boldsymbol{y}|\boldsymbol{\alpha}, \bar{\gamma}) \cdot p(\boldsymbol{\alpha}|\bar{\boldsymbol{\alpha}}) \cdot p(\bar{\boldsymbol{\alpha}}) \cdot p(\bar{\gamma}). \end{aligned} \tag{7.24}$$

The product on the right-hand side is proportional to $\exp\left\{-\frac{1}{2}Q\right\}$, where

$$Q = \frac{1}{\sigma^2}\left(\boldsymbol{y} - \boldsymbol{X}\boldsymbol{\alpha} - \tilde{\boldsymbol{Z}}\bar{\gamma}\right)'\left(\boldsymbol{y} - \boldsymbol{X}\boldsymbol{\alpha} - \tilde{\boldsymbol{Z}}\bar{\gamma}\right) + (\boldsymbol{\alpha} - \boldsymbol{A}_1\bar{\boldsymbol{\alpha}})'\boldsymbol{C}_2^{-1}(\boldsymbol{\alpha} - \boldsymbol{A}_1\bar{\boldsymbol{\alpha}}). \tag{7.25}$$

Rearranging Q in terms of the quadratic forms of $\boldsymbol{\alpha}$, $\bar{\gamma}$, $\bar{\boldsymbol{\alpha}}$, and $\boldsymbol{y}$ and noting that $\boldsymbol{X}'\tilde{\boldsymbol{Z}} = \boldsymbol{0}$, we have

$$
\begin{aligned}
Q &= \boldsymbol{\alpha}'\left[\frac{1}{\sigma^2}\boldsymbol{X}'\boldsymbol{X} + \boldsymbol{C}_2^{-1}\right]\boldsymbol{\alpha} - 2\boldsymbol{\alpha}'\left[\frac{1}{\sigma^2}\boldsymbol{X}'\boldsymbol{y} + \boldsymbol{C}_2^{-1}\boldsymbol{A}_1\bar{\boldsymbol{\alpha}}\right] \\
&\quad + \frac{1}{\sigma^2}\bar{\gamma}'\tilde{\boldsymbol{Z}}'\tilde{\boldsymbol{Z}}\bar{\gamma} - \frac{2}{\sigma^2}\bar{\gamma}'\tilde{\boldsymbol{Z}}'\boldsymbol{y} + \bar{\boldsymbol{\alpha}}'\boldsymbol{A}_1'\boldsymbol{C}_2^{-1}\boldsymbol{A}_1\bar{\boldsymbol{\alpha}} + \frac{1}{\sigma^2}\boldsymbol{y}'\boldsymbol{y} \\
&= Q_1 + Q_2 + \bar{\boldsymbol{\alpha}}'\boldsymbol{A}_1'\left[\boldsymbol{C}_2^{-1} - \boldsymbol{C}_2^{-1}\left(\frac{1}{\sigma^2}\boldsymbol{X}'\boldsymbol{X} + \boldsymbol{C}_2^{-1}\right)^{-1}\boldsymbol{C}_2^{-1}\right]\boldsymbol{A}_1\bar{\boldsymbol{\alpha}} \\
&\quad - 2\bar{\boldsymbol{\alpha}}'\boldsymbol{A}_1'\boldsymbol{C}_2^{-1}\left(\frac{1}{\sigma^2}\boldsymbol{X}'\boldsymbol{X} + \boldsymbol{C}_2^{-1}\right)^{-1}\left(\frac{1}{\sigma^2}\boldsymbol{X}'\boldsymbol{y}\right) \\
&\quad - \frac{1}{\sigma^4}\boldsymbol{y}'\boldsymbol{X}\left(\frac{1}{\sigma^2}\boldsymbol{X}'\boldsymbol{X} + \boldsymbol{C}_2^{-1}\right)^{-1}\boldsymbol{X}'\boldsymbol{y} - \frac{1}{\sigma^2}\boldsymbol{y}'\tilde{\boldsymbol{Z}}\left(\tilde{\boldsymbol{Z}}'\tilde{\boldsymbol{Z}}\right)^{-1}\tilde{\boldsymbol{Z}}'\boldsymbol{y} \\
&\quad + \frac{1}{\sigma^2}\boldsymbol{y}'\boldsymbol{y} \\
&= Q_1 + Q_2 + Q_3 + Q_4
\end{aligned}
\tag{7.26}
$$

where

$$
\begin{aligned}
Q_1 &= \left\{\boldsymbol{\alpha} - \left(\frac{1}{\sigma^2}\boldsymbol{X}'\boldsymbol{X} + \boldsymbol{C}_2^{-1}\right)^{-1}\left(\frac{1}{\sigma^2}\boldsymbol{X}'\boldsymbol{y} + \boldsymbol{C}_2^{-1}\boldsymbol{A}_1\bar{\boldsymbol{\alpha}}\right)\right\}'\left(\frac{1}{\sigma^2}\boldsymbol{X}'\boldsymbol{X} + \boldsymbol{C}_2^{-1}\right) \\
&\quad \cdot\left\{\boldsymbol{\alpha} - \left(\frac{1}{\sigma^2}\boldsymbol{X}'\boldsymbol{X} + \boldsymbol{C}_2^{-1}\right)^{-1}\left(\frac{1}{\sigma^2}\boldsymbol{X}'\boldsymbol{y} + \boldsymbol{C}_2^{-1}\boldsymbol{A}_1\bar{\boldsymbol{\alpha}}\right)\right\},
\end{aligned}
\tag{7.27}
$$

$$
Q_2 = \left[\bar{\gamma} - \left(\tilde{\boldsymbol{Z}}'\tilde{\boldsymbol{Z}}\right)^{-1}\tilde{\boldsymbol{Z}}'\boldsymbol{y}\right]'\left(\frac{1}{\sigma^2}\tilde{\boldsymbol{Z}}'\tilde{\boldsymbol{Z}}\right)\left[\bar{\gamma} - \left(\tilde{\boldsymbol{Z}}'\tilde{\boldsymbol{Z}}\right)^{-1}\tilde{\boldsymbol{Z}}'\boldsymbol{y}\right],
\tag{7.28}
$$

$$
\begin{aligned}
Q_3 &= \left\{\bar{\boldsymbol{\alpha}} - \left(\boldsymbol{A}_1'\left[\boldsymbol{C}_2 + \sigma^2(\boldsymbol{X}'\boldsymbol{X})^{-1}\right]^{-1}\boldsymbol{A}_1\right)^{-1}\boldsymbol{A}_1'\boldsymbol{C}_2^{-1}\right. \\
&\quad \left.\cdot\left(\frac{1}{\sigma^2}\boldsymbol{X}'\boldsymbol{X} + \boldsymbol{C}_2^{-1}\right)^{-1}\left(\frac{1}{\sigma^2}\boldsymbol{X}'\boldsymbol{y}\right)\right\}' \\
&\quad \cdot\left\{\boldsymbol{A}_1'\left[\boldsymbol{C}_2 + \sigma^2(\boldsymbol{X}'\boldsymbol{X})^{-1}\right]^{-1}\boldsymbol{A}_1\right\} \\
&\quad \cdot\left\{\bar{\boldsymbol{\alpha}} - \left(\boldsymbol{A}_1'\left[\boldsymbol{C}_2 + \sigma^2(\boldsymbol{X}'\boldsymbol{X})^{-1}\right]^{-1}\boldsymbol{A}_1\right)^{-1}\right. \\
&\quad \left.\cdot\boldsymbol{A}_1'\boldsymbol{C}_2^{-1}\left(\frac{1}{\sigma^2}\boldsymbol{X}'\boldsymbol{X} + \boldsymbol{C}_2^{-1}\right)^{-1}\left(\frac{1}{\sigma^2}\boldsymbol{X}'\boldsymbol{y}\right)\right\},
\end{aligned}
\tag{7.29}
$$

and

$$
\begin{aligned}
Q_4 &= \boldsymbol{y}'\left[\frac{1}{\sigma^2}I - \frac{1}{\sigma^4}\boldsymbol{X}\left(\frac{1}{\sigma^2}\boldsymbol{X}'\boldsymbol{X} + \boldsymbol{C}_2^{-1}\right)^{-1}\boldsymbol{X}' - \frac{1}{\sigma^2}\tilde{\boldsymbol{Z}}\left(\tilde{\boldsymbol{Z}}'\tilde{\boldsymbol{Z}}\right)^{-1}\tilde{\boldsymbol{Z}}'\right. \\
&\quad - \frac{1}{\sigma^4}\boldsymbol{X}\left(\frac{1}{\sigma^2}\boldsymbol{X}'\boldsymbol{X} + \boldsymbol{C}_2^{-1}\right)^{-1}\boldsymbol{C}_2^{-1}\boldsymbol{A}_1 \\
&\quad \cdot\left\{\boldsymbol{A}_1'\left[\boldsymbol{C}_2 + \sigma^2(\boldsymbol{X}'\boldsymbol{X})^{-1}\right]^{-1}\boldsymbol{A}_1\right\}^{-1}
\end{aligned}
\tag{7.30}
$$

$$\cdot A_1' C_2^{-1} \left(\frac{1}{\sigma^2} X'X + C_2^{-1} \right)^{-1} X' \Big] y \quad .$$

The form of Q_3 and Q_4 are derived by using the identity relation

$$(D + BFB')^{-1} = D^{-1} - D^{-1}B \left(B'D^{-1}B + F^{-1} \right)^{-1} B'D^{-1} \tag{7.31}$$

and its special case

$$(D + F)^{-1} = D^{-1} - D^{-1} \left(D^{-1} + F^{-1} \right)^{-1} D^{-1} \quad . \tag{7.32}$$

To derive results (i) - (iii), we note that as far as the joint distribution of $\alpha, \bar{\gamma}$, and $\bar{\alpha}$ given y is concerned, Q_4 is a constant. Therefore, Q_4 can be ignored. Furthermore, because of (7.11) and $X'\tilde{Z} = 0$, $p(\alpha, \bar{\alpha}, \bar{\gamma}|y) = p(\bar{\gamma}|y) \cdot p(\alpha, \bar{\alpha}|y) = p(\bar{\gamma}|y) \cdot p(\alpha|\bar{\alpha}, y) \cdot p(\bar{\alpha}|y)$. Since $p(\bar{\gamma}|y)$ is proportional to $\exp\{-\frac{1}{2}Q_2\}$, (iii) holds. Conditional on $\bar{\alpha}$ and y, Q_3 is a constant, $p(\alpha|\bar{\alpha}, y)$ is proportional to $\exp\{-\frac{1}{2}Q_1\}$, therefore (i) holds. To derive (ii), we note that $p(\bar{\alpha}|y) = \int p(\alpha, \bar{\alpha}|y)d\alpha = \int p(\alpha|\bar{\alpha}, y)d\alpha \cdot p(\bar{\alpha}|y)$, which is proportional to exp $\{-\frac{1}{2}Q_3\}$.

To derive (iv), we need to know

$$p(\alpha|y) = \int p(\alpha, \bar{\alpha}|y)d\bar{\alpha} \tag{7.33}$$

Equation (7.33) can be easily derived by considering $p(\alpha, \bar{\alpha}|y)$ as the product of $p(\bar{\alpha}|\alpha, y)$ and $p(\alpha|y)$. Therefore, we rewrite Q as

$$\begin{aligned} Q &= Q_3^* - \alpha' C_2^{-1} A_1 (A_1' C_2^{-1} A_1)^{-1} A_1' C_2^{-1} \alpha \\ &\quad + \alpha' [\frac{1}{\sigma^2} X'X + C_2^{-1}] \alpha - \frac{2}{\sigma^2} \alpha' X' y + \frac{1}{\sigma^2} (y - \tilde{Z}\bar{\gamma})' y - \tilde{Z}\bar{\gamma}) \\ &= Q_1^* + Q_2 + Q_3^* + Q_4^* \end{aligned} \tag{7.34}$$

where

$$\begin{aligned} Q_1^* &= \{\alpha - [\frac{1}{\sigma^2} X'X + C_2^{-1} - C_2^{-1} A_1 (A_1' C_2^{-1} A_1)^{-1} \\ &\quad \cdot A_1' C_2^{-1}]^{-1} (\frac{1}{\sigma^2} X' y)\}' \\ &\quad \cdot [\frac{1}{\sigma^2} X'X + C_2^{-1} - C_2^{-1} A_1 (A_1' C_2^{-1} A_1)^{-1} A_1' C_2^{-1}] \\ &\quad \cdot \{\alpha - [\frac{1}{\sigma^2} X'X + C_2^{-1} - C_2^{-1} A_1 (A_1' C_2^{-1} A_1)^{-1} \\ &\quad \cdot A_1' C_2^{-1}]^{-1} (\frac{1}{\sigma^2} X' y)\}, \end{aligned} \tag{7.35}$$

$$\begin{aligned} Q_3^* &= [\bar{\alpha} - (A_1' C_2^{-1} A_1)^{-1} A_1' C_2^{-1} \alpha]' (A_1' C_2^{-1} A_1) \\ &\quad \cdot [\bar{\alpha} - (A_1' C_2^{-1} A_1)^{-1} A_1' C_2^{-1} \alpha], \end{aligned} \tag{7.36}$$

and

$$\begin{aligned} Q_4^* &= y' \{\frac{1}{\sigma^2} I - \frac{1}{\sigma^2} \tilde{Z}(\tilde{Z}'\tilde{Z})^{-1}\tilde{Z}' - \frac{1}{\sigma^4} X [\frac{1}{\sigma^2} X'X + C_2^{-1} \\ &\quad - C_2^{-1} A_1 (A_1' C_2^{-1} A_1)^{-1} A_1' C_2^{-1}]^{-1} X'\} y. \end{aligned} \tag{7.37}$$

Integrating $\exp\{-\frac{1}{2}Q\}$ with respect to $\bar{\alpha}$ and $\bar{\gamma}$, we have

$$p(\alpha|y) \propto \exp\left\{-\frac{1}{2}Q_1^*\right\}, \tag{7.38}$$

hence (iv).

We now consider the general case where $C_1 \neq \sigma^2 I$ and $X'\tilde{Z} \neq 0$.

THEOREM 2: Suppose that given α and $\bar{\gamma}$,

$$y \sim \mathcal{N}(X\alpha + \tilde{Z}\bar{\gamma}, C_1), \tag{7.39}$$

and $\alpha, \bar{\gamma}$ satisfy (7.11) – (7.14). Then: (i) The posterior distribution of α given $\bar{\alpha}$ and y is $\mathcal{N}(\alpha^{**}, D_1^{**})$, where

$$\begin{aligned} \alpha^{**} &= \{X'[C_1^{-1} - C_1^{-1}\tilde{Z}(\tilde{Z}'C_1^{-1}\tilde{Z})^{-1}\tilde{Z}'C_1^{-1}]X + C_2^{-1}\}^{-1} \\ &\quad \cdot \{X'[C_1^{-1} - C_1^{-1}\tilde{Z}(\tilde{Z}'C_1^{-1}\tilde{Z})^{-1}\tilde{Z}'C_1^{-1}]y + C_2^{-1}A_1\bar{\alpha}\}, \end{aligned} \tag{7.40}$$

$$D_1^{**} = \{X'[C_1^{-1} - C_1^{-1}\tilde{Z}(\tilde{Z}'C_1^{-1}\tilde{Z})^{-1}\tilde{Z}'C_1^{-1}]X + C_2^{-1}\}^{-1} \tag{7.41}$$

(ii) The (unconditional) posterior distribution of α given y is $\mathcal{N}(\tilde{\alpha}^{**}, \tilde{D}_1^{**})$, where

$$\tilde{\alpha}^{**} = \tilde{D}_1^{**}\{X'[C_1^{-1} - C_1^{-1}\tilde{Z}(\tilde{Z}'C_1^{-1}\tilde{Z})^{-1}\tilde{Z}'C_1^{-1}]y\}, \tag{7.42}$$

$$\begin{aligned} \tilde{D}_1^{**} &= \{X'[C_1^{-1} - C_1^{-1}\tilde{Z}(\tilde{Z}'C_1^{-1}\tilde{Z})^{-1}\tilde{Z}'C_1^{-1}]X + C_2^{-1} \\ &\quad - C_2^{-1}A_1(A_1'C_2^{-1}A_1)^{-1}A_1'C_2^{-1}\}^{-1}. \end{aligned} \tag{7.43}$$

(iii) The posterior distribution of $\bar{\alpha}$ given y is $\mathcal{N}(\bar{\alpha}^{**}, \tilde{D}_2)$, where

$$\begin{aligned} \bar{\alpha}^{**} &= \tilde{D}_2\{A_1'C_2^{-1}(X'[C_1^{-1} - C_1^{-1}\tilde{Z}(\tilde{Z}'C_1^{-1}\tilde{Z})^{-1}\tilde{Z}'C_1^{-1}] \cdot X + C_2^{-1})^{-1} \\ &\quad \cdot X'[C_1^{-1} - C_1^{-1}\tilde{Z}(\tilde{Z}'C_1^{-1}\tilde{Z})^{-1}\tilde{Z}'C_1^{-1}]y\}, \end{aligned} \tag{7.44}$$

$$\begin{aligned} \tilde{D}_2 &= \{A_1'[C_2^{-1} - C_2^{-1}(X'C_1^{-1}X - X'C_1^{-1}\tilde{Z}(\tilde{Z}'C_1^{-1}\tilde{Z})^{-1} \\ &\quad \cdot \tilde{Z}'C_1^{-1}X + C_2^{-1})^{-1}C_2^{-1}]A_1\}^{-1} \\ &= \{A_1'(C_2 + [X'(C_1^{-1} - C_1^{-1}\tilde{Z}(\tilde{Z}'C_1^{-1}\tilde{Z})^{-1} \\ &\quad \cdot \tilde{Z}'C_1^{-1})X]^{-1})^{-1}A_1\}^{-1}. \end{aligned} \tag{7.45}$$

(iv) The posterior distribution of $\bar{\gamma}$ given y and α is $\mathcal{N}(\hat{\bar{\gamma}}, \hat{D}_3)$, where

$$\hat{\bar{\gamma}} = (\tilde{Z}'C_1^{-1}\tilde{Z})^{-1}\tilde{Z}'C_1^{-1}(y - X\alpha), \tag{7.46}$$

$$\hat{D}_3 = (\tilde{Z}'C_1^{-1}\tilde{Z})^{-1}. \tag{7.47}$$

(v) The posterior distribution of $\bar{\gamma}$ given y and $\bar{\alpha}$ is $\mathcal{N}(\hat{\bar{\gamma}}^*, \hat{D}_3^*)$, where

$$\begin{aligned} \hat{\bar{\gamma}}^* &= \hat{D}_3^*\tilde{Z}'\{[C_1^{-1} - C_1^{-1}X(X'C_1^{-1}X + C_2^{-1})^{-1}X'C_1^{-1}]y \\ &\quad - C_1^{-1}X(X'C_1^{-1}X + C_2^{-1})^{-1}C_2^{-1}A_1\bar{\alpha}\}, \end{aligned} \tag{7.48}$$

$$\hat{D}_3^* = \{\tilde{Z}'[C_1^{-1} - C_1^{-1}X(X'C_1^{-1}X + C_2^{-1})^{-1}X'C_1^{-1}]\tilde{Z}\}^{-1}. \tag{7.49}$$

(vi) The (unconditional) posterior distribution of $\bar{\gamma}$ given $\boldsymbol{y}$ is $\mathcal{N}(\bar{\gamma}^{**}, \boldsymbol{D}_3^{**})$, where

$$\begin{aligned} \bar{\gamma}^{**} &= \boldsymbol{D}_3^{**}\tilde{\boldsymbol{Z}}'\{\boldsymbol{C}_1^{-1} - \boldsymbol{C}_1^{-1}\boldsymbol{X}[\boldsymbol{X}'\boldsymbol{C}_1^{-1}\boldsymbol{X} + \boldsymbol{C}_2^{-1} \\ &\quad - \boldsymbol{C}_2^{-1}\boldsymbol{A}_1(\boldsymbol{A}_1'\boldsymbol{C}_2^{-1}\boldsymbol{A}_1)^{-1}\boldsymbol{A}_1'\boldsymbol{C}_2^{-1}]^{-1}\boldsymbol{X}'\boldsymbol{C}_1^{-1}\}^{-1}\boldsymbol{y}. \quad (7.50) \\ \boldsymbol{D}_3^{**} &= [\tilde{\boldsymbol{Z}}'\{\boldsymbol{C}_1^{-1} - \boldsymbol{C}_1^{-1}\boldsymbol{X}[\boldsymbol{X}'\boldsymbol{C}_1^{-1}\boldsymbol{X} + \boldsymbol{C}_2^{-1} - \boldsymbol{C}_2^{-1}\boldsymbol{A}_1(\boldsymbol{A}_1'\boldsymbol{C}_2^{-1}\boldsymbol{A}_1)^{-1} \quad (7.51) \\ &\quad \boldsymbol{A}_1'\boldsymbol{C}_2^{-1}]^{-1}\boldsymbol{X}'\boldsymbol{C}_1^{-1}\}\tilde{\boldsymbol{Z}}]^{-1}. \end{aligned}$$

(vii) The (unconditional) joint posterior distribution of $\boldsymbol{\alpha}$ and $\bar{\gamma}$ given $\boldsymbol{y}$ is

$$\mathcal{N}\left(\begin{pmatrix} \tilde{\boldsymbol{\alpha}}^{**} \\ \bar{\boldsymbol{y}}^{**} \end{pmatrix}, \boldsymbol{V}\right) \quad (7.52)$$

where

$$\boldsymbol{V}^{-1} = \begin{pmatrix} \boldsymbol{X}'\boldsymbol{C}_1^{-1}\boldsymbol{X} + \boldsymbol{C}_2^{-1} - \boldsymbol{C}_2^{-1}\boldsymbol{A}(\boldsymbol{A}_1'\boldsymbol{C}_2^{-1}\boldsymbol{A}_1)^{-1}\boldsymbol{A}_1'\boldsymbol{C}_2^{-1} & \boldsymbol{X}'\boldsymbol{C}^{-1}\tilde{\boldsymbol{Z}} \\ \tilde{\boldsymbol{Z}}'\boldsymbol{C}_1^{-1}\boldsymbol{X} & \tilde{\boldsymbol{Z}}'\boldsymbol{C}_1^{-1}\tilde{\boldsymbol{Z}} \end{pmatrix} \quad (7.53)$$

PROOF: Let

$$\begin{aligned} \boldsymbol{y}^* &= \boldsymbol{C}^{-\frac{1}{2}}\boldsymbol{y},\ \boldsymbol{X}^* = \boldsymbol{C}_1^{-\frac{1}{2}}\boldsymbol{X},\ \tilde{\boldsymbol{Z}}^* = \boldsymbol{C}_1^{-\frac{1}{2}}\tilde{\boldsymbol{Z}}, \\ \boldsymbol{u}^* &= \boldsymbol{C}_1^{-\frac{1}{2}}\boldsymbol{u}, \end{aligned} \quad (7.54)$$

and

$$\tilde{\boldsymbol{X}}^* = [\boldsymbol{I} - \tilde{\boldsymbol{Z}}^*(\tilde{\boldsymbol{Z}}^{*\prime}\tilde{\boldsymbol{Z}}^*)^{-1}\tilde{\boldsymbol{Z}}^{*\prime}]\boldsymbol{X}^*. \quad (7.55)$$

We can rewrite (7.10) as

$$\boldsymbol{y}^* = \tilde{\boldsymbol{X}}^*\boldsymbol{\alpha} + \tilde{\boldsymbol{Z}}^*\tilde{\gamma} + \boldsymbol{u}^*, \quad (7.56)$$

where

$$\tilde{\boldsymbol{Z}}^{*\prime}\tilde{\boldsymbol{X}}^* = \boldsymbol{0}, \quad (7.57)$$

$$\tilde{\gamma} = \bar{\gamma} + (\tilde{\boldsymbol{Z}}^{*\prime}\tilde{\boldsymbol{Z}}^*)^{-1}\tilde{\boldsymbol{Z}}^{*\prime}\boldsymbol{X}^*\boldsymbol{\alpha}, \quad (7.58)$$

and

$$E\boldsymbol{u}^* = \boldsymbol{0},\ E\boldsymbol{u}^*\boldsymbol{u}^{*\prime} = \boldsymbol{I}. \quad (7.59)$$

Substituting (7.54) and (7.55) into (7.14) – (7.17), (7.20), and (7.21), we have Theorem 2(i)-(iii).

By Theorem 1 (iii), the posterior distribution of $\tilde{\gamma}$ is $\mathcal{N}(\tilde{\gamma}^*, \tilde{\boldsymbol{D}}_3)$, where

$$\begin{aligned} \tilde{\gamma}^* &= (\tilde{\boldsymbol{Z}}^{*\prime}\tilde{\boldsymbol{Z}}^*)^{-1}\tilde{\boldsymbol{Z}}^{*\prime}\boldsymbol{y}^*, \quad (7.60) \\ \tilde{\boldsymbol{D}}_3 &= (\tilde{\boldsymbol{Z}}^{*\prime}\tilde{\boldsymbol{Z}}^*)^{-1}. \quad (7.61) \end{aligned}$$

Theorem 2(iv) follows from (7.58).

Substituting α^{**} from (7.40) into (7.56), we have

$$\hat{\hat{\gamma}}^* = \tilde{\gamma}^* - (\tilde{Z}^{*\prime}\tilde{Z}^*)^{-1}\tilde{Z}^{*\prime}X\alpha^{**}. \tag{7.62}$$

We derive theorem 2(v) by substituting (7.40), (7.41), and (7.60) into (7.62).

Similarly, theorem 2(vi) is derived by substituting (7.44), (7.45), and (7.60) into (7.58).

Theorem 2(vii) follows from the definition of the joint distribution as the product of the conditional and marginal distributions.

Many of the widely used estimators for various panel data models can be obtained by using theorems 1 and 2. In the following sections we shall relate these theorems to the commonly known estimators for various formulations that have been suggested to model unobserved heterogeneity, including a common model for all cross-sectional observations, a different model for each cross-sectional unit, variable intercept models, error components models, random coefficients models, and mixed fixed and random coefficients models.

3 Two Extreme Cases — All Cross-Sectional Units Have the Same Behavioral Pattern versus Different Units Have Different Behavioral Patterns

3.1 A Common Model for All Cross-Sectional Units

If there are no interindividual differences in behavioral patterns, we may assume that the regression coefficients are identical for all cross-sectional units at all times. Thus, we have

$$y_{it} = z_{it}'\bar{\gamma} + u_{it}, \qquad \begin{array}{l} i = 1, \ldots, N, \\ t = 1, \ldots, T. \end{array} \tag{7.63}$$

This model can be obtained from (7.5) by letting $X = \mathbf{0}$ and $A_2 = (I_{k_2}, \ldots, I_{k_2})'$, where I_{k_2} is a $k_x \times k_2$ identity matrix. Substituting these into (7.48) and (7.49), (or (7.46) and (7.47) the Bayes estimator of $\bar{\gamma}$ is the generalized least-squares estimator of (7.63),

$$\hat{\hat{\gamma}} = (\tilde{Z}'C_1^{-1}\tilde{Z})^{-1}(\tilde{Z}'C_1^{-1}y), \tag{7.64}$$

where

$$\tilde{Z} = (Z_1', Z_2', \cdot, Z_N')'.$$

3.2 Different Models for Different Cross-Sectional Units

When each individual unit is viewed as different, we can write the regression model in the form

$$y_{it} = z_{it}'\gamma_i + u_{it}, \qquad \begin{array}{l} i = 1, 2, \ldots, N, \\ t = 1, 2, \ldots, T, \end{array} \tag{7.65}$$

with γ_i different for different individuals. Model (7.65) can be obtained from (7.5) by letting $\boldsymbol{X} = \mathbf{0}$ and $\boldsymbol{A}_2$ be an $Nk_2 \times Nk_2$ identity matrix. Substituting $\boldsymbol{X} = 0$ and $\boldsymbol{A}_2 = \boldsymbol{I}_{Nk_2}$ into (7.50) and (7.51), we can see that $\bar{\gamma}^{**}$ is identical to Zellner's (1962) seemingly unrelated regression estimator,

$$\bar{\gamma}^{**} = (\boldsymbol{Z}'\boldsymbol{C}_1^{-1}\boldsymbol{Z})^{-1}(\boldsymbol{Z}'\boldsymbol{C}_1^{-1}\boldsymbol{y}). \tag{7.66}$$

In the case where C_1 is diagonal with

$$E\boldsymbol{u}_i\boldsymbol{u}_j' = \begin{cases} \sigma_i^2 \boldsymbol{I}_T & , \text{if } i = j, \\ \mathbf{0} & , \text{otherwise}, \end{cases} \tag{7.67}$$

or the case where $\boldsymbol{Z}_i = \boldsymbol{Z}_j$ for all i, Zellner's seemingly unrelated estimator for $\bar{\gamma} = (\gamma_1', \ldots, \gamma_N')$ is equivalent to applying least squares separately to the time series observations of each cross-sectional unit.

To assume that regression coefficients are identical over i and t may be too restrictive and often cannot be supported by the information contained in the data. On the other hand, to assume that γ_i are different across i may be too general. There is not much of a gain in efficiency by pooling the temporal cross-sectional data relative to the cost involved. Therefore, researchers have often resorted to some intermediate cases which we shall consider in the following sections.

4 Variable Intercept Model

In panel data analysis there are typically three types of variables: individual time-invariant, period individual-invariant, and individual time-varying variables. The individual time-invariant variables are variables that are the same for a given cross-sectional unit through time but that vary across cross-sectional units. Examples of these are attributes of individual-firm management, sex, and socioeconomic-background variables. The period individual-invariant variables are variables that are the same for all cross-sectional units at a given point in time but that vary through time. Examples of these variables are prices, interest rates, and widespread optimism or pessimism. The individual time-varying variables are variables that vary across cross-sectional units at a given point in time and also exhibit variations through time. Examples of these variables are firm profits, sales, and capital stock. The variable-intercept models assume that conditional on the observed explanatory variables, the effects of all omitted (or excluded) variables are also driven by these three types of variables in which the effects of the numerous omitted individual time-varying variables are each individually unimportant but are collectively significant and possess the property of a random variables that is uncorrelated with (or independent of) all other included variables. On the other hand, because the effects of all omitted individual time invariant or period individual-invariant variables either stay constant through time for a given cross-sectional unit or are the same for all cross-sectional units at a given point in time, or a combination of both, they can be absorbed into the intercept term of a regression model as a means to explicitly allow for the individual and/or time heterogeneity contained in the temporal cross-sectional data (e.g. Kuh 1963; Mundlak 1978).

Thus, we assume that

$$y_{it} = \gamma_{1i} + \bar{\gamma}_2 z_{2_{it}} + \ldots + \bar{\gamma}_{k_2} z_{k_2 it} + u_{it}, \quad i = 1, \ldots, N. \tag{7.68}$$

Model (7.68) can be obtained from (7.5) - (7.9) by letting $\boldsymbol{X} = \mathbf{0}$, the first column of $\boldsymbol{Z}_i$ be a $T \times 1$ vector of ones, $\boldsymbol{e}_T$,

$$\boldsymbol{A}_2 = (\boldsymbol{I}_N \otimes \boldsymbol{i}_{k_2}, \boldsymbol{e}_N \otimes \boldsymbol{I}^*_{k_2-1}), \quad \bar{\gamma}' = (\gamma_{11}, \ldots, \gamma_{1N}, \bar{\gamma}_2, \ldots, \bar{\gamma}_{k_2}),$$

where

$$\underset{k_2 \times 1}{\boldsymbol{i}_{k_2}} = \begin{pmatrix} 1 \\ 0 \\ \vdots \\ 0 \end{pmatrix}, \quad \underset{k_2 \times (k_2-1)}{\boldsymbol{I}^*_{k_2-1}} = \begin{pmatrix} \mathbf{0}' \\ \boldsymbol{I}_{k_2-1} \end{pmatrix}.$$

Again, the efficient estimator of $\bar{\gamma}$ is the one given by (7.50) after the substitution of $\boldsymbol{X} = \mathbf{0}$ into the formula. In the case that $\boldsymbol{C}_1$ is proportional to an identity matrix, (7.50) becomes the least squares dummy variable (LSDV) estimator of γ_{1i} and $\bar{\gamma}_2' = (\bar{\gamma}_2, \ldots, \bar{\gamma}_{k_2})$:

$$\hat{\bar{\gamma}}_{2w} = \left[\sum_{i=1}^{N}\sum_{t=1}^{T}(\boldsymbol{z}^*_{it} - \bar{\boldsymbol{z}}^*_i)(\boldsymbol{z}^*_{it} - \bar{\boldsymbol{z}}^*_i)'\right]^{-1} \cdot \left[\sum_{i=1}^{N}\sum_{t=1}^{T}(\boldsymbol{z}^*_{it} - \bar{\boldsymbol{z}}^*_i)(y_{it} - \bar{y}_i)\right] \tag{7.69}$$

$$\hat{\bar{\gamma}}_{1i} = \bar{y}_i - \hat{\bar{\gamma}}'_{2w}\bar{\boldsymbol{z}}^*_i, \quad i = 1, \ldots, N, \tag{7.70}$$

where $\boldsymbol{z}^{*\prime}_{it} = (z_{2it}, \ldots, z_{k_2 it})$, $\bar{\boldsymbol{z}}^*_i = \frac{1}{T}\sum_{t=1}^{T}\boldsymbol{z}^*_{it}$, and $\bar{y}_i = \sum_{t=1}^{T} y_{it}/T$.

The LSDV estimator of $\bar{\gamma}$ is also called the *within-group or covariance* estimator because it is the least-squares (LS) estimator of the transformed model.

$$y_{it} - \bar{y}_i = \bar{\gamma}_2'(z^*_{it} - \bar{z}^*_i) + (u_{it} - \bar{u}_i), \quad i = 1, \ldots, N, \ t = 1, \ldots, T, \tag{7.71}$$

where $\bar{u}_i = \sum_{t=1}^{t} u_{it}/T$. The transformation of the data attained by subtracting from each observation the time series mean for the corresponding cross-sectional unit is called the *covariance transformation.* Such a transformation eliminates the need to include dummy variables to handle the individual effects in the matrix of explanatory variables. Hence, we only need to invert a matrix of order $(k_2 - 1) \times (k_2 - 1)$, not a matrix of order $(N + k_2 - 1) \times (N + k_2 - 1)$.

The estimator (7.69) does not depend on the parameters γ_{1i} and is consistent when either N or T or both tend to infinity. The estimator $\hat{\bar{\gamma}}_{1i}$ (7.70) is consistent only when T tends to infinity. The LSDV estimator is the best linear unbiased estimator (BLUE) if the u_{it} is iid with mean zero and variance σ_u^2. Furthermore, if u_{it} is normally distributed, it is also the maximum likelihood estimator (MLE).

5 Error Components Models

When the effects of the omitted variables that reflect individual time-invariant differences are treated as random variables (just like the assumption on other components of the random disturbance term of a regression model), we have the error components model (e.g. Balestra and Nerlove 1966; Wallace and Hussain 1969).

$$y_{it} = \boldsymbol{z}'_{it}\bar{\gamma} + \alpha_{1i} + u_{it}, \tag{7.72}$$

with

$$E\alpha_{1i} = \bar{\alpha}_1, \tag{7.73}$$

if $\boldsymbol{z}_{it}$ does not include the common intercept term for all cross-sectional units, and

$$E\alpha_{1i} = 0, \tag{7.74}$$

if $\boldsymbol{z}_{it}$ includes the common intercept term, and

$$E(\alpha_{1i} - E\alpha_{1i})(\alpha_{1j} - E\alpha_{1j}) = \begin{cases} \sigma^2_{\alpha_1} & , \text{ if } i = j, \\ 0 & , \text{ otherwise.} \end{cases} \tag{7.75}$$

We can obtain (7.72) from (7.5)-(7.8) by letting $\boldsymbol{X}_i = \boldsymbol{e}_T, \boldsymbol{\alpha}' = (\alpha_{11}, \ldots, \alpha_{1N}), \boldsymbol{A}_1 = \boldsymbol{e}_N$,

$$\boldsymbol{C}_2 = E\boldsymbol{\epsilon\epsilon}' = \sigma^2_{\alpha_1}\boldsymbol{I}_N, \tag{7.76}$$

and $\bar{\alpha}_1 = 0$ if $\boldsymbol{z}_{it}$ contains the intercept term; or otherwise if $\boldsymbol{z}_{it}$ does not contain the intercept term and $\bar{\alpha}_1$ is an unknown constant. If we let $\bar{\alpha}_1 = 0$, the estimator of $\bar{\gamma}$ is given by (7.48). If $\bar{\alpha}_1$ is an unknown constant, the estimators are given by (7.50). However, the difference is only in notation. Both formulas give the same generalized least-squares (GLS) estimator.

In the special case when u_{it} is assumed to be iid over i and t, that is,

$$Eu_{it}u_{js} = \begin{cases} \sigma^2_u & , \text{ if } i = j \text{ and } t = s, \\ 0 & , \text{ otherwise,} \end{cases} \tag{7.77}$$

we have $C_1 = \sigma^2_u I_{NT}$. Assuming there that $\boldsymbol{z}_{it}$ does not contain the constant term and substituting (7.76), (7.77), $X_i = \boldsymbol{e}_T$, $A_1 = \boldsymbol{e}_N$ into (7.50), and (7.51), we have

$$\bar{\gamma}^{**} = \left[\sum_{i=1}^N \boldsymbol{Z}'_i\boldsymbol{V}^{-1}\boldsymbol{Z}_i\right]^{-1}\left[\sum_{i=1}^N \boldsymbol{Z}'_i\boldsymbol{V}^{-1}\boldsymbol{y}_i\right], \tag{7.78}$$

$$\bar{\alpha}_1^{**} = \bar{y} - \bar{\gamma}^{**\prime}\bar{\boldsymbol{z}}, \tag{7.79}$$

where $\bar{y}$ and $\bar{z}$ are overall means of y_{it} and z_{it}, and

$$\boldsymbol{V} = \sigma^2_u\boldsymbol{I}_T + \sigma^2_{\alpha_1}\boldsymbol{e}_T\boldsymbol{e}'_T. \tag{7.80}$$

Making use of the special pattern of $\boldsymbol{V}$ in (7.31), it is easy to verify,

$$\boldsymbol{V}^{-1} = \frac{1}{\sigma^2_u}\left[\boldsymbol{I}_T - \frac{\sigma^2_{\alpha_1}}{\sigma^2_u + T\sigma^2_{\alpha_1}}\boldsymbol{e}_T\boldsymbol{e}'_T\right] = \frac{1}{\sigma^2_u}\left[\boldsymbol{Q} + \psi\frac{1}{T}\boldsymbol{e}_T\boldsymbol{e}'_T\right], \tag{7.81}$$

where $Q = I_T - \frac{1}{T}e_T e_T'$, $\psi = \frac{\sigma_u^2}{\sigma_u^2 + T\sigma_{\alpha_1}^2}$. Substituting (7.81) into (7.78), the GLS estimator can be rewritten as a matrix weighted average of the between- and within-group estimators (Maddala 1971; also see Hsiao 1986, sec. 3.3),

$$\bar{\gamma}^{**} = \Delta\hat{\bar{\gamma}}_b + (I_{k_2} - \Delta)\hat{\bar{\gamma}}_w, \tag{7.82}$$

where $\hat{\bar{\gamma}}_w$ and $\hat{\bar{\gamma}}_b$ denote the within-group and the between-group estimator of $\bar{\gamma}$ with

$$\hat{\bar{\gamma}}_w = \left[\sum_{i=1}^{N} Z_i' Q Z_i\right]^{-1}\left[\sum_{i=1}^{N} Z_i' Q y_i\right], \tag{7.83}$$

and

$$\begin{aligned}
\hat{\bar{\gamma}}_b &= \left[\sum_{i=1}^{N}(\bar{z}_i - \bar{z})(\bar{z}_i - \bar{z})'\right]^{-1}\left[\sum_{i=1}^{N}(\bar{z}_i - \bar{z})(\bar{y}_i - \bar{y})\right], \\
\bar{z} &= \sum_{i=1}^{N}\sum_{t=1}^{T} z_{it}/NT = \sum_{i=1}^{N} \bar{z}_i/N, \\
\bar{y} &= \sum_{i=1}^{N}\sum_{t=1}^{T} y_{it}/NT = \sum_{i=1}^{N} \bar{y}_i/N.
\end{aligned}$$

The weighting matrix Δ is given by

$$\begin{aligned}
\Delta &= \psi T\left[\sum_{i=1}^{N}\sum_{t=1}^{T}(z_{it} - \bar{z}_i)(z_{it} - \bar{z}_i)' + \psi T\sum_{i=1}^{N}(\bar{z}_i - \bar{z})(\bar{z}_i - \bar{z})'\right]^{-1} \\
&\quad \cdot \left[\sum_{i=1}^{N}(\bar{z}_i - \bar{z})(\bar{z}_i - \bar{z})'\right].
\end{aligned}$$

As $\psi \to 1$, $\bar{\gamma}^{**}$ converges to the OLS estimator of (7.72). As $\psi \to 0$, $\bar{\gamma}^{**}$ converges to the within-group estimator (LSDV). In essence, ψ measures the weight given to the between-group variation. In the LSDV procedure (or fixed effect model), this source of variation is completely ignored. The OLS procedure corresponds to $\psi = 1$. The between-group and within-group variations are just added up. Thus, one may view the OLS and LSDV as somewhat all-or-nothing ways of utilizing the between-group variation. The procedure of treating α_i as random provides a solution intermediate to treating them all as different and treating them all as equal. However, when T tends to infinity, $\psi \to 0$, hence the GLS converges to the LSDV. This is because when T goes to infinity, we have an infinite number of observations for each i. Therefore, we can consider each α_i as a random variable which has been drawn once and forever so that for each i we can pretend that they are just like fixed parameters.

Computationally, the GLS estimator is fairly simple to implement. Noting the special form of V^{-1} from (7.81), we let $P = [I_T - (1-\psi^{\frac{1}{2}})\frac{e_T e_T'}{T}]$, whence $V^{-1} = PP'$. Therefore, we only need to transform the data by subtracting a fraction $(1 - \psi^{\frac{1}{2}})$ of individual means $\bar{y}_i$ and $\bar{z}_i$ from their corresponding y_{it} and z_{it} values, then regress $[y_{it} - (1 - \psi^{\frac{1}{2}})\bar{y}_i]$ on $[z_{it} - (1 - \psi^{\frac{1}{2}})\bar{z}_i]$.

Combining (7.40), (7.44), and (7.50), we obtain the best predictor formula of Taub (1979) and Wansbeek and Kapteyn (1978) for the individual outcome

$$\hat{y}_{i,T+1} = \bar{\alpha}_1^{**} + z'_{iT+1}\bar{\gamma}^{**} + \frac{T\sigma_{\alpha_1}^2}{T\sigma_{\alpha_1}^2 + \sigma_u^2}\bar{\hat{v}}_i, \tag{7.84}$$

where

$$\bar{\hat{v}}_i = \sum_{t=1}^{T}(y_{it} - \bar{\alpha}^{**} - z'_{it}\bar{\gamma}^{**})/T.$$

The fundamental assumption we made with regard to the error-components model is that the error term is serially uncorrelated, conditional on the individual effects α_{1i}. But there are cases in which the effects of unobserved variables last for more than one period. Anderson and Hsiao (1982) have considered the MLE of the model (7.72) with u_{it} following a first-order autoregressive process

$$u_{it} = \rho u_{i,t-1} + \eta_{it}, \tag{7.85}$$

where η_{it} are iid with zero mean and variance σ_η^2. However, computation of the MLE is very complicated. But if we treat the first-period time-series observations ($t = 1$) as given, and ρ is known, we can apply the usual Cochrane-Orcutt (1949) transformation to the data and obtain

$$y^* = y_{it} - \rho y_{i,t-1}, \tag{7.86}$$

$$\tilde{z}^* = \tilde{z}_{it} - \tilde{z}_{i,t-1}\rho, \quad i = 1, \ldots, N, \ t = 2, \ldots, T. \tag{7.87}$$

The transformed model

$$\boldsymbol{y}_i^* = \tilde{\boldsymbol{Z}}_i^*\bar{\gamma} + \alpha_{1i}(1-\rho)\boldsymbol{e}_{T-1} + \boldsymbol{\eta}_i \tag{7.88}$$

will have the usual error-components structure, with

$$\boldsymbol{V} = (1-\rho)^2\sigma_{\alpha_1}^2\boldsymbol{e}_{T-1}\boldsymbol{e}'_{T-1} + \sigma_\eta^2\boldsymbol{I}_{T-1}, \tag{7.89}$$

where the $(t-1)$th element of $u_{it}^* = u_{it} - \rho u_{i,t-1} = \eta_{it}$. The difference between (7.72) under (7.77) and (7.88) is that in the former case, the dimension is T and in the latter case, the dimension is $T - 1$. Hence the same estimation formulas (7.80) and (7.81) can be applied to (7.88). That is, we can first transform the data as before with ψ now equal to $\sigma_\eta^2/[\sigma_\eta^2 + (T-1)(1-\rho)^2\sigma_{\alpha_1}^2]$, then apply least squares to the transformed data.

The discussion above is based on the assumption that the first time-period observations are treated as fixed constants. In general, there is no reason to treat y_{i1} to be different from y_{it} for $t > 1$. When y_{i1}, just like any other y_{it}, are treated as random variables, then through the successive substitution of (7.85), it can be shown that

$$Eu_{i1}^2 = \frac{\sigma_\eta^2}{1-\rho^2} = Eu_{it}^2. \tag{7.90}$$

If the variance of u_{i1} is assumed to be equal to (7.90), we can first apply a Prais-Winston (1954) type of transformation to the data. That is, we transform the first observations by

$$y_{i1}^* = \sqrt{1-\rho^2} y_{i1} \tag{7.91}$$

and

$$\tilde{\boldsymbol{z}}_{i1}^* = \sqrt{1-\rho^2} \boldsymbol{z}_{i1} \quad , \tag{7.92}$$

and the remaining observations by (7.86). Then

$$\underset{NT\times 1}{\boldsymbol{y}^*} = \begin{pmatrix} \boldsymbol{y}_1^* \\ \vdots \\ \boldsymbol{y}_N^* \end{pmatrix} = \tilde{\boldsymbol{Z}}^* \bar{\gamma} + (1-\rho) \begin{pmatrix} \boldsymbol{e}_T^\rho & 0 & \dots & 0 \\ 0 & \boldsymbol{e}_T^\rho & & \vdots \\ \vdots & 0 & & \vdots \\ 0 & 0 & & \boldsymbol{e}_T^\rho \end{pmatrix} \begin{pmatrix} \alpha_{11} \\ \vdots \\ \vdots \\ \alpha_{1N} \end{pmatrix} + \begin{pmatrix} \boldsymbol{u}_1^* \\ \vdots \\ \vdots \\ \boldsymbol{u}_N^* \end{pmatrix}, \tag{7.93}$$

where $\boldsymbol{e}_T^\rho = [(1+\rho)^{\frac{1}{2}}/(1-\rho)^{\frac{1}{2}}, \boldsymbol{e}_{T-1}']'$ and $\boldsymbol{u}_i^* = [(1-\rho^2)^{\frac{1}{2}} u_{i1}, \eta_{i2}, \dots, \eta_{iT}]$.

We have

$$\begin{aligned} & E \quad \{(1-\rho)(\boldsymbol{I}_N \otimes \boldsymbol{e}_T^\rho)\boldsymbol{\alpha}_1 + \boldsymbol{u}^*\}\{(1-\rho)\boldsymbol{I}_N \otimes \boldsymbol{e}_T^\rho)\boldsymbol{\alpha}_1 + \boldsymbol{u}^*\}' \\ = & \quad (1-\rho)^2 \sigma_{\alpha_1}^2 [\boldsymbol{I}_N \otimes \boldsymbol{e}_T^\rho \boldsymbol{e}^{\rho\prime}{}_T] + \sigma_\eta^2 (\boldsymbol{I}_N \otimes \boldsymbol{I}_T) \\ = & \quad d^2(1-\rho^2)\sigma_{\alpha_1}^2 [\boldsymbol{I}_N \otimes \bar{\boldsymbol{J}}^\rho] + \sigma_\eta^2 (\boldsymbol{I}_N \otimes \boldsymbol{I}_T) \\ = & \quad \boldsymbol{V}^*, \end{aligned} \tag{7.94}$$

where $d^2 = \boldsymbol{e}_T^{\rho\prime} \boldsymbol{e}_T^\rho = T - 1 + \frac{1+\rho}{1-\rho}$ and $\bar{\boldsymbol{J}}_T^\rho = \boldsymbol{e}_T^\rho \boldsymbol{e}^{\rho\prime}{}_T / d^2$. Letting $\boldsymbol{Q}_T^\rho = \boldsymbol{I}_T - \bar{\boldsymbol{J}}_T^\rho$ and following the approach of Wansbeek and Kapteyn (1982, 1983), Baltagi and Li (1990) replace $\boldsymbol{I}_T$ by $\boldsymbol{Q}_T^\rho + \bar{\boldsymbol{J}}_T^\rho$ and obtain the spectral decomposition of $\boldsymbol{V}^*$,

$$\boldsymbol{V}^* = \boldsymbol{I}_N \otimes (\sigma_{\alpha_1}^{*2} \bar{\boldsymbol{J}}_T^\rho + \sigma_\eta^2 \boldsymbol{Q}_T^\rho) = \sigma_{\alpha_1}^{*2} (\boldsymbol{I}_N \otimes \bar{\boldsymbol{J}}_T^\rho) + \sigma_\eta^2 (\boldsymbol{I}_N \otimes \boldsymbol{Q}_T^\rho), \tag{7.95}$$

where $\sigma_{\alpha_1}^{*2} = d^2(1-\rho)^2 \sigma_{\alpha_1}^2 + \sigma_\eta^2$ is the first characteristic root of $\boldsymbol{V}^*$ of multiplicity N and σ_η^2 is the second characteristic root of $\boldsymbol{V}^*$ of multiplicity $N(T-1)$. Therefore

$$\begin{aligned} \sigma_\eta \boldsymbol{V}^{*-\frac{1}{2}} &= \frac{\sigma_\eta}{\sigma_\alpha^*} (\boldsymbol{I}_N \otimes \bar{\boldsymbol{J}}_T^\rho) + (\boldsymbol{I}_N \otimes \boldsymbol{Q}_T^\rho) \\ &= \boldsymbol{I}_N \otimes \boldsymbol{I}_T - \theta_\rho (\boldsymbol{I}_N \otimes \bar{\boldsymbol{J}}^\rho), \end{aligned} \tag{7.96}$$

where $\theta_\rho = 1 - (\frac{\sigma_\eta}{\sigma_{\alpha_1}^*})$. Thus, we can estimate $\bar{\gamma}$ by the following multistep procedure:

Step 1. Apply the Prais-Winston transformation [(7.86), (7.91), and (7.92)] to $\boldsymbol{y}$ and $\boldsymbol{z}$ to obtain $\boldsymbol{y}^*$ and $\tilde{\boldsymbol{z}}^*$.

Step 2. Premultiply $\boldsymbol{y}^*$ and $\tilde{\boldsymbol{z}}^*$ by $\sigma_\eta \boldsymbol{V}^{*-\frac{1}{2}}$ and obtain $\boldsymbol{y}^{**} = \sigma_\eta \boldsymbol{V}^{*-\frac{1}{2}} \boldsymbol{y}^*$ and $\tilde{\boldsymbol{z}}^{**} = \sigma_\eta \boldsymbol{V}^{*-\frac{1}{2}} \tilde{\boldsymbol{z}}^*$.

Step 3. Do least squares of $\boldsymbol{y}^{**}$ on $\tilde{\boldsymbol{z}}^{**}$.

The above estimation methods are based on the assumption that $\sigma_{\alpha_1}^2$, σ_u^2, and ρ are known. If they are unknown, we can replace them with their consistent estimators. When u_{it} is iid, under the representation of (7.72) and (7.73), one method is to use the estimators,

$$\hat{\sigma}_u^2 = \frac{\sum_{i=1}^N \sum_{t=1}^T \hat{u}_{it}^2}{N(T-1)-k_2}, \tag{7.97}$$

$$\hat{\sigma}_{\alpha_1}^2 = \frac{\sum_{i=1}^N (\bar{y}_i - \hat{\bar{\alpha}}_1 - \hat{\bar{\gamma}}_b' \bar{\bar{z}}_i)^2}{N-(k_2+1)} - \frac{1}{T}\hat{\sigma}_u^2, \tag{7.98}$$

where $\hat{\bar{\alpha}}_1$ and $\hat{\bar{\gamma}}_b$ are the between-group estimators of $\bar{\alpha}_1$ and $\bar{\gamma}$ obtained by regressing $\bar{y}_i$ on $(1, \bar{\bar{z}})$ and $\hat{u}_{it} = (y_{it} - \bar{y}_i) - \hat{\bar{\gamma}}_w'(z_{it} - \bar{z}_i)$. When u_{it} follows a first-order autoregressive process, we may estimate ρ by

$$\hat{\rho} = \left(\sum_{i=1}^N \sum_{t=2}^T \hat{u}_{it}\hat{u}_{i,t-1}\right) / \left(\sum_{i=1}^N \sum_{t=2}^T \hat{u}_{i,t-1}^2\right) \tag{7.99}$$

and σ_η^2 and $\sigma_{\alpha_1}^2$ by

$$\hat{\sigma}_\eta^2 = \sum_{i=1}^N \sum_{t=2}^T \left[\dot{y}_{it}^* - \dot{z}_{it}^{*\prime}\hat{\bar{\gamma}}_w\right]^2 / N(T-2), \tag{7.100}$$

$$\hat{\sigma}_{\alpha_1}^{*2} = \left\{\left[\sum_{i=1}^N \bar{y}_i^* - \hat{\bar{\alpha}}_1(1-\hat{\rho}) - \bar{\bar{z}}_i^{*\prime}\hat{\bar{\gamma}}_b\right]^2 / N(1-\hat{\rho})^2\right\} - \left[\hat{\sigma}_\eta^2/(T-1)(1-\hat{\rho})^2\right], \tag{7.101}$$

where

$$\dot{y}_{it}^* = y_{it}^* - \bar{y}_i^*, \quad \dot{z}_{it}^* = \bar{z}_{it}^* - \bar{\bar{z}}_i^*,$$

$$y_{it}^* = y_{it} - \hat{\rho} y_{i,t-1}, \quad \bar{z}_{it}^* = \bar{z}_{it} - \hat{\rho}\bar{z}_{i,t-1},$$

$$\bar{y}_i^* = \frac{1}{T-1}\sum_{t=2}^T y_{it}^*, \quad \bar{\bar{z}}_i^* = \frac{1}{T-1}\sum_{t=2}^T \bar{z}_{it}^*.$$

After we obtain the estimates of $\bar{\gamma}$, just like in the case of serially uncorrelated u_{it} (7.84), we can predict the future value of y_{it} by combining (7.42) and (7.50). Hence a predictor of $y_{i,T+1}$ is given by (Lee 1979; Baltagi and Li 1990)

$$\begin{aligned}\hat{y}_{i,T+1} &= \bar{z}_{i,T+1}'\bar{\gamma}^{**} + \rho u_{i,T} \\ &+ \left[\frac{(1-\rho)^2\sigma_{\alpha_1}^2}{\sigma_{\alpha_1}^{*2}}\right]\{[T-(T-2)\rho]\tilde{\alpha}_{1i}^{**} + \sum_{t=1}^T u_{it} \\ &- \rho\sum_{t=2}^T u_{i,t-1} + \rho u_{i1}\}\quad .\end{aligned} \tag{7.102}$$

Of course, before one performs transformations of the type (7.85), (7.91) and (7.92), one should test for $\rho = 0$. One such test is the generalized Durbin-Watson test, based on the test statistic

$$d_p = \sum_{i=1}^N \sum_{t=2}^T (\hat{u}_{it} - \hat{u}_{i,t-1})^2 / \sum_{i=1}^N \sum_{t=1}^T \hat{u}_{it}^2 \tag{7.103}$$

suggested by Bhargava, Franzini, and Narendranathan (1982).

6 Random Coefficients Models

When the heterogeneity among cross-sectional units cannot be captured completely by the time-invariant individual varying constants α_{1i}, a natural generalization would be to let the parameters vary across cross-sectional units. In Section 3.2 we considered the case where individual coefficients are treated as fixed and different. In this section we consider the case where the regression coefficients are viewed as random variables with a probability distribution(e.g. Hsiao 1974, 1975; Lindley and Smith 1971; Swamy 1970). The random coefficient specification reduces the number of parameters to be estimated substantially, while still allowing the coefficients to differ from unit to unit (and/or from time to time).

Under the assumption that the explanatory variables and coefficients are independent, the Swamy (1970)-type random coefficients formulation assumes that

$$y_{it} = \boldsymbol{x}'_{it}\boldsymbol{\alpha}_i + u_{it}, \tag{7.104}$$

$$E\boldsymbol{\alpha}_i = \bar{\boldsymbol{\alpha}}, \tag{7.105}$$

$$E(\boldsymbol{\alpha}_i - \bar{\boldsymbol{\alpha}})(\boldsymbol{\alpha}_j - \bar{\boldsymbol{\alpha}})' = \begin{cases} \boldsymbol{\Delta} & , \text{ if } i = j, \\ \boldsymbol{0} & , \text{ otherwise,} \end{cases} \tag{7.106}$$

and

$$E\boldsymbol{x}_{it}(\boldsymbol{\alpha}_i - \bar{\boldsymbol{\alpha}})' = 0. \tag{7.107}$$

The Swamy-type random coefficients models assume that individual differences satisfy De Finetti's (1964) exchangeability criterion. That is, the likelihood for the ith individual to have a particular realized value $\boldsymbol{\alpha}_i$ is the same as the jth individual. The individual $\boldsymbol{\alpha}_i$ all have the same expected value. The subscript i is purely a labeling device with no substantive content. The observed difference between $\boldsymbol{\alpha}_i$ and $\boldsymbol{\alpha}_j$ stems solely from the operation of chance mechanisms.

We can obtain model (7.104) – (7.107) from (7.10) and also using (7.7) and (7.8) by letting $\boldsymbol{Z}_i = 0$, $\boldsymbol{A}_1 = \boldsymbol{e}_N \otimes \boldsymbol{I}_{k_1}$, $\boldsymbol{\alpha} = (\boldsymbol{\alpha}'_1, \ldots, \boldsymbol{\alpha}'_N)$, $\boldsymbol{e}' = (\boldsymbol{\alpha}_1 - \bar{\boldsymbol{\alpha}}, \ldots, \boldsymbol{\alpha}_N - \bar{\boldsymbol{\alpha}})$, $\boldsymbol{C}_2 = \boldsymbol{I}_N \otimes \boldsymbol{\Delta}$. Hence the Bayes estimator of $\bar{\boldsymbol{\alpha}}$ is given by (7.44), which is identical to the GLS estimator of the model,

$$\boldsymbol{y} = \boldsymbol{X}\boldsymbol{A}_1\bar{\boldsymbol{\alpha}} + \boldsymbol{\eta}, \tag{7.108}$$

where

$$\boldsymbol{\eta} = \boldsymbol{X}\boldsymbol{e} + \boldsymbol{u}, \tag{7.109}$$

with

$$E\boldsymbol{\eta} = \boldsymbol{0} \tag{7.110}$$

and

$$E\boldsymbol{\eta}\boldsymbol{\eta}' = \boldsymbol{X}(\boldsymbol{I}_N \otimes \boldsymbol{\Delta})\boldsymbol{X}' + \boldsymbol{C}_1. \tag{7.111}$$

In the case that $E\boldsymbol{u}_i\boldsymbol{u}_i' = \sigma_i^2\boldsymbol{I}_T$ and $E\boldsymbol{u}_i\boldsymbol{u}_j' = \boldsymbol{0}$, by repeatedly using (7.31), (7.44), or (7.22) can be simplified as

$$\bar{\boldsymbol{\alpha}}^{**} = \sum_{i=1}^{N} \boldsymbol{W}_i\hat{\boldsymbol{\alpha}}_i \tag{7.112}$$

where

$$\boldsymbol{W}_i = \left\{\sum_{i=1}^{N}[\boldsymbol{\Delta} + \sigma_i^2(\boldsymbol{X}_i'\boldsymbol{X}_i)^{-1}]^{-1}\right\}^{-1} \cdot [\boldsymbol{\Delta} + \sigma_i^2(\boldsymbol{X}_i'\boldsymbol{X}_i)^{-1}]^{-1}, \tag{7.113}$$

and

$$\hat{\boldsymbol{\alpha}}_i = (\boldsymbol{X}_i'\boldsymbol{X}_i)^{-1}\boldsymbol{X}_i'\boldsymbol{y}_i. \tag{7.114}$$

Equation (7.112) shows that the Bayes or GLS estimator of $\bar{\boldsymbol{\alpha}}$ is a matrix-weighted average of the least-squares estimator for each cross-sectional unit, with the weights inversely proportional to their covariance matrices. It also shows that the GLS estimator requires only a matrix inversion of order k_1, and so it is not much more complicated to compute than the simple least-squares estimator.

When σ_i^2 and $\boldsymbol{\Delta}$ are unknown, Swamy (1970) proposes using the least-squares estimator $\hat{\boldsymbol{\alpha}}_i$ and residuals $\hat{\boldsymbol{u}}_i = \boldsymbol{y}_i - \boldsymbol{X}_i\hat{\boldsymbol{\alpha}}_i$ to obtain the unbiased estimators of σ_i and $\boldsymbol{\Delta}$,

$$\hat{\sigma}_i^2 = \frac{\hat{\boldsymbol{u}}_i'\hat{\boldsymbol{u}}_i}{T-k_1} = \frac{1}{T-k_1}\boldsymbol{y}_i'[\boldsymbol{I} - \boldsymbol{X}_i(\boldsymbol{X}_i'\boldsymbol{X}_i)^{-1}\boldsymbol{X}_i']\boldsymbol{y}_i, \tag{7.115}$$

$$\hat{\boldsymbol{\Delta}} = \frac{1}{N-1}\sum_{i=1}^{N}(\hat{\boldsymbol{\alpha}}_i - N^{-1}\sum_{i=1}^{N}\hat{\boldsymbol{\alpha}}_i)(\hat{\boldsymbol{\alpha}}_i - N^{-1}\sum_{i=1}^{N}\hat{\boldsymbol{\alpha}}_i)' - \frac{1}{N}\sum_{i=1}^{N}\hat{\sigma}_i^2(\boldsymbol{X}_i'\boldsymbol{X}_i)^{-1}. \tag{7.116}$$

Equation (7.116), just like (7.98) in the scalar case or (7.101), does not necessarily yield a nonnegative definite matrix. In this situation, instead of using the unbiased estimator for $\boldsymbol{\Delta}$, we can use the Bayes mode estimator suggested by Lindley and Smith (1971) and Smith (1972):

$$\boldsymbol{\Delta}^* = \left\{\boldsymbol{R} + \sum_{i=1}^{N}(\hat{\boldsymbol{\alpha}}_i - N^{-1}\sum_{i=1}^{N}\hat{\boldsymbol{\alpha}}_i)(\hat{\boldsymbol{\alpha}}_i - N^{-1}\sum_{i=1}^{N}\hat{\boldsymbol{\alpha}}_i)'\right\} \cdot (N + w - k_1 - 2)^{-1}, \tag{7.117}$$

where $\boldsymbol{R}$ and w are parameters, assuming that $\boldsymbol{\Delta}^{-1}$ has a Wishart distribution with w degrees of freedom and matrix $\boldsymbol{R}$.

While both the random coefficients model (7.104) and fixed coefficients model (7.64) are similar in the sense that both models allow the coefficients to be different across cross-sectional units, the random coefficients model (7.104) has imposed additional prior information, namely the coefficients are randomly distributed with a common mean. Therefore, in the random coefficients model formulation, our interest centers on the estimation of the common mean $\bar{\alpha}$ and the dispersion Δ, while in the fixed coefficients formulation (7.64), our interest centers on estimating individual response coefficients. For model (7.64), the best linear predictor for an individual outcome is to substitute the best linear unbiased estimator of the individual coefficients into the individual equation. But for model (7.104), because of the additional information that α_i has mean $\bar{\alpha}$ and variance covariance matrix Δ, the minimum mean square predictor of α_i is no longer the (generalized) least-squares estimator of α_i. In fact, it can be shown from (7.39) that the Bayes predictor for α_i is a weighted average of the (generalized) least-squares estimator of α_i and $\bar{\alpha}$ [see Section 7, eq. (7.129)]. In the case of the Swamy type model (7.104) – (7.108), the best predictor formula is simplified to [Lindley and Smith (1972)]

$$\begin{aligned} \hat{\alpha}_i^{**} &= \left(\frac{1}{\sigma_i^2}X_i'X_i + \Delta^{-1}\right)^{-1}\left(\frac{1}{\sigma_i^2}X_i'y_i + \Delta^{-1}\bar{\alpha}^{**}\right) \\ &= \left(\frac{1}{\sigma_i^2}X_i'X_i + \Delta^{-1}\right)^{-1}\left(\frac{1}{\sigma_i^2}X_i'X_i\hat{\alpha}_i + \Delta^{-1}\bar{\alpha}^{**}\right). \end{aligned} \tag{7.118}$$

7 Mixed Fixed and Random Coefficients Models

When some of the coefficients are treated as fixed constants and some are treated as random coefficients, we have the mixed fixed and random coefficients model (7.10) (e.g. Hsiao 1990; Hsiao et al. 1989). In this case we assume that responses to changes in certain conditions are similar, but responses to changes in other conditions may be individual specific.

Substituting (7.7) into (7.10), we have

$$y = XA_1\bar{\alpha} + \tilde{Z}\bar{\gamma} + \tilde{u}, \tag{7.119}$$

where

$$\tilde{u} = u + Xe \tag{7.120}$$

with $E\tilde{u} = \mathbf{0}$, and

$$E\tilde{u}, \tilde{u}' = C_1 + XC_2X'. \tag{7.121}$$

The Bayes estimators of $\bar{\alpha}$ ((7.44)) and $\bar{\gamma}$ ((7.50)) are the generalized least-squares (GLS) estimators of (7.119). To see this, we note that by repeatedly using the identities (7.31) and (7.32), we can show that

$$\begin{aligned} &[X'C_1^{-1}X + C_2^{-1} - X'C_1^{-1}\tilde{Z}(\tilde{Z}'C_1^{-1}\tilde{Z})^{-1}\tilde{Z}'C_1^{-1}X]^{-1} \\ &= (X'C_1^{-1}X + C_2^{-1})^{-1}\{I + X'C_1^{-1}\tilde{Z} \\ &\quad \cdot[\tilde{Z}'C_1^{-1}\tilde{Z} - \tilde{Z}'C_1^{-1}X(X'C_1^{-1}X + C_2^{-1})^{-1}X'C_1^{-1}\tilde{Z}]^{-1} \\ &\quad \cdot \tilde{Z}'C_1^{-1}X(X'C_1^{-1}X + C_2^{-1})^{-1}\} \\ &= [C_2 - C_2X'(XC_2X' + C_1)^{-1}XC_2] \\ &\quad \cdot\{I + X'C_1^{-1}\tilde{Z}[\tilde{Z}'(XC_2X' + C_1)^{-1}\tilde{Z}]^{-1} \\ &\quad \cdot\tilde{Z}'C_1^{-1}X(X'C_1^{-1}X + C_2^{-1})^{-1}\}. \end{aligned} \tag{7.122}$$

Substituting (7.122) into (7.45) and (7.44), we obtain

$$\tilde{D}_2 = \{A_1'X'[(XC_2X'+C_1)^{-1} - (XC_2X'+C_1)^{-1}\tilde{Z} \cdot\{\tilde{Z}'(XC_2X'+C_1)^{-1}\tilde{Z}\}^{-1}\tilde{Z}'(XC_2X'+C_1)^{-1}XA_1\}^{-1}, \quad (7.123)$$

and

$$\begin{aligned} & A_1'C_2^{-1}(X'[C_1^{-1} - C_1^{-1}\tilde{Z}(\tilde{Z}'C_1^{-1}\tilde{Z})^{-1}\tilde{Z}'C_1^{-1}]X + C_2^{-1})^{-1} \\ & \cdot X'[C_1^{-1} - C_1^{-1}\tilde{Z}(\tilde{Z}'C_1^{-1}\tilde{Z})^{-1}\tilde{Z}'C_1^{-1}]y \\ & = A_1'\{X' - X' + X'(XC_2X'+C_1)^{-1}C_1 \\ & -X'(XC_2X'+C_1)^{-1}\tilde{Z} \\ & \cdot[\tilde{Z}'(XC_2X'+C_1)^{-1}\tilde{Z}]^{-1}\tilde{Z}'(XC_2X'+C_1)^{-1}C_1 \qquad (7.124) \\ & \cdot[C_1^{-1} - C_1^{-1}\tilde{Z}(\tilde{Z}'C_1^{-1}\tilde{Z})^{-1}\tilde{Z}'C_1^{-1}]y \\ & = A_1'X'\{(XC_2X'+C_1)^{-1} - (XC_2X'+C_1)^{-1} \\ & \cdot\tilde{Z}[\tilde{Z}'(XC_2X'+C_1)^{-1}\tilde{Z}]^{-1}\cdot\tilde{Z}'(XC_2X'+C_1)^{-1}\}y. \end{aligned}$$

Combining (7.123) and (7.124), we establish the equivalence of $\bar{\alpha}^{**}$ (7.44) and the GLS estimator of $\bar{\alpha}$ for the model (7.119).

To show that the (unconditional) Bayes estimator of $\bar{\gamma}$ from (7.50) is the GLS estimator of $\bar{\gamma}$ for the model (7.119), again we repeatedly use the identities (7.31) and (7.32). We first note that

$$\begin{aligned} & X[X'C_1^{-1}X + C_2^{-1} - C_2^{-1}A_1(A_1'C_2^{-1}A_1)^{-1}A_1'C_2^{-1}]^{-1}X \\ = \; & X[C_2 - C_2X'(XC_2X'+C_1)^{-1}XC_2]\{I + C_2^{-1}A_1 \\ & \cdot[A_1'X'(XC_2X'+C_1)^{-1}XA_1]^{-1} \\ & \cdot A_1'C_2^{-1}[C_2 - C_2X'(XC_2X'+C_1)^{-1}XC_2]\}X \\ = \; & C_1 - C_1(XC_2X'+C_1)^{-1}C_1 + C_1(XC_2X'+C_1)^{-1} \\ & \cdot XA_1[A_1'X(XC_2X'+C_1)^{-1}XA_1]^{-1}\cdot A_1'X(XC_2X'+C_1)^{-1}C_1. \qquad (7.125) \end{aligned}$$

Substituting (7.125) into (7.50) and (7.51), we obtain

$$\begin{aligned} \bar{\gamma}^{**} = \; & \{\tilde{Z}'(XC_2X'+C_1)^{-1} - (XC_2X'+C_1)^{-1}XA_1 \\ & \cdot[A_1'X'(XC_2X'+C_1)^{-1}XA_1]^{-1}A_1'X(XC_2X'+C_1)\cdot\tilde{Z}\}^{-1} \\ & \tilde{Z}'\{(XC_2X'+C_1)^{-1} - (XC_2X'+C_1)^{-1}XA_1 \qquad (7.126) \\ & \cdot[A_1'X'(XC_2X'+C_1)^{-1}XA_1]^{-1}A_1'X'\cdot(XC_2X'+C_1)^{-1}y\}. \end{aligned}$$

On the other hand, the Bayes estimator of α is not the GLS estimator for the model (7.10). The Bayes estimator (7.42) is the "weighted average" of the GLS estimator of α for the model (7.10) with the weights proportional to the inverse of the precisions of respective estimates. In other words, the Bayes estimator of the individual coefficients α_i

"shrinks" the GLS estimator of α_i toward the grand mean $\bar{\alpha}$. The reason for doing so stems from the di Finneti's exchangeability assumption. When there are not enough time-series observations to allow for precise estimation of individual α_i (namely, T is small), additional information about α_i may be obtained by examining the behavior of others because the expected response is assumed the same and the actual differences in response among individuals are the work of a chance mechanism.

To establish this result, let $D = \tilde{X}^{*\prime}\tilde{X}^{*}+C_2^{-1}$, $B = C_2^{-1}A_1$, and $F = -(A_1'C_2^{-1}A_1)^{-1}$, where $\tilde{X}^{*}$ is defined as in (7.55). Using (7.31), we have

$$\begin{aligned}\tilde{D}_1^{**} &= (\tilde{X}^{*\prime}\tilde{X}^{*}+C_2^{-1})^{-1}\\ &\quad\cdot[I+C_2^{-1}A_1\{A_1'C_2^{-1}A_1 - A_1'C_2^{-1}(\tilde{X}^{*\prime}\tilde{X}^{*}+C_2^{-1})^{-1}C_2^{-1}A_1\}^{-1}\\ &\quad\cdot A_1'C_2^{-1}(\tilde{X}^{*\prime}\tilde{X}^{*}+C_2^{-1})^{-1}] \qquad (7.127)\\ &= (\tilde{X}^{*\prime}\tilde{X}^{*}+C_2^{-1})^{-1}[I+C_2^{-1}A_1\{A_1'\\ &\quad\cdot[C_2+(\tilde{X}^{*\prime}\tilde{X}^{*})^{-1}]^{-1}A_1\}^{-1}\cdot A_1'C_2^{-1}(\tilde{X}^{*\prime}\tilde{X}^{*}+C_2^{-1})^{-1}].\end{aligned}$$

Therefore

$$\tilde{\alpha}^{**} = (\tilde{X}^{*\prime}\tilde{X}^{*}+C_2^{-1})^{-1}\{(\tilde{X}^{*\prime}\tilde{X}^{*})\hat{\alpha}+C_2^{-1}A_1\bar{\alpha}^{**}\}, \qquad (7.128)$$

where

$$\hat{\alpha} = (\tilde{X}^{*\prime}\tilde{X}^{*})^{-1}\tilde{X}^{*\prime}y^{*}. \qquad (7.129)$$

8 Random or Fixed Effects (Parameters)

When differences among cross-sectional units are treated as fixed constants, the model is referred to as a fixed effects model. When such differences are treated as random variables, it is called a random effects model. The issue of whether to treat unobserved heterogeneity as fixed or random has paramount importance in panel data modeling (e.g. Ashenfelter 1978; Hausman 1978; Kiefer 1979). In this section we shall first give an example which illustrates the relevance of the fixed versus random effects formulation. Then we discuss some basic considerations for fixed versus random effects modeling and issues concerning correlation between the effects and the included explanatory variables. Finally we will discuss hypothesis testing and model selection for fixed versus random effects formulations.

8.1 An Example

In a study of Ontario regional electricity demand, Hsiao et al. (1989) use the model

$$q_{it} = a_i q_{i,t-1} + b_i' w_{it} + \delta_i' d_i + u_{it}, \qquad (7.130)$$

where q_{it} denotes the logarithm of monthly kilowatt-hour or kilowatt demand for region i at time t, w_{it} consists of climatic factors and the logarithm of income, own price, and price of its close substitute, all measured in real terms, d_i consists of twelve monthly dummies, and u_{it} denotes the error term. To allow for different patterns of heterogeneity across regions, four different specifications are considered:

Table 1. Root Mean Square Error of ln Kilowatt-Hours (One-Period-Ahead Forecast)

Municipality	Regional specific	Pooled	Random coefficients	Mixed
Hamilton	0.0865	0.0535	0.0825	0.0830
Kitchener-Waterloo	0.0406	0.0382	0.0409	0.0395
London	0.0466	0.0494	0.0467	0.0464
Ottawa	0.0697	0.0523	0.0669	0.0680
St. Catharines	0.0796	0.0724	0.0680	0.0802
Sudbury	0.0454	0.0857	0.0454	0.0460
Thunder Bay	0.0468	0.0615	0.0477	0.0473
Toronto	0.0362	0.0497	0.0631	0.0359
Windsor	0.0506	0.0650	0.0501	0.0438
Unweighted average	0.0558	0.0586	0.0568	0.0545
Weighted average[a]	0.0499	0.0525	0.0628	0.0487

[a]The weight is kilowatt-hours of the municipality in June 1985.
Source: Hsiao et al. (1989, p. 584).

1. The coefficients $\boldsymbol{\theta}_i' = (a_i, \boldsymbol{b}_i', \boldsymbol{\delta}_i')$ are fixed and different for different regions.

2. The coefficients $\boldsymbol{\theta}_i' = \boldsymbol{\theta}' = (a, \boldsymbol{b}', \boldsymbol{\delta}')$ for all i.

3. The coefficient vectors $\boldsymbol{\theta}_i$ are randomly distributed with common mean $\boldsymbol{\theta}$ and variance-covariance matrix

$$E(\boldsymbol{\theta}_i - \boldsymbol{\theta})(\boldsymbol{\theta}_j - \boldsymbol{\theta})' = \begin{cases} \boldsymbol{\Delta} & \text{if } i = j, \\ \mathbf{0} & \text{otherwise.} \end{cases}$$

4. The coefficients $(a_i, \boldsymbol{b}_i')$ are randomly distributed with common mean $(a, \boldsymbol{b}')$ and finite variance-covariance matrix $\boldsymbol{\Delta}_{11}$, and δ_i are fixed and different for different i.

Monthly data for Hamilton, Kitchener-Waterloo, London, Ottawa, St. Catharines, Sudbury, Thunder Bay, Toronto, and Windsor from January 1967 to December 1982 are used to estimate these four different specifications. Comparisons of the one-period-ahead prediction from January 1983 to December 1986 are summarized in Tables 1 and 2. As one can see from these tables, the simple pooling (model 2) and random coefficients (model 3) formulations on average yield less precise prediction for regional demand. The mixed fixed and random coefficients model (model 4) performs the best. It is interesting to note that

Table 2. Root Mean Square Error of ln Kilowatts
(One-Period-Ahead Forecast).

Municipality	Regional specific	Pooled	Random coefficients	Mixed
Hamilton	0.0783	0.0474	0.0893	0.0768
Kitchener-Waterloo	0.0873	0.0440	0.0843	0.0803
London	0.0588	0.0747	0.0639	0.0586
Ottawa	0.0824	0.0648	0.0846	0.0768
St. Catharines	0.0531	0.0547	0.0511	0.0534
Sudbury	0.0607	0.0943	0.0608	0.0614
Thunder Bay	0.0524	0.0597	0.0521	0.0530
Toronto	0.0429	0.0628	0.0609	0.0421
Windsor	0.0550	0.0868	0.0595	0.0543
Unweighted average	0.0634	0.0655	0.0674	0.0619
Weighted average[a]	0.0558	0.0623	0.0673	0.0540

[a]The weight is kilowatt-demand of the municipality in June, 1985.
Source: Hsiao et. al. (1989, p. 584).

combining information across regions together with a proper account of regional-specific factors is capable of yielding better predictions for regional demand than the approach of simply using regional-specific data (model 1). However, had Hsiao et. al. (1989) only considered all regional differences as random draws from a common population (the random coefficients approach), the prediction for regional demand would be worse than the results obtained by not pooling the data. These results appear to suggest that we should neither pool the data without taking account of heterogeneity across cross-sectional units (e.g. model 2), nor simply treat all regional heterogeneity as random draws from a common population (e.g. model 3).

8.2 Some Basic Considerations

The example above demonstrates that the way in which individual heterogeneity is taken into account makes a difference in the accuracy of inference. To present a unified framework for fixed versus random effects formulations, we shall assume from the outset that the effects are random. The fixed effects approach is viewed as one where investigators make inferences conditional on the effects that are in the sample. The random effects approach is viewed as one where investigators make unconditional or marginal inferences with respect to the population of all effects. There is really no distinction in the "nature of the effect". It is up to the investigator to decide whether he wants to make inferences with respect to the population characteristics or only with respect to the effects that are in the sample.

In general, whether one wishes to consider the conditional likelihood function or the marginal likelihood function depends on (Box and Tiao 1973):

(a) the objectives of the study, and

(b) the context of the data, the manner in which they were gathered, and the cnvironment from which they came.

When the objective is to make inferences about the *population* characteristics (say mean and variance) and the sample observations are *random* selections from the relevant population, a random effects model is appropriate. By contrast, the fixed effects setup implies that our interest centers on the outcome of an *individual* unit. Hence, whether the particular sample can be realistically considered as a random sample from the population is irrelevant. In still another situation, the objective may be to learn about the mean outcome of a specific factor over the population of other factors. In this case, the analysis is more appropriately conducted in terms of a mixed fixed and random effects framework, and the factors that were thought of as contributing random effects should be drawn randomly from the relevant populations.

To be more specific, consider an experiment in which there are six technicians caring for eight machines. Questions concerning the machine-technician populations which this kind of experiment might answer are:

1. What is the average output of a machine and how large a variance might be experienced by different technicians?

2. What is the mean output of each of the eight machines in the hands of the six particular technicians?

3. What is the mean output and variance of each of the eight particular machines as it performs in the hands of a population of technicians? Or what are the mean performance levels and variances of each of the six particular technicians when operating a machine?

In the first case, the objective is to use the individual machines and technicians to make inferences about the populations of machines and/or technicians. The individual machines and technicians in the experiment are of no interest in themselves. Here, both the machines and the technicians should be randomly selected from relevant and definable populations.

In the second case, interest centers on the mean performance of each of the eight particular machines in the hands of each of the six particular technicians, so a fixed effects model is appropriate. Whether the particular technicians, or machines can be realistically considered as relevant random samples from specific populations is of no particular concern.

In the third case, attention focuses on the individual mean performance of a machine when it is operated by a relevant population of technicians or the mean performance of a particular technician when operating a machine. In this case, the relevant analysis is a mixed fixed (machine or technician) and random (technician or machine) effects model. The factors contributing random effects should be random samples from the relevant populations.

In the example of regional electricity demand, the mixed effects model assumes that there are regional specific effects, and these effects are more appropriately captured by fixed and different coefficients for regional-seasonal dummy variables. The regional coefficients for economic variables (e.g. income, price variables) are considered random draws from a common population with constant mean and variance-covariance matrix. If a Marshallian representative consumer is a plausible approximation, then it is not unreasonable to assume that consumers respond more or less the same way toward changes in economic conditions, other things being held constant. The realized differences in regional response coefficients for economic variables can be viewed as outcomes of random draws from a common population. The subscript i is purely a labeling device with no substantive content.

In short, the situation to which a model applies and the inferences based on it are the deciding factors in determining whether we should treat effects as random or fixed. When inferences are going to be confined to the effects in the model, the effects are more appropriately considered fixed. When inferences will be made about a population of effects from which those in the data are considered to be a random sample, then the effects should be considered random.

8.3 Correlations between Effects and Included Explanatory Variables

Closely related to the question of fixed effects or random effects inference is the impact of nonorthogonality between the effects α_i and the included explanatory variables, $\boldsymbol{x}_i$ and $\boldsymbol{z}_i$, on the appropriateness of either approach [e.g. Mundlak (1978)], where $\boldsymbol{x}_i' = (\boldsymbol{x}_{i1}', \boldsymbol{x}_{i2}', \ldots, \boldsymbol{x}_{iT}')$ and $\boldsymbol{z}_i' = (\boldsymbol{z}_{i1}', \ldots, \boldsymbol{z}_{iT}')$. In this paper, when the effects are treated as random variables, we have assumed that the α_i are independent of the explanatory variables. This, as explained earlier, implies that the individual differences satisfy De Finetti's (1964) exchangeability criterion. The prior mean of α_i is the same as that of α_j, or any other α_i. The actual differences are the work of chance mechanisms. It does not allow α_i to depend on $\boldsymbol{x}_i$ or $\boldsymbol{z}_i$.

However, it is possible that the random effects and the included explanatory variables are correlated. For instance, consider a panel of random farms observed over several years. Suppose that y_{it} is a measure of the output of the ith farm in the t-th season, $\boldsymbol{z}_{it}$ are measured inputs, α_i represents the input reflecting soil quality, and u_{it} reflects unmeasured inputs which are not under the farmer's control (e.g. rainfall). We assume that α_i is known to the farmer but unknown to the investigating researcher. For profit-maximizing farmers, the factor-input decisions will in general be made conditional on α_i, although typically before knowing u_{it}. Hence, the assumption of independence between u_{it} and $\boldsymbol{z}_i$ may be valid, but certainly not the independence assumption between α_i and $\boldsymbol{z}_i$.

When the effects α_i are treated as unknown parameters, it is equivalent to adding the associated factors $\boldsymbol{x}_i$ to the set of the included explanatory variables $\boldsymbol{z}_i$. Therefore, if correlations exist between $\boldsymbol{\alpha}_i$ and $\boldsymbol{z}_i$, using a fixed effects model can capture such correlations, hence eliminate the bias arising from the correlation between the unobserved individual effects and the included explanatory variables. On the other hand, a random effects inference ignoring the correlation between the effects and explanatory variables can lead to biased estimation. However, this is a consequence of misspecification rather than the inappropriateness of the random effects inference.

To gain some intuition about the role of independence assumptions between the effects and the included explanatory variables within a conditional and unconditional inference framework, let us consider the following two experiments. Suppose a population is made up of a certain composition of red and black balls. The first experiment consists of N individuals, each picking a fixed number of balls randomly from this population to form his person-specific urn. Each individual then makes T independent trials of drawing a ball from his specific urn and putting it back. The second experiment assumes that individuals have different preferences for the composition of red and black balls from their specific urns and allows these personal attributes to affect the compositions of these urns. Specifically, prior to making T independent trials with replacement from their respective urns, individuals are allowed to take any number of balls from the population until their composition reaches a desired proportion.

If the researcher's interest lies in making inferences on an individual urn's compositions of red and black balls, a fixed effects model should be used whether the sample comes from the first or the second experiment. If his interest is in the population composition, a marginal or unconditional inference should be used. However, in the first experiment, differences in individual urns are outcomes of random sampling. The subscript i is purely a labeling device with no substantive content. A conventional random effects model assuming independence between α_i and $\boldsymbol{z}_{it}$ would be appropriate because the conditional distribution of α_i given $\boldsymbol{z}_i$, $f(\alpha_i|\boldsymbol{z}_i)$ is equal to the marginal distribution of α_i, $f(\alpha_i)$. In the second experiment, the differences in individual urns reflect differences in personal attributes. A proper marginal inference has to allow for these nonrandom influences. In other words, we have to derive the marginal distribution of α_i by integrating the conditional distribution of α_i given $\boldsymbol{z}_i$ over $\boldsymbol{z}_i$, $\int f(\alpha_i|\boldsymbol{z}_i)f(\boldsymbol{z}_i)d\boldsymbol{z}_i$.

Formally, let u_{it} and α_i be independent normal processes and be mutually independent. In the case of the first experiment, the α_i are independently distributed and independent of individual attributes, $\boldsymbol{z}_i = (\boldsymbol{z}'_{i1}, \ldots, \boldsymbol{z}'_{iT})'$, the distribution of α_i must be expressible as random sampling from a univariate distribution with mean $\bar{\alpha}$ and variance σ^2_α (Box and Tiao 1973; Chamberlain 1980). Thus the conditional distribution of $\{(\boldsymbol{u}_i + \boldsymbol{e}_T\alpha_i)', \alpha_i|\boldsymbol{z}_i\}$ is identical to the marginal distribution $\{\boldsymbol{u}_i + \boldsymbol{e}_T\alpha_i)', \alpha_i\}$,

$$f\left(\begin{bmatrix} u_{i1} + \alpha_i \\ \vdots \\ u_{iT} + \alpha_i \\ \cdots \\ \alpha_i \end{bmatrix}\right) = f\left(\begin{bmatrix} u_{i1} + \alpha_i & \vdots & \\ \vdots & & \\ u_{iT} + \alpha_i & & \boldsymbol{z}_i \\ \cdots & \vdots & \\ \alpha_i & & \end{bmatrix}\right)$$

$$\sim \mathcal{N}\left\{\begin{bmatrix} e\bar{\alpha} \\ \cdots \\ \bar{\alpha} \end{bmatrix}, \begin{bmatrix} \sigma_u^2 I_T + \sigma_\alpha^2 e_T e_T' & \vdots & \sigma_\alpha^2 e_T \\ \cdots & \cdots & \cdots \\ \sigma_\alpha^2 e_T' & \vdots & \sigma_\alpha^2 \end{bmatrix}\right\} \tag{7.131}$$

where $\boldsymbol{u}_i' = (u_{i1}, \ldots, u_{iT})$ and $\boldsymbol{e}_T$ is a $T \times 1$ vector of ones.

In the second experiment, α_i is correlated with the variables generating the process. Suppose that conditional on $\boldsymbol{z}_i$, $E(\alpha_i|\boldsymbol{z}_i) = a_i^* = \boldsymbol{a}'\boldsymbol{z}_i$ and var $(\alpha_i|\boldsymbol{z}_i) = \sigma_\omega^2$. Then the conditional distribution of $\{(\boldsymbol{u}_i + \boldsymbol{e}_T\alpha_i)' \vdots \alpha_i|\boldsymbol{z}_i\}$ is

$$f\left(\left[\begin{matrix} u_{i1} + \alpha_i \\ \vdots \\ u_{iT} + \alpha_i \\ \cdots \\ \alpha_i \end{matrix} \middle| \, \boldsymbol{z}_i \right]\right)$$

$$\sim \mathcal{N}\left[\begin{bmatrix} ea_i^* \\ \cdots \\ a_i^* \end{bmatrix}, \begin{bmatrix} \sigma_u^2 I_T + \sigma_w^2 e_T e_T' & \vdots & \sigma_w^2 e_T \\ \cdots & \cdots & \cdots \\ \sigma_w^2 e_T' & \vdots & \sigma_w^2 \end{bmatrix}\right]. \tag{7.132}$$

In both cases, the conditional density of $\boldsymbol{u}_i + \boldsymbol{e}_T\alpha_i$ given α_i is

$$(2\pi\sigma_u^2)^{-\frac{T}{2}} \exp\left\{-\frac{1}{2\sigma_u^2}\boldsymbol{u}_i'\boldsymbol{u}_i\right\}. \tag{7.133}$$

But the conditional densities of $\boldsymbol{u}_i + \boldsymbol{e}_T\alpha_i$ given $\boldsymbol{z}_i$ are different [(7.131) and (7.132), respectively]. Under the independence assumption, $\{\boldsymbol{u}_i + \boldsymbol{e}_T\alpha_i|\boldsymbol{z}_i\}$ has common mean $\boldsymbol{e}_T\bar{\alpha}$ for $i = 1, \ldots, N$. Under the assumption that α_i and $\boldsymbol{z}_i$ are correlated, $\{\boldsymbol{u}_i + \boldsymbol{e}_T\alpha_i|\boldsymbol{z}_i\}$ has a different mean $\boldsymbol{e}_T a_i^*$ for each different i. It is the mistaken use of (7.131) as the conditional density of $\{\boldsymbol{u}_i + \boldsymbol{e}_T\alpha_i|\boldsymbol{z}_i\}$ which creates the bias in the estimates. It is not in making inferences about the characteristics of a population that we should assume a fixed-effects model.

Although the unconditional or random effect inference in general will be more efficient than the conditional or fixed effect inference, the conditional density (7.133) demonstrates that the structural parameters of the fixed effect model does not suffer from the bias due to the omission of the relevant individual attributes. Furthermore, if we condition on α_i, there is no need to postulate the distribution of α_i. Thus, the fixed effect model has assumed paramount importance in empirical studies (e.g., Ashenfelter 1978; Hausman 1978; Kiefer 1979).

However, a typical panel contains a large number of cross-sectional units and only a small number of time-series observations. If we treat α_i as parameters to be estimated, there is not enough information available to determine α_i. Namely we have the classical incidental parameters problem in which the number of parameters increases with the number of observations (Neyman and Scott 1948). Whether the inconsistency in estimating the fixed effects will give rise to inconsistency for estimators of the structural parameters of interest depends on whether the estimators of γ satisfy the Neyman-Scott principle. That is, it depends on whether there exist functions of observables $\boldsymbol{y}_i = (y_{i1}, \ldots, y_{iT})'$,

$$\psi_{Nj}(\boldsymbol{y}_1, \ldots, \boldsymbol{y}_N | \gamma), \quad j = 1, \ldots, m, \tag{7.134}$$

which are independent of the incidental parameters such that when γ are the true values, $\psi_{Nj}(\boldsymbol{y}_1, \ldots, \boldsymbol{y}_N | \gamma)$ converge to zero in probability as N tends to infinity. If such functions exist, then an estimator $\hat{\gamma}$ derived by solving $\psi_{Nj}(\boldsymbol{y}_1, \ldots, \boldsymbol{y}_N | \hat{\gamma}) = 0, \ j = 1, \ldots, m$ is consistent under suitable regularity conditions. However, whether such functions exist depends on the type of model being analyzed. For nonlinear models, except for special cases (e.g. logit models), in general it is not possible to separate the estimation of common parameters and individual specific effects (e.g. see Hsiao 1986).

If the effects are treated as random, it is typical to assume they possess a probability density function characterized by a finite number of parameters. Hence, there is no incidental parameter problem. However, the random effect specification circumscribes the incidental parameter problem by making specific distributional assumptions. How restrictive such assumptions need to be again depend on the type of the model being investigated. For the linear models we have considered here, when the random effects $\boldsymbol{\alpha}_i$ are correlated with included explanatory variables, we assume that the conditional means of $\boldsymbol{\alpha}_i$ given the included explanatory variables are linear in the included explanatory variables. Therefore, we can decompose the random effects $\boldsymbol{\alpha}_i$ into two orthogonal components; one component is composed of a fixed parameters formulation, a_i^*, and the other is composed of a random effects formulation with the effects $\boldsymbol{\alpha}_i^*$ subject to

$$E\boldsymbol{\alpha}_i^* \boldsymbol{x}_i' = \boldsymbol{0} \quad , \quad E\boldsymbol{\alpha}_i^* \boldsymbol{z}_i' = \boldsymbol{0} \quad . \tag{7.135}$$

By regrouping the fixed effects terms and random effects terms, we have a model of the form (7.10).

8.4 Hypothesis Testing or Model Selection

In the error-components formulation, when $\boldsymbol{\alpha}$ is not orthogonal to Z, the LSDV estimator of γ is consistent but the GLS estimator is not. Let $\hat{\boldsymbol{q}}$ denote the difference between the LSDV estimator $\hat{\hat{\gamma}}_{2w}$, (7.69), and the GLS estimator, $\hat{\gamma}_{GLS}$, (7.77). The null of $E(\alpha|Z) = 0$ implies $H_0 : \boldsymbol{q} = \text{plim}_{N\to\infty} \hat{\boldsymbol{q}} = \boldsymbol{0}$. The alternative of $E(\alpha|\boldsymbol{Z}) \neq 0$ implies that $H_1 : \boldsymbol{q} \neq \boldsymbol{0}$. Therefore, Hausman (1978) suggests testing fixed versus random effects formulations, using the statistic

$$m = \hat{\boldsymbol{q}}' \, \hat{\text{Var}}(\hat{\boldsymbol{q}})\hat{\boldsymbol{q}}, \tag{7.136}$$

where $\hat{\text{Var}}(\hat{\boldsymbol{q}}) = \text{Var}(\hat{\hat{\gamma}}_{2w}) - \text{Var}(\hat{\hat{\gamma}}_{GLS})$. Similarly, Breush and Pagan (1979) have exploited the fact that under a random coefficients formulation, the error terms will have heteroscedastic variances, and they suggested a Lagrange multiplier test.

However, all these tests are *indirect* in nature. Rejection of the null does not automatically imply the acceptance of the specific alternative. In fact, it would be more useful to view fixed effects and random effects as *different models*. When we are confronted with different models, the classical sampling theory approach does not appear to be very satisfactory because the distribution of a test statistic is derived under an assumed true null hypothesis. It would appear more appropriate to view the choice of fixed versus random effects formulations as an issue of model selection.

In choosing among alternative formulations, we may follow a Bayesian approach. In the Bayesian approach, prior probabilities are assigned to hypotheses or models that reflect the degree of confidence associated with them, and Bayes's theorem is employed to compute and consider a posterior odds ratio. Therefore, for two hypotheses H_0 or H_1 with prior odds π_0 and π_1, the posterior odds ratio is defined as

$$K_{01} = \frac{\text{Prob}\,(H_0|\boldsymbol{y})}{\text{Prob}\,(H_1|\boldsymbol{y})} = \frac{\pi_0}{\pi_1} \cdot \frac{f(\boldsymbol{y}|H_0)}{f(\boldsymbol{y}|H_1)} \quad . \tag{7.137}$$

In computing K_{01}, because the likelihood function is employed to reflect the sample information, if $\frac{\pi_0}{\pi_1} = 1$, the posterior odds ratio will in general favor the model that has higher likelihood value. If the only difference between H_0 and H_1 is in random versus fixed effects (parameters), the value of the likelihood function is in general larger for the fixed effects model when diffuse priors are used because the fixed effects model contains more parameters to fit the data. In other words, with diffuse priors, the posterior odds ratio will in general favor the fixed effects model. On the other hand, if an informative prior is used, the posterior odds ratio will be heavily dependent on the prior.

An alternative to the use of the posterior odds ratio is to use the predictive density ratio (e.g. Min and Zellner 1990). In this framework, we can first divide the time series observations into two periods, 1 to T_1, denoted by $\boldsymbol{y}_1$, and T_1+1 to T, denoted by $\boldsymbol{y}_2$, and use the first T_1 observations to obtain the probability distribution of the parameters associated with H_0 and H_1, say $P(\theta_0|\boldsymbol{y}_1)$ and $P(\boldsymbol{\theta}_1|\boldsymbol{y}_1)$. We then construct the predictive density ratio as

$$\frac{\int f(\boldsymbol{y}_2|\boldsymbol{\theta}_0, \boldsymbol{y}_1)P(\boldsymbol{\theta}_0|\boldsymbol{y}_1)d\boldsymbol{\theta}_0}{\int f(\boldsymbol{y}_2|\boldsymbol{\theta}_1, \boldsymbol{y}_1)P(\boldsymbol{\theta}_1|\boldsymbol{y}_1)d\boldsymbol{\theta}_1}. \tag{7.138}$$

The advantage of using (7.138) is that the choice of the model is not dependent on the prior. It is also consistent with the theme that "a severe test for an economic theory, the only test and the ultimate test is its ability to predict" (Klein 1988, p. 21), and Zellner's (1988, p. 31) view of the "predictive principle that predictive performance is central in evaluating hypotheses" (also see Friedman 1953 and Geisser 1980). The disadvantage is that if $\boldsymbol{y}_2$ only contains a limited number of observations, the choice of the model becomes heavily sample dependent. If the number of observations in $\boldsymbol{y}_2$ is large, then a lot of sample information is not utilized to estimate unknown parameters. One remedy to this is to modify (7.138) by recursively updating the estimates when constructing the predictive density ratio:

$$\frac{\int f(y_T|\boldsymbol{\theta}_0, \boldsymbol{y}^{T-1})P(\boldsymbol{\theta}_0|\boldsymbol{y}^{T-1})d\boldsymbol{\theta}_0 \cdot \int f(y_{T-1}|\boldsymbol{\theta}_0, \boldsymbol{y}^{T-2}) \cdot P(\boldsymbol{\theta}_0|\boldsymbol{y}^{T-2})d\boldsymbol{\theta}_0}{\int f(y_T|\boldsymbol{\theta}_1, \boldsymbol{y}^{T-1})P(\boldsymbol{\theta}_1|\boldsymbol{y}^{T-1})d\boldsymbol{r}_1 \cdot \int f(y_{T-1}|\boldsymbol{r}_1, \boldsymbol{y}^{T-2}) \cdot P(\boldsymbol{\theta}_1|\boldsymbol{y}^{T-2})d\boldsymbol{\theta}_1} \tag{7.139}$$
$$\frac{\ldots \int f(y_{T_1+1}|\boldsymbol{\theta}_0, \boldsymbol{y}_1)P(\boldsymbol{\theta}_0|\boldsymbol{y}_1)d\boldsymbol{\theta}_0}{\ldots f(y_{T_1+1}|\boldsymbol{\theta}_1, \boldsymbol{y}_1)P(\boldsymbol{\theta}_1|\boldsymbol{y}_1)d\boldsymbol{\theta}_1},$$

where $P(\boldsymbol{\theta}|\boldsymbol{y}^T)$ denotes the posterior distribution of $\boldsymbol{\theta}$ given observations from 1 to T. Of course, (7.139) can be very laborious to compute.

While both the posterior odds ratio and posterior predictive density ratio can be computationally very laborious, the large sample approximation of the logarithm of the posterior odds ratio has been derived by Jeffreys (1967), Lindley (1961), and Schwarz (1978), among others. For instance, Schwarz (1978) recommends selecting the model that maximizes log $f(\boldsymbol{y}|H_i) - \frac{1}{2}m_i \log NT$, where m_i denotes the number of parameters for the model H_i. However, in the event that T is small, and the models under consideration involve choices between fixed versus random individual specific effects, it would be desirable to derive alternative approximations for various model selection criteria that balance the within-sample fit and post-sample prediction performance.

9 Conclusion

In this paper we have discussed various advantages of using panel data and surveyed some of the commonly used linear panel data models for drawing inferences while allowing for heterogeneity of individual sampling responses. Issues of fixed versus random effects formulations are also discussed. The formulas provided here are simple manipulations of matrix addition, subtraction, multiplication, and inversion. Therefore they can be easily programmed by using some well-known econometrics software such as GAUSS and SAS (for an evaluation of software that is useful for panel data analysis, see Blanchard 1992). However, the power of these models depends critically on the conformity of the statistical methods with the information content of the data. Otherwise, we may solve one problem but aggravate another. It is imperative that researchers explicitly recognize the limitations of the data and develop models and estimation methods based on what is known and available, rather than arbitrarily imposing assumptions on their models and data.

The discussion of this paper was based on the assumption that complete panels are available. In fact, one of the often confronted issues in using panel data is the occurrence of missing observations. With each new measurement over time, the collection of panel data places an additional burden on the respondents. Thus, attrition can increase over time. Furthermore, if the number of times respondents have been exposed to a survey gets large, the data may be affected and even behavioral changes may be induced (e.g. Binder and Hidiroglou 1988). In order to avoid these problems, rotating panels, that is, panels where part of the sample is replaced in each period and every individual is included in the panel for a limited number of periods, are often used in practice. If the observations are randomly missing, inference procedures discussed here can be modified as in Biorn (1982) and Hsiao (1986) (also see Nijman, Verbeek, and Van Soest 1991 for discussions on optimal rotating panel design). If the response probabilities are not independent of the endogenous variables, it creates the classical sample selection problem (e.g. Amemiya 1984; Heckman 1976, 1979). Results obtained from analyzing the complete observations may be subject to a severe bias (e.g. Hausman and Wise 1979; Hsiao 1986; Maddala 1978; Meghir and Saunders 1987; Ridder 1990; Verbeek 1991).

Another frequently occurring problem that is ignored here is the issue of measurement errors. When variables are subject to systematic measurement errors, their impact can be captured through the specification of auxiliary equations. If the variables are subject to random measurement errors, the application of the estimation methods discussed here will yield biased results. Measurement error raises complicated identification and estimation problems (e.g. Aigner, Hsiao, Kapteyn, and Wansbeek 1984; Hsiao 1976, 1977, 1979). However, panel data provide the possibility of identifying the model and obtaining consistent estimates of the unknown parameters even when variables are subject to random measurement errors (e.g. Biorn 1991; Griliches and Hausman 1986; Hsiao and Taylor 1991).

REFERENCES

Aigner, D. J., Hsiao, C., Kapteyn, A., and Wansbeek, T. (1985), "Latent Variable Models in Econometrics," in *Handbook of Econometrics*, Vol. 2, eds. Z. Griliches and M. D. Intriligator, Amsterdam: North-Holland, pp. 1322–1391.

Amemiya, T. (1984), "Tobit Models: A Survey," *Journal of Econometrics*, 24, 3–61.

Anderson, T. W. and Hsiao, C. (1981), "Estimation of Dynamic Models With Error Components," *Journal of the American Statistical Association*, 76, 598–606.

——— (1982), "Formulation and Estimation of Dynamic Models Using Panel Data," *Journal of Econometrics*, 18, 47–82.

Appelbe, T., Dineen, C., Solvanson, D. L., and Hsiao C. (1992), "Econometric Modeling of Canadian Long Distance Calling: A Comparison of Aggregate Time Series Versus Point-to-Point, Panel Data Approaches," *Empirical Economics* 17, 125–140.

Ashenfelter, O. (1978), "Estimating the Effect of Training Programs on Earnings," *Review of Economics and Statistics*, 60, 47–57.

Ashenfelter, O., and Solon, G. (1982), "Longitudinal Labor Market Data-Sources, Uses and Limitations," in *What's Happening to American Labor Force and Productivity Measurements*, W. E. Upjohn Institute for Economic Research, pp. 109–126.

Balestra, P., and Nerlove, M. (1966), "Pooling Cross-Section and Time Series Data in the Estimation of a Dynamic Model: The Demand for Natural Gas," *Econometrica*, 34, 585–612.

Baltagi, B. H. and Li, Qi (1990), "A Transformation That Will Circumvent the Problem of Autocorrelation in an Error Component Model," *Journal of Econometrics*, 48, 385–393.

Bhargava, A., Franzini, L., and Narendranathan, W. (1982), "Serial Correlation and the Fixed Effects Model," *Review of Economic Studies*, 49, 533–549.

Binder, D., and Hidiroglou, M. (1988), "Sampling in Time," in *Handbook of Statistics*, Vol. 6, eds. P. R. Krisnaiah and C. R. Rao, Amsterdam; Elsevier.

Biorn, E. (1992), "Econometrics of Panel Data With Measurement Errors," in *Econometrics of Panel Data: Theory and Applications*, eds. L. Màtyàs and P. Sevestre, Kluwer (forthcoming).

Blanchard, P. (1992), "Softwares," in *Econometrics of Panel Data: Theory and Application*, eds. L. Màtyàs and P. Sevestre, Kluwer (forthcoming).

Box, G. E. P. and Tiao, G. C. (1973), *Bayesian Inference in Statistical Analysis*, Menlo Park, CA: Addison-Wesley.

Breusch, T. S., and Pagan, A. R. (1980), "The Lagrange Multiplier Test and Its Application to Model Specification in Econometrics," *Review of Economic Studies*, 47, 239–254.

Chamberlain, G. (1978), "Omitted Variable Bias in Panel Data: Estimating the Returns to Schooling," *Annales de l'INSEE*, 30–31, 49–82.

Cochrane, D., and Orcutt, G. H. (1949), "Application of Least Squares Regression to Relationships Containing Autocorrelated Error Terms," *Journal of the American Statistical Association*, 44, 32–61.

De Finetti, B. (1964), "Foresight: Its Logical Laws, Its Subjective Sources," in *Studies in Subjective Probability*, eds. H. E. Kyburg, Jr., and H. E. Smokler, New York: Wiley, 93–158.
Geisser, S. (1980), "A Predictivistic Primer," in *Bayesian Analysis in Econometrics and Statistics: Essays in Honor of Harold Jeffreys*, Amsterdam: North Holland, pp. 363–382.
Griliches, Z. (1979), "Sibling Models and Data in Economics: Beginning of a Survey," *Journal of Political Economy*, 87, Suppl. 2, pp. S37-S64.
Griliches, Z., and J. A. Hausman (1986), "Errors-in-Variables in Panel Data," *Journal of Econometrics*, 31, 93–118.
Hausman, J. A. (1978), "Specification Tests in Econometrics," *Econometrica*, 46, 1251–1271.
Hausman, J. A., and Wise, D. A. (1979), "Attrition Bias in Experimental and Panel Data: The Gary Income Maintenance Experiment," *Econometrica*, 47, 455–473.
Heckman, J. J. (1976), "The Common Structure of Statistical Models of Truncation, Sample Selection, and Limited Dependent Variables and a Simple Estimator for Such Models," *Annals of Economic and Social Measurement*, 5, 475–492.
——(1978), "Simple Statistical Models for Discrete Panel Data Developed and Applied to Test the Hypothesis of True State Dependence Against the Hypothesis of Spurious State Dependence," Annales de l'INSEE, 30–31, 227–269.
——(1979), "Sample Selection Bias as a Specification Error," *Econometrica*, 47, 153–161.
——(1981a), "Statistical Models for Discrete Panel Data," in *Structural Analysis of Discrete Data with Econometric Applications*, eds. C. F. Manski and D. McFadden, Cambridge, MA: MIT Press, 114–178.
——(1981b), "Heterogeneity and State Dependence," in *Studies in Labor Markets*, ed. S. Rosen, Chicago: University of Chicago Press, 91–139.
Hsiao, C. (1974), "Statistical Inference for a Model With Both Random Cross-Sectional and Time Effects," *International Economic Review*, 15, 12–30.
——(1975), "Some Estimation Methods for a Random Coefficients Model," *Econometrica*, 43, 305–325.
——(1976), "Identification and Estimation of Simultaneous Equation Models with Measurement Error," *International Economic Review*, 17, 319–339.
——(1977), "Identification for a Linear Dynamic Simultaneous Error-Shock Model," *International Economic Review*, 18, 181–194.
——(1979), "Measurement Error in a Dynamic Simultaneous Equation Model with Stationary Disturbances," *Econometrica*, 47, 475–494.
——(1985), "Benefits and Limitations of Panel Data," *Econometric Review*, 4, 121–174.
——(1986), *Analysis of Panel Data*, New York: Cambridge University Press.
——(1990), "A Mixed Fixed and Random Coefficients Framework for Pooling Cross-Section and Time Series Data," paper presented at the Third Conference on Telecommunication Demand Analysis With Dynamic Regulation, Hilton Head, SC.
Hsiao, C. and Taylor, G. (1991), "Some Remarks on Measurement Errors and the Identification of Panel Data Models," *Statistica Neerlandica*, 45, 187–194.

Hsiao, C., Mountain, D. C., Tsui, K. Y., and Chan, Luke M. W. (1989), "Modeling Ontario Regional Electricity System Demand Using a Mixed Fixed and Random Coefficients Approach," *Regional Science and Urban Economics*, 19, 567–587.

James, W., and Stein, C. (1961), "Estimation with Quadratic Loss," in *Proceedings of the Fourth Berkeley Symposium on Mathematical Statistics and Probability*, ed. J. Neyman, 1, 361–379, Berkeley: University of California Press.

Jeffreys, H. (1967), *Theory of Probability*, London: Oxford University Press.

Judge, G., Griffiths, W., Hill, R., Lütkepohl, H., and Lee, T. (1985), *The Theory and Practice of Econometrics*, 2nd ed. New York: Wiley.

Kiefer, N. M. (1979), "Population Heterogeneity and Inference From Panel Data on the Effects of Vocational Education," *Journal of Political Economy*, 87, S213-S226.

Klein, L. R. (1988), "The Statistical Approach to Economics," *Journal of Econometrics*, 37, 7–26.

Kuh, E. (1963), *Capital Stock Growth: A Micro-Econometric Approach*, Amsterdam: North Holland.

Lee, L. F. (1979), "Estimation of Autocorrelated Error Components Model With Panel Data," mimeographed.

Lindley, D. V. (1961), "The Use of Prior Probability Distributions in Statistical Inference and Decision," in *Proceedings of the Fourth Berkeley Symposium on Mathematical Statistics and Probability*, ed. J. Neyman, Berkeley: University of California Press, Vol. 1, 453–468.

Lindley, D. V., and Smith, A. F. M. (1972), "Bayes Estimates for the Linear Model," *Journal of the Royal Statistical Society*, Ser. B, 34, 1–41.

Maddala, G. S. (1971), "The Use of Variance Components Models in Pooling Cross Section and Time Series Data," *Econometrica*, 39, 341–358.

——— (1978), "Selectivity Problems in Longitudinal Data," *Annales de l'INSEE*, 30–31, 423–450.

Meghir, C. and Saunders, M. (1987), "Attrition in Company Panels and the Estimation of Investment Equations," working paper, University College, London.

Mundlak, Y. (1978), "On the Pooling of Time Series and Cross Section Data," *Econometrica*, 46, 69–85.

Min, C. K., and Zellner, A. (1993), "Bayesian and Non-Bayesian Methods for Combining Models and Forecasts With Applications to Forecasting International Growth Rates," *Journal of Econometrics*, 59, 63–86.

Neyman, J., and Scott, E. L. (1948), "Consistent Estimates Based on Particularly Consistent Observations," *Econometrica*, 16, 1–32.

Nijman, T. H. E., Verbeek, M., and van Soest, A. (1991), "The Efficiency of Rotating Panel Designs in an Analysis of Variance Model," *Journal of Econometrics*, 49, 373–399.

Prais, S. J., and Winston, C. B. (1954), "Trend Estimators and Serial Correlation," Coweles Commission Discussion Paper No. 383, Chicago.

Ridder, G. (1990), "Attrition in Multi-Wave Panel Data," in *Panel Data and Labor Market Studies*, eds. J. Hartog, G. Ridder and J. Theeuwes, Amsterdam: North Holland.

Schwarz, G. (1978), "Estimating the Dimension of a Model," *Annals of Statistics*, 6, 461–464.

Smith, A. F. M. (1973), "A General Bayesian Linear Model," *Journal of The Royal Statistical Society*, Ser. B, 35, 67–75.

Swamy, P. A. V. B. (1970), "Efficient Inference in a Random Coefficient Regression Model," *Econometrica*, 38, 311–323.

Verbeek, M. (1991), "The Design of Panel Surveys and the Treatment of Missing Observations," unpublished Ph.D. dissertation, Tilburg University.

Wallace, T. D., and Hussain, A. (1969), "The Use of Error Components Models in Combining Cross-Section with Time Series Data," *Econometrica*, 37, 55–72.

Wansbeek, T., and Kapteyn, A. (1978), "The Separation of Individual Variation and Systematic Change in the Analysis of Panel Data," *Annales de l'INSEE*, 30–31, 659–680.

——(1982), "A Simple Way to Obtain the Spectral Decomposition of the Variance Components Models for Balanced Data," *Communications in Statistics*, A11, 2105–2112.

——(1983), "A Note on Spectral Decomposition and Maximum Likelihood Estimation of ANOVA Models With Balanced Data," *Statistics and Probability Letters*, 1, 213–215.

——(1989), "Estimation of the Error Components Model With Incomplete Panels," *Journal of Econometrics*, 41, 341–361.

Zellner, A. (1962), "An Efficient Method of Estimating Seemingly Unrelated Regressions and Tests for Aggregation Bias," *Journal of the American Statistical Association*, 57, pp. 348–368.

Zellner, A. (1988), "Bayesian Analysis in Econometrics," *Journal of Econometrics*, 37, 27–50.

Chapter 8

Panel Analysis for Qualitative Variables

ALFRED HAMERLE AND GERD RONNING

1 Introduction

In this chapter we consider models which take the special structure of qualitative panel data into consideration. Models for qualitative dependent variables in cross-section analysis are discussed by Arminger in Chapter 3. Hsiao, in Chapter 7, presents panel models for metric dependent variables. Therefore, these two chapters are closely related to the present one.

The growing availability of panel data for individuals, households, and firms has created demand for adequate models. For example, the British Consumer Expenditure Survey (CES) contains information about expenditures and the state of employment. Demand analysis by means of individual expenditures uses methods described by Hsiao in Chapter 7, whereas this chapter presents methods suitable for the analysis of individual labor supply and its determinants. Of course, joint analysis of consumption and labor supply is also of interest. The resulting mixed (discrete/continuous) models use elements from both chapters.

Our usage of the term "qualitative" follows the usual loose interpretation which subsumes the following types of variables:

- unordered categorical data (nominal scale)
- ordered categorical data (ordinal scale)
- integer-valued data ("count data")
- censored data ("limited dependent" variables)

ALFRED HAMERLE • Lehrstuhl für Statistik, Universität Regensburg, Universitätsstr. 31, D–93053 Regensburg, Germany. GERD RONNING • Abteilung Statistik und Ökonometrie I, Department of Economics, Eberhard-Karls-Universität, Mohlstr. 36, D–72074 Tübingen, Germany. • We thank Gerhard Arminger and Michael Sobel for helpful comments on an earlier version of this chapter.

Handbook of Statistical Modeling for the Social and Behavioral Sciences, edited by Gerhard Arminger, Clifford C. Clogg, and Michael E. Sobel. Plenum Press, New York, 1995.

Note that the last item characterizes a variable which is of a continuous nature. For example, Tobin (1958) considered expenditures for some durable goods like television sets or dish washers. Some households show "zero expenditure" in some periods whereas others spend various amounts of money for these goods. Therefore, both the qualitative aspect of consumption/no consumption and the quantitative aspect of the level of consumption must be considered. Such models are called Tobit models.

Section 2 reviews the main results for regression models with binary (cross-section) outcomes including estimation by maximum likelihood and—in the case of repeated observations—by generalized least squares. Section 3 considers the case of binary panel data. We present a fixed effects logit model and a random effects probit model and compare their advantages and disadvantages. Some generalizations are also treated. We then move to dynamic specifications: First we consider the case of autocorrelated errors in probit models. Proper dynamic structures are then introduced by considering autoregressive probit models which contain "lagged endogenous" variables among the explanatory variables. Both types of models are no longer estimable by standard estimation methods. Therefore, some alternative methods are proposed and illustrated by simulation experiments.

In Section 4 we give some results for Markov chain models which consider the transition between states over time and note the relationship to autoregressive probit models and discrete time duration models. Section 5 treats a random effects Tobit model which parallels to some extent the specification of the random effects probit model.Finally, Section 6 considers panel models for count data which usually are modeled by either the Poisson or the negative binomial distribution. Since the latter is a (gamma) mixture of Poisson distributions, it can be interpreted as introducing "heterogeneity" into the model. Both a fixed effects specification and a random effects specification are presented.

2 Some Regression Models for Binary Outcomes

2.1 Probit Model, Logit Model, Linear Probability Model, and Maximum Likelihood Estimation

We only consider the simplest of the discrete regression models in which the dependent variable y is binary; y takes the value 1 if an event occurs and 0 if it does not. Examples include participation in the labor force, union membership, purchases of durables in a given period, acceptance of loan applications, the decision to marry, and the like.

Let us assume that the Bernoulli random variable y_i has expectation $E(y_i) = P_i$ where P_i is the probability that the event occurs. Economists and social scientists are typically interested in the problem of studying how various explanatory variables affect P_i. Therefore, we assume that P_i is a function of explanatory variables $\boldsymbol{x}_i$, and a vector of unknown parameters $\boldsymbol{\beta}$. Then, we assume

$$P(y_i = 1 \mid \boldsymbol{x}_i) = E(y_i \mid \boldsymbol{x}_i) = F(\boldsymbol{x}_i' \boldsymbol{\beta})$$

where $F(\cdot)$ usually has the properties of a cumulative distribution function. In the following, all models are considered as conditional on observed values of $\boldsymbol{x}$.

A way to motivate such discrete regression models is based on the maximization of expected utility by an individual choosing between two alternatives. Let the utility connected with a choice be based on the attributes, as perceived by the individual, denoted by $\boldsymbol{v}_{i1}$ and $\boldsymbol{v}_{i0}$, the individual's socioeconomic characteristics (denoted by $\boldsymbol{z}_i$), and a random disturbance, e_{i1} or e_{i0} say. Let u_{i1} and u_{i0} denote the (random) utilities of the two choices. Then, we assume the linear relationships

$$\begin{aligned} u_{i0} &= \alpha_0 + \boldsymbol{z}_i' \gamma_0 + \boldsymbol{v}_{i0}' \boldsymbol{\delta} + e_{i0}, \\ u_{i1} &= \alpha_1 + \boldsymbol{z}_i' \gamma_1 + \boldsymbol{v}_{i1}' \boldsymbol{\delta} + e_{i1}. \end{aligned}$$

If $u_{i1} > u_{i0}$, alternative 1 is chosen, and, if $u_{i1} < u_{i0}$, the second alternative is chosen by the decision maker. Consequently,

$$\begin{aligned} P(y_i = 1) &= P(u_{i1} > u_{i0}) \\ &= P(e_{i0} - e_{i1} < (\alpha_1 - \alpha_0) + \boldsymbol{z}_i'(\gamma_1 - \gamma_0) + (\boldsymbol{v}_{i1} - \boldsymbol{v}_{i0})'\boldsymbol{\delta}). \end{aligned}$$

Defining $\boldsymbol{x}_i' = (1, \boldsymbol{z}_i', (\boldsymbol{v}_{i1} - \boldsymbol{v}_{i0})')$ and $\boldsymbol{\beta}' = (\alpha_1 - \alpha_0, (\gamma_1 - \gamma_0)', \boldsymbol{\delta}')$ we obtain

$$P(y_i = 1) = F(\boldsymbol{x}_i' \boldsymbol{\beta})$$

where F is the cumulative distribution function of $(e_{i0} - e_{i1})$.

The type of model obtained depends on the choice of F. The most frequently used choices in economics and social sciences are the linear probability model:

$$F(\boldsymbol{x}_i' \boldsymbol{\beta}) = \boldsymbol{x}_i' \boldsymbol{\beta}, \tag{8.1}$$

the probit model:

$$F(\boldsymbol{x}_i' \boldsymbol{\beta}) = \int_{-\infty}^{\boldsymbol{x}_i' \boldsymbol{\beta}} \frac{1}{\sqrt{2\pi}} \exp\left\{-\frac{t^2}{2}\right\} dt, \tag{8.2}$$

the logit model:

$$F(\boldsymbol{x}_i' \boldsymbol{\beta}) = \frac{\exp(\boldsymbol{x}_i' \boldsymbol{\beta})}{1 + \exp(\boldsymbol{x}_i' \boldsymbol{\beta})}. \tag{8.3}$$

In the linear probability model the expectation $E(y_i \mid \boldsymbol{x}_i)$ is equal to $\boldsymbol{x}_i' \boldsymbol{\beta}$ as in the usual regression framework. However, $E(y_i \mid \boldsymbol{x}_i)$ is the probability that the event will occur, so $E(y_i \mid \boldsymbol{x}_i)$ must fall in the interval $[0, 1]$. However, $\boldsymbol{x}_i{}'\boldsymbol{\beta}$ need not fall in the unit interval, revealing a deficiency of the linear probability model. While the linearity assumption may be appropriate over a range of values of the explanatory variables, it is certainly not appropriate for either extremely large or small values. As an alternative to the linear probability model, $P_i = E(y_i \mid \boldsymbol{x}_i)$ may be assumed to be a nonlinear function of the explanatory variables as in the probit and in the logit models. The functions F for the probit and logit models are the distribution functions of the standard normal distribution and the logistic distribution, respectively. Because they are distribution functions, they are bounded between 0 and 1.

Sometimes it is more convenient to assume that there is an underlying continuous response variable y_i^* defined by the regression relationship

$$y_i^* = \boldsymbol{x}_i' \boldsymbol{\beta} + u_i \tag{8.4}$$

(see Goldberger 1964; Heckman 1981). In practice, y_i^* is not observable. What we observe is a dummy variable y_i defined by

$$y_i = \begin{cases} 1 & \text{if } y_i^* > 0, \\ 0 & \text{otherwise.} \end{cases} \tag{8.5}$$

In this formulation, $\boldsymbol{x}_i' \boldsymbol{\beta}$ is not $E(y_i \mid \boldsymbol{x}_i)$ as in the linear probability model, it is $E(y_i^* \mid \boldsymbol{x}_i)$. According to Heckman (1981), introducing a latent continuous variable that crosses thresholds (and thus generates discrete outcomes) simplifies the analysis, and provides a natural framework for formulating choice theoretic econometric models. The probit and logit models correspond to the cumulative distribution of u_i being standard normal and logistic.

In particular for the probit model we use var$(u_i) = 1$. This is in contrast to the linear regression model where the error variance can vary freely. In order to understand why here the variance has to be restricted, consider the probability of the events $\{y = 1\}$ and $\{y = 0\}$ for arbitrary error variance σ^2. Disregarding the subscript i we obtain

$$P(y=0) = P\left(\frac{y^* - \boldsymbol{x}_i'\boldsymbol{\beta}}{\sigma} \leq \frac{-\boldsymbol{x}_i'\boldsymbol{\beta}}{\sigma}\right) = \Phi\left(\frac{-\boldsymbol{x}_i'\boldsymbol{\beta}}{\sigma}\right) \tag{8.6}$$

and

$$P(y=1) = P\left(\frac{y^* - \boldsymbol{x}_i'\boldsymbol{\beta}}{\sigma} > \frac{-\boldsymbol{x}_i'\boldsymbol{\beta}}{\sigma}\right) = 1 - \Phi\left(\frac{-\boldsymbol{x}_i'\boldsymbol{\beta}}{\sigma}\right) \quad , \tag{8.7}$$

where Φ denotes the distribution function of the standard normal. In these two probabilities the vector $\boldsymbol{\beta}$ could be arbitrarily rescaled when doing the same with the parameter σ. This means that only the ratio $\boldsymbol{\beta}/\sigma$ is identified. Therefore we set $\sigma = 1$.

If N observations are available, the likelihood function for the three models can be written as

$$L = \prod_{i=1}^{N} [F(\boldsymbol{x}_i'\boldsymbol{\beta})]^{y_i} \, [1 - F(\boldsymbol{x}_i' \boldsymbol{\beta})]^{1-y_i} .$$

The log-likelihood function is

$$l(\boldsymbol{\beta}) = \ln L = \sum_{i=1}^{N} \{y_i \ln F(\boldsymbol{x}_i' \boldsymbol{\beta}) + (1 - y_i) \ln [1 - F(\boldsymbol{x}_i' \boldsymbol{\beta})]\} . \tag{8.8}$$

The log-likelihood function can be rewritten as

$$l(\boldsymbol{\beta}) = \sum_{i=1}^{N} \left\{ y_i \ln \frac{F(\boldsymbol{x}_i' \boldsymbol{\beta})}{1 - F(\boldsymbol{x}_i' \boldsymbol{\beta})} + \ln(1 - F(\boldsymbol{x}_i' \boldsymbol{\beta})) \right\} . \tag{8.9}$$

Differentiating (8.9) with respect to $\boldsymbol{\beta}$ yields the score vector

$$\boldsymbol{s}(\boldsymbol{\beta}) = \frac{\partial\, l(\boldsymbol{\beta})}{\partial\, \boldsymbol{\beta}} = \sum_{i=1}^{N} \frac{f(\boldsymbol{x}_i' \boldsymbol{\beta})}{F(\boldsymbol{x}_i' \boldsymbol{\beta})\,(1 - F(\boldsymbol{x}_i' \boldsymbol{\beta}))}\,(y_i - F(\boldsymbol{x}_i' \boldsymbol{\beta}))\, \boldsymbol{x}_i \tag{8.10}$$

where $f(\cdot)$ is the density function corresponding to F and MLE $\hat{\boldsymbol{\beta}}$ is a solution (if it exists) of $\boldsymbol{s}(\boldsymbol{\beta}) = \mathbf{0}$. The matrix of second-order derivatives is

$$\begin{aligned}\frac{\partial^2\, l(\boldsymbol{\beta})}{\partial\, \boldsymbol{\beta}\, \partial\, \boldsymbol{\beta}'} &= -\sum_{i=1}^{N} \left[\frac{y_i - F(\boldsymbol{x}_i' \boldsymbol{\beta})}{F(\boldsymbol{x}_i' \boldsymbol{\beta})(1 - F(\boldsymbol{x}_i' \boldsymbol{\beta}))}\right]^2 [f(\boldsymbol{x}_i' \boldsymbol{\beta})]^2\, \boldsymbol{x}_i\, \boldsymbol{x}_i' \\ &\quad + \sum_{i=1}^{N} \frac{y_i - F(\boldsymbol{x}_i' \boldsymbol{\beta})}{F(\boldsymbol{x}_i' \boldsymbol{\beta})\,(1 - F(\boldsymbol{x}_i' \boldsymbol{\beta}))}\, f'(\boldsymbol{x}_i' \boldsymbol{\beta})\, \boldsymbol{x}_i\, \boldsymbol{x}_i',\end{aligned} \tag{8.11}$$

where $f'(\boldsymbol{x}_i'\boldsymbol{\beta})$ is the first derivative of the density function evaluated at $\boldsymbol{x}_i'\boldsymbol{\beta}$.

Since the expectation of the second sum in (8.11) is zero, we obtain a simpler expression for the Fisher information. It is given by

$$\mathrm{I}(\boldsymbol{\beta}) = -\,E\left(\frac{\partial^2\, l(\boldsymbol{\beta})}{\partial\, \boldsymbol{\beta}\, \partial\, \boldsymbol{\beta}'}\right) = \sum_{i=1}^{N} \frac{(f(\boldsymbol{x}_i' \boldsymbol{\beta}))^2}{F(\boldsymbol{x}_i' \boldsymbol{\beta})\,(1 - F(\boldsymbol{x}_i' \boldsymbol{\beta}))}\, \boldsymbol{x}_i\, \boldsymbol{x}_i'. \tag{8.12}$$

Iterative methods are necessary to solve for the first-order conditions $s(\boldsymbol{\beta}) = \mathbf{0}$. For the probit and logit models, iteration is simple because the likelihood is globally concave (see, e.g., Amemyia 1985, p. 273; or Ronning 1991, p. 33, p. 46). A method of scoring iteration is given by

$$\hat{\boldsymbol{\beta}}^k = \hat{\boldsymbol{\beta}}^{k-1} - \left[E\left(\frac{\partial^2\, l(\boldsymbol{\beta})}{\partial\, \boldsymbol{\beta}\, \partial\, \boldsymbol{\beta}'}\right)\right]^{-1}_{\boldsymbol{\beta}=\hat{\boldsymbol{\beta}}^{k-1}} \boldsymbol{s}(\hat{\boldsymbol{\beta}}^{k-1}) \tag{8.13}$$

where $\hat{\boldsymbol{\beta}}^k$ denotes the value of $\hat{\boldsymbol{\beta}}$ at the kth iteration. The Newton–Raphson iterative procedure can also be used with iteration

$$\hat{\boldsymbol{\beta}}^k = \hat{\boldsymbol{\beta}}^{k-1} - \left[\frac{\partial^2\, l(\boldsymbol{\beta})}{\partial\, \boldsymbol{\beta}\, \partial\, \boldsymbol{\beta}'}\right]^{-1}_{\boldsymbol{\beta}=\hat{\boldsymbol{\beta}}^{k-1}} \boldsymbol{s}(\hat{\boldsymbol{\beta}}^{k-1}).$$

Under various regularity conditions (see, for example, Amemyia 1985, sect. 9.2.2), the MLE $\hat{\boldsymbol{\beta}}$ is consistent and asymptotically normal

$$\sqrt{N}\,(\hat{\boldsymbol{\beta}} - \boldsymbol{\beta}) \xrightarrow{d} \mathcal{N}\left(0\,;\, \lim\left[-\,N^{-1}\, E\left(\frac{\partial^2\, l(\boldsymbol{\beta})}{\partial\, \boldsymbol{\beta}\, \partial \boldsymbol{\beta}'}\right)\right]^{-1}\right).$$

Here $\xrightarrow{d}$ denotes convergence in distribution. If the Newton–Raphson method is used, the asymptotic covariance matrix of $\hat{\beta}$ can be approximated by

$$\hat{\mathbf{V}}(\hat{\boldsymbol{\beta}}) = -\left[\frac{\partial^2\, l(\boldsymbol{\beta})}{\partial\, \boldsymbol{\beta}\, \partial\, \boldsymbol{\beta}'}\right]^{-1}_{\boldsymbol{\beta}=\hat{\boldsymbol{\beta}}}$$

for finite samples. If the method of scoring is used, the asymptotic covariance matrix of $\hat{\beta}$ can be estimated by

$$[\boldsymbol{I}(\boldsymbol{\beta})]^{-1}_{\boldsymbol{\beta}=\hat{\boldsymbol{\beta}}} \quad .$$

For the probit model where F and f are the standard normal cumulative distribution and probability density function,

$$\frac{\partial\, l(\boldsymbol{\beta})}{\partial\, \boldsymbol{\beta}} = \sum_{i=1}^{N} \frac{\varphi(\boldsymbol{x}_i'\, \boldsymbol{\beta}) \cdot \boldsymbol{x}_i}{\Phi(\boldsymbol{x}_i'\, \boldsymbol{\beta})\,(1 - \Phi(\boldsymbol{x}_i'\, \boldsymbol{\beta}))} (y_i - \Phi(\boldsymbol{x}_i'\, \boldsymbol{\beta})) \tag{8.14}$$

where $\varphi(t) = \frac{1}{\sqrt{2\pi}} \exp\left(-\frac{t^2}{2}\right)$ and $\Phi(t) = \int_{-\infty}^{t} \varphi(s)\,ds$.
Also note that

$$\varphi'(t) = -t \cdot \varphi(t) \quad \text{and} \quad \Phi(-t) = 1 - \Phi(t) \quad .$$

Then

$$\begin{aligned} \frac{\partial^2\, l(\boldsymbol{\beta})}{\partial\, \boldsymbol{\beta}\, \partial\, \boldsymbol{\beta}'} = & - \sum_{i=1}^{N} \left\{ \varphi(\boldsymbol{x}_i'\, \boldsymbol{\beta}) \left[\frac{y_i - \Phi(\boldsymbol{x}_i'\, \boldsymbol{\beta})}{\Phi(\boldsymbol{x}_i\, \boldsymbol{\beta})\,(1 - \Phi(\boldsymbol{x}_i'\, \boldsymbol{\beta}))} \right]^2 \right. \\ & \left. + \frac{[y - \Phi(\boldsymbol{x}_i'\, \boldsymbol{\beta})]\; (\boldsymbol{x}_i'\, \boldsymbol{\beta})}{\Phi(\boldsymbol{x}_i'\, \boldsymbol{\beta})\,(1 - \Phi(\boldsymbol{x}_i'\, \boldsymbol{\beta}))} \right\} \varphi(\boldsymbol{x}_i'\, \boldsymbol{\beta})\, \boldsymbol{x}_i \boldsymbol{x}_i' \end{aligned} \tag{8.15}$$

and the information matrix is

$$\boldsymbol{I}(\boldsymbol{\beta}) = -E \left(\frac{\partial^2\, l(\boldsymbol{\beta})}{\partial\, \boldsymbol{\beta}\, \partial\, \boldsymbol{\beta}'} \right) = \sum_{i=1}^{N} \frac{\varphi^2(\boldsymbol{x}_i'\boldsymbol{\beta})}{\Phi(\boldsymbol{x}_i'\boldsymbol{\beta})(1 - \Phi(\boldsymbol{x}_i'\boldsymbol{\beta}))} \boldsymbol{x}_i \boldsymbol{x}_i' \,. \tag{8.16}$$

For the logit model $P_i = F(\boldsymbol{x}_i'\boldsymbol{\beta})$ where

$$F(t) = \frac{\exp(t)}{1 + \exp(t)} \quad \text{and} \quad f(t) = \frac{\exp(t)}{(1 + \exp(t))^2} = F(t)\,(1 - F(t)) \quad . \tag{8.17}$$

From (8.17) it follows that

$$\begin{aligned} \frac{f(t)}{F(t)} &= 1 - F(t), \\ \frac{f'(t)}{f(t)} &= -F(t)(1 - \exp(-t)), \\ F(-t) &= 1 - F(t). \end{aligned}$$

Using these relations it is not difficult to show that for the logit model

$$\frac{\partial l(\boldsymbol{\beta})}{\partial \boldsymbol{\beta}} = \sum_{i=1}^{N} (y_i - F(\boldsymbol{x}_i'\boldsymbol{\beta}))\boldsymbol{x}_i$$

and

$$\frac{\partial^2 l(\boldsymbol{\beta})}{\partial \boldsymbol{\beta} \partial \boldsymbol{\beta}'} = -\sum_{i=1}^{N} f(\boldsymbol{x}_i'\boldsymbol{\beta})\boldsymbol{x}_i \boldsymbol{x}_i'.$$

The matrix of second partials no longer depends on y_i and, therefore, coincides with its expectation.

The simple expressions for the score function and the Hessian matrix follow from the theory of generalized linear models and exponential families (see, e.g., McCullagh and Nelder 1989). In a generalized linear model with "canonical link" the score function is particularly simple and the Hessian matrix is nonstochastic. If the distribution of the Bernoulli variables y_i is rewritten in an exponential family representation, we obtain

$$P(y_i) = \exp\{y_i \ln \frac{P_i}{1 - P_i} + \ln(1 - P_i)\},$$

and $\theta_i = \ln[P_i/(1 - P_i)]$ is the canonical parameter. It is easily seen that θ_i is the inverse function of the logistic function $P_i = \exp(\boldsymbol{x}_i'\boldsymbol{\beta})/(1 + \exp(\boldsymbol{x}_i'\boldsymbol{\beta}))$. Hence the logit model is given by $\ln[P_i/(1 - P_i)] = \boldsymbol{x}_i'\boldsymbol{\beta}$ (the "linear predictor"; see, e.g., McCullagh and Nelder 1989), and corresponds to the "natural parameterization" where the canonical parameter θ_i of the exponential family distribution is equal to the linear predictor $\boldsymbol{x}_i'\boldsymbol{\beta}$. Note that in the probit case we do not have the "canonical link", and obtain more complicated expressions for the derivatives of the log–likelihood function with respect to $\boldsymbol{\beta}$.

For both the probit and the logit models, the matrix of second partials is negative definite (the information matrix is positive definite) at each stage of iteration, because the log-likelihood function is strictly concave. Hence the iterative procedure will converge to a maximum of the likelihood function, no matter what the starting value is. The negative of the inverse of the Hessian matrix or the inverse of the information matrix, evaluated at the final converged estimate $\hat{\boldsymbol{\beta}}$, is an estimate of the asymptotic covariance matrix. These estimated variances and covariances will enable us to test hypotheses about the different elements of $\boldsymbol{\beta}$.

2.2 Generalized Least Squares Estimation When There Are Repeated Observations

The data we have considered thus far are individual; each observation consists of $(y_i, \boldsymbol{x}_i)$, the actual response of an individual and an associated regressor vector. In the case where repeated observations (or grouped data, respectively) are available, we observe the response of n_i individuals, all of whom have the same regressor vector $\boldsymbol{x}_i$. The observed dependent variable will consist of the proportion, p_i , of the n_i individuals who respond with $y = 1$. This proportion is an estimate of the corresponding probability, $P(y_i = 1|\boldsymbol{x}_i) = F(\boldsymbol{x}_i\prime\boldsymbol{\beta})$. Since the function $F(\boldsymbol{x}_i\prime\boldsymbol{\beta})$ is assumed strictly monotonic, it has an inverse. Let the inverse be denoted by h. Using the estimate p_i we can formulate a regression model

$$h(p_i) = \boldsymbol{x}_i'\boldsymbol{\beta} + \varepsilon_i \quad . \tag{8.18}$$

For the linear probability model we have

$$h(p) = p,$$

for the logit model we have

$$h(p) = \ln \frac{p}{(1-p)},$$

and for the probit model we have

$$h(p) = \Phi^{-1}(p)$$

where Φ^{-1} is the inverse function of the standard normal distribution function.

The regression model (8.18) is heteroskedastic. The variance of p_i is

$$\text{var}(p_i) = \frac{P_i(1-P_i)}{n_i}$$

where n_i is the number of repeated observations connected with the regressor vector $\boldsymbol{x}_i$. Using the first derivative $dh(p)/dp$, we can define a generalized least-squares principle where the residuals $h(p_i) - \boldsymbol{x}_i'\boldsymbol{\beta}$ are multiplied by the weights $\left[\text{var}(p_i)\left(\frac{dh(p_i)}{dp}\right)^2\right]^{-1}$, yielding generalized least-squares estimates $\hat{\boldsymbol{\beta}}$. The matrix diag $\left\{\text{var}(p_i)\left(\frac{dh(p_i)}{dp}\right)^2\right\}$ is the asymptotic covariance matrix of the vector $h(p)$. If h is a linear transformation, this matrix is the exact covariance matrix in finite samples. For the linear probability model, we have for a single element p

$$\frac{dh(p)}{dp} = 1,$$

for the logit model, we have

$$\frac{dh(p)}{dp} = \frac{1}{p(1-p)},$$

and for the probit model, we obtain

$$\frac{dh(p)}{dp} = [\varphi(p)]^{-1}$$

where $\varphi(\cdot)$ is the density function of the standard normal. The resulting estimator, which is generally referred to as the minimum-chi-square estimator, has the same asymptotic efficiency as the maximum likelihood estimator. In finite samples, the minimum-chi-square estimator may even have a smaller mean squared error than the maximum likelihood estimator as shown in some simulation experiments (see, e.g., Amemyia 1981; Berkson 1955, 1980; Kritzer 1978). However, the analysis of panel models is often based on survey data where many explanatory variables are continuous. Then, it will often be the case that the number n_i of outcomes y_i observed for each set of explanatory variables will be just one. That is, we only observe one value of the random variable y_i for each $\boldsymbol{x}_i$, so that $n_i = 1$. The minimum-chi-square method is not defined in this case, since its application requires repeated observations for each value of the vector of explanatory variables. For this reason, we shall confine our attention to the maximum likelihood method. It should be noted, however, that in experimental design and laboratory settings, the minimum-chi-square method is attractive and computationally simpler than maximum likelihood estimation.

2.3 A Note on Interpretation

Finally, we note the interpretation of the estimated coefficients in discrete choice models. We look at the derivatives of the probability of the event occuring with respect to a particular independent variable. Let x_{ik} be the kth element of the vector of explanatory variables $\boldsymbol{x}_i$, and let $\boldsymbol{\beta}_k$ be the kth element of $\boldsymbol{\beta}$. The derivatives for the probabilities given by the linear probability model, probit model, and logit model are, respectively,

$$\begin{aligned}
\frac{\partial}{\partial x_{ik}}(\boldsymbol{x}_i'\boldsymbol{\beta}) &= \boldsymbol{\beta}_k, \\
\frac{\partial}{\partial x_{ik}}\Phi(\boldsymbol{x}_i'\boldsymbol{\beta}) &= \varphi(\boldsymbol{x}_i'\boldsymbol{\beta})\,\boldsymbol{\beta}_k \\
\frac{\partial}{\partial x_{ik}}\frac{\exp(\boldsymbol{x}_i'\boldsymbol{\beta})}{1+\exp(\boldsymbol{x}_i'\boldsymbol{\beta})} &= \frac{\exp(\boldsymbol{x}_i'\boldsymbol{\beta})}{(1+\exp(\boldsymbol{x}_i'\boldsymbol{\beta}))^2}\,\boldsymbol{\beta}_k.
\end{aligned}$$

These derivatives will be needed for predicting the effects of changes in one of the independent variables. In the linear probability model, these derivatives are constant, and estimated coefficients indicate the increase in the probability of the event occuring given a one unit increase in the corresponding independent variable. For the probit and logit models, the coefficients reflect the effect of a one unit change in an independent variable upon the corresponding inverse distribution functions. These inverse distribution functions are $\Phi^{-1}(P_i)$ for the probit model and $\ln[P_i/(1-P_i)]$ for the logit model. In both cases the amount of the increase in the probability depends on the original probability and thus on the values of all the independent variables and their coefficients. Hence, we need to calculate the derivatives at different levels of the independent variables to get an idea of the range of variation of the resulting changes in the probabilities.

2.4 Models for Limited Dependent Variables

Tobin (1958) proposed a realistic model for household expenditures on consumer durables: households either purchase or do not purchase the good during a given time period, but among those who purchase the good, the amount spent increases with household income. Tobin called the variable which explains this mixed discrete/continuous behavior "limited dependent"; Goldberger (1964) coined the term "Tobit model". However, subsequent research has made apparent that there are in fact two types of Tobit models: the censored Tobit model and the truncated Tobit model. We first explain the terms "truncation" and "censoring" and then present the two versions of the model.

Consider a variable y^*, for which we observe the constant c whenever the variable takes on values smaller than c and for which we observe the values themselves otherwise. More formally, we define a (left) censored variable y by

$$y = \begin{cases} y^* & \text{if } \; y^* > c \\ c & \text{if } \; y^* \leq c \end{cases}$$

Such a situation arises, for example, in the case already described above with $c = 0$. For values of c different from zero some care must be taken in interpreting the model: Since

c is merely a threshold it should not be regarded as minimal consumption. Rather the event $\{y = c\}$ should be read as the event "no consumption." We shall see later on when introducing explanatory variables like, for example, income that the threshold will always be set equal to zero in order to identify the remaining parameters.

Alternatively we now consider a random variable which takes on only values greater than c. This case arises in demand analysis if only the buying households are observed whereas no information is available for those not buying. We call such a random variable (or rather the underlying distribution) "truncated (from below)." Technically, the density function $f(y)$ of a truncated distribution is derived from the density function $g(y)$ of the same random variable y when facing no such restriction. In the above example, g would describe the buying behavior when all households would be observed. The density f of the truncated random variable is then defined by

$$f(y) = \frac{g(y)}{\int_c^\infty g(t)\,dt} \quad , \tag{8.19}$$

which is merely a renormalization of the density function g.

The "Standard Tobit model" considers the case of censored variables and is specified as follows:

$$\begin{aligned} y^* &= \boldsymbol{x}'\boldsymbol{\beta} + \varepsilon \\ \varepsilon &\sim N(0, \sigma^2) \\ y &= \begin{cases} y^* & \text{if} \quad y^* > 0 \\ 0 & \text{if} \quad y^* \leq 0 \end{cases} \end{aligned} \tag{8.20}$$

Here we implicitly assume that all households have the same minimal consumption. Why can we specify minimal consumption by the value 0 ? Considering the case of just one explanatory variable z, we see from

$$\begin{aligned} P(y^* \leq y_0) &= P\left(\frac{y^* - \mu}{\sigma} \leq \frac{y_0 - \mu}{\sigma}\right) \\ &= \Phi\left(\frac{y_0 - \beta_1 - \beta_2 z}{\sigma}\right) \end{aligned} \tag{8.21}$$

that only the difference $y_0 - \beta_1$ is identified. Therefore we normalize β_1 such that $y_0 = 0$.

For the nonbuying households we get information about the unknown parameters from equation (8.21), which we rewrite as

$$\Phi\left(\frac{-\boldsymbol{x}_t'\boldsymbol{\beta}}{\sigma}\right) \quad .$$

For the buyers we observe the expenditures, which are modeled by the density

$$\frac{1}{\sigma}\varphi\left(\frac{y_t - \boldsymbol{x}_t'\boldsymbol{\beta}}{\sigma}\right) \quad .$$

If we now use the notation 0 for the nonbuyers and 1 for the buyers, we obtain the following likelihood function of the censored Tobit model:

$$L = \prod_0 \Phi\left(\frac{-\boldsymbol{x}_t'\boldsymbol{\beta}}{\sigma}\right) \prod_1 \frac{1}{\sigma}\varphi\left(\frac{y_t - \boldsymbol{x}_t'\boldsymbol{\beta}}{\sigma}\right) \quad . \tag{8.22}$$

By comparing the probabilities entering the likelihood with probabilities (8.6) and (8.7) in case of the probit model, we note that in this model the standard deviation σ is identified. Technically identification is achieved by the variable y_t, which describes expenditures of buyers and which for this model appears in the likelihood (8.22).

The log-likelihood function of the standard Tobit model is globally concave with respect to the parameters $\boldsymbol{\alpha}$ and h (see Olson 1978), where

$$\boldsymbol{\alpha} = \frac{\boldsymbol{\beta}}{\sigma} \quad \text{and} \quad h = \frac{1}{\sigma} \quad .$$

This makes maximum likelihood estimation of this model easy.

We now turn to the case of the truncated standard Tobit model, which considers only observations from purchasing households. This situation is described by a truncated distribution which starts from the conditional density of the observed random variable y given that $y^* > 0$. For this density we obtain from (8.19) under the normality assumption:

$$f(y) = \frac{1}{1-\Phi\left(-\frac{\boldsymbol{x}'\boldsymbol{\beta}}{\sigma}\right)} \frac{1}{\sigma} \varphi\left(\frac{y-\boldsymbol{x}'\boldsymbol{\beta}}{\sigma}\right) = \frac{1}{\sigma} \frac{\varphi\left(\frac{y-\boldsymbol{x}'\boldsymbol{\beta}}{\sigma}\right)}{\Phi\left(\frac{\boldsymbol{x}'\boldsymbol{\beta}}{\sigma}\right)} , \tag{8.23}$$

which leads to the likelihood function

$$L = \prod_{1} \frac{1}{\sigma} \frac{\varphi\left(\frac{y_t - \boldsymbol{x}_t'\boldsymbol{\beta}}{\sigma}\right)}{\Phi\left(\frac{\boldsymbol{x}_t'\boldsymbol{\beta}}{\sigma}\right)} . \tag{8.24}$$

3 Binary Regression Models for Panel Data

In the discrete regression models discussed in the last section the basic assumption is that the regression coefficients are the same for all individuals at all times. These models ignore individual differences, and the overall homogeneity assumption may not be a realistic one. Ignoring such parameter heterogeneity among cross-sections or time-series units could lead to inconsistent or meaningless estimates of structural parameters. Let us consider an example. As pointed out by Ben-Porath (1973), a discovery of a group of married women having an average yearly labor participation rate of 50% could lead to diametrically opposite inferences. At one extreme this might be interpreted as impyling that each woman in a homogeneous population has a 50% chance of being in the labor force in any given year, while at the other extreme it might imply that 50% of the women in a heterogeneous population always work and 50% never work. In the first case, each woman would be expected to spend half of her married life in the labor force and half out of the labor force, and job turnover would be expected to be frequent, with an average job duration of two years. In the second case, there is no turnover, and current information about work status is a perfect predictor of future work status. Either explanation is consistent with the finding relying on cross-sectional data. To discriminate among the many possible models, we need information on individual labor-force histories in different subintervals of the life cycle. This is possible only if we observe a number of individuals over time.

The main advantage of panel data is that models for the analysis of panel data can take into account unobserved heterogeneity across individuals and/or through time. Unobserved heterogeneity may be due to omitted variables. Conditional on the observed explanatory variables , the effects of omitted or excluded variables are in general driven by three types of variables (Hsiao 1986, p. 25):

- Individual time-invariant variables; these are variables that are the same for a given cross-sectional unit through time but that vary across cross-sectional units. Examples are sex, ability, attributes of firm-specific management.
- Period individual-invariant variables; these are variables that are the same for all individuals at a given point in time but that vary through time. Examples are prices and interest rates.
- Individual time-varying variables; these are variables that vary across cross-sectional units at a given point in time and also exhibit variations through time.

It turns out that with longitudinal data we have a much better chance to control for individual heterogeneity than with cross-sectional data or repeated cross-sections. For a more detailed discussion of the advantages of panel data, see, for example, Chamberlain (1984) and Hsiao (1985, 1986). Controlling for heterogeneity is in most applications a means to obtain consistent estimates of the systematic part of the model. A convenient starting point to illustrate the issues is the latent continuous variable model with a threshold model generalized for panel data. For the latent continuous variable y_{it}^* we assume the linear regression model

$$y_{it}^* = \boldsymbol{x}_{it}'\boldsymbol{\beta} + u_{it},$$

$i = 1, \ldots, N;\ t = 1, \ldots, T$. In practice, y_{it}^* is not observable. What we observe is a binary variable y_{it} defined by

$$y_{it} = \begin{cases} 1 & \text{if } y_{it}^* > 0, \\ 0 & \text{otherwise.} \end{cases} \tag{8.25}$$

In the most restrictive model that corresponds to cross-sectional analysis, the errors $\{u_{it}\}$ are assumed to be serially independent. Then, we have a quantal response model with NT observations. However, very often the serial independence assumption cannot be supported empirically. According to the remarks given above we introduce individual-specific and period-specific effects by decomposing the error term u_{it} into

$$u_{it} = \alpha_i + \lambda_t + \varepsilon_{it}. \tag{8.26}$$

If α_i and λ_t are treated as parameters that have to be estimated along with the structural parameters, the model is referred to as the "fixed effects" model. This model is likely to present some difficult estimation problems. First, in the typical panel data set T is small, often less than 10. Thus, considering the λ_t's as parameters does not cause problems. On the other hand, if there are a large number of cross-sectional units, say thousands, the number of unknown parameters α_i, that must be estimated will quickly become intractable.

Second, in this nonlinear setting it is not possible to eliminate heterogeneity α_i by taking differences of successive observations. Whereas in the case of continuous outcomes and no autoregressions the fixed effects model gives consistent estimates of the slope parameters $\boldsymbol{\beta}$ (see, e.g., Hsiao 1986), this is not the case when the dependent variable y_{it} is observed only as a qualitative variable and there are only a few time-series observations per individual. Andersen (1973) and Chamberlain (1980) demonstrate this for the logit model and suggest a conditional likelihood approach. The idea is to consider the likelihood function conditional on sufficient statistics for the incidental parameters α_i. This conditional likelihood approach only works in the logit model and in the log-linear model. Thus, in applying the fixed effects models to qualitative dependent variables based on panel data, the logit model and the log-linear models seem to be the only choices. We shall present the fixed effects logit model in Section 3.1.

If α_i (and λ_t) are treated as random variables, the model is referred to as the "random effects" model. The arguments for using random effects models instead of fixed effects models are several. One argument used in the analysis of variance literature is that if we want to make inferences only about the set of cross-section units in the sample, then we should treat α_i as fixed. On the other hand, if we want to make inferences about the population from which these cross-section data came, we should treat α_i as random. If we have a large number of cross-section units, instead of estimating N parameters α_i as in the fixed effects models, we estimate in the random effects models only the mean and variance. Another argument is the one given in Maddala (1971) that the α_i measure individual specific effects that we are ignorant about just in the same way that u_{it} measures effects for the ith cross-section unit in the t-th period that we are ignorant about. Thus, if u_{it} is treated as a random variable, then there is no reason why α_i should not also be treated as random. However, an advantage of the fixed effects model is that it does not require assumptions about the correlation between the individual-specific effects and the included explanatory variables $\boldsymbol{x}_{it}$. We shall discuss the random effects model in some detail in Section 3.2.

3.1 The Fixed Effects Logit Model

A fixed effects logit model that accounts for heterogeneity is

$$P(y_{it} = 1 \mid \boldsymbol{x}_{it}) = \frac{\exp(\alpha_i + \boldsymbol{x}'_{it}\boldsymbol{\beta})}{1 + \exp(\alpha_i + \boldsymbol{x}'_{it}\boldsymbol{\beta})}. \tag{8.27}$$

The outcomes y_{it} are assumed to be independent given α_i and $\boldsymbol{x}_i$. Assuming the individual specific effect α_i to be fixed requires the estimation of both $\boldsymbol{\beta}$ and α_i, $i = 1, \ldots, N$. When T tends to infinity, the ML estimator of α_i is consistent. However, usually there are only a few time-series observations per individual. Thus there is only a limited number of observations to estimate α_i, and we have the familiar incidental-parameter problem (see Neyman and Scott 1948).

A simple version of (8.27) was used for about 30 years in item response theory in psychology. Item response theory is the collective name for a number of models that describe the subject's response as an explicitly specified probability function containing one or more subject and item parameters. The logistic item response model was first formulated by

Rasch in 1961. The α_i's are subject parameters, for example representing "ability" in an intelligence test. The Rasch model also contains an item parameter per test item characterizing the difficulty of the items. T is the number of items instead of the number of observation times in panel analysis. In psychological testing, a main goal is the estimation of the subject parameters, whereas in the analysis of panel data interest centers on estimation of the structural parameter $\boldsymbol{\beta}$.

Unfortunately, contrary to the linear-model case, the ML estimators $\boldsymbol{\beta}$ and α_i cannot be separated in the nonlinear models for qualitative outcomes. When T is fixed, the inconsistency of $\hat{\alpha}_i$ is transmitted into the ML estimator for $\boldsymbol{\beta}$. Even if N tends to infinity, the ML estimator of $\boldsymbol{\beta}$ remains inconsistent. Andersen (1973) (see also Chamberlain 1980; Hsiao 1986) demonstrates this for the logit model with one regressor and $T = 2$. It turns out that the ML estimator $\hat{\boldsymbol{\beta}}$ converges in probability to $2\boldsymbol{\beta}$ which is not consistent. Unfortunately, the result could not be extended for the case $T > 2$. Several simulation studies of the Rasch model indicate that $\hat{\boldsymbol{\beta}}$ seems to converge to $T/(T-1) \cdot \boldsymbol{\beta}$.

Andersen (1970, 1973) proposed a conditional likelihood approach to estimate $\boldsymbol{\beta}$. To illustrate this conditional ML method, we consider the joint probability for a reponse pattern $\boldsymbol{y}_i' = (y_{i1}, \ldots, y_{iT})$. For simplicity, we assume that the explanatory variables are nonstochastic constants.

In general, the joint probability of $\boldsymbol{y}_i$ is (using the independence assumption)

$$P(\boldsymbol{y}_i) = \prod_{t=1}^{T} \pi_{it}^{y_{it}} (1 - \pi_{it})^{1-y_{it}}$$

where $\pi_{it} = P(y_{it} = 1), t = 1, \ldots, T$. The distribution of the Bernoulli variable y_{it} is a special case of the binomial distribution. The binomial distribution is a member of the exponential family. This can be seen, if $P(\boldsymbol{y}_i)$ is rewritten as

$$P(\boldsymbol{y}_i) = \exp\{\sum_{t=1}^{T} y_{it} \ln\left(\frac{\pi_{it}}{1 - \pi_{it}}\right) + \sum_{t=1}^{T} \ln(1 - \pi_{it})\} \quad . \tag{8.28}$$

The natural parameter in (8.28) is $\rho_{it} = \ln[\pi_{it}/(1 - \pi_{it})]$, and only in the logit model, ρ_{it} is a linear function of individual specific effects and regression parameters. We have

$$\ln\left(\frac{\pi_{it}}{1 - \pi_{it}}\right) = \alpha_i + \boldsymbol{x}_{it}'\boldsymbol{\beta}. \tag{8.29}$$

Inserting this into (8.28) we obtain

$$P(\boldsymbol{y}_i) = \exp\{\alpha_i \sum_{t=1}^{T} y_{it} + \boldsymbol{\beta}' \sum_{t=1}^{T} \boldsymbol{x}_{it} y_{it} + a(\alpha_i, \boldsymbol{\beta})\} \tag{8.30}$$

where

$$a(\alpha_i, \boldsymbol{\beta}) = \ln \prod_{t=1}^{T} (1 + \exp(\alpha_i + \boldsymbol{x}_{it}'\boldsymbol{\beta})) \quad .$$

From the theory for distributions from the exponential family it follows immediately that $\sum_{t=1}^{T} y_{it}$ is a (minimal) sufficient statistic for the individual specific parameter α_i. From the definition of a sufficient statistic follows that the conditional probability of $\boldsymbol{y}_i$, given $\sum_{t=1}^{T} y_{it}$, does not depend on α_i. So we have found a conditional distribution that does not contain the incidental parameters, and estimation of the structural parameters $\boldsymbol{\beta}$ can be based on this conditional distribution.

In the logit model, (8.30) can be rewritten as

$$P(\boldsymbol{y}_i) = \frac{\exp\left[\alpha_i \sum_{t=1}^{T} y_{it} + \boldsymbol{\beta}' \sum_{t=1}^{T} \boldsymbol{x}_{it} y_{it}\right]}{\prod_{t=1}^{T}(1 + \exp(\alpha_i + \boldsymbol{x}'_{it}\boldsymbol{\beta}))} \quad . \tag{8.31}$$

The probability distribution of the statistic $\sum_{t=1}^{T} y_{it}$ is given by

$$P(\sum_{t=1}^{T} y_{it} = \tau_i) = \sum_{\sum_{t=1}^{T} y_{it} = \tau_i} \frac{\exp\left[\alpha_i \tau_i + \boldsymbol{\beta}' \sum_{t=1}^{T} \boldsymbol{x}_{it} y_{it}\right]}{\prod_{t=1}^{T}(1 + \exp(\alpha_i + \boldsymbol{x}'_{it}\boldsymbol{\beta}))}, \tau_i = 0, 1, \ldots, T. \tag{8.32}$$

Dividing (8.31) by (8.32) we obtain the conditional probability for $\boldsymbol{y}_i$, given $\sum_{t=1}^{T} y_{it} = \tau_i$:

$$P(\boldsymbol{y}_i | \sum_{t=1}^{T} y_{it} = \tau_i) = \frac{\exp(\boldsymbol{\beta}' \sum_{t=1}^{T} \boldsymbol{x}_{it} y_{it})}{\sum_{\boldsymbol{d}_i \epsilon B(\tau_i)} \exp(\boldsymbol{\beta}' \sum_{t=1}^{T} \boldsymbol{x}_{it} d_{it})} \tag{8.33}$$

where $B(\tau_i) = \{\boldsymbol{d}_i = (d_{i1}, \ldots, d_{iT}) : d_{it} = 0 \text{ or } 1, \text{and} \sum_{t=1}^{T} d_{it} = \tau_i\}$. This conditional probability in (8.33) is independent of α_i. Moreover, it is in a conditional logit form (McFadden 1974) with the alternative sets $B(\tau_i)$ varying across individuals. Hence the conditional ML estimator of $\boldsymbol{\beta}$ can be obtained by using standard programs for ML estimation in logit models provided that they allow for individual specific alternative sets. However, the number of elements of $B(\tau_i)$ is $\binom{T}{\tau_i}$ and increases rapidly if the number of observation times increases. Since the response patterns of different individuals are independent, the total conditional likelihood is the product of the N individual contributions (8.33). There are $T + 1$ distinct alternative sets B(τ_i) corresponding to $\tau_i = 0, 1, \ldots, T$. Individuals for whom $\tau_i = 0$ or T do not contribute to the conditional likelihood function, because the corresponding probability in this case is equal to 1. So only $T - 1$ alternative sets are relevant. Consequently, there may be considerable loss of data, in particular when there are only a few panel waves and many individuals never change states.

Let us now demonstrate the conditional likelihood approach in the simple case when $T = 2$. The only relevant case is $\tau_i = y_{i1} + y_{i2} = 1$. Define $z_i = 1$ if $\boldsymbol{y}_i = (0, 1)$, and $z_i = 0$ if $\boldsymbol{y}_i = (1, 0)$. The corresponding conditional probability in (8.33) becomes

$$\begin{aligned} P(z_i = 1 | \tau_i = 1) &= \frac{P(z_i = 1)}{P(z_i = 1) + P(z_i = 0)} = \frac{1}{1 + P(z_i = 0)/P(z_i = 1)} \\ &= \frac{1}{1 + \exp(-\boldsymbol{\beta}'(\boldsymbol{x}_{i2} - \boldsymbol{x}_{i1}))} \quad . \end{aligned} \tag{8.34}$$

It is easily seen that equation (8.34) is in the form of a binary logit function with response z_i and explanatory variables $(\boldsymbol{x}_{i2} - \boldsymbol{x}_{i1})$. In this model changes in the $\boldsymbol{x}_{it}$'s are used to explain changes in the dichotomous dependent variables (0, 1) and (1, 0). The conditional log-likelihood function is

$$\log L_C = \sum_{i \in I} \{z_i(\boldsymbol{x}_{i2} - \boldsymbol{x}_{i1})'\boldsymbol{\beta} - \log(1 + \exp((\boldsymbol{x}_{i2} - \boldsymbol{x}_{i1})'\boldsymbol{\beta}))\} \qquad (8.35)$$

where $I = \{i : \tau_i = 1\}$. From (8.34) and (8.35) it becomes apparent that the effects of time-invariant explanatory variables cannot be estimated in fixed-effects models.

For general T, one has to consider the sets $\tau = 1, 2, \ldots, (T-1)$, and computations may become more cumbersome. For the case of $T = 3$ see Maddala (1987). Andersen (1970, 1973) has shown that the inverse of the information matrix based on the conditional likelihood function provides an asymptotic covariance matrix for the conditional ML estimator of the remaining parameters of the conditional likelihood under mild regularity conditions. He applied these results to the estimation of the item difficulty parameters in the Rasch models. Chamberlain (1980) extended the results to allow for the inclusion of stochastic regressor variables. Andersen (1973) and Chamberlain (1980) also showed that the conditional logit model can be extended to the multinomial logit model as well as the log-linear model.

Let us now consider an example. Cecchetti (1986) (see also Greene 1993) applied the fixed effects logit model in a study of the frequency of the price changes for a sample of 38 magazines observed over a 27-year period. The constant term for each magazine was allowed to change every 3 years, implying a total of 318 free parameters. The data were analyzed in 9-year subperiods, so we have $T = 3$. Since a price change could occur in any period, the sums of y_{it} were zero, one, two, or three. The model included as regressors (1) the time since the last price change, (2) inflation since the last price change, (3) the size of the last price change, (4) current inflation, (5) industry sales growth, and (6) a measure of sales volatility. Among his results, Cecchetti reports the regression coefficients and z-ratios listed in Table 1. The model suggests that the most important determinant of the probability of a price increase is the path of inflation.

Table 1. CML Estimates of the Fixed Effects Logit Model

	Coefficient	z-ratio
Time since last change	0.79	(2.68)
Inflation since last change	24.18	(2.94)
Previous change	-6.33	(1.50)
Current inflation	-27.60	(1.97)
Industry sales growth	7.93	(2.23)
Sales volatility	0.15	(1.45)

Finally, a further remark is necessary. From the theory of exponential families it follows that only for the natural parameterization a sufficient statistic exists. In the case of

Bernoulli variables, this natural parameterization is given by the logit specification. For the probit model no sufficient statistic for the individual specific effect exists, and the conditional ML method does not produce computational simplifications because the individual specific effects do not cancel out. This implies that all N individual specific effects must be estimated as part of the estimation procedure. As in the logit model, unconditional ML estimation in the probit model gives inconsistent estimates for both α_i and the structural parameters $\boldsymbol{\beta}$.

Heckman (1981b) has conducted a limited set of Monte Carlo experiments to get some idea of the order of the bias of the ML estimator for the fixed effects probit models. In Heckman's experiment with the fixed effects ML estimator of the probit model, the bias is never more than 10% and is always toward zero. Wright and Douglas (1976) obtained similar results for the unconditional ML estimator of the fixed effects logit model. They found that for $T \geq 20$ the ML estimator of the structural parameters is virtually unbiased, and its distribution is well described by a normal distribution, with the covariance matrix based on the inverse of the estimated information matrix. In summary we see that applying the fixed effects models to qualitative dependent variables based on panel data, for small T the logit model and the log-linear models seem to be appropriate. If the number of panel waves increases, unconditional ML estimation of both logit and probit models can be used as well.

3.2 Random Effects Models

In the random effects approach, the person effects α_i are assumed random, and we have to specify a distribution of the α_i's. If the person effects are correlated with the included explanatory variables, we have to specify a conditional distribution for α, given $\boldsymbol{x}$. Thus this approach resolves the "incidental parameters" problem introduced by α_i at the cost of making an additional distributional assumption. By specifying the conditional distribution of α given $\boldsymbol{x}$, and estimating the fixed number of parameters of that conditional distribution along with structural model parameters, we avoid the problem of having the number of parameters increase as the number of individuals in the sample increases. The drawback of this approach is the need to assume a functional form for the distribution of α. If the distributional assumption is incorrect, then the estimates may be inconsistent. We again use the latent continuous variable model crossing thresholds. For the latent continuous variable we assume the regression relationship

$$y_{it}^* = \boldsymbol{x}_{it}'\boldsymbol{\beta} + u_{it} ,$$

$i = 1, \ldots, N; t = 1, \ldots, T$. We only observe a binary variable y_{it} defined by

$$y_{it} = \begin{cases} 1 & \text{if} \quad y_{it}^* > 0 \\ 0 & \text{otherwise} \end{cases} \tag{8.36}$$

In the decomposition of the error term u_{it},

$$u_{it} = \alpha_i + \lambda_t + \epsilon_{it} ,$$

the λ_t are treated as parameters. Since the length of the panel is typically short (while the number of individuals is large), only a few additional parameters have to be estimated. By an appropriate extension of the vector $\boldsymbol{x}_{it}$ of regressor variables, the time-specific effects can be absorbed in the parameter vector $\boldsymbol{\beta}$. For identification reasons only $T-1$ new parameters $\lambda_1, \ldots, \lambda_{T-1}$ are introduced, and we define $\tilde{\boldsymbol{\beta}}' = (\lambda_1, \ldots, \lambda_{T-1}, \boldsymbol{\beta}')$. Analogously, $\boldsymbol{x}_{it}$ is extended by introducing $T-1$ dummy variables with a one at the tth position and zeros otherwise. In the following we denote the possibly extended vector of explanatory variables also by $\boldsymbol{x}_{it}$, and the possibly extended parameter vector by $\boldsymbol{\beta}$ so that the linear combination $\boldsymbol{x}_{it}'\boldsymbol{\beta}$ may contain the time specific effect λ_t if necessary. With random effects the composite error term

$$u_{it} = \alpha_i + \epsilon_{it}$$

is correlated within persons across time. With the logit model, where the errors are assumed to have a logistic distribution, we need to use the multivariate logistic distribution. This distribution is described by Johnson and Kotz (1972, pp. 291–294). In the specific case where the ϵ_{it} are independently and identically distributed (iid), only the marginal distributions are needed, and, for given α_i, we obtain the familiar logit model. However, in more general cases we need to use the joint T-variate distribution.

The multivariate logistic distribution has the disadvantage that the correlations are all constrained to be 0.5. Although some generalizations are possible, the multivariate logistic distribution does not permit much flexibility. On the other hand, with the probit model we use the multivariate normal distribution, and general correlation structures can be modeled. So, in the case of random effects models for qualitative outcomes it is the probit model that is computationally tractable rather than the logit model.

Maximum Likelihood Estimation of the Random Effects Probit Model with Heterogeneity

In this section we consider the model

$$\begin{aligned} y_{it}^* &= \boldsymbol{x}_{it}'\boldsymbol{\beta} + \alpha_i + \epsilon_{it} \\ y_{it} &= \begin{cases} 1 & \text{if } y_{it}^* > 0 \\ 0 & \text{otherwise} \end{cases} \end{aligned} \tag{8.37}$$

where $\alpha_i \sim N(0, \sigma_\alpha^2)$, $\epsilon_{it} \sim N(0, \sigma_\epsilon^2)$, $E(\alpha_i \epsilon_{it}) = 0$. We also assume that the individual specific effects, α_i, are independent of $\boldsymbol{x}_i$ and for identification reasons, $\sigma_\epsilon^2 = 1$. For the joint probability of $\boldsymbol{y}_i$, given α_i and $\boldsymbol{x}_i$, we obtain

$$P(\boldsymbol{y}_i|\alpha_i, \boldsymbol{x}_i) = \prod_{t=1}^{T} \Phi\left(\boldsymbol{x}_{it}'\boldsymbol{\beta} + \alpha_i\right)^{y_{it}} \left[1 - \Phi\left(\boldsymbol{x}_{it}'\boldsymbol{\beta} + \alpha_i\right)\right]^{1-y_{it}}.$$

For the joint probability of $\boldsymbol{y}_i$, (given $\boldsymbol{x}_i$) we obtain

$$P(\boldsymbol{y}_i|\boldsymbol{x}_i) = \int \prod_{t=1}^{T} \Phi\left(\boldsymbol{x}_{it}'\boldsymbol{\beta} + \sigma_\alpha^2\, \tilde{\alpha}_i\right)^{y_{it}} \cdot \left[1 - \Phi\left(\boldsymbol{x}_{it}'\boldsymbol{\beta} + \sigma_\alpha^2\, \tilde{\alpha}_i\right)\right]^{1-y_{it}} \cdot \varphi(\tilde{\alpha}_i) d\tilde{\alpha}_i \tag{8.38}$$

where $\alpha_i = \sigma_\alpha \tilde{\alpha}_i$ and φ denotes the density function of the standard normal distribution. Equation (8.38) replaces the probability function for $\boldsymbol{y}_i$ conditional on α_i by a probability function that is marginal with respect to α_i. It is a function of a finite number of parameters, $\boldsymbol{\beta}$ and σ_α^2. Thus, maximizing the log-likelihood

$$\log L = \sum_{i=1}^{N} \log P(\boldsymbol{y}_i | \boldsymbol{x}_i), \tag{8.39}$$

under weak regularity conditions, will give consistent estimators for $\boldsymbol{\beta}$ and σ_α^2 as N tends to infinity. The log-likelihood (8.39) requires a numerical integration per observation of products of cumulative normal error functions. In the case of normally distributed individual specific effects, α_i, Gaussian quadrature can be used, which is computationally efficient and easy to implement (see Butler and Moffit 1982). Using an M-point Gaussian quadrature, an integral of the form $\int f(x)\varphi(x)d(x)$ is approximated by a weighted sum

$$\sum_{m=1}^{M} w_m f(x_m)$$

where x_m are the Gaussian quadrature points and w_m the associated weights. The terms $w_m\sqrt{\pi}$ and $w_m\sqrt{2}$ are given in Abramowitz and Stegun (1965, p. 924). The use of the probit model with the entire set of NT observations ignoring the correlations will give an initial consistent estimate of $\boldsymbol{\beta}$. The first application of the random effects probit model is due to Heckman and Willis (1976). An efficient algorithm that uses analytical second derivatives of the log-likelihood function was developed by Spiess (1993). A set of Monte Carlo experiments shows that the algorithm converges quickly and that the variance estimates calculated from the matrix of second derivatives are closer to the true parameter values than the estimates derived from the outer product of the first derivatives of the log-likelihood.

Note finally that is not necessary to assume that α_i or ϵ_{it} are normal variates. For example, assuming the ϵ_{it} logistic leads to the logit model for given α_i. This model is considered in the next section, but other choices for the distributions of ϵ_{it} and α_i are possible as well. Moreover, it is possible to permit the error variances to differ among time periods and to estimate the ratio among error variances in different periods.

A Simple Test for Heterogeneity

Hamerle (1990) proposes a simple test for the detection of unmeasured heterogeneity in regression models for panel data. Conditionally on the individual-specific effects and the regressor variables, the distribution of the dependent variable is assumed to belong to the linear exponential family which includes the binomial regression model. The null hypothesis is that there is no neglected heterogeneity. This can be expressed as $\sigma_\alpha^2 = 0$. Since the distribution of α is in general not known, we need a test statistic that does not require knowledge about the form of the distribution of α. This is achieved by an appropriate approximation of the likelihood function with a Taylor expansion neglecting terms of order higher than two. Then, the proposed test is a score test for the null hypothesis $\sigma_\alpha^2 = 0$. In the cases studied here, the outcomes y_{it} are Bernoulli variables, and conditionally on the unobservable individual specific effect α_i and the regressor variables, we have the response probabilities

$$P(y_{it}|\boldsymbol{x}_{it}, \alpha_i) = F(\boldsymbol{x}'_{it}\boldsymbol{\beta} + \alpha_i)$$

where F is an appropriate distribution function. Furthermore conditionally on α_i, the outcomes are assumed serially independent. Hamerle (1990) conducted a Monte Carlo experiment for the binary logistic regression model. Here we have

$$P(y_{it} = 1|\boldsymbol{x}_{it}, \alpha_i) = L(\boldsymbol{x}'_{it}\boldsymbol{\beta} + \alpha_i)$$

where $L(z) = \exp(z)/(1 + \exp(z))$. In this case the numerator of the test statistic is given by (see Hamerle 1990)

$$U = \frac{1}{2}\sum_{i=1}^{N}\left(\left[\sum_t (y_{it} - \hat{\mu}_{it})\right]^2 - \sum_t \hat{\mu}_{it}(1 - \hat{\mu}_{it})\right)$$

where $\hat{\mu}_{it} = L(\boldsymbol{x}'_{it}\hat{\boldsymbol{\beta}})$, and $\hat{\boldsymbol{\beta}}$ denotes the maximum likelihood estimator of $\boldsymbol{\beta}$ under the null hypothesis, that is when σ_α^2 is known to be zero. Note that U can be calculated with available computer programs. Hamerle (1990) uses the variance estimate

$$s_U^2 = \frac{1}{4}\sum_{i=1}^{N}(s_i - \bar{s})^2,$$

where $s_i = \sum_t (y_{it} - \hat{\mu}_{it})^2 - \sum_t \hat{\mu}_{it}(1 - \hat{\mu}_{it})$ and $\bar{s}$ is the sample mean of s_i. This is the conditional estimate of the variance of U given the estimates of the regression coefficients.

However, this choice of the variance estimator ist not optimal because it ignores a possible asymptotic correlation between the parameter estimates and the test statistics and may render the test conservative (Orme 1992, personal communication). Commenges et al. (1993) also discuss this issue and propose an improved variance estimator.

$$\boldsymbol{I} = \boldsymbol{I}_{\sigma^2\sigma^2} - \boldsymbol{I}_{\sigma^2\boldsymbol{\beta}}\,\boldsymbol{I}_{\boldsymbol{\beta\beta}}^{-1}\,\boldsymbol{I}_{\sigma^2\boldsymbol{\beta}} \tag{8.40}$$

where

$$\begin{aligned}
\boldsymbol{I}_{\sigma^2\sigma^2} &= \sum_{i=1}^{N} \mathrm{E}(\partial l_i/\partial\sigma_\alpha^2)^2 \\
\boldsymbol{I}_{\boldsymbol{\beta\beta}} &= \sum_{i=1}^{N} \mathrm{E}(\partial l_i/\partial\boldsymbol{\beta})(\partial l_i/\partial\boldsymbol{\beta})' \\
\boldsymbol{I}_{\sigma^2\boldsymbol{\beta}} &= \sum_{i=1}^{N} \mathrm{E}(\partial l_i/\partial\sigma_\alpha^2)(\partial l_i/\partial\boldsymbol{\beta})'
\end{aligned}$$

where

$$l_i = \log \int \prod_{t=1}^{T} \mu_{it}(\alpha_i)^{y_{it}}(1 - (\mu_{it}(\alpha_i))^{1-y_{it}} dG(\alpha_i) \tag{8.41}$$

$\mu_{it} = \mathrm{L}(\boldsymbol{x}'_{it}\boldsymbol{\beta} + \alpha_i)$ and $G(\alpha_i)$ denotes the distribution function of α_i. Both the scores $\partial l_i/\partial\sigma^2_\alpha$, $\partial l_i/\partial\boldsymbol{\beta}$ and the expectations are calculated under H_0, that is at $\sigma^2_\alpha = 0$, followed by replacing $\boldsymbol{\beta}$ by $\hat{\boldsymbol{\beta}}$. It can be shown (see Commenges et al. 1993) that

$$\begin{aligned}
\boldsymbol{I}_{\sigma^2\sigma^2} &= \frac{1}{4}\sum_{i=1}^{N}\left[\sum_t \mu_{it}(1-\mu_{it})(1-6\mu_{it}(1-\mu_{it})) + 2\left(\sum_t \mu_{it}(1-\mu_{it})\right)^2\right] \\
\boldsymbol{I}_{\boldsymbol{\beta}\boldsymbol{\beta}} &= \sum_{i=1}^{N}\sum_t \mu_{it}(1-\mu_{it})\,\boldsymbol{x}_{it}\boldsymbol{x}'_{it} \\
\boldsymbol{I}_{\sigma^2\boldsymbol{\beta}} &= \sum_{i=1}^{N}\sum_t \mu_{it}(1-\mu_{it})\left(\frac{1}{2}-\mu_{it}\right)\boldsymbol{x}_{it}
\end{aligned}$$

Commenges et al. (1993) (see also Orme 1993, and personal communication) show in a simulation study that this variance estimator provides a more powerful test.

General One-Factor Schemes

Heckman (1981a) has generalized the random effects model (8.37) by allowing an error process of the form

$$u_{it} = \gamma_t\alpha_i + \epsilon_{it}, \quad i = 1,\ldots,N;\ t = 1,\ldots,T. \tag{8.42}$$

The following assumptions are made:

(1) α_i is distributed independently of ϵ_{it} and of the included exogenous variables,

(2) $\mathrm{E}\alpha_i = \mathrm{E}\epsilon_{it} = 0$, and $\mathrm{E}\epsilon^2_{it} = \sigma^2_{\epsilon,t}$, $\mathrm{E}\alpha^2_i = \sigma^2_\alpha$,

(3) ϵ_{it} are serially independent, and in the probit specification all random variables are normally distributed.

The random effects model with error process (8.42) is referred to as a "one-factor scheme". The γ_t are parameters that can be estimated, and the error variances $\sigma^2_{\epsilon,t}$ are allowed to vary across time. The simple components of variance model $u_{it} = \alpha_i + \epsilon_{it}$ is a special case with $\gamma_t = 1$ and $\sigma^2_{\epsilon,t} = \sigma^2_\epsilon$ for all t.

The probability that $y_{it} = 1$ given $\boldsymbol{x}_{it}$ and α_i is

$$\begin{aligned}
P(y_{it} = 1|\boldsymbol{x}_{it},\alpha_i) &= P(\epsilon_{it} \geq -\boldsymbol{x}'_{it}\boldsymbol{\beta} - \gamma_t\alpha_i|\boldsymbol{x}_{it},\alpha_i) \\
&= \Phi(\boldsymbol{x}'_{it}\tilde{\boldsymbol{\beta}}_t + \frac{\gamma_t\sigma_\alpha}{\sigma_{\epsilon,t}}\tilde{\alpha}_i)
\end{aligned} \tag{8.43}$$

where $\tilde{\alpha}_i$ is the standardized α_i variable, and $\tilde{\boldsymbol{\beta}}_t = \boldsymbol{\beta}/\sigma_{\epsilon,t}$.

Since the ϵ_{it} are serially independent, the joint probability of $\boldsymbol{y}'_i = (y_{i1},\ldots,y_{iT})$, given $\boldsymbol{x}_i$ and α_i is given by the product of the corresponding probabilities in (8.43) (and $P(y_{it} = 0|\boldsymbol{x}_{it},\alpha_i)$ resp.) for $t = 1,\ldots,T$. For the joint probability of $\boldsymbol{y}_i$ given $\boldsymbol{x}_i$ we obtain

$$P(\boldsymbol{y}_i|\boldsymbol{x}_i) = \int_{-\infty}^{\infty}\prod_{t=1}^{T}\Phi\{[\boldsymbol{x}'_{it}\tilde{\boldsymbol{\beta}}_t + \frac{\gamma_t\sigma_\alpha}{\sigma_{\epsilon,t}}\tilde{\alpha}][2y_{it}-1]\}\varphi(\tilde{\alpha})d\tilde{\alpha} \tag{8.44}$$

For identification reasons a normalization restriction is necessary. For example, we can set $\sigma^2_{\epsilon,1}$ equal to 1. Using (8.44), it is possible to estimate $\boldsymbol{\beta}, \gamma_t\sigma_\alpha$, and $\sigma^2_{\epsilon,t}$, $t = 2, \ldots, T$, by the method of maximum likelihood. Other normalizations are possible as well. For more details the reader is referred to Heckman (1981a). It must be noted, however, that one-factor schemes have not received much attention in applications.

3.3 Random Coefficients Models

A further extension which is well known for linear random effects models is the incorporation of several random effects. Let the latent variable y^*_{it} now be given by

$$y^*_{it} = \boldsymbol{x}'_{it}\boldsymbol{\beta} + \alpha_i + \boldsymbol{z}'_{it}\boldsymbol{a}_i + \epsilon_{it} \tag{8.45}$$

and

$$y_{it} = \begin{cases} 1 & \text{if } y^*_{it} > 0 \\ 0 & \text{otherwise} \end{cases} \tag{8.46}$$

where $\boldsymbol{z}_{it}$ is a $q \times 1$ vector of covariates whose coefficients vary across individuals. $\boldsymbol{z}_{it}$ is in general a subset of $\boldsymbol{x}_{it}$ so we can adopt the convention $E(\boldsymbol{a}_i) = \boldsymbol{0}$.

Let us consider the special case $\boldsymbol{z}_{it} = \boldsymbol{x}_{it}$. Then we have

$$y^*_{it} = \boldsymbol{x}'_{it}(\boldsymbol{\beta} + \boldsymbol{a}_i) + \alpha_i + \epsilon_{it}$$

where the slopes of $\boldsymbol{x}_{it}$ are random. This may arise, for example, if there are interactions of included covariates and unobserved variables. Consider the covariates sex and age. If both covariates are included we have the model

$$y^*_{it} = \beta_0 + \beta_1 \cdot \text{sex}_i + \beta_2 \cdot \text{age}_i + \beta_3 \cdot \text{sex}_i \cdot \text{age}_i + \alpha_i + \epsilon_{it}$$

where β_3 is the weight of the interaction between sex and age. Rearranging terms we obtain

$$y^*_{it} = \beta_0 + (\beta_1 + \beta_3 \cdot \text{age}_i) \cdot \text{sex}_i + \beta_2 \cdot \text{age}_i + \alpha_i + \epsilon_{it}$$

Now assume that the variable age is not observed. If the interaction term is negligible, then $\beta_2 \cdot \text{age}_i$ is absorbed in the random effect α_i and we obtain again a regression model with a random intercept. However, if the interaction is important and cannot be neglected, the coefficient of the variable sex is $a_i = \beta_1 + \beta_3 \cdot \text{age}_i$ which varies across individuals. A random coefficients model like (8.45) may be appropriate in this case.

As in one-factor models, the correlation among responses for an individual is assumed to arise completely from natural heterogeneity in regression coefficients across people. Given one person's coefficients, the responses are assumed to be independent observations. Setting $\boldsymbol{b}'_i = (\alpha_i, \boldsymbol{a}'_i)$ and $\boldsymbol{z}^{*\prime}_{it} = (1, \boldsymbol{z}'_{it})$ we have

$$P(\boldsymbol{y}_i | \boldsymbol{x}_i, \boldsymbol{b}_i) = \prod_{t=1}^{T} F(\boldsymbol{x}'_{it}\boldsymbol{\beta} + \boldsymbol{z}^{*\prime}_{it}\boldsymbol{b}_i)^{y_{it}} \left(1 - F(\boldsymbol{x}'_{it}\boldsymbol{\beta} + \boldsymbol{z}^{*\prime}_{it}\boldsymbol{b}_i)\right)^{1-y_{it}}$$

It should be noted again, that in general it is not necessary that all individuals have the same number of observations. This number T_i may vary across individuals. We made the assumption $T_1 = \ldots = T_N = T$ only for notational convenience.

Finally we have to assume a specific form for the distribution of $\boldsymbol{b}_i$. A traditional choice is the $q + 1$ dimensional Gaussian distribution with mean zero and unknown covariance matrix $\boldsymbol{C}$.

Only a few articles are concerned with random coefficients models for qualitative responses. Random coefficients models are rather difficult to fit because evaluation of the likelihood requires numerical methods in most problems. The problems arise because high dimensional integral expressions need to be calculated repeatedly. For example, if a probit specification in (8.45) in used, the likelihood function involves the multivariate normal integral which in intractable even for moderate dimensions. One approach to avoid the numerical integrals is to approximate the integrals with simple expansions whose integrals have closed forms. This was the approach proposed from a Bayesian perspective by Stiratelli, Laird, and Ware (1984). Hennevogl (1991) and Fahrmeir and Tutz (1994) discuss in detail maximum likelihood estimation of the structural model parameters (i.e., $\boldsymbol{\beta}$ and the elements of $\boldsymbol{C}$) based on the EM algorithm. Their approach is general by imbedding model (8.45) into the framework of multivariate generalized linear models. The integrals appearing in the E step are approximated by Monte Carlo methods or Gauss-Hermite quadrature in the case of normally distributed random effects. A Bayesian approach for fitting random coefficients models for qualitative outcomes using Gibbs sampling has been proposed by Zeger and Karim (1991). Other simulation-based numerical integration methods have been developed by McFadden (1989), Pakes and Pollard (1989), and others. An overview of these simulation techniques for the estimation of limited dependent variable models is given by Hajivassiliou (1992). Another approach is to fit "marginal models" where the regression and within-subject correlation are modeled separately. The correlation over time is considered as nuisance and "maximum quasi-likelihood estimation" or "generalized estimating equations" techniques (Liang and Zeger 1986) are used for the estimation of the structural regression parameters, which is the main interest. These methods are briefly described in the next section.

3.4 Probit Models With Autocorrelated Errors

The random effects model (8.37) assumes that for the latent continuous process the correlation between successive disturbances for the same individual unit is a constant. This specification is often referred to as the specification of "equicorrelation" or "compound symmetry." It should be noted that this correlation structure across time does not carry over to the observable outcomes y_{it}, due to the nonlinear relationship between the latent continuous variable and the observable outcome. If we relax this assumption and introduce, for instance, an autocorrelated error process

$$\epsilon_{it} = \rho\epsilon_{i,t-1} + v_{it} \tag{8.47}$$

where $|\rho| < 1$, and the $\{v_{it}\}$ are iid normal, the likelihood function contains a T-fold normal integral whose evaluation is computationally infeasible for more than five or six observation times. In this section we shall discuss some estimation procedures of the regression coefficients that avoid the calculation of T-variate integrals.

For simplicity we always assume the error process (8.47), but the methods can be generalized to more general error processes. We assume $\mathrm{E}v_{it} = 0$, and for identification we assume $\mathrm{Var}(u_{it}) = \mathrm{Var}(\alpha_i + \epsilon_{it}) = 1$. So we have

$$\sigma_\alpha^2 = 1 - \sigma_\epsilon^2$$

and assuming stationarity,

$$\sigma_\epsilon^2 = \frac{\sigma_v^2}{1-\rho^2}.$$

For the covariance between u_{it} and $u_{it'}$ we obtain

$$\mathrm{cov}(u_{it}, u_{it'}) = (1 - \sigma_\epsilon^2) + \sigma_\epsilon^2 \rho^{|t-t'|}. \tag{8.48}$$

Because $\sigma_u^2 = 1$ this is also the correlation between u_{it} and $u_{it'}$.

We also make the assumption

$$\mathrm{E}(\boldsymbol{u}_i|\boldsymbol{x}_i) = 0$$

where $\boldsymbol{u}_i' = (u_{i1}, \ldots, u_{iT})$ and $\boldsymbol{x}_i' = (x_{i1}', \ldots, x_{iT}')$. This implies that the x's are strictly exogenous, that is that u_{it} is uncorrelated with functions of past, current, and future x's. For the observable binary outcomes we obtain

$$\mu_{it}(\boldsymbol{\beta}) = E(y_{it}|\boldsymbol{x}_{it}) = \Phi(\boldsymbol{x}_{it}'\boldsymbol{\beta}) \quad , \quad t = 1, \ldots, T, \quad i = 1, \ldots, N.$$

Note that other distributions than the normal may be used as well. In such cases we have $E(y_{it}|\boldsymbol{x}_{it}) = F(\boldsymbol{x}_{it}'\boldsymbol{\beta})$ where F denotes the distribution function of the error u_{it}.

Defining

$$\boldsymbol{y}_i' = (y_{i1}, \ldots, y_{iT}), \; \boldsymbol{\mu}_i(\boldsymbol{\beta})' = (\Phi(\boldsymbol{x}_{i1}'\boldsymbol{\beta}), \ldots, \Phi(\boldsymbol{x}_{iT}'\boldsymbol{\beta})),$$

we obtain the nonlinear regression model

$$\boldsymbol{y}_i = \boldsymbol{\mu}_i(\boldsymbol{\beta}) + \boldsymbol{d}_i \quad , \qquad i = 1, \ldots, N \tag{8.49}$$

where $\boldsymbol{d}_i' = (d_{i1}, \ldots, d_{iT})$, and $d_{it} = y_{it} - \mu_{it}(\boldsymbol{\beta})$, $t = 1, \ldots, T$.

The "true" parameter vector will be denoted by $\boldsymbol{\beta}_0$, and interest centers on obtaining a reasonable estimate of $\boldsymbol{\beta}_0$. Throughout, we assume that the "true" expectation of $\boldsymbol{y}_i$ given $\boldsymbol{x}_i$ is $\boldsymbol{\mu}_i(\boldsymbol{\beta}_0)$, and that its variance $\boldsymbol{\Omega}_0(\boldsymbol{x}_i)$ exists for all $\boldsymbol{x}_i$. We always assume for simplicity that the dependent variable y_{it} depends on the exogenous variables $\boldsymbol{x}_{it}$ only through the linear combination $\boldsymbol{x}_{it}'\boldsymbol{\beta}$. However, the results can be extended to allow also nonlinear relationships. The observable outcomes y_{it}, $t = 1, \ldots, T$, for each individual are Bernoulli variables. The specification (8.47) for the error process together with the normality assumption leads to a multivariate probit model that involves a T-variate normal integral. Its estimation is computationally infeasible for more than five or six observation times. Other specifications may lead to even more complicated distributions. In general, the functional form of the joint T-variate distribution of $\boldsymbol{y}_i$ may be not completely known. In the following we shall discuss some approaches that avoid the complicated evaluation of the T-variate integral.

Maximum Quasi-likelihood Estimation

The "quasi-likelihood" approach was initiated by Wedderburn (1974). Much literature, both of theory and applications, has followed; see, for example, McCullagh (1983), McCullagh and Nelder (1989), and Diggle, Liang, and Zeger (1993) for an overview.

One approach to maximum quasi-likelihood estimation is via least-squares. For our model (8.49), the least-squares (or weighted least-squares) estimator $\hat{\boldsymbol{\beta}}$ is obtained by minimizing $\sum_{i=1}^{N} Q_i$ where

$$Q_i = (\boldsymbol{y}_i - \boldsymbol{\mu}_i(\boldsymbol{\beta}))'\boldsymbol{\Omega}_i^{-1}(\boldsymbol{y}_i - \boldsymbol{\mu}_i(\boldsymbol{\beta})) : \tag{8.50}$$

For known $\boldsymbol{\Omega}_i$, $\hat{\boldsymbol{\beta}}$ satisfies the first-order conditions

$$\frac{\partial}{\partial \boldsymbol{\beta}} \sum_i Q_i = \sum_{i=1}^{N} \frac{\partial \boldsymbol{\mu}_i(\boldsymbol{\beta})'}{\partial \boldsymbol{\beta}} \boldsymbol{\Omega}_i^{-1}(\boldsymbol{y}_i - \boldsymbol{\mu}_i(\boldsymbol{\beta})) = \boldsymbol{0}, \tag{8.51}$$

obtained by differentiating $\sum_{i=1}^{N} Q_i$ with respect to $\boldsymbol{\beta}$. Here $\partial \boldsymbol{\mu}_i(\boldsymbol{\beta})/\partial \boldsymbol{\beta}$ is a $(T \times p)$-matrix with the (t, j)th element $\partial \mu_{it}(\boldsymbol{\beta})/\partial \beta_j$. Since we always assume that $\mu_{it}(\boldsymbol{\beta})$ depends on $\boldsymbol{\beta}$ through the linear combination $\boldsymbol{x}_{it}'\boldsymbol{\beta}$, we have

$$\frac{\partial \boldsymbol{\mu}_i(\boldsymbol{\beta})}{\partial \boldsymbol{\beta}} = \boldsymbol{X}_i \operatorname{diag}(\varphi(\boldsymbol{x}_{i1}'\boldsymbol{\beta}), \ldots, \varphi(\boldsymbol{x}_{iT}'\boldsymbol{\beta})) = \boldsymbol{X}_i \boldsymbol{D}_i$$

where $\varphi(z)$ is the density function of the standard normal distribution, evaluated at $z = \boldsymbol{x}_{it}'\boldsymbol{\beta}$ and $\boldsymbol{X}_i = (\boldsymbol{x}_{i1}, \ldots, \boldsymbol{x}_{iT})$.

However, in practice $\boldsymbol{\Omega}_i$ is not known. If $\boldsymbol{\Omega}_i$ equals $E(\boldsymbol{d}_i\boldsymbol{d}_i')$, (8.50) corresponds to GLS estimation. The equation (8.51) is now used to define, as its solution, the maximum quasi-likelihood estimator which is also denoted by $\hat{\boldsymbol{\beta}}$. Note that this is no longer equivalent to the minimization of $\sum Q_i$ since differentiation would now introduce extra terms involving $\partial \boldsymbol{\Omega}_i/\partial \boldsymbol{\beta}$. For the univariate case, (8.51) can often be identified with the likelihood equation corresponding to an exponential-family distribution for the observations. This is often taken to be the primary derivation for (8.51). Liang and Zeger (1986) propose partial specification of the model in a generalized linear model framework. Only the marginal moments of y_{it}, $t = 1, \ldots, T$, the means and variances, are explicitly modeled. The correlation over time is taken as a nuisance.

Liang and Zeger (1986) denote (8.51) as "Generalized Estimating Equations (GEE)", using the choice

$$\boldsymbol{\Omega}_i = \boldsymbol{V}_i \boldsymbol{R}(\boldsymbol{\theta}) \boldsymbol{V}_i \tag{8.52}$$

where

$$\boldsymbol{V}_i = \operatorname{diag}\left\{[\mu_{i1}(\boldsymbol{\beta})(1 - \mu_{i1}(\boldsymbol{\beta}))]^{1/2}, \ldots, [\mu_{iT}(\boldsymbol{\beta})(1 - \mu_{iT}(\boldsymbol{\beta}))]^{1/2}\right\}$$

$\boldsymbol{R}(\boldsymbol{\theta})$ is a "working" correlation matrix depending on a (possibly unknown) parameter vector $\boldsymbol{\theta}$. $\boldsymbol{R}(\boldsymbol{\theta})$ may be misspecified. If $\boldsymbol{R}(\boldsymbol{\theta})$ is taken as $\boldsymbol{I}_T$, the $T \times T$ identity matrix, (8.51) reduces to the "independence equations", that is the set of equations which would be correct if the responses were serially uncorrelated and $\boldsymbol{\Omega}_i$ were truly diagonal. In a more general case where the responses are permitted to be freely correlated over time, T contains $T(T-1)/2$ additional parameters. Other possible choices for $\boldsymbol{R}(\boldsymbol{\theta})$ are suggested in Liang and Zeger (1986). If $\boldsymbol{\theta}$ is known, (8.51) is sufficient for estimation of $\boldsymbol{\beta}$. Otherwise $\boldsymbol{\theta}$ itself must be estimated. We shall give a brief description of some possible estimation procedures for $\boldsymbol{\theta}$ later on.

In order to derive the asymptotic properties of $\hat{\boldsymbol{\beta}}$, we can rely on results that establish the asymptotic properties of estimators obtained by maximizing a function; see, for example, Huber (1967), White (1982), Burguete, Gallant, and Souza (1982), Amemyia (1985), chap. 4, Fahrmeir (1987, 1990), or Li and Duan (1989). Here we want to estimate $\boldsymbol{\beta}_0$ by minimizing (8.50) with the first-order conditions (8.51) where the correlation matrix $\boldsymbol{R}(\boldsymbol{\theta})$ may be misspecified. It can be shown that under mild regularity conditions, the estimator $\hat{\boldsymbol{\beta}}_N$ obtained by minimizing (8.50) with the first-order conditions (8.51) is strongly consistent for $\boldsymbol{\beta}_0$. Moreover, $\hat{\boldsymbol{\beta}}_N$ is such that

$$\sqrt{N}(\hat{\boldsymbol{\beta}}_N - \boldsymbol{\beta}_0) \xrightarrow{d} \mathcal{N}(\mathbf{0}, \boldsymbol{A}^{-1}\boldsymbol{B}\boldsymbol{A}^{-1})$$

where $\boldsymbol{A}$ and $\boldsymbol{B}$ can be consistently estimated by

$$\begin{aligned}\hat{\boldsymbol{A}} &= \frac{1}{N}\sum_{i=1}^{N} \boldsymbol{X}_i'\boldsymbol{D}_i\boldsymbol{V}_i^{-1}\boldsymbol{R}(\hat{\boldsymbol{\theta}})^{-1}\boldsymbol{V}_i^{-1}\boldsymbol{D}_i\boldsymbol{X}_i \\ \hat{\boldsymbol{B}} &= \frac{1}{N}\sum_{i=1}^{N} \boldsymbol{X}_i'\boldsymbol{D}_i\boldsymbol{V}_i^{-1}\boldsymbol{R}(\hat{\boldsymbol{\theta}})^{-1}\boldsymbol{V}_i^{-1}\hat{\boldsymbol{d}}_i\hat{\boldsymbol{d}}_i'\boldsymbol{V}_i^{-1}\boldsymbol{R}(\hat{\boldsymbol{\theta}})\boldsymbol{V}_i^{-1}\boldsymbol{D}_i\boldsymbol{X}_i\end{aligned} \tag{8.53}$$

and $\boldsymbol{\beta}$ is replaced by the estimate $\hat{\boldsymbol{\beta}}_N$.

Therefore, even in the presence of misspecification of the correlation structure across time, the regression coefficients can be consistently estimated if the regression is correctly specified, that is the data are generated by the model $y_{it} = \Phi(\boldsymbol{x}_{it}'\boldsymbol{\beta}_0) + d_i$. Moreover, robust variance estimators are available. The results also hold for more general error processes than (8.47).

From the first-order conditions (8.51) a modified scoring algorithm can be derived. Given a current estimate $\tilde{\boldsymbol{\theta}}_k$ of the nuisance parameter $\boldsymbol{\theta}$, we use the following iterative procedure

$$\begin{aligned}\hat{\boldsymbol{\beta}}_{k+1} &= \hat{\boldsymbol{\beta}}_k + \left[\sum_{i=1}^{N} \boldsymbol{X}_i'\boldsymbol{D}_i(\hat{\boldsymbol{\beta}}_k)\boldsymbol{V}_i(\hat{\boldsymbol{\beta}}_k)^{-1}\boldsymbol{R}(\tilde{\boldsymbol{\theta}}_k)^{-1}\boldsymbol{V}_i(\hat{\boldsymbol{\beta}}_k)^{-1}\boldsymbol{D}_i(\hat{\boldsymbol{\beta}}_k)\boldsymbol{X}_i\right]^{-1} \\ &\quad \cdot\sum_{i=1}^{N} \boldsymbol{X}_i'\boldsymbol{D}_i(\hat{\boldsymbol{\beta}}_k)\boldsymbol{V}_i(\hat{\boldsymbol{\beta}}_k)^{-1}\boldsymbol{R}(\tilde{\boldsymbol{\theta}}_k)^{-1}\boldsymbol{V}_i(\hat{\boldsymbol{\beta}}_k)^{-1}[\boldsymbol{y}_i - \boldsymbol{\mu}_i(\hat{\boldsymbol{\beta}}_k)] \quad .\end{aligned} \tag{8.54}$$

An initial estimate of $\boldsymbol{\beta}$ may be obtained from the "independence model" where the off-diagonal elements of $\boldsymbol{R}$ are constrained to be zero.

In each iteration step in estimating $\boldsymbol{\beta}$, a further estimation procedure is implemented to estimate the nuisance parameter. A simple specification is to constrain the off-diagonal terms of $\boldsymbol{R}$ to be equal. Define

$$c_{tt'} = \sum_{i=1}^{N}(y_{it} - \Phi(\boldsymbol{x}_{it}'\hat{\boldsymbol{\beta}}))(y_{it'} - \Phi(x_{it'}'\hat{\boldsymbol{\beta}}))$$

and

$$c^*_{tt'} = \frac{c_{tt'}}{\sqrt{c_{tt} \cdot c_{t't'}}}.$$

Then, the off-diagonal estimates of $\boldsymbol{R}$ can be estimated in each iteration step by

$$\tilde{\theta} = \sum_{t'>t} c^*_{tt'}/[T(T-1)/2]\,. \tag{8.55}$$

Note that this correlation structure is misspecified. It arises for the latent continuous variables if the error process $\{\epsilon_{it}\}$ is not autoregressive, that is if $\rho = 0$ in (8.47). Moreover, even if $\rho = 0$ in (8.47), it is only the true correlation structure for the latent variables, and by no means for the observable binary outcomes. However, simulation studies indicate that maximum quasilikelihood estimation with the specification (8.55) performs well and thus efficiency of the regression parameter estimates is high. In our simulation experiments more complex procedures to estimate the nuisance parameters σ^2_v and ρ were also investigated. We also considered a nonlinear least squares procedure, in particular a Gauss-Newton algorithm, using the residuals $y_{it} - \Phi(\boldsymbol{x}'_{it}\hat{\boldsymbol{\beta}}_k)$ (in the kth iteration step). It turns out, however, that the efficiency gain in the simulation studies was negligible compared with specification (8.55), in particular if T is small, a typical case in panel studies. For further estimation procedures for $\boldsymbol{\theta}$, see Diggle, Liang, and Zeger (1993).

In summary, the quasi-likelihood approach can be highly recommended for the estimation of panel models for qualitative outcomes, when the likelihood function becomes intractable because complicated multivariate integrals appear. If the regression, that is the marginal mean given the regressors, is correctly specified, the maximum quasi-likelihood estimator of the regression parameter is consistent and asymptotically normally distributed regardless whether the covariance structure across time is correctly specified or not. The simulation experiments show that there is only a small loss of efficiency in most cases, in particular if there are only a few panel waves. A further advantage is that the scoring algorithm for the maximum quasi-likelihood estimator is easy to implement, for example by using a matrix language program such as SAS/IML or GAUSS. Even the "independence model"which completely ignores the dependence across time may perform well in many cases. The estimator $\hat{\beta}$ of this model can be computed with existing software, such as SAS or GLIM. So the "independence model" can be used as a basis model to start with.

Simulation-Based Inference

Recently, simulation methods have been shown to be useful to approximate objective functions in which integrals appear. Therefore, they seem to be particularly useful in panel models for discrete data involving multivariate probit specifications. Recent developments of these inference methods are described by several authors including McFadden (1989), Pakes and Pollard (1989), Börsch-Supan and Hajivassiliou (1990), and Gourieroux and Monfort (1991).

There are several simulation-based methods available, including the simulated generalized methods of moments, the simulated maximum likelihood method, and the simulated pseudo-maximum likelihood methods. For a survey of the recent developments of these inference methods, see Gourieroux and Monfort (1991) or Hajivassiliou, McFadden, and Ruud (1992). Hajivassiliou (1992) surveys simulation estimation methods for limited dependent variable models. Here we shall only give a brief description of the simulated maximum likelihood method and how it applies to the discrete panel data models.

For the discrete panel data model considered here, the maximum likelihood estimator is obtained from the maximization of the log-likelihood function

$$\log L = \sum_{i=1}^{N} \log P(\boldsymbol{y}_i|\boldsymbol{x}_i, \boldsymbol{\theta})$$

where $\boldsymbol{y}_i' = (y_{i1}, \ldots, y_{iT}), \boldsymbol{x}_i' = (x_{i1}', \ldots, x_{iT}')$, and the parameter vector $\boldsymbol{\theta}$ contains all relevant model parameters, in particular the regression coefficients $\boldsymbol{\beta}$. We have seen that for the model with autocorrelated errors, T-fold normal integrals appear in the likelihood function.

In general, when $P(\boldsymbol{y}_i|\boldsymbol{x}_i; \boldsymbol{\theta})$ does not possess a closed form it may be possible to use an "unbiased simulator," that is a function $\tilde{P}(\boldsymbol{y}_i, \boldsymbol{x}_i, \zeta_i, \boldsymbol{\theta})$ where ζ is an artificial random variable whose distribution is known and such that

$$E_\zeta \tilde{P}(\boldsymbol{y}_i, \boldsymbol{x}_i, \zeta_i; \boldsymbol{\theta}) = P(\boldsymbol{y}_i|\boldsymbol{x}_i; \boldsymbol{\theta}).$$

If such an unbiased simulator is available, $P(\boldsymbol{y}_i|\boldsymbol{x}_i; \boldsymbol{\theta})$ may be approximated by

$$\frac{1}{H}\sum_{k=1}^{H} \tilde{P}(\boldsymbol{y}_i, \boldsymbol{x}_i, \zeta_{ki}; \boldsymbol{\theta})$$

where the ζ_{ki}'s are independent drawings from the distribution of ζ. The simulated log-likelihood function is now given by

$$\log L_s = \sum_{i=1}^{N} \log[\frac{1}{H}\sum_{k=1}^{H} \tilde{P}(\boldsymbol{y}_i, \boldsymbol{x}_i, \zeta_{ki}; \boldsymbol{\theta})]$$

The maximization of $\log L_s$ provides the simulated maximum likelihood estimator $\hat{\boldsymbol{\theta}}^S_{N,H}$.

It can be shown (see Gourieroux and Monfort 1991) that the simulated maximum likelihood method has the same asymptotic properties as the exact ML method if not only the number of simulations goes to infinity (at least as fast as the square root of the number of observations); if the number of simulations is fixed, the asymptotic bias of the simulated ML method can be evaluated. Especially for the problem of evaluating multivariate normal probabilities and the derivatives, several simulation techniques are now available and are discussed in the literature [see Hajivassiliou (1992) or Hajivassiliou et al. (1992)]. However, in applications further research is needed. For example, it would be important to have a more precise idea of the performance of the simulators in finite samples. Some applications that are already available are promising (see e.g. Hajivassiliou and McFadden 1990, or Börsch-Supan and Hajivassiliou 1993), but further experience is needed.

The MECOSA Approach

A further approach that can be utilized for the analysis of discrete panel data based on normality was developed by Muthén (1984) and extended by Küsters (1987) and Arminger (1994). The general model is discussed by Browne and Arminger in Chapter 4. In this approach the covariances over time are estimated using the bivariate marginal distributions and polychoric correlation coefficients. Hence, estimation only requires the integration of bivariate normal distributions instead of the T-variate normal integral. The approach is quite general and is not restricted to the analysis of panel data or binary response variables, but includes ordinal and censored metric dependent variables. For a technical description of the corresponding GAUSS computer program, see Schepers and Arminger (1992).

3.5 Autoregressive Probit Models

The random effects probit models that we considered in the previous section can be generalized to the case that include "state dependence" or lagged values of y. We shall analyze a model that is simple enough to be computationally feasible and yet general enough to contain most of the interesting features of this type of model. Let

$$y_{it}^* = \boldsymbol{x}_{it}'\boldsymbol{\beta} + \gamma y_{i,t-1} + u_{it} \qquad (8.56)$$

$$y_{it} = \begin{cases} 1 & \text{if } \; y_{it}^* > 0 \\ 0 & \text{otherwise} \end{cases} \qquad (8.57)$$

$i = 1, \ldots, N; t = 1, \ldots, T$. The error term is assumed to be independent of $\boldsymbol{x}_i$ and is independently distributed over i. "True state dependence" is defined to mean $\gamma \neq 0$, whereas heterogeneity means serial correlation of $\{u_{it}\}$.

The static models in the previous sections assume that the probability of moving or staying in or out of a state is independent of the occurrence or nonoccurrence of the event in the past. It should be noted, however, that also in this case when $\gamma = 0$ but there is unobserved heterogeneity and the $\{u_{it}\}$ are serially correlated, we have

$$P(y_{it} = 1 \,|\, y_{i,t-1} = 1) \geq P(y_{it} = 1)$$

(see, e.g., Amemiya 1985, p. 349). This is referred to as "spurious state dependence" by some authors. If there are unobservable variables that affect people differently in regard to their tendency to experience the event, then integrating out these variables will result in a nonnegative correlation between the events $\{y_t = 1\}$ and $\{y_{t-1} = 1\}$. However, conditional on the person-specific effects, y_t is independent of y_{t-1}. On the other hand, in the autoregressive model the conditional probability that an individual will experience the event in the future is a function of past experience. In this case actual experience of an event has modified individual behavior.

A new problem with the autoregressive model is how to treat the initial observations. In the dynamic linear-regression model for panel data, Balestra and Nerlove (1966) encountered problems with maximum likelihood estimation treating y_{i0} as fixed. Anderson and Hsiao (1982) and Bhargava and Sargan (1983) showed the importance of specifying the distribution of y_{i0}. Hsiao (1986, p. 170), points out that the assumption that initial conditions are fixed constants is valid only if the disturbances that generate the process are serially independent and if a genuinely new process is fortuitously observed at the beginning of the sample. However, if the process has been in operation prior to the starting date of the sample, the presample history and hence the initial conditions are not exogenous. Therefore, we assume that the initial state for individual i, y_{i0}, is determined by the process generating the panel sample. Furthermore, we assume the variance components model

$$u_{it} = \alpha_i + \varepsilon_{it}$$

where the $\{\varepsilon_{it}\}$ are iid normal, and the $\{\alpha_i\}$ are treated as random. We assume that α_i is distributed independently of ε_{it} with $E(\alpha_i) = 0$. The total error variance σ_u^2 is normalized to be equal to 1. Let $f(y_{i0}|\,\alpha_i)$ denote the marginal probability of y_{i0}, given α_i. Then, the sample likelihood for this random effects model is

$$L = \prod_{i=1}^{N} \int_{-\infty}^{\infty} \prod_{t=1}^{T} \Phi(\psi_{it} + \alpha)^{y_{it}} \left[1 - \Phi(\psi_{it} + \alpha)\right]^{1-y_{it}} f(y_{i0}|\,\alpha) dG(\alpha) \tag{8.58}$$

where $\psi_{it} = \boldsymbol{x}'_{it}\boldsymbol{\beta} + \gamma y_{i,t-1}$ and $G(\alpha)$ denotes the cumulative distribution function of α.

In practice the marginal distribution $f(y_{i0}|\,\alpha)$ is not easy to derive. Moreover, maximizing (8.58) is considerably more involved. A less preferable but more tractable procedure is to assume that the $\{y_{i0}\}$ are random variables with a probability distribution

$$P(y_{i0} = 1\,|\,\boldsymbol{x}_{i0}) = F(\boldsymbol{x}'_{i0}\boldsymbol{\delta})$$

where $\boldsymbol{\delta}$ is a set of unknown parameters to be estimated. This is the assumption used by Heckman (1981b).

In the fixed-effect probit model treating $\{\alpha_i\}$ as unknown constants, the maximum likelihood estimators of $\boldsymbol{\beta}$, γ, and σ_α^2 are consistent only when T tends to infinity. If T is finite, the MLE is biased. Heckman's (1981b) Monte Carlo experiments show that the fixed-effect probit estimator exhibits a downward bias in γ, as in the Balestra-Nerlove model. Moreover, the greater the variance of the individual effects σ_α^2, the greater the bias.

If y_{it}^* is observable, (8.56) corresponds to Balestra and Nerlove's model, where it is well-known that the ordinary least-squares estimator of γ is inconsistent. It overestimates γ, and adding exogenous variables does not alter the direction of bias. Using the quasi-likelihood approach discussed in Section 3.3.1, some simulation experiments showed a similar result for the discrete model, if the "independence model" is used, that is if we specify a diagonal matrix for the covariance matrix of y_{it} , $t = 1, \ldots, T$. The true autocorrelation coefficient was always overestimated. However, our simulation results also show that the autocorrelation coefficient γ, and the other regression parameters as well, can be consistently estimated if we use a nondiagonal matrix for $\sum_i$ in the estimating equations. For a theoretical justification of these results further research is needed.

The maximum quasi-likelihood approach is also useful in solving the initial conditions problem. Using Heckman's approach, $F(\boldsymbol{x}'_{i0}\boldsymbol{\delta})$ is substituted by $P(y_{i0} = 1\,|\boldsymbol{x}_{i0})$ where F is the distribution function of the random variable

$$y_{i0}^* = \boldsymbol{x}'_{i0}\boldsymbol{\delta} + u_{i0} \quad .$$

The error term u_{i0} is allowed to be correlated with α_i and $\varepsilon_{it}, t = 1, \ldots, T$. Including $y_{i0}^{\star}$ in the vector of dependent latent variables, we obtain the augmented covariance matrix of $\tilde{\boldsymbol{u}}_i = (u_{i0}, u_{i1}, \ldots, u_{it})'$

$$\operatorname{cov}(\tilde{\boldsymbol{u}}_i) = \begin{pmatrix} \sigma_{u0}^2 & & & \\ \sigma_{01} & \sigma_\alpha^2 + \sigma_\varepsilon^2 & & \\ \sigma_{02} & \sigma_\alpha^2 & \sigma_\alpha^2 + \sigma_\varepsilon^2 & \\ \vdots & \vdots & & \ddots \\ \sigma_{0T} & \sigma_\alpha^2 & \sigma_\alpha^2 & & \sigma_\alpha^2 + \sigma_\varepsilon^2 \end{pmatrix}.$$

In practice, further restrictions must be imposed in order to identify the parameters of the model, for example $\sigma_{u0}^2 = 1, \sigma_\varepsilon^2 = 1$. Note that $\operatorname{cov}(\tilde{\boldsymbol{u}}_i)$ is the covariance matrix of the latent variables $y_{i0}^{\star}, y_{i1}^{\star}, \ldots, y_{iT}^{\star}$. However, as described in the previous section, this covariance matrix may also be used for $y_{i0}, y_{i1}, \ldots, y_{iT}$ although this is not the correct specification for the observables. The "working augmented correlation matrix" contains T additional parameters in order to capture the correlation between the initial value and the subsequent states y_{it}. In this case, the correlation matrix $\tilde{\boldsymbol{R}}$ is given by

$$\begin{pmatrix} 1 & & & & & \\ \rho_{01} & 1 & & & & \\ \rho_{02} & \rho & 1 & & & \\ \vdots & \vdots & \vdots & \ddots & & \\ \rho_{0T} & \rho & \rho & \rho & \rho & 1 \end{pmatrix}$$

where ρ and $\rho_{0t}, t = 1, \ldots, T$, are estimated from appropriate residuals. In particular, the correlations can be estimated from the residuals without any restrictions. Note, however, that this leads to $T(T-1)/2$ additional parameters to be estimated. Our simulation experiments show that in this approach much efficiency may be lost in estimating the regression parameters $\boldsymbol{\beta}$. Therefore, in most cases, it is useful to restrict the "working correlation matrix."

Maddala (1987) also discusses autoregressive fixed-effects logit models, and a conditional maximum likelihood estimation procedure for estimation of γ. It has been used by Chamberlain (1978) and Corcoran (1982) to study labor force participation. A first limitation of the autoregressive fixed-effects logit model is that the use of the conditional estimation method might involve discarding a large number of sample observations and using a very small portion of the data, as in the paper by Corcoran and Hill (1985). The second limitation is that the model does not permit the use of exogenous variables, which is a computational problem with this technique. Consequently, there are not many empirical applications for these models.

3.6 Panel Models for Ordinal Data

Only a few articles are concerned with regression models for panel data where the dependent variable consists of ordered categories (see Harville and Mee 1984; Stram, Wei, and Ware 1988; Jansen, 1990). Recently, Tutz and Hennevogl (1993) considered the general case of an ordinal cumulative model including random effects. The cumulative model is embedded into the framework of multivariate generalized linear models. Then, the authors give three alternative estimation procedures, two of them based on numerical integration, and the last one being a variant of the EM algorithm.

In the present paper we only give a brief description of a straightforward extension of the random effects model to the case of ordered categorical responses. We shall restrict ourselves to the most popular cumulative-type model (McCullagh 1980). In general, without referring to panel models, the existence of a latent continuous variable $y^\star$ is assumed, which is connected to the covariate vector $\boldsymbol{x}$ by

$$y^\star = -\boldsymbol{x}'\boldsymbol{\beta} + \epsilon$$

where ϵ is an iid error variable with distribution function F. The observable ordinal response variable $y \in \{1, ..., k\}$ is considered as a coarser version of $y^\star$ and is defined by the relationship

$$y = r \Leftrightarrow \theta_{r-1} < y* \leq \theta_r \tag{8.59}$$

where $-\infty = \theta_o < \theta_1 < ... < \theta_k = \infty$ are thresholds or cutpoints. The cumulative model following from (8.59) is given by

$$P(y \leq r|\boldsymbol{x}) = F(\theta_r + \boldsymbol{x}'\boldsymbol{\beta}) \quad . \tag{8.60}$$

A traditional choice of F is the logistic distribution $F(z) = 1/(1 + \exp(-z))$ yielding the cumulative logit or proportional odds model. In general, the thresholds θ_r are unknown parameters that have to be estimated along with the regression coefficients.

Lets us now consider the model for ordinal panel data. The latent variable $y^\star_{it}$ associated with individual i and observation t is now given by

$$y'_{it} = -\boldsymbol{x}'_{it}\boldsymbol{\beta} - \alpha_i + \epsilon_{it} \ , \tag{8.61}$$

$i = 1, ..., N; t = 1, ..., T.$ ϵ_{it} has distribution function F, and α_i is an individual specific random variable which is usually assumed to be normally distributed. Again, let α_i, ϵ_{it} be independent. The threshold approach (8.59) is assumed to work conditionally, that is for given α_i the observable variable $y_{it} \in (1, ..., h)$ is defined by

$$y_{it} = r \Leftrightarrow \theta_{r-1} < y^\star_{it} \leq \theta_r$$

yielding

$$P\left(y_{it} \leq r \mid \boldsymbol{x}_{it}, \alpha_i\right) = F\left(\theta_r + \boldsymbol{x}'_{it}\boldsymbol{\beta} + \alpha_i\right) \quad . \tag{8.62}$$

Jansen (1990) considered this model. Tutz and Hennevogl (1993) discuss a more complex model with random coefficients

$$P\left(y_{it} \leq r \,|\boldsymbol{x}_{it}, \alpha_i, \boldsymbol{b}_i\right) = F\left(\theta_r + \boldsymbol{x}'_{it}\boldsymbol{\beta} + \alpha_i + \boldsymbol{v}'_{it}\boldsymbol{b}_i\right)$$

where $\boldsymbol{v}_{it}$ denotes a subvector of $\boldsymbol{x}_{it}$ and $\boldsymbol{b}_i$ is an individual specific random vector with $\boldsymbol{b}_i \sim \mathcal{N}(\boldsymbol{0}; \boldsymbol{C})$. They use several variants of the EM algorithm for the estimation of the structural parameter β and the unknown elements of C. In addition, Tutz and Hennevogl (1993) discuss the case of random thresholds where the thresholds are assumed to be normally distributed. The estimation procedures are illustrated and compared by two examples.

There are only a few applications of ordinal cumulative models including random effects. One might also extend the quasi-likelihood or generalized estimating equations approach to panel models for ordinal data where the covariance structure over time may be misspecified. However, further research is needed on this topic.

4 Markov Chain Models

Here we shall only consider a two-state Markov model with exogenous variables that accounts for heterogeneity and nonstationarity of the data. This model is closely related to the models considered in Section 3.4. The model easily generalizes to the case where there are more than two states. For Markov models with exogenous variables, only a few applications have appeared in the literature. On the other hand, there is much literature, both theoretical and applied, for Markov chain models without the inclusion of explanatory variables. The reader who wishes to pursue this topic in greater detail should consult the textbooks by Bartholomew (1982), Feller (1968), or Howard (1971), among others.

In order to make the discussion comparable to the discussion of Section 3.4, we shall use the same notation. Let the states be denoted by 0 and 1 (rather than 1 and 2), and let $y_{it} = 1$ if individual i is in state 1 at time t and $y_{it} = 0$ otherwise. Markov chain models specify the probability distribution of y_{it} as a function of y_{is}, $s = t-1, t-2, \ldots$, as well as possibly exogenous variables. Therefore, Markov chain models can be regarded as generalizations of quantal response models. In fact, Markov chain models reduce to qualitative response models if the y_{it} are independent over t.

Models where the distribution of y_{it} does not depend on $y_{i,t-2}, y_{i,t-3}, \ldots$ are called first-order Markov models. We shall focus on such models, although most of the results readily generalize to higher-order Markov models. First-order Markov models are completely characterized if we specify the transition probabilities $P^i_{jk}(t)$, defined as the probability that individual i is in state k at time t given that he was in state j at time $t-1, j, k = 0, 1$, and the distribution of y_{i0}, the initial conditions.

A Markov model where $P^i_{jk}(t) = P_{jk}$ for all t is called a stationary Markov model. If $P^i_{jk}(t) = P_{jk}(t)$ for all i, the model is called homogeneous (see Amemiya 1985, p. 414). Most applications in the social sciences deal with stationary, homogeneous Markov chains. Here we focus on Markov models that are both nonstationary and heterogeneous.

If we define $P_{i0} = P(y_{i0} = 1)$, the likelihood function of the first-order Markov model can be written as

$$\begin{aligned} L &= \prod_i \prod_t P^i_{01}(t)^{(1-y_{i,t-1})y_{it}} \cdot P^i_{10}(t)^{y_{i,t-1}(1-y_{it})} \\ &\quad \left(1 - P^i_{01}(t)\right)^{(1-y_{i,t-1})\cdot(1-y_{it})} \left(1 - P^i_{10}(t)\right)^{y_{i,t-1}y_{it}} \\ &\quad P_{i0}^{y_{i0}} \left(1 - P_{i0}\right)^{1-y_{i0}} . \end{aligned} \tag{8.63}$$

Next, we have to specify the transition probabilities $P_{01}^i(t)$ and $P_{10}^i(t)$ dependent upon the exogenous variables. A concise specification [presented by Amemiya (1985, p. 422)] is to assume

$$P(y_{it} = 1 \mid y_{i,t-1}) = F(\boldsymbol{x}_{it}' \boldsymbol{\beta} + \boldsymbol{x}_{it}' \boldsymbol{\alpha} y_{i,t-1}) \tag{8.64}$$

where F is a certain distribution function. Formula (8.64) is equivalent to

$$\begin{aligned} P_{01}^i(t) &= F(\boldsymbol{x}_{it}'\boldsymbol{\beta}) \\ P_{11}^i(t) &= F(\boldsymbol{x}_{it}'(\boldsymbol{\alpha} + \boldsymbol{\beta})) \quad . \end{aligned}$$

Formula (8.64) makes clear the similarity of this model to a quantal response model. The likelihood function for this model conditional on y_{i0} can be written as

$$\begin{aligned} L &= \prod_i \prod_t F(\boldsymbol{x}_{it}'\boldsymbol{\beta})^{(1-y_{i,t-1})y_{it}} \cdot (1 - F(\boldsymbol{x}_{it}'(\boldsymbol{\alpha} + \boldsymbol{\beta})))^{y_{i,t-1}(1-y_{it})} \\ &\quad \cdot (1 - F(\boldsymbol{x}_{it}'\boldsymbol{\beta})^{(1-y_{i,t-1})(1-y_{it})} \cdot F(\boldsymbol{x}_{it}'(\boldsymbol{\alpha} + \boldsymbol{\beta}))^{y_{i,t-1}y_{it}} . \end{aligned}$$

The similarity to the quantal response model becomes even clearer if we write it alternatively as

$$L = \prod_i \prod_t F(\boldsymbol{x}_{it}'\boldsymbol{\beta} + y_{i,t-1}\boldsymbol{x}_{it}'\boldsymbol{\alpha})^{y_{it}} \cdot (1 - F(\boldsymbol{x}_{it}'\boldsymbol{\beta} + y_{i,t-1}\boldsymbol{x}_{it}'\boldsymbol{\alpha}))^{1-y_{it}} . \tag{8.65}$$

There are only few applications of Markov models with exogenous variables in the literature. Boskin and Nold (1975) assumed

$$\begin{aligned} P_{01}^i(t) &= L(\boldsymbol{x}_i'\boldsymbol{\beta}) \\ P_{10}^i(t) &= L(\boldsymbol{x}_i'\boldsymbol{\alpha}) \end{aligned}$$

where $L(z) = (1 - e^{-z})^{-1}$, the distribution function of the logistic distribution. They applied the model to data on 440 households during a 60-month period. State 1 represents the state of an individual being on welfare, and state 0 an individual being off welfare. The exogenous variables are dummy variables characterizing the economic and demographic characteristics of the individuals. They do not depend on time, indicating that the model is heterogeneous but stationary. As a consequence, it is interesting to evaluate the equilibrium probability – the probability that a given individual is on welfare after a suffiently long time has elapsed since the initial time. Toikka (1976) considered a three-state Markov model of labor market decisions in which the three states are employed, unemployed and actively looking for a job, and out of the labor force.

Finally we note that there is also a similarity of the Markov models to discrete-time duration or event history models. Whereas common event history models assume that time is measured as a continuous variable, discrete-time event history models assume that the data are not available as exact response times but record only the particular interval of time in which the event or failure occurs. The duration T of a spell is then a discrete random variable, and $\{T = t\}$ means that a transition has occured in the interval $(a_{t-1}, a_t]$. Let us consider a single-spell model where state 0 represents the initial state and state 1 the termination state. Then, only $P^i_{01}(t)$ is of interest. The probability of duration t of the spell, that is the individual stays t periods or time intervals in state 0 until a transition occurs, is given by

$$P_i(t) = \prod_{k=1}^{t-1} \left(1 - P^i_{01}(k)\right) \quad P^i_{01}(t) \quad .$$

The likelihood function can be expressed in terms of duration as well. This is especially useful in the statistical analysis of duration or event history models. For a detailed treatment of event history models in continuous time is given by Petersen in Chapter 9.

5 Tobit Models for Panel Data

Consider the situation where a market research institute analyzes the consumption of a durable good and the data are taken from a panel of households. This calls for a Tobit model. Let y_{it} denote *observed* consumption of household i in period t, which is related to the latent variable y^*_{it} by equation (8.20) in Section 2.4. The linear regression model relating this latent variable to the set of explanatory variables is now more explicitly written as

$$y^*_{it} = \boldsymbol{x}'_{it}\boldsymbol{\beta} + u_{it} \ , \quad i = 1, \ldots, N \ , \ t = 1, \ldots, T \quad , \tag{8.66}$$

where $\boldsymbol{x}_{it}$ denotes a vector of household characteristics for household i in period t. We also assume that the error term u_{it} has the following structure:

$$u_{it} = \alpha_i + \varepsilon_{it} \quad . \tag{8.67}$$

This corresponds to the special case of the random effects model (8.45) which we discussed in Section 3.3. (A more general specification would include interactions between error components and explanatory variables.) Note that both α_i and ε_{it} are random variables which satisfy $E(\alpha_i) = E(u_{it}) = 0$, $Var(\alpha_i) = \sigma^2_\alpha$, and $Var(\varepsilon_{it}) = \sigma^2_\varepsilon$. Moreover, all α_i and ε_{it} are mutuallly uncorrelated, i.e. $E(\alpha_i \alpha_j) = 0$ and $E(\varepsilon_{it}\varepsilon_{js}) = 0$, if $i \neq j$ or $s \neq t$. The total variance of u_{it} is given by

$$\sigma^2 = \sigma^2_\alpha + \sigma^2_\varepsilon \quad . \tag{8.68}$$

In the following we additionally assume that both error components are normally distributed:

$$\varepsilon_{it} \sim N(0, \sigma^2_\varepsilon) \quad , \quad \alpha_i \sim N(0, \sigma^2_\alpha) \quad .$$

This implies that for *fixed* time period t, the random variables y^*_{it} are independent, whereas for fixed index i, the y^*_{it} are correlated over time with correlation parameter $\gamma = \sigma^2_\alpha/\sigma^2$. Note that this correlation is constant and does not depend on the distance between two periods. Correlation is higher when the "individual effect" α_i shows more variation, that is when the (unobserved) "heterogeneity" among households is greater.

We now derive the likelihood function for this model. First, we observe that for *given* α_i and $\boldsymbol{x}_{it}$ we have

$$y_{it}^* \sim N(\alpha_i + \boldsymbol{x}_{it}'\boldsymbol{\beta}\ ,\ \sigma_\varepsilon^2) \quad .$$

Therefore we obtain for the conditional probability that *zero* consumption will be observed:

$$\begin{aligned} P(y_{it} = 0 \mid \alpha_i, \boldsymbol{x}_{it}) &= P\left(\frac{y_{it}^* - \alpha_i - \boldsymbol{x}_{it}'\boldsymbol{\beta}}{\sigma_\varepsilon} \leq \frac{-\alpha_i - \boldsymbol{x}_{it}'\boldsymbol{\beta}}{\sigma_\varepsilon} \middle| \alpha_i, \boldsymbol{x}_{it} \right) \\ &= \Phi\left(\frac{-\alpha_i - \boldsymbol{x}_{it}'\boldsymbol{\beta}}{\sigma_\varepsilon} \right) \equiv \Phi_{it} \quad . \end{aligned} \tag{8.69}$$

We can also write for the conditional density of y_{it}^* (given α_i and $\boldsymbol{x}_{it}$):

$$\varphi(y_{it}^* | \alpha_i, \boldsymbol{x}_{it}) = \frac{1}{\sigma_\varepsilon \sqrt{2\,\pi}} \exp\left\{ -\frac{1}{2\,\sigma_\varepsilon^2} (y_{it} - \alpha_i - \boldsymbol{x}_{it}'\boldsymbol{\beta})^2 \right\} \equiv \varphi_{it} \quad . \tag{8.70}$$

We now recall that the (time-independent) individual effect α_i was responsible for correlation over time; that is for *given* α_i and $\boldsymbol{x}_{it}$ the random variables

$$y_{i1}^*, \ldots, y_{iT}^*$$

are independent. Let $\mathcal{N}_i$ and $\mathcal{K}_i$ denote, respectively, the set of periods for which zero consumption was observed for household i, and the set of periods for which positive consumption occurred. Then, for given α_i the likelihood of $y_{i1}, \ldots, y_{iT}$ for household i can be written as

$$L_{i|\alpha} = \prod_{t \in \mathcal{N}_i} \Phi_{it} \prod_{t \in \mathcal{K}_i} \varphi_{it} \tag{8.71}$$

where Φ_{it} and φ_{it} are given by (8.70) and (8.69), respectively.

We now use the assumption that α_i is normally distributed and therefore has density function

$$h(\alpha_i) = \frac{1}{\sigma_\alpha \sqrt{2\,\pi}} \exp\left(-\frac{\alpha_i^2}{2\,\sigma_\alpha^2} \right) \quad .$$

Multiplying the conditional likelihood $\mathcal{L}_{i|\alpha}$ by this density and then "integrating out" leads to the unconditional likelihood

$$L_i = \int_{-\infty}^{\infty} \prod_{t \in \mathcal{N}_i} \Phi_{it} \prod_{t \in \mathcal{K}_i} \varphi_{it}\, h(\alpha_i)\, d\alpha_i \quad ,$$

which no longer can be written as an explicit function. The full likelihood for all N households is then given by the product of the individual likelihoods:

$$L = \prod_{i=1}^{N} \left\{ \int_{-\infty}^{\infty} \left[\prod_{t \in \mathcal{N}_i} \Phi_{it} \prod_{t \in \mathcal{K}_i} \varphi_{it} \frac{1}{\sigma_\alpha \sqrt{2\,\pi}} \exp\left(-\frac{\alpha_i^2}{2\,\sigma_\alpha^2} \right) \right] d\,\alpha_i \right\} . \tag{8.72}$$

Since our model includes only one error component, evaluation of the (one-dimensional) integral by the method proposed by Butler and Moffitt (1982) gives a good approximation. See Maddala (1987, p. 318–319). For more general specifications of random coefficients models, the remarks apply which were made in connection with the closely related probit model (cf. Section 3.3).

6 Models for Count Data

In this section we consider an integer-valued random variable y, that is $y \in \{0, 1, \dots\}$. Such situations arise if, for example, the number of unemployment spells or the number of patents issued during some time interval is considered. Data of this type are called *count data* which can be modeled by the Poisson distribution or the negative binomial distribution. However both distributions are restrictive in the sense that their mean and variance are not varying independently: For the Poisson distribution mean and variance are identical, whereas in case of the negative binomial distribution the variance is proportional to its mean. These restrictions are reflected by the assumptions concerning the data-generating process.

6.1 Poisson Distribution and Negative Binomial Distribution

For the well-known Poisson distribution, probabilities

$$P\{y = j\} = \frac{e^{-\lambda}\lambda^j}{j\,!}, \quad j = 0, 1, 2, \dots \tag{8.73}$$

with $E(y) = \text{Var}(y) = \lambda > 0$ are derived under the assumption that the number of observed units is proportional to the length of the time unit, and that occurrences of events in different (nonoverlapping) time intervals are independent of one other. Therefore this model is not adequate if—in case of unemployment—the unemployment registrations cluster during winter time or—in the case of modeling issuance of patents over the year—firms register all patents at the end of the year. In both cases the probability varies over time, violating assumptions for the Poisson process.

The negative binomial distribution can be derived from the Poisson distribution by (continuous) mixing, as described below. Its probabilities are given by

$$P\{y = j\} = \frac{\Gamma(\theta + j)}{\Gamma(\theta)\, j\,!}\, p^{\theta}\,(1 - p)^j\,, \quad j = 0, 1, 2, \dots \tag{8.74}$$

with parameters $\theta > 0$ and $0 < p < 1$. For the mean and variance of the distribution we obtain

$$E(y) = \frac{\theta p}{1 - p} \quad \text{and} \quad \text{Var}(y) = \frac{\theta p}{(1 - p)^2} > E(y). \tag{8.75}$$

Note that the variance exceeds the mean; this is called *overdispersion*.

If θ is an integer then we call (8.74) a *Pascal distribution* and write it as follows:

$$P\{y = j\} = \frac{(\theta + j - 1)\,!}{(\theta - 1)\,!\, j!}\, p^{\theta}\,(1 - p)^j\,, \quad j = 0, 1, 2, \dots \quad . \tag{8.76}$$

This distribution can be interpreted as follows: In a Bernoulli experiment with success probability p, equation (8.76) is the probability that in $(\theta + j)$ trials, j *failures* have been observed before the first success occurs. This explains the name of this distribution; in contrast to the binomial distribution, the number of trials is not fixed in advance. Finally, for the special case $\theta = 1$ we obtain the *geometric* distribution.

6.2 Mixtures of Poisson Distributions

We now describe how the Poisson distribution is related to the negative binomial distribution through the mixing procedure: Let us assume that the parameter λ of the Poisson distribution is not exactly known. Therefore we attach highest weights to those values which are most plausible or, more generally, attach a weight to each parameter value. Since λ is nonnegative a convenient weight function is provided by the density function of the gamma distribution (with parameters κ and θ). Interpreting λ now as a random variable, we find a new marginal distribution for y by multiplying the Poisson probabilities (8.73) by the gamma density and then integrating out with respect to λ: This leads to the following probability density function:

$$\begin{aligned} P\{y = j \mid \kappa, \theta\} &= \int_0^\infty \frac{e^{-\lambda}\lambda^j}{j\,!} \frac{\kappa^\theta}{\Gamma(\theta)} \lambda^{(\theta-1)} e^{-\kappa\lambda}\, d\lambda \\ &= \frac{\Gamma(\theta+j)}{\Gamma(\theta)\, j\,!} \frac{\kappa^\theta}{(\kappa+1)^{(\theta+j)}} \\ &= \frac{\Gamma(\theta+j)}{\Gamma(\theta)\, j\,!} \left(\frac{\kappa}{\kappa+1}\right)^\theta \left(\frac{1}{\kappa+1}\right)^j \quad . \end{aligned} \tag{8.77}$$

Clearly this is the probability density function of the negative binomial distribution with parameters θ and $p = \kappa/(\kappa+1)$. In deriving the result we made use of the fact that

$$\frac{\Gamma(a)}{t^a} = \int_0^\infty y^{a-1} e^{-ty}\, dy \tag{8.78}$$

holds. If we use θ/κ and θ as parameters for the (mixing) gamma distribution, we obtain

$$\begin{aligned} P\{y = j \mid \kappa, \theta\} &= \frac{\Gamma(\theta+j)}{\Gamma(\theta)\, j\,!} \left(\frac{\frac{\theta}{\kappa}}{\frac{\theta}{\kappa}+1}\right)^\theta \left(\frac{1}{\frac{\theta}{\kappa}+1}\right)^j \\ &= \frac{\Gamma(\theta+j)}{\Gamma(\theta)\, j\,!} \left(\frac{\theta}{\theta+\kappa}\right)^\theta \left(\frac{\kappa}{\theta+\kappa}\right)^j \quad . \end{aligned} \tag{8.79}$$

This is the probability density function of the negative binomial distribution with parameters θ and $p = \theta/(\theta+\kappa)$. This formulation leads to the expectation $E(y) = \theta(1-p)/p = \kappa$, which is particularly simple.

6.3 The Poisson Model

We assume that the count data (number of unemployment spells, number of patents registered) is generated by a Poisson distribution. Furthermore, we assume that a vector x of explanatory variables influences the probability of a certain realization.

Since the Poisson parameter λ is nonnegative, a convenient parameterization is given by

$$\lambda(\boldsymbol{x}) = \exp(\boldsymbol{x}'\boldsymbol{\beta}). \tag{8.80}$$

Assuming a random sample with n observations, the following likelihood function is obtained:

$$L = \prod_{i=1}^{n} \frac{e^{-\exp(\boldsymbol{x}_i'\boldsymbol{\beta})} \, [\exp(\boldsymbol{x}_i'\boldsymbol{\beta})]^{y_i}}{y_i\,!} \quad , \tag{8.81}$$

and the log-likelihood function is given by

$$\ell = -\sum_i \exp(\boldsymbol{x}_i'\boldsymbol{\beta}) + \sum_i y_i\,(\boldsymbol{x}_i'\boldsymbol{\beta}) - \sum_i \log(y_i!) \quad . \tag{8.82}$$

For the case of K explanatory variables it can be shown that the Hessian matrix of this function is negative definite if the K vectors $\boldsymbol{x}_k = (x_{1k}, \ldots x_{nk})'$ are linearly independent. See Maddala (1983, p. 52). Therefore the log-likelihood function is globally concave and the maximum likelihood estimates are unique.

6.4 A Model with Overdispersion

In empirical research one often encounters situations where the restriction "mean = variance" inherent in the Poisson distribution will be violated and overdispersion is observed. This also applies to the *conditional* model discussed in the last subsection. We also consider the model of the negative binomial distribution, which features overdispersion.

First of all, for this model we have to discuss the way in which we want to specify the influence of the explanatory variable x.

Table 2. Alternative Parameterizations
for the Model of the Negative Binomial Distribution

	Parameterization	
	(a) $p = \kappa/(1+\kappa)$	(b) $p = \theta/(\theta+\kappa)$
$E(y)$	θ/κ	κ
$V(y)$	$(\theta/\kappa)\,(1+1/\kappa)$	$\kappa\,(1+\kappa/\theta)$
$V(y)/E(y)$	$1+1/\kappa$	$1+\kappa/\theta$

Table 2 shows the mean, the variance, and their ratio for the two parameterizations

$$\text{(a)} \quad p = \kappa/(1+\kappa) \quad \text{and} \quad \text{(b)} \quad p = \theta/(\theta+\kappa)$$

where the first is related to the probability density function (8.77) and the latter to (8.79). The relation with $\boldsymbol{x}$ could now be specified, for example, by

$$\kappa(\boldsymbol{x}) = \exp(\alpha + \boldsymbol{x}'\boldsymbol{\beta}) \tag{8.83}$$

or

$$\theta(\boldsymbol{x}) = \exp(\alpha + \boldsymbol{x}'\boldsymbol{\beta}) \quad . \tag{8.84}$$

One could also relate the variance/mean ratio linearly to $\boldsymbol{x}$. Note that under specification (8.83), the alternatives (a) and (b) are equivalent except for a reinterpretation of parameters. This can be seen as follows: Under specification (8.83) we have

$$\begin{aligned}
p^{\theta}(1-p)^{y} &= \left[\frac{\kappa}{1+\kappa}\right]^{\theta}\left[\frac{1}{1+\kappa}\right]^{y} \\
&= \left[\frac{\exp(\alpha+\boldsymbol{x}'\boldsymbol{\beta})}{1+\exp(\alpha+\boldsymbol{x}'\boldsymbol{\beta})}\right]^{\theta}\left[\frac{1}{1+\exp(\alpha+\boldsymbol{x}'\boldsymbol{\beta})}\right]^{y} \\
&= \left[\frac{\exp(\alpha)}{\exp(\alpha)+\exp(-\boldsymbol{x}'\boldsymbol{\beta})}\right]^{\theta}\left[\frac{\exp(-\boldsymbol{x}'\boldsymbol{\beta})}{\exp(\alpha)+\exp(-\boldsymbol{x}'\boldsymbol{\beta})}\right]^{y} \\
&= \left[\frac{\exp(c+\alpha)}{\exp(c+\alpha)+\exp(c-\boldsymbol{x}'\boldsymbol{\beta})}\right]^{\theta}\left[\frac{\exp(c-\boldsymbol{x}'\boldsymbol{\beta})}{\exp(c+\alpha)+\exp(c-\boldsymbol{x}'\boldsymbol{\beta})}\right]^{y} \\
&= \left[\frac{\theta}{\theta+\kappa^{*}}\right]^{\theta}\left[\frac{\kappa^{*}}{\theta+\kappa^{*}}\right]^{y}
\end{aligned}$$

with $\theta \equiv \exp(c+\alpha)$ und $\kappa^{*}(\boldsymbol{x}) = \exp(c-\boldsymbol{x}'\boldsymbol{\beta})$ from which c may be determined. The left-hand side of the first line pertains to alternative (a) whereas the last line gives the corresponding result for alternative (b). Note the reversed sign of $\boldsymbol{\beta}$. An example will illustrate this later on in Section 6.6.

Maximum likelihood estimation of the various specifications proceeds in the usual manner. For illustrative purpose we briefly describe estimation for specification (8.83) in connection with alternative (a). The likelihood is

$$L = \prod_{i=1}^{n} \frac{\Gamma(\theta+y_i)}{\Gamma(\theta)\, y_i!} \frac{\exp(\alpha+\boldsymbol{x}_i'\boldsymbol{\beta})^{\theta}}{[1+\exp(\alpha+\boldsymbol{x}_i'\boldsymbol{\beta})]^{\theta+y_i}} \tag{8.85}$$

and the log-likelihood function

$$\begin{aligned}
\ell &= \sum_i [\log(\Gamma(\theta+y_i)) - \log(\Gamma(\theta)) - \log(y_i!) + \theta(\alpha+\boldsymbol{x}_i'\boldsymbol{\beta}) \\
&\quad -(\theta+y_i)\log(1+\exp(\alpha+\boldsymbol{x}_i'\boldsymbol{\beta}))] \quad .
\end{aligned} \tag{8.86}$$

From this we obtain the following first-order derivatives:

$$\frac{\partial \ell}{\partial \alpha} = n\,\theta - \sum_i (\theta + y_i)\,\frac{\exp(\alpha + \boldsymbol{x}_i'\boldsymbol{\beta})}{1 + \exp(\alpha + \boldsymbol{x}_i'\boldsymbol{\beta})} \tag{8.87}$$

$$\frac{\partial \ell}{\partial \boldsymbol{\beta}} = \theta \sum_i \boldsymbol{x}_i - \sum_i (\theta + y_i)\,\boldsymbol{x}_i\,\frac{\exp(\alpha + \boldsymbol{x}_i'\boldsymbol{\beta})}{1 + \exp(\alpha + \boldsymbol{x}_i'\boldsymbol{\beta})} \tag{8.88}$$

$$\begin{aligned}\frac{\partial \ell}{\partial \theta} &= \sum_i \psi(\theta + y_i) - n\,\psi(\theta) + \sum_i (\alpha + \boldsymbol{x}_i'\boldsymbol{\beta}) \\ &\quad - \sum_i \log\left(1 + \exp(\alpha + \boldsymbol{x}_i'\boldsymbol{\beta})\right)\end{aligned} \tag{8.89}$$

where ψ denotes the digamma function, that is $\psi(x) = \frac{d}{dx}\log(\Gamma(x))$. Using $p_i = \exp(\alpha + \boldsymbol{x}_i'\boldsymbol{\beta})/[(1 + \exp(\alpha + \boldsymbol{x}_i'\boldsymbol{\beta})]$, $\boldsymbol{z}_i = (1, \boldsymbol{x}_i')$ and $\boldsymbol{\gamma} = (\alpha\,,\,\boldsymbol{\beta})'$ we then obtain the following second-order derivatives:

$$\frac{\partial^2 \ell}{\partial \boldsymbol{\gamma}\,\partial \boldsymbol{\gamma}'} = -\sum_i (\theta + y_i)\,p_i\,(1 - p_i)\,\boldsymbol{z}_i\,\boldsymbol{z}_i' \tag{8.90}$$

$$\frac{\partial^2 \ell}{\partial \boldsymbol{\gamma}\,\partial \theta} = \sum_i (1 - p_i)\,\boldsymbol{x}_i \tag{8.91}$$

$$\frac{\partial^2 \ell}{(\partial \theta)^2} = \sum_i \psi_1(\theta + y_i) - n\,\psi_1(\theta) \tag{8.92}$$

where ψ_1 denotes the trigamma function, that is $\psi_1(x) = \frac{d^2}{(dx)^2}\log(\Gamma(x))$. To the best of our knowledge no analytical proof of negative definiteness of this Hessian matrix exists. However, our own experience shows that the log-likelihhod function seems to be well-behaved. See also the example in Section 6.6.

6.5 Maximum Quasi-likelihood Estimation Under Overdispersion

In Section 3.4 we considered the possibility of obtaining reasonable estimates of the unknown parameters of the probit model even when the stochastic model is not properly specified. We present this method of maximum quasilikelihood estimation here also for models with discrete dependent variables. Again we have in mind the case of neglected heterogeneity, that is, in the present context "overdispersion." To see this connection, consider the conditional distribution of y given $\boldsymbol{x}$ and u (the latter variable characterizing unobserved heterogeneity) with mean

$$E(y\,|\,\boldsymbol{x}, u) = \exp(\boldsymbol{x}'\boldsymbol{\beta} + u) = \exp(\boldsymbol{x}'\boldsymbol{\beta})\,\exp(u) \tag{8.93}$$

Note that under the Poisson specification we have $V(y\,|\,\boldsymbol{x}, u) = E(y\,|\,\boldsymbol{x}, u)$. Now assume that $\exp(u)$ is a random variable with expectation 1. Also, let $V(\exp(u)) = \sigma_u^2$. Furthermore, let $\boldsymbol{x}$ and u be uncorrelated. Then the mean and variance of the random variable y for given

$\boldsymbol{x}$ only can be written as follows:

$$\begin{aligned} E(y|\boldsymbol{x}) &= E\,E(y|\boldsymbol{x},u) = E(\exp(\boldsymbol{x}'\boldsymbol{\beta})\,\exp(u)) = \exp(\boldsymbol{x}'\boldsymbol{\beta}) \\ V(y|\boldsymbol{x}) &= E\,[V(y|\boldsymbol{x},u)] + V\,[E(y|\boldsymbol{x},u)] \\ &= E[\exp(\boldsymbol{x}'\boldsymbol{\beta})\,\exp(u)] + V[\exp(\boldsymbol{x}'\boldsymbol{\beta})\,\exp(u)] \\ &= \exp(\boldsymbol{x}'\boldsymbol{\beta}) + \sigma_u^2\{\exp(\boldsymbol{x}'\boldsymbol{\beta})\}^2 \quad . \end{aligned} \tag{8.94}$$

This shows that unobserved heterogeneity implies overdispersion in the conditional distribution of $y|\boldsymbol{x}$. In particular we obtain this structure for the negative binomial distribution with $\lambda(\boldsymbol{x}) = \kappa(\boldsymbol{x}) = \exp(\boldsymbol{x}'\boldsymbol{\beta})$ and $V(\varepsilon) = 1/\theta$, the latter being the variance of the gamma mixing distribution. See Table 2.

We now consider the estimator $\hat{\boldsymbol{\beta}}$ that maximizes the log-likelihood function (8.82), whereas the true log-likelihood function would satisfy $V(\exp(u)) > 0$. Then under mild regularity conditions, this estimator satisfies

$$\sqrt{N}(\hat{\boldsymbol{\beta}}_N - \boldsymbol{\beta}_0) \xrightarrow{d} \mathcal{N}(0, N\,\boldsymbol{A}^{-1}\boldsymbol{B}\boldsymbol{A}^{-1}) \tag{8.95}$$

with $\boldsymbol{A}$ equal to the information matrix and $\boldsymbol{B}$ equal to the expected value of the outer product of the score vector, both with respect to the "wrong" log-likelihood function (8.82). For the Poisson model we obtain the following expressions:

$$\begin{aligned} \boldsymbol{A} &= \sum_i \exp(\boldsymbol{x}_i'\boldsymbol{\beta})\,\boldsymbol{x}_i\boldsymbol{x}_i' \;, \\ \boldsymbol{B} &= \sum_i V(\boldsymbol{y}_i|\boldsymbol{x}_i,\boldsymbol{\beta})\,\boldsymbol{x}_i\boldsymbol{x}_i' \end{aligned} \tag{8.96}$$

In practice we substitute $\hat{\boldsymbol{\beta}}$ for $\boldsymbol{\beta}$ and estimate $V(\boldsymbol{y}_i|\boldsymbol{x}_i,\boldsymbol{\beta})$ by its sample estimate $\frac{1}{N-1}\sum_i(\boldsymbol{y}_i - \exp(\boldsymbol{x}_i'\hat{\boldsymbol{\beta}}))^2$. Winkelmann and Zimmermann (1991) present Monte Carlo results which show that use of the Poisson model under overdispersion will produce consistent estimates in samples of the modest size of 100 observations. Moreover, and more importantly, the standard errors from the covariance matrix $\boldsymbol{A}^{-1}\boldsymbol{B}\boldsymbol{A}^{-1}$ will give standard errors of the estimates which provide reliable statistical inference, whereas the standard errors computed from the Poisson model are severely biased toward zero. See also the example in the next subsection.

One could also test for heterogeneity using the Lagrange multiplier test as proposed by Hamerle (1990). However, this test is only useful when panel data are available. See Section 6.7, and see also the corresponding test for probit models in Section 3.2.

6.6 An Example with Cross-Sectional Data

Applied econometricians have often considered research and development (R & D) efforts of enterprises. One way of measuring this activity is by the number of patents issued to the enterprises and to relate this variable to explanatory variables like profit, firm size, or market structure. See, for example, Hausman, Hall and Griliches (1984). In the following we consider a sample of 143 German firms whose patent activity was analysed already by Zimmermann and Schwalbach (1991). Table 3 shows the empirical distribution of number of patents. The large share of firms that has not issued any patent and the very large number of patents issued by a minority of firms should be noted. Therefore, a poor fit of the Poisson distribution is to be expected, unless explanatory variables can take account of this heterogeneity.

Table 3. Empirical Distribution of Patents

Number of Patents	Frequency
0	43
1 - 10	36
11 - 100	33
101 - 500	18
501 - 1000	6
1000 <	7
maximum:	7390
minimum:	0
average:	193.1
standard deviation:	743.1

In Table 4 the parameter estimates are given for a regression model with the explanatory variable RISK; this variable is the empirical covariance of firm-specific profit rate and the average rate of growth of the German economy during the years 1963 to 1975.

One should note the drastic reduction of the value of the log-likelihood when comparing results from Poisson and negative binomial model. The likelihood ratio test as well as t-values seem to indicate that the Poisson model has explanatory power. However t-values as given for the negative binomial model are more reliable since they account for heterogeneity.

The equivalence of the two alternative parameterizations (a) and (b) under specification $\kappa = \exp(\alpha + \beta x)$ as discussed in Section 6.4 can be checked from Table 4: Under (a) we have $\alpha = -6.986$ and $\theta = 0.155$ whereas under alternative (b) $c = 5.124$ and $\theta = \exp(\alpha + c) = \exp(-1.862) = 0.115$. Moreover we have already remarked while discussing results of Table 2 that the sign of β changes when switching from (a) to (b). Therefore the parameterization must be stated before the sign of the vector β can be interpreted meaningfully.

Table 4. Estimation Results for Explanatory Variable RISK

Parameterization		Poisson distribution			
		estimate		t-value	
$\lambda = \exp(\alpha + \beta\, x)$	α	5.212		801.395	
	β	0.187		28.211	
	L	-56556.9			
	LR	747.8			
		Negative binomial distribution			
		model			
		(a) $p = \kappa/(1+\kappa)$		(b) $p = \theta/(\theta+\kappa)$	
		estimate	t-value	estimate	t-value
$\kappa = \exp(\alpha + \beta\, x)$	α	-6.986	-27.465	5.124	22.381
	β	-0.469	-1.305	0.469	1.305
	θ	0.155	9.026	0.155	9.023
	L	-646.89		-646.89	
	LR	1.50		1.50	
		estimate	t-value	estimate	t-value
$\theta = \exp(\alpha + \beta\, x)$	α	-1.893	-16.415	-1.893	- 16.175
	β	0.094	0.877	0.082	0.635
	κ	0.001	4.057	196.338	4.673
	L	-647.27		-647.44	
	LR	0.74		0.40	

NOTE: L = log-likelihood, LR = likelihood ratio statistic.

6.7 Panel Models for Count Data

Let us assume that for T subsequent years the innovational activity for a panel of N firms is observed. Let y_{it} be the number of patents issued by firm i in year t. If we assume these observations are generated by Poisson distribution, then we obtain

$$P(y_{it}) = \frac{e^{-\lambda}\, \lambda^{y_{it}}}{y_{it}!} \quad \text{with} \left\{ \begin{array}{rcl} y_{it} & = & 0,1,2,\ldots \\ i & = & 1,\ldots,N \\ t & = & 1,\ldots,T \end{array} \right. . \tag{8.97}$$

The influence of the vector $\boldsymbol{x}$ of explanatory variables and the individual-specific effect α_i is specified by

$$\lambda(\boldsymbol{x}_{it}, \alpha_i) = \exp(\alpha_i + \boldsymbol{x}_{it}' \boldsymbol{\beta}) \quad , \tag{8.98}$$

which leads to

$$P(y_{it} \mid \boldsymbol{x}_{it}, \alpha_i) = \frac{\exp\left(-\exp(\alpha_i + \boldsymbol{x}_{it}'\boldsymbol{\beta})\right)\, \exp(\alpha_i + \boldsymbol{x}_{it}'\boldsymbol{\beta})^{y_{it}}}{y_{it}!} \quad . \tag{8.99}$$

Let us assume that the y_{it} are independently distributed for *given* α_i. (The same assumption has already been used for the logit and probit panel models. See Section 3.2.) We then obtain

$$\begin{aligned} & P(y_{i1}, \ldots, y_{iT} \mid \alpha_i, \boldsymbol{x}_i) \\ &= \prod_{t=1}^{T} \frac{\exp(-\exp(\alpha_i + \boldsymbol{x}'_{it}\boldsymbol{\beta}))\ \exp(\alpha_i + \boldsymbol{x}'_{it}\boldsymbol{\beta})^{y_{it}}}{y_{it}!} \\ &= \left[\prod_{t=1}^{T} \frac{\bar{\lambda}_{it}^{y_{it}}}{y_{it}\,!}\right] \exp\left(-\exp(\alpha_i) \sum_t \bar{\lambda}_{it}\right) \exp\left(\alpha_i \sum_t y_{it}\right) \end{aligned} \quad (8.100)$$

where

$$\bar{\lambda}_{it} = \exp(\boldsymbol{x}'_{it}\boldsymbol{\beta}) \quad . \quad (8.101)$$

This result can be used to formulate a fixed effects Poisson model. We proceed in a manner analogous to that for the fixed effects logit model (see Section 3.1). First, note that $\sum_t y_{it}$ is Poisson distributed with parameter $\sum_t \lambda_{it}$. The corresponding probability is given by

$$\begin{aligned} & P(\sum_t y_{it}) \\ &= \frac{1}{(\sum_t y_{it})!} \exp(-\exp(\alpha_i) \sum_t \exp(\boldsymbol{x}'_{it}\boldsymbol{\beta}))[\exp(\alpha_i) \sum_t \exp(\boldsymbol{x}'_{it}\boldsymbol{\beta})]^{\sum_t y_{it}} \\ &= \frac{1}{(\sum_t y_{it})!} \exp(-\exp(\alpha_i) \sum_t \exp(\boldsymbol{x}'_{it}\boldsymbol{\beta}) + \alpha_i \sum_t y_{it})[\sum_t \exp(\boldsymbol{x}'_{it}\boldsymbol{\beta})]^{\sum_t y_{it}} \quad . \end{aligned} \quad (8.102)$$

From (8.100) and (8.102) we then get for individual i the following conditional probability:

$$\begin{aligned} & P(y_{i1}, \ldots, y_{iT} \mid \alpha_i, \boldsymbol{x}_{it}, \sum_t y_{it}) \\ &= (\sum_t y_{it})! \prod_t \frac{\exp(\boldsymbol{x}'_{it}\boldsymbol{\beta})^{y_{it}}}{y_{it}!\ [\sum_t \exp(\boldsymbol{x}'_{it}\boldsymbol{\beta})]^{y_{it}}} = (\sum_t y_{it})! \prod_t \frac{\pi_{it}^{y_{it}}}{y_{it}!} \end{aligned} \quad (8.103)$$

where

$$\pi_{it} = \frac{\exp(\boldsymbol{x}'_{it}\boldsymbol{\beta})}{\sum_t \exp(\boldsymbol{x}'_{it}\boldsymbol{\beta})} \quad (8.104)$$

As in the fixed effect logit model, this conditional probability no longer involves the individual-specific effects α_i. Therefore, the structural parameter vector $\boldsymbol{\beta}$ can be estimated from the product of this probability for all N individuals. Note that (8.104) has the form of the multinomial logit model whose log-likelihood function (with respect to $\boldsymbol{\beta}$) is globally concave. Therefore computation of maximum-likelihood estimates is easy. See Hausman et al. (1984, p. 919) for an application of this approach.

We now turn to an alternative approach that assumes random effects. If we substitute $\exp(\alpha_i)$ by ε_i then the conditional probability (8.100) can also be written as follows:

$$P(y_{i1}, \ldots, y_{iT} \mid \alpha_i, \boldsymbol{x}_{it})$$

$$= \left[\prod_{t=1}^{T} \frac{(\bar{\lambda}_{it}\,\varepsilon_i)^{y_{it}}}{y_{it}\,!}\right] \exp\left(-\sum_t (\bar{\lambda}_{it}\,\varepsilon_i)\right)$$

$$= \left[\prod_{t=1}^{T} \frac{\bar{\lambda}_{it}^{y_{it}}}{y_{it}\,!}\right] \exp\left(-\varepsilon_i \sum_t \bar{\lambda}_{it}\right) \varepsilon_i^{\sum_t y_{it}} \qquad (8.105)$$

where, as before, $\bar{\lambda}_{it} = \exp(\boldsymbol{x}_{it}'\boldsymbol{\beta})$. We now assume that ε_i follows the gamma distribution with expected value equal to 1, that is

$$f(\varepsilon_i) = \frac{\theta^\theta}{\Gamma(\theta)}\, \varepsilon_i^{\theta-1} \exp(-\theta\,\varepsilon_i)$$

and ε_i acts a a "mixing variable" (see Section 6.2). Then we obtain for the marginal probability

$$P\,(y_{i1}\,,\,\ldots\,,\,y_{iT}\,|\,\boldsymbol{x}_i)$$

$$= \frac{\theta^\theta}{\Gamma(\theta)} \left[\prod_{t=1}^{T} \frac{\bar{\lambda}_{it}^{y_{it}}}{y_{it}\,!}\right] \int_0^\infty \exp\left(-\varepsilon_i \sum_t \bar{\lambda}_{it}\right) \varepsilon_i^{\sum_t y_{it}}\, \varepsilon_i^{\theta-1} \exp(-\theta\,\varepsilon_i)\; d\,\varepsilon_i$$

$$= \frac{\theta^\theta}{\Gamma(\theta)} \left[\prod_{t=1}^{T} \frac{\bar{\lambda}_{it}^{y_{it}}}{y_{it}\,!}\right] \int_0^\infty \exp\left(-\varepsilon_i\,(\theta + \sum_t \bar{\lambda}_{it})\right) \varepsilon_i^{(\theta + \sum_t y_{it} - 1)}\; d\,\varepsilon_i$$

$$= \left[\prod_{t=1}^{T} \frac{\bar{\lambda}_{it}^{y_{it}}}{y_{it}\,!}\right] \frac{\Gamma\,(\theta + \sum_t y_{it})}{\Gamma\,(\theta)} \left(\frac{\theta}{\theta + \sum_t \bar{\lambda}_{it}}\right)^\theta \left(\frac{1}{\theta + \sum_t \bar{\lambda}_{it}}\right)^{\sum_t y_{it}}. \qquad (8.106)$$

The last line is derived by using (8.78). From this we obtain the likelihood function by considering the product of probabilities (8.106) for all firms:

$$L = \prod_{i=1}^{N} \left\{ \left[\prod_{t=1}^{T} \frac{\bar{\lambda}_{it}^{y_{it}}}{y_{it}\,!}\right] \frac{\Gamma\,(\theta + \sum_t y_{it})}{\Gamma\,(\theta)} \left(\frac{\theta}{\theta + \sum_t \bar{\lambda}_{it}}\right)^\theta \left(\frac{1}{\theta + \sum_t \bar{\lambda}_{it}}\right)^{\sum_t y_{it}} \right\} \qquad (8.107)$$

This likelihood function cannot be factorized into expressions related to the different points of time, which is due to the *intertemporal correlation* in this model.

An alternative approach is to assume that α_i follows a normal distribution with zero mean and variance σ^2. Then we obtain for the marginal probabilities

$$\mathrm{P}(y_{i1}, \ldots, y_{iT}) = \frac{1}{\prod_t y_{it}!} \int_{-\infty}^{+\infty} \prod_t \exp\left(-\exp\left(\sigma_\alpha \tilde{\alpha}_i + \boldsymbol{x}_{it}'\boldsymbol{\beta}\right)\right)$$
$$\times\; \exp\left(\sigma_\alpha \tilde{\alpha}_i + \boldsymbol{x}_{it}'\boldsymbol{\beta}\right)^{y_{it}} \varphi\,(\tilde{\alpha}_i)\, d\tilde{\alpha}_i \qquad (8.108)$$

where $\varphi(.)$ denotes the density of the standard normal distribution. For the involved numerical integration, Gaussian quadrature can be used.

Example

The data are taken from the German Socioeconomic Panel, and contain information on the number of consultations with a doctor or specialist within one-year periods from 1984 to 1989. We do not discuss the substantive economic issues but confine discussion to matters of statistical methodology. The sample consists of 2420 women over 16 years old who answered all the questions essential to our analysis. For a similar study with cross-sectional data see Cameron and Trivedi (1986) and the literature cited therein. The dependent variable has a range from 0 to 100, and the empirical frequencies show that the Poisson model appears adequate. The explanatory variables in the model are Age, Chronic disease (1 = yes/0 = no), Participation in sport (every week, every month, seldom, never), and Life style (0 = not at all content ... 10 = very content). In addition, the regression model includes five (fixed) time effects for 1984 to 1988. The last year, 1989, is the reference category. The (marginal) maximim likelihood estimation method uses Gaussian quadrature and analytical second derivatives of the log-likelihood function. See Spiess (1993) for details. Table 5 contains the parameter estimates of the regression coefficients and the parameter σ_α together with their standard deviations.

Table 5. Estimation Results for the Poisson Panel Model

	Parameter estimate	Standard error
Constant	0.6014	0.0344
Time 1	0.1631	0.0146
Time 2	0.0541	0.0147
Time 3	0.1986	0.0143
Time 4	0.1617	0.0143
Time 5	-0.0431	0.0149
Age	0.0187	0.0007
Chronic disease	0.0493	0.0116
Sport 1	-0.0239	0.0196
Sport 2	-0.0956	0.0245
Sport 3	-0.0614	0.0168
Contentment	-0.0588	0.0029
σ_α	0.5852	0.0053

The results show that the estimate of σ_α is highly significant (t-value=109.8) so that there seems to be heterogeneity among the women that is not accounted for by the explanatory variables.

REFERENCES

Abramowitz, M., and Stegun, I. A. (1965). *Handbook of Mathematical Functions*. New York: Dover.

Amemiya, T. (1981). "Qualitative Response Models: A Survey." *Journal of Economic Literature*, 19, 483–536.

Amemiya, T. (1985). *Advanced Econometrics*. Oxford: Basil Blackwell.

Andersen, E. B. (1970). "Asymptotic properties of conditional maximum likelihood estimators." *Journal of the Royal Statistical Society*, B32, 283–301.

———(1973). *Conditional Inference and Models for Measuring*. Copenhagen: Mentalhygieniejuisk Forlag.

Anderson, T. W., and Hsiao, C. (1982). "Formulation and estimation of dynamic models using panel data." *Journal of Econometrics*, 18, 47–82.

Arminger, G. (1994). "Probit-Models for the Analysis of Non-Metric Panel Data," *Allgemeines Statistisches Archiv*, 78, 125–140.

Balestra, P., and Nerlove, M. (1966). "Pooling cross-section and time series data in the estimation of a dynamic model: The demand for natural gas." *Econometrica*, 34, 585–612.

Bartholomew, D. J. (1982). *Stochastic Models for Social Processes*. 3rd ed. New York: Wiley.

Berkson, J. (1955). "Maximum likelihood and minimum χ^2 estimates of the logistic function." *Journal of the American Statistical Association*, 50, 130–162.

Berkson, J. (1980). "Minimum chi square, not maximum likelihood." *Annals of Statistics*, 8, 457–487.

Ben-Porath, Y. (1973). "Labor force participation rates and the supply of labor." *Journal of Political Economy*, 81, 697–704.

Bhargava, A., and Sargan, J. D. (1983). "Estimating dynamic random effects models from panel data covering short time periods." *Econometrica*, 51, 1635–1659.

Börsch-Supan, A., and Hajivassiliou, V. (1993). "Smooth Unbiased Multivariate Probability Simulators for Maximum Likelihood Estimation of Limited Dependent Variable Models." *Journal of Econometrics*, 58, 347–368.

Boskin, M. J., and Nold, F. C. (1975). "A Markov model of turnover in Aid to Families with Dependent Children." *Journal of Human Resources*, 10, 476–481.

Butler, J. S., and Moffitt, R. (1982). "A computationally efficient quadrature procedure for the one-factor multinomial probit model." *Econometrica*, 50, 761–764.

Burguete, J. F., Gallant, A. R., and Souza, G. (1982). "On unification of the asymptotic theory of nonlinear econometric models." *Econometric Review*, 1, 151–190.

Cecchetti, S. (1986). "The frequency of price adjustment: A study of the newsstand prices of magazines." *Journal of Econometrics*, 31, 355–274.

Chamberlain, G. (1978). "Omitted variable bias in panel data: Estimating the returns to schooling." *Annales de l'INSEE*, 30–1, 49–82.

———(1980). "Analysis of Covariance with Qualitative Data." *Review of Economic Studies*, 47, 225–238.

———(1984). "Panel Data." In: Z. Griliches, and M. D. Intriligator. *Handbook of Econometrics, Vol. 2*. Amsterdam: North-Holland, pp. 1247–1318.

Commenges, D., Letenneur, L., Jacquin, H., and Dartigues, J. F. (1993). "Test of homogeneity of binary data with explanatory variables." *Biometrics*, forthcoming.
Corcoran, M. (1982). "The employment, wage and fertility consequences of teenage women's nonemployment. In: R. B. Freeman and D. A. Wise eds., *The Youth Labor Market Problem: Its Nature, Causes, and Consequences*. Chicago: University of Chicago Press.
Corcoran, M., and Hill, M. (1985). "Reoccurence of unemployment among young adult men." *The Journal of Human Resources*, 20, 165–183.
Diggle, P. J., Liang, K. Y., and Zeger, S. L. (1993). *Analysis of Longitudinal Data* (in press).
Fahrmeir, L. (1987). "Asymptotic likelihood inference for nonhomogeneous observations." *Statistical Papers*, 28, 81–116.
Fahrmeir, L. (1990). "Maximum likelihood estimation in misspecified generalized linear models." *Statistics*, 21, 487–502.
Fahrmeir, L., and Tutz, G. (1994). *Multivariate Statistical Modeling based on Generalized Linear Models*, New York: Springer-Verlag.
Feller, W. (1968). *Probability Theory and Its Application*. Vol. 1, New York: Wiley.
Goldberger, A. S. (1964). *Econometric Theory*. New York: Wiley.
Gourieroux, C., and Monfort, A. (1990). "Simulation based influence: A survey with special reference to panel data models," *Mimeographed*. l'INSEE.
Greene, W. H. (1993). *Econometric Analysis*, Second edition. New York: Macmillan.
Hajivassiliou, V. A. (1992). "Simulation estimation methods for limited dependent variable models," In G. S. Maddala, C. R. Rao, and H. D. Vinod eds., *Handbook of Statistics*, Vol. 11 (Econometrics). Amsterdam: North-Holland.
Hajivassiliou, V. A., and McFadden, D. (1990). "The method of simulated scores for the estimation of LDV models," *Cowles Foundation Discussion Paper*, No. 967, Yale University.
Hajivassiliou, V., McFadden, D., and Ruud, P. (1992). "Simulation of multivariate normal orthant probabilities: Theoretical and computational results," *mimeographed*, Yale University.
Hamerle, A. (1990). "On a Simple Test for Neglected Heterogeneity in Panel Studies." *Biometrics*, 46, 193–199.
Hausman, J. A., Hall, B., and Griliches, Z. (1984). "Econometric Models for Count Data with an Application to the Patents - R & D Relationship." *Econometrica*, 52, 909–938.
Heckman,J. J. (1981a). "Statistical Models for Discrete Panel Data." In: C. F. Manski, and D. McFadden, eds., *Structural Analysis of Discrete Data With Econometric Applications*. Cambridge, MA: The MIT Press, pp. 114–178.
———(1981b). "The Incidental Parameters Problem and the Problem of Initial Conditions in Estimating a Discrete Time-Discrete Data Stochastic Process." In: C. F. Manski, and D. McFadden, eds., *Structural Analysis of Discrete Data With Econometric Applications*. Cambridge, MA: MIT Press, pp. 179–195.
Heckman,J. J., and Willis,R. (1976). "Estimation of a Stochastic Model of Reproduction: An Econometric Approach." In: N. Terleckyj, ed., *Household Production and Consumption*. New York: National Bureau of Economic Rese

Hennevogl, W. (1991). "Schätzung generalisierter Regressions- und Zeitreihenmodelle mit variierenden Parametern," Ph.D. Thesis, Regensburg University.

Howard, R. A. (1971). *Dynamic Probabilistic Systems*, Vol. 1: Markov Models, New York: Wiley.

Hsiao, C. (1985). *Benefits and Limitations of Panel Data*. Cambridge: Cambridge University Press.

——(1986). *Analysis of Panel Data*. Cambridge: Cambridge University Press.

Huber, P. (1967). "The Behavior of Maximum Likelihood Estimates Under Nonstandard Conditions," *Proceedings of the Fifth Berkeley Symposiums*, Berkeley: University of California Press.

Jansen, J. (1990). "On the statistical analysis of ordinal data when extravariation is present," *Applied Statistics*, 39, 74–85.

Johnson, N. L., and Kotz, S. (1972). *Distributions in Statistics: Continuous Multivariate Distributions*. New York: Wiley.

Kritzer, H. M. (1978). "An introductuion to multivariate contingency table analysis," *American Journal of Political Science*, 22, 187–226.

Küsters, U. (1987). *Hierarchische Mittelwert- und Kovarianzstrukturmodelle mit nichtmetrischen endogenen Variablen*, Heidelberg: Physica Verlag.

Li, K. C., and Duan, N. (1989). "Regression analysis under link violation," *The Annals of Statistics*, 17, 1009–1052.

Liang, K. Y., and Zeger, S. L. (1986). "Longitudinal data analysis using generalized linear models," *Biometrica*, 73, 13–22.

Maddala, G. S .(1971). "The Use of Variance Components Models in Pooling Cross Section and Time Series Data." *Econometrica*, 39, 341–358.

——(1983). *Limited-dependent and Qualitative Variables*. Cambridge: Cambridge University Press.

——(1987). "Limited Dependent Variable Models Using Panel Data." *Journal of Human Resources*, 22, 307–338.

McCullagh, P. (1980). "Regression models for ordinal data (with discussion)," *Journal of the Royal Statistical Society*, B42, 109–142.

——(1983). "Quasi-likelihood functions," *Annals of Statistics*, 11, 59–67.

McCullagh, P., and Nelder, J. A. (1989). *Generalized linear models*, 2nd ed., New York: Chapman and Hall.

McFadden, D. (1974). "The Measurement of Urban Travel Demand." *Journal of Public Economics*, 3, 303–328.

——(1989). "A Method of Simulated Moments for Estimation of Multinomial Probits without Numerical Integration." *Econometrica*, 57, 995–1226.

Neyman, J., and Scott, E. L. (1948). "Consistent estimates based on partially consistent observations," *Econometrica*, 16, 1–32.

Olson, R. J. (1978). "Note on the Uniqueness of the Maximum Likelihood Estimator in the Tobit Model." *Econometrica*, 46, 1211–1215.

Orme, C. (1993). "A comment on 'A simple test for neglected heterogeneity in panel studies'," *Biometrics*, 49, 665–667.

Pakes, A., and Pollard, D. (1989). "Simulation and the Asymptotics of Optimization Estimators." *Econometrica*, 57, 1027–1057.

Ronning, G. (1991). *Mikroökonometrie*. Berlin: Springer-Verlag.
Schepers, A., and Arminger, G. (1992). *MECOSA: A Program for the Analysis of General Mean and Covariance Structures with Non-Metric Variables, User Guide*, Frauenfeld, Switzerland: SLI-AG.
Spiess, M. (1993). Parameterschätzung in Regressionsmodellen mit diskreten, korrelierten endogenen Variablen, Dissertation, Konstanz University.
Stiratelli, R., Laird, N., and Ware, J. H. (1984). "Random effects models for serial observations with binary responses," *Biometrics*, 40, 961–971.
Tobin, J. (1958). "Estimation of Relationships for Limited Dependent Variables." *Econometrica*, 26, 24–36.
Toikka, R. S. (1976). "A Markovian model of labor market decisions by workers," *American Economic Review*, 66, 821–834.
Tutz, G., and Hennevogl, W. (1993). *Random effects in ordinal regression models*, Preprint, University of Munich.
Wedderburn, R. W.M. (1974). "Quasi-likelihood functions, generalized linear models and the Gauss-Newton method," *Biometrika*, 61, 439–447.
White, H. (1982). "Maximum likelihood estimation of misspecified models," *Econometrica*, 50, 1–25.
Winkelmann, R., and K. F. Zimmermann (1991). "Robust Poisson Regression," in L. Fahrmeir, and B. Francis eds., *Advances in GLIM. Statistical Modelling*. Proceedings of the GLIM 92 Workshop on Statistical Modelling. Munich, 13–17 June 1991. Berlin: Springer-Verlag, pp. 201–206.
Wright, B., and Douglas, G. A. (1976). "Better procedures for sample-free item analysis." Research Memorandum no. 20, Statistical Laboratory, Dept. of Education, University of Chicago.
Zeger, S. L., and Karim, M. R. (1991). "Generalized linear models with random effects: a Gibbs sampling approach," *Journal of the American Statistical Association*, 86, 79–86.
Zimmermann, K. F., and Schwalbach, J. (1991). "Determinanten der Patentaktivität." *Ifo-Studien*, 37 (1991), 201–227.

Chapter 9

Analysis of Event Histories

TROND PETERSEN

1 Introduction

Event histories are generated by so-called failure-time processes and take the following form. The dependent variable—for example, some social state—is discrete or continuous. Over time it evolves as follows. For finite periods of time (from one calendar date to another) it stays constant at a given value. At a later date, which is a random variable, the dependent variable jumps to a new value. The process evolves in this manner from the calendar date, when one change occurs, to a later date, when another change occurs. Between the dates of the changes, the dependent variable stays constant.

Data on such processes typically contain information about the date a sample member entered a social state, for example, an employment state, the date the state was subsequently left or the date the person was last observed in the state, and if the state was left, the value of the next state entered, and so on.

In analyzing such data the foci are on what determines the amount of time spent in each state and on what determines the value of the next state entered. Typically, one would like to assess the effects of covariates on the amount of time spent in a state and on the value of the next state entered.

TROND PETERSEN • Walter A. Haas School of Business, University of California, Berkeley, CA 94720, USA. • I thank the handbook editors (Gerhard Arminger, Clifford Clogg, and Michael Sobel), George Kephart, and Anand Swaminathan for comments on an earlier version of the article, and Peggy Anne Davis for careful copy editing of parts of the manuscript. For research assistance I thank Kenneth Koput. Many of the materials in this paper appeared in Petersen (1990a, 1991a). However, there are new materials in Sections 3, 4, 7, and 9, and all of the materials in Sections 10, 15, and 16, as well as the empirical examples, are new to this paper. The research was supported by the U.S. National Institute of Aging, grant AG04367, and by the Institute of Industrial Relations at the University of California, Berkeley. The opinions expressed are those of the author.

Handbook of Statistical Modeling for the Social and Behavioral Sciences, edited by Gerhard Arminger, Clifford C. Clogg, and Michael E. Sobel. Plenum Press, New York, 1995.

This paper discusses three types of failure-time or jump processes. The first and simplest type, called a single-state nonrepeatable event process, obtains when there is a single state that can be occupied only once. A person currently in the state may leave it or not. If it is left, one does not distinguish among different reasons for leaving or different destination states. The state, once left, cannot be reentered. An example is entry into first marriage, provided one makes no distinction between religious and secular marriages (Hernes 1972). Another example is mortality (Vaupel, Manton, and Stallard 1979). Being alive is a state that cannot be reentered, and typically one does not distinguish between different destination states, heaven, purgatory, or hell.

The second type is the so-called multistate process. In such a process, the state currently occupied can be left for several distinct reasons. For example, in AIDS research, an AIDS victim may die from Kaposi's sarcoma, pneumonia, or some other complication. One might be interested in detecting which complication is most deadly, which in turn would allow one to make judgments about the most effective alleviating measures. In most case the number of states is finite, that is, the state space is discrete, although in some instances it is continuous, as, for example, in the analysis of individual earnings histories.

Finally, I consider the repeatable-event process. In this process a person can occupy a state several times. Job histories fall within this class of process (Tuma 1976). The researcher focuses on the amount of time spent in each job. Each sample member may contribute more than one job. Typically, such processes also have multiple states: employed, unemployed, and out of the labor force (Flinn and Heckman 1983).

In all three types of failure-time processes the objective of the empirical analysis is, as stated above, to analyze the determinants of the amount of time that elapses between changes and the value of the destination state once a change occurs. An exception to this characterization is the counting-process framework (see Andersen and Borgan 1985), where the concept of a failure time plays only a marginal role.

The remainder of the paper is organized in 16 sections. Section 2 discusses why ordinary regression methods are not suited for analyzing event histories. Section 3 outlines the basic strategy for analyzing event histories by means of hazard-rate models. Section 4 explains how time-independent explanatory variables can be introduced into the hazard rate. Section 5 explains how time-dependent explanatory variables can be introduced into the hazard rate. In Section 6 comparisons to more familiar regression-type models are made. Section 7 discusses repeatable event processes. Section 8 discusses multistate processes with a discrete state space. Section 9 discusses multistate processes with a continuous state space. Section 10 discusses methods of estimation and closes with a discussion of computer software for estimating hazard-rate models. Section 11 addresses the issue of unobserved variables that influence the rate. Section 12 discusses time-aggregation bias. Section 13 discusses the use of continuous- versus discrete-time methods. Section 14 discusses briefly some theoretical models that may generate event histories. Section 15 discusses at some length problems that arise in connection with a common sampling plan for event histories, where one samples spells that are in progress at a given point in time. Section 16 discusses how to deal with left censoring. Section 17 concludes the paper. Many of the sections include empirical examples, using real-life as well as simulated data. The appendix provides a proof of one of the results stated in Section 3.

2 Motivation

Suppose a researcher has collected career histories on employees in a hierarchically organized company. The researcher might be interested in analyzing the determinants of promotion or the amount of time that elapses before a promotion occurs.

Let t_k be the amount of time that elapsed from the date an employee entered a certain level in the organization before he or she was promoted, or before he or she left the company without having been promoted, or before he or she was last observed in the company without having received a promotion. The use of the subscript k to t will be explained in Section 3. Let $\boldsymbol{x}$ denote the vector of explanatory variables, for example, race, sex, and marital status.

One may formulate a linear regression model

$$\ln t_k = \boldsymbol{\beta x} + \epsilon, \tag{9.1}$$

where $\ln t_k$ is the natural logarithm of t_k, $\boldsymbol{\beta}$ the effect parameters pertaining to $\boldsymbol{x}$, and ϵ a stochastic error term.

There are at least two problems with the approach in (9.1). *First*, it treats employees who were never promoted or who left the company without receiving a promotion in the same way as those who did experience a promotion. The former cases are referred to as right-censored. We know only that they had not experienced the event of interest when they were last observed in the company. A *second* problem with the formulation in (9.1) arises when the covariates in $\boldsymbol{x}$ change over time. The number of patterns that $\boldsymbol{x}$ may take over time can be very large, and to account for all of these on the right-hand side of (9.1) may be close to impossible; it would be hard if not impossible to derive a general formulation.

One response to the censoring problem is to restrict analysis to those employees who were promoted. This solution, however, generates other problems. For one, there may be systematic differences between those who were promoted and those who were not. If the research interest is to assess the determinants of promotions, the bias introduced by excluding those who were not promoted may be severe. We will only learn about the amount of time that elapsed before a promotion occurred among those who were promoted in the data set.

Another response to the problem of right-censoring would be to define a dummy variable C that is equal to one if a promotion occurred and zero otherwise and then estimate a logit (or probit) model predicting the probability of having been promoted, as follows (in the logit case):

$$P(C = 1 \mid \boldsymbol{x}) = \exp(\boldsymbol{\beta x})/[1 + \exp(\boldsymbol{\beta x})]. \tag{9.2}$$

However, this procedure ignores the amount of time that elapsed before a promotion or censoring occurred. Being promoted after six months is a different career trajectory from being promoted after six years, but (9.2) does not distinguish the two cases. Introducing t_k on the right-hand side would not help. It would be tantamount to treating a dependent variable as an independent, or using what is to be predicted as one of the predictors, which would make matters worse. Also, (9.2) cannot account for time-varying covariates, unless one defines the probability in (9.2) separately for each observed time unit, say, each week or month (see Allison 1982). In the latter case, one has defined a discrete-time model. It will be treated in more detail below, first in Section 3.2 and then in Section 13.

In the next three sections I show how the hazard-rate framework can be used to solve the two problems associated with the regression framework.

3 The Hazard-Rate Framework

3.1 Basic Concepts

The solution to the problems of right-censoring and time-dependent covariates is now described. Instead of focusing on the entire duration t_k spent in a state, proceed in a more step-wise manner. The central idea is to divide the duration t_k into several segments. Set t_0=0. Then divide t_k into k segments of time from duration 0 to duration t_k. The first segment covers the interval 0 to t_1, the second covers $t_1$to t_2, and so on up until t_{k-1} to t_k, where $0=t_0 < t_1 < t_2 < \ldots < t_{k-1} < t_k$. Each segment has length $\Delta t = t_{j+1} - t_j$.

Now let T be the random variable denoting the amount of time spent in a state before a transition or censoring occurs. The hazard-rate framework proceeds by specifying the probability that the state is left during the duration interval t_j to t_{j+1}, given that it was not left before t_j,

$$P(t_j \leq T < t_j + \Delta t \mid T \geq t_j), \quad \text{where } t_j + \Delta t = t_{j+1}. \tag{9.3}$$

There are two outcomes in a given time interval: the state is left or is not left. The probability that the state is not left in the duration interval t_j to t_{j+1}, given that it was not left before t_j, is therefore just one minus the probability that it was left, given in (9.3), namely

$$P(T \geq t_{j+1} \mid T \geq t_j) = 1 - P(t_j \leq T < t_{j+1} \mid T \geq t_j), \text{where } t_{j+1} = t_j + \Delta t. \tag{9.4}$$

By means of these two probabilities, that is, (9.3) and (9.4), defined for each small segment of time t_j to t_{j+1}, one can derive the probability that the state was not left before duration t_k as

$$P(T \geq t_k) = \prod_{j=0}^{k-1} P(T \geq t_{j+1} \mid T \geq t_j), \quad \text{where } t_0=0 \text{ and } t_{j+1} = t_j + \Delta t, \tag{9.5}$$

which follows from rules for conditional probabilities.

The interpretation of (9.5) is this: the probability of not having an event before duration t_k equals the probability of surviving beyond duration t_1, times the probability of surviving beyond duration t_2, given survival at t_1, and so on up until the probability of surviving beyond duration t_k, given survival at t_{k-1}.

Similarly, the probability that the state was left between duration t_k and $t_k + \Delta t$ follows as

$$P(t_k \leq T < t_k + \Delta t) = P(T \geq t_k) \times P(t_k \leq T < t_k + \Delta t \mid T \geq t_k) \tag{9.6}$$

$$= \prod_{j=0}^{k-1} P(T \geq t_{j+1} \mid T \geq t_j) \times P[t_k \leq T < t_k + \Delta t \mid T \geq t_k],$$

which again follows from rules for conditional probabilities, as in (9.5).

The interpretation of (9.6) is this: the probability of leaving the state in the duration interval t_k to $t_k + \Delta t$ equals the probability of not leaving it before duration t_k, that is (9.5), times the probability of leaving the state between duration t_k and $t_k + \Delta t$, given that it was not left before duration t_k.

In conclusion, if one can specify the probability of a transition in a small time interval, that is, (9.3), given no transition before entry into the interval, then one can derive (9.5) and (9.6).

3.2 Discrete-Time Formulations

In the so-called discrete-time formulation for analyzing event histories, one proceeds by specifying the probability in (9.3) directly, for the given value of Δt, say, Δt=1. To see how this is done, let C_j be a dummy variable that is equal to 1 if the state was left in the duration interval t_j to t_{j+1} and zero otherwise. One may then specify (9.3) using a model for binary outcomes, say, a logit or probit model. The logit model yields

$$P(t_j \leq T < t_{j+1} \mid T \geq t_j) = P(C_j = 1 \mid T \geq t_j) = \frac{\exp(\alpha + \gamma t_j)}{1 + \exp(\alpha + \gamma t_j)}, \tag{9.7}$$

where γ measures the effect of time already spent in the state on the probability of a transition between t_j and t_{j+1} and α is a constant. How to introduce covariates is discussed in Section 4.

This procedure has been used by many (see, e.g., Massey 1987; Morgan and Rindfuss 1985). In order to use the discrete-time framework one needs to insert the right-hand side of (9.7) into (9.5) and (9.6) above. For each time unit where no transition occurs, one specifies the probability of no transition occurring, namely $1 - P(C_j = 1 \mid T \geq t_j)$. For the time units where a transition occurs, one specifies the probability of a transition occurring, namely $P(C_j = 1 \mid T \geq t_j)$. A censored observation would be specified by (9.5), where each time unit gets specified by $1 - P(C_j = 1 \mid T \geq t_j)$. A noncensored observation would be specified by (9.6), where each time unit before the last gets specified by $1 - P(C_j = 1 \mid T \geq t_j)$, while the last time unit is specified as $P(C_j = 1 \mid T \geq t_j)$. There would be one logit equation (or some other probability model) specified per time unit that an individual was observed. Only in the last time unit may the variable C_j be equal to 1, if a transition occurs in that time unit. Otherwise the individual is censored and C_j is equal to 0 for all time units.

The discrete-time formulation as applied to continuous-time processes has two drawbacks. *First*, the estimated coefficients will depend on the length of the time interval Δt. That means that results will not be comparable across studies that vary in the lengths of the time intervals for which the probabilities are specified. Fortunately, in most models (e.g., the logit model) only the constant term will be severely affected, whereas the coefficients of explanantory variables tend to be less affected by the length of the time interval. Some further comments on this are made in Section 13.

Second, in software packages, say, for logit or probit models, the discrete-time formulation requires the researcher to create one record of data per observed time unit on an individual. For a model without duration dependence, this data organization is not required if one has access to a software package that can handle grouped data [e.g., GLIM, see Baker and Nelder (1978)]. All records that do not end in a transition are coded as censored, C_j=0, while the last record on a case, if it ends in a transition, is coded as noncensored, C_j=1. So there will be many more records of data than there are individuals. This imposes burdens on the researcher in terms of data management and increases computation time. Note, however, that multiple records per observation and different number of records across observations *do not deflate standard errors or induce other biases* (Petersen 1986a, pp. 229–233). That is, observations do not become differentially weighted by this procedure. This is a feature that arises out of the hazard-rate framework and is an issue that I will return to later.

3.3 Continuous-Time Formulations

In this subsection I derive the continuous-time formulation in terms of the hazard-rate framework, the formulation used by most researchers. To do this, one step remains. The choice of Δt, that is, the length of the time interval for which (9.3) is specified, is arbitrary. One would like to obtain a formulation that yields consistency across studies. The convention, therefore, since time is continuous, is to let Δt approach zero, namely

$$\lim_{\Delta t \downarrow 0} P(t_j \leq T < t_j + \Delta t \mid T \geq t_j), \quad \text{where } t_j + \Delta t = t_{j+1}. \tag{9.8}$$

Since duration T is taken to be an absolutely continuous variable, the probability of any specific realization of T is zero, and hence (9.8) is also equal to zero. Therefore one divides the probability in (9.3) by Δt, as with probabilities for other continuous variables, which yields a probability per time unit divided by the time unit itself. Then one takes the limit of this ratio as the time unit Δt goes to zero. This operation yields the central concept in event-history analysis, the *hazard rate*

$$\lambda(t_j) \equiv \lim_{\Delta t \downarrow 0} P(t_j \leq T < t_j + \Delta t \mid T \geq t_j)/\Delta t, \tag{9.9}$$

which is a conditional density function: the density that the state is left at duration t_j, given that it was not left before duration t_j.

From (9.9) we find that for small Δt the probability of a transition, initially specified in (9.3), becomes

$$P(t_j \leq T < t_j + \Delta t \mid T \geq t_j) \approx \lambda(t_j)\Delta t, \quad \text{when } \Delta t \text{ is small.} \tag{9.10}$$

Then, inserting (9.10) into (9.4), yields

$$P(T \geq t_{j+1} \mid T \geq t_j) \approx 1 - \lambda(t_j)\Delta t, \quad \text{where } t_{j+1} = t_j + \Delta t. \tag{9.11}$$

Next, insert (9.11) into (9.5),

$$P(T \geq t_k) \approx \prod_{j=0}^{k-1}[1 - \lambda(t_j)\Delta t]. \tag{9.12}$$

Inserting (9.12) and (9.10) into (9.6) yields

$$P(t_k \leq T < t_k + \Delta t) \approx \prod_{j=0}^{k-1}[1 - \lambda(t_j)\Delta t] \times \lambda(t_k)\Delta t. \tag{9.13}$$

Equations (9.12) and (9.13) have the same interpretations as (9.5) and (9.6), but the right-hand sides are now expressed exclusively in terms of the hazard rate in (9.9).

The approximations in equations (9.10) – (9.13) can be made exact by replacing Δt on the right-hand sides with Δt^*, where $\Delta t^* < \Delta t$. This gives

$$\begin{aligned} P(t_j \leq T < t_j + \Delta t \mid T \geq t_j) &= \lambda(t_j)\Delta t^*, \\ P(T \geq t_{j+1} \mid T \geq t_j) &= 1 - \lambda(t_j)\Delta t^*, \\ P(T \geq t_k) &= \prod_{j=0}^{k-1}[1 - \lambda(t_j)\Delta t^*], \\ P(t_k \leq T < t_k + \Delta t) &= \prod_{j=0}^{k-1}[1 - \lambda(t_j)\Delta t^*] \times \lambda(t_k)\Delta t^*. \end{aligned}$$

If Δt=1 in (9.10) – (9.13) and $\lambda(t_j)$ stays constant from t_j to t_j+1, then the value of Δt^* needed to make the approximation exact depends only on the rate $\lambda(t_j)$. Set Δt=1, that is, $t_j + 1 = t_{j+1}$. Then, when the rate is .05, Δt^* is .975, while when the rate is .50, Δt^* is .787. The smaller the rate, the closer Δt^* is to 1.0. The relationship between the rate $\lambda(t_j)$ and Δt^*, when $\lambda(t_j)$ stays constant in the interval t_j to t_j+1 and Δt=1, is $\Delta t^* = \{1 - \exp[-\lambda(t_j)]\}/\lambda(t_j)$, which is less than 1.

Since duration is absolutely continuous the expressions in (9.12) and (9.13) must, as in (9.9), be evaluated as Δt goes to zero. When Δt goes to zero, the number of segments k goes to infinity, since $k = t_k/\Delta t$. To proceed, set $t_k = t_e$. Then, t_k is treated as a fixed duration, set equal to t_e, while k and Δt are variable, so that the duration t_k remains constant as the number of time units into which it gets divided goes to infinity. Computing the limit of (9.12) as $\Delta t \downarrow 0$ and $k \to \infty$ yields the famous expression for the probability of surviving beyond duration t_k

$$\begin{aligned} P(T \geq t_e = t_k) &= \lim_{\substack{\Delta t \downarrow 0 \\ k \to \infty}} \prod_{j=0}^{k-1}[1 - \lambda(t_j)\Delta t] \\ &= \exp[-\int_0^{t_e} \lambda(s)ds], \qquad \text{where } t_e = t_k, \end{aligned} \tag{9.14}$$

known as the *survivor function*, where lim denotes the limit. A proof of (9.14) is given in the appendix. In the last equality of (9.14), s denotes duration in state.

Here I consider a proof of (9.14) in the special case when the hazard rate equals a constant θ for all t. The integral of $\lambda(t)=\theta$ from 0 to t_k then equals θt_k and the survivor function in (9.14) is hence $\exp(-\theta t_k)$. In order to show this, using the limit operations on the right-hand side of the first equality in (9.14), set first $\Delta t = t_k/k$. Then let $k \rightarrow \infty$ (i.e., $\Delta t \downarrow 0$), which, for a fixed t_k $(=t_e)$, yields

$$\begin{aligned} P(T \geq t_e = t_k) &= \lim_{k\rightarrow\infty} \prod_{j=0}^{k-1}(1-\theta t_e/k) \\ &= \lim_{k\rightarrow\infty}(1-\theta t_e/k)^k \qquad (9.15) \\ &= \exp(-\theta t_e), \qquad \text{where } t_e = t_k, \end{aligned}$$

where the last equality follows from a well-known fact in calculus (see Apostol 1967, eq. [10.13], p. 380). The last equality in (9.15) is often taken to be the definition of the exponential function itself.

Finally, consider the limit of (9.13) as Δt goes to zero and k goes to infinity, but now first dividing by Δt on both sides of (9.13), for the same reason as in (9.9). We obtain the *density function* for the duration t_k as

$$\begin{aligned} f(t_e = t_k) &\equiv \lim_{\Delta t \downarrow 0} P(t_k \leq T < t_k + \Delta t)/\Delta t \\ &= \lim_{\substack{\Delta t \downarrow 0 \\ k\rightarrow\infty}} \prod_{j=0}^{k-1}[1-\lambda(t_j)\Delta t] \times \lambda(t_k)\Delta t/\Delta t \qquad (9.16) \\ &= \exp[-\int_0^{t_e}\lambda(s)ds] \times \lambda(t_e), \qquad \text{where } t_e = t_k. \end{aligned}$$

The key point of all this is that by specifying the hazard rate, as in (9.9), one can derive the survivor function and the density function for the duration t_k. The survivor function, equation (9.14), accounts for right-censored observations, those that did not experience a transition. The density function, equation (9.16), accounts for observations that did experience a transition. Once one has specified the hazard rate, the survivor and density functions follow directly from (9.14) and (9.16). So the researcher needs first to choose a particular hazard rate that makes sense for describing the process. Thereafter, the rest follows.

How this framework can be used to account for time-dependent covariates is the subject of Section 5. Before proceeding to this task it may be instructive to consider some specifications of the hazard rate. Perhaps the most famous specification of (9.9) is the exponential model

$$\lambda(t) = \exp(\alpha), \qquad (9.17)$$

where the rate at duration t is independent of t. Exponentiation of α is done to ensure non-negativity of the rate, an issue that is of importance when covariates are introduced, but which does not arise in the rate above.

A straightforward extension of the exponential rate, which allows the rate to depend on duration t, is the so-called piecewise constant rate

$$\lambda(t) = \exp[\sum_{j=0}^{k} \alpha_j D_j(t)], \tag{9.18}$$

where $D_j(t)$=1 when $t \in (t_j, t_{j+1}]$ and 0 otherwise. The rate stays constant at $\exp(\alpha_j)$ within each period t_j to t_{j+1}, but varies arbitrarily between periods. The length of each period is chosen by the researcher. When α_j=α for all j the constant rate model in (9.17) obtains.

Explicit expressions exist for the estimators of the rates in equations (9.17) and (9.18). I consider the former. Let C=0 if an observation is right censored and 1 if not. The likelihood, log-likelihood, and gradient of an observation is then

$$\begin{aligned} \mathcal{L} &= [\exp(\alpha)]^C \times \exp[-t\exp(\alpha)] \\ L &= C \times \alpha - t \times \exp(\alpha) \\ \partial L/\partial\alpha &= C - t\exp(\alpha). \end{aligned} \tag{9.19}$$

Let π denote the proportion of noncensored observations in a sample and let $\bar{t}$ denote the average duration of the observations in the sample, where the average is computed across censored and noncensored observations. The maximum likelihood (ML) estimators of α and of the rate itself follow from (9.19) above, as

$$\hat{\alpha} = \ln(\pi/\bar{t}), \tag{9.20}$$

and

$$\hat{\lambda}(t) = \pi/\bar{t}. \tag{9.21}$$

As (9.21) shows, for a fixed proportion of noncensored observations, the higher the average duration the lower the estimated rate. This makes sense, since the lower the rate the more time will on average be spent in the state. Conversely, for a fixed average duration $\bar{t}$, the higher the proportion of noncensored observations the higher the estimated rate. This also makes sense. For a fixed average duration in the sample, which often means a fixed observation period for the durations, the higher the rate the higher will be the proportion of observations that experience a transition.

Several other specifications allow the rate to depend on duration t. A simple but general specification would be

$$\lambda(t) = \exp(\alpha + \gamma_1 t + \gamma_2 \ln t), \tag{9.22}$$

in which case the exponential model obtains as a special case when γ_1=γ_2=0; the Gompertz when γ_2=0; and the Weibull when γ_1=0 and $\gamma_2 > -1$. In Figure 1 the Weibull specification is plotted for various parameter values.

In the Weibull model the survivor function has the form

$$P(T \geq t_k) = \exp[-\frac{1}{\gamma_2 + 1} t_k^{\gamma_2+1} \exp(\alpha)]. \tag{9.23}$$

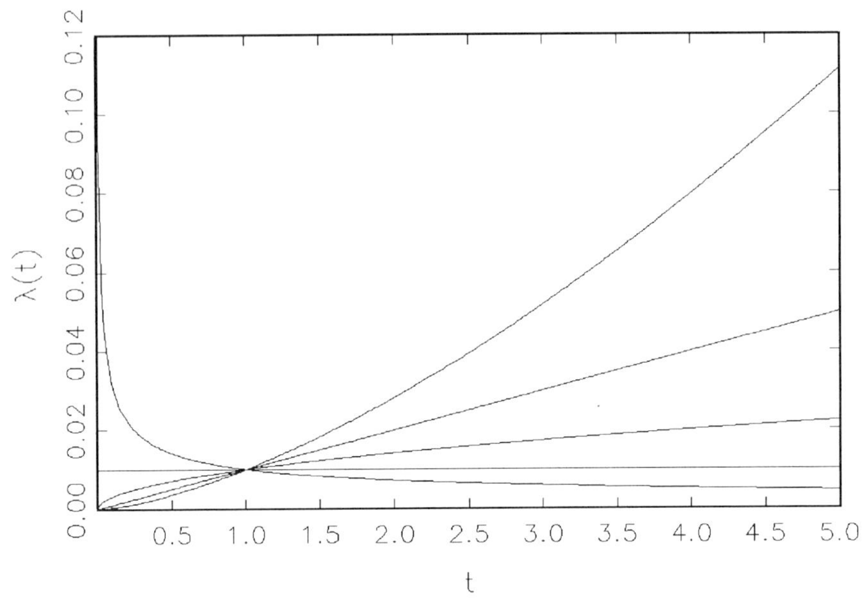

FIGURE 1. The Weibull rate for five values of γ_2 (γ_2=–.5, 0.0, .5, 1.0, 1.5) and for α=–4.61 from equation (9.22) when γ_1=0. Starting from the right vertical axis of the figure, the curve at the bottom of the figure corresponds to γ_2=–.50, while the curves above it correspond to increasing values of γ_2.

In the case of the Weibull model, no analytic solution exists for the estimates of α and γ_2. They must be obtained by means of iterative procedures. Some details on the likelihood, loglikehood and gradient for the Weibull model are given in Section 10.

I present an example of the rates in equations (9.17), (9.18), and (9.22). Data are taken from the personnel records of a large U.S. insurance company. For each employee in the company, we know the date he or she entered the company and the dates of all movements within the company until the end of the study or the date the person quit the company (end of study is December 1978). For further descriptions of the data see Petersen and Spilerman (1990). In this section I present estimates of the rates of departure from the company.

I restrict the analysis to lower-level clerical employees in the company, all of whom are employed in salary grade levels 1 to 6. I use data on a 50% random sample of the employees, leaving data on 10,850 employees.

In this analysis, the duration t is the seniority of the employee in the company. It is measured in months. For the piecewise constant rate, I report estimates for 10 seniority groups. For seniority of nine years or less, there are nine groups, one for each year of seniority less than or equal to nine, where each group is captured by the coefficient α_j, where j refers to the number of years of seniority (j=0,. . .,8). For seniority greater than nine years, there is one group, nine-plus years of seniority, captured by the coefficient α_9. The parameter estimates are given in Table 1.

Table 1. Estimates of the Rate of Departure from Company
(Estimated Standard Errors in Parentheses)

	Constant rate Eq. (9.17)	Piecewise constant rate[a] Eq. (9.18)	Gompertz Eq. (9.22) $\gamma_2=0$	Weibull Eq. (9.22) $\gamma_1=0$
β_0	−3.747 (.011)		−3.215 (.012)	−2.829 (.023)
γ_1			−.015 (.000)	
γ_2				−.330 (.008)
α_0		−3.019 (.014)		
α_1		−3.684 (.026)		
α_2		−4.082 (.038)		
α_3		−4.319 (.048)		
α_4		−4.674 (.065)		
α_5		−4.759 (.078)		
α_6		−4.699 (.087)		
α_7		−4.985 (.115)		
α_8		−4.987 (.131)		
α_9		−4.850 (.056)		
L[b]	−37,695	−35,852	−36,511	−36,388
N	10,089	10,089	10,089	10,089
Number of events	7,947	7,947	7,947	7,947

NOTE: Data are taken from the personnel records of a large U.S. insurance company. The dependent duration variable is the number of months since entry into the company (i.e., seniority) before a departure or censoring (end of study) occurs. See section 3.3 for further description of the data. The estimation routine is described in Petersen (1986b). See Blossfeld, Hamerle, and Mayer (1989, chap. 6) for an extensive discussion of the BMDP implementation of the routine. All computations are done in BMDP (1985). All coefficients are significantly different from zero at the 5% level (two-tailed tests). A 50% random sample of all individuals about which we have data are used in this analysis. For spells that were started prior to January 1, 1970 and that were still in progress at that date, the likelihood contribution is given by equation (9.112) in Section 15.1 below. Spells that started and ended prior to January 1, 1970 are not included in our sample.

[a]For the piecewise constant rate, the duration is divided into 10 groups of seniority. For seniority of less or equal to nine years, there are nine groups, one for each year of seniority, where each group is captured by the coefficient α_j, where j refers to the number of years of seniority (j=0,. . .,8) at the beginning of the period to which α_j pertains. For seniority greater than nine years, there is one group, nine-plus years of seniority, captured by the coefficient α_9.

[b]This is the log-likelihood of the model.

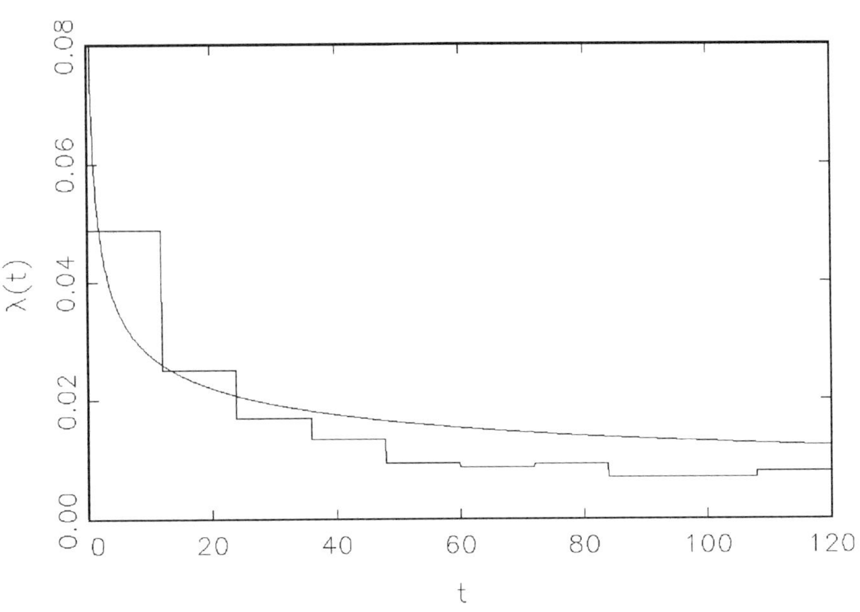

FIGURE 2. Plots of the rates estimated in columns 2 and 4 of Table 1. The trapezoidal curve is the piecewise constant rate in equation (9.18) and the smooth curve is the Weibull rate in equation (9.22), when $\gamma_1=0$. Duration (t) is measured in months.

The constant rate model gives an estimate of β_0 of –3.747, which means that the estimated expected time before a departure is 42 months among lower-level clerical employees, computed by the formula for the expected duration in the exponential model [see eq. (9.59) in Section 6], namely $1/\lambda(t)$, which in this case yields $1/\exp(-3.747)$. From the Gompertz and Weibull models we see that seniority has a significant negative effect on the rate of departure. The longer someone has been with the company the less likely he or she is to leave. As discussed in Section 9, this may reflect unobserved heterogeneity in the rates due to omitted variables in Table 1. The piecewise constant rate confirms the findings of the Gompertz and Weibull models. The rate declines with every year of seniority, until it reaches a low level after about eight years.

Estimates of the rates for the piecewise constant and the Weibull model are plotted in Figure 2. We see that the Weibull model, relative to the piecewise constant-rate model, overestimates the departure rate somewhat for employees with four to five years or more seniority, but that the agreement between the models is quite close for seniority of less than five years (60 months).

It is useful to consider in more detail the accuracy of the approximation $\Delta t=1$ in (9.12). Assume that the rate is

$$\lambda(t) = \lambda_j \quad \text{if } t_j < t \leq t_j+1. \tag{9.24}$$

The probability of no event between duration t_j and duration t_j+1 , given no event prior to t_j, is then given by

$$P(T > t_j+1 \mid T > t_j) = \exp(-\lambda_j), \tag{9.25}$$

and the probability of an event between duration t_j and duration t_j+1 , given no event prior to t_j, is

$$P(t_j < T \leq t_j+1 \mid T > t_j) = 1 - \exp(-\lambda_j). \tag{9.26}$$

Note now that for small λ_j,

$$\lambda_j \approx 1 - \exp(-\lambda_j). \tag{9.27}$$

For λ_j=0, the approximation is exact. But λ_j=0, is not permissible. Note also that

$$\lambda_j > 1 - \exp(-\lambda_j) \quad \text{for all } \lambda_j > 0, \tag{9.28}$$

which means that the rate λ_j always is larger than the probability of an event in the relevant time interval.

Figure 3 plots the relationship between λ_j and the probability of an event between duration t_j and duration t_j+1 , given no event prior to t_j, for values of λ_j in the interval 0 to .25. It shows that the discrepancy between λ_j and the probability of an event in the interval increases with λ_j and that the rate is larger than the true probability. For small rates, say, $\lambda_j \leq .10$, the two are very close. So, the approximation Δt=1 is quite accurate when the rate is low.

4 Time-Independent Covariates

The hazard rate at duration t may depend not only on t, as in (9.22), but also on explanatory variables. Explanatory variables can be grouped broadly into two types, which I treat separately: those that stay constant over time and those that change or may change over time. Examples of the former are sex, race, and birthplace. Examples of the latter are marital status, number of children, and socioeconomic status. This section describes the simpler case where the covariates stay constant over time; Section 5 discusses the more complicated case where they depend on time.

Let $\boldsymbol{x}$ denote the set of time-constant covariates. The hazard rate at duration t, given the covariates $\boldsymbol{x}$, is now defined as

$$\lambda(t \mid \boldsymbol{x}) \equiv \lim_{\Delta t \downarrow 0} P(t \leq T < t + \Delta t \mid T \geq t, \boldsymbol{x})/\Delta t, \tag{9.29}$$

giving the rate at which a transition occurs at duration t, given no transition before t, and given the covariates $\boldsymbol{x}$.

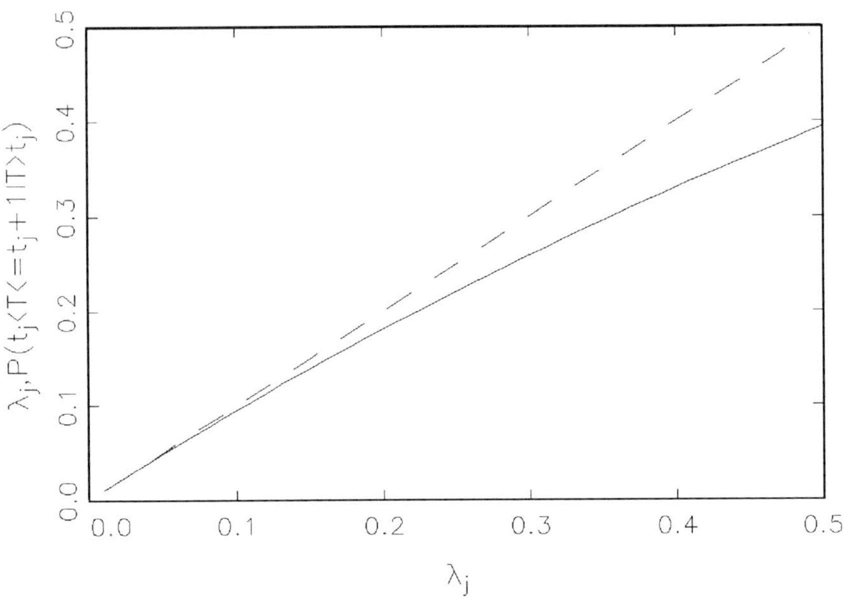

FIGURE 3. Plots of the relationship between the rate λ_j and the probability of an event between duration t_j and duration $t_j + 1$, given no event prior to t_j, $P(t_j < T \leq t_j+1 \mid T > t_j)$, for values of λ_j in the interval 0 to .25, when the rate stays constant at λ_j in the interval t_j to t_{j+1}. The horizontal axis gives λ_j, while the vertical axis gives λ_j (the 45-degree dashed line) and $P(t_j < T \leq t_j+1 \mid T > t_j)$ (the full-drawn curve).

The survivor function, given the covariates $\boldsymbol{x}$, is

$$P(T \geq t_k \mid \boldsymbol{x}) = \exp\left[-\int_0^{t_k} \lambda(s \mid \boldsymbol{x})ds\right] . \tag{9.30}$$

It might be instructive to consider some specific examples of the hazard rate. This can easily be done within the framework of (9.17) and (9.22) of Section 3. In the case of the exponential model, the approach is to say that the parameter α differs between groups in the sample, so that individual i has parameter, say, $\alpha_i = \boldsymbol{\beta x}_i$, where $\boldsymbol{x}_i$ are the covariates for individual i, and $\boldsymbol{\beta}$ is a vector of effect parameters conforming to $\boldsymbol{x}_i$, where $\boldsymbol{x}_i$ usually contains the constant 1, and the first element of $\boldsymbol{\beta}$ usually is the constant term. The rate then becomes

$$\lambda(t \mid \boldsymbol{x}_i) = \exp(\boldsymbol{\beta x}_i). \tag{9.31}$$

The covariates shift the rate up and down. Therefore, differences may occur in the rates between individuals due to differences in the covariates. If $\beta_h > 0$, the corresponding covariate increases the rate; if $\beta_h = 0$, the covariate has no effect on the rate; and if $\beta_h < 0$, the covariate lowers the rate.

If $\boldsymbol{x}_i$ only contains a constant 1 and a single categorical covariate (e.g., ethnicity), that is, $\boldsymbol{x}_i = (1, x_{1i})$, then the ML estimator of $\boldsymbol{\beta}$ in (9.31) has an analytic expression. It is informative to consider this case. In the more general case where x_{1i} contains two or more explanatory variables (e.g., education and age), no explicit expressions exist for the parameter estimates. They must be obtained by means of iterative procedures.

Suppose x_{1i} is a dummy variable equal to 0 or 1, in which case $\boldsymbol{\beta}=(\beta_0, \beta_1)$. Let π_0 denote the proportion of noncensored observations for cases with x_{1i}=0 and let π_1 denote the proportion of noncensored observations for cases with x_{1i}=1. Further, let $\bar{t}_0$ denote the average duration for cases with x_{1i}=0 and let $\bar{t}_1$ denote the average duration for cases with x_{1i}=1. The ML estimators of β_0, $\beta_0+\beta_1$, and β_1 are

$$\begin{aligned}\hat{\beta}_0 &= \ln(\pi_0/\bar{t}_0),\\ \widehat{\beta_0+\beta_1} &= \ln(\pi_1/\bar{t}_1),\\ \hat{\beta}_1 &= \ln[(\bar{t}_0/\bar{t}_1)(\pi_1/\pi_0)],\end{aligned} \tag{9.32}$$

and for the two rates

$$\begin{aligned}\hat{\lambda}(t \mid x_{1i}=0) &= \exp(\hat{\beta}_0) = \pi_0/\bar{t}_0,\\ \hat{\lambda}(t \mid x_{1i}=1) &= \exp(\hat{\beta}_0+\hat{\beta}_1) = \pi_1/\bar{t}_1.\end{aligned} \tag{9.33}$$

As the first equation of (9.33) shows, the estimate of the rate for men equals the proportion of noncensored cases among men divided by the average duration for men, with the same relationship holding for the estimate of the rate for women. This is no different from the case without covariates in equation (9.21).

We see from (9.32) that the estimate of the effect β_1 of being female depends on two ratios, $\bar{t}_0$ to $\bar{t}_1$ and π_1 to π_0. Fixing $\pi_0=\pi_1$, the effect parameter β_1 for females is positive if the average duration for men $\bar{t}_0$ is larger than that for women, $\bar{t}_1$. This makes sense. The higher the rate, the lower the average duration. A positive value for β_1 indicates a higher rate for women. If, in contrast, the opposite holds, $\bar{t}_0$ is less than $\bar{t}_1$, the estimate of β_1 will be negative. If $\bar{t}_0$ equals $\bar{t}_1$, then the estimate of β_1 will be zero: there are no differences in the rates between the two groups.

Conversely, fixing $\bar{t}_0=\bar{t}_1$, then, if the proportion of noncensored cases is larger for men than for women, the estimate of β_1 is positive. Women have a higher rate. The opposite occurs when the proportion of noncensored cases is smaller for men than for women. When the two proportions are equal, the effect of being female is zero. This makes sense. Fixing the average durations to be equal, the group with the lower rate will experience a lower proportion of noncensored cases and vice versa.

In the case of the more general rate in (9.22), the approach is the same. Typically, one assumes that γ_1 and γ_2 do not vary among groups in the sample, but that α does, yielding

$$\lambda(t \mid \boldsymbol{x}_i) = \exp(\boldsymbol{\beta}\boldsymbol{x}_i + \gamma_1 t + \gamma_2 \ln t). \tag{9.34}$$

Now, differences may occur in the rates between individuals at a given duration, due to differences in the covariates, and intraindividual differences may occur in the rate over time due to the effect of duration itself.

In the case of the Weibull model, where γ_2=0 in (9.34), the survivor function becomes

$$P(T \geq t_k \mid \boldsymbol{x}_i) = \exp\left[-\frac{1}{\gamma_2+1}\, t_k^{\gamma_2+1} \exp(\boldsymbol{\beta}\boldsymbol{x}_i)\right]. \tag{9.35}$$

Further details on the likelihood for the Weibull model are given in Section 10.

The size of the coefficients in (9.31) and (9.34) will depend on (a) the units in which duration is measured, that is, days, weeks, and so on; (b) the units in which $\boldsymbol{x}$ is measured, as always; and (c) how often transitions occur, that is, the rate at which changes occur.

Regarding (a), if duration is measured in months, the estimated rates per month will be roughly four times bigger than the estimated rates per week had durations been measured in weeks. Except for the constant term, the coefficients in $\boldsymbol{\beta}$ will not be much affected by the units in which duration is measured. The reason for this is that a coefficient can be roughly interpreted as the percentage deviation in the rate from the baseline group captured by the constant term. This percentage deviation will be unaffected by the units in which duration is measured. These properties are easily illustrated by the ML estimators in equation (9.32). For the constant term, the estimate from the first equation of (9.32) clearly depends on the units in which durations are measured. If duration is measured in years and fractions of years, the denominator on the right side of the first equation of (9.32) will be twelve times smaller than when durations are measured in months and fractions of months. This means that the constant term with durations measured in yearly units will be equal to ln 12=2.48 plus the constant term when durations are measured in monthly units. This follows because $\hat{\beta}_0 = \ln(\pi_0/\bar{t}_0/12) = \ln(\pi_0/\bar{t}_0) + \ln 12$. So the constant term merely gets adjusted by a constant that depends on 12, namely $\ln 12$. For the effect parameter β_1, we see that the estimate in fact is independent of the units in which durations are measured, because, when one goes from monthly to yearly measurement, the monthly measurements in the first term on the right-hand side of the third equation of (9.32) are divided by 12 in both the numerator and denominator, thus cancelling each other. In the rate considered in (9.31), the rate itself is twelve times larger for yearly than monthly measurements, so a monthly rate of .10 translates into an annual rate of 1.2, as can be seen from (9.33). But in terms of the survivor function as well as all other relevant measures that can be derived from the rate, the meaning of the two numbers .10 and 1.2 are the same. For example, the probability of surviving the first year will be given by $\exp(-.10 \times 12)$=.30 and $\exp(-1.2 \times 1)$=.30 in the case of monthly and annual measurements of durations respectively.

Not much can be said about the number of observations needed in order to estimate the parameters of a rate. Estimation procedures are usually nonlinear. My experience is that hazard-rate models typically can be estimated from the same number of observations as binary logit and probit models. Furthermore, parameter estimates are usually stable with respect to where one starts the iteration routine. For the model in (9.31), final estimates are indeed independent of initial guesses, since the model contains no local maxima of the likelihood.

An illustration of these procedures is found in Table 2, column 1. I use the same data as in Table 1 (see Section 3.3) and estimate the same rate—the rate of departure—but add the covariates sex and ethnicity/race (white, black, Hispanic, or Asian). The Weibull model is used, namely

$$\lambda_d(t \mid \boldsymbol{x}_i) = \exp(\boldsymbol{\beta}_d \boldsymbol{x}_i + \gamma_d \ln t), \tag{9.36}$$

where the subscript d denotes that this is the departure rate and that the coefficients $\boldsymbol{\beta}_d$ and

γ_d pertain to that rate, so as to distinguish it from the promotion rate considered in Sections 7 and 8.

The seniority effect (i.e., γ_d) is much the same as in Table 1. Men and whites have higher departure rates than the other groups.

To illustrate the meaning of the size of the coefficients, it is useful to consider the constant rate model in equation (9.31). That model allows one to easily calculate several measures. If we estimate this model, using the same data and same variables as in column 1 of Table 2, except that we exclude duration, we get the following estimates of the constant and the sex effect: –3.642 and –.125. The estimated rate for white males is then $\hat{\lambda}(t \mid \text{Sex}=0, \text{Race}=0)=\exp(-3.642)=.026$ and for white women it is $\hat{\lambda}(t \mid \text{Sex}=1, \text{Race}=0)=\exp(-3.642 - .125)=.023$. Thus, the monthly probability of leaving the company is approximately .026 for white men and .023 for white women.

Since every employee sooner or later will leave the company, these rates can also be used straightforwardly to compute estimates of the expected time before a departure. For white men it is $\hat{E}(T \mid \text{Sex}=0, \text{Race}=0)=1/.026=38.4$ months, while for white women it is $\hat{E}(T \mid \text{Sex}=1, \text{Race}=0)=1/.023=43.4$ months [see eq. (9.59) of Section 6). Similarly, the survivor function can be used to compute the probability of no departure before, say, 12 months. For white men, this gives .730, using the standard expression for the survivor function in (9.14), or .727 using the approximation Δt=1 in (9.12). For white women, this gives .757, using (9.14), or .755, using the approximation Δt=1 in (9.12). As we see, the approximation Δt=1 in (9.12) is accurate.

5 Time-Dependent Covariates

In this section I treat, drawing heavily on Petersen (1992), the considerably more difficult case where the hazard rate depends on covariates that may change over time, so-called time-dependent covariates. For example, in the analysis of departure rates, as in Sections 3.3 and 4, some of the covariates on which the departure rate depends may change over time. This typically will be the case for salaries, for position within the company, perhaps for work location, and the like. In the analysis of the rate, one would generally like to take account of these changes in the explanatory variables.

Time-dependent covariates are often grouped into three classes (Kalbfleisch and Prentice 1980, pp. 122–127). First, there are the deterministic time-dependent covariates, such as calendar time or any function of time that is prespecified. Second, there are stochastic covariates that are generated by a stochastic mechanism external to the process being studied. An example may be fluctuations in interest rates which may influence the behavior of an individual but which themselves are not influenced by his or her behavior. Below, these types of covariates are referred to as exogenous. Third, there are stochastic covariates that are generated by a stochastic mechanism that is internal to the process being studied. An example might be how the number of children a couple has depends on whether they remain married or not. Now, whether the couple remains married or not may also depend on the number of children they acquire. Hence, marital status is a covariate that is partially determined by the dependent fertility process. The latter types of covariates are referred

Table 2. Estimates of Effect Parameters on the Rates of Departure From and Promotion Within Company (Estimated Standard Errors in Parentheses)

	Departure		Promotion	
Constant	−2.739 (.043)	−1.610 (.073)	−.054* (.057)	1.415 (.108)
Constant of hazard[a]			−6.121 (.056)	−6.760 (.062)
Duration[b]	−.335 (.008)	−.234 (.011)	1.033 (.033)	.930 (.027)
Seniority[c]				−.002 (.000)
Time in grade[d]		−.003 (.001)		
Age[e]		−.013 (.001)		−.018 (.001)
Sex[f]	−.061* (.039)	−.229 (.040)	−.297 (.034)	−.245 (.035)
Ethnicity/Race[g]				
Black	−.057* (.029)	−.054* (.030)	.016* (.027)	−.199 (.027)
Asian	−.023* (.096)	−.027* (.094)	.097* (.088)	.065* (.088)
Hispanic	−.182 (.048)	−.186 (.048)	.114 (.040)	−.085* (.041)
Education[h]		−.422 (.048)		.220 (.051)
Company Location[i]		−.480 (.028)		.239 (.024)
Salary grade level[j]				
2		−.148 (.034)		−.236 (.035)
3		−.258 (.035)		−.393 (.035)
4		−.461 (.040)		−.513 (.038)
5		−.734 (.052)		−.561 (.043)
6		−.711 (.085)		−.879 (.051)
L[k]	−36,368	−35,904	−39,949	−39,322
	10,089	10,089	10,089	10,089
Number of events	7,947	7,947	32,815	32,815

*Not significantly different from zero at the .05 level (two-tailed tests).

NOTE: Data and estimation procedures are described in the note to Table 1 and in section 4 for the numbers in column 1, section 5 for the numbers in column 2, and section 7 for the numbers in columns 3 and 4. For the departure rate the Weibull model is used; see equation (9.36) for the estimates in column 1 and equation (9.51) for the estimates in column 2. For the promotion rate the proportional-hazards version of the log-logistic model is used; see equation (9.64).

[a] This is the γ_0 parameter in the log-logistic model in equation (9.64).

[b] For departures, duration is measured as months since employment in the company started (i.e., seniority). For promotions, duration is measured as months since the currently occupied salary grade level was entered.

[c] Seniority is measured as months of employment in the company. In the promotion rate its path is approximated by a step function, updated as a time-dependent covariate every 12 months.

[d] Time in grade is measured as months since the currently occupied salary grade level was entered. In the departure rate its path is approximated by a step function, updated as a time-dependent covariate every 12 months.

[e] This is the age of the employee measured in years. In both rates its path is approximated by a step function, updated as a time-dependent covariate every 12 months.

[f] Reference category: male.

[g] Reference category: white.

[h] High school education or more = 1; less than high school education = 0.

[i] Home office branch = 1; branch in another city = 0.

[j] Reference category: salary grade level 1,

[k] This is the loglikelihood of the model.

to as endogenous, that is, dependent on the failure-time process, the primary dependent process.

The first type of covariates does not create any specific conceptual problems. The second and third do. The point is that although estimation is not affected by whether the covariates are deterministic or stochastic (and if stochastic, whether they are endogenous or exogenous), the construction of the survivor function is.

Let $\boldsymbol{x}(t)$ be the vector of explanatory variables at duration t, where $\boldsymbol{x}(t)$ may include lagged values of the explanatory variables.

Let the sample path of covariates from duration 0 to t_k, which usually is a sequence such that $\boldsymbol{x}$ changes a finite number of times, be denoted by

$$\boldsymbol{X}(t_k) \equiv \{\boldsymbol{x}(s)\}_{s=0}^{s=t_k}. \tag{9.37}$$

Let further $\boldsymbol{x}(t^-)$ denote the value of $\boldsymbol{x}$ when evaluated immediately prior to t. Similarly, let $\boldsymbol{X}(t^-)$ denote the history of the process up to but not including t. Formally, $\boldsymbol{x}(t^-)$ is defined as

$$\boldsymbol{x}(t^-) = \lim_{s \uparrow t} \boldsymbol{x}(s), \tag{9.38}$$

using a notation from Self and Prentice (1982, p. 1122). The definition is similar for $\boldsymbol{X}(t^-)$. If $\boldsymbol{X}$ is a failure-time or set of failure-time processes, often referred to as jump processes, the limit in (9.38) will, for some values of t, depend on whether it is taken from below or above t, that is, as $s \downarrow t$ or $s \uparrow t$.

It will differ at those points in time when changes in $\boldsymbol{X}$ occur. This means that $\boldsymbol{x}(t) - \boldsymbol{x}(t^-)$ will differ from zero for some t. In most applications $\boldsymbol{X}$ is a failure-time process, as opposed to a diffusion process. In the latter case $\boldsymbol{X}$ would change gradually but never jump [for comparisons of the two processes, see Petersen (1988), sec. 2.4].

The hazard rate at duration t can now be defined as

$$\lambda[t \mid \boldsymbol{X}(t^-)] \equiv \lim_{\Delta t \downarrow 0} P[t \leq T < t + \Delta t \mid T \geq t, \boldsymbol{X}(t^-)]/\Delta t, \tag{9.39}$$

giving the rate at which a transition occurs at duration t, given no transition before t, and given the covariates up until but not including t. The reason that one conditions on the covariates up to but not including t is that the "cause" must preceed the effect in time. Andersen and Gill (1982) provide additional technical justifications for this type of specification. Petersen (1992) defines (9.39) in a more rigorous manner, using a slightly different notation.

Operationally, the specification in (9.39) means that in specifying the hazard rate at t one uses lagged rather than contemporaneous values of the covariates. In principle the lag should be infinitesimally small, using values of $\boldsymbol{x}$ right before t. In practice, the length of the lag depends on the frequency with which measures are available. If the process is measured down to monthly intervals, then one would, in specifying the hazard rate in month t, condition on the covariate process $\boldsymbol{X}$ up to and including month $t-1$, but not up to and including month t.

Suppose now that instead of conditioning on $\boldsymbol{X}$ only up to but not including t, one conditions all the way up to and including t. One will then get a quantity that differs from the rate in (9.39). In order to discuss the ensuing quantity I will assume that all elements of $\boldsymbol{X}$ are failure-time processes. In the case when $\boldsymbol{x}(t)=\boldsymbol{x}(t^-)$, the rate is defined (Petersen 1991a, 1992) as

$$\lambda[t \mid \boldsymbol{X}(t)] \equiv \lim_{\Delta t \downarrow 0} P[t \leq T < t+\Delta t \mid T \geq t, \boldsymbol{X}(t+\Delta t), \boldsymbol{x}(t+\Delta t)=\boldsymbol{x}(t^-)]/\Delta t. \quad (9.40)$$

Here, one conditions explicitly on the fact of no change in $\boldsymbol{X}(t)$ at t. The case where $\boldsymbol{x}(t) \neq \boldsymbol{x}(t^-)$ requires a different probability statement. It is treated in Petersen (1992).

The rate in (9.40) will in general be different from the rate in (9.39) even though $\boldsymbol{x}(t)=\boldsymbol{x}(t^-)$ in (9.40). It is useful to state this inequality

$$\lambda[t \mid \boldsymbol{X}(t)] \neq \lambda[t \mid \boldsymbol{X}(t^-)], \quad \text{where } \boldsymbol{x}(t)=\boldsymbol{x}(t^-). \quad (9.41)$$

The reason for this inequality is rather straightforward and is best explained by a simple example. Suppose $\boldsymbol{x}(t)$ consists of a single dummy variable. If we condition on the $\boldsymbol{X}$ process up to but not including t, that is, on $\boldsymbol{X}(t^-)$, then the rate of change in the dependent failure-time process at t equals the rate at which the process changes alone without $\boldsymbol{X}$ changing at t plus the rate at which a change occurs in both the dependent failure-time process and in $\boldsymbol{X}$ at t. This is a feature of multistate processes, processes in which several types of changes may occur. It is explained in more detail in Section 8 on multistate processes [esp. eq. (9.66)]. If we condition on the fact that no change occurred in $\boldsymbol{X}(t)$ at t, then we have already ruled out the possibility of the dependent failure-time processes and $\boldsymbol{X}$ changing at the same time. Hence the rate must be different from the rate where we do not condition on $\boldsymbol{X}$ at t, only at t^-. Therefore, the inequality in (9.41). The case where $\boldsymbol{x}(t) \neq \boldsymbol{x}(t^-)$ requires a different probability statement. It is treated in Petersen (1992).

There is one exception to the inequality in (9.41). When the dependent failure-time process and the covariate process $\boldsymbol{X}$ *cannot* change at the same time, one finds that

$$\begin{aligned} \lambda[t \mid \boldsymbol{X}(t)] &= \lambda[t \mid \boldsymbol{X}(t^-)], && \text{for the case where } \boldsymbol{x}(t)=\boldsymbol{x}(t^-), \\ \lambda[t \mid \boldsymbol{X}(t)] &= 0, && \text{for the case where } \boldsymbol{x}(t) \neq \boldsymbol{x}(t^-). \end{aligned} \quad (9.42)$$

Here, the second equation of (9.42) is quite straightforward to explain. If a change occurred in $\boldsymbol{X}$ at t, no change can occur in the dependent failure-time process at t, because the two processes cannot change at the same time, by assumption above. Hence the rate is zero as stated. The equality in the first equation of (9.42) is harder to explain, but it shows that when the dependent failure-time process and the $\boldsymbol{X}$ process cannot change at the same time, then the two rates, conditioning on $\boldsymbol{X}(t)$ and $\boldsymbol{X}(t^-)$ respectively, are equal, provided that no change occurred in $\boldsymbol{X}$ at t. It is discussed in Petersen (1992).

As mentioned above, covariates that change over time may either be endogenous or exogenous relative to the dependent failure-time process. The covariates are exogenous when they influence the probability of a failure, but are themselves not influenced by the failure-time process. Otherwise they are endogenous. The relevant exogeneity condition, in the case when $\boldsymbol{x}(t)=\boldsymbol{x}(t^-)$, is

$$\lambda[t \mid \boldsymbol{X}(t_k)] = \lambda[t \mid \boldsymbol{X}(t)] \quad \text{for all } t_k > t, \text{ when } \boldsymbol{x}(t)=\boldsymbol{x}(t^-), \quad (9.43)$$

which is an extension to continuous-time processes of Chamberlain's (1982) generalization of Sims's (1972) exogeneity condition for time-series data (see Petersen 1992). On the right-hand side of (9.43) the conditioning on $\boldsymbol{X}$ is all the way up to and including t, for the reason explored above, namely the inequality in (9.41). Petersen (1992) provides a more extensive treatment of exogeneity conditions in hazard-rate models. These issues are difficult and no comprehensive treatment exists in the literature. Lancaster (1990, esp. chap. 2) gives a partial treatment, drawing on an earlier (1986) version of Petersen (1992).

The exogeneity condition in (9.43) says that when the covariates are exogenous to the dependent failure-time process, future values of the covariates are not informative with respect to the probability of a present failure. In contrast, when the covariates are endogenous to the dependent failure-time process, in which case they are outcomes of the failure-time process, their future values will add information about the probability of a current failure.

The condition makes sense only when the covariates are stochastic. That is, covariates may exist whose future values will influence the likelihood of a present failure, but which are not endogenous, because they are nonstochastic. An example would be an inheritance determined at birth that is to be received at the age of 20. It may influence behavior before age 20, but is itself not influenced by that behavior.

One should also note that (9.43) does not preclude expectations about the future influencing the probability of a transition at t, but expectations are to be distinguished from realizations of the future.

To obtain the survivor function, I consider the case where $\boldsymbol{X}$ and the dependent failure-time process cannot change at the same time. In that case, the relationship in (9.42) holds between the two rates in (9.39) and (9.40). The survivor function, irrespective of whether the covariates are exogenous or not, given the covariates from 0 to t_k, is then

$$P[T \geq t_k \mid \boldsymbol{X}(t_k)] = \exp\{-\int_0^{t_k} \lambda[s \mid \boldsymbol{X}(t_k)]ds\}. \tag{9.44}$$

As stated on the right-hand side of (9.44), one conditions at each $s < t_k$, not only on the covariates up to s but also on future values of the covariates (up until t_k).

Under the assumption of exogeneity of the covariates, that is (9.43), the survivor function, given the sequence of covariates from 0 to t_k, becomes

$$\begin{aligned} P[T \geq t_k \mid \boldsymbol{X}(t_k)] &= \exp\{-\int_0^{t_k} \lambda[s \mid \boldsymbol{X}(s)]ds\} \\ &= \exp\{-\int_0^{t_k} \lambda[s \mid \boldsymbol{X}(s^-)]ds\}. \end{aligned} \tag{9.45}$$

In the integral on the right-hand side of the first equality in (9.45), one conditions at each $s < t_k$, on the history of the covariates up until and including s. This is justified by the exogeneity condition in (9.43). The conditioning on $\boldsymbol{X}$ up to and including s, rather than just up to but not including s (i.e., to s^-), follows from the inequality in (9.41). However, I have assumed that $\boldsymbol{X}$ and the dependent failure-time process cannot change at the same time. Then the equality in (9.42) holds. From that, the expression on the right-hand side of the second equality in (9.45) follows, where the conditioning is on $\boldsymbol{X}$ up to but not including s.

In most applications, the covariates in $\boldsymbol{x}$ change according to step-functions of time. That is, the covariates stay constant at, say, $\boldsymbol{x}(t_j)$, from duration t_j to t_{j+1}, at which time they jump to $\boldsymbol{x}(t_{j+1})$, and so on. Suppose the covariates stay constant for k such periods of time. Suppose also that the exogeneity condition in (9.43) holds. In that case the survivor function in (9.44) or (9.45) reduces to

$$P[T \geq t_k \mid \boldsymbol{X}(t_k)] = \exp\{-\sum_{j=0}^{k-1} \int_{t_j}^{t_{j+1}} \lambda[s \mid \boldsymbol{X}(s^-)]ds\} \text{ where } t_0{=}0. \tag{9.46}$$

Note that since $\boldsymbol{x}(t)$ stays constant at $\boldsymbol{x}(t_j)$ in the duration interval t_j to t_{j+1}, $\boldsymbol{X}$ poses no difficulty of integration in such an interval. Each term on the right-hand side of (9.46) has the interpretation

$$P[T \geq t_{j+1} \mid T \geq t_j, \boldsymbol{X}(t_{j+1})] = \exp\{-\int_{t_j}^{t_{j+1}} \lambda[s \mid \boldsymbol{X}(s^-)]ds\}, \tag{9.47}$$

giving the probability of surviving beyond duration t_{j+1}, given survival at t_j and given the covariates from 0 to t_{j+1}. In this integral one needs only to use $\boldsymbol{X}$ from t_j to t_{j+1}^-, because $\boldsymbol{X}$ jumps at t_{j+1}, in which case $\lambda[t_{j+1} \mid \boldsymbol{X}(t_{j+1})]{=}0$, due to (9.42). Suppose the rate at t depends on $\boldsymbol{X}$ only through the values of the covariates at t^-, that is, $\boldsymbol{x}(t^-)$, where $\boldsymbol{x}(t^-)$ may include elements of the past history of $\boldsymbol{X}$, but not on, say, time since the last change in $\boldsymbol{X}$. Then the integral from t_j to t_{j+1} in (9.47) is easy to evaluate. One just uses the covariates as evaluated at t_j, since $\boldsymbol{X}$ stays constant from t_j to t_{j+1}. For estimating the survivor function this is tantamount to using $\boldsymbol{X}$ only up to t_j, since $\boldsymbol{x}(t_{j+1}^-){=}\boldsymbol{x}(t_s){=}\boldsymbol{x}(t_j)$ for all s in the interval t_j to t_{j+1}^-. Note however that in the interpretation of the survivor function (9.47), the conditioning is on $\boldsymbol{X}$ in the entire interval up to t_{j+1}.

Again, some specific examples of the hazard rate might be instructive to consider. This can easily be done within the framework of (9.17) and (9.22) of Section 3. In the case of the exponential model, the approach is to say that the parameter α differs between groups in the sample, so that individual i has parameter, say, $\alpha_i{=}\boldsymbol{\beta x}_i(t)$, where $\boldsymbol{x}_i(t)$ are the covariates at duration t for individual i, and $\boldsymbol{\beta}$ is a vector of effect parameters conforming to $\boldsymbol{x}_i$. The rate then becomes

$$\lambda[t \mid \boldsymbol{x}(t^-)] = \exp[\boldsymbol{\beta x}_i(t^-)], \tag{9.48}$$

where only the value of the covariates immediately prior to t enter. $\boldsymbol{x}(t^-)$ may of course include elements of the past history of $\boldsymbol{X}$. The covariates shift the rate up and down. There may be differences in the rates between individuals at a given duration t, due to differences in the covariates, and there may be intraindividual differences in the rate over time, due to intraindividual changes in $\boldsymbol{x}$ over time. Again, if $\beta_h > 0$, the corresponding covariate increases the rate; if $\beta_h{=}0$, the covariate has no effect on the rate; and if $\beta_h < 0$, the covariate lowers the rate.

With the more general rate in (9.22), the approach is the same. Typically, one assumes that γ_1 and γ_2 do not vary between groups in the sample, but that α does, yielding

$$\lambda[t \mid \boldsymbol{x}_i(t^-)] = \exp[\boldsymbol{\beta x}_i(t^-) + \gamma_1 t + \gamma_2 \ln t]. \tag{9.49}$$

With respect to estimation, much can be said, but the central results are these. If the covariates are exogenous, just use (9.47) for each period within which the covariates X stayed constant, or more generally the expression in (9.46) covering all periods. At the points in time when a failure occurs, the contribution is just the hazard rate. If the covariates are endogenous, the same procedure can be used, as Kalbfleisch and Prentice (1980, pp. 121–127) have shown, but this is a topic that requires separate treatment. The central difference then is that when X is endogenous to the dependent failure-time process, the expressions in (9.46) and (9.47) no longer have interpretations as survivor functions. Petersen (1992) treats these issues in considerable detail and discusses alternative approaches.

Consider further a specific example of the likelihood used in the presence of time-dependent covariates. Focus on a Weibull model, which obtains from (9.49) when γ_1=0. For a duration interval t_j to t_{j+1}, where the covariates stay constant at $\boldsymbol{x}(t_j)$, the likelihood part used in estimation is

$$\begin{aligned}\mathcal{L}^* &= \exp[\boldsymbol{\beta x}(t_j) + \gamma_2 \ln t_{j+1}]^C \times \exp\{-\int_{t_j}^{t_{j+1}} \exp[\boldsymbol{\beta x}(t_j) + \gamma_2 \ln s]ds\} \\ &= \exp[\boldsymbol{\beta x}(t_j) + \gamma_2 \ln t_{j+1}]^C \times \exp\{-\frac{1}{\gamma_2 + 1}(t_{j+1}^{\gamma_2+1} - t_j^{\gamma_2+1}) \times \exp[\boldsymbol{\beta x}(t_j)]\}.\end{aligned} \tag{9.50}$$

where C is a censoring indicator equal to 1 if a transition occurs at t_{j+1} and 0 otherwise. covariate and the dependent When the covariates X are exogenous and X and the dependent failure-time process cannot change simultaneously, the likelihood in (9.50) has a standard likelihood interpretation. For example, the second term on the right-hand side has the interpretation of a survivor function, as given on the left-hand side of (9.47). When the covariates X are endogenous or when X and the dependent failure-time process can change simultaneously, this likelihood can still be used to estimate the rate in (9.49). However, the likelihood expression then no longer has a standard likelihood interpretation. For example, the second term on the right-hand side cannot be interpreted as a survivor function.

Data management is cumbersome in the presence of time-dependent covariates. In most applications the covariates change according to step-functions of time, and if they do not, their paths are approximated by step-functions of time. The typical strategy is then to create a new record of data each time a change in one of the covariates occurs. There will be as many records of data as there are periods within which the covariates stayed constant. Each record will cover a period in which the covariates stayed constant. Each record then contains information about duration in the focal state (the dependent failure-time process) at the beginning of the period the record covers, duration of the period the record covers, and whether a transition occurred or not at the end of the period the record covers, as well as the values of the covariates during the period the record covers (i.e., equal to their values at the beginning of the period). Justification for this procedure is found in the survivor function in (9.46) or (9.47) above, where each piece of the survivor function pertains to a period in which the covariates stayed constant. This does not deflate standard errors or inflate statistical significance (see Petersen 1986a, pp. 229–233). Blossfeld et al. (1989, pp. 199–205) provide a detailed description of this procedure for data management in the presence of time-dependent covariates.

An illustration of the use of time-dependent covariates is found in column 2 in Table 2. Added to the variables in column 1 are the following time-dependent covariates: the person's age (years), the time (months) spent in the currently occupied salary grade level, his or her education (high school versus less), the salary grade level currently occupied, and the work location (home office versus branch in other city). The variables age of employee and time spent in the currently occupied salary grade level, change continuously with time. In this analysis I approximate their continuous change by a step function. I let their values change every 12 months. The alternative solutions are either to let their values change every month or to let them change continuously. The first alternative is computer-intensive. The second requires numerical integration of the hazard rate and is, therefore, also computer-intensive. The rate is

$$\lambda_d[t \mid \boldsymbol{x}_i(t^-)] = \exp[\boldsymbol{\beta}_d \boldsymbol{x}_i(t^-) + \gamma_d \ln t], \tag{9.51}$$

where the subscript d denotes departure rate and that the coefficients $\boldsymbol{\beta}_d$ and γ_d pertain to that rate—to distinguish it from the promotion rate considered in Sections 7 and 8.

As in the results in Table 1 and in column one of Table 2, the departure rate declines with seniority. Time since last promotion has a negative effect on the rate of leaving the company. The longer one has waited without having received a promotion the less likely one is to leave. Age (years) has a negative effect, about half the effect of time in grade (months), when the latter is multiplied by 12. Employees with a high school diploma have lower rates of departure than those without. In Petersen (1991a, Table 2) the education variable was incorrectly coded so that the reported estimate of having a high school diploma was positive. The rate of departure is lower in the home office than elsewhere. The rate of departure declines with the salary grade level among these lower-level clerical employees.

6 Observability of the Dependent Variable

The dependent variable in hazard-rate models is often said to be an unobservable quantity—that is, the hazard rate (see, e.g., Allison 1984, p. 23). This is incorrect. The dependent variable in hazard-rate models is not the hazard rate, but one of the two following quantities, depending on one's point of view.

According to the first view, which I will call the event-history formulation, the dependent variable is whether or not an event takes place in a small time interval t to $t + \Delta t$, which is closed on the left, that is, it includes t (now dropping the subscripts to periods of time used in Section 3). The dependent variable is then a zero-one variable that takes the value of 1 if an event takes place in the small time interval and 0 if not. We need as many such zero-one variables as there are observed time intervals for an individual.

According to the second view, which I will call the duration formulation, the dependent variable is the amount of time that elapses before an event or censoring occurs.

Both ways of viewing the dependent variable are equally valid, and they amount to the same specification, estimation, and interpretation of the models. Now I explore both viewpoints.

Let $D(t^-)$ be equal to 0 at t^-, with no transition prior to t. Assuming that at most one transition can occur in a small time interval t to $t + \Delta t$, then

$$D(t+\Delta t) = \begin{cases} 1 & \text{if a transition occurs in the time interval } t \text{ to } t+\Delta t, \\ 0 & \text{if no transition occurs in the time interval } t \text{ to } t+\Delta t. \end{cases} \tag{9.52}$$

From (9.52),

$$P[D(t+\Delta t) = 1 \mid T \geq t, \boldsymbol{x}] \approx \lambda(t \mid \boldsymbol{x})\Delta t, \quad \text{for small } \Delta t, \tag{9.53}$$

which can be made exact by replacing Δt on the right-hand side with Δt^*, where $\Delta t^* < \Delta t$, as in Section 3.

If, at most, one change in D can occur between t and $t + \Delta t$, it follows that

$$D(t+\Delta t) = \lambda(t \mid \boldsymbol{x})\Delta t + \epsilon(t), \quad \text{when } T \geq t \text{ and } \Delta t \text{ is small}, \tag{9.54}$$

where $\epsilon(t)$ is a stochastic error term with expectation 0, conditional on $T \geq t$, and $\boldsymbol{x}$. Further, from (9.54),

$$E[D(t+\Delta t) \mid T \geq t, \boldsymbol{x}] \approx \lambda(t \mid \boldsymbol{x})\Delta t \quad \text{for small } \Delta t, \tag{9.55}$$

which also can be made exact by replacing Δt on the right-hand side with Δt^*, as above.

The point here is that in (9.54), which captures the event-history formulation, the dependent variable is whether an event takes place between t and $t + \Delta t$, given no event prior to t. This dependent variable takes the value of 0 in all time units in which no event takes place. Only in the last time unit may it take the value of 1, if the observation is noncensored. Since Δt goes to zero, there will be infinitely many such zero-one variables that will account for the entire duration in a state. *We can conclude that the dependent variable in event-history analysis is observable.*

Once the hazard rate has been specified, the survivor function, $P(T \geq t \mid \boldsymbol{x})$, and the probability density function, $f(t \mid \boldsymbol{x})$, follow, by (9.14) and (9.16). The mean value of the duration T can be derived from the probability density function, as

$$E(T \mid \boldsymbol{x}) = \int_0^\infty s f(s \mid \boldsymbol{x}) ds, \tag{9.56}$$

from which it follows that

$$T = E(T \mid \boldsymbol{x}) + \epsilon = \int_0^\infty s f(s \mid \boldsymbol{x}) ds + \epsilon, \tag{9.57}$$

where ϵ is a stochastic error term with mean zero, conditional on $\boldsymbol{x}$.

For example, if the rate is

$$\lambda(t \mid \boldsymbol{x}) = \exp(\boldsymbol{\beta x}), \tag{9.58}$$

then

$$E(T \mid \boldsymbol{x}) = \exp(-\boldsymbol{\beta x}), \tag{9.59}$$

and hence

$$T = \exp(-\boldsymbol{\beta x}) + \epsilon, \tag{9.60}$$

where we can estimate $\boldsymbol{\beta}$ by nonlinear least squares (henceforth NLLS). However, since we have specified a hazard rate, it is preferable to compute the ML rather than the NLLS estimates. The hazard rate uniquely defines the survivor and density functions. NLLS relies on specification of the hazard rate and hence on the entire probability distribution of the duration. But then ML is more efficient while NLLS has no gains in terms of being more robust.

Again, *the central point is that the dependent variable is not some unobservable instantaneous rate*. In the representation in (9.57), which corresponds to the duration formulation, the dependent variable is the amount of time that elapses before an event takes place. We focus on one aspect of this amount of time, the hazard rate, and we try to estimate the parameters of this rate.

7 Repeated Events

I consider the case of job mobility. Each person in the sample has held at least one job and some have held two or more jobs. The focus of the analysis will still be on the determinants of the amount of time spent in each job. A straightforward extension of the framework developed in the earlier sections will accomplish this.

Consider a person who when last observed had held m jobs with durations $t_1, t_2, \ldots, t_m$, where the last duration may be censored. Note that t_j now refers to the amount of time spent in job j, not to the duration at which period j within a job was entered, as in Section 3. Let C_m=0 if the last job was censored and C_m=1 if not.

Within the ML framework we need to derive the probability density of the entire job history of the person, which now is the unit of the analysis and which may consist of more than one job. Define

$$H_{j-1} \equiv \{t_g\}_{g=1}^{j-1} \quad \text{for } j \geq 2, \tag{9.61}$$

which gives the sequence of job durations for job 1 through job $j-1$.

The probability density of the entire job history of a person with m jobs can now be written

$$f(t_1, \ldots, t_m) =$$

$$f(t_1)\prod_{j=2}^{m-1} f(t_j \mid H_{j-1}) \times [\lambda(t_m \mid H_{m-1})]^{C_m} P(T_m \geq t_m \mid H_{m-1}), \tag{9.62}$$

where $f(t_j \mid H_{j-1})$ gives the density of the duration in job j, given the sequence of previous jobs 1 through $j - 1$. The specification allows for full dependence of the duration in, say, job j on the previous job history. Here, $f(t_1)$ gives the probability density of the duration in the first job. Then, each of the $m - 2$ terms within the product, gives the probability density of the duration in job j, given the prior history of jobs up until job j, namely H_{j-1}. Finally, the last two terms give the probability contribution of the last job, be it censored or noncensored, given the history of jobs prior to the last, namely H_{m-1}. So, the probability density of the entire job sequence obtains as the product of the density of first job, density of second job, and so forth until the last job. Covariates can be introduced into the hazard rate in the manner discussed in Sections 4 and 5.

Taking the logarithm of (9.62) yields the log-likelihood of the job history of the individual as

$$\begin{aligned} L &= \ln f(t_1) + \sum_{j=2}^{m-1} \ln f(t_j \mid H_{j-1}) + C_m \ln \lambda(t_m \mid H_{m-1}) \\ &\quad + \ln P(T_m \geq t_m \mid H_{m-1}). \end{aligned} \tag{9.63}$$

We see that the log-likelihood of the entire job history consists of the sum of the log-likelihoods of each job.

In specifying the hazard rate of leaving, say, job j, two procedures are common. In the first, one assumes that the shape of the rate and the parameters of the rate are the same for all jobs. Dependence on previous history may be captured through explanatory variables. In the second procedure one assumes that the rate and its parameters differ from job to job, or at least between subsets of jobs [say, early and late jobs; see, e.g., Blossfeld and Hamerle (1989)]. Using the same data as in Tables 1 and 2 to analyze promotion processes, Petersen et al. (1989) employ the first procedure, while a variant of the second is used in Petersen and Spilerman (1990).

When the form for the hazard rate and its parameters are common to all jobs, one just pools all the jobs on each individual, and estimates the parameters from the data on all the jobs. When the hazard rate and its parameters vary between jobs, depending on, say, the job number, one estimates the parameters separately for each job number. One would specify separate hazard rates for each job, say, $\lambda^j(t \mid H_{j-1})$ for job j, and each $\lambda^j(\cdot)$ would depend on a separate set of parameters. The parameters for job j are estimated using only information on the durations in job j, and so on. Of course, in analyzing the rate in job j one may condition on information on the prior history of jobs (i.e., H_{j-1}). This conditioning can, for example, be on the number of prior jobs, the average duration in prior jobs, and the types of prior jobs.

In both cases, one creates one record of data for each job a person held. Justification for this can be seen from (9.63), where the log-likelihood of a person's job history is the sum of the log-likelihoods of each job. This sum can be computed from m different records of data on a person who held m jobs. It is important to note that this procedure for arranging the data makes no assumption about independence between the jobs on the same individual.

The procedure is valid if the rate in each job does not depend on unobserved variables that are common to or correlated across jobs within an individual's job history. Under this assumption each job can be treated as a separate observation, provided that we condition correctly on the past history of the process.

Note also that even if a possibly unobserved variable is neither common to nor correlated across jobs, ignoring it will still create biases. The bias created does not arise because each job is treated separately, but because the unobservable is not taken into account in deriving the likelihood, a problem that arises even if each sample member held only one job (see Section 11). Thus, restricting the analysis to only first jobs or spells will still yield inconsistent estimates if there are unobservables and these are not taken into account in the likelihood.

An analysis of repeated events is presented in columns 3 and 4 in Table 2, using the same data as in Sections 3–5. The focus is now on the promotion process, and estimates of the rate of promotion in the company are presented. The dependent duration variable is the number of months, measured as time spent in the currently occupied salary grade level, that elapses before a promotion, departure, or censoring occurs. A person can be promoted several times. Hence, it is a repeatable-event process. The variables are the same as in columns 1 and 2, which were discussed in Sections 4 and 5. The rate is specified as a proportional-hazards log-logistic model

$$\lambda_p(t \mid \boldsymbol{x}(t^-)) = \{\exp(\gamma_0 + \gamma_p \ln t)/[1 + \exp(\gamma_0 + (\gamma_p + 1)\ln t)]\} \times \exp[\boldsymbol{\beta}_p \boldsymbol{x}(t^-)], \qquad (9.64)$$

where $\gamma_p > -1$ and $\boldsymbol{\beta}_p$ is a vector of parameters conforming to $\boldsymbol{x}(t)$ (for a use of this model, see Petersen et al. 1989). When $\gamma_p < 0$, the rate declines monotonically with duration t. When $\gamma_p > 0$, the rate first increases with duration in grade, it then reaches a peak, whereafter it declines. That is, the rate is a bell-shaped function of time, which seems reasonable in the context of promotion processes (see, e.g., Petersen et al. 1989). In the specification it is assumed that the shape of the hazard and its parameters are the same for all repetitions of the process.

Focusing on column 4, we see that the rate of promotion declines with the salary grade level. The higher up in the hierarchy a lower-level clerical employee is, the less likely he or she is to get promoted. The sex and race effects are as one would expect. Female and black employees are less likely to be promoted. The effects of age and seniority on the promotion rate are negative. Employees in the home office have higher promotion rates than those employed elsewhere. Since $\gamma_p > 0$, the promotion rate is a bell-shaped function of time in grade, low during the initial months, then rising to a peak, whereafter it declines.

In Figure 4 the promotion rate, as a function of time in a grade level, is plotted for white men and white women.

The plots are based on the estimates in column 3 of Table 2. The plots show that the promotion rate reaches its peak after about 20 months in a grade and that the rate for women is substantially lower than the rate for men.

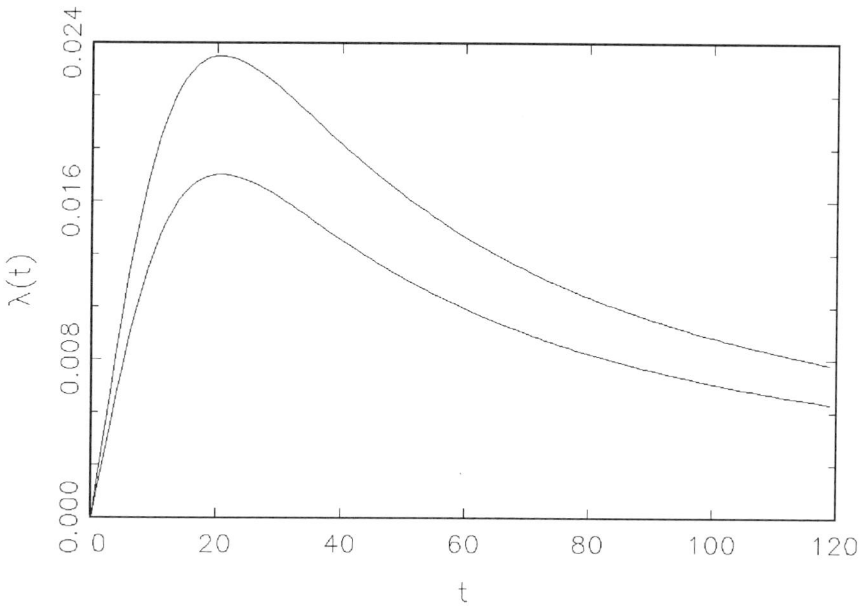

FIGURE 4. Plots of the estimated promotion rates in column 3 of Table 2, from the log-logistic model in equation (9.64). Upper curve is the rate for white men and lower curve is the rate for white women. Duration (t) is measured in months.

8 Multistate Processes: Discrete State Space

In most applications, when a failure or transition occurs, the person enters a new state or the transition occurs for a specific reason. Sometimes, the number of states that can be entered is finite. The state space is then referred to as discrete, an example of which is labor force transitions between being employed and unemployed or out of the labor force. There may also be a continuum of states, in which case the state space is referred to as continuous. Examples of the latter are individual-level socioeconomic status and earnings histories.

In this section I discuss discrete state space processes. In the next section I discuss continuous state space processes.

Let Z be a random variable denoting the state entered when a transition occurs, and let z denote a specific realization of Z, where Z is categorical, that is, has a finite number of values. *The destination-specific* rate of transition, $\lambda_z(t \mid z_j)$, where z_j denotes the state occupied immediately prior to t, is defined as

$$\lambda_z(t \mid z_j) \equiv \lim_{\Delta t \downarrow 0} P(t \leq T < t + \Delta t, Z = z \mid T \geq t, z_j)/\Delta t, \tag{9.65}$$

again dropping the subscripts to subperiods of time t_j. Equation (9.65) gives the rate at which a transition to state z occurs at duration t, given no transition prior to t and given that state z_j was occupied immediately prior to t. For z=z_j, the rate is zero. Covariates can be introduced in the same manner as in Sections 4 and 5 and the rates may depend on the entire past history of the process, including the nature of previous transitions and durations.

Let z' denote the number of possible destination states. The overall rate of transition at duration t, irrespective of the destination state, follows by straightforward probability calculus as

$$\lambda(t \mid z_j) = \sum_{z=1}^{z'} \lambda_z(t \mid z_j), \tag{9.66}$$

because the z' states are mutually exclusive.

Let $P(Z = z \mid T = t, z_j)$ denote the probability that state z was entered, given that a transition occurred at t, and given that state z_j was occupied before the transition. It is defined as

$$P(Z = z \mid T = t, z_j) = \lim_{\Delta t \downarrow 0} P(Z = z \mid t \leq T < t + \Delta t, z_j). \tag{9.67}$$

Since the states are mutually exclusive and exhaust the possible transitions, we get

$$\sum_{z \neq z_j} P(Z = z \mid T = t, z_j) = 1. \tag{9.68}$$

Now, using probability calculus, one can decompose the destination-specific rate of transition as follows:

$$\lambda_z(t \mid z_j) = \lambda(t \mid z_j) \times P(Z = z \mid T = t, z_j), \tag{9.69}$$

that is, into the overall rate of transition times the probability of the destination state, given that a transition occurred.

The survivor function follows as

$$P[T \geq t \mid z_j] = \exp[-\int_0^t \sum_{z=1}^{z'} \lambda_z(s \mid z_j) ds], \tag{9.70}$$

which is obtained by inserting the overall rate of transition in (9.66) into the general expression for the survivor function in equation (9.14) of Section 3.

In analyzing discrete state space processes one can either specify the destination-specific rate of transition directly, as in (9.65), or the overall rate of transition and the probability of the destination state, given a transition, as on the right-hand side of (9.69). In the first case one estimates the destination-specific rates directly. In the second case, one estimates first the overall rate of transition, using a hazard rate routine, and then the probabilities of the destination states, given a transition, using, for example, a multinomial logit model.

Focusing on the destination-specific rate as in (9.65), one can, for purposes of estimation, use a hazard rate routine for estimating single-state space processes. Estimates of each of the destination-specific rates can be obtained by separate analyses. In estimating, say, $\lambda_1(t \mid z_j)$, each transition that occurred for reason 1 is treated as noncensored, all the other observations, that is, transitions to other states or censored observations, are treated as censored. To estimate all the z' different rates, just perform z' separate estimations, one for each state. This procedure is valid provided there are no restrictions on the parameters across the destination-specific rates and no unobserved variables common to, or correlated across, the rates. Each of the destination-specific rates may be given a separate functional form—Weibull, Gompertz, and so forth—and may depend on different explanatory variables.

An example of analysis of a multiple state process is given in Table 2. Getting promoted and leaving the company are the two states. If a departure occurs, a promotion cannot. However, if one is promoted, which is a repeatable-event process, one is still at risk for departing as well as at risk for further promotions.

In the departure analysis the dependent duration variable is the number of months that elapses from the time a person enters the company until he or she leaves or until censoring (end of study) occurs. In the promotion analysis, the dependent duration variable is the number of months that elapses in a given salary grade before a promotion, departure, or censoring occurs.

Both sets of estimates were discussed in Sections 4 and 7. I stress here what can be learned additionally from considering the two-state model.

First, we see that both the rates of departure and of promotion decline strongly with the salary grade level occupied. The higher up in the company, the less likely an employee is to leave and the longer it takes to get promoted. This probably means that the benefits accruing from being in the upper echelons of the salary grade levels for lower-level clerical employees must outweigh the drawback of the lower promotion rates once these grades have been reached. Otherwise, one would expect departure rates to increase with salary grade level.

Second, the rate of promotion is higher in the home office than elsewhere, while the rate of departure is lower. When opportunities for advancement are high, quit rates are lower, given the level of already obtained achievement, that is, the salary grade level. Thus, the two-state model gives insight into how employees respond in terms of quit rates to the opportunity structure of the company.

9 Multistate Processes: Continuous State Space

For some processes the state space is continuous. Examples arise in analysis of intragenerational mobility studies, where one focuses on changes in socioeconomic status, and in analysis of individual-level wage and earnings dynamics. If the state space is continuous, the framework of (9.65) must be modified correspondingly (see Petersen 1988, 1990b).

Let Y be an absolutely continuous random variable and let y denote a specific realization of Y. In specifying the destination-specific rate of transition, focus on the probability density of y being entered in a small time interval, given what has happened up to the start of the interval (see Petersen 1988, p. 144). In order to specify the rate at duration t, let y_j denote the state occupied immediately prior to t, where the subscript "j" may denote that we are looking at the jth transition in a repeated events framework, while it does not indicate that there is a countable number of states on y. The destination-specific rate of transition is defined as

$$\lambda(t, y \mid y_j) \equiv \lim_{\substack{\Delta t \downarrow 0 \\ \Delta y \downarrow 0}} P[t \leq T < t + \Delta t, y \leq Y < y + \Delta y \mid T \geq t, y_j]/\Delta t \Delta y. \tag{9.71}$$

The definition of the rate in (9.71) differs from the definition of the rate in the discrete state space case in (9.65) in that one divides by and takes the limit with respect to both Δy and Δt, whereas in (9.65) one divides by and takes the limit only with respect to Δt. In (9.71) one needs to take the limit also with respect to Δy because y is continuous and probability measures for continuous variables are defined in terms of the limits that give densities.

Covariates can be introduced into the rate in (9.71) in the same manner as in Section 4 and the rates may depend on the entire past history of the process, including the nature of previous transitions and durations (as discussed in Petersen 1988).

The overall rate of transition follows, in a manner analogous to the discrete state space framework in (9.66), by integrating over all the destination-specific rates, namely

$$\lambda(t \mid y_j) = \int_{D(y)} \lambda(t, y \mid y_j) dy, \tag{9.72}$$

where $D(y)$ denotes the domain of y.

Define, in a manner analogous to (9.67), the density of the destination state, given a transition at duration t and given that state y_j was occupied prior to t, as

$$g(y \mid T = t, y_j) \equiv \lim_{\substack{\Delta t \downarrow 0 \\ \Delta y \downarrow 0}} P[y \leq Y < y + \Delta y \mid t \leq T < t + \Delta t, y_j]/\Delta y. \tag{9.73}$$

Since $g(y \mid T = t, y_j)$ is a density function for y, given that a change in y occurred at t, we find, analogously to (9.68) in the discrete case, that

$$\int_{D(y)} g(y \mid T = t, y_j) dy = 1. \tag{9.74}$$

The destination-specific rate of transition can be decomposed into the overall rate of transition times the probability density of the destination state, analogously to (9.69) in the discrete state space framework, namely

$$\lambda(t, y \mid y_j) = \lambda(t \mid y_j) \times g(y \mid T = t, y_j). \tag{9.75}$$

For estimation one can either focus on the destination-specific rate directly (see Petersen 1990b), as in (9.71), or on its decomposition into the overall rate of transition times the probability density of the destination state, given a transition (see Petersen 1988), as on the right-hand side of (9.75).

The survivor function follows in complete analogy to (9.70) in the discrete state space case, as

$$P(T \geq t \mid y_j) = \exp[-\int_0^t \int_{D(y)} \lambda(s, y \mid y_j) dy ds], \tag{9.76}$$

which one may obtain either by specifying $\lambda(t, y \mid y_j)$ directly as in (9.71) or by using the decomposition in (9.75). In the latter case, one first integrates $g(y \mid T = t, y_j)$ over the domain of y, which yields 1. Thereafter, one integrates the overall rate of transition in (9.72) from 0 to t.

I present a brief example of this framework using an empirical study by Visher (1984), as reported in Petersen (1988, pp. 157–161), using the two-step procedure where one specifies the overall rate of transition and the density of the destination state given that a transition occurred, as in (9.7). The data were taken from the Norwegian Life History Study for Men, which was directed by Natalie Rogoff Ramsøy at the Institute of Applied Social Research in Oslo and is described in detail in Rogoff Ramsøy (1977, pp. 43-60). The Norwegian Central Bureau of Statistics collected and organized the data. A representative sample of 3,470 Norwegian men born in 1921, 1931, and 1941 were interviewed retrospectively on their life histories from age 14 up to the date of interview in 1971. Detailed month-by-month employment histories as well as histories in other life spheres were collected.

The analysis focuses on the rate of upward shifts in socieconomic status and on the value of socioeconomic status after an upward shift occurred. An upward shift is defined as a job change that results in an increase in socioeconomic status over the highest level previously attained. Almost all changes in socioeconomic status in this data set are upward. Jobshifts leading to either no change or a downward change in socioeconomic status are treated as if no change occurred, since theories of intragenerational status attainment are primarily about gains in attainment and have little to say about downward and lateral changes in socioeconomic status (see Sørensen 1984, pp. 91-93, 97). If a person holds more than one job before improving his attainment over the previous highest level, the duration before the upward shift is the sum of the durations in the jobs held since the previous highest level of attainment was reached. The measure of socioeconomic status (see Skrede 1971) runs from a low of 3 to a high of 52 and can for all practical purposes be considered continuous.

The rate at which upward shifts occur depends on the sector in which the person works (private or public), on the highest level of socioeconomic status previously attained (i.e., y_j), educational attainment (junior high school or less, or high school or more), on occupational position (manager, professional, craftsman, or manual worker), on labor force experience and on duration since the last upward shift. Other than y_j, all variables are treated as time-dependent, including labor force experience. The latter is allowed to vary continuously with time since the last upward shift (as detailed in Petersen 1986, pp. 231-232). Visher (1984, p. 123) specifies the rate of upward shifts as (suppressing subscripts to individual observations)

$$\lambda_1(t \mid y_j, \boldsymbol{x}(t), L_j) = \exp[\boldsymbol{\beta}\boldsymbol{x}(t) + \rho y_j + \alpha(L_j + t) + \gamma t], \tag{9.77}$$

where $\boldsymbol{\beta}$ is a vector of parameters giving the effects of the covariates in $\boldsymbol{x}$, which includes a constant 1, education, sector, and occupation, measured as of the job held immediately prior to duration t; L_j is the employee's labor force experience (measured in months) at the date the last upward shift occurred; L_j+t is the labor force experience at duration t after the last shift occurred, with effect α; y_j is the highest socioeconomic status previously reached and ρ its effect; and γ is the effect of duration since the last shift.

The specification for the new value of socioeconomic status, given that an upward shift occurred, is

$$y_{j+1} = \boldsymbol{\theta}_1 \boldsymbol{x}_j + \delta_1 y_j + \epsilon, \tag{9.78}$$

where $\boldsymbol{\theta}_1$ is a vector of parameters giving the effects of the covariates in $\boldsymbol{x}_j$, which includes a constant 1, education, sector and occupation; δ_1 gives the effect of the highest level of socioeconomic status previously attained and ϵ is a stochastic error term (see Visher 1984, p. 158). Sector and occupation are measured as of the job held immediately prior to the change in Y (if that job differs from the job held when status y_j was entered). It is assumed that the parameters pertaining to the new value of Y, given a shift in direction d, differ for upward and downward shifts. Hence, we can correctly estimate (9.78) on the basis of upward shifts alone, with no correction for truncation, because there is no truncation problem, as discussed in Petersen [1988, eq.(19)].

In Visher's specification, therefore, the instantaneous rate of an upward change in Y depends on its highest value previously reached, on the time since that value was obtained, and on the exogenous variables, as seen from (9.77). The density of the new value of Y, given that an upward shift occurred, depends on the highest value of Y prior to the change and on the exogenous variables, but not on time since y_j was reached, as seen from (9.78). *There is, however, nothing in the general model specification that prevents one from entering the time elapsed since status y_j was achieved as a predictor in (9.78).*

Assuming that there is no autocorrelation in the ϵ's and that the expectation of ϵ, conditional on an upward shift and on the right-hand-side variables in (9.78), is zero, the parameters of (9.78) can be consistently estimated by linear least squares. No distribution needs to be imposed on the error term. If the latter is normal, least squares and ML coincide, and if not, least squares still yields consistent estimates, under the usual assumptions. The parameters of the hazard rate were estimated by ML (e.g., Tuma and Hannan 1984, chap. 5).

Table 3 gives the estimates of equations (9.77) and (9.78) (taken from Visher 1984, Table 5.2, col. 1 and Table D, panel B, col. 1). I will not comment on every number in the table. Instead, I will focus on the conclusions from this analysis that one could not obtain solely from analyses of the rate of upward shifts or of the size of shifts. In the first column we see that managers have a lower rate of upward shifts than the other occupational groups. That is, on the average they wait longer before experiencing an upward shift (net of the other variables).

From the analysis of upward shifts alone, as in Sørensen and Tuma (1981), one would conclude that managers are the most constrained in their opportunities for increasing rewards, a conclusion that seems plausible in light of their already high rewards and the ceiling effects that may set in. In the second column, we see that managers on the average make the largest jumps, given that an upward shift occurred. From the analysis of the size

Table 3. Estimates of the Effects on the Rate of Upward Shifts in Socioeconomic Status and of the Density of the New Socioeconomic Status Given That an Upward Shift Occurred (Standard Errors in Parentheses)

Variables	Equation (9.77)[a]	Equation (9.78)[b]
Constant	– 3.642 (.042)	10.780 (.205)
Duration (in months), γ	0.001*(.000)	
Labor force experience (in months), α	– 0.005 (.000)	
Socioeconomic status before shift	– 0.078 (.003)	0.635 (.018)
Education (1=high school or more)[c]	0.663 (.040)	5.098 (.227)
Sector (1=public, 0=private)	0.096*(.052)	– 0.247*(.289)
Occupation[d]		
Manager	– 0.797 (.256)	5.356 (1.37)
Professional	– 0.361 (.134)	– 0.812*(.717)
Craftsman	– 0.108 (.046)	1.177 (.263)
Log-likelihood[e]	– 22727.4	
Number of events[f]	6523	3730

*Not significantly different from zero at the .05 level, two-tailed tests.

SOURCE: Visher (1984, Table 5.2 and Table D). For exact definitions of the sample and variables, see Visher (1984, chaps. 5-6). The data were taken from the Norwegian Life History Study for Men (see Rogoff Ramsøy 1977, pp. 43-60; Visher 1981).

[a]These are ML estimates of the rate of upward shifts in socioeconomic status (from Visher 1984, Table 5-2. For estimation procedures see the note to Table 1.

[b]These are estimates of $\boldsymbol{\theta}_1$ and δ_1 in the density for the new value of socioeconomic status, given that an upward shift occurred (from Visher 1984, Table D, panel B, col. 1). The estimates were obtained by least squares, which coincide with the ML estimates when the error term in (9.78) is normally distributed.

[c]The reference category is educational attainment equal to junior high school, its equivalent (in years)

[d] The reference category is manual workers.

[e]Using a likelihood ratio test, we can reject the constant rate model, $\lambda(t) = \lambda$, against the model in (9.77) at any reasonable level of significance.

[f]In column 1, the number of observed spells is 6523, out of which 3730 are noncensored. In column 2, the number of observed upward shifts is 3730.

of the gain alone, as in Sørensen's (1974) difference equation model approach, one would conclude that managers are the least constrained in their opportunities to get ahead. Considering equations (9.77) and (9.78) simultaneously yields a more nuanced picture. The process of intragenerational mobility appears to differ between managers and the reference group, manual workers, in the following way. The former wait longer before they experience upward shifts, but once they shift, they also jump farther. Managers climb in few, but long, steps, whereas manual workers climb in many, but correspondingly shorter, steps. The approach taken here to the study of continuous state space failure time processes allows us to characterize the difference in the processes in this way.

10 Estimation Procedures

So far I have not discussed estimation of parameters in much detail. I have specified the hazard rate, sometimes the corresponding likelihood, and occasionally made a comment on estimation. For models where the hazard rate depends on explanatory variables in a parametric manner, three estimation procedures are common. Below, I elaborate on each of these.

The first is the standard ML procedure. For a hazard-rate model with time-constant covariates $\boldsymbol{x}_i$ and hazard rate $\lambda(t \mid \boldsymbol{x}_i)$, the likelihood of observation i with elapsed duration t_i is

$$\mathcal{L}_i = [\lambda(t_i \mid \boldsymbol{x}_i)]^{C_i} \exp[-\int_0^{t_i} \lambda(s \mid \boldsymbol{x}_i) ds], \tag{9.79}$$

where as before C_i is a censoring indicator equal to 1 if a transition occurred and 0 otherwise. The ML estimates are obtained by maximizing the sum of the natural logarithm of the likelihood across individuals i with respect to parameters that are to be estimated.

In the case of a Weibull model [see (9.22) with γ_1=0], one gets the following expressions for the likelihood, log-likelihood, and gradient of observation i:

$$\begin{aligned}
\mathcal{L}_i &= [\exp(\boldsymbol{\beta x}_i + \gamma_2 \ln t_i)]^C \cdot \exp[-\frac{1}{\gamma_2+1} t_i^{\gamma_2+1} \exp(\boldsymbol{\beta x}_i)] \\
L_i &= C \cdot (\boldsymbol{\beta x}_i + \gamma_2 \ln t_i) - \frac{1}{\gamma_2+1} t_i^{\gamma_2+1} \cdot \exp(\boldsymbol{\beta x}_i) \\
\partial L_i / \partial \beta_j &= C \cdot x_{ij} - \frac{1}{\gamma_2+1} t_i^{\gamma_2+1} \cdot \exp(\boldsymbol{\beta x}_i) \cdot x_{ij} \\
\partial L_i / \partial \gamma_2 &= C \cdot \ln t_i + \frac{1}{\gamma_2+1} t_i^{\gamma_2+1} \cdot \exp(\boldsymbol{\beta x}_i) (\frac{1}{\gamma_2+1} - \ln t_i),
\end{aligned} \tag{9.80}$$

where β_j is the coefficient in $\boldsymbol{\beta}$ that pertains to variable j in $\boldsymbol{x}_i$, x_{ij}.

In the Weibull model, as with most models, no analytic solution exists for the estimates of $\boldsymbol{\beta}$ and γ_2. They must be obtained by means of iterative procedures.

The second procedure is known as the Partial Likelihood (henceforth PL) principle. It can be used for so-called proportional-hazard models, models that have the form

$$\lambda(t \mid \boldsymbol{x}_i) = h(t) \exp(\boldsymbol{\beta x}_i), \tag{9.81}$$

where $h(t)$ gives the effect of duration in a state on the rate. The central feature of this specification is that $h(t)$ enters multiplicatively with the covariates $\exp(\boldsymbol{\beta x}_i)$. Examples of proportional-hazards model are the exponential, Weibull, Gompertz, and log-logistic, given respectively by

$$h(t) = \begin{cases} 1 \\ \exp(\gamma \ln t) \\ \exp(\gamma t) \\ \exp(\gamma_0 + \gamma \ln t)/[1 + \exp(\gamma_0 + (\gamma+1)\ln t)] \end{cases} \tag{9.82}$$

Each of these models can be estimated by ML, as has been done for the exponential, Weibull, and log-logistic in Tables 1 and 2. However, Cox (1975) proposed an ingenious principle for estimating all proportional-hazards models. Using this principle one estimates the effects $\boldsymbol{\beta}$ of the covariates $\boldsymbol{x}_i$ but not the effect of duration $h(t)$. In fact, one estimates $\boldsymbol{\beta}$ and at the same time lets $h(t)$ be totally free. The only restriction on $h(t)$ is that it enters multiplicatively with $\exp(\boldsymbol{\beta x}_i)$.

This principle is very attractive and is often used in social science research. I will, somewhat unusually, use a multistate framework to derive and motivate the principle. First, let there be J noncensored durations. Order them from t_1 to t_J, that is, $t_1 < t_2 < \ldots < t_j < \ldots < t_{J-1} < t_J$. Let $R(t_j)$ denote the set of individuals who are at risk of having an event at immediately prior to the duration t_j at which event j happened. These are the individuals who did not have an event prior to t_j. In continuous time, at most one event can happen at any point in time. If an event occurs at duration t_j, that event happens to one of the individuals in the risk-set $R(t_j)$ at duration t_j. The overall rate at which an event occurs at t_j is now given as the sum of the individual rates for those in the risk set at t_j, namely

$$\begin{aligned} \lambda[t_j \mid \{\boldsymbol{x}_\ell\}_{\ell\in R(t_j)}] &= \sum\nolimits_{\ell\in R(t_j)} \lambda(t_j \mid \boldsymbol{x}_\ell) \\ &= h(t_j) \sum\nolimits_{\ell\in R(t_j)} \exp(\boldsymbol{\beta x}_\ell), \end{aligned} \tag{9.83}$$

where $\{\boldsymbol{x}_\ell\}_{\ell\in R(t_j)}$ denotes the sequence of covariates for those who are at risk at duration t_j.

This overall rate at duration t_j factors into the product of $h(t)$, which is common to all individuals, and the sum of the individual-specific contributions $\exp(\boldsymbol{\beta x}_i)$. This is an important property of the framework. The overall rate gives the rate at which an event happens at duration $t{=}t_j$, irrespective of which individual experiences the event. It corresponds to the overall rate of transition in (9.66) in the multistate framework, except that in (9.83) it is the entire set of individuals that survived to t_j that is at risk, not a single individual as in (9.66). The rate of a particular individual i having an event at duration t is given by individual-specific rate in (9.81). Each individual is described by the same set of parameters $\boldsymbol{\beta}$, but the covariates $\boldsymbol{x}_i$ vary among individuals.

Next, let I be the random variable denoting which particular individual had an event at a given duration and let i denote its realization. In the PL principle one now asks this question: Given that an individual had an event at duration t_j, what is the probability that that event happened to individual i ? This question can be answered by the same formula as in equation (9.69) in the multistate framework, after having divided by the overall rate on both sides of (9.69), namely

$$\begin{aligned}\mathcal{L}_j &= P[I = i \mid T_j = t_j, \{\boldsymbol{x}_\ell\}_{\ell \in R(t_j)}, i \in R(t_j)] \\ &= \lambda(t_j \mid \boldsymbol{x}_i)/\lambda[t_j \mid \{\boldsymbol{x}_\ell\}_{\ell \in R(t_j)}] \\ &= \lambda(t_j \mid \boldsymbol{x}_i)/\sum_{\ell \in R(t_j)} \lambda(t_j \mid \boldsymbol{x}_\ell) \\ &= \exp(\boldsymbol{\beta x}_i)/\sum_{\ell \in R(t_j)} \exp(\boldsymbol{\beta x}_\ell).\end{aligned} \tag{9.84}$$

Here, comparing (9.84) to the sets of equations in the discrete state space framework in Section 8, the right-hand side of the first equation of (9.84) corresponds to (9.67), whereas the right-hand side of the second equation of (9.84) corresponds to (9.69), after having divided by the overall rate on both sides of (9.69).

As one can see from (9.84), all individuals that have not experienced an event by the time immediately prior to duration t_j contribute to the denominator. This number becomes smaller and smaller as duration t grows, as more and more individuals will have experienced an event and will hence be withdrawn from the risk set. The individual who experiences an event at duration t_j also contributes to the numerator.

Further, one sees from the fourth equation of (9.84) that the duration-dependence term $h(t)$ drops out of the likelihood once this probability is formed. Hence, one can use (9.84) as a basis for estimating $\boldsymbol{\beta}$ without having to worry about the form of the duration-dependence term $h(t)$. An interesting property of the probability in (9.84) is that it has the same functional form as the probability of choosing response i in a multinomial logit model where the response probabilities depend on a common parameter $\boldsymbol{\beta}$ and the alternatives depend on characteristics $\boldsymbol{x}_i$ that vary among the $R(t_j)$ alternatives, known as McFadden's conditional logit model (see, e.g., McFadden 1984).

In order to form the likelihood of the data, one just needs to specify (9.84) for each of the durations at which an event happened. The total likelihood becomes

$$\mathcal{L} = \prod_j^J \mathcal{L}_j. \tag{9.85}$$

So, in this likelihood one gets the product of J terms, one term per event. Each term consists of the probability in (9.84). The estimates of the parameters $\boldsymbol{\beta}$ are obtained by maximizing (9.85) with respect to $\boldsymbol{\beta}$.

In forming this likelihood the only information needed on the durations is their rank ordering, not their actual values. Some information is therefore lost in using this likelihood. Nevertheless, this estimator is asymptotically equivalent in distribution to the ML estimator (Efron 1977). Additionally, it does not suffer from the potential inconsistency of the ML estimator due to misspecification of $h(t)$.

The PL framework is easily extended to the cases with repeated events, multiple states, and time-dependent covariates. For the latter, in (9.84) one just replaces $\boldsymbol{x}_i$ and $\boldsymbol{x}_\ell$ with their values at duration t_j. Otherwise, there are no changes to the likelihood. The framework can also be extended to the case where two or more events happen at a given duration t_j. Developing the exact PL estimator in that case is more cumbersome, but good approximations exist (see, e.g., Kalbfleisch and Prentice 1980, chap. 3).

The third procedure is related to the regression-type model discussed in Section 6 [esp. eq. (9.57)]. Instead of specifying the expectation of the completed duration t_i, as in Section 6, one specifies the expectation of the logarithm of t_i, $\ln t_i$. For some models, for example the exponential and the Weibull, this expectation has a linear form, yielding the linear regression model

$$\ln t_i = \boldsymbol{\beta}^* \boldsymbol{x}_i + \sigma \epsilon_i, \tag{9.86}$$

where $\boldsymbol{\beta}^*$ will be explained below and ϵ_i is a random error term and σ a so-called scale parameter. This model is often referred to as the *accelerated* failure-time model.

The distribution of ϵ_i depends on which hazard-rate model one believes governs the process. For a Weibull model, ϵ_i has an extreme-value distribution (see, e.g., Kalbfleisch and Prentice 1980, p. 24). The parameters $\boldsymbol{\beta}^*$ and σ are estimated by ML, which is straightforward once the distribution for ϵ_i is chosen. Note that specifying the probability distribution for ϵ_i is equivalent to specifying the probability distribution for $\ln t_i$, as in a linear regression model. So, in the framework in (9.86) one specifies the density function for $\ln t_i$ as opposed to t_i, as in the standard ML framework.

Each parameter in the $\boldsymbol{\beta}^*$ vector gives roughly the percentage change in the expected duration resulting from a unit increase in the corresponding explanatory variable in $\boldsymbol{x}_i$. The parameter $\boldsymbol{\beta}^*$ is related to the parameter $\boldsymbol{\beta}$ in the hazard-rate framework by $\boldsymbol{\beta}=-\boldsymbol{\beta}^*/\sigma$. In the Weibull model, σ is related to the duration-dependence parameter γ_2 in equation (9.22) by $\gamma_2=1/\sigma$, while in the exponential model, $\sigma=1$ and the duration-dependence parameter is zero. Estimates of $\boldsymbol{\beta}^*$ have the opposite sign of the estimates of the hazard-rate parameters $\boldsymbol{\beta}$ and may differ from it by a constant multiplicative factor $1/\sigma$.

Procedures relying on the specification in (9.86) are quite popular, mainly because they are available in some standard statistical software packages. They have at least two drawbacks. First, if one is primarily interested in the impact of variables on the rate, one has to translate the estimates of $\boldsymbol{\beta}^*$ and σ into estimates of $\boldsymbol{\beta}$ $(=-\boldsymbol{\beta}^*/\sigma)$. This requires also translating the estimated standard errors of $\boldsymbol{\beta}^*$ and σ into those of $\boldsymbol{\beta}$, which can be cumbersome. Second, the specification apparently does not allow for time-dependent covariates nor multiple states. Researchers occassionally use procedures based on (9.86) for dealing with time-dependent covariates (e.g., Hannan and Freeman 1988, chap. 10). Doing so requires various shortcuts, and the results obtained are at best approximations to the results from a correct likelihood specification. To my knowledge, the accuracy of these approximate results has not been investigated.

In conclusion, the first two procedures, ML and PL, seem preferable in research. The choice between ML and PL is not easy. If one specifies a proportional-hazards model, for example, a Weibull model, and has no intrinsic interest in the duration-dependence parameters, then the PL procedure seems preferable. If one either has a direct interest in the duration-dependence parameters or one specifies a nonproportional-hazards model, only the ML procedure is available. The third procedure dominates the two first only if one is solely interested in the percentage impact on the expected duration of changes in explanatory variables x_i, not in the hazard-rate parameters directly, and one studies a process with no time-dependent covariates and no multiple states.

In terms of software for estimating these models, several special-purpose routines are available. Unfortunately, the situation is grimmer with respect to standard widely dispersed mainframe packages. BMDP (1990) and SAS (1991) provide procedures for PL estimation with and without time-dependent covariates, although the latter can be cumbersome to implement. LIMDEP (see Greene 1991, chap. 50) claims to do ML estimation with and without time-dependent covariates for several models, based on the results in Petersen (1986a,b), to a large extent utilizing the routine developed by Petersen which was documented in Blossfeld et al. (1989, chap. 6 and app. 2), but the extent to which this is the case I have not been able to verify. LIMDEP appears to be the only widely-dispersed general-purpose program that does ML estimation of several hazard-rate models. LIMDEP and SAS can both estimate the accelerated failure-time specification in (10.8). The use of several of these programs, as well as other programs such as GLIM (see Baker and Nelder 1978), is described in Blossfeld et al. (1989, chaps. 5, 6).

11 Unobserved Heterogeneity

Unobserved heterogeneity refers to the situation where some of the covariates on which the hazard rate depends are not observed. The issue is distinct from the existence of an error term in hazard-rate models. As discussed in Section 6, the dependent variable in analysis of event histories can always be expressed as a sum of its mean and an error term.

To fix ideas, consider the hazard rate

$$\lambda(t \mid \boldsymbol{x}, \mu) = \exp(\boldsymbol{\beta x} + \delta\mu), \tag{9.87}$$

where μ is the unobserved variable and δ its effect; μ may for example capture whether a person is a mover or a stayer (see Spilerman 1972).

Continuing the example of equation (9.54) in Section 6, the value of the dependent 0–1 variable $D(t + \Delta t)$ corresponding to (9.87) can be written as

$$D(t + \Delta t) = \exp(\boldsymbol{\beta x} + \delta\mu)\Delta t + \epsilon(t), \quad \text{when } T \geq t \text{ and } \Delta t \text{ is small}, \tag{9.88}$$

where $\epsilon(t)$ is an error term with mean of zero, conditional on $\boldsymbol{x}$, μ, and $T \geq t$. Equation (9.88) shows that μ is distinct from an error term $\epsilon(t)$.

The unmeasured variable μ presents the problem that even if its mean, conditional on $\boldsymbol{x}$, is zero, that is, $E(\mu \mid \boldsymbol{x})$=0, the mean of $\exp(\delta\mu)$, conditional on $\boldsymbol{x}$ and t, will typically be different from zero (by Jensen's inequality; see, e.g., DeGroot 1970, p. 97). For example, if μ is gamma distributed with mean 1 and variance $1/\sigma$ and we normalize δ=1, then the average rate, unconditional on μ, for those who had not experienced an event by duration t becomes

$$\lambda(t \mid \boldsymbol{x}) = \sigma \cdot \exp(\boldsymbol{\beta x})/[t \cdot \exp(\boldsymbol{\beta x}) + \sigma] \tag{9.89}$$

which depends on t (see, e.g., Tuma and Hannan 1984, p. 179). The rate in (9.87) is independent of t, but when μ is not observed, the rate, unconditional on μ, ends up depending on t.

Therefore, when an unobserved variable enters the hazard rate, as in (9.87), it typically does not cancel out when it comes to estimation, as the error term $\epsilon(t)$ in (9.88) does. To avoid biases in the estimates, one must take account of the unobservable in the estimation procedure.

The most well-known problem arising from not controlling for unobserved variables is that the estimated hazard rate becomes biased toward negative duration dependence (see, e.g., Heckman and Singer 1984b, pp. 77–78). Intuitively, this occurs because individuals with high death rates on the average leave the population at risk before those with low death rates, so that the rate in the population, not controlling for the individual frailty levels, declines over time. That this is the case can easily be seen from (9.89). By taking the derivative with respect to t one sees that the hazard rate declines with t.

Several solutions to the problem of unobserved heterogeneity have been proposed, but two are prominent in the literature (see Yamaguchi 1986).

The first solution assumes, as above, that μ is a random variable, commonly referred to as the random-effects procedure. The main approach in that case is to impose a specific distribution on μ, say, normal, log-normal, or gamma. In terms of estimation, one first derives the likelihood for the observed history on an individual, conditional on observed and unobserved variables. Thereafter, one uses the imposed distribution of the unobservable to compute the mean of the likelihood when the unobserved variable is not taken into account. That is, one computes the average value of the likelihood, where the averaging is done over all possible values of the unobserved variable. This average depends on the observed variables, on the parameters of the hazard rate, and on the parameters of the distribution of the unobservable. This procedure is repeated for each individual in the sample. Thereafter, the likelihood of the sample is maximized by standard procedures (see Lancaster 1979). A variant of this procedure is the so-called EM algorithm (see Dempster, Laird, and Rubin 1977).

An example of this approach obtains if we assume μ in (9.87) to be binomially distributed with parameter q, which gives the probability of a person being, for example, a mover or a stayer. The likelihood of an observation, after having "expected out" the unobservable, then is

$$\begin{aligned}\mathcal{L} = {} & q[\exp(\boldsymbol{\beta x} + \delta)]^C \exp[-t \cdot \exp(\boldsymbol{\beta x} + \delta)] \\ & +(1-q)[\exp(\boldsymbol{\beta x})]^C \exp[-t \cdot \exp(\boldsymbol{\beta x})],\end{aligned} \tag{9.90}$$

where C, as before, is a censoring indicator. The objective of the analysis based on this likelihood is to estimate the structural parameters $\boldsymbol{\beta}$ and δ, where δ is the effect of the unobservable μ, as well as the parameter q of the distribution of the unobservable μ.

Neither theory nor data give much guidance when choosing the distribution for μ. Heckman and Singer (1984a) therefore developed an estimator of the hazard rate that is nonparametric with respect to the distribution of μ. There is not much experience with this estimator (see, however, Hoem 1989; Trussel and Richards 1985). The central idea in their estimator is the same as in (9.90). First, they impose a multinomial distribution on the unobservable

μ. Then the objective of the estimation procedure is to estimate (a) the number of categories on μ [i.e., two in (9.90)], (b) the effect of each category [i.e., δ in (9.90)], and (c) the probability of each category [i.e., q in (9.90)]. It generalizes the idea in (9.90) by letting the number of categories on μ being a quantity that is to be estimated rather than assumed a priori.

The second main solution is to treat μ as a fixed variable (Chamberlain 1985), referred to as the fixed-effects procedure. While appealing in that few assumptions need to be imposed on μ, this procedure has the drawback that it applies only to processes where the event is repeated over time and where one has observed at least two transitions on some of the individuals in the sample. Furthermore, and perhaps more restrictive, only the effects of covariates that change over time can be estimated. For example, one cannot estimate the effects of sex and race (for technical details, see Chamberlain 1985). This parallels the identical result in the panel-data setting (e.g., Chamberlain 1985).

It is instructive to consider how the random-effects procedure works. Continuing the example in the previous sections, I conduct a Monte Carlo experiment and consider the case of departures from a corporation. Suppose that there are two types of individuals, movers and stayers, as in equation (9.90). The movers have a high rate of quitting, whereas the stayers have a low rate. In the experiment I first specify a true underlying model, namely the rate

$$\lambda(t \mid \mu) = \exp(\beta_0 + \delta\mu), \tag{9.91}$$

which is a special case of (9.87), where μ is a dummy variable equal to 0 or 1, indicating whether a person is a mover (=1) or a stayer (=0), and δ is its effect. I set β_0=–2.302 and δ=1.609. The rates are then $\lambda(t \mid \mu=0)=\exp(\beta_0)=.10$ and $\lambda(t \mid \mu=1)=\exp(\beta_0+\delta)=.50$.

With these death rates I assume that 100 persons on each value on μ enter the organization each year—for simplicity, at the beginning of the year. With these departure rates and the entry process I used the program language GAUSS (Edlefsen and Jones 1988) to generate a data set. More specifically, I used the random number generator in GAUSS (see Edlefsen and Jones 1988, p. 424), as here programmed with the probability integral transformation for going from a uniform to an exponential random variable (see, e.g., DeGroot 1986, pp. 154–156).

I record the time of departure for each person within a cohort of entrants, in double precision (i.e., with 16 decimals), so there is no time-aggregation bias in the ensuing estimates (see, e.g., Petersen 1991b). I let the process run in this way for a total of 21 years. Observations that died during or after the twenty-first year of the process, as measured in calendar time (not individual duration), were treated as right-censored in this study, in order to make it comparable to real-life data sets, where right-censoring is ever present. Altogether, I generated data for 4,200 observations. A similar process is investigated in detail in Petersen and Koput (1991).

With the assumed entry process, the distribution of μ is binomial with $P(\mu=1)\equiv q$=.50. With these data I report estimates of four different versions of this rate

$$\lambda(t \mid \mu) = \exp(\beta_0 + \delta\mu + \gamma \ln t). \tag{9.92}$$

In the first specification I control for μ by using its values in the data (0 or 1). In the second specification I do the same, but control for duration (i.e., $\ln t$) as well. The true value of the duration effect γ is zero. In the third specification there is no control for μ (i.e., δ=0), but I control for duration ($\ln t$). That specification corresponds to the rate with unobserved heterogeneity where no attempt is made to correct for it. In the fourth specification I exclude μ and control for duration ($\ln t$), but I try to correct for the unobserved heterogeneity induced by the exclusion of μ by using the random-effects procedure whose likelihood is given in (9.90). The objective is to see if I can recover the structural parameters β_0=–2.302, δ=1.608, and γ=0, as well as the distribution of μ, q=.50, even when μ is not observed.

Table 4 gives the results. The first column gives the estimates of the true underlying model when μ is observed. The structural parameters are quite well recovered. The second column gives the estimates of the true model with duration added. Again, the structural parameters are well recovered. In particular, the duration effect, whose true value is zero, is not significantly different from zero. The third column gives the estimates of the model excluding the variable μ. This is the model with an unobservable and no attempt to control for it. The duration dependence parameter is negative and significantly different from zero. Thus, the results confirm the well-known fact that unobserved heterogeneity tends to produce negative duration dependence, because as time passes, more and more low-death-rate observations will be left in the sample. The duration dependence effect picks up this phenomenon when one does not control for the frailty levels of the observations. The fourth column reports estimates of a model with an effect of duration but with no *measured* controls for μ. However, I attempt to control for the unobserved variable μ by means of the random-effects procedure in (9.90). The column shows: (1) the structural parameters β_0 and δ are well recaptured; (2) there is no spurious duration effect, that is, the parameter γ is not significantly different from zero; and (3) the parameter q for the distribution for μ is well recaptured with an estimate of .508; its true value is .50.

Finally, note in Table 4 that the model where an attempt is made to correct for the unobserved heterogeneity by using the random-effects procedure produces a considerably higher fit than the model where no such attempt is made (compare the log-likelihoods in columns 3 and 4). This correctly indicates that there is unobserved heterogeneity in the data. As one would expect, both models with measured controls for μ (columns 1 and 2) produce higher fits than the model where μ is controlled for by the random-effects procedure.

To conclude, the random-effects procedure yields good results in this case. However, note that the case studied was simple: I knew the correct specification of the hazard rate as well as the correct distribution of the unobservable μ. In more complicated models with less a priori knowledge, it appears that these techniques for dealing with unobserved heterogeneity are less reliable (e.g., Hoem 1989).

12 Time-Aggregation Bias

The estimators used to estimate the parameters of the hazard rate typically assume that the available measures of time are continuous.The assumption that time is exactly or continuously measured is rarely met. Instead, researchers typically know that the amount of time

Table 4. Estimated Death Rates for Data With and Without Unobserved Heterogeneity Using Models With and Without Correction for Unobserved Heterogeneity (Estimated Standard Errors in Parentheses)

	No Unobserved Heterogeneity		Unobserved Heterogeneity (μ is Unobserved)		
	No Age Effect	Age Effect	No Control for Unobservables	Random-Effects Procedure	True Parameters
β_0	−2.305	−2.315	−1.505*	−2.384	−2.303
	(.0281)	(.009)	(.019)	(.072)	
δ	1.630	1.640		1.739	1.609
	(.036)	(.040)		(.090)	
γ		.009	−.200*	.034	.000
		(.014)	(.0130)	(.025)	
q				.509	.500
				(.030)	
L^a	−7,466	−7,466	−8,356	−8,299	
N	4,200	4,200	4,200	4,200	

*Significantly different from the *true* value of the coefficient at the .05 level (two-tailed tests). For the true values of the parameters see column 5 above.

NOTE: Data were generated by the random number generator in GAUSS (see Edlefsen and Jones 1988, p. 424) programmed here with the probability integral transformation for going from a uniform to an exponential random variable (see, e.g., DeGroot 1986, pp. 154–155). The model used to generate the censored and noncensored failure times is the hazard rate in equation (9.91), using the parameters reported in column 5 above. See section 11 for further details. The estimation routine is described in Petersen (1986b). See Blossfeld, Hamerle, and Mayer (1989, chap. 6) for an extensive discussion of the BMDP implementation of the routine. All computations are done in BMDP (1985).

[a]This is the log-likelihood of the model.

spent in a state lies between, say, $j-1$ and j months, but not the exact time within that window. Researchers then customarily set the duration in the state equal to j and treat this as the exact duration. Or, if censoring occurred between $j-1$ and j, the censoring time is set equal to j. These practices give rise to some bias in the estimates. This bias is called *time-aggregation bias*—a problem that arises in virtually all analyses of event history data. Rarely is something done to deal with it. In the remainder of this section I draw heavily on Petersen (1991b).

The most natural solution to the problem of time aggregation is to develop estimators that take account of the grouping of the duration measurements. These estimators are derived from the assumption that the process evolves continuously in time, but they adjust for the fact that measurements of time are not continuous (see, e.g., Cox and Oakes 1984, pp. 56–60; Heijtan 1989, pp. 172–174). Specifically, for an observation that experienced a transition between $j-1$ and j, the likelihood contribution becomes the *probability* of experiencing a transition between $j-1$ and j. Each term in this probability can be derived from the hazard rate. The likelihood based on this probability yields *consistent* estimates of the parameters of the hazard rate. If, in contrast, the researcher assumes that the exact duration is j, the likelihood contribution of the observation becomes the *probability density* of experiencing a transition at duration j. The likelihood based on this probability density yields *inconsistent* estimates of the parameters of the hazard rate. In sociological research, the most popular solution to time aggregation has been to employ discrete-time methods. These are discussed briefly in Section 13.

One reason why many researchers may choose not to implement the estimator that adjusts for the grouping of the duration measurements is that it becomes cumbersome to do so in multistate models. To illustrate, consider the simplest of cases: Transitions can be made to two states, and the rates are constant, r_1 and r_2. Let Z be the random variable denoting the state entered upon a transition and let z denote its realization. The probability that a transition occurs between j_i-1 and j_i and that it is to state z then becomes

$$\begin{aligned} P(j_i-1 < T_i \leq j_i, Z=z) &= \\ \exp[-(r_1+r_2)(j_i-1)] \cdot \{1-\exp[-(r_1+r_2)]\} &\cdot \frac{r_z}{r_1+r_2} \end{aligned} \tag{9.93}$$

[for the general formula, see, e.g., Elandt-Johnson and Johnson 1980, p. 274, eq. (9.15)]. The second and third terms on the right-hand side of (9.93) will, after taking logarithms, not factor into separate components for each of the destination-specific rates. Therefore, joint maximization of the rates is required.

It would always be preferable to use the estimator that adjusts for the grouping of the duration measurements. But researchers almost never do so, in part because of convention, and in part for the following four reasons. First, it is often impractical to develop the relevant estimator. In many cases it would require writing special-purpose estimation routines. Second, as discussed above, in the case with more than one state that can be entered when a transition occurs, the parameters of all the relevant rates must be estimated simultaneously when using the estimator that adjusts for the grouping of the duration measurements. Using the estimator that assumes exact measurements of durations allows one to estimate first the rate to state 1, then the rate to state 2, and so on. Third, the estimator that adjusts for the grouping of the duration measurements requires the researcher to specify a model for the censoring process and to estimate its parameters along with the parameters of the failure-time process. This is needed in order to deal with censored cases. This is also the case for the discrete-time formulation. But researchers rarely do this, assuming instead that censoring occurred at the end of the interval in question. Unfortunately, little is usually known about the censoring process and the researcher rarely has an intrinsic interest in learning about it. Fourth, even though the problem of time-aggregation bias is known, the researcher may believe it to be negligible, in the sense that the inconsistent estimates may not be far off from the consistent estimates.

For these four reasons, the estimator based on the assumption that durations are exactly measured is often used and seems, from a practical point of view, preferable, in spite of its bias. Petersen (1991b) considered a constant rate model with no covariates and no censoring and addressed two issues in connection with time-aggregation bias. First, Petersen (1991b) discussed the size of the bias when an estimator based on exact measurements of durations is applied to grouped measurements of durations. That discussion provided guidelines for when the problem is likely or unlikely to affect severely the conclusions reached. Two conclusions were reported. (1) For a given window (i.e., number of time units) within which the duration is known to lie, the higher the rate the higher the bias. (2) For a given rate, the wider the window within which the durations are known to lie, the higher the bias.

Second, Petersen (1991b) addressed how one can minimize the time aggregation bias when using an estimator based on the assumption of exact measurements of durations. It was shown, in the case of a constant rate, that the assigned duration ought to lie in the first half of the window within which the event happened. The higher the rate, the closer the assigned duration is to the beginning point of the window. The lower the rate, the closer the optimal duration is to the midpoint of the window within which the event happened. Petersen and Koput (1992) extended these results to a constant rate model with a single categorical covariate with several values and to the case of a fixed right-censoring scheme.

As a simple rule of thumb, one might therefore assign the duration that lies at the midpoint of the interval within which the event happened, a rule that has been used extensively in estimation routines based on the life table (see, e.g., Breslow and Crowley 1974; Cox and Oakes 1984, pp. 56–60).

13 Continuous- Versus Discrete-Time Models

It is important to distinguish between processes that are continuous in time and those that occur discretely in time. This chapter treats the former. In the latter type of process, events can only happen at specific prespecified times. Examples are presidential elections (once every four years) and graduating from major universities (at the end of each semester or term). As discussed in the previous section, measures of most continuous-time processes occur only at discrete time intervals, or duration measures are rounded up to, say, nearest integer. Nevertheless, the process itself is continuous in time. It just happens that our measures of it are not, a problem that may cause time-aggregation bias. If a process is continuous in time, it is clearly preferable to use continuous-time methods. If there is time aggregation in the duration measures, just adjust for these, as in equation (9.93). If the process is discrete in time, one clearly should use discrete-time methods.

Since discrete-time methods as applied to continuous-time processes nevertheless are popular in sociology (see, e.g., Allison 1982), a few comments are in order (see Petersen 1991b, Section 5). The main drawback of the discrete-time approach is that the coefficients estimated, say, from a logit or a probit model for each time interval, need not be entirely comparable to coefficients obtained from a continuous-time hazard-rate model. However, if the probability of an event in each time interval is small, then the coefficients obtained from the discrete-time specification for most models will be quite close to those obtained from the continuous-time specification. For example, the coefficients from a logit model will approximate the coefficients from a proportional hazards model (e.g., Weibull or Gompertz) quite well, as long as the probability of an event in each time interval is small. And only the constant term, which captures the rate for the baseline group, will depend strongly on the length of the time interval (see Arjas and Kangas 1988), whereas the other coefficients, which have the interpretations as relative risks, tend to be unaffected by changes in the time unit, as is also the case in a continuous-time proportional-hazards model [see, e.g., Section 4, (9.32)]. An often-used specification is the complementary log-log function available in *GLIM* (see Baker and Nelder 1978). It provides consistent estimates of the proportional-hazards parameters (e.g., a piecewise constant rate), regardless of the interval length or (equivalently) the size of the rate (see, e.g., Allison 1982, pp. 72–73; Arjas and Kangas 1988, p. 5; Prentice and Gloeckler 1978, p. 58). This property is not shared by the logistic model. If the objective is to obtain estimates of the parameters of an underlying continuous-time hazard-rate model, the complementary log-log function has the advantage of doing so (see the discussion in Allison 1982, pp. 72–73).

To see the point that only the constant term is severely affected by the length of the time interval, consider, without loss of generality, a single-state process. In the discrete-time framework, the probability of a transition in the next time interval, between, say, month $j-1$ and month j, may be specified by a logit model

$$\begin{aligned} P(j-1 < T \leq j \mid T > j-1, \boldsymbol{x}_{j-1}) &= 1/\{1+\exp[-\boldsymbol{\beta x}_{j-1} - g(j,\gamma)]\} \\ &\equiv p(j \mid \boldsymbol{x}_{j-1}), \end{aligned} \tag{9.94}$$

where $\boldsymbol{x}_{j-1}$ are the covariates evaluated at entry into period $j-1$ to j and $g(j,\gamma)$ is the duration dependence term and γ the coefficients pertaining to j.

When $p(j \mid x_{j-1})$ is small, we easily find the approximation

$$\exp[\boldsymbol{\beta x}_{j-1} + g(j,\gamma)] = p(j \mid \boldsymbol{x}_{j-1})/[1 - p(j \mid \boldsymbol{x}_{j-1})] \approx p(j \mid \boldsymbol{x}_{j-1}). \tag{9.95}$$

Hence, $\exp[\boldsymbol{\beta x}_{j-1}+g(j,\gamma)]$ roughly has the interpretation as the probability of a transition in the duration interval $j-1$ to j, when that probability is small. But this term has the form of a proportional-hazards model, say, a Weibull, Gompertz, or piecewise constant rate. It is well known that the instantaneous rate of transition from a continuous-time model approximates well the probability of a transition in the next time unit, say, week or month, when that transition probability is small (see Sections 3 and 4 above). Therefore, for small transition probabilities per time unit, the parameters from a discrete-time logit and a continuous-time proportional-hazards specification are close. So, in these cases, even when one uses an inappropriate discrete-time model, it enables us to recapture quite well the underlying continuous-time parameters.

14 Structural Models for Event Histories

Often, the researcher starts with a theoretical model for how the data were generated. The objective of the empirical analysis is then to use the data and the statistical techniques to estimate the parameters of the underlying theoretical model. In this section, I consider a simple example of such a situation, drawn from sociological research on intragenerational mobility (Sørensen 1979).

A person with socioeconomic status (or earnings) y_j receives job offers at a rate $\lambda(t \mid y_j)$, where t is the time elapsed since y_j was entered. The offer y comes from a distribution with density $g(y \mid t, y_j)$.

Assume that the worker maximizes socioeconomic status and that there are no costs of changing jobs (an assumption that is straightforward to relax). Assume also that all job changes are voluntary. If so, a person will accept any offer for which $y > y_j$, and otherwise reject. Let $D = 1$ denote that the offer is acceptable, that is, $y > y_j$, and $D = 0$ otherwise.

The rate $\lambda_1(t \mid y_j)$ at which an upward change in socioeconomic status occurs then equals

$$\hat{\lambda}_1(t \mid y_j) = \lambda(t \mid y_j) \times P[D = 1 \mid T = t, y_j]. \tag{9.96}$$

That is, the rate of an upward shift equals the overall rate at which offers arrive times the probability that the offer is acceptable. The researcher might be interested in recovering (i.e., estimating) both the overall rate and the distribution of the offers.

A specific example where the researcher can achieve both objectives may help clarify the ideas. Let the arrival rate of offers be given by

$$\lambda(t \mid y_j) = \exp(\beta_0 + \beta_1 y_j), \tag{9.97}$$

and let the density of an offer y, given an arrival at duration t, be

$$g(y \mid t, y_j) = \xi \times \exp(-\xi y), \tag{9.98}$$

where ξ is a parameter to be estimated.

The rate at which an upward shift occurs then equals

$$\begin{aligned} \lambda_1(t \mid y_j) &= \exp(\beta_0 + \beta_1 y_j) \times \exp(-\xi y_j) \\ &= \exp[\beta_0 + (\beta_1 - \xi) y_j], \end{aligned} \tag{9.99}$$

and the density of an accepted offer, y_{j+1}, at duration t_{j+1}, is

$$g(y_{j+1} \mid T = t_{j+1}, y_j, y_{j+1} > y_j) = \xi \cdot \exp[-\xi(y_{j+1} - y_j)]. \tag{9.100}$$

From data on the durations before an upward shift, it is clear that we can identify β_0 and a reduced form of β_1, namely $\rho \equiv \beta_1 - \xi$, but not the structural coefficient β_1, as is seen from (9.99).

However, from data on the values of socioeconomic status before and after an upward shift, we can identify ξ, as is seen from (9.100). And then, using the estimates of ξ and of the reduced form parameter ρ, we can identify the structural parameter β_1, from $\beta_1 \equiv \rho + \xi$.

The example shows that it may be useful to consider some underlying process that generates the data. The estimates obtained from the data will have interpretations relative to the underlying model. Specifically, in the example given, focusing only on the rate of upward shifts, that is, equation (9.99), will not enable one to recover the parameters of interest. The reduced form estimates mix the parameters of the rate at which offers arrive and the parameters of the distribution of the offers.

15 Sampling Plans

In many samples of event histories, the sampling plan itself gives rise to problems in the analyses. Suppose one samples only the last two events on each individual in a target sample. In principle one should then construct an estimator that adjusts for the fact that only portions of the life histories are observed (see Hoem 1985). Or, suppose one samples individuals who are in a certain state at calendar time τ_0, say, the state of being employed in a federal bureaucracy in January 1980. This is often called sampling from the stock, as opposed to sampling from the flow, where entrants into a state are sampled. Even when one knows the dates at which the people in the sample as of τ_0 entered that state, so there is no left censoring, this type of data creates problems. The standard estimators of the rate are no longer consistent. To obtain consistent estimates one must adjust for having sampled from the stock of those who survived in the state up until τ_0. If in addition, one does not know the time at which those in the stock as of τ_0 entered the state, the problem of left censoring remains. In that case the problem is twofold: first, the need to adjust for having sampled from the stock; second, the need to adjust for the left censoring of durations among those sampled from the stock.

In this section I treat the case where one samples from the stock but where there is no left censoring, drawing heavily on Heckman and Singer (1984b, sec. 8), Lancaster (1990, chap. 5), and Hamerle (1991). In the next section I extend the analysis to taking account of left censoring. The materials below are considerably more difficult than those of the preceeding sections. To some extent this is so because the issues themselves are more difficult. To some extent it is due to the lack of other easily accessible treatments of these topics. Although the present discussion is not easy to read, it does, unlike other discussions, detail more carefully the various steps taken in order to derive the relevant likelihoods. For the reader interested merely in the conclusions not in how they were reached, go to the end of Section 15.1 as well as to the end of the entire section. A shorter version, emphasizing the conclusions not the technical details, can be found in Petersen (1993).

In order to get started, some additional notation is needed. Let $Z_{\tau_0}=z_{\tau_0}$ denote the event that someone is in the stock, that is, in state z at calendar time τ_0. Let t_b denote the amount of time spent in the state at τ_0, where the subscript *b* denotes that this is the amount of time spent in the state *before* the sample was taken at τ_0. It is often referred to as the *backward recurrence time*. We may observe t_b or it may be left censored, in which case it is not observed. Further, let t_a denote the amount of time spent in the state after the sample was taken at τ_0, where the subscript *a* denotes that it is the amount of time that elapses in the state *after* τ_0. It is often referred to as the *forward recurrence time*. The duration t_a may be right censored or not. When it is right censored, C_a=0, otherwise, C_a=1. The total duration in the state is denoted $t_c = t_b + t_a$. If t_c is right censored, C_c=0, otherwise, C_c=1. Finally, let $e_{\tau_0 - t_b}$ denote the event that the focal state z was entered at calendar time $\tau_0 - t_b$.

To discuss estimators in the presence of stock sampling, three notions will frequently be used: the superpopulation, the population, and the sample. The superpopulation consists of everyone who either is in the focal state or who is at risk of entry into the focal state. The population consists of everyone who entered the focal state. The sample consists of those members of the population who survived in the state until τ_0, that is, of those who entered the state at τ_0 or some time prior to τ_0 and that were still in the state at τ_0.

The central objective of the statistical analysis is to estimate the population exit rate from state z at duration t

$$\lambda(t \mid \boldsymbol{x}, e_{\tau_0 - t_b}) \equiv \lim_{\Delta t \downarrow 0} P(t \leq T < t + \Delta t \mid T \geq t, \boldsymbol{x}, e_{\tau_0 - t_b})/\Delta t. \tag{9.101}$$

Note in (9.101) that I condition explicitly on the fact of entry into the focal state. The population consists of everyone who entered the state at some time. I consider only dependence on time-constant covariates.

In order to estimate the parameters of the population hazard rate we need a few quantities. First, we need the entry density into state z. The entry density often gets referred to as the entry rate (e.g., Hamerle 1991, p. 398), but that is misleading. Equation (9.102) below defines a density not a rate. Let $N(\tau_0 - t_b)$ be the number of entries into the state of interest from calendar time 0 to $\tau_0 - t_b$. The *entry density* is defined as

$$e(\tau_0 - t_b \mid \boldsymbol{x}) \equiv \lim_{\Delta t \downarrow 0} P[N(\tau_0 - t_b + \Delta t) - N(\tau_0 - t_b) \mid \boldsymbol{x}]/\Delta t. \tag{9.102}$$

Next, we shall need a subtle property of renewal processes, that is, of processes where individuals at risk of entry into or currently in the focal state may, after having left the state, reenter it at a later point in time. Let T_r be a random variable denoting the time spent from the time of entry into the focal state before it is reentered. It equals the amount of time spent in the focal state plus the amount of time spent in other states before the focal state is reentered. As shown in Cox (1962, chap. 7), for such processes it is then the case that

$$\lim_{\tau_0 \to \infty} e(\tau_0 - t_b) \mid \boldsymbol{x}) = 1/E(T_r \mid \boldsymbol{x}), \tag{9.103}$$

where the denominator on the right-hand side is the expected time that a case spends before reentering the focal state z, given $\boldsymbol{x}$. The idea here is that as the process has been running for a very long time, that is, is far away from its origin at calendar time 0, then the entry density becomes a constant.

Finally, we shall have use for the following expression for the expected time spent in a state

$$E(T \mid \boldsymbol{x}, e_{\tau 0-t_b}) = \int_0^{\infty} P(T \geq t \mid \boldsymbol{x}, e_{\tau 0-t_b}), \tag{9.104}$$

which says that the expected time in a state equals the integral of the survivor function from 0 to ∞. Equation (9.104) is a property that holds for all nonnegative random variables for which the expectation exists (see, e.g., Billingsley 1979, p. 239).

With these quantities in hand, I will, in what follows, derive the likelihood expressions first for t_a, given t_b; next for t_b; then for t_a and t_b; and finally for t_a, t_b and $\boldsymbol{x}$. In each of the likelihoods I condition on the fact that the observation survived in the focal state until τ_0. In the three first likelihoods I also condition on $\boldsymbol{x}$. Note that the only sample information available about these durations is conditional on survival at τ_0, therefore, the conditioning on survival to τ_0. However, our interest is in recovering the distribution of these quantities (i.e., the hazard rate) in the population that entered the focal state at some time, not only the distribution among the survivors. The objective, then, is to express the densities for the sample quantities exclusively in terms of the population densities, that is, the population hazard rate.

In the derivations below, we shall make use of the fact that the event of being in the sample at τ_0, namely $z_{\tau 0}$, is a subset of the event of having entered the sample at $\tau_0 - t_b$ and having survived in the state for t_b periods. I state this formally

$$z_{\tau 0} \subset (e_{\tau 0-t_b}, t_b) \quad \text{or} \quad (e_{\tau 0-t_b}, t_b) \Rightarrow z_{\tau 0}. \tag{9.105}$$

In a notation encompassing censored and noncensored observations, the population density, corresponding to the population hazard rate that we are interested in estimating, is

$$f(t_c \mid \boldsymbol{x}, e_{\tau 0-t_b}) = [\lambda(t_c \mid \boldsymbol{x}, e_{\tau 0-t_b})]^{C_c} \exp[-\int_0^{t_c} \lambda(s \mid \boldsymbol{x}, e_{\tau 0-t_b}) ds], \tag{9.106}$$

where

$$P(T \geq t_c \mid \boldsymbol{x}, e_{\tau 0-t_b}) = \exp[-\int_0^{t_c} \lambda(s \mid \boldsymbol{x}, e_{\tau 0-t_b}) ds]. \tag{9.107}$$

Note that

$$f(t_a \mid \boldsymbol{x}, z_{\tau 0}, t_b) = f(t_c \mid \boldsymbol{x}, z_{\tau 0}, t_b), \tag{9.108}$$

which will be used later. The equality follows because $t_c = t_b + t_a$.

Unfortunately, one cannot consistently estimate the rate in (9.101) using the density in (9.106), as in the procedures in the preceeding sections (see, e.g., Section 10), because one must first adjust for the fact that one's sample consists of those who survived in state z until τ_0. Intuitively, one will on the average observe more persons with low exit rates from z when one's sample consists of those who survived in that state until τ_0, since those with high exit rates are more likely to have left before τ_0. So, estimating the rate in (9.101) directly on the basis of t_b and t_a using (9.106) will lead to biased estimates. The estimates will be biased downwards.

We shall also have use for the density of t_c, conditional on survival at t_b, namely

$$f(t_c \mid \boldsymbol{x}, e_{\mathcal{T}0-t_b}, t_b) = [\lambda(t_c \mid \boldsymbol{x}, e_{\mathcal{T}0-t_b})]^{C_c} \exp[-\int_{t_b}^{t_c} \lambda(s \mid \boldsymbol{x}, e_{\mathcal{T}0-t_b}) ds], \tag{9.109}$$

where

$$\begin{aligned} P(T \geq t_c \mid \boldsymbol{x}, e_{\mathcal{T}0-t_b}, T \geq t_b) &\equiv P(T \geq t_c \mid \boldsymbol{x}, e_{\mathcal{T}0-t_b}, t_b) \\ &= \exp[-\int_{t_b}^{t_c} \lambda(s \mid \boldsymbol{x}, e_{\mathcal{T}0-t_b}) ds]. \end{aligned} \tag{9.110}$$

Finally note that

$$P(T \geq t_c \mid \boldsymbol{x}, e_{\mathcal{T}0-t_b}) = P(T \geq t_c \mid \boldsymbol{x}, e_{\mathcal{T}0-t_b}, t_b) \times P(T \geq t_b \mid \boldsymbol{x}, e_{\mathcal{T}0-t_b}), \tag{9.111}$$

by the chain-rule for probabilities (see eq. 9.5).

15.1 A Conditional Likelihood for t_a, given t_b

Lancaster (1979) developed an ingenious approach for dealing with the peculiarities of stock samples. He specified the density of t_a, given t_b (i.e., $T \geq t_b$), $\boldsymbol{x}$, and $z_{\mathcal{T}0}$. This is

$$\begin{aligned} f(t_a \mid \boldsymbol{x}, z_{\mathcal{T}0}, t_b) &= f(t_a \mid \boldsymbol{x}, e_{\mathcal{T}0-t_b}, t_b) \\ &= f(t_c \mid \boldsymbol{x}, e_{\mathcal{T}0-t_b}, t_b) \\ &= f(t_c \mid \boldsymbol{x}, e_{\mathcal{T}0-t_b}) / P(T \geq t_b \mid \boldsymbol{x}, e_{\mathcal{T}0-t_b}). \end{aligned} \tag{9.112}$$

The first equality in (9.112) follows from (9.105), because

$$(z_{\mathcal{T}0}, t_b) \Longleftrightarrow (e_{\mathcal{T}0}, t_b). \tag{9.113}$$

The second equality follows due to (9.108) and the third due to (9.111). As we see, the density $f(t_c \mid \boldsymbol{x}, z_{\mathcal{T}0}, t_b)$ for t_c in the sample has been expressed in terms of the population density $f(t_c \mid \boldsymbol{x}, e_{\mathcal{T}0-t_b}, t_b)$ for t_c, on the right-hand sides of (9.112). When one conditions on t_b, the two coincide.

In this approach, the only difference between the sample and population likelihoods is that in the survivor function in the sample the integral is computed from t_b to t_c, rather than from 0 to t_c. That is, one computes the survivor function conditional on survival to t_b.

The approach is very appealing. All one needs to specify is the population hazard and then to make the minor adjustment to the likelihood in the survivor function. This can be done with any standard hazard-rate program that handles time-dependent covariates. One specifies the likelihood contribution in the duration interval t_b to t_a, given survival at t_b.

There are two drawbacks to this likelihood. First, it treats t_b as a parameter rather than as a random variable whose distribution is to be estimated. This leads to some loss in efficiency of estimates. Second, it does not specify the distribution of the covariates $\boldsymbol{x}$.

As we shall see, this leads to some further inefficiency, because the marginal distribution of the covariates among those who survived to τ_0, in general is informative with respect to the parameters of the hazard rate. The approach in the next subsection removes the first inefficiency, while the approach in the final subsection removes both sources of inefficiency.

In spite of these two inefficiencies, Lancaster's (1979) approach is the one I recommend researchers use, for reasons explicated in the next two subsections.

15.2 Likelihood for t_b and Joint Likelihood for t_a and t_b

The second approach removes the inefficiency caused by treating t_b as a parameter rather than as the realization of a random variable. In removing this inefficiency, one specifies the joint distribution of t_b and t_a, given z_{τ_0} and $\boldsymbol{x}$. In this approach, t_b and t_a are treated as separate variables, as opposed to specifying the distribution of their sum, $t_c{=}t_b{+}t_a$.

First, decompose the joint distribution of t_b and t_a

$$\begin{aligned} f(t_b, t_a \mid \boldsymbol{x}, z_{\tau_0}) &= f(t_b \mid \boldsymbol{x}, z_{\tau_0}) \times f(t_a \mid \boldsymbol{x}, z_{\tau_0}, t_b) \\ &= f(t_b \mid \boldsymbol{x}, z_{\tau_0}) \times f(t_a \mid \boldsymbol{x}, e_{\tau_0 - t_b}). \end{aligned} \tag{9.114}$$

We have already derived the distribution of $f(t_a \mid \boldsymbol{x}, z_{\tau_0}, t_b)$. It is given by (9.112). It remains to derive the distribution of t_b.

The first term on the right-hand side of (9.114) can be decomposed as follows:

$$f(t_b, \mid \boldsymbol{x}, z_{\tau_0}) = f(t_b, z_{\tau_0} \mid \boldsymbol{x})/P(z_{\tau_0} \mid \boldsymbol{x}), \tag{9.115}$$

which gives the density function for the backward recurrence time in the sample.

Both terms on the right-hand side of (9.115) can be developed further. For the first term we can write

$$f(t_b, z_{\tau_0} \mid \boldsymbol{x}) = e(\tau_0 - t_b \mid \boldsymbol{x}) \times P(T \geq t_b \mid \boldsymbol{x}, e_{\tau_0 - t_b}). \tag{9.116}$$

This shows that the joint density of the backward recurrence time and the event that the person is in the state as of τ_0 equals the density of entering the state at calendar time $\tau_0 - t_b$ times the probability of surviving in the state for t_b periods, given entry at $\tau_0 - t_b$.

For the second term we can write

$$P(z_{\tau_0} \mid \boldsymbol{x}) = \int_0^\infty e(\tau_0 - t_b \mid \boldsymbol{x}) \times P(T \geq t_b \mid \boldsymbol{x}, e_{\tau_0 - t_b}) dt_b. \tag{9.117}$$

Here the backward recurrence time t_b is integrated out of the joint density for t_b and z_{τ_0} from (9.116).

Collecting terms, by first inserting (9.116)–(9.117) into (9.115) and then inserting (9.115) and (9.112) into (9.114), we now get

$$\begin{aligned} f(t_b, t_a \mid \boldsymbol{x}, z_{\tau_0}) &= f(t_b \mid \boldsymbol{x}, z_{\tau_0}) \times f(t_a \mid \boldsymbol{x}, z_{\tau_0}, t_b) \\ &= f(t_a \mid \boldsymbol{x}, e_{\tau_0 - t_b}) \times f(t_b, z_{\tau_0} \mid \boldsymbol{x})/P(z_{\tau_0} \mid \boldsymbol{x}) \\ &= \frac{e(\tau_0 - t_b \mid \boldsymbol{x}) \times f(t_c \mid \boldsymbol{x}, e_{\tau_0 - t_b})}{\int_0^\infty e(\tau_0 - t_b \mid \boldsymbol{x}) \times P(T \geq t_b \mid \boldsymbol{x}, e_{\tau_0 - t_b}) dt_b}, \end{aligned} \tag{9.118}$$

where $f(t_c \mid \boldsymbol{x}, e_{\tau_0 - t_b})$ is given in (9.106). The relationship in (9.111) is used to obtain $f(t_c \mid \boldsymbol{x}, e_{\tau_0 - t_b})$. This follows from (9.111) because $f(t_c \mid \boldsymbol{x}, e_{\tau_0 - t_b}) = f(t_c \mid \boldsymbol{x}, e_{\tau_0 - t_b}, t_b) \times P(T \geq t_b \mid \boldsymbol{x}, e_{\tau_0 - t_b})$.

The central feature of this density, be it right censored or not, is that it looks quite different from the population density $f(t_c \mid \boldsymbol{x}, e_{\tau_0 - t_b})$ in (9.106). In the latter, all spells are sampled from their beginnings instead of only sampling spells that survived until τ_0. The second term in the numerator of (9.118) coincides with the population density. The other terms in the density adjust for the sampling plan.

In general, (9.118) is hard to use in research. It is too complicated to compute. In one case it simplifies considerably, namely when

$$e(\tau \mid \boldsymbol{x}) = e(\boldsymbol{x}), \tag{9.119}$$

that is, when the entry density is time homogenous. There are two ways to justify a time-homogenous entry density. First and obviously, the entry density might be time homogenous over the entire time axis. This might occur, for example, when an organization hires a fixed number of new people each year or when a country admits a fixed number of immigrants each year. Second, it can be justified if the process has been running for a very long time, in which case the entry density becomes time homogenous, as stated in (9.103).

Under the assumption of a time-homogenous entry density, the density in (9.118) reduces to

$$f(t_b, t_a \mid \boldsymbol{x}, z_{\tau_0}) = \frac{f(t_c \mid \boldsymbol{x}, e_{\tau_0 - t_b})}{\int_0^\infty P(T \geq t_b \mid \boldsymbol{x}, e_{\tau_0 - t_b}) dt_b} = \frac{f(t_c \mid \boldsymbol{x}, e_{\tau_0 - t_b})}{E(T \mid \boldsymbol{x}, e_{\tau_0 - t_b})}. \tag{9.120}$$

Here, the denominator on the right-hand side of the second equality, the expected time spent in the focal state, follows from (9.104).

This density also differs from the standard population density had we sampled all spells from their beginnings. So, if we sample from the stock at τ_0, observe t_b and t_a, and the entry density is time homogenous, then the sample density of t_b and t_a equals the population density divided by the expected time spent in the state, provided the expectation exists.

The density in (9.118) is more efficient than the density in (9.112). However, it requires that one specifies the entry density $e(\tau \mid \boldsymbol{x})$. Furthermore, to make (9.118) feasible it requires the assumption of a time-homogenous entry density, $e(\tau \mid \boldsymbol{x}) = e(\boldsymbol{x})$. The central drawback of this procedure is the danger of misspecifying the entry density. From a practical viewpoint, the density in (9.118) or even its special case in (9.120), is computationally more cumbersome than the conditional density in (9.112). No standard software exists for computing estimates based on (9.118) or (9.120).

As a piece of advice, therefore, I propose that researchers use the density in (9.112). It is more robust and simpler computationally. It seems sensible to strive for consistency at some loss in efficiency, rather than trying to obtain both consistency and efficiency at the risk of sacrificing both due to potential misspecification of the entry density. Additionally, the density in (9.112) is computationally simple. All it requires is a standard hazard-rate program that handles time-dependent covariates.

An example of (9.120) is useful to consider. Let the rate be

$$\lambda(t \mid \boldsymbol{x}, e_{\tau_0 - t_b}) = \exp(\boldsymbol{\beta x}). \tag{9.121}$$

Then (9.120) reduces to

$$f(t_b, t_a \mid \boldsymbol{x}, z_{\tau_0}) = \lambda(t_c \mid \boldsymbol{x}, e_{\tau_0 - t_b}) \times f(t_c \mid \boldsymbol{x}, e_{\tau_0 - t_b}). \tag{9.122}$$

In the sample, therefore, with an exponential rate, censored observations contribute with the density of noncensored observations in the population. Noncensored observations contribute with the same density multiplied by the hazard rate evaluated at the time of failure, so that the density becomes the survivor function times the square of the hazard rate.

In (9.114) I specified the joint distribution of t_b and t_a, given $\boldsymbol{x}$ and survival to τ_0. An alternative is to specify the density for the sum t_c=t_b+t_a, given $\boldsymbol{x}$ and survival to τ_0. This can be done by noting that

$$f(t_b, t_c \mid \boldsymbol{x}, z_{\tau_0}) = f(t_b, t_a \mid \boldsymbol{x}, z_{\tau_0}), \tag{9.123}$$

as can be seen from (9.108). From this density we can integrate out t_b. The reason we need to integrate out t_b is that once we specify the density of the sum t_c=t_b+t_a, we do not use the information on t_b and t_a separately. Due to the decomposition of $f(t_b, t_a \mid \boldsymbol{x}, z_{\tau_0})$ in (9.114) we get:

$$\begin{aligned} f(t_c \mid \boldsymbol{x}, z_{\tau_0}) &= \int_0^{t_c} f(t_b, t_c \mid \boldsymbol{x}, z_{\tau_0}) dt_b \\ &= \int_0^{t_c} f(t_b \mid \boldsymbol{x}, z_{\tau_0}) \times f(t_c \mid \boldsymbol{x}, z_{\tau_0}, t_b) dt_b \end{aligned} \tag{9.124}$$

The integral goes from 0 to t_c because the backward recurrence time, t_b=$t_c - t_a$, must be equal to or less than the total time t_c, that is, t_b=$t_c - t_a \leq t_c$.

We have already computed the expressions for $f(t_b, t_c \mid \boldsymbol{x}, z_{\tau_0})$ within the integral, in (9.118). So, we only need to insert the right-hand side of (9.118) into the integrand, and then integrate the resulting expression from 0 to t_c, noting that the denominator in (9.118) is given by the left-hand side of (9.117), which is independent of t_b. Thus we get

$$\begin{aligned} f(t_c \mid \boldsymbol{x}, z_{\tau_0}) &= \int_0^{t_c} f(t_b, t_c \mid \boldsymbol{x}, z_{\tau_0}) dt_b \\ &= \int_0^{t_c} [e(\tau_0 - t_b \mid \boldsymbol{x}) \times f(t_c \mid \boldsymbol{x}, e_{\tau_0 - t_b})] dt_b / P(z_{\tau_0} \mid \boldsymbol{x}) \\ &= f(t_c \mid \boldsymbol{x}, e_{\tau_0 - t_b}) \times [\int_0^{t_c} e(\tau_0 - t_b \mid \boldsymbol{x}) dt_b] / P(z_{\tau_0} \mid \boldsymbol{x}). \end{aligned} \tag{9.125}$$

This is not an easy expression to work with in its general form. As we see, the density for t_c or (9.125) is quite different from the joint density for t_b and t_a in (9.118).

In the special case where the entry density is time homogenous, as in (9.121), we get

$$f(t_c \mid \boldsymbol{x}, z_{\tau 0}) = t_c f(t_c \mid \boldsymbol{x}, e_{\tau 0 - t_b})/E(T \mid \boldsymbol{x}, e_{\tau 0 - t_b}). \tag{9.126}$$

Again the density for t_c or (9.126) is quite different from the joint density for t_b and t_a in (9.120).

The literature appears to give no advice for choosing between the two formulations, specifying the density for t_b and t_a or specifying the density for their sum t_c. My advice is to use the first formulation. From a computational viewpoint both are equally cumbersome. The former is, I believe, more efficient because it uses more features of the data, both the pre- and post-sample durations t_b and t_a, whereas the latter uses only their sum t_c.

15.3 Full Likelihood in t_b, t_a, and x

The likelihoods in the two subsections above were all specified conditional on the covariates $\boldsymbol{x}$ and $z_{\tau 0}$. In now derive the full likelihood for t_b, t_a, and $\boldsymbol{x}$, given $z_{\tau 0}$. This likelihood reveals that the marginal distribution of the covariates $\boldsymbol{x}$ in the sample in general is informative with respect to the parameters of the hazard rate. That this is so can be seen from a simple argument. Specify the marginal distribution of the covariates $\boldsymbol{x}$ in the superpopulation, consisting of both those that entered the focal state as well as those at risk of entering the state, by

$$f(\boldsymbol{x}) \equiv \lim_{\Delta \boldsymbol{x} \downarrow 0} P(\boldsymbol{x} \leq \boldsymbol{X} < \boldsymbol{x} + \Delta \boldsymbol{x})/\Delta \boldsymbol{x}, \tag{9.127}$$

where $\boldsymbol{x}$ may be a vector, in which case $\Delta \boldsymbol{x} = (\Delta x_1, \ldots, \Delta x_r)'$ when there are r covariates and $\Delta x_j = 1$ if x_j is categorical.

Now, the marginal distribution of $\boldsymbol{x}$ among entrants into the state will in general differ from the marginal distribution $\boldsymbol{x}$ in the superpopulation. This will be so if the entry density into the state depends on the covariates $\boldsymbol{x}$. The marginal distribution of the covariates among entrants into the state obtains as

$$f(\boldsymbol{x} \mid e_{\tau 0 - t_b}) = f(\boldsymbol{x}) e(\tau_0 - t_b \mid \boldsymbol{x})/e(\tau_0 - t_b). \tag{9.128}$$

As we see, the marginal distribution of the covariates among entrants equals the marginal distribution in the superpopulation only if the entry density is independent of the covariates. Hence, the marginal distribution of $\boldsymbol{x}$ among the entrants is in general informative about the parameters of the entry density. So will also the marginal distribution of $\boldsymbol{x}$ in the sample of survivors to τ_0 be.

The joint distribution of t_b and t_a in (9.118) involves the entry density. As just discussed, the marginal distribution of $\boldsymbol{x}$ in the sample is informative with respect to the parameters of the entry density. It will therefore also be informative with respect to the parameters of the hazard rate, even if it does not carries information about the hazard-rate parameters directly. This is so because the marginal distribution of $\boldsymbol{x}$ in the sample carries information about the entry-density parameters, and the latter carries information about the hazard-rate parameters.

In what follows, I show that the marginal distribution of $\boldsymbol{x}$ in the sample in general is also directly informative with respect to the parameters of the hazard rate, not only with respect to the parameters of the entry density.

We need to specify the joint density for t_b, t_a, and $\boldsymbol{x}$. It can be written as

$$f(t_b, t_a, \boldsymbol{x} \mid z_{\tau 0}) = f(t_b, t_a \mid \boldsymbol{x}, z_{\tau 0}) \times f(\boldsymbol{x} \mid z_{\tau 0}). \tag{9.129}$$

We have already derived the distribution of $f(t_b, t_a \mid \boldsymbol{x}, z_{\tau 0})$ in (9.118). So we need to concentrate on the second term. Using Bayes' rule it can be rewritten as follows:

$$f(\boldsymbol{x} \mid z_{\tau 0}) = f(\boldsymbol{x}) \times P(z_{\tau 0} \mid \boldsymbol{x}) / P(z_{\tau 0}). \tag{9.130}$$

We already have derived the expressions for the two terms in the numerator, for $P(z_{\tau 0} \mid \boldsymbol{x})$ in (9.117) and for $f(\boldsymbol{x})$ from the definition in (9.127). From (9.130) we see that the sample distribution of $\boldsymbol{x}$ equals the population distribution only when

$$P(z_{\tau 0} \mid \boldsymbol{x}) = P(z_{\tau 0}) \,. \tag{9.131}$$

The equality in (9.131) occurs only when both the entry density $e(\tau_0 - t_b \mid \boldsymbol{x})$ and the hazard rate $\lambda(t \mid \boldsymbol{x}, e_{\tau 0 - t_b})$ are independent of $\boldsymbol{x}$, as can be seen from (9.117). That is, the covariates must have effects neither on the entry density nor on the hazard rate. Formally, one can see this from

$$\begin{aligned} P(z_{\tau 0}) &= \int_{D(\boldsymbol{x})} P(z_{\tau 0} \mid \boldsymbol{x}) \times f(\boldsymbol{x}) d\boldsymbol{x} \\ &= \int_{D(\boldsymbol{x})} \int_0^\infty e(\tau_0 - t_b) \mid \boldsymbol{x}) P(T \geq t_b \mid \boldsymbol{x}, e_{\tau 0 - t_b}) f(\boldsymbol{x}) dt_b d\boldsymbol{x}, \end{aligned} \tag{9.132}$$

where $D(\boldsymbol{x})$ denotes the domain of $\boldsymbol{x}$ and where (9.117) is used on the right-hand side of the second equality. As we see, the equality in (9.131) only obtains when both the entry density and the hazard rate are independent of $\boldsymbol{x}$.

To derive the expression for the joint density for t_b, t_a, and $\boldsymbol{x}$ in (9.129), start with (9.130). Insert first (9.117) and (9.132) into the right-hand side of (9.130). Then insert the resulting expression and the right-hand side of (9.118) into (9.129). We get the joint distribution of t_b, t_a, and $\boldsymbol{x}$, given $z_{\tau 0}$, as

$$\begin{aligned} f(t_b, t_a, \boldsymbol{x} \mid z_{\tau 0}) &= f(t_b, t_a \mid \boldsymbol{x}, z_{\tau 0}) \times f(\boldsymbol{x}) \times P(z_{\tau 0} \mid \boldsymbol{x}) / P(z_{\tau 0}) \\ &= \frac{f(\boldsymbol{x}) \times e(\tau_0 - t_b \mid \boldsymbol{x}) \times f(t_c \mid \boldsymbol{x}, e_{\tau 0 - t_b})}{\int_{D(\boldsymbol{x})} \int_0^\infty e(\tau_0 - t_b \mid \boldsymbol{x}) P(T \geq t_b \mid \boldsymbol{x}, e_{\tau 0 - t_b}) f(\boldsymbol{x}) dt_b d\boldsymbol{x}}. \end{aligned} \tag{9.133}$$

As the joint likelihood shows, if either the entry density or the hazard rate or both depend on the covariates $\boldsymbol{x}$, full maximum-likelihood estimation of the parameters of the hazard rate requires specifying both the entry density and the marginal distribution of the covariates in the superpopulation.

The likelihood in (9.133) is fully efficient. However, the efficiency comes at two costs. First, it is computationally complicated. Second, it need not be very robust. The reason is that one has to specify both the entry density and the distribution of the covariates in the superpopulation. Both are subject to possible misspecification which in turn usually will lead to inconsistent estimates of the parameters of the hazard rate. My advice, then, is the same as earlier: Use the first approach in (9.112). It is more robust and is computationally exceedingly simple.

A specific example might be useful. Let the entry density be time homogenous and independent of x; $e(\tau_0 - t_b \mid x) = c$. Let the hazard rate be

$$\lambda(t \mid x, e_{\tau_0 - t_b}) = \exp(\beta_0 + \beta_1 x), \tag{9.134}$$

where x is a dummy variable equal to 0 or 1. The marginal distribution of x in the superpopulation, and in this case also among entrants into the state, since $e(\tau_0 - t_b \mid x) = c$, is given by

$$f(x = 1) = p. \tag{9.135}$$

Then we get for (9.133)

$$f(t_b, t_a, x \mid z_{\tau_0}) = \frac{p \times [\exp(\beta_0 + \beta_1 x)]^{C_c} \exp[-t_c \exp(\beta_0 + \beta_1 x)]}{p/\exp(\beta_0 + \beta_1)] + (1 - p)/\exp(\beta_0)]}. \tag{9.136}$$

This density is quite different from the density conditional on x, which is given in (9.122) in the case of the rate in (9.134).

The marginal distribution of x, given survival at τ_0, is now

$$f(x \mid z_{\tau_0}) = \frac{p/\exp(\beta_0 + \beta_1 x)}{p/\exp(\beta_0 + \beta_1) + (1 - p)/\exp(\beta_0)}. \tag{9.137}$$

Only when $\beta_1 = 0$, when the covariate has no effect on the hazard rate, do we get

$$f(x \mid z_{\tau_0}) = f(x) = p, \tag{9.138}$$

in which case the sample distribution of the covariate coincides with the population and superpopulation distributions.

Much more can be said about this topic. For example, when the covariates depend on time, the issue becomes even more complicated [see, e.g., Heckman and Singer (1984b, sec. 8)].

Since this section has been quite technical, I reiterate, as the final remark, the conclusion that has been running through the entire section: In the presence of stock sampling, Lancaster's (1979) approach of (9.112) seems preferable to the two other approaches, partly because it is more robust and partly because it is computationally simple.

16 Left Censoring

For left-censored observations one knows that someone is in a state as of a given date τ_0, but one does not know the date prior to τ_0 that the state was entered, that is, one does not know the duration in the state as of τ_0. Thus, left-censoring usually arises when one samples from the stock of those who survived in a state up to τ_0 and one does not know the amount of time spent in the state prior to the sampling date. The problem is then twofold: First, the need to adjust for having sampled from the stock, which I treated in the previous section; second, the need to adjust for the left-censoring of durations among those sampled from the stock. In this section I extend the analysis to taking account of left censoring.

Using the notation from Section 14, I will focus on the likelihood for t_a, conditioning on x and z_{τ_0}, rather than on the joint likelihood for t_a and $\boldsymbol{x}$. The latter will be more efficient but less robust. The point of departure can be taken either in the approach in Section 15.1 or in Section 15.2. Both lead to the same results. The idea in both is to integrate out the unobservable t_b. From the approach in Section 15.1 we get

$$\begin{aligned} f(t_a \mid \boldsymbol{x}, z_{\tau_0}) &= \int_0^\infty f(t_b, t_a \mid \boldsymbol{x}, z_{\tau_0}) dt_b \\ &= \int_0^\infty f(t_a \mid \boldsymbol{x}, z_{\tau_0}, t_b) \times f(t_b \mid \boldsymbol{x}, z_{\tau_0}) dt_b. \end{aligned} \tag{9.139}$$

Each of the two terms within the integral (i.e., the integrands) have already been derived, in (9.112) and (9.115) respectively.

The rate depends on the fact of entry at calendar time $\tau_0 - t_b$ through $e_{\tau_0 - t_b}$. Assume now that the dependence on $e_{\tau_0 - t_b}$ only is through the amount of time elapsed since the state was entered, that is, only through the backward recurrence time, not on calendar time or historical circumstances prior to τ_0 per se. We can then write the rate as

$$\lambda(t \mid \boldsymbol{x}, e_{\tau_0 - t_b}) = \lambda(t \mid \boldsymbol{x}, e), \tag{9.140}$$

where the conditioning on e is to indicate that this is the population rate pertaining to all entrants into the state, not the sample rate pertaining to those that survived in the state until τ_0. In this case the density for t_c, covering right-censored and noncensored observations, is

$$f(t_c \mid \boldsymbol{x}, e_{\tau_0 - t_b}) = f(t_c \mid \boldsymbol{x}, e) = [\lambda(t_c \mid \boldsymbol{x}, e)]^{C_a} \times \exp[-\int_0^{t_c} \lambda(s \mid \boldsymbol{x}, e) ds], \tag{9.141}$$

where C_a=0 if t_a or equivalently t_c is right censored, otherwise, C_a=1.

If $\lambda(t \mid \boldsymbol{x}, e_{\tau_0 - t_b})$ also depended on $e_{\tau_0 - t_b}$, that is, on the historical circumstance that entry occurred at calendar time $\tau_0 - t_b$, we would need to integrate out the conditioning on this historical circumstance in addition to integrating out the backward recurrence time t_b.

Using (9.108), the expression for $f(t_b, t_a \mid \boldsymbol{x}, z_{\tau_0})$ from (9.118), and the equality in (9.140), the expression in (9.139) becomes

$$\begin{aligned} f(t_a \mid \boldsymbol{x}, z_{\tau_0}) &= \int_0^\infty f(t_b, t_a \mid \boldsymbol{x}, z_{\tau_0}) dt_b \\ &= \int_0^\infty [f(t_b + t_a \mid \boldsymbol{x}, e) \times e(\tau_0 - t_b \mid \boldsymbol{x})] dt_b / P(z_{\tau_0} \mid \boldsymbol{x}), \end{aligned} \tag{9.142}$$

where the expression for the denominator is given in (9.117). In its general form, the expression in (9.142) is quite unwieldy. The special case of a constant entry density, $e(\tau_0 - t_b \mid \boldsymbol{x}) = e(\boldsymbol{x})$, gives

$$f(t_a \mid \boldsymbol{x}, z_{\tau_0}) = P(T \geq t_a \mid \boldsymbol{x}, e) \times [E(T - t_a \mid \boldsymbol{x}, e, T \geq t_a)]^{1-C_a} / E(T \mid \boldsymbol{x}, e). \qquad (9.143)$$

A more complete discussion of (9.143) is given in Hamerle (1991, pp. 408–409). The numerator here follows because

$$\int_0^\infty f(t_b + t_a \mid \boldsymbol{x}, e) dt_b = P(T \geq t_a \mid \boldsymbol{x}, e) \times \int_0^\infty f(t_b + t_a \mid \boldsymbol{x}, e, T \geq t_a) dt_b, \qquad (9.144)$$

where

$$\int_0^\infty f(t_b + t_a \mid \boldsymbol{x}, e, T \geq t_a) dt_b = \begin{cases} 1 & \text{when } C_a = 1, \\ [E(T - t_a \mid \boldsymbol{x}, e, T \geq t_a)] & \text{when } C_a = 0, \end{cases} \qquad (9.145)$$

where the right-hand side of the second equation of (9.145) follows due to (9.104).

The expression in (9.143) shows that when the entry density is constant and covariates are time independent, the density of a noncensored forward recurrence time t_a is the same as the density of the backward recurrence time t_b. This can be seen from comparing the density for the forward recurrence time (9.143) and the density for the backward recurrence time (9.115), inserting (9.116)–(9.117) in (9.115), while utilizing the fact about expectations in (9.104).

If the distribution of T is exponential, as in (9.121), then (9.143) reduces to the population density for a complete duration equal to t_a. Hence, assuming an exponential distribution for T allows one to treat t_a as if it were the total duration in the state. Neither the sampling plan nor the left censoring creates a problem in that case.

Were one to use the full likelihood for t_a and $\boldsymbol{x}$, one would have to integrate out t_b from the expression in (9.133). This would obviously be a very difficult task in the general case.

17 Conclusion

I have presented an overview of some central themes in analyzing event histories. Several topics have been omitted, three of which, in my opinion, are important.

First, I have not discussed one important formulation found in the statistical literature, the counting-process framework (see, e.g., Andersen and Borgan 1985; Andersen et al. 1992; Fleming and Harrington 1991).

Second, I have not discussed nonparametric estimators of a single hazard-rate without covariates but with duration dependence (see, e.g., Wu 1989) or local hazard-rate models (see Wu and Tuma 1990). In the latter, the effect parameters of covariates may depend in rather arbitrary ways on the amount of time spent in a state.

Third, I have not discussed in detail problems that arise when the hazard rate depends on covariates that are endogenous to the dependent failure-time process. These issues were touched upon in Section 5. I have tried to develop them in more detail in Petersen (1992).

Appendix: Proof of Equation (9.14)

We shall prove the survivor function in equation (9.14):

$$\begin{aligned} P(T \geq t_e = t_k) &= \lim_{\substack{\Delta t \downarrow 0 \\ k \to \infty}} \prod_{j=0}^{k-1} [1 - \lambda(t_j)\Delta t] \\ &= \exp[-\int_0^{t_e} \lambda(s)ds], \qquad \text{where } t_e = t_k. \end{aligned} \tag{A1}$$

Before proceeding, two facts (Petersen 1990a, pp. 284–285) will be useful in the proof, (one)

$$1 - f(\tau_j + \Delta\tau)\Delta\tau \approx \exp[-f(\tau_j + \Delta\tau)\Delta\tau], \quad \text{i.e., } 1 - x \approx \exp(-x), \tag{A2}$$

when $f(\tau_j + \Delta\tau)\Delta\tau$ is positive and small; (two)

$$\lim_{\substack{\Delta t \downarrow 0 \\ k \to \infty}} \sum_{j=0}^{k-1} f(\tau_j + \Delta t)\Delta t \equiv \int_{\tau_0}^{\tau_e} f(\tau)d\tau, \tag{A3}$$

where by construction $\tau_k = \tau_e$ and where the left-hand side is nothing but the definition of a Riemann-Stieltjes integral (see, e.g., Kolmogorov and Fomin 1970, pp. 367–368). Here, τ_k is treated as fixed, set equal to τ_e, so that when k goes to infinity and Δt goes to zero, τ_k still stays fixed at τ_e.

Using fact (A2) on the right-hand side of the first equality in (A1) we get

$$1 - \lambda(t_j)\Delta t \approx \exp[-\lambda(t_j)\Delta t] \quad \text{for } \lambda(t_j)\Delta t \text{ positive and small.} \tag{A4}$$

Insert then the right-hand side of (A4) into each term on the right-hand side of the first equality in (A1). This yields

$$\begin{aligned} \lim_{\substack{\Delta t \downarrow 0 \\ k \to \infty}} \prod_{j=0}^{k-1} [1 - \lambda(t_j)\Delta t] &= \lim_{\substack{\Delta t \downarrow 0 \\ k \to \infty}} \prod_{j=0}^{k-1} \exp[-\lambda(t_j)\Delta t] \\ &= \exp[-\lim_{\substack{\Delta t \downarrow 0 \\ k \to \infty}} \sum_{j=0}^{k-1} \lambda(t_j)\Delta t] \\ &= \exp[-\int_0^{t_e} \lambda(s)ds], \qquad \text{with } t_e = t_k, \end{aligned} \tag{A5}$$

where the right-hand side of the first equality follows from (A4); the right-hand side of the second equality follows by rules for products and exponents; while the right-hand side of the last equality follows from the definition in (A3). The interchange of the limit and the exponentiation operator implicit on the right-hand side of the second equality follows because the exponentiation operator is continuous. The latter preserves convergence (see Buck, 1978, chap. 6). Thus, we have proved (9.14).

REFERENCES

Allison, P. D. (1982), "Discrete-time methods for the analysis of event histories," in *Sociological Methodology*, eds. S. Leinhardt, San Francisco: Jossey-Bass, pp. 61–98.

——(1984), *Event History Analysis. Regression for Longitudinal Event Data*, Beverly Hills: Sage.

Andersen, P. K., and Borgan, Ø. (1985), "Counting process models for life history data: A review (with discussion)," *Scandinavian Journal of Statistics*, 12, 97–158.

Andersen, P. K., and Gill, R. D. (1982), "Cox regression model for counting processes: A large sample study," *Annals of Statistics*, 10, 1100–1120.

Andersen, P. K., Borgan, Ø, Gill, R. D., and Keiding, N. (1992), *Statistical Models Based on Counting Process*. New York: Springer-Verlag.

Apostol, T. M. (1967), *Calculus Volume 1*. 2nd ed., New York: Wiley.

Arjas, E., and Kangas, P. (1988), "A Discrete Time Method for the Analysis of Event Histories," *Stockholm Research Reports in Demography No. 49*, Stockholm: University of Stockholm.

Baker, R. J., and Nelder, J. A. (1978), *The GLIM System*, Oxford: Numerical Algorithms Group.

Billingsley, P. (1979), *Probability and Measure*, New York: Wiley.

Blossfeld, H.-P., and Hamerle, A. (1989), "Using Cox models to study multiepisode processes," *Sociological Methods and Research*, 17, 432–448.

Blossfeld, H.-P., Hamerle, A., and Mayer, K. U. (1989), *Event History Analysis*, Hillsdale, NJ: Erlbaum.

BMDP (1990), *BMDP Statistical Software*, Berkeley, CA: University of California Press.

Breslow, N., and Crowley, J. (1974), "A Large Sample Study of the Life Table and Product Limit Estimates Under Random Censorship," *Annals of Statistics*, 2, 437–453.

Buck, R. C. (1978), *Advanced Calculus*, New York: McGraw-Hill.

Chamberlain, G. (1982), "The general equivalence of Granger and Sims causality," *Econometrica*, 50, 569–581.

——(1985), "Heterogeneity, omitted variable bias, and duration dependence," in *Longitudinal Analysis of Labor Market Data*, eds. J. J. Heckman, and B. Singer, New York: Cambridge University Press, pp. 3–38.

Cox, D. R. (1962), *Renewal Theory*, London: Chapman and Hall.

——(1975), "Partial likelihood," *Biometrika*, 62, 269–276.

Cox, D. R., and Oakes, D. (1984), *Survival Analysis*, London: Chapman and Hall.

DeGroot, M. H. (1970), *Optimal Statistical Decisions*, New York: McGraw-Hill.

——(1986), *Probability and Statistics*, Reading, MA: Addison-Wesley.

Dempster, A. P., Laird, N., and Rubin, D. B. (1977), "Maximum likelihood from incomplete data via the EM algorithm," *Journal of the Royal Statistical Society*, Ser. B, 39, 1–38.

Edlefsen, L. E., and Jones, S. D. (1988), *GAUSS Manual*, Kent, WA: Aptech Systems.

Efron, B. (1974), "Efficiency of Cox's likelihood function for censored data," *Journal of the American Statistical Association*, 53, 557–565.

Elandt-Johnson, R. C., and Johnson, N. L. (1980), *Survival Models and Data Analysis*, New York: Wiley.

Fleming, T. R., and Harrington, D. P. (1991), *Counting Processes and Survival Analysis*, New York: Wiley.

Flinn, C. J., and Heckman, J. J. (1983), "Are unemployment and out of the labor force behaviorally distinct labor force states," *Journal of Labor Economics*, 1, 28–42.

Greene, W. H. (1991), *LIMDEP. User's Manual and Reference Guide*, Version 6.0, New York: Econometric Software.

Hamerle, A. (1991), "On the treatment of interrupted spells and initial conditions in event history analysis," *Sociological Methods and Research*, 19, 388–414.

Hannan, M. T., and Freeman, J. (1988), *Organizational Ecology*, Cambridge, MA: Harvard University Press.

Heckman, J. J., and Singer, B. (1984*a*), "A method for minimizing the impact of distributional assumptions in econometric models for duration data," *Econometrica*, 52, 271–320.

———(1984*b*), "Econometric duration analysis," *Journal of Econometrics*, 24, 63–132.

Heijtan, D. F. (1989), "Inference from Grouped Continuous Data: A Review," *Statistical Science*, 4, 164–183.

Hernes, G. (1972), "The process of entry into first marriage," *American Sociological Review*, 37, 173–182.

Hoem, J. M. (1985), "Weighting, misclassification, and other issues in the analysis of survey samples of life histories," in *Longitudinal Analysis of Labor Market Data*, eds. J. J. Heckman, and B. Singer, New York: Cambridge University Press, pp. 249–283.

———(1989), "Limitations of a Heterogeneity Technique: Selectivity Issues in Conjugal Union Disruption at Parity Zero in Contemporary Sweden," in *Convergent Issues in Genetics and Demography*, eds. J. Adams, O. A. Lam, A. Hermalin, and P. C. Smouse, Oxford: Oxford University Press, pp. 133–153.

Kalbfleisch, J. D., and Prentice, R. L. (1980), *The Statistical Analysis of Failure Time Data*, New York: Wiley.

Kolmogorov, A. N., and Fomin, S. V. (1970), *Introductory Real Analysis*, New York: Dover.

Lancaster, T. (1979), "Econometric methods for the duration of unemployment," *Econometrica*, 47, 939–956.

———(1990), *The Econometric Analysis of Transition Data*, New York: Cambridge University Press.

McFadden, D. L. (1984), "Econometric Analysis of Qualitative Response Models." Chap. 22 (pp. 1395–1457) in Z. Griliches and M. D. Intriligator (eds.), *Handbook of Econometrics, Volume 2* New York: Cambridge University Press.

Massey, D. S. (1987), "Understanding Mexican Migration to the United States," *American Journal Sociology*, 92, 1372–1403.

Morgan, S. P., and Rindfuss, R. R. (1985), "Marital disruption: Structural and temporal dimensions," *American Journal of Sociology*, 90, 1055–1077.

Petersen, T. (1986a), "Fitting parametric survival models with time-dependent covariates," *Journal of the Royal Statistical Society*, Ser. C, 35, 281–288.

———(1986b), "Estimating fully parametric hazard rate models with time-dependent covariates. Use of Maximum Likelihood," *Sociological Methods and Research*, 14, 219–246.

———(1988), "Analyzing change over time in a continuous dependent variable: Specification and estimation of continuous state space hazard rate models," in *Sociological Methodology*, ed. C. C. Clogg, Washington, DC: American Sociological Association, pp. 137–164.

———(1990a), "Analyzing event histories," in *New Statistical Methods in Longitudinal Research Volume 2*, ed. A. von Eye, Orlando, FL: Academic Press, pp. 258–288.

———(1990b), "Analyzing continuous state space failure time processes: Two further results," *Journal of Mathematical Sociology*, 10, 247–256.

———(1991a), "The Statistical Analysis of Event Histories," *Sociological Methods and Research*, 19, 270–323.

———(1991b), "Time-aggregation bias in continuous-time hazard-rate models," in *Sociological Methodology Volume 21*, ed. P. V. Marsden, Cambridge, MA: Basil Blackwell, pp. 263–290.

———(1992), "Simultaneous equations models for analysis of event-history data," Unpublished manuscript, Walter A. Haas School of Business, University of California, Berkeley.

———(1993), "Recent Advances in Longitudinal Methodology," in *Annual Review of Sociology*, 19, pp. 425–454, Palo Alto, CA: Annual Reviews.

Petersen, T., and Koput, K. W. (1991), "Dependence in Organizational Mortality Processes: Legitimacy or Unobserved Heterogeneity?," *American Sociological Review*, 56 (3), 399–409.

———(1992), "Time-aggregation bias in hazard-rate models with covariates," *Sociological Methods and Research*, 21(1), 25–51.

Petersen, T., and S. Spilerman (1990), "Job-quits from an internal labor market," in *Applications of Event History Analysis in Life Course Research*, Chap. 4, eds. K. U. Mayer, and N. B. Tuma, Madison: University of Wisconsin Press.

Petersen, T., Spilerman, S., and Dahl, S.–Å. (1989), "The structure of employment termination among clerical employees in a large bureaucracy," *Acta Sociologica*, 32, 319–338.

Prentice, R. L., and Gloeckler, L. A. (1978), "Regression Analysis of Grouped Survival Data With Application to Breast Cancer Data," *Biometrics*, 34, 57–67.

Ridders, G. (1984), "The distribution of single–spell duration data," in *Studies in Labor Market Dynamics*, eds. G. R. Neumann, and N. C. Westergard-Nielsen, New York: Springer-Verlag, pp. 45–73.

Rogoff Ramsøy, N. (1977), *Sosial Mobilitet i Norge* (Social Mobility in Norway), Oslo: Tiden Norsk Forlag.

SAS (1991), *SAS/STAT User's Guide, Release 6.04*, Cary, NC: SAS Institute.

Self, S. G., and Prentice, R. L. (1982), "Commentary on Andersen and Gill's 'Cox regression model for counting processes: A large sample study'," *Annals of Statistics*, 10, 1121–1124.

Sims, C. A. (1972), "Money, income, and causality," *American Economic Review*, 62, 540–552.

Skrede, K. (1971), *Sosioøkonomisk Klassifisering av Yrker i Norge, 1960* (Socioeconomic classification of Occupations in Norway, 1960), Report 71–1, Oslo: Institute of Applied Social Research.

Sørensen, A. B. (1974), "A Model for Occupational Careers," *American Journal of Sociology*, 80, 44–57.

———(1979), "A model and a metric for the analysis of the intragenerational status attainment process," *American Journal of Sociology*, 85, 361–384.

———(1984), "Interpreting Time Dependency in Career Processes," in *Stochastic Modelling of Social Processes*, eds. A. Diekman, and P. Mitter, New York: Academic Press, pp. 89–122.

Sørensen, A. B., and Tuma, N. B. (1981), "Labor Market Structure and Job Mobility," in *Research in Social Stratification and Mobility*, eds. D. Treiman, and R. V. Robinson, Vol. 1, Greenwich, CT: JAI Press, pp. 69–81.

Spilerman, S. (1972), "Extensions of the mover-stayer model," *American Journal of Sociology*, 78, 599–627.

Trussel, J., and Richards, T. (1985), "Correcting for unmeasured heterogeneity using the Heckman-Singer procedure," in *Sociological Methodology*, ed. N. B. Tuma, San Francisco: Jossey-Bass, pp. 242–276.

Tuma, N. B. (1976), "Rewards, resources and the rate of mobility: a nonstationary multivariate stochastic model," *American Sociological Review*, 41, 338–360.

Tuma, N. B., and Hannan, M. T. (1984), *Social Dynamics. Models and Methods*, Orlando, FL: Academic Press.

Vaupel, J. W., Manton, K. G., and Stallard, E. (1979), "The impact of heterogeneity in individual frailty on the dynamics of mortality," *Demography*, 16, 439–454.

Visher, M. G. (1984), "The Workers of the State and the State of State Workers: A Comparison of Public and Private Employment in Norway," Ph.D. dissertation, Department of Sociology, University of Wisconsin, Madison.

Wu, L. L. (1989), "Issues in Smoothing Empirical Hazard Rates," in *Sociological Methodology*, ed. C. C. Clogg, Washington, DC: American Sociological Association, pp. 127–159.

Wu, L. L., and Tuma, N. B. (1990), "Local Hazard Models," in *Sociological Methodology*, ed. C. C. Clogg, Washington, DC: American Sociological Association, pp. 141–180.

Yamaguchi, K. (1986), "Alternative approaches to unobserved heterogeneity in the analysis of repeatable events," in *Sociological Methodology*, ed. N. B. Tuma, San Francisco: Jossey-Bass, pp. 213–249.

Chapter 10

Random Coefficient Models

NICHOLAS T. LONGFORD

1 Introduction

Quantitative social research relies heavily on data that originate either from surveys with sampling designs that depart from simple random sampling, or from observational studies with no formal sampling design. Simple random sampling is often not feasible, or its use would yield data with less information about certain features of interest, and it is often economically prohibitive. For example, in studies of school effectiveness it may be difficult to secure the cooperation of a school or a classroom. Therefore it would be rather wasteful to collect data from a small number of students in such a classroom. Data from a larger proportion, or from all the students, could be collected at a small additional expense, thus reducing the number of classrooms required for a sample to contain sufficient information for the intended purposes. Similarly, in household surveys, having contacted a selected individual, it would make sense to collect data from the rest of the members of the household at the same time. When this is done, we usually end up with data for which the standard assumptions of independence (such as in ordinary regression) are inappropriate.

Individuals within a cluster, e.g., members of a family, may share the same characteristics or the extent to which they do so may be of interest. In various social, political, and economic systems, such as education, government, or corporate business, we can recognize hierarchical (nested) structures, such as district, school, classroom, and student, or central, regional, and local government, and citizen. These structures are used as natural clusters in clustered sampling designs, and often the purpose of the study is inference about the differences among these clusters. When there are a large number of clusters, pairwise

---

NICHOLAS T. LONGFORD • Educational Testing Service, Princeton, New Jersey 08541, USA. • I wish to acknowledge M. Sobel's and G. Arminger's thorough reviews of earlier versions of the chapter. Most of the work on the chapter was conducted while I was a visiting associate professor in the Department of Mathematics, University of California, Los Angeles, and a visiting lecturer at the Institute for Electronic Systems, Aalborg University, Aalborg, Denmark.

Handbook of Statistical Modeling for the Social and Behavioral Sciences, edited by Gerhard Arminger, Clifford C. Clogg, and Michael E. Sobel. Plenum Press, New York, 1995.

comparisons of the clusters are much less informative than a suitable "global" measure for between-cluster differences. Pairwise comparisons are even more problematic when the clusters contain small numbers of observations.

Consider the problem of estimating the mean of a normally distributed random variable in a clustered sample of N observations from M $(< N)$ sampled clusters. Within-cluster homogeneity of the observations implies a correlation of any two observations in the same cluster. The observations can be represented by the simple variance-component model

$$y_{ij} = \mu + \delta_j + \varepsilon_{ij}\,, \tag{10.1}$$

where y_{ij} is the observed value of the variable y for the ith subject in the jth cluster, and $\{\delta_j\}_j$ and $\{\varepsilon_{ij}\}_{i,j}$ are mutually independent random samples from centered normal distributions with respective variances σ_2^2 and σ_1^2. The within-cluster correlation is

$$\rho = \operatorname{corr}(y_{ij}, y_{i'j}) = \frac{\sigma_2^2}{\sigma_1^2 + \sigma_2^2}\,.$$

If $\rho = 1$, then $\sigma_1^2 = 0$ and all observations from a cluster are identical. Any estimator of the mean would be based, essentially, only on M observations, one from each cluster. If $\rho = 0$, or equivalently, $\sigma_2^2 = 0$, then the observations are independent, and the arithmetic average of the N observations is a suitable estimator of the population mean. For intermediate values of the correlation ρ it appears that within a cluster each observation after the first one partially replicates some information about the mean μ contained in the first observation; therefore N observations from M clusters $(M < N)$ contain less information about the mean than N observations from N clusters (one from each cluster).

More specifically, the variance of the sample mean $\bar{y} = \sum_j \sum_i y_{ij}/N$ is equal to

$$\begin{aligned}\operatorname{var}(\bar{y}) &= \sum_j \frac{n_j^2\sigma_2^2}{N^2} + \frac{\sigma_1^2}{N}\\ &= \left\{1 + \rho\sum_j \frac{n_j(n_j-1)}{N}\right\}\frac{\operatorname{var}(y)}{N}\,,\end{aligned}$$

where n_j is the size of cluster j. Thus, for a given nesting design the variance of the estimator of the population mean $\bar{y}$ is an increasing function of the within-cluster correlation.

Note, however, that the sampling design with N observations from N clusters contains no information about the correlation ρ. In practice, when selecting a clustered sampling design it is often necessary to compromise between the conflicting goals of optimizing information about the population mean and the within-cluster correlation. Given a total sample size N, information about the mean is optimized by selecting not more than one observation from a cluster, but information about the correlation is optimized by sampling a certain number of subjects from each selected cluster. The optimal cluster sample size for the correlation depends on the (unknown) correlation.

Information about the within-cluster correlation can substantially enhance estimation of the within-cluster means. The arithmetic average $\bar{y}_j = \sum_i y_{ij}/n_j$ is the minimum variance unbiased estimator of the mean for cluster j (based solely on data from the cluster). Its (conditional) variance is equal to σ_1^2/n_j. In the extreme case of no within-cluster correlation $\sigma_2^2 = 0$, and $\bar{y} = \sum_j \sum_i y_{ij}/N$ is a much more efficient estimator for the mean in every cluster j because its mean squared error is σ_1^2/N.

In general, when ρ is known, the inefficient estimator $\bar{y}_j$ and the estimator of the population mean $\bar{y}$, biased for the mean of each cluster j, can be combined to produce the *shrinkage* estimator of the cluster mean,

$$\bar{y}_{s,j} = \frac{n_j \rho}{1 + (n_j - 1)\rho} \bar{y}_j + \frac{1 - \rho}{1 + (n_j - 1)\rho} \bar{y}\,.$$

The mean squared error of this estimator is approximately equal to $1/(\sigma_2^{-2} + n_j \sigma_1^{-2})$, which is smaller than both σ_1^2/n_j and σ_2^2.

This chapter takes up random coefficient models. These can be described as various generalizations of the simple variance component model (10.1). First, we consider adaptations of the model (10.1) that incorporate linear regression with a single variable (Section 2) and several variables (Section 3). Section 4 deals with maximum likelihood estimation for these models. Next, in Section 5 we discuss models for data with multiple layers of nesting (e.g., individuals in groups within areas, etc.). Third, in Section 6 we allow for departures from the assumption of normality featured in Sections 1–5. Fourth, Section 7 extends factor analysis and measurement error models to the case where observations are clustered.

Section 8 contains an example about wage inflation in engineering firms in Britain. Each of a large number of companies has provided very modest information (12 records on average) about the wages of their employees. The main issue is comparison of wage increases across the firms. If analysis of covariance were applied a large number of estimates, most of them with very large standard errors, would be obtained, and they would have to be subjected to a secondary analysis to generate a meaningful summary of the data. The random coefficient approach is superior to the analysis of covariance because it provides a flexible framework for description of the differences among the companies' wage trends by parsimonious models. Software for fitting random coefficient models is reviewed in Section 9.

Random coefficient models have a long history. The astronomer Airy made a sequence of telescopic observations of a phenomenon over several nights, and, to control the measurement error, he made several observations each night. In Airy (1861) he considers the within- and between-night sources of variation, and the importance of their separation. Galton identified the problem of variance component estimation when he considered the variability of brothers' statures within families. Stigler (1986) gives a detailed account of his approach.

Random coefficient models have also been considered in agricultural and animal-breeding applications (Crump 1951; Eisenhart 1947). Henderson (1953) discussed an application to artificial insemination: An animal breeding cooperative has approximately 60 bulls in service, and it is believed that the allocation of the bulls' semen to the cows in the cooperative is random. The focus of the study is on milk production of the daughters sired by each bull. The records span several years and contain information about several thousand daughters from about 2000 herds and 100 sires. Any form of the classical analysis of variance would be meaningless due to the large number of categories of several factors, and because most cells in the multiway table associated with the factors are empty (most herd-by-sire combinations are not realized in the data). Disentagling the various sources of variation (herd, bull, sire, etc.) is important for the design of efficient testing and selection programs. An important feature of the study is that selection of the most (least) "productive" bulls among those included in the study is of lesser importance than the global assessment of variability in the genetic characteristics of the bulls.

The similarity of the issues above to the inferential problems encountered in social sciences has been realized and exploited only recently. Methodological impetus in quantitative educational research was provided by the focus on within-classroom and within-school context in educational research by Cronbach and Webb (1975), Burstein, Linn, and Capell (1978), Firebaugh (1978), and Burstein (1980).

Developments in statistical applications were set in motion by the seminal paper of Dempster, Laird, and Rubin (1977) on the Expectation Maximization algorithm. Rubin, Laird, and Tsutakawa (1981) and Aitkin, Anderson, and Hinde (1981) presented the first applications of the EM algorithm in educational research, and they were followed by Raudenbush and Bryk (1986), Aitkin and Longford (1986), Goldstein (1987), and others. For a comprehensive review, see Raudenbush (1988). The EM methodology has provided a new computational and conceptual approach to estimation with random coefficient models, in addition to already established direct maximum likelihood methods. For a review of the latter, see Searle (1971) and Harville (1977).

1.1 An Illustration

The purpose of this illustration is to highlight the problematic nature of formal statistical inference from observational studies. We will also refer to this example subsequently to motivate some of the models introduced.

One focus of school effectiveness research is on assessing the importance of selection of the school or classroom for a student. Say, Jim attends school A, and his grades are G. But his parents may conjecture that if he attended school A', his grades would be higher. Such a judgment might be based on the comparison of the students with characteristics and attributes similar to Jim's, who attend the school A'.

It is important to distinguish two modes of inference. First, inference about Jim's grades in a school other than A would be impossible even in a rigorous experimental design. What we need are two identical "copies" of Jim, one enrolled in school A, the other one in school A'. This is a standard problem in design of experiments, e.g., when comparing a set of treatments. See Holland (1986) for a theoretical framework and a detailed discussion.

The second mode of inference is concerned with the within-school averages of the grades. In an experiment, a pool of students would be randomly allocated to the selected schools, the school means could be estimated straightforwardly, and they could be the basis for comparing the schools. In an observational study, students are "allocated" to schools by a selection process which renders such a comparison problematic. A school may rank high because it attracts high-ability students (from its catchment area, by its own efforts, or similar), not because it provides a high-quality educational "treatment".

Random allocation of students to schools could ensure an equitable comparison of the schools, but in most contexts that is not a realistic proposition. Even if random allocation were arranged, any "school-effect" comparisons would still be contentious because the school does not represent the sole "treatment" in this fictitious experiment; the effects of parental care, neighborhoood, and the like, cannot be controlled for.

Attempts to take account of selection bias have led educational researchers to application of regression methods. In a general context we consider for each school a hypothetical formula relating the educational performance (grades) to the background characteristics of the students. Differences among the schools would then be adequately described by the differences in these formulas and a suitable summary of the educational system (the population of schools) would be provided by the (multivariate) distribution of these formulas. Such formulas are usually linear regressions, and so the distribution of their coefficients would provide a suitable description.

In view of unexplored selection processes that, say, result in Jim living in a neighborhood near school A, in him being enrolled in school A, and so on, the usual comparison of "effectiveness" of the schools is contentious, even if students' grades in the two schools were comparable (i.e., if the grades had a common scale), and if the regression formulas for each school were known exactly.

1.2 Clustered Design

The simplest nontrivial data structure with hierarchies is that of a sample of groups with a sample of individuals within each sampled group. We emphasize at the outset that random coefficient models, as applied in the context considered here, rely on the following experimental design-type assumptions:

A. Groups are a simple random sample from a population of groups.

B. Individuals are a simple random sample from a population of individuals.

C. Individuals are allocated to groups by a random procedure.

Assumptions A and B are standard for a clustered design and assumption C is essential for equitable comparisons among the clusters.

Assumption A is that of *exchangeability*, or "anonymity" of the groups, and it is motivated by the absence of any a priori information about unique features of each group. This assumption is usually relaxed to that of conditional exchangeability, given some explanatory variables; we assume that after taking account of these explanatory variables, the groups are exchangeable. It is convenient to adopt the assumption of exchangeability when the sample studied consists of a large number of groups, and the analyst's interest is on no particular group but on the differences, or *variation*, among the groups in general.

In the experimental-design terminology we associate each group with an 'effect' (the result of a *treatment*), and differences among these effects represent the between-group differences.

If there are only a small number of groups and their identification is essential (say parents would contemplate enrolling Jim in only one of the three schools in their town) then specific pairwise differences among the group effects may be of interest. The classical analysis of covariance is often an appropriate method for analysis in this context. If the assumption of exchangeability is adopted, analysis of covariance remains applicable, but the lack of any parsimonious description for the between-group differences becomes a drawback. Random coefficient models are better suited for such a situation because they have a more flexible parametric structure for the between-group variation.

Assumptions A–C are often not satisfied, or they cannot be formally tested in many contexts encountered in social science research, especially when the data at hand do not contain any information about the processes of allocation of individuals into groups, and of individuals and groups being selected into (or omitted from) the study. Assumptions A and B can be replaced by the assumptions that the groups are a *representative* sample from the population of groups, and the individuals sampled within a group are a representative sample from that group. Inferences that fail to adjust for violations of these assumptions are subject to inaccuracies. The size of these is difficult to evaluate without information about the form and magnitude of these violations. An informal assessment of the biases due to departure from assumptions A–C should accompany every study that does not employ a formal experimental design.

2 Models With a Single Explanatory Variable

To illustrate a simple random coefficient model, we consider a skeletal version of the school-effectiveness problem of Section 1.1. For the moment, we ignore issues related to the choice of the explanatory variables, those related to the relationship of the variables with the characteristics they represent, the issues of finiteness of the respective populations of the individuals, as well as issues related to experimental design.

For each group, $j = 1, 2, \ldots, M$, we consider a simple regression with group-specific intercept a_j and slope b_j:

$$y_{ij} = a_j + x_{ij} b_j + \varepsilon_{ij}, \tag{10.2}$$

where x_{ij} is the value of the univariate background characteristic x for the ith element of the jth group, and ε_{ij} is a random term (error), assumed to be a random draw from $\mathcal{N}(0, \sigma^2)$. For the *within-group coefficients* we adopt the following distributional assumption:

$$(a_j, b_j) \sim \mathcal{N}\{(\alpha, \beta), \Sigma\} \qquad \text{(iid)}, \tag{10.3}$$

that is, the coefficients are a random sample from a bivariate normal distribution. We will use the familiar subscript notation for the elements of the covariance matrix Σ, that is, Σ_{kh} will stand for the element of Σ in row k and column h.

The parameters (α, β) represent the average intercept and slope, or the intercept and slope for an average group, while $\boldsymbol{\Sigma}$ is a measure of the between-group variation. The model described is essentially an analysis-of-covariance model (ANACOVA) with the additional stochastic assumption (10.3). This model is sometimes referred to as seemingly unrelated regressions (Zellner 1962), or as mixed model of analysis of (co-)variance (Hartley and Rao 1967). The former terminology is partially motivated by (10.2), which on its own suggests that the coefficients a_j and b_j should be estimated by applying ordinary least squares (OLS) to the observations from group j only.

The OLS estimators for the vectors $\boldsymbol{\beta}_j = (a_j, b_j)$ are unbiased, and among the unbiased estimators have minimum variances. But this estimator for each vector $\boldsymbol{\beta}_j$ is based only on the observations from group j, which may constitute only a small proportion of the data. Thus we have a large number of estimated pairs of coefficients, $(\hat{a}_j, \hat{b}_j)$, each subject to large sampling variation. If the covariance matrix $\boldsymbol{\Sigma}$ were available these estimators could be improved upon. For example, in the extreme case of no between-group variation, $\boldsymbol{\Sigma} = \mathbf{0}$, the OLS estimator for the entire set of observations (all groups) is unbiased for each pair of coefficients $\boldsymbol{\beta}_j$, and this estimator is superior to the OLS estimator based solely on the data from group j. Thus, it seems that when $\boldsymbol{\Sigma}$ is small the ANACOVA estimator for $\boldsymbol{\beta}_j$ should be improved by combining it with the OLS estimator. On the other hand, when the covariance matrix $\boldsymbol{\Sigma}$ is very large, estimation of the coefficients $\boldsymbol{\beta}_j$ should be based solely on the data from group j. It turns out that, in general, a suitable estimator for the within-group coefficients is constructed by a compromise (a matrix weighted average) of the two estimators for the extreme values of $\boldsymbol{\Sigma}$:

$$\hat{\boldsymbol{\beta}}_j = \boldsymbol{G}_j^{-1}\hat{\boldsymbol{\beta}}^* + (\boldsymbol{I} - \boldsymbol{G}_j^{-1})\hat{\boldsymbol{\beta}}_j^{(OLS)} , \tag{10.4}$$

where $\boldsymbol{G}_j$ is a 2×2 weight matrix (a function of σ^2 and $\boldsymbol{\Sigma}$), $\hat{\boldsymbol{\beta}}^*$ is a pooled estimator for the mean regression parameter $\boldsymbol{\beta}$, and $\hat{\boldsymbol{\beta}}_j^{(OLS)}$ is the OLS estimator for the coefficients $\boldsymbol{\beta}_j$ based on the data from group j. Full details (in a more general context) are given in Section 4.

As we saw in Section 1, estimation of the within-group coefficients is affected by between-group variation. It would appear that a crucial issue is whether to treat the coefficients (a_j, b_j) as parameters, as in the classical ANACOVA, or as random variables (vectors). Certainly, if the groups can be regarded as a sample from a population of groups, the stochastic assumption (10.3) is a natural one. In classical ANACOVA the (*unbiased*) OLS estimator is routinely used. Random coefficient models lead to the (*biased*) estimator (10.4). Lindley and Smith (1972) present arguments for biased estimation, of the form (10.4) in the case of ANACOVA, irrespective of the assumptions about the coefficients (a_j, b_j). They argue that in situations with large numbers of parameters certain biased estimators are preferable because they have smaller mean-square errors than their (uniformly minimum variance) unbiased counterparts.

2.1 Patterns of Variation

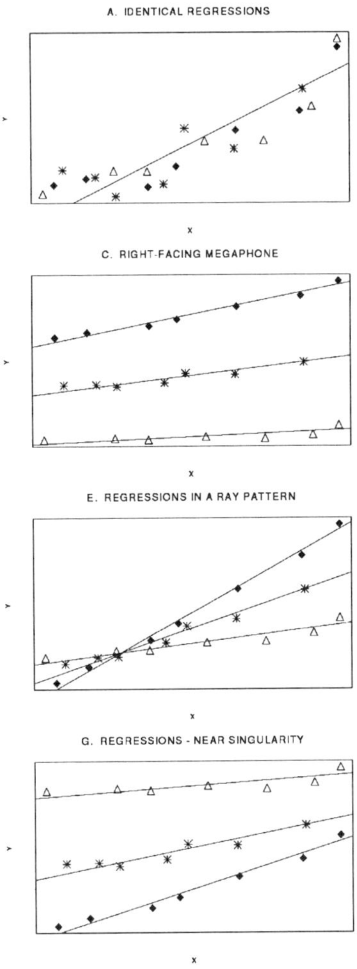

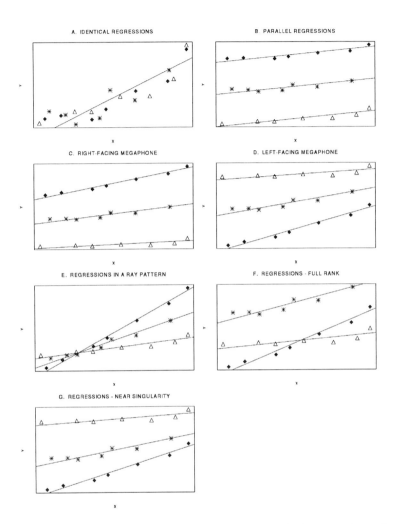

FIGURE 1. Patterns of Between-Group Variation. A. Identical regressions; B. Parallel regressions; C. Right-facing megaphone; D. Left-facing megaphone; E. Ray pattern; F. Full rank variation; G. Near-singularity of variation. The symbols identify three different groups.

The covariance matrix $\boldsymbol{\Sigma}$ can be effectively used to describe the between-group variation. First, the case $\boldsymbol{\Sigma} = \mathbf{0}$ corresponds to no between-group variation, and equal within-group regressions, as depicted in panel A of Figure 1. If the slope-variance vanishes, $\Sigma_{22} = 0$, but the intercept-variance Σ_{11} is positive, then the within-group regressions (10.2) are parallel ($b_j \equiv \beta$); see panel B. In this case there is a natural ranking for the groups. For any value x_0 of the explanatory variable x, the group with the highest intercept also has the highest expected outcome

$$E(y \mid x = x_0;\ a_j, \beta) = a_j + \beta x_0.$$

If Σ_{22} is positive, we distinguish two cases:

A. $\boldsymbol{\Sigma}$ is singular ($\Sigma_{12}^2 = \Sigma_{11}\Sigma_{22}$)

B. $\boldsymbol{\Sigma}$ is of full rank 2

In case A the within-group regressions display a "megaphone" or a "ray" pattern, as depicted in panels C – E. Their common feature is that the regression lines intersect at a single point, say, $(x^*,\ a + bx^*)$. The megaphone pattern arises when the value of the regressor $x = x^*$ is outside the range of the realized values of x. In the case of the megaphone pattern we can still define a meaningful ranking, but for the ray pattern we would have to define two mutually reverse rankings, one applicable for the values of x smaller than x^* and the other for the values greater than x^*.

Case B represents the complement of all the special cases described above (see panel F). In general, no meaningful ranking for the groups can be defined, unless no two within-group regressions intersect in the range of the observed values of x. This is likely to happen only when the covariance matrix $\boldsymbol{\Sigma}$ is close to a matrix of deficient rank (panel G).

The conditional (within-group) variance of an observation is equal to σ^2, whereas the unconditional variance is

$$\text{var}(y) = \sigma^2 + (1,\ x)\boldsymbol{\Sigma}(1,\ x)' = \sigma^2 + \Sigma_{11} + 2x\Sigma_{12} + x^2\Sigma_{22}, \tag{10.5}$$

which is a quadratic function of the explanatory variable x. Since the matrix $\boldsymbol{\Sigma}$ is nonnegative definite, the variance (10.5) is greater than or equal to σ^2. It reduces to the constant $\sigma^2 + \Sigma_{11}$ when the regression slopes are constant, i.e., when $\Sigma_{22} = 0$. Otherwise, the minimum of (10.5) is achieved for the value $x^* = -\Sigma_{12}/\Sigma_{22}$, and the corresponding variance is equal to $\sigma^2 + \Sigma_{11} - \Sigma_{12}^2/\Sigma_{22}$. This minimum variance is equal to σ^2 if and only if $\boldsymbol{\Sigma}$ is of rank 1. For nonconstant variance (10.5) it may be useful to distinguish the following three situations: The variance var(y) is either decreasing in x ($x^* > x$ for all realized x), increasing in x ($x^* < x$ for all realized x), or it first decreases and then increases (when x^* is within the range of realized values of x). Note that this definition of x^* coincides with the earlier definition for case A.

The unconditional covariance of two observations from the same group is

$$(1,\ x_1)\boldsymbol{\Sigma}(1,\ x_2)' = \Sigma_{11} + (x_1 + x_2)\Sigma_{12} + x_1x_2\Sigma_{22},$$

where x_1 and x_2 are the values of the explanatory variable for the two observations.

Up to this point we have tacitly assumed that the explanatory variable x is not constant within each group; otherwise the within-group regression slopes on x would not be identifiable. In the corresponding analysis of covariance we make the same assumption, although in the random coefficient model (10.2)–(10.3) it suffices to assume that the explanatory variable be nonconstant within at least two groups.

Suppose each group is represented by a single observation ($n_j = 1$ for all j). Then (10.2) – (10.3) is a regression model which specifies the expectation of an observation as a linear function of the regressor x, and the variance as a quadratic function of x. Aitkin (1987) describes a class of models with an alternative parametrization for the variance as a function of regressor(s) and unknown parameter(s). Even with clustered observations ($n_j > 1$ for several groups), we may consider a variable defined for groups (constant within groups) in (10.2), with positive variance Σ_{22}, to allow the within-group covariance to depend on an explanatory variable. Thus, random-coefficent models can be used for modeling variance heterogeneity.

Returning to the dilemma of Jim's parents about his hypothetical performance in a different school (and ignoring the numerous caveats stated in Section 1.1), the variance component $(1,\ x)\boldsymbol{\Sigma}(1,\ x)'$ provides a description of the risk associated with Jim's placement in one of the schools. For each school we may consider the conditional expectation $\hat{\boldsymbol{\beta}}_j$ of its "effect," given by (10.4). The expected performance $\hat{y}_j$ of a student with the value x of the explanatory variable is $\hat{y}_j(x) = (1,\ x)\hat{\boldsymbol{\beta}}_j$. Students with value x tend to do best in the school(s) with highest $\hat{y}_j(x)$. We emphasize the dependence of $\hat{y}_j$ on x because, as discussed earlier, the relative position (rank) of $\hat{y}_j$ among the schools j may vary with x.

In many contexts we may be interested in the proportion of the groups that have a slope exceeding a certain benchmark (e.g., have a positive slope). Counting the number of groups with positive conditional expectation of the slope, (10.4), has a more elegant alternative–determining the proportion of groups from the population distribution of the slopes, $\mathcal{N}(b,\ \Sigma_{22})$. The latter has a lesser chance of capitalizing on chance, and the effects of shrinkage are also negated.

In the social sciences we frequently encounter variables defined on arbitrary scales, such as scales for standardized educational tests, IQ, and the like. Therefore, a minimal requirement for any statistical model is that it have certain invariance properties with respect to changes of the scales. Here we consider only linear transformations of the explanatory variable. Suppose we replace the variable x by its linear transform $x' = (x - A)/B$, where A and $B > 0$ are some constants. Then the model corresponding to (10.2) is

$$\begin{aligned} y_{ij} &= a_j + b_j x_{ij} + \varepsilon_{ij} \\ &= a_j + Ab_j + Bb_j x'_{ij} + \varepsilon_{ij} \quad (= a'_j + b'_j x'_{ij} + \varepsilon_{ij}), \end{aligned}$$

so that the individual-level random term ε remains unaltered, the "new" within-group slope is the B-multiple of the "old" one, and the "new" within-group intercept is related to the "old" one by the linear formula $a'_j = a_j + Ab_j$. Thus the "new" covariance matrix $\boldsymbol{\Sigma}'$ is related to the "old" one by the quadratic form

$$\boldsymbol{\Sigma}' = \boldsymbol{S}\boldsymbol{\Sigma}\boldsymbol{S}',$$

where

$$\boldsymbol{S} = \begin{pmatrix} 1 & A \\ 0 & B \end{pmatrix}.$$

In particular, the meaning and interpretation of the sizes of the variance of the intercept, Σ_{11}, and of the slope-by-intercept covariance, Σ_{12}, are tied to the scale for the explanatory variable. However, the slopes and their distribution are interpretable, since they can be easily converted from one scale to another: the slopes $\{b_j\}$ have the $\mathcal{N}(b, \Sigma_{22})$ distribution, and the slopes $\{b'_j\}$ have the $\mathcal{N}(Bb, B^2\Sigma_{22})$ distribution. These invariance properties also extend to the maximum likelihood estimators for the means and variances of the slopes.

From panels C – G in Figure 1, or from the formula (10.5), it is easy to see that as the origin of the explanatory variable is moved farther and farther to the left (or right) the intercept-variance increases beyond all bounds, and the correlation of the intercept and slope converges to +1 or −1. The analogous discussion of linear transformations of the outcome variable is trivial.

Of course, for nonlinear monotone transformations such invariance properties do not hold; in fact, finding a nonlinear transformation of the data that improves their description by a linear model is often of substantive interest. These issues carry over directly from ordinary regression.

2.2 Contextual Models

Substantive researchers often argue that context variables, such as certain characteristics of the composition of the group, have an effect on the outcomes, and modeling of these "contextual" effects has gained considerable importance. A *context variable* can be represented by a variable defined for groups (group-level variable), and a natural approach is to augment the basic regression formula (10.2) by the context variable

$$y_{ij} = a_j + b_j x_{ij} + c z_j + \varepsilon_{ij} \, , \tag{10.6}$$

where z_j is the value of the context variable z in group j, and c is an unknown (context) parameter. Here, in the absence of any group-level observed attributes, an appealing choice for z_j is that of the within-group mean of the explanatory variable x, $\bar{x}_j$. The problematic nature of this choice is discussed at length by Aitkin and Longford (1986). A more general (though still problematic) framework for incorporating "contextual effects" involves additional regression equations for the random terms $(a_j, \; b_j)$ in (10.2):

$$\begin{aligned} a_j &= \Gamma_{a1} + \Gamma_{a2} z_j + \delta_{aj} \\ b_j &= \Gamma_{b1} + \Gamma_{b2} z_j + \delta_{bj} \, , \end{aligned} \tag{10.7}$$

where $\Gamma_{a1}, \; \ldots, \; \Gamma_{b2}$ are unknown constants and the random vectors $\{(\delta_{aj}, \; \delta_{bj})\}_j$ are assumed to be a random sample from the $\mathcal{N}(\mathbf{0}, \; \boldsymbol{\Sigma})$ distribution. In the econometric literature equations (10.2) and (10.7) are often referred to as *micro* and *macro* model equations, respectively. Disregarding the problems with experimental design-type inference from observational data, the model equations (10.7) are often referred to as "explanation" of the between-group differences (effects). Direct substitution of (10.7) into (10.2) yields

$$y_{ij} = g_{aj} + g_{bj} x_{ij} + \Gamma_{a2} z_j + \Gamma_{b2} x_{ij} z_j + \varepsilon_{ij} \, , \tag{10.8}$$

where $g_{aj} = \Gamma_{a1} + \delta_{aj}$ and $g_{bj} = \Gamma_{b1} + \delta_{bj}$. This is essentially the model (10.6) augmented by the interaction term $\Gamma_{b2}x_{ij}z_j$. The model (10.8) may be further extended by allowing the regression coefficient Γ_{b2} to vary across the groups. Often there are no a priori reasons for assuming the same form for the group-level equations (10.7). For example, we could assume that $\Gamma_{b2} = 0$ (no "explanation" for variation of b_j), or even replace the regressor z associated with the parameter Γ_{b2} by a different group-level variable.

Of course, in the contextual model (10.6) with $z_j = \bar{x}_j$ we can replace the regressors x_{ij} and $\bar{x}_j$ by their "centered" versions $x_{ij} - \bar{x}_j$ and $\bar{x}_j - \bar{x}^*$, where $\bar{x}^*$ is the sample mean, or any other constant (e.g., the known population mean). The model (10.6) with these centered variables,

$$y_{ij} = \{a_j + (c + b_j)\bar{x}^*\} + b_j(x_{ij} - \bar{x}_j) + (c + b_j)(\bar{x}_j - \bar{x}^*) + \varepsilon_{ij},$$

is not equivalent to the original model (10.6), because the 'new' context variable $\bar{x}_j - \bar{x}^*$ is now associated with variance $\mathrm{var}(c + b_j) = \Sigma_{22}$.

The *Multilevel Modelling Newsletter* (1989, 1990) contains a discussion of the relevance of centering in multilevel models. See also Boyd and Iversen (1979).

2.3 Terminology: A Review

Historically, the development of random coefficient models proceeded in close connection with applications in mutually secluded areas, and this has resulted in usage of a number of different terms for very similar classes of models, reflecting primarily the approach and motivation, rather than methodological differences.

Early applications in agricultural and animal-breeding experiments focused on the basic analysis of variance model (10.1), with the standard assumptions of normality and independence of the random terms δ_j and ε_{ij}. Of primary interest was the decomposition of the variance of the outcome y into its within- and between-group components; hence the term *variance component analysis*. The model (10.2) – (10.3) is an obvious extension of this "raw" model incorporating an explanatory variable. The term "variance component model" remains meaningful; in (10.2)–(10.3) the residual variance of the outcome also has two additive components, although the between-group component is no longer constant, but becomes a quadratic function; see (10.5). Rubin et al. (1981) refer to models of the form (10.2)–(10.3), and extensions thereof, as covariance component models (the description for the between-group variation involves a covariance, Σ_{12}).

If the global parameters β and Σ are known (which is unlikely), we can regard the $\mathcal{N}(\beta, \Sigma)$ distribution as the (Bayesian) prior distribution for the within-group coefficients. In practice we often have no prior information about the global parameters, and so they would be estimated from the data, that is, from the same dataset which would be used for inference about the within-group coefficients. The term empirical Bayes analysis is motivated by this approach. For a more detailed discussion see Deeley and Lindley (1981) and Morris (1983). A comprehensive Bayes analysis would assume a joint prior distribution for the global parameters β and Σ. Within a Bayesian framework we can consider a hierarchy of parameters: The vectors of regression coefficients $\{\beta_j\}_j$ are a set of parameters which are further modeled by the set of *hyperparameters* β and Σ.

Before the potential of random coefficients for constructing models that mix levels of analysis was recognized, the choice between aggregated and disaggregated data was a key issue in quantitative social science research. Analysis of group-aggregate data is referred to as group-level analysis, and analysis of the disaggregated (individual-level) data as individual-level analysis. Extensive discussion of the differences between these two kinds of analyses has resulted in the realization that neither of the analyses is suitable, and some form of simultaneous analysis that recognizes the levels (hierarchy) of clustering is necessary. Adaptations of random coefficient models for these problems have been termed multilevel models (Burstein 1980; Goldstein 1987) and hierachical linear models (Raudenbush and Bryk 1986).

The term seemingly unrelated regressions (SUR) is generally attributed to Zellner (1962), and has been used in a variety of applications, especially in longitudinal analysis (see, e.g., Stanek and Koch 1985). Models of the form (10.2) are also referred to as mixed models (when they feature a combination of "fixed" and "random" coefficients), or random models (when all coefficients are random).

2.4 Applications

Random coefficient models are relevant in a variety of situations that give rise to clustered observations. Clustering of students within classrooms and schools, clustering of individuals within families or households, paired organs within individuals, production units within factories, respondents within interviewers, medical cases within hospitals, offspring within parents, plots within a field, and geographical subdivisions are some typical examples. Usually the cluster is associated with a treatment (in the sense of an experimental design); description of the sample of these treatments is then an important inferential problem. See Aitkin, Anderson, and Hinde (1981); Anderson and Aitkin (1985); Longford (1985); Rosner (1984); Strenio, Weisberg, and Bryk (1983); and Zelen (1957) for examples.

Surveys covering a large country often involve formidable expenditures that can be justified only if information about a variety of issues is provided. National totals for variables of interest may be one focus of such a survey, but subtotals for smaller subdivisions of the country (regions, districts) may also be desired. Of course, small districts may be very sparsely represented in the survey sample. Any estimator for a district subtotal based only on the data from that district would have a very large sampling variation; it would appear that the neighboring districts, or the rest of the districts in the country, also contain information about the district in question, due to similarity of the subtotals (or means) across the districts. Thus information contained in the district's data can be supplemented by that from other districts ("borrowing strength"). Such applications arise in demography, epidemiology, economics, in studies of voting behavior, and other areas where surveys with a geographical region and its subdivisions are used as a sampling frame. The associated inferential problems are referred to in survey literature as *small-area statistics*; see Battese, Harter, and Fuller (1988) for an application and Ghosh and Rao (1994) for a review.

In many areas of research, experiments can be conducted only with a small number of subjects, so that a single experiment is not sufficient for testing research hypotheses with desired confidence. Thus it is necessary to combine information from experiments with the same (or similar) protocol. Each experiment/study involves slightly different circumstances (physical, psychological, and other setting), and so within-experiment relationships of the variables of interest, such as regressions, may vary. The observations within experiments represent the clustered data structure, and random coefficient models can be easily formulated. Between-experiment variation is a measure of importance of the uncontrolled settings of the experiments. The corresponding analysis requires certain within-experiment data summaries which may not be available (especially when information about the experiments is available only from publications). Meta-analysis (Hedges and Olkin, 1985) is a collection of methods for combining within-experiment estimators.

In many situations measurements or observations are subject to substantial errors, and a common remedy for the lack of reliability of a measurement is its replication. In schemes where measurements on subjects are repeated we can regard the observations on an individual as a cluster of correlated data. In a regression model systematic differences among the replications can be accounted for by a variable defined for the elementary level (observation), for example, a categorical variable indicating the "type" of measurement.

In multivariate regression analysis we have multiple observations (outcomes) for each individual, and therefore a clustered structure of the data. As an alternative to multivariate analysis we may consider random coefficient models with a suitable pattern of variation of the observations within an individual. An unrestricted pattern of variation among the components can be modeled by associating the categorical variable indicating the component of the outcome vector with variation (see Section 3.2). The main advantage of the random coefficient models is in their flexibility; data missing at random are straightforward to accommodate, because within the random coefficient model the missing data merely produces an unbalanced design. Additionally, random coefficient models contain a rich class of alternatives that are more restricted than the multivariate regression analysis model, and more general than the trivial model of independence of the components of the outcome vector.

Random coefficient models are also applicable in many longitudinal studies. In a typical experiment we consider a number of exchangeable (or conditionally exchangeable, given some explanatory variables) subjects. For each subject the value of the outcome variable is observed/measured at designated time points, not necessarily identical for each subject. The basic structure of the resulting dataset is that of sets of observations within subjects. Formally, we can regard data from each subject as a small experiment carried out in a different context of the unique characteristics of the subject.

If the relationship of the outcome on time is linear for each subject, and the appropriate assumptions of normality and within-group homoscedasticity are satisfied, the model (10.2)–(10.3) provides a suitable description. The added significance of the exchangeability assumption A, expressed by (10.3), is that in situations where only a small number of observations per subject is collected, the OLS estimates of within-group regressions would both have extremely poor resampling properties and would be of marginal interest; in most applications of longitudinal analysis the substantive focus is on description of growth for the population represented by the subjects involved in the study. Whereas each subject's growth curve is very poorly estimated when data solely from that subject are used, the pooled dataset may contain abundant information about the distribution of growth across the relevant population. Estimation of each subject's growth curve can be improved upon by application of the shrinkage estimator (10.4). This can be interpreted as "borrowing strength", i.e., using information from other subjects to estimate within-subject coefficients. Borrowing strength is an important consideration in applications of the model (10.2)–(10.3) whenever aggregate units exhibit substantial similarity and contain small numbers of elementary observations.

3 The General Two-Level Model

We assume that there are N_2 aggregate units (groups, or units at level 2), with each unit j containing n_j elementary units (individuals, units at level 1). The outcome variable y is defined for the elementary units, and the explanatory variables $\boldsymbol{x}$ may contain variables defined at either level, say, $\boldsymbol{x} = \left(1,\ \boldsymbol{x}^{(1)},\ \boldsymbol{x}^{(2)}\right)$, where 1 is the constant variable which represents the intercept, and the vector of variables $\boldsymbol{x}^{(h)}$, of length p_h, is defined at level h $(= 1, 2)$. We assume a linear regression model

$$y_{ij} = \boldsymbol{x}_{ij}\boldsymbol{\beta}_j + \varepsilon_{ij}\ , \qquad (10.9)$$

where $\boldsymbol{\beta}_j$ is a vector of regression coefficients, of length $p = 1 + p_1 + p_2$, assumed to be a random sample from a multivariate normal distribution $\mathcal{N}(\boldsymbol{\beta}, \boldsymbol{\Sigma})$ and the elementary-level random terms $\left(\varepsilon_{1j}\ , \varepsilon_{2j}\ , \ldots, \varepsilon_{n_j j}\right)$ form mutually independent random samples from $\mathcal{N}(0, \sigma_j^2)$. The mean regression vector $\boldsymbol{\beta}$ is an arbitrary p-dimensional vector, and the between-group covariance matrix $\boldsymbol{\Sigma}$ is nonnegative definite. Usually we assume that the within-group variance σ_j^2 is constant across the groups, $\sigma_j^2 \equiv \sigma^2$. In situations where we have no prior information about the pattern of variation, or the functional form of the within-group covariance, we seek a model description as simple as possible which nevertheless adequately captures the salient features exhibited by the data.

The variance of an observation is equal to

$$\mathrm{var}(y_{ij}) = \sigma^2 + \boldsymbol{x}_{ij}\boldsymbol{\Sigma}\boldsymbol{x}_{ij}'\ , \qquad (10.10)$$

and the covariance of two observations in the same group is equal to

$$\mathrm{cov}(y_{ij}, y_{i'j}) = \boldsymbol{x}_{ij}\boldsymbol{\Sigma}\boldsymbol{x}_{i'j}'\ ; \qquad (10.11)$$

two observations from different groups are independent.

We will use the following matrix notation: let $\boldsymbol{y}_j = (y_{1j}, y_{2j}, \ldots, y_{n_j j})'$ be the vector of outcomes for group j, $\boldsymbol{X}_j = (\boldsymbol{x}_{1j}', \boldsymbol{x}_{2j}', \ldots, \boldsymbol{x}_{n_j j}')'$ the $n_j \times p$ "design" matrix for group j, $\boldsymbol{V}_j$ the corresponding covariance matrix, $\boldsymbol{y} = (\boldsymbol{y}_1', \boldsymbol{y}_2', \ldots, \boldsymbol{y}_{N_2}')'$ the vector of all outcomes, and $\boldsymbol{V}$ the corresponding covariance matrix. Since the clusters are mutually independent, $\boldsymbol{V}$ is a block-diagonal matrix, with blocks $\boldsymbol{V}_j$.

We will use the Kronecker product notation for block-diagonal matrices and for "stacking" vectors and matrices. Let $\boldsymbol{A}$ be a $K \times H$ matrix and $\{\boldsymbol{v}_{kh}\}_{k=1,\dots,K,\, h=1,\dots,H}$ a two-dimensional array of matrices of the sizes $m_k \times p_h$. The Kronecker product $\boldsymbol{A} \otimes \{\boldsymbol{v}_{kh}\}$ is defined as the $\sum_k m_k \times \sum_h p_h$ matrix consisting of the blocks $A_{kh}\boldsymbol{v}_{kh}$:

$$\boldsymbol{A} \otimes \{\boldsymbol{v}_{kh}\} = \begin{pmatrix} A_{11}\boldsymbol{v}_{11} & A_{12}\boldsymbol{v}_{12} & \cdots & A_{1H}\boldsymbol{v}_{1H} \\ A_{21}\boldsymbol{v}_{21} & \cdot & \cdots & A_{2H}\boldsymbol{v}_{2H} \\ \cdot & \cdot & & \cdot \\ \cdot & \cdot & & \cdot \\ \cdot & \cdot & & \cdot \\ A_{K1}\boldsymbol{v}_{K1} & A_{K2}\boldsymbol{v}_{K2} & \cdots & A_{KH}\boldsymbol{v}_{KH} \end{pmatrix}.$$

We denote by $\mathbf{1}_n$ the column vector of ones of length n, by $\boldsymbol{I}_N$ the $N \times N$ identity matrix, and by $\boldsymbol{J}_N$ the $N \times N$ matrix of ones ($\boldsymbol{J}_N = \mathbf{1}_N\mathbf{1}'_N$). With this notation we have

$$\boldsymbol{y}_j = \mathbf{1}_{n_j} \otimes \{y_{ij}\}, \quad \boldsymbol{y} = \mathbf{1}_{N_2} \otimes \{\boldsymbol{y}_j\}, \quad \boldsymbol{X}_j = \mathbf{1}_{n_j} \otimes \{\boldsymbol{x}_{ij}\}.$$

We will slightly abuse this notation by writing $\boldsymbol{V} = \boldsymbol{I}_{N_2} \otimes \{\boldsymbol{V}_j\}$ for the block- diagonal matrix $\boldsymbol{V}$ with blocks $\boldsymbol{V}_j$, and $\boldsymbol{J}_N \otimes \boldsymbol{B}$ for the matrix consisting of $N \times N$ identical blocks each equal to the matrix $\boldsymbol{B}$.

We denote by $\mathbf{0}_n$ the column vector of zeros of length n. When the dimensions are obvious from the context, we drop the subscripts for $\mathbf{1}$, $\mathbf{0}$, $\boldsymbol{I}$ and $\boldsymbol{J}$. We will use the symbol $\mathbf{0}$ also for matrices of zeros. We define the (sample) design matrix as $\boldsymbol{X} = \mathbf{1}_{N_2} \otimes \{\boldsymbol{X}_j\}$, that is, $\boldsymbol{X} = (\boldsymbol{X}'_1, \boldsymbol{X}'_2, \dots, \boldsymbol{X}'_{N_2})'$.

For the model (10.9) we have

$$\boldsymbol{V}_j = \sigma^2\boldsymbol{I} + \boldsymbol{X}_j\boldsymbol{\Sigma}\boldsymbol{X}'_j\,. \tag{10.12}$$

An alternative formulation of the model (10.9) is

$$y_{ij} = \boldsymbol{x}_{ij}\boldsymbol{\beta} + \gamma_{ij}\,, \tag{10.13}$$

where $\boldsymbol{\beta}$ is a vector of parameters and the random terms γ_{ij} are normally distributed with zero means and variances, and covariances for pairs of observations within the same group, given by a parametric formula. The simplest choice of the within-group covariance structure for both models (10.9) and (10.13) is that of constant within-group covariance (compound symmetry),

$$\text{cov}(y_{ij}, y_{i'j}) = \theta\,,$$

which, for $\theta \geq 0$, corresponds to (10.11) with $\Sigma_{11} = \theta$ and $\Sigma_{kh} = 0$ for all the other elements of $\boldsymbol{\Sigma}$;

$$y_{ij} = \boldsymbol{x}_{ij}\boldsymbol{\beta} + \delta_j + \varepsilon_{ij}\,. \tag{10.14}$$

Thus the apparent generality of the model (10.13), in this special case, is in allowing negative within-group covariances. It can be shown by a simple eigenvalue analysis that $\boldsymbol{V}_j$

is nonnegative definite if and only if $\theta \geq -\frac{1}{n_j-1}$ for all j. Hence it is meaningful to consider a negative covariance θ only in cases when there is an upper bound on the number of elementary observations within a group.

In most applications it turns out that for a general specification of the variance-covariance structure, the parameterization implied by the model (10.9) is sufficiently flexible; in fact, selection among the alternative models for description of between-group variation usually poses a more serious problem. Therefore we restrict our discussion to the model (10.9). The number of variance and covariance parameters grows quadratically with the number of explanatory variables, p, and so for practical purposes it becomes necessary to assume that some of the variances in Σ are equal to zero, that is, to assume a common slope across the groups for each of a subset of the explanatory variables.

On the one hand, the more variances are equal to zero, the simpler the model description; on the other hand, too simple a model may not provide an adequate description of the pattern of variation. Note that assuming a zero variance, $\Sigma_{kk} = 0$, implies that all the covariances in the same row and column of Σ are also equal to zero, $\Sigma_{kh} = 0$ for all $h = 1, \ldots, p$. Additionally we may assume that certain covariances are equal to zero, say $\Sigma_{kh} = 0$, even though the *constituent* variances, Σ_{kk} and Σ_{hh}, are positive; in other words, that the corresponding within-group coefficients are uncorrelated.

The discussion in Section 2.1 implies that if we insist on invariance with respect to linear rescaling, we cannot assume that an intercept-by-slope covariance, Σ_{1k}, vanishes unless the slope variance Σ_{kk} also vanishes. To show this, suppose we change the origin of the $k$thvariable, i.e., we replace the variable x_k with $x_k - c_k$ where c_k is an arbitrary constant. Then the intercept for the model (10.9) becomes

$$\beta_{j1}^* = \beta_{j1} + c_k \beta_{jk}$$

and all the other regression coefficients are unchanged ($\beta_{jk}^* = \beta_{jk}$, $k = 2, \ldots, p$). Hence the covariance matrix Σ^* for the "new" parametrization differs from the "old" covariance matrix Σ only in the following two terms:

- the intercept-variance $\Sigma_{11}^* = \Sigma_{11} + 2c_k\Sigma_{1k} + c_k^2 \Sigma_{kk}$
- the intercept-by-slope covariance $\Sigma_{1k}^* = \Sigma_{1k} + c_k \Sigma_{kk}$.

If variables have arbitrary origins, as is often the case, then the intercept variance and the intercept-by-slope covariances (the first row and column of Σ) depend on the choices of origins of the explanatory variables (k) *associated* with positive variances (Σ_{kk}). The variances of the other coefficients are independent of the choice of origin.

We shall say that the explanatory variable x_k is associated with (between-group) variation if the corresponding variance Σ_{kk} is a free parameter. As discussed in Section 2, association of a group-level variable with variation does not have an interpretation in terms of varying regression coefficients; the corresponding variances are usually assumed to be equal to zero [as in (10.6) and (10.8)]. Nevertheless, association of group-level variables with variation may still be useful in modeling between-group variation.

In most applications it suffices to consider only a small number of variables associated with variation, and, if there is no information guiding the choice among the variables to be associated with variation, elementary-level variables should take precedence because of easier interpretation.

3.1 Categorical Variables and Variation

A categorical explanatory variable with K categories is represented in the models (10.9) and (10.13) by $K-1$ $0/1$ dummy variables. In a standard parameterization the first (reference) category corresponds to the vector $(1, 0, \ldots, 0)$ and the kth category to $(1, 0, \ldots, 0, 1, 0, \ldots, 0)$ (ones in the first and kth positions). Association of the categorical variable, that is, of the dummy variables, with variation corresponds to varying within-group differences between pairs of categories.

As an example consider a categorical variable with three categories, A, B, and *not known*. Suppose each subject belongs to the category A or B, but the record of this variable has been lost for some subjects. Suppose the within-cluster means are constant for both categories A and B. The *not known* category may have an uneven representation of subjects from categories A and B, in which case the differences between the categories *not known* and A would vary across the clusters.

An important generalization of the model (10.9) involves a more general form of the within-group variance, σ^2, such as unrelated within-group variances σ_j^2 $(j = 1, \ldots, N_2)$, or different within-group variances for each of a number of categories defined for groups. More generally, we may consider the within-group variance as a function of some of the explanatory variables, in the same fashion as we consider the between-group variance as a function of some of the explanatory variables. These generalizations are closely related to variance heterogeneity modelling.

3.2 Multivariate Regression as a Random Coefficient Model

The standard multivariate regression can be expressed as a random coefficient model using the device of associating a categorical variable with variation. We demonstrate this by an example, and discuss the advantages of this representation at the end of this section.

We consider a K-variate regression model with independent outcome vectors $\{\boldsymbol{y}_j\}$, $j = 1, 2, \ldots, N_2$, and an arbitrary dependence structure of the components,

$$\boldsymbol{y}_j = \boldsymbol{\Gamma} \boldsymbol{v}_j + \boldsymbol{\delta}_j^* \quad , \tag{10.15}$$

where $\boldsymbol{\Gamma}$ is a matrix of parameters, $\boldsymbol{v}_j$ is the (column) vector of regressors for the jth vector of outcomes, and $\{\boldsymbol{\delta}_j^*\}_j$ a random sample from $\mathcal{N}_K(\mathbf{0}, \boldsymbol{\Sigma})$. To obtain the componentwise version of (10.15), we consider $\boldsymbol{y} = \mathbf{1}_{N_2} \otimes \{\boldsymbol{y}_j\}$ as a vector of *univariate* outcomes and define the nesting structure of *components* within *subjects* (vectors). When there are no missing observations, each cluster contains K observations.

We define the categorical component-level variable $c_{ij} = i$ which indicates the position of the observation in the original vector $\boldsymbol{y}_j$. Let $\boldsymbol{z}_{ij}$ be the corresponding vector of dummy variables (with standard parameterization), and let $\boldsymbol{x}_{ij}$ be the set of all interactions of c_{ij} ($\boldsymbol{z}_{ij}$) with the regressors $\boldsymbol{v}_j$. Then

$$y_{ij} = \boldsymbol{x}_{ij} \boldsymbol{\beta} + \boldsymbol{z}_{ij} \boldsymbol{\delta}_j , \tag{10.16}$$

which is the random coefficient model (10.9) with $\sigma^2 = 0$. There is a one-to-one correspondence between the parameter vector $\boldsymbol{\beta}$ and the matrix $\boldsymbol{\Gamma}$ [$\boldsymbol{\beta}$ is a linear nonsingular transformation of vec($\boldsymbol{\Gamma}$)]. The random terms $\boldsymbol{\delta}_j^*$ and $\boldsymbol{\delta}_j$ are related by

$$\delta_{j1} = \delta_{j1}^* \quad \text{and} \quad \delta_{jk} = \delta_{jk}^* - \delta_{j1} \quad (k = 2, \ldots, K),$$

and so var($\boldsymbol{\delta}_j$) can be an arbitrary covariance matrix.

While in the model (10.15) dealing with missing data (that is, missing at random) can be a serious problem, sometimes dealt with by listwise deletion, in (10.16) missing observations merely cause a second level unit to have fewer than K observations. Even when a large proportion of observations is missing (because of imperfect data collection, or by design), no available observations have to be discarded in the random coefficient model. An additional advantage is that with the random coefficient models, submodels of (10.16) with restricted pattern of the covariance matrix $\boldsymbol{\Sigma}$ can be considered.

3.3 Contextual Models

An apparent generalization of the model (10.9) involves separate equations for each level. The development has been outlined in Section 2.2; here we present the extension for multiple regression. We supplement the *elementary-level model* (10.9) by a set of ordinary regression models for the random coefficients:

$$\boldsymbol{\beta}_j = \boldsymbol{\Gamma} \boldsymbol{U}_j + \boldsymbol{\delta}_j , \qquad (10.17)$$

where $\boldsymbol{\Gamma}$ is a matrix of parameters, $\boldsymbol{U}_j$ is a vector of group-level explanatory variables, and $\{\boldsymbol{\delta}_j\}_j$ is a random sample from $\mathcal{N}(\mathbf{0}, \boldsymbol{\Sigma})$. However, (10.17) is only an alternative formulation of the general model (10.9), since by direct substitution of (10.17) into (10.9) we obtain another random coefficient model of the form (10.9),

$$\boldsymbol{y}_j = \boldsymbol{X}_j \boldsymbol{\Gamma} \boldsymbol{U}_j + \boldsymbol{X}_j \boldsymbol{\delta}_j + \boldsymbol{\varepsilon}_j \qquad (10.18)$$

($\boldsymbol{X}_j = \mathbf{1}_{n_j} \otimes \{\boldsymbol{x}_{ij}\}$), which includes *cross-level* interactions (interactions between variables defined at different levels) that are not associated with variation. Of course, such constructed variables can be included in the model directly, without reference to a group-level model (10.17).

The two-stage formulation, (10.9) and (10.17), is quite appealing because it appears to separate processes at elementary and group levels. It should be noted, however, that a group-level variable can be "disguised" as an elementary-level one, and then the model equations (10.9) and (10.17) are changed substantially. This can create problems in interpreting estimated regression parameters.

The two-stage model formulation also has a long tradition in longitudinal research, see, for instance, Grizzle and Allen (1969). See also Chi and Reinsel (1987), and Ware (1985) for a review.

3.4 Random Polynomials

Nonlinear regression, such as polynomial regression, can be accomodated in (10.9) by inclusion of nonlinear functions of some of the regressors. These "constructed" regressors can be associated with variation. Care should be exercised in the choice of constraints for the covariances that correspond to positive variances. Suppose variables x_k and x_h are both associated with variation, and we would consider a rotation of these two regressors as meaningful, that is, the pair of variables (x_h, x_k) could be replaced by a nonsingular rotation $(x'_h, x'_k)' = \boldsymbol{A}(x_h, x_k)'$, where $\boldsymbol{A}$ is a nonsingular matrix. For example, replacing the linear and quadratic terms $(t,\ t^2)$ in quadratic regression with t and $t^2 - 2t$ corresponds to the rotation matrix

$$\boldsymbol{A} = \begin{pmatrix} 1 & 0 \\ -2 & 1 \end{pmatrix}.$$

Let the submatrix of $\boldsymbol{\Sigma}$ associated with the "old" variables (t, t^2) be $\boldsymbol{\Theta}$. Then the corresponding submatrix for the "new" variables $(t, t^2 - 2t)$ is $\boldsymbol{A}^{-1}\boldsymbol{\Theta}(\boldsymbol{A}^{-1})'$. The covariance Σ_{hk} is not invariant with respect to rotation of the constituent variables, and therefore it can be assumed to be equal to zero only at the expense of model invariance with respect to rotation. In general, it is meaningful to assume zero covariances in $\boldsymbol{\Sigma}$ only for pairs of variables for which linear combinations are not meaningful; for example, sex and socio-economic background.

3.5 Fixed and Random Parts

Another formulation of the random coefficient model (10.9),

$$\boldsymbol{y}_j = \boldsymbol{X}_j\boldsymbol{\beta} + \boldsymbol{Z}_j\boldsymbol{\delta}_j + \boldsymbol{\varepsilon}_j, \tag{10.19}$$

emphasizes the following three model components:

A. Expectation (regression or fixed part), $\boldsymbol{x\beta}$

B. Group-level random part, $\boldsymbol{z\delta}$

C. Individual-level random part ε.

Here $\boldsymbol{X}_j$ is the $n_j \times p$ matrix of the values of the explanatory variables for the observations in group j, and $\boldsymbol{Z}_j$ is the $n_j \times r$ matrix of the values of the explanatory variables associated with variation, and $\boldsymbol{\varepsilon}_j = \mathbf{1}_{n_j} \otimes \{\varepsilon_{ij}\}$. Although we have indicated that the variables associated with variation are selected from those represented in the regression formula, $\boldsymbol{X\beta}$, other variables, used solely for description of variation, can also be considered. In fact, they can be formally included in the regression part, with the associated regression parameters assumed equal to zero. In most situations we wish to adhere to the conventions of the analysis of covariance, where an interaction is considered only if all the 'main effects' and all the (sub-) interactions are represented in the model by free parameters.

3.6 Model Identification

In a more compact matrix notation we have

$$y = X\beta + Z\delta + \varepsilon,$$

where $X = \mathbf{1}_{N_2} \otimes \{X_j\}$, $Z = I_{N_2} \otimes \{Z_j\}$, $\delta = \mathbf{1}_{N_2} \otimes \{\delta_j\}$, and $\varepsilon = \mathbf{1}_{N_2} \otimes \{\varepsilon_j\}$. Note that δ and ε are column vectors of respective lengths rN_2 and $N = \sum_{j=1}^{N_2} n_j$.

If a variable included in the design matrix Z is associated with variance (diagonal element of Σ) equal to zero, then this variable (column of data) can be deleted from the matrices Z_j without affecting the distribution of (10.19). To avoid this ambiguity we will adopt for the rest of the chapter the following convention: The matrix Z will always stand for the *minimal* design matrix, that is, the corresponding covariance matrix Σ will not contain any variances equal to zero. A similar convention is usually assumed for the regression parameters: A variable with corresponding regression coefficients equal to zero is assumed to be excluded from the design matrix X. The added importance of the convention for Z (Σ) is that it is often of interest whether Σ is of full rank (after deleting rows and columns corresponding to zero variances).

The joint distribution of the outcomes is

$$y \sim \mathcal{N}_N\{X\beta,\ \sigma^2 I_N + Z(I_{N_2} \otimes \Sigma)Z'\}. \tag{10.20}$$

Instead of the model description with random terms we may consider random coefficient models merely as prescribing a specific pattern for the mean and the covariance matrix of the observations.

The formulation (10.20) implies that full rank of the design matrix X is a necessary condition for uniqueness of β. The information matrix associated with the vector β is equal to $X'V^{-1}X$, where $V = \text{var}(y)$ (see Section 4). Therefore full rank of X is also a sufficient condition for uniqueness of β. In practice, if the matrix X is of full rank, but close to a matrix of deficient rank, estimates of β are subject to substantial sampling variation (are "poorly identified"). This is a problem analogous to model identification in ordinary regression. Proximity of X to a matrix of deficient rank can be assessed by the *condition number* of X, defined as the ratio of the largest and the smallest eigenvalues of $X'X$. This condition number is a suitable criterion for model identifiability for the fixed part of random coefficient models, although the eigenvalues of $X'V^{-1}X$ should actually be used instead. The disadvantage of the latter is that it depends on the (usually unknown) covariance matrix Σ.

Problems of identification are equally applicable to the parameters for the covariance matrix Σ, but there are no established methods for their exploration. Obviously, if each group has not more than s individuals, then there is no scope for modeling the between-group variation by more than s variables in the random part. In general, the number of variables that can be considered for the random part depends on the variables in the fixed part, on the pattern of variation of the random part variables within and between groups, and on the covariance matrix Σ, and so the only feasible approach rests on evaluation of the information matrix for the elements of Σ. If several (say, s) variables are included in the random part this information matrix has large dimensions ($t \times t$, where $t = \frac{s^2+s}{2}$). Formulas for the elements of the information matrix are given in Section 4.

In practice, when too rich a pattern of between-group variation is declared, the estimate of the covariance matrix $\boldsymbol{\Sigma}$ is usually of deficient rank. Deficient rank of $\boldsymbol{\Sigma}$ indicates that the components of the random terms $\boldsymbol{\delta}_j$ are linearly dependent. Therefore deficient rank of the estimate $\hat{\boldsymbol{\Sigma}}$ implies that the model for the random part is probably overparameterized. Note, however, that singularity of $\boldsymbol{\Sigma}$ is not related in any obvious way to singularity of the information matrices for the parameters in $\boldsymbol{\Sigma}$ or $\boldsymbol{\beta}$.

4 Estimation

The first comprehensive approach to estimation with random coefficient models is due to Henderson (1953). The main virtues of the three algorithms described in Henderson (1953) are simplicity (the algorithms could in principle be executed on a calculator), intuitive appeal, and reasonable resampling properties. With the advent of the computer the emphasis has shifted from relative simplicity to exact maximum likelihood methods. The first such solution, for simultaneous estimation of the regression and variance-covariance parameters, was given by Hartley and Rao (1967). In ordinary regression the maximum likelihood estimator (MLE) of the residual variance is known to be biased. For the random coefficient models, the MLE is also biased; the estimator for the variances and covariances does not take into account the uncertainty about the regression parameters $\boldsymbol{\beta}$. This problem has been rectified by Patterson and Thompson (1971), who proposed to maximize the likelihood for a set of error contrasts. The connection between the two approaches (referred to as *full* and *restricted* maximum likelihood estimation, respectively) is clarified and comprehensively discussed by Harville (1974, 1977). The adjectives *full* and *restricted* do not imply superiority of the approach associated with the former. We will refer to the two approaches, corresponding estimators, methods, and so on, as $\mathrm{MLE_F}$ and $\mathrm{MLE_R}$, respectively.

The computational approaches of Hartley and Rao (1967) and Patterson and Thompson (1971) are essentially straightforward algorithms for maximization of the likelihood as a nonlinear function of several parameters, using the Newton-Raphson or Fisher scoring methods. The log-likelihood associated with the model (10.19) is

$$\log \lambda_F = -\frac{1}{2} N \log(2\pi) - \frac{1}{2}\log(\det \boldsymbol{V}) - \frac{1}{2}\mathrm{tr}(\boldsymbol{V}^{-1}\boldsymbol{S}), \tag{10.21}$$

where $\boldsymbol{S} = (\boldsymbol{y} - \boldsymbol{X}\boldsymbol{\beta})(\boldsymbol{y} - \boldsymbol{X}\boldsymbol{\beta})'$ is the matrix of cross-products of the deviations ("errors"), and $\boldsymbol{V} = \mathrm{var}(\boldsymbol{y})$ is the covariance matrix for the observations. The log-likelihood corresponding to a set of error contrasts for $\mathrm{MLE_R}$ is (apart from an additive constant dependent on the choice of the set of contrasts)

$$\log \lambda_R = -\frac{1}{2}(N-p)\log(2\pi) - \frac{1}{2}\log\{\det \boldsymbol{V}\ \det(\boldsymbol{X}'\boldsymbol{V}^{-1}\boldsymbol{X})\} - \frac{1}{2}\mathrm{tr}(\boldsymbol{V}^{-1}\boldsymbol{S}) \tag{10.22}$$

(Harville 1974). Note that $\log \lambda_R$ differs from $\log \lambda_F$ only by the additive term $C + \frac{1}{2}\log\{\det(\boldsymbol{X}'\boldsymbol{V}^{-1}\boldsymbol{X})\}$, where C is a constant not depending on the model parameters.

Newton-Raphson and Fisher scoring methods lead directly to the generalized least-squares solution for the regression parameters $\boldsymbol{\beta}$,

$$\hat{\boldsymbol{\beta}} = (\boldsymbol{X}'\boldsymbol{V}^{-1}\boldsymbol{X})^{-1}\boldsymbol{X}'\boldsymbol{V}^{-1}\boldsymbol{y}\,, \tag{10.23}$$

for both $\mathrm{MLE_F}$ and $\mathrm{MLE_R}$ approaches, regardless of whether the variance-covariance parameters are known or not. If they are not known, they are replaced in the formula (10.23) by their estimates. The matrix $\boldsymbol{X}'\boldsymbol{V}^{-1}\boldsymbol{X}$ is the information matrix for the parameters $\boldsymbol{\beta}$.

The first-order partial derivatives with respect to the variance and covariance parameters can be derived using formal matrix differentiation formulas; for example, for an arbitrary parameter $\boldsymbol{\theta}$ involved in $\boldsymbol{\Sigma}$ (or σ^2) we have

$$\frac{\partial \log(\det \boldsymbol{V}_j)}{\partial \boldsymbol{\theta}} = \operatorname{tr}\left(\boldsymbol{V}_j^{-1}\frac{\partial \boldsymbol{V}_j}{\partial \boldsymbol{\theta}}\right) \tag{10.24}$$

$$\frac{\partial \operatorname{tr}(\boldsymbol{V}^{-1}\boldsymbol{S})}{\partial \boldsymbol{\theta}} = -\operatorname{tr}\left(\boldsymbol{V}^{-1}\frac{\partial \boldsymbol{V}}{\partial \boldsymbol{\theta}}\boldsymbol{V}^{-1}\boldsymbol{S}\right) \tag{10.25}$$

$$\frac{\partial \boldsymbol{V}_j}{\partial \sigma^2} = \boldsymbol{I}_{n_j} \tag{10.26}$$

$$\frac{\partial \boldsymbol{V}_j}{\partial \boldsymbol{\theta}} = \boldsymbol{Z}_j\frac{\partial \boldsymbol{\Sigma}}{\partial \boldsymbol{\theta}}\boldsymbol{Z}'_j. \tag{10.27}$$

The expected information submatrix (the negative of the expectation of the matrix of second-order partial derivatives) corresponding to the regression and variance-covariance parameters is equal to zero:

$$-\operatorname{E}\left\{\frac{\partial^2 \log \lambda_F}{\partial \boldsymbol{\beta}\,\partial \boldsymbol{\theta}}\right\} = -\operatorname{E}\left\{\frac{\partial^2 \log \lambda_R}{\partial \boldsymbol{\beta}\,\partial \boldsymbol{\theta}}\right\} = \mathbf{0}, \tag{10.28}$$

and an element of the information matrix corresponding to a pair of variance-covariance parameters has the form

$$-\operatorname{E}\left\{\frac{\partial^2 \log \lambda_F}{\partial \boldsymbol{\theta}\,\partial \boldsymbol{\theta}'}\right\} = \frac{1}{2}\operatorname{tr}\left(\boldsymbol{V}^{-1}\frac{\partial \boldsymbol{V}}{\partial \boldsymbol{\theta}}\boldsymbol{V}^{-1}\frac{\partial \boldsymbol{V}}{\partial \boldsymbol{\theta}'}\right), \tag{10.29}$$

which can be expressed as a group-wise sum of cross-products of the elements of the matrices $\boldsymbol{Z}'_j\boldsymbol{V}_j^{-1}\boldsymbol{Z}_j$ (see Jennrich and Schluchter 1986, or Longford 1987).

The block-diagonal form of the information matrix allows each iteration to be split into two parts: estimation of the regression parameters, and estimation of the variance and covariance parameters. For estimation of the within-group variance σ^2, it is advantageous to consider the reparameterization $(\sigma^2, \boldsymbol{\Sigma}) \to (\sigma^2, \boldsymbol{\Psi})$, where $\boldsymbol{\Psi} = \sigma^{-2}\boldsymbol{\Sigma}$. It is easy to see that in this parameterization the variance σ^2 can be partialed out from both log-likelihoods (10.21) and (10.22), and by setting the first-order partial derivative with respect to σ^2 to zero we obtain the respective solutions for $\mathrm{MLE_F}$ and $\mathrm{MLE_R}$,

$$\hat{\sigma}^2_F = \frac{\mathrm{tr}(\boldsymbol{W}^{-1}\boldsymbol{S})}{N} \tag{10.30}$$

$$\hat{\sigma}^2_R = \frac{\mathrm{tr}(\boldsymbol{W}^{-1}\boldsymbol{S})}{N-p}, \tag{10.31}$$

where $\boldsymbol{W} = \sigma^{-2}\boldsymbol{V} = \boldsymbol{I}_N + \boldsymbol{I}_{N_2} \otimes \{\boldsymbol{Z}_j\boldsymbol{\Psi}\boldsymbol{Z}'_j\}$. We see that MLE_R can be interpreted as taking into account the degrees of freedom lost due to the p regression parameters. The matrix derivative $\partial\boldsymbol{\Sigma}/\partial\boldsymbol{\theta}$ is equal to the $r \times r$ matrix of elementwise derivatives, that is, for a variance or covariance parameter $\boldsymbol{\theta}$ it is equal to a matrix containing one(s) in the position(s) where $\boldsymbol{\theta}$ appears, and zeros elsewhere.

If the group sizes are substantial, it becomes paramount for any computationally efficient algorithm to avoid numerical inversion of the within-group variance matrices $\boldsymbol{V}_j$. The pattern of these matrices allows analytical expressions for their inverses and determinants in terms of inverses and determinants of small matrices. We have

$$\log(\det \boldsymbol{V}_j) = n_j \log(\sigma^2) + \log(\det \boldsymbol{G}_j) \tag{10.32}$$

$$\boldsymbol{V}_j^{-1} = \sigma^{-2}\boldsymbol{I}_{n_j} - \sigma^{-4}\boldsymbol{Z}_j\boldsymbol{\Sigma}\boldsymbol{G}_j^{-1}\boldsymbol{Z}'_j, \tag{10.33}$$

where $\boldsymbol{G}_j = (\boldsymbol{I}_r + \sigma^{-2}\boldsymbol{Z}'_j\boldsymbol{Z}_j\boldsymbol{\Sigma})$ is an $r \times r$ matrix. Note that $\boldsymbol{Z}'_j\boldsymbol{V}_j^{-1}\boldsymbol{Z}_j = \sigma^{-2}\boldsymbol{G}_j^{-1}\boldsymbol{Z}'_j\boldsymbol{Z}_j$.

The conditional distribution of the random terms $\boldsymbol{\delta}_j$ is

$$(\boldsymbol{\delta}_j \mid \boldsymbol{\beta}, \sigma^2, \boldsymbol{\Sigma}; \boldsymbol{X}, \boldsymbol{Z}, \boldsymbol{y}) \sim \mathcal{N}(\boldsymbol{\Sigma}\boldsymbol{Z}'_j\boldsymbol{V}_j^{-1}\boldsymbol{e}_j,\ \boldsymbol{\Sigma} - \boldsymbol{\Sigma}\boldsymbol{Z}'_j\boldsymbol{V}_j^{-1}\boldsymbol{Z}_j\boldsymbol{\Sigma}) \tag{10.34}$$

($\boldsymbol{e} = \boldsymbol{y} - \boldsymbol{X}\boldsymbol{\beta}$), which is equivalent to $\mathcal{N}(\sigma^{-2}\boldsymbol{\Sigma}\boldsymbol{G}_j^{-1}\boldsymbol{Z}'_j\boldsymbol{e}_j,\ \boldsymbol{\Sigma}\boldsymbol{G}_j^{-1})$. The proof of (10.32)–(10.34) is given in the appendix.

Maximum likelihood estimates of the variance and covariance parameters can also be obtained by the generalized least-squares method, as demonstrated by Goldstein (1986). Suitable linear combinations of the pairwise products of residuals, matched with their expectations, yield equations for these parameters. The advantage of this method is its simple description and straightforward generalizability. An economic organization of the computations is paramount (note that the number of cross-products of residuals is $\frac{N(N+1)}{2}$).

The Expectation Maximization (EM) algorithm gives a general approach for maximizing the likelihood in complex problems. A comprehensive description of the EM algorithm, with several examples, is provided by Dempster et al. (1977), although the general idea had been widely applied earlier. The observed data ($\boldsymbol{y}$) are assumed to be a subset of a dataset $(\boldsymbol{y}, \boldsymbol{\gamma})$; whereas the joint density (likelihood) for the observed data $\boldsymbol{y}$, referred to in this context as the *incomplete data*, has a complex form, the maximization of the likelihood for the *complete data* $(\boldsymbol{y}, \boldsymbol{\gamma})$ would represent a simpler problem. Any choice for the complement of the incomplete data, $\boldsymbol{\gamma}$ (referred to as the *missing data*), that renders the complete data analysis a simple one (e.g., an ordinary regression), is suitable. For the random coefficient model (10.19) the obvious choice for $\boldsymbol{\gamma}$ are the random vectors $\{\boldsymbol{\delta}_j\}$. If they were observed/known, the constructed outcomes $y_j^* = y_j - Z_j\delta_j$ would satisfy the ordinary regression model

$$y^* = X\beta + \varepsilon,$$

($y^* = \mathbf{1}_{N_2} \otimes \{y^*\}$) and OLS would yield the maximum likelihood estimate of the regression parameters. The between-group covariance matrix would be estimated from the vectors $\{\delta_j\}$ in an obvious way.

The EM algorithm is an iterative procedure, with each iteration consisting of two parts: In the first part (*Expectation*, or E step) the conditional expectations of the complete-data sufficient statistics are estimated, based on the current estimates of all the model parameters. The second stage (*Maximization*, or M step) involves maximization of the complete-data likelihood, with functions of the complete data replaced by their conditional expectations obtained in the E step. The iterations of the EM algorithm are carried out until the corrections for the estimated parameters are smaller than a prescribed tolerance. Note that, in general, the M step of an EM procedure may itself involve an iterative procedure.

An application of the EM algorithm to random coefficient models is outlined by Dempster et al. (1977), and described in detail by Rubin et al. (1981); see also Raudenbush (1988). The EM algorithm tends to have very poor convergence properties, especially for complex problems, and when the likelihood has its maximum for a singular matrix $\hat{\Sigma}$ (Lindstrom and Bates 1989; Thompson and Meyer 1986). Each iteration of the EM algorithm requires calculation of the conditional expectations and variances of the group-level random terms, as given in (10.34). However, the Fisher scoring, the Newton-Raphson, and moment methods require similar expressions. The convergence properties of the EM algorithm can be improved substantially by implementing simple acceleration routines, such as the Aitken accelerator. Each iteration of the EM algorithm results in a nonnegative definite estimate of the covariance matrix $\hat{\Sigma}$, and in an increased value of the likelihood; the algorithm converges to a local maximum under very mild regularity conditions (Dempster et al. 1977).

The Fisher scoring, the Newton-Raphson, and moment-type methods have, under normal circumstances, quadratic convergence properties, and they usually provide a clear indication of confounding of the estimated parameters (singularity of the matrix of second-order partial derivatives, or of the information matrix), but their iterations may result in decreased values of the likelihood, or in estimated variance matrices with negative eigenvalues. While the former does not constitute a serious problem (it usually occurs only in earlier iterations, and the losses are quickly recovered), the latter is a nontrivial problem. A general solution, involving damping of the corrections, is proposed by Longford (1989). Negative estimated variances, and, by extension, estimated variance matrices with negative eigenvalues, provide an important diagnostic tool. A negative eigenvalue implies that the solution, restricted to the parameter space, is a singular matrix, and as such it may indicate overparametrization for the covariance matrix Σ. The standard solution by Lagrange multipliers is often not practicable. Lindstrom and Bates (1989) propose to estimate the LR decomposition of the covariance matrix Σ (a decomposition to a product of two triangular matrices), which leads to an unconstrained maximization problem. However, their method cannot be directly applied to patterned variance matrices (e.g., certain covariances constrained to be zero).

4.1 The Fisher Scoring Algorithm

This section contains computational details of the Fisher scoring algorithm for the random coefficient model (10.19). Readers who are unfamiliar with the algorithm can find a general exposition in the appendix on numerical analysis.

Owing to (10.28), an iteration of the Fisher scoring algorithm consists of two parts: updating of the regression parameters, and updating of the covariance structure parameters. The former is done using the generalized least squares formula (10.23) with the covariance matrix $\boldsymbol{V}$ evaluated for the current values of the covariance structure parameters. Equation (10.33) implies that

$$\boldsymbol{x}_1'\boldsymbol{V}^{-1}\boldsymbol{x}_2 = \sigma^{-2}(\boldsymbol{x}_1'\boldsymbol{x}_2 - \sigma^{-2}\sum_j \boldsymbol{x}_{1j}'\boldsymbol{Z}_j\boldsymbol{\Sigma}\boldsymbol{G}_j^{-1}\boldsymbol{Z}_j'\boldsymbol{x}_{2j}), \tag{10.35}$$

where $\boldsymbol{x}_1$ and $\boldsymbol{x}_2$ are arbitrary $N \times 1$ vectors, and $\boldsymbol{x}_{1j}$ and $\boldsymbol{x}_{2j}$ are their respective subvectors corresponding to group j.

For the first-order partial derivatives with respect to an element of $\boldsymbol{\Sigma}$ expressions of the form (10.24) and (10.25) have to be evaluated. By substituting (10.27) in (10.24) we obtain

$$\operatorname{tr}\left(\boldsymbol{V}_j^{-1}\frac{\partial \boldsymbol{V}_j}{\partial \boldsymbol{\theta}}\right) = \operatorname{tr}\left(\boldsymbol{Z}_j'\boldsymbol{V}_j^{-1}\boldsymbol{Z}_j\frac{\partial \boldsymbol{\Sigma}}{\partial \boldsymbol{\theta}}\right) = \sigma^{-2}\operatorname{tr}\left(\boldsymbol{G}_j^{-1}\boldsymbol{Z}_j'\boldsymbol{Z}_j\frac{\partial \boldsymbol{\Sigma}}{\partial \boldsymbol{\theta}}\right), \tag{10.36}$$

which is equal to the kth diagonal element of $\sigma^{-2}\boldsymbol{G}_j^{-1}\boldsymbol{Z}_j'\boldsymbol{Z}_j$ for $\boldsymbol{\theta} = \Sigma_{kk}$, and to twice the element (k, h) of $\sigma^{-2}\boldsymbol{G}_j^{-1}\boldsymbol{Z}_j'\boldsymbol{Z}_j$ for the covariance $\boldsymbol{\theta} = \Sigma_{kh}$. Further, (10.25) can be rewritten as

$$\begin{aligned}\operatorname{tr}\left(\boldsymbol{V}^{-1}\frac{\partial \boldsymbol{V}}{\partial \boldsymbol{\theta}}\boldsymbol{V}^{-1}\boldsymbol{S}\right) &= \sum_j \operatorname{tr}\left(\boldsymbol{e}_j'\boldsymbol{V}_j^{-1}\boldsymbol{Z}_j\frac{\partial \boldsymbol{\Sigma}}{\partial \boldsymbol{\theta}}\boldsymbol{Z}_j'\boldsymbol{V}_j^{-1}\boldsymbol{e}_j\right) \\ &= \sigma^{-4}\sum_j \boldsymbol{e}_j'\boldsymbol{Z}_j(\boldsymbol{G}_j^{-1})'\frac{\partial \boldsymbol{\Sigma}}{\partial \boldsymbol{\theta}}\boldsymbol{G}_j^{-1}\boldsymbol{Z}_j'\boldsymbol{e}_j\end{aligned} \tag{10.37}$$

($\boldsymbol{e}_j = \boldsymbol{y}_j - \boldsymbol{X}_j\boldsymbol{\beta}$). Group j contributes to (10.37) with the product of two elements of the vector $\boldsymbol{G}_j^{-1}\boldsymbol{Z}_j'\boldsymbol{e}_j$. Finally, an element of the information matrix (10.29) is equal to a constant multiple of

$$\begin{aligned}&\sum_j \operatorname{tr}\left(\boldsymbol{V}_j^{-1}\frac{\partial \boldsymbol{V}_j}{\partial \boldsymbol{\theta}}\boldsymbol{V}_j^{-1}\frac{\partial \boldsymbol{V}_j}{\partial \boldsymbol{\theta}'}\right) = \sum_j \operatorname{tr}\left(\boldsymbol{Z}_j'\boldsymbol{V}_j^{-1}\boldsymbol{Z}_j\frac{\partial \boldsymbol{\Sigma}}{\partial \boldsymbol{\theta}}\boldsymbol{Z}_j'\boldsymbol{V}_j^{-1}\boldsymbol{Z}_j\frac{\partial \boldsymbol{\Sigma}}{\partial \boldsymbol{\theta}'}\right) = \\ &\sigma^{-4}\sum_j \operatorname{tr}\left(\boldsymbol{G}_j^{-1}\boldsymbol{Z}_j'\boldsymbol{Z}_j\frac{\partial \boldsymbol{\Sigma}}{\partial \boldsymbol{\theta}}\boldsymbol{G}_j^{-1}\boldsymbol{Z}_j'\boldsymbol{Z}_j\frac{\partial \boldsymbol{\Sigma}}{\partial \boldsymbol{\theta}'}\right),\end{aligned} \tag{10.38}$$

and each summand is a product of two elements of the matrix $\boldsymbol{G}_j^{-1}\boldsymbol{Z}_j'\boldsymbol{Z}_j$, or a sum of two such products. Expressions for a different parameterization are obtained by application of the chain rule.

In summary, the Fisher scoring algorithm consists of the following steps:

1. Ordinary regression; starting values for β, σ^2 and Σ

2. Fisher scoring iteration (use the chain rule if applicable)

3. Adjust the new solution, if necessary to ensure nonnegative definiteness of Σ, and decide whether to terminate iterations or return to step 2.

The second step consists of calculation of the contributions to the various elements of the scoring vector and the information matrix. It is useful to organize these calculations in such a way that only a single loop over the groups is needed, for example:

1. Calculate G_j, $\det G_j$, G_j^{-1}, ΣG_j^{-1}

2. Calculate and accumulate the contributions to $(X, y)'V^{-1}(X, y)$

3. Calculate $G_j^{-1}Z_j'Z_j$ and accumulate all cross-products of its elements.

Step 2 assumes that the sample totals of squares and cross-products $(X, y)'(X, y)$ are stored throughout the iterations (they are required first for the OLS). The term $e'V^{-1}e$ in the log-likelihood is obtained using the identity

$$e'V^{-1}e = y'V^{-1}y - 2\beta'X'V^{-1}y + \beta'X'V^{-1}X\beta.$$

The iterations can be terminated when the vector of corrections for all the estimated parameters has a norm smaller than a prescribed tolerance (which may depend on the number of parameters), and/or when the difference of two consecutive values of the log-likelihood is smaller than another prescribed threshold.

An important advantage of such a method is that it can be relatively easily adapted to become the M step of an EM algorithm, so that the random coefficient model can itself be considered as the complete-data problem. Conjugate gradient methods (see, e.g., McIntosh 1982 or Luenberger 1984) can be used with advantage for models with large numbers of parameters. This avenue for improving the computational efficiency of algorithms for random coefficient models appears not to have been fully explored.

Lange and Laird (1989) describe a class of balanced designs for which maximum likelihood solutions can be obtained by noniterative procedures.

A set of minimal sufficient statistics for the model (10.9) is:

A. Sample totals of cross-products $y'y$ and $y'X$

B. Within-group totals of cross-products $X_j'y_j$, $j = 1, 2, \ldots, N_2$.

Note that the random coefficient model (10.19) and the corresponding analysis of covariance model require the same set of minimal sufficient statistics. In addition, the likelihood function depends on the data via the cross-products $X'X$, $\{X_j'Z_j\}_j$, and $\{Z_j'Z_j\}_j$.

Minimum variance quadratic estimation and minimum norm quadratic estimation are two nonlikelihood-based methods for estimation with random coefficient models. See Rao (1971a,b) for details.

4.2 Diagnostics

Diagnostic procedures for ordinary regression are based principally on the residuals, $e = y - \boldsymbol{x}\hat{\boldsymbol{\beta}}$. Clearly the first step in extending these procedures for the random coefficient models consists of separation of the group- and individual-level components of the residual e.

The natural analogue to the OLS residuals is the conditional expectation of the random terms. In addition to formula (10.34) for the group-level random terms, we have

$$\left(\boldsymbol{\varepsilon} \mid \boldsymbol{\beta}, \sigma^2, \boldsymbol{\Sigma}; \boldsymbol{X}, \boldsymbol{y}\right) \sim N\left(\sigma^2 \boldsymbol{V}^{-1}\boldsymbol{e},\ \sigma^2 \boldsymbol{I} - \sigma^4 \boldsymbol{V}^{-1}\right); \tag{10.39}$$

see the appendix for proof. Note that

$$\mathrm{E}\left(\varepsilon_{ij} \mid \boldsymbol{\beta}, \sigma^2, \boldsymbol{\Sigma}; \boldsymbol{X}, \boldsymbol{y}\right) + \boldsymbol{x}'_{ij}\mathrm{E}\left(\boldsymbol{\delta}_j \mid \boldsymbol{\beta}, \sigma^2, \boldsymbol{\Sigma}; \boldsymbol{X}, \boldsymbol{y}\right) = y_{ij} - \boldsymbol{x}_{ij}\boldsymbol{\beta}\,.$$

We will refer to the conditional expectations of the random terms, (10.34) and (10.39), as the group-level and the individual-level residuals, and denote them by $\hat{\boldsymbol{\delta}}_j$ and $\hat{\varepsilon}_{ij}$, respectively. It is now natural to define a pair of diagnostic procedures, one for each level of nesting, to see if the residuals conform with the assumptions of normality, homoscedasticity, and independence of the corresponding random terms. For the individual level we have the univariate residuals $\{\hat{\varepsilon}_{ij}\}_{i,j}$, and their inspection for normality and absence of stochastic dependence is analogous to the corresponding procedures in OLS.

Although such procedures perform satisfactorily in most problems, they disregard the fact that under the assumptions of the model the residuals are neither independent nor have equal variances. The residuals are normally distributed, though, and can be "standardized" by a suitable linear transformation. See Cook and Weisberg (1982) for details.

In addition, the assumption of equal within-group variances has to be considered, especially for data with large groups. A trivial method for checking this assumption would consist of generating summaries, such as the within-group means of squares of the residuals, and using graphical displays to search for extreme values of these summaries, or for some pattern among them. Such a procedure could also be enhanced by taking into account the distribution of the residuals.

Apart from the compound symmetry models, similar diagnostic procedures for the group-level random terms involve inspection for multivariate normality, and are therefore more complex. Componentwise diagnostics are an obvious solution, but for small dimensions of $\boldsymbol{\Sigma}$ it is more natural to plot the linear functions $\boldsymbol{z}\hat{\boldsymbol{\delta}}_j$ against $\boldsymbol{z}$, in the range of values of $\boldsymbol{z}$ occurring in the data for group j. Such plots can be much more readily used for identification of exceptional groups. See Longford (1985), Fieldsend, Longford, and McLeay (1987), or Section 8 (Figure 3) for examples. A more formal approach that takes into account the distribution of the conditional expectations is given by Lange and Ryan (1989).

4.3 Model Selection

Another aspect of model diagnostics involves assessment of the parametric structure for the mean and for the variances and covariances. The general principles of the likelihood ratio test can be applied for pairwise comparisons of models. Suppose that in the model (10.19) the parameters $(\boldsymbol{\beta}, \sigma^2, \boldsymbol{\Sigma})$ belong to the space $\boldsymbol{\Omega}$, and a null hypothesis,

$$\left(\boldsymbol{\beta}, \sigma^2, \boldsymbol{\Sigma}\right) \in \boldsymbol{\Omega}_0 \subset \boldsymbol{\Omega},$$

is specified. Let λ_0 and λ_1 be the values of the likelihood at the respective maximum likelihood solutions for the *constrained* model $(\boldsymbol{\Omega}_0)$ and the unconstrained model $(\boldsymbol{\Omega})$. Then under the null hypothesis, and under certain regularity conditions, the likelihood ratio statistic

$$2 \log \left(\frac{\lambda_1}{\lambda_0}\right)$$

has an asymptotically χ^2_s distribution, where s is the number of functionally independent constraints imposed on the parameters in the null hypothesis, in addition to the constraints in the parameter space $\boldsymbol{\Omega}$. Intuitively, the asymptotics in the context of individuals within groups refer to increasing number of groups, with the average within-group information converging to a positive definite matrix. See Miller (1977) for a formal treatment. The likelihood ratio statistic has an asymptotically χ^2 distribution only if the (unknown) true parameter vector lies in the interior of the parameter space $\boldsymbol{\Omega}_0$. This assumption is particularly problematic when the estimated variance matrix $\hat{\boldsymbol{\Sigma}}$ is nonsingular. The full likelihood λ_F can be used for both model selection for the regression and random parts, whereas the use of the restricted likelihood λ_R is appropriate only with a fixed regression part (no constraints on the regression parameters in $\boldsymbol{\Omega}_0$), because different regression parts lead to different sets of error contrasts, and hence the corresponding likelihoods involve different datasets.

5 Multiple Levels of Nesting

The random coefficient model (10.9) involves a single grouping factor. In numerous applications we encounter data structures that involve several grouping factors, for example, students within departments within colleges in educational research, or observations within subjects, with interviewer as another factor. The former example involves nested classification since students within a department also belong to the same college, while the latter, depending on the allocation of interviewers to respondents, may be an example of cross-classification–a respondent may be interviewed by several interviewers, and each interviewer contacts several respondents.

Extension of the model (10.9) to multiple classifications presents no conceptual problems; for illustration we discuss in detail only the nested classification models. Suppose we have elementary units (e.g., individuals), $i = 1, 2, \ldots, n_{2,jh}$, within level 2 units (say, groups), $j = 1, 2, \ldots, n_{3,h}$, and level 2 units within level 3 units (say, areas), $h = 1, 2, \ldots, N_3$. The sample sizes of elementary and level 2 units are $N_1 = \sum_h \sum_j n_{2,jh}$, and $N_2 = \sum_h n_{3,h}$, respectively. Suppose we have an outcome variable y and a set of p explanatory variables $\boldsymbol{x}$, consisting of the intercept and the variables defined at the three specified levels of nesting, $\boldsymbol{x} = (1,\ \boldsymbol{x}_1,\ \boldsymbol{x}_2,\ \boldsymbol{x}_3)$. We assume a model with linear within-group and within-area regressions:

$$y_{ijh} = \boldsymbol{x}_{ijh}\boldsymbol{b}_{jh} + \varepsilon_{ijh}\,, \tag{10.40}$$

where $\{\varepsilon_{ijh}\}$ form a random sample from $\mathcal{N}(0, \sigma^2)$, and the random coefficients $\boldsymbol{b}_{jh}$ have decompositions

$$\boldsymbol{b}_{jh} = \boldsymbol{\beta} + \boldsymbol{\delta}_{3,h} + \boldsymbol{\delta}_{2,jh} \tag{10.41}$$

into the vector of the population average regression parameters ($\boldsymbol{\beta}$), the deviations of the *area-average regressions* from the population average ($\boldsymbol{\delta}_{3,h}$), and the deviations of the *group-level regressions* from the area-average regressions ($\boldsymbol{\delta}_{2,jh}$). The area-level deviations $\boldsymbol{\delta}_{3,h}$ are assumed to be a random sample from a centered normal distribution, $\mathcal{N}(\mathbf{0},\ \Sigma_3)$ and the group-level deviations $\boldsymbol{\delta}_{2,jh}$ are assumed to be another random sample from a centered normal distribution, $\mathcal{N}(\mathbf{0},\ \Sigma_2)$. The three random samples, $\{\varepsilon_{ijh}\}$, $\{\boldsymbol{\delta}_{2,jh}\}$, and $\{\boldsymbol{\delta}_{3,h}\}$, are mutually independent.

The terminology of "association with variation" introduced for the two-level model in Section 3 can be extended for the general random coefficient model. For the model (10.40)–(10.41) we will say that a variable x is associated with group-level variation if the corresponding components of the random vectors $\boldsymbol{\delta}_{2,jh}$ are not constrained to zero (i.e., their variance is not constrained to zero). Similarly, a variable is said to be associated with area-level variation if the corresponding components of the random vectors $\boldsymbol{\delta}_{3,h}$ are not constrained to zero. If we want to adhere to the interpretation in terms of varying regression slopes, then we do not associate area-level variables with either group- or area-level variation, and associate group-level variables only with area-level variation.

The variance of an observation decomposes into three parts:

$$\mathrm{var}(y) = \sigma^2 + \boldsymbol{x}\Sigma_2\boldsymbol{x}' + \boldsymbol{x}\Sigma_3\boldsymbol{x}'. \tag{10.42}$$

The covariance of a pair of observations from the same group is

$$\mathrm{cov}(y_{ijh}\,,\ y_{i'jh}) = \boldsymbol{x}_{ijh}\Sigma_2\boldsymbol{x}'_{i'jh} + \boldsymbol{x}_{ijh}\Sigma_3\boldsymbol{x}'_{i'jh}\,, \tag{10.43}$$

and the covariance of a pair of observations from the same area, but from different groups, is equal to

$$\mathrm{cov}(y_{ijh}\,,\ y_{i'j'h}) = \boldsymbol{x}_{ijh}\Sigma_3\boldsymbol{x}'_{i'j'h}\,. \tag{10.44}$$

Thus we can use the model (10.40)–(10.41) to describe the variation of regression coefficients, within-unit covariance, or between-unit heterogeneity.

5.1 Estimation

The methods for maximum likelihood estimation with random coefficient models discussed in Section 4 have conceptually straightforward extensions for the multilevel models. In the EM approach the random samples $\{\boldsymbol{\delta}_{2,jh}\}$ and $\{\boldsymbol{\delta}_{3,h}\}$ form the missing data, so that the complete data method is ordinary regression. Direct likelihood maximization procedures require computationally efficient formulas for inversion of the variance matrices for the observations.

The notation for two-level data, introduced in Section 4, can be extended for multiple levels. Data for group j in area h are

$$\boldsymbol{y}_{jh} = \mathbf{1}_{n_{2,jh}} \otimes \{y_{ijh}\} \quad \text{and} \quad \boldsymbol{X}_{jh} = \mathbf{1}_{n_{2,jh}} \otimes \{\boldsymbol{x}_{ijh}\} \quad ,$$

and the data for area h are formed by stacking the data for its groups:

$$\boldsymbol{y}_{h} = \mathbf{1}_{n_{3,h}} \otimes \{\boldsymbol{y}_{jh}\} \quad \text{and} \quad \boldsymbol{X}_{h} = \mathbf{1}_{n_{3,h}} \otimes \{\boldsymbol{X}_{jh}\} \quad .$$

The area vectors of outcomes $\boldsymbol{y}_h$ are mutually independent and have the variance matrices

$$\text{var}(\boldsymbol{y}_h) \;=\; \boldsymbol{I}_{n_{3,h}} \otimes \{\boldsymbol{V}_{2,jh}\} \;+\; \boldsymbol{X}_h \boldsymbol{\Sigma}_3 \boldsymbol{X}'_h \,, \tag{10.45}$$

where $\boldsymbol{J}_n$ is the $n \times n$ matrix of ones, and the matrices $\boldsymbol{V}_{2,jh}$ are equal to

$$\boldsymbol{V}_{2,jh} \;=\; \sigma^2 \boldsymbol{I}_{n_{2,jh}} \;+\; \boldsymbol{X}_{jh} \boldsymbol{\Sigma}_2 \boldsymbol{X}'_{jh} \,. \tag{10.46}$$

The pattern of the covariance matrix (10.45) can be exploited to obtain formulas for its inverse and determinant. The inverse of the covariance matrix (10.45) can be expressed in terms of the inverse of the variance matrices (10.46) (see appendix), and for these the formula (10.33) is applicable.

The formulas (10.21)–(10.31) carry over to the multilevel case. The complexity of these formulas is not affected by the numbers of elementary observations within groups or within areas, but it depends somewhat on the number of groups within areas.

5.2 Proportion of Variation Explained in Multilevel Models

In ordinary regression "the proportion of variation explained" (R^2) is often used as a measure of reduction of the residual variance attributed to the explanatory variables. For independent outcomes $\{y_i\}$ we consider the raw model

$$y \;=\; \mu \;+\; \varepsilon, \tag{10.47}$$

with variance $\text{var}(\varepsilon) \;=\; \sigma_0^2$, and the *adopted model* involving a selected list of explanatory variables,

$$y \;=\; \boldsymbol{x}\boldsymbol{\beta} \;+\; \varepsilon,$$

with residual variance var(ε) = σ^2. The proportion of variation explained is defined as

$$R^2 = 1 - \frac{\sigma^2}{\sigma_0^2};$$

the variances in this formula are routinely replaced by their estimates.

An extension of the R^2 for two-level models has to acknowledge that explanatory variables can "explain" either only the elementary-level or the group-level variation, or contribute to the reduction of both variances.

In order to parallel this definition for ordinary regression, for a specific two-level dataset we consider the *raw* model

$$y_{ij} = \mu + \delta_j + \varepsilon_{ij},$$

with variance components Σ_0^2 = var(δ_j) and σ_0^2 = var(ε_{ij}), and the *adopted* model

$$y_{ij} = \boldsymbol{x}_{ij}\boldsymbol{\beta} + \boldsymbol{z}_{ij}\boldsymbol{\delta}_j + \varepsilon_{ij},$$

with elementary-level variance σ^2 and group-level variance $\boldsymbol{z}_{ij}\boldsymbol{\Sigma}_2\boldsymbol{z}'_{ij}$; the variables $\boldsymbol{z}_{ij}$ are defined at individual level. We define the elementary- and group-level R^2's as

$$R_1^2 = 1 - \frac{\sigma^2}{\sigma_0^2} \tag{10.48}$$

and

$$R_2^2 = 1 - \frac{\boldsymbol{z}_{ij}\boldsymbol{\Sigma}_2\boldsymbol{z}'_{ij}}{\Sigma_0^2}, \tag{10.49}$$

respectively. The elementary-level R^2 is always in the interval $[0, 1]$. Note that the group-level R^2 depends on the explanatory variables, and since the numerator in (10.49) is a quadratic function, R_2^2 may attain negative values, in particular for values of $\boldsymbol{z}$ outside the range of the data. The proportion R_2^2 is constant only for the compound symmetry model (10.14) (when all slope-variances vanish). In practice the quantities (10.48) and (10.49) are replaced by their naive estimates. Just as in ordinary regression, a high value of estimated R^2 (close to 1) at either level may be a sign of having indiscriminantly used an unduly large number of explanatory variables, or having selected the variables from a large pool.

In analyses with human subjects as the elementary-level units, reseachers have typically observed much higher values of R_2^2 than of R_1^2 (e.g., Aitkin and Longford 1986; Lockheed and Longford 1989); it appears that it is much easier to find a description of the group-level variation than of the subject-level variation.

In compound symmetry models inclusion of a group-level variable (i.e., a variable constant within groups) will result in an increment to R_2^2, whereas R_1^2 will remain unchanged. For other variables the relative sizes of reductions of R_2^2 and R_1^2 depend on the decomposition of variation of the explanatory variable to its within- and between-group components. Inclusion of a variable that is balanced across the groups (no between-group component) will result in an increment to R_1^2 only. The more unbalanced the explanatory variable (larger between-group component) the larger the relative increment to R_2^2.

Extension of the definition of R^2 to multiple levels of nesting is straightforward.

6 Generalized Linear Models

There are many settings in which data structures with multiple layers of nesting arise, where the observed outcomes have a distinctly nonnormal distribution, e.g., counts, dichotomous (0/1) data, or small positive quantities. In this section we discuss extensions of the random coefficient models to nonnormally distributed data, or equivalently, extensions of the generalized linear model (GLM) for independent observations to correlated observations.

Ordinary regression models rely on the assumption of normality of the outcome variable y. The class of GLMs is a natural extension of the ordinary regression models that accomodates a variety of distributional assumptions without any loss of flexibility of modeling of the dependence of the outcome variable on the explanatory variables.

We assume that $\{y_i\}_i$ are independent observations with a (discrete or absolutely continuous) density

$$f(y;\theta,\phi) = \exp\left\{\frac{y\theta - b(\theta)}{a(\phi)} + c(y,\phi)\right\}, \tag{10.50}$$

where a, b, and c are some functions and ϕ is a parameter (the *dispersion parameter*, or the *scale*); the role of θ is elaborated below.

The mean and variance of an observation are

$$\begin{aligned} \mathrm{E}(y) &= b'(\theta) \\ \mathrm{var}(y) &= b''(\theta)a(\phi), \end{aligned}$$

respectively, where b' and b'' denote the first- and second-order derivatives of b with respect to θ.

We assume that the expectation $b'(\theta)$ is related to the explanatory variables through a *link* function η:

$$\eta\{\mathrm{E}(y)\} = \boldsymbol{x}\boldsymbol{\beta},$$

where $\boldsymbol{\beta}$ is a vector of (regression) parameters.

Usually we set $a(\phi) = \phi$ (or ϕ/w, when considering a prior weight w). Note that if ϕ is known, (10.50) belongs to the *exponential family* of distributions. If we disregard the scale factor $a(\phi)$, the mean and the variance of an observation depend only on θ. The relationship of the variance of an observation to its mean is described by the *variance function* $b''(\theta)$, which we write as a function of the mean, $V(\mu)$.

For example, the Bernoulli distribution corresponds to (10.50) with

$$a(\phi) = 1, \quad b(\theta) = \log(1+\exp\,\theta), \quad c(y,\theta) = 0,$$

where θ is the logit of the probability of a successful outcome, $\theta = \log\left(\frac{p}{1-p}\right)$. The most commonly used link functions for the Bernoulli distribution are:

- the logit link:

$$\log\left(\frac{p}{1-p}\right), \tag{10.51}$$

- complementary log-log link:

$$\log\{-\log(1-p)\}, \tag{10.52}$$

- probit link:

$$\Phi^{-1}(p), \tag{10.53}$$

where Φ is the distribution function of the standard normal distribution. For each distribution (10.50) there is a unique *canonical* link function for which the linear predictor $\boldsymbol{x\beta}$ is equal to θ. The canonical link for the binary distribution is the logit function. For the Poisson distribution,

$$f(y;\lambda,\phi) = \frac{e^{-\lambda}\lambda^{y}}{y!} = \exp\{y\log\lambda - \lambda - log(y!)\},$$

it is the logarithm, $\theta = \log\lambda$. Ordinary (normal) regression corresponds to the identity link function, with constant variance function, and scale $\phi = \sigma^2$.

In the GLM we assume a set of independent outcomes $\{y_i\}_i$ and an associated set of vectors $\{\boldsymbol{x}_i\}_i$ of values of explanatory variables. The outcome y_i has the distribution with density $f(y_i;\theta_i,\phi)$ given by (10.50), and its expectation $b'(\theta_i)$ is related to the explanatory variables through the link function η:

$$\eta\{b'(\theta_i)\} = \boldsymbol{x}_i\boldsymbol{\beta};$$

$\boldsymbol{\beta}$ a vector of regression parameters. The scale $a(\phi)$ is assumed constant over i.

6.1 Estimation

Maximum likelihood estimation of the regression parameters $\boldsymbol{\beta}$ is accomplished by the method of iteratively reweighted least squares. The log-likelihood for the observations $\{y_i\}$ is

$$\log l = \sum_i \left\{\frac{y_i\theta_i - b(\theta_i)}{a(\phi)} + c(y_i,\phi)\right\} \tag{10.54}$$

and the first-order partial derivatives with respect to $\boldsymbol{\beta}$ are equal to

$$\frac{\partial \log l}{\partial\boldsymbol{\beta}} = \sum_i \frac{y_i-\mu_i}{a(\phi)}\frac{\partial\mu_i}{\partial\eta_i}\{V(\mu_i)\}^{-1}\boldsymbol{x}_i, \tag{10.55}$$

where μ_i is the expectation of the observation i and the partial derivatives are evaluated for $\eta_i = \boldsymbol{x}_i\boldsymbol{\beta}$.

For the canonical link, $\frac{\partial\mu_i}{\partial\eta_i} = V(\mu_i)$, which simplifies (10.55) somewhat. Then the vectors of totals of cross-products $\sum_i y_i\boldsymbol{x}_i$ are a set of sufficient statistics for the parameters in the linear predictor $\boldsymbol{x\beta}$.

The expected value of the matrix of second-order partial derivatives in the general case is equal to

$$\mathrm{E}\left\{\frac{\partial^2 \log l}{\partial \boldsymbol{\beta}\, \partial \boldsymbol{\beta}'}\right\} = -\{a(\phi)\}^{-1} \sum_i \left(\frac{\partial \mu_i}{\partial \eta_i}\right)^2 \{V(\mu_i)\}^{-1} \boldsymbol{x}_i' \boldsymbol{x}_i. \tag{10.56}$$

We define the *residual function*

$$r_i = (y_i - \mu_i)\frac{d\eta_i}{d\mu_i}, \tag{10.57}$$

and the *weight function*

$$W_i = \{V(\mu_i)\}^{-1}\left(\frac{d\mu_i}{d\eta_i}\right)^2. \tag{10.58}$$

An iteration of the Fisher scoring algorithm for maximization of the log-likelihood (10.54) is essentially a weighted least-squares procedure, with the residual function and the weight function in the respective roles of the residuals and the weights. Since the residual and weight functions depend on the linear predictor which is adjusted at each iteration, (10.57) and (10.58) have to be recalculated at each iteration.

A GLM has a number of alternative specifications. Instead of the log-likelihood (distribution) and the link function, the link and variance functions, or the residual and weight functions, can be used to specify a GLM. There are combinations of link and variance functions which do not correspond to any proper distribution in the exponential family. This can be regarded as an advantage—we may wish to define a GLM without reference to a distribution, but merely by a functional relationship of the variance to the mean. Additionally, of course, the form of the linear predictor, $\boldsymbol{x}\boldsymbol{\beta}$, has to be specified, in complete analogy with ordinary regression.

For a detailed background on the GLM we refer the reader to McCullagh and Nelder (1989). Dobson (1983) and Aitkin et al. (1989) contain more elementary introductions to the GLM.

6.2 Quasi-likelihood

Definition of a GLM without explicit reference to the log-likelihood motivated the original definition of the quasi-likelihood by Wedderburn (1974). The quasi-likelihood for an observation y with mean μ and variance function $V(\mu)$ is defined by the partial derivative

$$\frac{\partial Q(y;\mu)}{\partial \mu} = \frac{y-\mu}{V(\mu)}, \tag{10.59}$$

so that the quasi-likelihood itself is given as the integral of (10.59), with an appropriate normalizing constant (if the integral converges). The scale $a(\phi)$ is absorbed in $V(\mu)$. See McCullagh and Nelder (1989, chap. 9) for further motivation.

In the GLM, testing hypotheses about nested subsets of explanatory variables in the linear predictor is accomplished by comparison of the likelihoods (the likelihood ratio test).

The *deviance function* is defined as the likelihood ratio test statistic for the hypothesis for a specific model against the alternative that the model fit is exact for each observation. For the quasi-likelihood this statistic has the form

$$D(y; \mu) = -2 \int_y^{\mu} \frac{y-u}{V(u)} du . \qquad (10.60)$$

Thus, when working with the quasi-likelihood, the *quasi-likelihood ratio* test statistic is calculated as the difference of the expressions (10.60) for two competing models. The quasi-likelihood can be used for comparison of models with different link functions, but not for comparisons involving different variance functions. Nelder and Pregibon (1987) defined an extended quasi-likelihood that eliminates this deficiency. This quasi-likelihood has a normal-lookalike form:

$$Q(y; \mu) = -\frac{1}{2} \log \{2\pi\phi V(y)\} - \frac{1}{2\phi} D(y; \mu) , \qquad (10.61)$$

where the deviance function D is defined by the integral (10.60). This quasi-likelihood is a saddlepoint approximation to the log-likelihood (10.50).

The GLIM package (Numerical Algorithmus Group, 1986) implements the Fisher scoring algorithm for fitting GLMs with facilities for user-defined link and variance functions, but the algorithm can also be easily programmed using a computer language for matrix operations, such as SPlus, Matlab, or GAUSS.

6.3 Extensions for Dependent Data

In analogy with the extension of the ordinary regression models to the random coefficient models, in the GLM we can also consider randomly varying regression coefficients for data involving one or several layers of nesting. A general formulation for a two-level model involves a link function η relating the *conditional* expectation for an observation to the conditional linear predictor, given the regression coefficients $\boldsymbol{\beta}_j$ for group j. The regression coefficients $\boldsymbol{\beta}_j$ are assumed to be a random sample, across the groups, from a multivariate normal distribution, $\mathcal{N}(\boldsymbol{\beta}, \boldsymbol{\Sigma})$. Thus $\boldsymbol{\beta}$ represents the mean regression coefficient vector, although, because of the nonlinear nature of the link, not the regression for the average group. The covariance matrix $\boldsymbol{\Sigma}$ provides a description for the between-group variation, on an underlying scale defined by the link function.

Alternatively, such a model can be specified by the fixed and random design matrices. For example (using the notation of Section 4), for a two-level dataset we assume that the conditional expectation of the outcome y in group j, given the random vector $\boldsymbol{\delta}_j$, is related to the "design" row vectors $\boldsymbol{x}$ and $\boldsymbol{z}$ by

$$\eta\{\mathrm{E}(y \mid \boldsymbol{\delta}_j)\} = \boldsymbol{x\beta} + \boldsymbol{z\delta}_j , \qquad (10.62)$$

where η is a link function and $\boldsymbol{\delta}_j \sim \mathcal{N}(\mathbf{0}, \boldsymbol{\Sigma})$.

6.4 Estimation for Models With Dependent Data

Whereas definition of the models extending the GLM to the case of dependent observations appears to be straightforward, extensions of the associated computational algorithms involve appreciable difficulty.

Williams (1982) considered a logistic model for clustered binary data, with within-cluster correlation, and constructed an algorithm for estimating this correlation based on the observed variation in excess of what would be expected if the data were independent. This algorithm is suitable only for situations with small between-group variation.

Stiratelli, Laird, and Ware (1984) and Anderson and Aitkin (1985) applied the EM algorithm to estimation with binary data and the commonly used link functions (10.51)–(10.53). Although this algorithm is easy to extend for other distributional assumptions and link functions, it is computationally intensive, because it involves numerical integration at the E step of each iteration, and the number of iterations tends to be very large. Nevertheless, the algorithm is relatively easy to adapt for other distributions and link functions.

Rosner (1984, 1989) developed a class of logistic models and an approximate maximum likelihood algorithm for their estimation, but the method cannot be adapted for other distributional assumptions.

Morton (1987) describes an algorithm for variance component estimation in Poisson data with logarithm link function, taking advantage of the multiplicative nature of the random terms.

Bonney (1987) and Connolly and Liang (1988) express the joint distribution of the binary outcomes for a group in terms of various conditional distributions. This leads to a tractable log-likelihood which is relatively easy to maximize.

The log-likelihood for a general random coefficient GLM is given by the integral of the conditional densities,

$$\sum_j \log\left[\int \cdots \int \prod_i f\left\{y_{ij}; \eta^{-1}(\boldsymbol{x}_{ij}\boldsymbol{\beta} + \boldsymbol{z}_{ij}\boldsymbol{\delta}_j), \phi\right\} g(\boldsymbol{\delta}_j; \mathbf{0}, \Sigma) d\boldsymbol{\delta}_j\right], \qquad (10.63)$$

where f is given by (10.50) and $g(\boldsymbol{\delta}; \mathbf{0}, \Sigma)$ is the density of $\mathcal{N}(\mathbf{0}, \Sigma)$. Of course, in (10.63) the density f can be replaced by the exponential of the quasi-likelihood (10.61):

$$\sum_j \log\left(\int \cdots \int \exp\left[\sum_i Q\left\{y_{ij}; \eta^{-1}(\boldsymbol{x}_{ij}\boldsymbol{\beta} + \boldsymbol{z}_{ij}\boldsymbol{\delta}_j)\right\}\right] g(\boldsymbol{\delta}_j; \mathbf{0}, \Sigma) d\boldsymbol{\delta}_j\right). \qquad (10.64)$$

Evaluation of the integrals in (10.63) and (10.64) may not always be feasible, especially with several explanatory variables in the random part. Gianola and Im (1988) give an application in animal-breeding research, and use a simplex method to maximize the log-likelihood (10.63) for binomial observations. The simplex method requires repeated evaluation of (10.63). The integrals in (10.63) and (10.64) can be approximated using a quadrature method.

Longford (1988a) derived an approximation to the integrated quasi-likelihood (10.64) that has a form similar to the normal likelihood (10.21). Based on this *approximation*, an approximate quasi-likelihood estimation procedure is constructed. The procedure is a combination of the Fisher scoring algorithm described in Section 4.1 with the generalized least-squares algorithm of Section 6.1. At each iteration of the algorithm the scoring vector and the information matrix are calculated as in Section 4.1 but the residuals are replaced by the residual function (10.57). Further, the matrices of cross-products $(\boldsymbol{e}_j, \boldsymbol{X}_j)'(\boldsymbol{e}_j, \boldsymbol{X}_j)$ are replaced by their weighted versions

$$(\boldsymbol{r}_j, \boldsymbol{X}_j)'\boldsymbol{W}_j(\boldsymbol{r}_j, \boldsymbol{X}_j),$$

where $\boldsymbol{r}_j = \mathbf{1}_{n_j} \otimes \{r_{ij}\}$ is the vector of residual function values for group j and $\boldsymbol{W}_j = \boldsymbol{I}_{n_j} \otimes \{w_{ij}\}$ is the diagonal matrix of values of the weight function for group j. The residual and weight functions are evaluated at the (unconditional) linear predictor $\boldsymbol{x}_{ij}\hat{\boldsymbol{\beta}}$. Since these quantities depend on the estimated linear predictor, the cross-products have to be recalculated at each iteration. It would appear that the order of the iterations should be carefully arranged. However, the GLM solution, adopted as a starting solution, usually provides values of the fitted mean and weight very close to those at the termination of the iterations, and therefore the organization of the iterations is not important. It suffices to carry out one iteration of the algorithm for random coefficient models after recalculating the residual and weight functions.

The software package VARCL (Longford 1988b) implements this algorithm in conjuction with the Fisher scoring algorithm for random coefficient models, but the adaptation can in principle be implemented using any other algorithm for normal random coefficient models. The resampling properties of the resulting estimators have not been widely studied, but it appears that the inverse of the estimated information matrix in random coefficient GLM models matches the observed resampling variation only for very large datasets and relatively simple models.

McCullagh and Nelder (1989) describe a general approach for estimation with GLM models with a limited pattern of within-group covariance.

A similar approach is discussed by Liang and Zeger (1986) in the context of longitudinal analysis; see also Zeger, Liang, and Albert (1988), and Zeger and Qadish (1988). They define the derivative of the quasi-likelihood by a direct extension of the matrix version of (10.60) to nondiagonal covariance matrix functions, and impose a suitable parameterization for this covariance matrix.

The methods of Liang and Zeger (1986), McCullagh and Nelder (1989), and Longford (1988a) share a common estimator for the regression parameters, the generalized least-squares estimator

$$\hat{\boldsymbol{\beta}} = (\boldsymbol{X}'\boldsymbol{V}^{-1}\boldsymbol{X})^{-1}\boldsymbol{X}'\boldsymbol{V}^{-1}\boldsymbol{y}, \tag{10.65}$$

where $\boldsymbol{V}$ is a (generalized) covariance matrix for the outcomes. The matrix $\boldsymbol{V}$ depends on the generalized variances of the outcomes, and on one or several parameters describing within-group correlation. For compound-symmetry-type models, Liang and Zeger (1986) propose models with constant within-group correlations in $\boldsymbol{V}$, whereas the development of Longford (1988a) leads to constant within-group covariances in $\boldsymbol{V}$. The latter approach has a natural extension for complex patterns of variation and yields approximate maximum likelihood estimators of the parameters involved in $\boldsymbol{V}$, and enables approximate likelihood ratio testing.

Issues concerning restricted and full maximum likelihood estimation carry over to models assuming nonnormal distributions. See Longford (1994) for an approach exploiting analogies with normal models.

7 Factor Analysis and Structural Equations

7.1 Factor Analysis

Decomposition of the total variance into its individual- and group-level components is an important issue not restricted to regression with clustered data, but equally relevant in other statistical methods. In this section we discuss an extension of factor analysis appropriate for clustered observations. When applying classical factor analysis (Lawley and Maxwell 1971; Bartholomew 1987) for clustered vectors of observations it may not be obvious whether the resulting factors reflect between-group or between-individual differences. A solution to this problem, analogous to the extension of the multiple regression model to the random coefficient regression model, is to allow for separate within- and between-group covariance structures. Such a *two-level factor-analysis* model is described in this section. The exposition is restricted to normally distributed outcomes.

Formally, we assume a one-way layout of individuals $i = 1, 2, \ldots, n_j$ in groups $j = 1, 2, \ldots, N_2$, and each individual has a $p \times 1$ vector of observations $\boldsymbol{y}_{ij}$. For the within-group factor structure we assume that (conditionally on the mean $\boldsymbol{m}_j$)

$$\boldsymbol{y}_{ij} = \boldsymbol{m}_j + \boldsymbol{\Lambda}_1\boldsymbol{\delta}_{1,ij} + \boldsymbol{\varepsilon}_{1,ij}\,, \qquad (10.66)$$

where $\boldsymbol{\Lambda}_1$ is a $p \times r_1$ matrix of constants, $\{\boldsymbol{\delta}_{1,ij}\}_{i,j}$ is a random sample from $\mathcal{N}(\mathbf{0}, \boldsymbol{\Theta}_1)$, $\{\boldsymbol{\varepsilon}_{1,ij}\}_{i,j}$ is a random sample from $\mathcal{N}(\mathbf{0}, \boldsymbol{\Psi}_1)$, and these two samples are mutually independent.

For the group-means $\boldsymbol{m}_j$ we impose a similar factor-analysis model,

$$\boldsymbol{m}_j = \boldsymbol{\mu} + \boldsymbol{\Lambda}_2\boldsymbol{\delta}_{2,j} + \boldsymbol{\varepsilon}_{2,j}\,, \qquad (10.67)$$

with assumptions analogous to those associated with (10.66): $\boldsymbol{\Lambda}_2$ is a $p \times r_2$ matrix of constants, $\{\boldsymbol{\delta}_{2,j}\}_j$ and $\{\boldsymbol{\varepsilon}_{2,j}\}_j$ are mutually independent random samples from $\mathcal{N}(\mathbf{0}, \boldsymbol{\Theta}_2)$ and $\mathcal{N}(\mathbf{0}, \boldsymbol{\Psi}_2)$, respectively, and they are in turn mutually independent from their counterparts in (10.66). Unconditionally, we have

$$\boldsymbol{y}_{ij} = \boldsymbol{\mu} + \boldsymbol{\Lambda}_1\boldsymbol{\delta}_{1,ij} + \boldsymbol{\varepsilon}_{1,ij} + \boldsymbol{\Lambda}_2\boldsymbol{\delta}_{2,j} + \boldsymbol{\varepsilon}_{2,j}\,. \qquad (10.68)$$

The conditional (within-group) covariance matrix for an observation is equal to

$$\mathrm{var}(\boldsymbol{y}_{ij} \mid \boldsymbol{m}_j) = \boldsymbol{\Lambda}_1\boldsymbol{\Theta}_1\boldsymbol{\Lambda}_1' + \boldsymbol{\Psi}_1\,, \qquad (10.69)$$

and the unconditional covariance matrix is

$$\mathrm{var}(\boldsymbol{y}_{ij}) = \boldsymbol{W}_1 + \boldsymbol{W}_2\,, \qquad (10.70)$$

where $\boldsymbol{W}_h = \boldsymbol{\Lambda}_h \boldsymbol{\Theta}_h \boldsymbol{\Lambda}_h' + \boldsymbol{\Psi}_h$, $h = 1, 2$. In complete analogy to classical factor analysis we can specify a variety of parametric models for the variance matrices $\boldsymbol{W}_h$, by constraining the factor loadings $\boldsymbol{\Lambda}_h$ and the covariance matrices $\boldsymbol{\Psi}_h$ and $\boldsymbol{\Theta}_h$. To ensure identifiability, the residual covariance matrices $\boldsymbol{\Psi}_h$ can be assumed to be diagonal, and in some settings it is meaningful to set $\boldsymbol{\Psi}_2 = \mathbf{0}$. No generality is lost by assuming the covariance matrices for the factors, $\boldsymbol{\Theta}_h$, to be correlation matrices.

For each level we can distinguish two modes: confirmatory and exploratory. In the exploratory mode we assume uncorrelated factors, $\boldsymbol{\Theta}_h = \boldsymbol{I}$, and an arbitrary set of mutually orthogonal factor loadings, whereas in the confirmatory mode a general form for the covariance matrix of the factors is assumed, but certain constraints on the factor loadings are imposed. If there are no *cross-level* constraints, the same conventional rules about the degrees of freedom can be established as in classical factor analysis. Additionally cross-level constraints can be considered, such as equal-variance structures, $\boldsymbol{W}_1 = \boldsymbol{W}_2$, equal-factor structures, $\boldsymbol{\Lambda}_1 = \boldsymbol{\Lambda}_2$, "parallel" covariance matrices, $\boldsymbol{\Psi}_1 = c\boldsymbol{\Psi}_2$, or the like. For further background and motivation we refer the reader to Muthén (1989).

In principle, exact maximum likelihood procedures for estimation with two-level factor-analysis models can be constructed by using the algorithms for random coefficient models, insofar as the two classes of models involve similar structures for the means and covariance matrices of the observations. Goldstein and McDonald (1988) indicate how the IRLS (moment matching) procedure can be directly applied to factor analysis, and McDonald and Goldstein (1989) describe a direct maximization algorithm for a class of models encompassing multilevel factor analysis.

The generalization for Heywood cases can now occur at both levels. In particular, the estimate of the diagonal matrix $\boldsymbol{\Psi}_2$ may have negative elements. If $\boldsymbol{\Psi}_2 = \mathbf{0}$ then the factor structure at level 2 coincides with that assumed in principal component analysis.

The log-likelihood associated with the observations $\boldsymbol{y}$, ($\boldsymbol{y} = \mathbf{1}_{N_2} \otimes \{\boldsymbol{y}_j\}$ and $\boldsymbol{y}_j = \mathbf{1}_{n_j} \otimes \{\boldsymbol{y}_{ij}\}$) is equal to

$$\log l = -\frac{1}{2}\left\{Np \log(2\pi) + \log(\det \boldsymbol{V}) + \boldsymbol{e}'\boldsymbol{V}^{-1}\boldsymbol{e}\right\}, \tag{10.71}$$

where $N = \sum_j n_j$ is the number of observations, $\boldsymbol{e} = \boldsymbol{y} - \mathbf{1}_N \otimes \boldsymbol{\mu}$ is the $Np \times 1$ vector of deviations from the mean vector, and $\boldsymbol{V} = \boldsymbol{I}_{N_2} \otimes \{\boldsymbol{V}_j\}$ is the (block-diagonal) covariance matrix for all the observations.

Evaluation of the log-likelihood (10.71) can be simplified by making use of the patterned form of the covariance matrix for the observations

$$\boldsymbol{V}_j = \boldsymbol{I}_{n_j} \otimes \boldsymbol{W}_1 + \boldsymbol{J}_{n_j} \otimes \boldsymbol{W}_2 . \tag{10.72}$$

We have

$$\boldsymbol{V}_j^{-1} = \boldsymbol{I}_{n_j} \otimes \boldsymbol{W}_1^{-1} - \boldsymbol{J}_{n_j} \otimes \left\{\boldsymbol{H}_{2j}^{-1}\boldsymbol{W}_2\boldsymbol{W}_1^{-1}\right\}$$

and

$$\det \boldsymbol{V}_j = \det(\boldsymbol{W}_1)^{n_j - 1} \det \boldsymbol{H}_{2j} ,$$

where $H_{2j} = W_1 + n_j W_2$. Thus the log-likelihood (10.71) can be evaluated without inversion of large matrices. Note that H_{2j} is the n_j-multiple of the covariance matrix for the within-group means, $H_{2j} = n_j \mathrm{var}(\bar{y}_j)$.

Let S be the matrix of totals of squares and cross-products,

$$S = \sum_j \sum_i (y_{ij} - \mu)(y_{ij} - \mu)'.$$

The matrix S can be decomposed into its within- and between-group components, $S = S_1 + \sum_j D_j$, where

$$S_1 = \sum_j \sum_i (y_{ij} - \bar{y}_j)(y_{ij} - \bar{y}_j)',$$

and

$$D_j = n_j(\bar{y}_j - \bar{y})(\bar{y}_j - \bar{y})' + n_j(\bar{y} - \mu)(\bar{y} - \mu)'.$$

For completeness, we denote the matrix of between-group totals of squares and cross-products by S_2 $(= \sum_j D_j)$.

It can be shown that

$$e'V^{-1}e = \mathrm{tr}(W_1^{-1}S_1) + \sum_j \mathrm{tr}(H_{2j}^{-1}D_j) \quad . \tag{10.73}$$

See Longford and Muthén (1992) for details. This further simplifies calculation of the log-likelihood (10.71). Note that for balanced data, $(n_j \equiv n)$ we have

$$\log l = -\frac{1}{2}\left\{N_2 np \log(2\pi) + N_2 \log(\det V_1) + \mathrm{tr}(W_1^{-1}S_1) + \mathrm{tr}(H_2^{-1}S_2)\right\}, \tag{10.74}$$

where V_1 and H_2 are the respective common values of V_j and H_{2j} for all j.

It is easy to see that the matrix S_1 and the vectors of means $\left\{\bar{y}_j\right\}_j$ form a set of minimal sufficient statistics. For balanced data, S_1 and S_2 (with μ replaced by the sample mean $\bar{y}$) are a set of minimal sufficient statistics. If the data contain a small number of different group sizes, then a suitable set of minimal sufficient statistics is formed by S_1 and the totals $\sum D_j$ over groups with the same group sizes.

The factor analysis model with no restrictions on the covariance structures is likely to be of interest in most applications, because it provides a benchmark for comparison of all the fitted models. This saturated model coincides with the three-level random coefficient model (10.40), with the nesting hierarchy of observations within individuals within groups, and with the categorical explanatory variable indicating the component of the vector of observations (p categories) included in the fixed part and both level 2 and level 3 random parts of the model. In this model the elementary-level variance (σ^2) is confounded with the level 2 covariance matrix; in model fitting the variance σ^2 has to be constrained to a small positive constant. See Longford (1990) for further details.

Although such a model can be fitted using standard software, maximum likelihood estimates of the variance matrices $\boldsymbol{W}_1$ and $\boldsymbol{W}_2$ can be obtained by matching the statistics $\boldsymbol{S}_1$ and $\boldsymbol{S}_2$ to their expectations. We have

$$E(\boldsymbol{S}_1 \mid \boldsymbol{\mu}, \boldsymbol{W}_1, \boldsymbol{W}_2) = (N - N_2)\boldsymbol{W}_1 \tag{10.75}$$

$$E(\boldsymbol{S}_2 \mid \boldsymbol{\mu}, \boldsymbol{W}_1, \boldsymbol{W}_2) = N_2\boldsymbol{W}_1 + \left(N - \sum_j \frac{n_j^2}{N}\right) \boldsymbol{W}_2 , \tag{10.76}$$

and these equations, with $\boldsymbol{S}_1$ and $\boldsymbol{S}_2$ in place of the respective expectations, are easy to solve for $\boldsymbol{W}_1$ and $\boldsymbol{W}_2$. We emphasize the dependence on the mean $\boldsymbol{\mu}$, because in analogy with the random coefficient models, restricted maximum likelihood [taking account of the degrees of freedom used to estimate $\boldsymbol{\mu}$ see (10.21) and (10.22)] can be considered as an alternative to full maximum likelihood (10.71). The moment matching equations for the restricted maximum likelihood equations are obtained from (10.76) by replacing the term $N_2\boldsymbol{W}_1$ with $(N_2 - 1)\boldsymbol{W}_1$; the corresponding log-likelihood differs from (10.71) by a constant and the term

$$-\frac{1}{2}\log\left\{\det\left(\sum_j n_j \boldsymbol{H}_{2j}^{-1}\right)\right\} . \tag{10.77}$$

The estimator for $\boldsymbol{W}_1$, common to the full and restricted maximum likelihood, has a Wishart distribution with $N - N_2$ degrees of freedom. If the between-group covariance matrix $\boldsymbol{W}_2$ were equal to zero, the MLE for $\boldsymbol{W}_1$ would have the Wishart distribution with $N - 1$ degrees of freedom. This indicates that the estimator for $\boldsymbol{W}_1$ based on (10.75) may be fairly efficient, especially when the number of groups N_2 is much smaller than the number of individuals N.

An approximate MLE of the factor structures in the two-level model (10.68) can be obtained by solving the moment matching equations (10.75) and (10.76), and then subjecting the resulting estimated matrices $\hat{\boldsymbol{W}}_1$ and $\hat{\boldsymbol{W}}_2$ to separate classical factor analyses. Exact maximum-likelihood estimation procedures involve iterative algorithms, and this "makeshift" estimator provides a suitable starting value for the iterations.

Formulas for the Fisher scoring algorithm are given in Longford and Muthén (1992), but the principles for their derivation are given in Section 4. We note that two-level factor analysis models are likely to contain a large number of parameters, and although calculation of all the items for the Fisher scoring algorithm is not time consuming, having to invert a large information matrix (or solve the associated linear system of equations) at each iteration may be a distinct disadvantage. This problem can be effectively resolved by application of conjugate gradient (quasi-Newton) methods, as indicated by McDonald and Goldstein (1989). Applications of conjugate gradient methods to classical factor analysis are reported in Jamshidian and Jennrich (1988).

Longford (1990) describes an application of two-level factor analysis in educational testing.

7.2 Structural Equation Models

Extensions of structural equations models to clustered observations are also of interest, particularly in educational research. In this section we describe a general measurement error model for clustered observations and outline some of the problems with model specification and estimation.

We assume that the univariate outcome y_{ij}, $i = 1, \ldots, n_j, j = 1, \ldots, N_2$, is related to a vector of known regressors $\mathbf{z}_{ij}$ and a vector of latent regressors $\mathbf{x}_{ij}$ by a random coefficient model,

$$y_{ij} = \mathbf{x}_{ij}\boldsymbol{\beta} + \mathbf{z}_{ij}\boldsymbol{\delta}_j + \varepsilon_{ij}, \tag{10.78}$$

where $\boldsymbol{\delta}_j \sim \mathcal{N}(\boldsymbol{\mu}_\delta, \boldsymbol{\Sigma}_\delta)$ and $\varepsilon_{ij} \sim \mathcal{N}(0, \sigma^2)$ are two random samples and $\boldsymbol{\beta}$ is a vector of (regression) parameters. Instead of $\mathbf{x}_{ij}$ a random vector $\mathbf{s}_{ij}$, related to $\mathbf{x}_{ij}$ by a measurement error model,

$$\mathbf{s}_{ij} = \mathbf{U}_{ij}\mathbf{x}_{ij} + \boldsymbol{\gamma}_{2,j} + \boldsymbol{\gamma}_{1,ij}, \tag{10.79}$$

is observed. In (10.79), $\mathbf{U}_{ij}$ is a matrix of known constants, and $\boldsymbol{\gamma}_{2,j} \sim \mathcal{N}(\mathbf{0}, \boldsymbol{\Sigma}_{\gamma,2})$ and $\boldsymbol{\gamma}_{1,ij} \sim \mathcal{N}(\mathbf{0}, \boldsymbol{\Sigma}_{\gamma,1})$ are two random samples. Further, we assume that

$$\mathbf{x}_{ij} = \boldsymbol{\mu}_x + \boldsymbol{\xi}_{2,j} + \boldsymbol{\xi}_{1,ij}, \tag{10.80}$$

where $\boldsymbol{\mu}_x$ is a vector of parameters, and $\boldsymbol{\xi}_{2,j} \sim \mathcal{N}(\mathbf{0}, \boldsymbol{\Sigma}_{\xi,2})$ and $\boldsymbol{\xi}_{1,ij} \sim \mathcal{N}(\mathbf{0}, \boldsymbol{\Sigma}_{\xi,1})$ are two random samples. The random samples $\{\varepsilon_{ij}\}$, $\{\boldsymbol{\gamma}_{2,j}\}$, $\{\boldsymbol{\gamma}_{1,ij}\}$, $\{\boldsymbol{\xi}_{2,j}\}$, and $\{\boldsymbol{\xi}_{1,ij}\}$ are assumed to be mutually independent.

The formulation (10.78)–(10.80) is quite general, and can be extended further to multivariate outcomes $\mathbf{y}_{ij}$. Specification of constraints that would assure identifiability of the covariance matrices $\boldsymbol{\Sigma}_\delta$, $\boldsymbol{\Sigma}_{\gamma,2}$, $\boldsymbol{\Sigma}_{\gamma,1}$, $\boldsymbol{\Sigma}_{\xi,2}$ and $\boldsymbol{\Sigma}_{\xi,1}$ is an unresolved problem.

We consider the random vectors $\{\mathbf{u}_j\}_j$, $j = 1, \ldots, N_2$, $\mathbf{u}_j = \mathbf{1}_{n_j} \otimes \{\mathbf{u}_{ij}\}$, where $\mathbf{u}_{ij} = (y_{ij}, \mathbf{s}'_{ij})'$. The joint distribution of $\mathbf{u}_{ij}$ is

$$\mathcal{N}\left\{\begin{pmatrix} \boldsymbol{\mu}_x\boldsymbol{\beta} + \mathbf{z}_{ij}\boldsymbol{\mu}_\delta \\ \mathbf{U}_{ij}\boldsymbol{\mu}_x \end{pmatrix}, \begin{pmatrix} \boldsymbol{\beta}'\boldsymbol{\Sigma}_\xi\boldsymbol{\beta} + \mathbf{z}_{ij}\boldsymbol{\Sigma}_\delta\mathbf{z}'_{ij} + \sigma^2 & \boldsymbol{\beta}'\boldsymbol{\Sigma}_\xi\mathbf{U}'_{ij} \\ \mathbf{U}_{ij}\boldsymbol{\Sigma}_\xi\boldsymbol{\beta} & \mathbf{U}_{ij}\boldsymbol{\Sigma}_\xi\mathbf{U}'_{ij} + \boldsymbol{\Sigma}_\gamma \end{pmatrix}\right\} \tag{10.81}$$

($\boldsymbol{\Sigma}_\xi = \boldsymbol{\Sigma}_{\xi,1} + \boldsymbol{\Sigma}_{\xi,2}$ and $\boldsymbol{\Sigma}_\gamma = \boldsymbol{\Sigma}_{\gamma,1} + \boldsymbol{\Sigma}_{\gamma,2}$), and the covariance of two vectors $\mathbf{u}_{ij}$ and $\mathbf{u}_{i'j}$ is equal to

$$\operatorname{cov}(\mathbf{u}_{ij}, \mathbf{u}_{i'j}) = \begin{pmatrix} \boldsymbol{\beta}'\boldsymbol{\Sigma}_{\xi,2}\boldsymbol{\beta} + \mathbf{z}_{ij}\boldsymbol{\Sigma}_\delta\mathbf{z}'_{i'j} & \boldsymbol{\beta}'\boldsymbol{\Sigma}_{\xi,2}\mathbf{U}'_{i'j} \\ \mathbf{U}_{ij}\boldsymbol{\Sigma}_{\xi,2}\boldsymbol{\beta} & \mathbf{U}_{ij}\boldsymbol{\Sigma}_{\xi,2}\mathbf{U}'_{i'j} + \boldsymbol{\Sigma}_{\gamma,2} \end{pmatrix}. \tag{10.82}$$

The log-likelihood for the random vector $\mathbf{1}_{N_2} \otimes \{\mathbf{u}_j\}$ has the form (10.71) with the covariance matrix $\mathbf{V}$ given by the blocks (10.81) and (10.82). Inversion and determinant formulas similar to (10.32) and (10.33) can be derived for this covariance matrix, and they facilitate efficient calculation of the log-likelihood and its partial derivatives. See Longford (1993) for details. The general treatment of the problem appears to be of formidable complexity.

Muthén (1989) provides an authoritative review of the subject area, and McDonald and Goldstein (1989) describe a comprehensive modeling framework and a computational algorithm.

8 Example: Wage Inflation in Britain

The following example illustrates the random coefficient methods for two-level data discussed in Sections 2–4. We demonstrate the unsatisfactory nature of the ordinary regression method for two-level (and by implication multilevel) data, and give a substantive interpretation for the inferred covariance structure. We fit an ordinary regression, a random effects, and a random coefficient model, each time substantially improving the model fit (as measured by the likelihood ratio criterion). Thus between-company variation is an important feature of the data; the random coefficient model provides a description for its pattern.

The example is drawn from a study of wage inflation in engineering firms in the northwest region of England. Weekly wages for nine occupational categories were obtained from 57 companies over six years (1979–1984). The data contain 679 records on the following variables:

- year (coded 1–6)
- occupational category (1–9)
- wages (per week, in pounds U.K.),

and each record is associated with a company. The minimum number of records per company in these data is 1, and the maximum is 40. Only ten companies have provided more than five records from at least one year, and none have provided more than eight from a year. For years 1–6 there are 126, 104, 99, 117, 108, and 125 records, respectively. The nine occupational categories are distributed evenly among the years and within the larger companies.

Our analysis focuses on between-company comparisons of wage increases over the six years of the study. Analysis of covariance is not feasible because all the company-by-year combinations are represented very sparsely in the data. Preliminary exploration of the data indicates that substantial wage increases were awarded between years 1 and 2, and thereafter the increases were at an approximately constant annual rate; the mean wages for years 1–6 are

72.38, 88.80, 95.97, 99.85, 105.17, 112.22,

respectively, corresponding to average annual wage increases of

22.7%, 8.1%, 4.0%, 5.3%, 6.7%.

We consider first a linear model for the structure of unconditional expectations of the logarithms of the wages (log-wages):

$$\mathrm{E}(lw) = C + f_t + p_s , \qquad (10.83)$$

where lw is the log-wage, C the intercept, f_t, $t = 1, \ldots, 6$, is the adjusted mean for the year t, and p_s, $s = 1, \ldots, 9$, represents the offset for occupational category s.

For the occupational categories we use the standard parameterization. We designate the first occupational category as reference and set $p_1 = 0$. Then p_s, $s = 2, \ldots, 9$, is the difference of the log-wages between occupational categories s and 1, adjusted for year. Note that the difference of log-wages is equal to the logarithm of the ratio of the wages. The occupational categories are not ordered in any meaningful way.

To decide on the functional form of f_t we may consider six unrelated values (though one of them is confounded with the intercept C), a smooth curve, or a specific increase from year 1 to year 2, and then constant (percentage) increases. We choose the latter alternative, and then check its appropriateness throughout the analysis. Specifically, we use the following parameterization: The mean for the first occupational category in the first year is $\beta_1 - \beta_2$, and in years 2–6 β_1, $\beta_1 + \beta_3$, $\beta_1 + 2\beta_3$, $\beta_1 + 3\beta_3$, and $\beta_1 + 4\beta_3$, respectively. Thus β_2 is the mean increase of log-wages from year 1 to year 2, and β_3 the mean annual increase during years 3–6.

The OLS fit for the model (10.83) is given in column A of Table 1. For orientation, the fitted mean wage for occupational category 1 in year 2 is exp(4.30) = 73.70 pounds, and for category 9 in year 4 is $\exp(4.300 - 0.314 + 2 \times 0.0571) = 60.35$ pounds. For small numbers, the logarithmic scale is very close to the proportion scale; for instance, $\hat{\beta}_3 = 0.0571$ corresponds to average annual wage increase of about 5.7%. Occupational categories 3–5 appear to have very similar average wages (p_3, p_4, and p_5 could be replaced by a single parameter). The estimate of the residual variance is 0.00897. Of course, based on ordinary regression it is not possible to apportion the residual variance to its two potential sources—employees and companies. Companies' pay awards reflect their (present or projected) financial performances as well as the situation in the local labor market, and these may vary from company to company. These factors contribute to between-company variation. Similarly, the wages of an employee in a company are determined by a number of factors besides the occupational category, such as age, experience, performance, company loyalty, and the like. Unfortunately, no data related to these factors are available. These factors contribute to within-company variation. Thus the sources of variation provide information about the strength of association of company- and employee-level factors, although, of course, the importance of none of the single factors can be assessed.

In order to separate the company and employee as sources of variation we consider the random effects model (10.14). This model assumes a constant within-company covariance of the log-wages. We use the same regression part as in (10.83). For the covariance structure we use the parameterization (σ^2, ψ), where ψ is the variance ratio (see Section 4). We set the starting solution for ψ to 0.1, and that for σ^2 to the residual variance from OLS. The first iteration yields the estimates $\hat{\sigma}^2 = 0.00621$ and $\hat{\psi} = 0.5174$, implying that the between-company variance is much higher than we anticipated. The deviance of the starting solution is equal to -1443.39, a substantial reduction from the OLS deviance of -1273.91. Further iterations yield respective deviances -1599.2, -1650.9, and -1659.5, with decreasing estimates of σ^2 and increasing estimates of the variance ratio ψ.

The iterations were terminated after nine cycles when the difference of consecutive deviances became smaller than 10^{-4} and each parameter estimate was altered by less than 10^{-6}. The maximum likelihood estimates are given in column B of Table 1.

The estimated between-company variance, $\hat{\psi}\hat{\sigma}^2 = 0.00705$, is almost twice as large

Table 1. Regression Parameter Estimates for Wage Inflation Data.

Parameter	Method A	Method B	Method C	
β_1	4.300 (0.0113)	4.563 (0.0117)	4.568 (0.0139)	
β_2	0.2001 (0.0111)	0.2153 (0.00680)	0.2174 (0.00796)	
β_3	0.0571 (0.00284)	0.0535 (0.00356)	0.05011 (0.00371)	
p_2	6.26×10^{-4} (0.0133)	-6.64×10^{-3} (0.00700)	$-5.98\text{E}\times10^{-3}$ (0.00864)	
p_3	0.0529 (0.0191)	0.0316 (0.0139)	0.0281 (0.0143)	
p_4	0.0595 (0.0193)	0.0335 (0.0123)	0.0329 (0.0131)	
p_5	0.0512 (0.0193)	0.0257 (0.0128)	0.0237 (0.0135)	
p_6	0.0156 (0.0150)	0.0196 (0.00890)	0.0214 (0.0104)	
p_7	-0.133 (0.0147)	-0.140 (0.00818)	-0.139 (0.00950)	
p_8	-0.168 (0.0128)	-0.170 (0.00666)	-0.170 (0.00855)	
p_9	-0.314 (0.0131)	-0.316 (0.00678)	-0.314 (0.00862)	
σ^2	0.00897	0.00404	0.00350	
ψ		1.747 (0.360)	1.917 (0.440)	-0.114 (0.088)
			-0.114 (0.088)	0.0899 (0.0315)
Deviance	−1273.91	−1660.74	−2120.29	

NOTE: The methods are: A—ordinary regression, B—random-effects model, C—random coefficient model (β_2 associated with variation). The parameters are introduced in the text. Standard errors are in parentheses.

as the within-company variance. The standard deviation corresponding to the former is 0.0840. This figure, 8.4%, can be interpreted as the risk associated with the choice of the company, or as a measure of difference in how companies reward employees. On the same scale, the within-company standard deviation of wages is 6.35%.

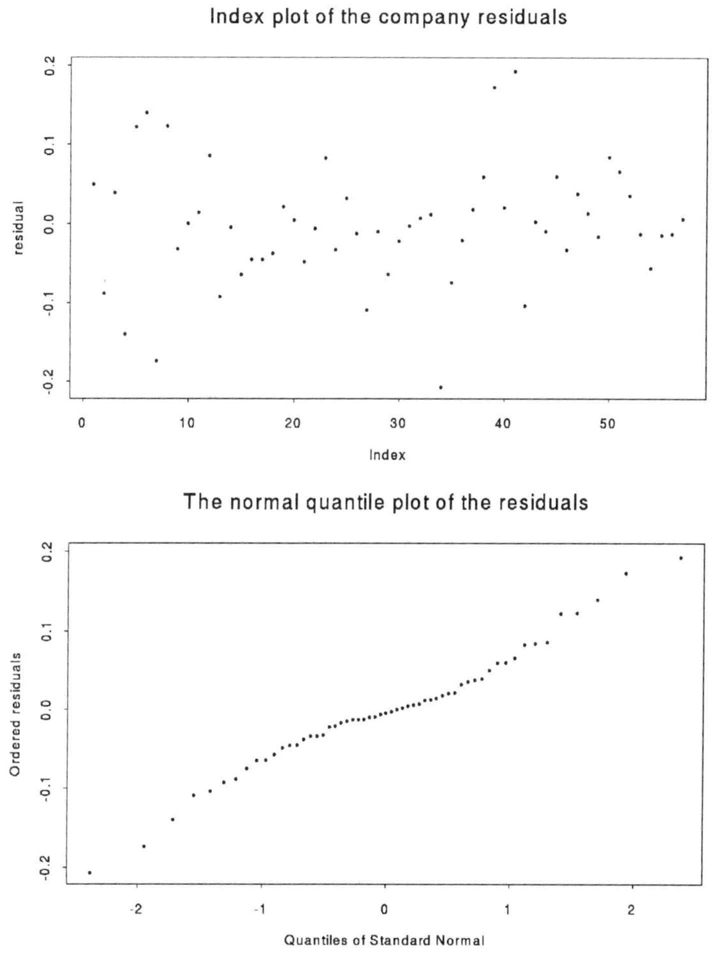

FIGURE 2. Wage Inflation in Britain. The index plot and the normal quantiles plot of the residuals from the random-effects model fit for log-wages.

We regard the conditional expectations of the company-level random terms as residuals, and subject them to a standard check for normality. The plots in Figure 2 indicate that the normality assumption is reasonable. The top panel contains a plot of the residuals against the order of the companies, and the bottom panel displays the normal quantile plot. Other plots of the residuals, such as against time, display no discernible patterns.

Of course, the model assumes the same percentage wage increase for all the companies within each year. Of interest is whether there are between-company differences in the rate of wage increases (differential wage inflation). We consider therefore the random coefficient model (10.19) with the fixed part (10.83), in which we allow the coefficients related to the year (β_2 and β_3) to vary across the companies.

The estimated scaled covariance matrix $\hat{\Psi}$ attains very large diagonal entries,

$$\hat{\Psi} = \begin{pmatrix} 20.335 & -21.627 & 0.119 \\ -21.627 & 25.190 & -0.182 \\ -0.119 & -0.182 & 0.0696 \end{pmatrix}, \qquad (10.84)$$

and the associated value of the deviance is –2124.84, a substantial reduction from the random effects model fit. Now the fitted variances of the intercept and of the year 2 – year 1 differences are unrealistically large, and the corresponding random terms are almost perfectly negatively correlated. The associated standard errors are also large, and the information matrix for the covariance structure parameters is extremely ill-conditioned. The ill-conditioning is caused principally by the top left-hand corner of Ψ. It seems that we cannot make inference about both variation of the year 2 – year 1 differences and the average differences over the six years. This is not surprising; the years 1 and 2 contain about 100 records each, from about 30 companies, and so a description of such delicate patterns of variation is not meaningful, especially in view of adjustment for other factors. Indeed, when the year 2 – year 1 contrast is excluded from the random part, the estimate of the scaled between-company covariance matrix becomes

$$\hat{\Psi} = \begin{pmatrix} 1.917 & -0.0114 \\ -0.0114 & 0.08994 \end{pmatrix}, \qquad (10.85)$$

with deviance –2120.29. Exclusion of the three covariance structure parameters involves only a small increase in the value of the deviance. The fitted variance for an observation is equal to $\hat{\sigma}^2\{1 + (1, t-2)\hat{\Psi}(1, t-2)'\}$, where $t = 1, \ldots, 6$ is the year of the observation. The relative changes (ratios) of the variance depend on $\hat{\Psi}$ only. For years 1 – 6 the fitted variance is equal to the following multiples of $\hat{\sigma}^2$:

3.03, 2.91, 2.98, 3.23, 3.66, 4.26,

that is, the variance of the log-wages has been increasing over the last two to three years of the study. Details of the model fit are given in column C of Table 1. The regression parameter estimates obtained for the three different model assumptions (ordinary regression, random effects, and random coefficients) differ only marginally; the latter two sets of estimates are somewhat more similar to each other than to the OLS estimates. There are substantial differences among the associated standard errors, though. Contrary to intuition, the standard errors for the random effects and random coefficient models are much smaller than the standard errors obtained by least squares method. For example, the OLS standard error for the parameter p_5 is 0.0193, respectively 1.5 and 1.4 times larger than its counterparts in the random effect and random coefficient models. Note that the corresponding t

ratios are 2.6, 2.0, and 1.75; different conclusions about significance of the parameter p_5 would be arrived at using the three models. Larger standard errors for the regression parameters in the random coefficient model are a consequence of substantially smaller fitted within-company variance than the residual variance estimated by OLS.

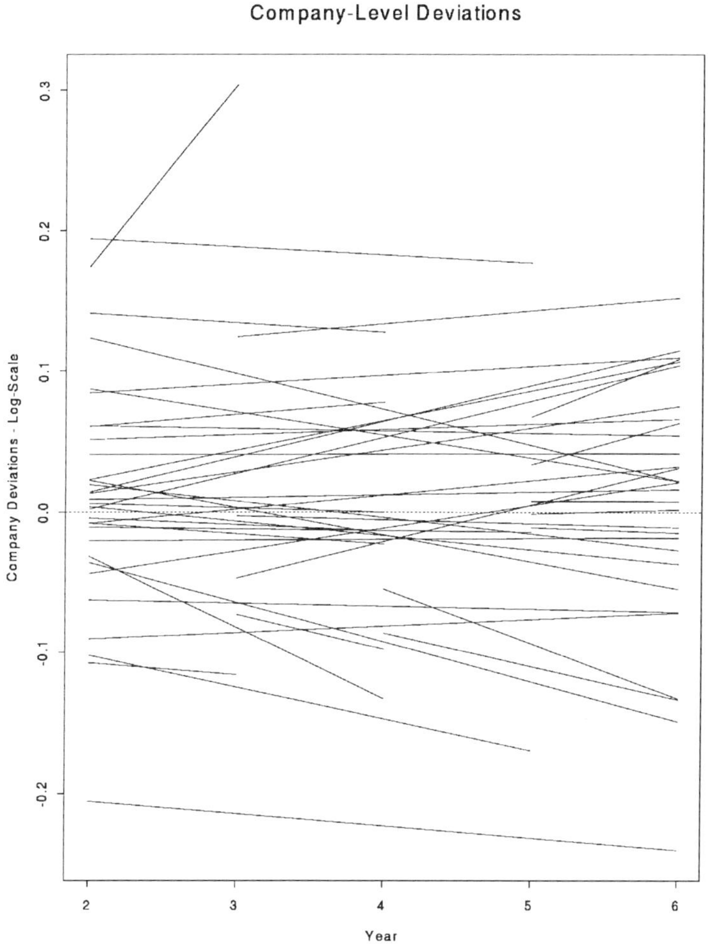

FIGURE 3. Wage Inflation in Britain. Fitted deviations of company wage increases from the average wage increases for years 2–6, on the log-scale. Each company is represented by a segment delineated by the years for which data were provided. Companies that provided data for only one of these years are omitted from the plot.

The average slope for the time variable (β_3) is 0.0501, and so the average wage increase over the years 2–6 is about 5%. But the standard deviation of the within-company wage increases is $\sqrt{0.08994} = 0.300$, and so there are a lot of companies where the wages have actually been decreasing. A fuller picture can be obtained from the plots of the linear combinations of the random coefficients, $z\hat{\delta}$ (see Section 4.2). These are displayed in Figure 3. The linear functions $z\hat{\delta}$ are plotted only in the range of years that occur in the data; the 14 companies that provided data for one year only are not represented in the plot. For such companies the conditional expectations of the company-deviations in that year are very close to zero because substantial shrinkage takes place.

The plot in Figure 3 has two complementary purposes. First, it provides a check on the normality assumption. There appears to be only one outlier, in the upper left-hand corner of the plot, a company that supplied data for years 2 and 3 only. In year 3 this company increased wages disproportionately more than other companies. To convert the residuals to the percentage scale, the values on the y-axis should be exponentiated, but for most purposes it is accurate enough to multiply them by 100, since $\exp(y) \approx 1 + y$ in the range of values of the fitted deviations.

Other elements of model checking involve exploration of alternatives to the adopted fixed part (10.83). The data contain very sparse information about interactions of the occupational categories with other variables, and so their modeling is not feasible. The linear inflation of log-wages during years 2 – 6 is appropriate, a quadratic term is ignorable and not significant at the 5% level. The nine occupational categories could be collapsed to fewer ones, although the aggregation is substantively not meaningful.

Serious substantive interpretation of these results is hampered by lack of experimental design in the study. The companies have essentially volunteered to participate in the study, and missingness may be informative. Allocation of employees to companies is not random; companies carefully select among the applicants, and job-seekers select among the companies that have offered them positions. Structural and personnel changes in the companies, combined with imperfect taxonomy of the occupations, are likely to distort the results further. In particular, they may be a substantial component of the between-company variation. Also, timing of the wage increases is relevant in this context. For instance, companies may award wage increases in periods different from exactly one year. And of course, wages are an imperfect measure of companies' expenditures on its employees, or of employee benefits. Nevertheless, it is plausible that the substantial differences in rates of wage increases reflect the differences in the fortunes of the companies in a harsh economic climate.

The analysis was replicated using REML. The same pattern of variation emerged, with a slight increase in all the estimated covariance structure parameters.

In conclusion, random coefficient models enable a more detailed examination of the data than ordinary regression models, provide a description for between-group variation and thus indicate the relevance of unavailable variables. The inferred pattern of variation is also of substantive interest. From the theoretical point of view it is important that the random coefficient models dispense with the assumption of independence of the observations.

9 Software

The use of random coefficient models, under a variety of names, has proliferated recently as suitable computational software has become available, such as HLM (based on Raudenbush and Bryk 1986), ML3 (based on Goldstein 1986), and VARCL (based on Longford 1987), as well as modules in BMPD, SAS, and other statistical packages. Specialized stand-alone software (HLM, ML3, and VARCL) did not fill a void in statistical methodology, but it has helped to bring these methods to the attention of social scientists. For example, the REML package (based on Patterson and Thompson 1971), designed primarily for agricultural and animal-breeding experiments, would, with minor modifications, be suitable for fitting of all the models described in this chapter.

This section contains a brief description of the programs ML3, VARCL, and HLM. A much more comprehensive review of these programs and of GENMOD (based on Mason, Wong, and Entwisle 1984) is contained in Kreft, DeLeeuw, and Kim (1990).

9.1 ML3

The program ML3 (Rasbash, Prosser, and Goldstein 1989) is distributed and mantained by the Multilevel Models Project at the Institute of Education, University of London, U.K.

The range of the models that can be fitted by the software is quite comprehensive. An arbitrary linear regression part and essentially an arbitrary pattern of variation at both levels 2 and 3 can be fitted. In addition the elementary-level variance can be modeled as constant, a quadratic function of a variable, or varying from group to group.

The user interface is unavoidably complex, though several reasonable defaults are implemented. The iterations can be controlled by the user; at prespecified iterations parameters can be "frozen" (constrained to their current value) or "set free." The results are displayed in the standard form of estimates and their standard errors, and the value of the fitted log-likelihood. Both ML and REML estimates can be obtained. A nonnegative definite estimated covariance matrix is ensured by setting certain variance(s) to zero; this may not be a satisfactory solution in some settings. The program stores the entire dataset in the core memory. Consequently, the capacity of the program is somewhat limited.

The computational core of the program is written in C and is merged with NANOSTAT (Healy 1989), a general-purpose package with graphics facilities. The Multilevel Models Project organizes frequent workshops where the software is demonstrated. Further development of the software is in progress. See Kreft and Kim (1991) for a detailed review.

9.2 VARCL

VARCL (Longford 1988b) is distributed by iec ProGamma in Groningen, the Netherlands. Two- and three-level random coefficient models can be fitted with arbitrary linear regression part and essentially arbitrary pattern of variation at both levels 2 and 3. The elementary-level variance is a constant, either known or estimated. The package is loosely modeled on GLIM (Numerical Algorithms Group 1986). It is written in Fortran. An independent Fortran program for fitting simple covariance structure models for data with up to nine levels of nesting is part of the software.

Each computer program consists of a computational core and a user interface. Several features of the software reflect the principal aim of fitting a set of related models for a large dataset. The interface is organized in such a way that the user has to specify the set of "candidate" models even before the data are input. When normality is assumed the data are passed through only once and their summary is generated. The data are not stored. Together with the use of dynamic storage allocation, this allows for utilization of large datasets, especially when mainframe computers are used.

Once the input has been completed, the user can fit any sequence of models for which the generated summaries are sufficient. VARCL is a purely computational package, with no data preprocessing or graphics facilities. The data have to be provided in a suitable format, together with a small control file, but all the model specifications are interactive. The residuals at levels 2 and 3 can be written into a file and then processed by a different program, if desired. The program outputs estimates and their standard errors and the value of the deviance. All model fits are maximum likelihood.

9.3 HLM

HLM (Bryk, Raudenbush, Seltzer, and Congdon 1988) implements the EM algorithm for REML estimation with random coefficient models. The software emphasizes contextual models and complex patterns of variation in two-level models. A three-level version is being developed. Data input and model specification are flexible. The main drawback of HLM is its slow convergence and lack of information about confounding of covariance structure parameters, especially in complex problems.

HLM defines a saturated model against which all fitted models (submodels of the saturated model) are compared. The program does not use the likelihood-ratio criterion. Estimates of the regression parameters, in the parameterization of micro and macro equations, are accompanied by their standard errors, but significance of the random terms and of the associated variances is assessed by certain χ^2-distributed reliability statistics.

9.4 Outlook

Recently, general-purpose statistical software packages (high-level programming languages), such as GAUSS, Matlab, and SPlus, have become popular among statisticians. Their syntax is very compact, and matrix operations in general are efficient. The main problem with their application seems to be the substantial statistical expertise required for their use. Many of the older programs (written in Fortran, and the like) are likely to be supplanted by routines written in this new generation of statistical software.

Appendix: Some Matrix Identities

Determinant of a Patterned Matrix

Theorem A1. Let V^* be a nonsingular matrix and Z and Σ any matrices such that the matrix $V^* + Z\Sigma Z'$ is well defined. Then

$$\det(V^* + Z\Sigma Z') = \det V^* \det G, \tag{10.A1}$$

where $G = I + Z'V^{*-1}Z\Sigma$.

Proof: Let L be the partitioned matrix

$$L = \begin{pmatrix} I & -Z' \\ Z\Sigma & V^* \end{pmatrix}. \tag{10.A2}$$

The determinant of L remains unchanged if we subtract the $(Z\Sigma)$-multiple of the blocks at the top, I and $-Z'$, from the corresponding blocks at the bottom, $Z\Sigma$ and V^*. The resulting matrix

$$L_1 = \begin{pmatrix} I & -Z' \\ \mathbf{0} & V^* + Z\Sigma Z' \end{pmatrix} \tag{10.A3}$$

has the determinant $\det(V^* + Z\Sigma Z')$. Similarly, if the $(V^{*-1}Z\Sigma)$-multiple of the right-hand-side blocks of L, $-Z'$, and V^*, are subtracted from the corresponding left-hand side blocks, we obtain the matrix

$$L_2 = \begin{pmatrix} G & -Z' \\ \mathbf{0} & V^* \end{pmatrix}, \tag{10.A4}$$

the determinant of which is $\det V^* \det G$. The two sides of the formula (10.A1) are the determinants of (10.A3) and (10.A4), and therefore (10.A1) holds. □

The formula (10.32) corresponds to (10.A1) with $V^* = \sigma^2 I_N$.

Inverse of a Patterned Matrix

Theorem A2. Let V^*, Z, Σ, and G be matrices as in Theorem A1. Then

$$(V^* + Z\Sigma Z')^{-1} = V^{*-1} - V^{*-1}Z\Sigma G^{-1}Z'V^{*-1}. \tag{10.A5}$$

Proof: The inversion formula (10.A5) is proved directly by multiplication:

$$\begin{aligned}(V^{*-1} - V^{*-1}Z\Sigma G^{-1}Z'V^{*-1})(V^* + Z\Sigma Z') &= \\ I + V^{*-1}Z\Sigma\{I - G^{-1}(I + Z'V^{*-1}Z\Sigma)\}Z' &= I.\end{aligned}$$

The matrix $Z'V^{*-1}Z$ is nonnegative definite, and therefore G is positive definite; its inverse is well defined. □

The identity (10.33) corresponds to (10.A5) with $V^* = \sigma^2 I_N$.

Conditional Distribution

Theorem A3. Suppose the random vector $\boldsymbol{y}$ has the normal distribution with mean $\boldsymbol{\mu}$ and positive definite covariance matrix $\boldsymbol{\Sigma}$, $\boldsymbol{y} \sim \mathcal{N}(\boldsymbol{\mu}, \boldsymbol{\Sigma})$. Let $\boldsymbol{y} = (\boldsymbol{y}_1', \boldsymbol{y}_2')'$, $\boldsymbol{\mu} = (\boldsymbol{\mu}_1', \boldsymbol{\mu}_2')'$, and

$$\boldsymbol{\Sigma} = \begin{pmatrix} \boldsymbol{\Sigma}_{11} & \boldsymbol{\Sigma}_{12} \\ \boldsymbol{\Sigma}_{21} & \boldsymbol{\Sigma}_{22} \end{pmatrix}$$

be compatible partitionings of $\boldsymbol{y}$, $\boldsymbol{\mu}$ and $\boldsymbol{\Sigma}$ (so that, for instance, $\boldsymbol{y}_h \sim \mathcal{N}(\boldsymbol{\mu}_h, \boldsymbol{\Sigma}_{hh})$ for $h = 1, 2$). Then the conditional distribution of $\boldsymbol{y}_1$ given $\boldsymbol{y}_2$ is

$$(\boldsymbol{y}_1 \mid \boldsymbol{y}_2) \sim \mathcal{N}\left\{\boldsymbol{\mu}_1 + \boldsymbol{\Sigma}_{12}\boldsymbol{\Sigma}_{22}^{-1}(\boldsymbol{y}_2 - \boldsymbol{\mu}_2),\ \boldsymbol{\Sigma}_{11} - \boldsymbol{\Sigma}_{12}\boldsymbol{\Sigma}_{22}^{-1}\boldsymbol{\Sigma}_{21}\right\}. \tag{10.A6}$$

Proof: Most textbooks of multivariate analysis give a proof by construction; see, for example, Mardia, Kent, and Bibby (1979, theorem 3.2.4). We present an alternative elementary proof.

We assume first that $\boldsymbol{\mu}_1 = \mathbf{0}$ and $\boldsymbol{\mu}_2 = \mathbf{0}$. The conditional density of $\boldsymbol{y}_1$ given $\boldsymbol{y}_2$ is equal to the ratio of the joint density of $\boldsymbol{y}$ and of the marginal density of $\boldsymbol{y}_2$:

$$f(\boldsymbol{y}_1 \mid \boldsymbol{y}_2) = (2\pi)^{-\frac{n_1}{2}} \left\{\frac{\det \boldsymbol{\Sigma}}{\det \boldsymbol{\Sigma}_{22}}\right\}^{-\frac{1}{2}} \exp\left\{-\frac{1}{2}(\boldsymbol{y}'\boldsymbol{\Sigma}^{-1}\boldsymbol{y} - \boldsymbol{y}_2'\boldsymbol{\Sigma}_{22}^{-1}\boldsymbol{y}_2)\right\}, \tag{10.A7}$$

where n_1 is the length of $\boldsymbol{y}_1$. Using the methods of proof in theorems A1 and A2, it can be shown that $\det \boldsymbol{\Sigma} = \det \boldsymbol{\Sigma}_{22} \det(\boldsymbol{\Sigma}_{11} - \boldsymbol{\Sigma}_{12}\boldsymbol{\Sigma}_{22}^{-1}\boldsymbol{\Sigma}_{21})$ (implying that all these determinants are positive), and

$$\boldsymbol{\Sigma}^{-1} = \begin{pmatrix} \boldsymbol{G}_1 & -\boldsymbol{G}_1\boldsymbol{\Sigma}_{12}\boldsymbol{\Sigma}_{22}^{-1} \\ -\boldsymbol{\Sigma}_{22}^{-1}\boldsymbol{\Sigma}_{21}\boldsymbol{G}_1 & \boldsymbol{G}_2 \end{pmatrix},$$

where $\boldsymbol{G}_h = (\boldsymbol{\Sigma}_{hh} - \boldsymbol{\Sigma}_{h,3-h}\boldsymbol{\Sigma}_{3-h,3-h}^{-1}\boldsymbol{\Sigma}_{3-h,h})^{-1}$, $h = 1, 2$. Now (10.A7) simplifies to

$$f(\boldsymbol{y}_1 \mid \boldsymbol{y}_2) = \tag{10.A8}$$

$$(2\pi)^{-\frac{n_1}{2}} (\det \boldsymbol{G}_1)^{\frac{1}{2}} \exp\left\{-\frac{1}{2}(\boldsymbol{y}_1'\boldsymbol{G}_1\boldsymbol{y}_1 - 2\boldsymbol{y}_1'\boldsymbol{G}_1\boldsymbol{\Sigma}_{12}\boldsymbol{\Sigma}_{22}^{-1}\boldsymbol{y}_2 + \boldsymbol{y}_2'(\boldsymbol{G}_2 - \boldsymbol{\Sigma}_{22}^{-1})\boldsymbol{y}_2)\right\} =$$

$$(2\pi)^{-\frac{n_1}{2}} (\det \boldsymbol{G}_1)^{\frac{1}{2}} \exp\left\{-\frac{1}{2}(\boldsymbol{y}_1 - \boldsymbol{\Sigma}_{12}\boldsymbol{\Sigma}_{22}^{-1}\boldsymbol{y}_2)'\boldsymbol{G}_1(\boldsymbol{y}_1 - \boldsymbol{\Sigma}_{12}\boldsymbol{\Sigma}_{22}^{-1}\boldsymbol{y}_2)\right\} \exp\{g(\boldsymbol{y}_2)\},$$

where $g(\boldsymbol{y}_2)$ is a function not involving $\boldsymbol{y}_1$. Since the conditional density (10.A8) integrates to 1 (over the range of $\boldsymbol{y}_1$), necessarily $g(\boldsymbol{y}_2) \equiv 0$, and so (10.A8) is the density of (10.A6) for $\boldsymbol{\mu}_h = \mathbf{0}$, $h = 1, 2$. The extension for nonzero means is proved directly by reference to $\boldsymbol{y}_h - \boldsymbol{\mu}_h$, $h = 1, 2$. □

Formulas (10.34) and (10.39) are obtained by direct application of (10.A6) to $(\boldsymbol{\delta}_j', \boldsymbol{y}')'$ and $(\boldsymbol{\varepsilon}', \boldsymbol{y}')'$, respectively.

REFERENCES

Airy, G. B. (1861), *On the Algebraical and Numerical Theory of Errors of Observations and the Combination of Observations*, London: McMillan.

Aitkin, M. (1987), "Modelling variance heterogeneity in normal regression using GLIM," *Applied Statistics*, 36, 332–339.

Aitkin, M., Anderson, D., and Hinde, J. (1981), "Statistical modelling of data on teaching styles." *Journal of the Royal Statistical Society*, Ser. A, 144, 148–161.

Aitkin, M. A., Anderson, D., Francis, B., and Hinde, J. (1989), *Statistical Modelling in GLIM*, Oxford: Oxford University Press.

Aitkin, M., and Longford, N. T. (1986), "Statistical modelling issues in school effectiveness studies." *Journal of the Royal Statistical Society*, Ser. A, 149, 1–43.

Anderson, D., and Aitkin, M. (1985), "Variance component models with binary response: Interviewer variability." *Journal of the Royal Statistical Society*, Ser. B, 47, 203–210.

Bartholomew, D. J. (1987), *Latent Variable Models and Factor Analysis*, London: C. Griffin & Co.; New York: Oxford University Press.

Battese, G. E., Harter, R. M., and Fuller, W. A. (1988), "An error components model for prediction of county crop areas using survey and satellite data." *Journal of American Statistical Association*, 83, 28–36.

Berk, K. (1987), "Computing for incomplete repeated measures." *Biometrics*, 43, 385–398.

Bock, R. D. (1989), *Multilevel Analysis of Educational Data*, San Diego: Academic Press.

Bonney, G. E. (1987), "Logistic regression for dependent binary observations." *Biometrics*, 43, 951–973.

Boyd, L. H., and Iversen, G. R. (1979), *Contextual Analysis: Concepts and Statistical Techniques*, Belmont, CA: Wadsworth.

Bryk, A. S., Raudenbush, S. W., Seltzer, M., and Congdon, R. T. (1988), *An Introduction to HLM: Computer Program and Users' Guide*, Chicago: University of Chicago Press.

Burstein, L. (1980), "The role of levels of analysis in the specification of educational effects." In *The Analysis of Educational Productivity*, Vol. 1, chap. 3, 11–90. Issues in Microanalysis, R. Dreeben and J. A. Thomas, eds. Cambridge, MA: Ballinger Publishing Co.

Burstein, L., Linn, R. L., and Capell, F. (1978), "Analyzing multilevel data in the presence of heterogeneous within-class regressions." *Journal of Educational Statistics*, 3, 347–383.

Chi, E. M., and Reinsel, G. C. (1987), "Models for longitudinal data with random effects and AR(1) errors." *Journal of the American Statistical Association*, 84, 452–459.

Connolly, M., and Liang, K.-Y. (1988), "Conditional logistic regression models for correlated binary data." *Biometrika*, 75, 501–506.

Cook, R. D., and Weisberg, S. (1982), *Residuals and Influence in Regression*, New York: Chapman and Hall.

Cronbach, L. J., and Webb, N. (1975), "Between-class and within-class effects in a reported aptitude-by-treatment interaction: Re-analysis of a study by G. L. Anderson." *Journal of Educational Psychology*, 67, 717–724.

Crump, S. L. (1951), "Present status of variance component analysis." *Biometrics*, 24, 527–540.

Deeley, J. J., and Lindley D. V. (1981), "Bayes Empirical Bayes." *Journal of the American Statistical Association*, 76, 833–841.

Dempster, A. P., Laird, N. M., and Rubin, D. B. (1977), "Maximum likelihood from incomplete data via the EM algorithm." *Journal of the Royal Statistical Society*, Ser. B, 39, 1–38.

Dobson, A. (1983), *An Introduction to Statistical Modelling*, London: Chapman and Hall.

Eisenhart (1947), "The assumptions underlying the analysis of variance." *Biometrics*, 33, 615–628.

Fieldsend, S., Longford, N. T., and McLeay, S. (1987), "Industry effects and the proportionality assumption in ratio analysis: A variance component analysis." *Journal of Business Finance and Accounting*, 14, 497–517.

Firebaugh, G. (1978), "A rule for inferring individual-level relationships from aggregate data." *American Sociological Review*, 4, 557–572.

Galton, F. (1886), "Family likeness in stature." *Proceedings of the Royal Society of London*, 40, 42–73.

Ghosh, M., and Rao, J. N. K. (1994), "Small area estimation: an appraisal," *Statistical Science*, 9, 55–93.

Goldstein, H. (1986), "Multilevel mixed linear model analysis using iterative generalized least squares." *Biometrika*, 73, 43–56.

——— (1987), *Multilevel Models in Educational and Social Research*, London: C. Griffin & Co.; New York: Oxford University Press.

Goldstein, H., and McDonald, R. P. (1988), "A general model for the analysis of multilevel data." *Psychometrika*, 53, 435–467.

Grizzle, J. E., and Allen, M. D. (1969), "Analysis of growth and dose response curves." *Biometrics*, 25, 357–381.

Hartley, H. O., and Rao, J. N. K. (1967), "Maximum likelihood estimation for the mixed analysis of variance model." *Biometrika*, 54, 93–108.

Harville, D. (1974), "Bayesian inference for variance components using only error contrasts." *Biometrika*, 61, 383–385.

——— (1977), "Maximum likelihood approaches to variance component estimation and to related problems." *Journal of the American Statistical Association*, 72, 320–340.

Healy, M. J. R. (1989), *NANOSTAT User Manual*, London: Alphabridge.

Hedges, L. V., and Olkin, I. (1985), *Statistical Methods for Meta-analysis*, New York: Academic Press.

Hemmerle, W. J., and Hartley, H. O. (1973), "Computing maximum likelihood estimates for the mixed A. O. V. model using the W transformation." *Technometrics*, 15, 819–831.

Henderson, C. R. (1953), "Estimation of variance and covariance components." *Biometrics*, 9, 226–252.

Holland, P. W. (1986), "Statistics and causal inference." *Journal of the American Statistical Association*, 81, 945 – 968.

Im, S., and Gianola, D. (1988), "Mixed models for binomial data with an application to lamb mortality." *Applied Statistics*, 37, 196–204.

Jamshidian, M., and Jennrich, R. I. (1988), "Conjugate gradient methods in confirmatory factor analysis." UCLA Statistics, Series 8.

Jennrich, R. I., and Sampson, P. F. (1976), "Newton-Raphson and related algorithms for maximum likelihood variance component estimation." *Technometrics*, 18, 11–17.

Jennrich, R. I., Schluchter, M. D. (1986), "Unbalanced repeated-measures models with structured covariance matrices." *Biometrics*, 42, 805–820.

Kreft, I. G. G., DeLeeuw, J., and Kim, K. S. (1990), "Comparing four different statistical packages for hierarchical linear regression: GENMOD, HLM, ML2, and VARCL." UCLA Statistics, Series 50.

Kreft, I. G. G., and Kim, K. S. (1991), "ML3." In *Statistical Software Reviews. Applied Statistics*, 40, 343–347.

Laird, N. M., Lange, N., and Stram, D. (1987), "Maximum likelihood computations with repeated measures: Application of the EM algorithm." *Journal of the American Statistical Association*, 82, 97–105.

Laird, N. M., and Ware, J. H. (1982), "Random-effects models for longitudinal data." *Biometrics*, 38, 963–974.

Lange, N., and Laird, N. M. (1989), "The effect of covariance structure on variance estimation in balanced growth-curve models with random parameters." *Journal of the American Statistical Association*, 84, 241–247.

Lange, N., and Ryan, L. (1989), "Assessing normality in random effects models." *Annals of Statistics*, 17, 624–643.

Lawley, D. N., and Maxwell, A. E. (1971), *Factor Analysis as a Statistical Method*, 2nd ed., London: Butterworth & Co.

Liang, K.-Y., and Zeger, S. L. (1986), "Longitudinal data analysis using generalized linear models." *Biometrika*, 73, 13–22.

Lindley, D. V., and Smith, A. M. F. (1972), "Bayes estimates for the linear model." *Journal of the Royal Statistical Society*, Ser. B, 34, 1–18.

Lindstrom, M. J., and Bates, D. M. (1989), "Newton-Raphson and EM algorithms for linear mixed-effects models for repeated measures data." *Journal of the American Statistical Association*, 84, 1014–1022.

Lockheed, M. E., and Longford, N. T. (1989), "A multilevel model of school effectiveness in a developing country." World Bank Discussion Papers 69, Policy Planning Research. Washington: World Bank.

Longford, N. T. (1985), "Mixed linear models and an application to school effectiveness." *Computational Statistics Quarterly*, 2, 109–117.

———(1987), "A fast scoring algorithm for maximum likelihood estimation in unbalanced mixed models with nested random effects." *Biometrika*, 74, 817–827.

———(1988a), "A quasi-likelihood adaptation for variance component analysis." Proceedings of the Section on Statistical Computing Section of the American Statistical Association. Alexandria, VA: American Statistical Association.

———(1988b), *VARCL Software for variance component analysis of data with hierarchically nested random effects (maximum likelihood)*, Princeton, NJ: Educational Testing Service.

———(1989), "Fisher scoring algorithm for variance component analysis of data with multilevel structure." In *Multilevel Analysis of Educational Data*, R. D. Bock, ed., 297–310. San Diego: Academic Press.

———(1990), "Multivariate variance component analysis: An application in test development." *Journal of Educational Statistics*, 15, 91–112.

———(1994), "Logistic regression with random coefficients." *Computational Statistics and Data Analysis*, 97, 1–15.

Longford, N. T., and Muthén, B. O. (1990), "Factor analysis for clustered observations." *Psychometrika*, 57, 581–597.

Luenberger, D. G. (1984), *Linear and Nonlinear Programming*, 2nd ed., Reading, MA: Addison-Wesley.

Mardia, K. V., Kent, J. T., and Bibby, J. M. (1979), *Multivariate Analysis*, London: Academic Press.

Mason, W. M., Wong, G. Y., and Entwisle, B. (1984), Contextual analysis through the multilevel linear model. In *Sociological Methodology* (S. Leinhardt, ed.), Jossey Bass, 72–103.

McCullagh, P., and Nelder, J. A. (1989), *Generalized Linear Models*, 2nd edition, London: Chapman and Hall.

McDonald, R. P., and Goldstein, H. (1989), "Balanced versus unbalanced designs for linear structural relations in two-level data." *British Journal of Mathematical and Statistical Psychology*, 42, 215–232.

McIntosh, A. (1982), *Fitting Linear Models: An Application of Conjugate Gradient Algorithms*, Lecture Notes in Statistics 10, New York: Springer Verlag.

Miller, J. J. (1977), "Asymptotic properties of maximum likelihood estimates in the mixed model of the analysis of variance." *The Annals of Statistics*, 5, 746–762.

Morris, C. N. (1983), "Parametric empirical Bayes inference: Theory and Applications." *Journal of the American Statistical Association*, 78, 47–65.

Morton, R. (1987), "A generalized linear model with nested strata of extra-Poisson variation." *Biometrika*, 74, 247–257.

Multilevel Modelling Newsletter (1989, 1990), London: Institute of Education, University of London.

Muthén, B. O. (1989), "Latent variable modelling in heterogeneous populations." *Psychometrika*, 54, 557–585.

Nelder, J. A., and Pregibon, D. (1987), "An extended quasi-likelihood function." *Biometrika*, 74, 221–232.

Numerical Algorithms Group (NAG) (1986), *The GLIM System. Release 3.77, Manual*, London: Royal Statistical Society.

Patterson, H. D., and Thompson, R. (1971), "Recovery of inter-block information when block sizes are unequal." *Biometrika*, 58, 545–554.

Rao, C. R. (1971a), "Estimation of variance and covariance components – MINQUE theory." *Journal of Multivariate Analysis*, 1, 257–275.

———(1971b), "Minimum variance quadratic unbiased estimation of variance components." *Journal of Multivariate Analysis*, 1, 445–456.

Rasbash, J., Prosser, R., and Goldstein, H. (1989), *Software for Two-level Analysis, Users' Guide*, London: Institute of Education.

Raudenbush, S. W. (1988), "Educational applications of hierarchical linear models: A review." *Journal of Educational Statistics*, 13, 85–116.

Raudenbush, S. W., and Bryk, A. S. (1986), "A hierarchical model for studying school effects." *Sociology of Education*, 59, 1–17.

Rosner, B. (1984), "Multivariate methods in opthalmology with application to other paired-data situations." *Biometrics*, 40, 1025–1035.

———(1989), "Multivariate methods for clustered binary data with more than one level of nesting." *Journal of the American Statistical Association*, 84, 373–380.

Rubin, D. B., Laird, N. M., and Tsutakawa, R. K. (1981), "Estimation in covariance component models." *Journal of the American Statistical Association*, 76, 341–353.

Searle, S. R. (1971), "Topics in variance component estimation." *Biometrics*, 27, 1–76.

Stanek, E. J., and Koch, G. G. (1985), "The equivalence of parameter estimates from growth curve models and seemingly unrelated regression models." *American Statistician*, 39, 149–152.

Stigler, S. M. (1986), *The History of Statistics. The Measurement of Uncertainty before 1900*, Cambridge, MA and London: The Belknap Press of Harvard University Press.

Stiratelli, R., Laird, N. M., and Ware, J. H. (1984), "Random effects models for serial observations with binary response." *Biometrics*, 40, 961–971.

Strenio J. F., Weisberg, H. I., and Bryk, A. S. (1983), "Empirical Bayes estimation of individual growth-curve parameters and their relationship to covariates." *Biometrics*, 39, 71–86

Thompson, R. (1979), "The estimation of variance and covariance components with an application when records are subject to culling." *Biometrics*, 29, 527–550.

Thompson, R., and Meyer, K. (1986), "Estimation of variance components: What is missing in the EM algorithm?" *Journal Statistical Computing and Simulation*, 24, 215–230.

Ware, J. H. (1985), "Linear models for the analysis of longitudinal studies." *The American Statistician*, 39, 95–101.

Williams, D. A. (1982), "Extra-binomial variation in logistic linear models." *Applied Statistics*, 31, 144–148.

Wedderburn, R. W. M. (1974), "Quasi-likelihood functions, generalized linear models and the Gauss-Newton method." *Biometrika*, 61, 439–447.

Zeger, S. L., Liang, K.-Y., and Albert, P. S. (1988), "Models for longitudinal data: A generalized estimating equation approach." *Biometrics*, 44, 1049–1060.

Zeger, S. L., and Qadish, B. (1988), "Markov regression models for time series: A quasi-likelihood approach." *Biometrics*, 44, 1019–1031.

Zelen, M. (1957), "The analysis of covariance for incomplete block designs." *Biometrics*, 13, 309–332.

Zellner, A. (1962), "An efficient method for estimating seemingly unrelated regression equations and test for aggregation bias. *Journal of the American Statistical Association*, 57, 348–368.

Index

Accelerated failure-time model, 491
Adjustment cells, 46
Airy, George Biddell, 521
Aliasing, 84
AMOS, 242
Analysis of covariance (ANACOVA), 524, 525
Analysis of variance (ANOVA), 252, 256, 282
Arbitrary heteroscedasticity, 127
Area-average regressions, 548
Aristotle, 4
Artificial insemination, 521–522
Asymptotically distribution-free (ADF) estimator, 191, 193, 229
Asymptotic covariance matrix, 99, 144
Attitudes over time, stability of, 210–214
Autocorrelated errors, probit models with, 423–429
Autoregressive fixed-effects logit models, 431
Autoregressive probit models, 429–431
Available-case (AC) analysis, 45, 50, 51
Average effects, 18, 19, 21–22, 24

Backward causation, 7
Backward recurrence time, 502, 505, 507, 511–512
BAN estimator, 87, 297
Basic models
- for panel analysis, metric data, 367–368
- for three-way tables, 282–285
- for two-way tables, 266–270

Bayesian methods, 366, 368–374, 394
- contingency table analysis and, 299
- event histories and, 509
- missing data and, 55, 59, 62–63, 65, 66–69
- mixed fixed and random coefficients models and, 384–386
- random coefficient models and, 382, 383, 384, 423, 530–531

Bernoulli distribution
- dichotomous outcome models and, 148–149, 150
- generalized linear models and, 551
- logit models and, 153, 154

Bernoulli distribution (*cont.*)
- missing data and, 58
- pseudo-maximum likelihood estimation and, 116, 117, 149

Bernoulli variables
- contingency table analysis and, 269
- fixed effects logit model and, 414, 417
- panel analysis, qualitative variables and, 402, 407
- random effects models and, 420

Best asymptotically normal estimator, *see* BAN estimator
Best linear unbiased estimator (BLUE), 376
Binary regression models
- examples of, 402–411
- for panel data, 411–433

Binomial distribution, 116, 154, 159, *see also* Negative binomial distribution
BMDP, 52
- contingency table analysis and, 301
- event histories and, 463, 492
- factor analysis models and, 569
- mean structures and, 177
- missing data and, 54

BMDPAM, 52
BMDP8D, 42, 43
Bootstrap samples, 55, 63–64
Borrowing strength, 531, 533
Boundary solutions, 335
Box–Cox transformation, 143
British Consumer Expenditure Survey (CES), 401
Broyden–Fletcher–Goldfarb–Shanno (BFGS) procedure, 125

CALIS, 242
CAN estimators, 297
Canonical link, 155, 407, 552
Canonical parameter
- generalized linear models and, 153–154, 158
- panel analysis, qualitative variables and, 407
- pseudo-maximum likelihood estimation and, 114–115

Categorical covariates, 340–343
Categorical variables/variation, 536
Causal analysis, 268–270
Causal effects, *see* Effects
Causal inference, 1–32
  in causal models, 4, 27–31
  deterministic, *see* Deterministic causation
  experimental approach and, 2, 17–26, 32
  probabilistic, *see* Probabilistic causation
  statistics and, 17–26
  structural-equation approach to, 28
Causal models, 4, 27–31
Causal priority, 14
  deterministic causation and, 7–8
  Granger causation and, 15, 16
  manipulative accounts and, 23
Causal sequences
  deterministic causation and, 6–7, 9
  probabilistic causation and, 10–11
Causes, *see* Factors
CDAS, 302, 312
Cell counts, 53
Censored data, 401
Censored outcomes, 150–153
  doubly, 79, 164–166
Censored Tobit model, 150–152, 409–411
Censoring, 479, 480, 483, 497
  left, 501, 511–512
  right, 455, 456, 498, 502, 506, 511
Chained data augmentation, 68
Chain rule, 125
Chi-squared distribution
  contingency table analysis and, 299
  covariance/mean structures and, 190, 193, 195, 198, 214, 229
  latent class models and, 319, 323, 336
  panel analysis, qualitative variables and, 408
Cholesky decomposition, 131, 137
Circumplex structures, 220
Classical scaling models, 344–348
Classified metric outcome models, 164–166
Clogg–Goodman formulation, 340, 341
Clusters
  latent class analysis and, 333–336
  random coefficient models and, 519–521, 523–524, 531–532
Cochrane–Orcutt transformation, 379
Coefficients of determination
  dichotomous outcomes and, 149–150
  generalized linear models and, 157–160
  linear regression models and, 89–90
Collapsibility, 285–287
Column effects model, 279–280, 291
Common cause, principle of the, 12
Compact Reticular Action Model (RAM), 207–208, 212
Complementary log-log link, 552
Complete-case (CC) analysis, 44–45, 50, 51
  weighting adjustments in, 46–48
Complete data, 542
Computer programs, *see* Software
Concomitants vs. factors (causes), 23, 25, 29
Conditional covariance structures, 221–223
Conditional distribution, 572
Conditional independence, 292–293, 325
Conditional likelihood, 504–505
Conditional likelihood ratio test, 271–272, 273, 277, 299
Conditional mean imputation, 60
Conditional mean structures, 221–223
Conservatism/liberalism, 255–259
Constant effect functions, 26
Contextual models, 529–530, 537
Contingency table analysis, 251–303
  estimation theory for, 293–298
  of higher-way tables, 291–293
  model-selection procedures in, 298–300
  residual analysis and, 298–300
  software for, 258, 300–301
  of three-way tables, *see* Three-way tables
  of two-way tables, *see* Two-way tables
  univariate distribution models and, 253–259
Continuous state space, 483–488
Continuous-time formulations, 457–465
  discrete-time formulations vs., 498–500
Continuous Work History Sample (CWHS), 361
Cook statistic, 93, 96, 156
  modified, 130–131, 138
Correlation structures
  minimum discrepancy estimation of, 193–194
  scaling considerations for, 187–194
COSAN, 242
Count data, 401, 402, 437–447
  negative binomial distribution and, 402, 437, 443
  overdispersion and, 437, 439–442
  panel models for, 444–447
  Poisson distribution and, 402, 437, 438, 441–442, 443, 444–445
  Poisson model and, 438–439, 447
  univariate nonlinear regression models and, 139–143
Counterfactual conditionals, 23, 25
Covariance estimator, *see* Least squares dummy variable estimator
Covariance matrix, 191–193
  asymptotic, 99, 144
Covariance structures, 1–2, 185–242
  background and notation on, 186–187
  computational aspects in, 200–203
  conditional, 221–223
  direct product, 219–220
  discrepancy functions and, 195–200

Covariance structures (*cont.*)
estimation of parameters in, 227–231
large sample properties of estimators, 194–200
multigroup analysis of, 232
with nonmetric dependent variables, 220–241
population drift assumption and, 195, 198, 200
reduced form parameters of, 228–229, 233–234
reference functions and, 195–200
scaling considerations for, 187–194
software for, 241–242
unconditional, 221–223
Covariance transformation, 376
Cressie–Read power-divergence family, 319
Cross-sectional data
count data and, 442–443
panel analysis, metric data and, 361–362, 368, 374–376, 377, 386, 392
Current Population Survey (CPS), 40–41, 46, 57, 59, 60–62, 68, 361
Curse of dimensionality, 78

Data augmentation, 67
chained, 68
Davidon–Fletcher–Powell (DFP) procedure, 125
De Finetti's exchangeability criterion, 382, 386, 390
Density function, 460
Dependent variable observability, 476–478
Destination-specific rate of transition, 481–484, 497
Determinant of patterned matrix, 571
Deterministic causation, 3, 13
in philosophy, 4–10
Deterministic time-dependent covariates, 469, 471
Deviance function, 554
Deviance residuals, 156, 299
Dichotomous outcomes
mean structures and, 146–150, 160
multivariate, 174
random utility maximization model for, 166–167
Direct effects, *see* Unit effects
Direct product covariance structures, 219–220
Direct Quartimin procedure, 205
Discrepancy functions, 195–200
Discrepancy measures, 157–160
Discrete data, 52–53
Discrete state space, 481–483
Discrete-time formulations, 457–458, 497, 498
continuous-time formulations vs., 498–500
Discrete-time Markov chains, 313
Dispersion parameter, 132
Distinctness, 4
Double-sided censored dependent variables, 224
Doubly censored outcomes, 79, 164–166
Durbin–Watson test, 381

ECM algorithm, 50, 53
Econometric approach to causal inference, 27–28
Economics, Granger causation in, 14–17
Education
attrition in, 40
income dependence on, 94–97
Effects, 4, 18
Efficient causes, 4
Employment status, 150, 151, 174–177
Endogenous time-dependent covariates, 471, 472–473, 475
Endogenous variables
in causal models, 27–29
Granger causation and, 14–15
Entry density, 502, 506–510, 512
EQS, 241, 352
Equilibrium multipliers, 28
Equiprobability hypotheses, 267
Error components model, 365, 376–381
E step
missing data and, 50, 53, 66
random coefficient models and, 543
Estimation
for contingency table analysis, 293–298
for covariance structure parameters, 227–231
for event histories, 488–492
for generalized linear models, 552–553
for generalized linear models with dependent data, 555–557
for random coefficient models, 540–547, 549
Event histories, 453–513
continuous state space and, 483–488
continuous-time formulations and, *see* Continuous-time formulation
dependent variable observability and, 476–478
discrete state space and, 481–483
discrete-time formulations and, *see* Discrete-time formulations
estimation procedures for, 488–492
hazard-rate model for, *see* Hazard-rate model
motivation in, 455–456
repeated events in, 478–480, 483
sampling plans of, 501–510
structural models and, 500–501
time-aggregation bias and, 495–498
time-dependent covariates and, *see* Time-dependent covariates
time-independent covariates and, 465–469
Exchangeability of groups, 523
Exogenous time-dependent covariates, 469–471, 472–473, 475
Exogenous variables
in causal models, 27–31
Granger causation and, 14–15, 23
Expectation, in random coefficient models, 538
Expectation maximization (EM) algorithm
event histories and, 493–495
factor analysis models and, 570

Expectation maximization (EM) algorithm (*cont.*)
generalized linear models and, 555
latent class models and, 312, 313, 333
missing data and, 49–50, 52, 53–54, 57, 66, 67
panel analysis, qualitative variables and, 432–433
random coefficient models and, 423, 522, 542–543, 545, 549
Expectation step, *see* E step
Expected-value parameterization, 114–115
Experimental approach, 2, 17–26, 32
Explicit imputation models, 60–61
Exponential curve, 217
Exponential model, 489
continuous-time formulations and, 460–461
time-dependent covariates and, 474
time-independent covariates and, 466
External validity, 23–24
Extrinsic aliasing, 84

Facets, 219–220
Factor analysis models, 187, 192, 202, 203–205, 222, 557–560
direct product covariance structures compared with, 219
latent curve models compared with, 217
matrix identities and, 571–572
software for, 568–570
two-level, 557–560
Factored likelihoods, 51
Factors (causes) vs. concomitants, 23, 25, 29
Failure-time processes, 453–454, 471, 472, 473, 475
Fatal Accident Reporting System (FARS), 41, 59, 61
Fertility study, 261–262, 284–285
Final causes, 4
Fisher scoring
contingency table analysis and, 300
factor analysis models and, 560
generalized linear models and, 553, 554, 556
latent class models and, 312, 323
maximum likelihood estimate and, 102, 104, 202
panel analysis, qualitative variables and, 405
pseudo-maximum likelihood estimators and, 124–127
quasi generalized PML and, 155
random coefficient models and, 540, 543, 544–545
Fisher's transformation, 229, 329
Five-way tables, 292
Fixed coefficients model, 383–384
Fixed effects
panel analysis, metric data and, 363, 366, 386–395
panel analysis, qualitative variables and, 412
unobserved heterogeneity and, 494–495
Fixed effects logit models, 413–417
autoregressive, 431
Fixed mean curves, with time series deviations, 218
Fixed parts, 538
Fixity, 7, 16
Formal causes, 4
Forward recurrence time, 502, 512
Full likelihood, 508–510
Full maximum likelihood estimation, 540–542

Galton, Francis, 521
Gamma distribution, 118, 154
GAUSS, 241
contingency table analysis and, 302
event histories and, 494
factor analysis models and, 570
generalized linear models and, 554
linear regression models and, 84
maximum likelihood estimation and, 110
mean structures and, 177
panel analysis, metric data and, 395
panel analysis, qualitative variables and, 429
probit models and, 427
Gauss–Hermite integration, 228
Gaussian (normal) distribution, 154
latent class models and, 351
pseudo-maximum likelihood estimation and, 117, 149
quasi generalized PML estimation and, 134
Gaussian quadrature
count data and, 446, 447
random effects models and, 419
Gauss–Jordan sweeps, 203
Gauss–Newton algorithm, 200–201, 202, 301, 427
GEM, 312
General homoscedasticity, 99
Generalized estimating equations (GEE)
for the first and second moments, 173
for mean structures, 79, 172–177
for missing data, 55
for probit models, 425
Generalized inverse, 85
Generalized least-squares discrepancy function, 189, 190, 194, 197, 200–201
Generalized least-squares (GLS) estimation, 393, 407–408
error components model and, 377–378
mixed fixed/random coefficients models and, 384–386
probit models and, 425
random coefficient models and, 382–383
repeated observations and, 407–408
Generalized linear models (GLM), 82, 132, 196, 407, 551–557
coefficients of determination and, 157–160
discrepancy measures and, 157–160
estimation for, 552–553
estimation for models with dependent data, 555–557
extensions for dependent data, 554

Generalized linear models (GLM) (*cont.*)
mean structures and, 137, 140, 153–160
missing data and, 55
parameterization of, 118
specification of, 153–155
univariate, 79
General location model, 53–54
General multivariate heteroscedasticity, 113
General one-factor schemes, 421–422
General Social Survey (GSS), 255–259
General two-level random coefficient models, 533–540
GENLOG, 302
GENMOD, 569
Geometric distribution, 437
German Socio-Economic Panel (GSOEP), 94, 150, 165, 174–177, 447
Gibbs sampler, 228
missing data and, 55, 67–68
random coefficient models and, 423
GLIM
contingency table analysis and, 270, 293, 300–301
event histories and, 458, 499
factor analysis models and, 569
generalized linear models and, 554
latent class models and, 313, 320
linear regression models and, 82, 84
mean structures and, 177
probit models and, 427
Gompertz curve, 217
Gompertz model, 489, 499
continuous-time formulations and, 461, 464
discrete state space and, 483
Goodman–Kruskal lambda measure of association, 337
Goodman's log-bilinear models, 257–258
Goodman's modification of Guttman's model, 344–346
Goodman's scaling model, 347–348, 351
Goodness-of-fit
contingency table analysis and, 253, 272–273, 298, 300
covariance/mean structures and, 230
latent class models and, 318–319, 321
structural equation models and, 214–216
Granger causation, 2, 3, 4, 14–17, 23
Graphical analysis method, 324–325
Graphical models, 292–293
Grouped metric variables, 79
Group-level random parts, 538
Group-level regressions, 548
Guest workers, attitude toward, 162–164
Guttman's scaling model
Goodman's modification of, 344–346
latent class models and, 348

Hazard-rate model, 456–465, 473, 474–475, 476, 499, 512
continuous state space and, 486
Hazard-rate model (*cont.*)
dependent variable observability and, 476–478
repeated events in, 479–480
sampling plans and, 504, 507, 508–510
software and, 492
time-aggregation bias and, 495
unobserved heterogeneity and, 492–494
Heckman's approach, 430
Hessian matrices
count data and, 439
covariance/mean structures and, 198–199, 201–202
latent class models and, 333
panel analysis, qualitative variables and, 407
pseudo-maximum likelihood estimation and, 125
Heterogeneity, 388–389, 412–413
fixed effects logit model with, 413–417
interindividual, 365
intertemporal, 365
random effects models with, 418–421
test for, 419–421
unobserved, *see* Unobserved heterogeneity
Heteroscedastic consistent (HC) covariance matrix, 89, 92
Heteroscedasticity
arbitrary, 127
general multivariate, 113
quasi generalized PML estimation and, 134
standard nonlinear regression models and, 144
Higher-way tables, 291–293
HILOGLINEAR, 302
HLM, 568–569, 570
Homoscedasticity, 90, 92–93, 95, 132
doubly censored/classic metric outcomes and, 164
general, 99
maximum likelihood estimation and, 86, 88–89, 112
multivariate, 98
Hot-deck procedures, 60–62, 63–64
Human capital theory, 94
Hume, David, 5–6, 11

Ignorable nonresponse models
imputation and, 61–63, 64
maximum likelihood estimation and, 48–55
IML, 427
Implicit imputation models, 60–61
Imputation, 59–66
conditional mean, 60
explicit vs. implicit models in, 60–61
multiple, *see* Multiple imputation
proper methods of, 63–64
stochastic regression, 60, 61
unconditional mean, 60
Imputation step, 67
IMS, 177
Income
as classified metric dependent variable, 165–166
education and, 94–97

Incomplete data, 542
Independence model, 427, 430
Independence of irrelevant alternatives (ILA), 169–170
Independent multinomial sampling, 251, 252, 253, 261, 262–263
Indirect effects, 29
Individual-level random parts, 538
Individual time-invariant variables, 375
Individual time-varying variables, 375
Influential points
- generalized linear models and, 156
- linear regression models and, 93
- pseudo-maximum likelihood estimation and, 130–131
- quasi generalized PML estimation and, 138

Information sandwich, 89, 173
Instantaneous causation, 16–17, 25, 26, 30
Integer-valued data, 401
Interactive forks, 12
Interative Bayesian simulation, 68–69
Intercept-by-slope-variance, 535
Intercept-variance, 535
Interindividual heterogeneity, 365
Internal validity, 23
Intertemporal heterogeneity, 365
Intrinsic underidentification, 333
Invariant under a constant scaling factor (ICSF), 192
Inverse-function theorem, 115
Inverse link function, 132, 153
Inverse of patterned matrix, 571
IRLS procedure, 558
Item nonresponse, 42
Item ordering, 345
Item-response theory (IRT) models, 313
- fixed effects logit model and, 413–414
- latent class models and, 348, 352

Iterative weighted least squares (IWLS) procedure
- pseudo-maximum likelihood estimation and, 127
- quasi generalized PML estimation and, 135, 155
- standard nonlinear regression models and, 143
- univariate nonlinear regression models and, 141

Jackknife samples, 55
Jacobian matrices
- covariance/mean structures and, 198, 201, 203
- latent class models and, 333

Jeffreys prior distribution, 59, 300
Jensen's inequality, 492
Joint likelihood, 505–508
Jump processes, 471

Kant, Immanuel, 6
Kronecker delta, 161
Kronecker products, 97, 534
Kullback–Leibler Information Criterion (KLIC), 101, 112
Kullback's inequality, 121

Lack of fit, 195
LACORD, 313
Lagged endogenous variables, 402
Lagrange multiplier (LM), 203, 230
- contingency table analysis and, 254
- maximum likelihood estimation and, 78, 100, 104–107
- panel analysis, metric data and, 393
- pseudo-maximum likelihood estimation and, 128–129
- quasi generalized PML estimation and, 135–136
- random coefficient models and, 543

Laplace approximations, 55
Large-sample methods, 66, 68
- covariance structure estimators as, 194–200

LAT, 312, 320, 333, 338, 349, 352
Latent budget analysis, 320, 324
Latent class analysis
- categorical covariates in, 340–343
- clustering and, 333–336

Latent class models (LCM), 311–353
- alternative forms of, 319–321
- basic concepts and notation for, 315–317
- latent structure models and, 313–315
- measuring fit with, 318–319, 328–330
- medical diagnosis and, 327–328
- missing data and, 332–333
- models related to, 324–326
- in multiple groups, 340–343
- predicting membership with, 336–339
- rater agreement measured with, 330–332
- scaling and, 343–352
- software for, 312–313
- *T* class, 318, 326, 343
- for two-way tables, 321–324

Latent curve models, 217
Latent distance model, 344
Latent Markov models, 317, 318, 353
Latent profile model, 351
Latent structure models, 313–315
LCAG, 312–313
Least squares dummy variable (LSDV) estimator, 376, 378, 393
Least-squares method, 144
- event histories and, 486
- generalized linear models and, 556
- missing data and, 57
- panel analysis, metric data and, 363

Left censoring, 501, 511–512
Leverage points
- generalized linear models and, 156
- linear regression model and, 93
- quasi generalized PML estimation and, 138

Liberalism/conservatism, 255–259
Likelihood, 505–508
- conditional, 504–505
- factored, 51

Likelihood (*cont.*)
full, 508–510
joint, 505–508
Likelihood function, 227–228
Likelihood ratio (LR), 230
conditional, 271–272, 273, 277, 299
contingency table analysis and, 255
generalized linear models and, 158–159, 553–554
latent class models and, 319
maximum likelihood estimation and, 78, 100, 104–107
pseudo-maximum likelihood estimation and, 130
quasi generalized PML estimation and, 135–136
Likert scales, 78, 232, 314
LIMDEP, 177, 492
Limited dependent variable models, 401, 409–411
LINCS, 242
Linear-by-linear association model, 279–280, 290–291
Linear exponential family, 114–121
multivariate, 119–120
with nuisance parameter, 114, 158–159
parameterized in the mean, 120–121
univariate, 115–118, 132
Linear predictor, 153
Linear probability model, 146
panel analysis, qualitative variables and, 402–408, 409
Linear regression models, 132, 221
arbitrary heteroscedasticity and, 127
causal regression models compared with, 27
event histories and, 455
matrix of, 80
mean structures and, 80–99
model specification in, 80–84
regression coefficient estimation in, 80–81, 84–89
regression diagnostics in, 89–97
multivariate, *see* Multivariate linear regression model
univariate, 78, 80, 177
Link function, 153, 154
inverse, 132, 153
LISCOMP, 241, 314, 352
LISREL, 205–209, 241
latent class models and, 314, 352
latent curve models and, 217
Listwise deletion, *see* Complete-case analysis
Local independence, axiom of, 317
Logistic curve, 217
Logit link, 551
Logit models
contingency table analysis and, 267–270, 274–278, 284, 287
event histories and, 457, 458, 499
fixed effects, 413–417, 431
mean structures and, 147–148, 149–150, 151, 153, 154, 164
Logit models (*cont.*)
multinomial, *see* Multinomial logit models
panel analysis, metric data and, 393
panel analysis, qualitative variables and, 402–407, 408, 409
random effects models and, 418
Log-likelihood function
contingency table analysis and, 254
factor analysis models and, 558–559, 560
in maximum likelihood estimation, 100, 101–102
LOGLINEAR, 302
Log-linear models, 153, 154
contingency table analysis and, 251–252, 253, 254–255, 256, 257, 258, 261–263, 266–270, 271, 279, 282–283, 287, 291–292, 300
count data and, 140, 141
missing data and, 53
saturated, 256, 263, 267, 282–283
Log-logistic model, 489
Log-nonlinear model of symmetry, 258
Longitudinal Study of American Youth (LSAY), 221, 232–241

Macro model equations, 529
Mahalanobis distance, 109
Manipulative accounts, 2, 3, 29, 32
of deterministic causation, 4
statistics and, 17, 23–26
Marginal homogeneity (MH), 270–272, 273, 289
MARKOV, 302
Markov chain models
discrete-time, 313
panel analysis, qualitative variables and, 402, 433–435
Markov imputation, 68
Markov latent models, 317, 318, 353
Material causes, 4
Matlab, 554, 570
Matrix identities, in factor analysis models, 571–572
Maximization step, *see* M step
Maximum likelihood discrepancy function, 189, 191, 197, 202
Maximum likelihood estimation (MLE), 122–123, 146, 148, 149, 150, 169
autoregressive probit models and, 430
censored outcomes and, 152
contingency table analysis and, 253, 254–255, 257, 261, 262–263, 266, 270, 274, 293–297
count data and, 141
of covariance structures, 223, 224, 225, 228–229
event histories and, 461, 466–467, 468, 478, 486, 488, 489, 490–492
factor analysis models and, 192, 205, 560
fixed effects logit model and, 413–417
full, 540–542
generalized linear models and, 154–155, 552
latent class models and, 321

Maximum likelihood estimation (MLE) (*cont.*)
of mean structures, 78–79, 86–89, 91, 95, 98, 99, 100–111, 143
conditional/unconditional, 223, 224, 225
log-likelihood function in, 100
properties of estimator in, 101–104
reduced form parameters in, 228–229
restrictions on parameters in, 108–111
missing data and, 42, 45, 66
ignorable nonresponse models in, 48–55
nonignorable nonresponse models in, 55–59
under misspecification, 111–113
ordered categorical variable models and, 161
panel analysis, metric data and, 376, 379
panel analysis, qualitative variables and, 402–407
quasi generalized PML estimation compared with, 134, 136, 138
random coefficient models and, 540–542, 549
of random effects probit models with heterogeneity, 418–419
restricted, 540–542
standard nonlinear regression models and, 143–144
theory of, 48–49
Maximum quasi-likelihood estimation
under overdispersion, 441–442
probit models and, 425–427, 430
McFadden's conditional logit model, 490
Mean structures, 78–79, 185–242
background and notation on, 186–187
computational aspects in, 200–203
conditional, 221–223
discrepancy functions and, 195–200
estimation of parameters in, 227–231
examples of, 203–220
generalized estimating equations for, 79, 172–177
large sample properties of estimators, 194–200
multigroup analysis of, 232
with nonmetric dependent variables, 220–241
with nuisance parameters, 88
population drift assumption and, 195
reduced form parameters of, 228–229, 233–234
reference functions and, 195–200
regression models and, 77–177, *see also* Linear regression models; Maximum likelihood estimation; Multivariate nonlinear regression models; Pseudo-maximum likelihood estimation; Quasi generalized pseudo-maximum likelihood estimation; Univariate nonlinear regression models
scaling considerations for, 187–194
software for, 177, 241–242
unconditional, 221–223
MECOSA, 233, 241–242
latent class models and, 314, 352
panel analysis, qualitative variables and, 429
Medical diagnosis, 327–328
Megaphone patterns, 527
Membership prediction, with latent class models, 336–339
Metrically classified dependent variables, 224
Metrically scaled dependent variables, 223
Metric data, panel analysis for, *see* Panel analysis, metric data
M-facets, 219–220
Micro model equations, 529
Minimum discrepancy estimation
of correlation structures, 193–194
of covariance matrix, 191–193
of covariance/mean fundamental parameters, 229–230
Minimum distance estimation (MDE), 79, 109–111
Minimum ignorance estimator, 112
Missing at random (MAR), 43–44
ignorable nonresponse and, 49
latent class models and, 332
maximum likelihood estimation and, 58
Missing completely at random (MCAR), 43–44
complete-case analysis and, 44
latent class models and, 332
maximum likelihood estimation and, 49, 56, 58
unconditional mean imputation and, 45
Missing data, 39–69, 232
completely at random, *see* Missing completely at random
examples of, 40–41
item nonresponse, 42
latent class models for, 332–333
maximum likelihood estimation in, *see* under Maximum liklihood estimation
mechanisms of, 43–44
monotone, 42, 51
multiple imputation and, 59–66, 67
multivariate, 43
naive approaches to, 44–46
patterns of, 42
at random, *see* Missing at random
random coefficient models and, 542
unit nonresponse, *see* Unit nonresponse
univariate nonresponse, *see* Univariate nonresponse
Misspecification
linear regression model under, 89, 90–92
maximum likelihood estimation under, 111–113
Mixed fixed and random coefficients models, 365, 384–386
ML3, 569
MLLSA, 312–313, 333, 334
Mobility table, 270
Modified Cook statistic, 130–131, 138
Moment methods, 543
Monotone missing data, 42, 51

Monte Carlo experiments, 241
autoregressive probit models and, 430
binary logicstic regression model and, 420
contingency table analysis and, 297
count data and, 442
fixed effects logit model and, 417
random coefficients models and, 423
random effects models and, 419
unobserved heterogeneity and, 494
M step
latent class models and, 312
missing data and, 50, 52, 53, 66
random coefficient models and, 543, 545
Multigroup analysis of covariance/mean structures, 232
Multilevel Models Project, 569
Multinomial distribution
contingency table analysis and, 251, 252, 255
pseudo-maximum likelihood estimation and, 119
Multinomial logit model, 171
independence of irrelevant alternatives and, 169–170
as random utility maximization model, 168–169
Multinomial models, 52–53
Multinomial probit models, 170–171
Multinomial sampling, 262–263, 266, 290
Multiple groups, latent class models in, 340–343
Multiple imputation
interative Bayesian simulation and, 68–69
latent class models and, 337–338
missing data and, 59–66, 67
Multivariate central limit theorem, 103
Multivariate delta method, 79, 108–109
Multivariate dichotomous outcomes, 174
Multivariate homoscedasticity, 98
Multivariate linear exponential family, 119–120
Multivariate linear regression models, 78, 97–99
Multivariate missing data, 43
Multivariate nonlinear regression models, 79, 160–177
for classified metric outcomes, 164–166
for doubly censored outcomes, 79, 164–166
for ordered categorical variables, 79, 160–164
software for, 177
for unordered categorical variables, 166–171
Multivariate normal model
missing data and, 51–52, 54
pseudo-maximum likelihood estimation and, 120
Multivariate regression, 532, 536–537
Mx, 242

Naive approaches to missing data, 44–46
NANOSTAT, 569
National Survey of Families and Households, 334
Negative binomial distribution
count data and, 402, 437, 443
pseudo-maximum likelihood estimation and, 117, 118
Nesting levels, in random coefficient models, 547–550
NEWTON, 312, 320, 333, 338, 349, 352
Newton algorithms, 54
Newton–Raphson algorithm
contingency table analysis and, 300, 301, 302
latent class models and, 312, 323, 333
missing data and, 50
panel analysis, qualitative variables and, 405
random coefficient models and, 540, 543
Neyman–Scott principle, 393
Nominal scale, 401
Noncausal sequences
deterministic causation and, 6–7, 9
probabilistic causation and, 10–11
Noncensored durations, 489
Noncensored observations, 511
Nonignorable nonresponse models
imputation and, 61–63
maximum likelihood estimation and, 55–59
Nonlinear least squares (NLLS), 478
Nonmetric dependent variables, 220–241
estimation of, 227–231
multigroup analysis of, 232
threshold models and, 223–226
Non-normal models for repeated measures, 55
Normal density with constant variance, 145–146
Normal distribution, *see* Gaussian (normal) distribution
Normality assumption
doubly censored/classic metric outcomes and, 164
linear regression models and, 90, 92–93
Normal models for repeated measures, 54
Normal pattern-mixture models, 58–59
Normal theory
generalized least squares discrepancy function of, 189, 190
maximum likelihood discrepancy function of, 189, 191
Norwegian Life History Study for Men, 485
Nuisance parameters, 48
correlation structures and, 194
doubly censored/classic metric outcomes and, 164–166
linear exponential family with, 114, 158–159
mean structures with, 88
pseudo-maximum likelihood estimation with, 132–134
univariate nonlinear regression models and, 141
Null hypothesis
contingency table analysis and, 251–252, 255, 262–263, 284–285
linear regression model and, 85, 87–88
of marginal homogeneity, 271–272, 273
maximum likelihood estimation and, 105, 106
pseudo-maximum likelihood estimation and, 128

Observability condition, 22–23, 24
Occupational codes, 41, 59, 62
Occupational mobility, 270, 272–274, 281–282, 295, 321–326

Odds ratios
posterior, 394–395
for three-way tables, 265–266
for two-by-two tables, 259–263
for two-way tables, 264
Omitted variable bias, 285–286
One-sided censored dependent variables, 224
One-way analysis of variance models, 19
Ordered categorical data, 401
Ordered categorical dependent variables, 224
Ordered categorical variables, 78, 79
models for, 160–164
Ordered variables, 289–291
Ordinal data, 431–433
Ordinal probit regression, 228
Ordinal Probit Relation, 224
Ordinal scale, 401
Ordinal variables, 274–282
Ordinal *X*, 343–344
Ordinary least squares (OLS) estimation
classified metric dependent variables and, 165
error components model and, 378
factor analysis models and, 566–567
linear regression models and, 84, 88, 91–92
multivariate linear regression models and, 99
random coefficient models and, 525, 532, 543, 545, 546, 563
Ordinary residuals
generalized linear models and, 156
linear regression model and, 90–92
pseudo-maximum likelihood estimation and, 129
Orthogonal rotation procedures, 205
Overdispersion, 437, 439–442

Panel analysis, metric data, 361–396
basic model for, 367–368
primary uses of, 362
random or fixed effects in, 386–395
variable types in, 375
Panel analysis, qualitative variables, 401–447
binary regression models for, 411–433
count data in, *see* Count data
Markov chain models and, 402, 433–435
ordinal data and, 431–433
simulation-based inference in, 427–428
Panel Study of Income Dynamics (PSID), 361
PANMARK, 313
Partial association models, 291
Partial Likelihood (PL) principle, 488–492
Pascal distribution, 117, 437
Path analysis, 1, 81
Patterned matrix, 571
Pattern-mixture models, 56
normal, 58–59
Pearson's goodness-of-fit statistic, 318–319
Pearson's residuals, 298, 300, 301, 302
Pearson's X^2 statistic, 159–160, 255, 272
Perfect mobility, 321
Period individual-invariant variables, 375
Philosophy
deterministic causation in, 4–10
probabilistic causation in, 10–14
Piecewise constant rate, 461, 462–465, 499
Poisson distribution
contingency table analysis and, 251, 252, 253, 266, 293–294
count data and, 402, 437, 438, 441–442, 443, 444–445
generalized linear models and, 159, 160, 552, 555
pseudo-maximum likelihood estimation and, 114, 115, 117, 122, 141–143, 146
univariate nonlinear regression models and, 140–141
Poisson error function, 300
Poisson log-linear models, 153, 154, 302
Poisson model, 438–439, 447
Polychoric correlation coefficients, 226–227, 229, 234
Polychoric covariance coefficients, 226–227
Polynomial regression, 538
Polyserial correlation coefficients, 226–227, 229, 234
Polyserial covariance coefficients, 226–227, 234
Population, 502, 504, 506, 509
Population drift assumption, 195, 198, 200
Posterior odds ratio, 394–395
Posterior step, 67
Post-stratification method, 47
Power divergence statistic, 255
Prais-Winston transformation, 380
Predictive density ratio, 394–395
Predictive mean matching, 61
PRELIS 2, 241
Probabilistic causation, 3, 10–17, 23, 32
philosophical treatments of, 10–14
Probability density, 497
Probit link, 552
Probit models, 221, 225, 417
with autocorrelated errors, 423–429
autoregressive, 429–431
event histories and, 457, 458, 499
generalized estimating equations and, 176
mean structures and, 147–148, 149–150, 151, 153, 162, 164
multinomial, 170–171
panel analysis, qualitative variables and, 402–407, 408, 409
random effects with heterogeneity, 418–419
univariate, 187
Probit Relation, Two-Limit, 224
Probit selection model, 56–58
Proctor–Goodman model, 347–348
Proctor's model, 346, 347
Product multinomial sampling, *see* Multinomial sampling

Proper imputation methods, 63–64
Proportional hazards model, 488–489, 491
Proportion of reduced deviance (PED), 159
Proportion of reduced error (PRE), 159
Pseudo-maximum likelihood (PML) estimation, 79, 144, 149
  contingency table analysis and, 293, 298
  count data and, 141–143
  generalized estimating equations and, 172, 173
  generalized linear models and, 153
  linear exponential family in, 114–121
  of mean structures, 89, 91, 95, 113–131
    Fisher scoring and, 124–127
    properties of estimator and, 121–124
    regression diagnostics in, 129–131
  under normal density with constant variance, 145–146
  with nuisance parameters, 132–134
  ordered categorical variable models and, 162
  software and, 177

Quadratic log-linear model of symmetry, 257
Qualitative variables, data analysis for, *see* Panel analysis, qualitative variables
Quantit models, 148, 150–153, 164
Quasi generalized pseudo-maximum likelihood (QGPML) estimation, 79, 149
  computation of, 135
  count data and, 141
  generalized estimating equations and, 173, 174
  generalized linear models and, 153, 154–155, 156
  mean and variance specification in, 131–132
  of mean structures, 88, 131–138
  ordered categorical variable models and, 162
  regression diagnostics under, 136–138
  software for, 177
  standard nonlinear regression models and, 144
Quasi-independence (QI) model, 271, 321–322
Quasi-likelihood estimation, 132
  contingency table analysis and, 297
  generalized linear models and, 156, 553–554, 555–556
Quasi-likelihood ratio test statistic, 554
Quasi maximum likelihood (QML), 79, 111–113
Quasi-Newton methods, 560
Quasi-symmetry (QS) model, 281, 288–289
  square tables and, 271–274
Quasi-uniform association (QUA) model, 281
Quasi-Wiener Simplex structures, 218

RAMONA, 242
Random coefficient models, 365, 382–384, 519–572
  applications of, 531–533
  categorical variables/variation and, 536
  contextual models and, 537
Random coefficient models (*cont.*)
  diagnostics for, 546
  estimation for, 540–547, 549
  general two-level, 533–540
  identification of, 539–540
  illustration of, 522–523
  model selection and, 546–547
  multivariate regression as, 532, 536–537
  nesting levels in, 547–550
  panel analysis, qualitative variables and, 422–423
  proportion of variation explained in, 549–550
  with single explanatory variable, 524–533
  terminology of, 530–531
Random effects models
  general one-factor schemes in, 421–422
  panel analysis, metric data and, 363, 366, 386–395
  panel analysis, qualitative variables and, 413, 417–422, 423
  unobserved heterogeneity and, 493–495
Randomized experiments, 23–24, 32
Random parts, 538
Random polynomials, 538
Random utility maximization (RUM) model, 79, 170
  for dichotomous outcomes, 166–167
  multinomial logit model as, 168–169
Rao's efficient score, 106
Rasch model
  fixed effects logit model and, 414, 416
  latent class models and, 312, 314, 330, 348–351, 352
Rater agreement, 330–332
Ray patterns, 527
R+C model, 280, 281
RC(1) model, 278–281, 289–291
RC association model, 322
RC(M) model, 278–279, 290
Reciprocal linear model, 154
Reference functions, 195–200
Regression analysis, 22–23
Regression coefficients, 80–81, 84–89
Regression diagnostics
  in linear regression models, 89–97
  in pseudo-maximum likelihood estimation, 129–131
  in quasi generalized PML estimation, 136–138
Regression models
  binary, *see* Binary regression models
  mean structures and, 77–177, *see also* under Mean structures
Regressor matrix, 80
Regularity theories, 4, 5, 6–7, 8
REML, 568, 569, 570
Renewal processes, 502
Repeated events, 478–480, 483
Repeated measurements
  non-normal models for, 55
  normal models for, 54
  over time, 216–220

Repeated observations, 407–408
RESET test, 91–92, 95–96
Residual analysis, 298–300
Residual function, 553
Response propensity stratification, 46–47
Restricted factor analysis, 205
Restricted maximum likelihood estimation, 540–542
Reticular Action Model (RAM), 207–216
    compact, 207–208, 212
    latent curve models and, 217
    software and, 242
Riemann–Stieltjes, 513
Right censoring, 455, 456, 498, 502, 506, 511
RMSEA, 230, 237
Row effects model, 279–280, 291
Russell, Bertrand, 6

S, 177
Sampling from the flow, 501
Sampling from the stock, 501–502, 504, 506, 510
Sampling plans, 501–510
SAS
    contingency table analysis and, 301
    event histories and, 492
    factor analysis models and, 569
    linear regression models and, 84
    mean structures and, 177
    panel analysis, metric data and, 395
    probit models and, 427
SAS Proc Mixed, 54
Saturated log-linear models, 256, 263, 267, 282–283
Scaled deviance, 158
Scaling models
    classical, 344–348
    as latent class models, 343–352
    for mean, covariance and correlation structures, 187–194
Schopenhauer, Arthur, 6
Seemingly unrelated regressions (SUR), 531
SEM algorithm, 50
SEPATH, 242
Shrinkage estimators, 521
Simulated generalized methods of moments, 427–428
Simulated maximum likelihood methods, 427–428
Simulated pseudo-maximum likelihood methods, 427–428
Simulation-based inference, 427–428
Simultaneous-equation models, 4, 14–15, 27, 31
Simultaneous latent structure analysis, 340
Single-equation models, 27
Single explanatory variable, 524–533
Slutsky's theorem, 103
Small-area statistics, 531
Software, *see also* specific programs
    for contingency table analysis, 258, 300–301
    for covariance structures, 241–242

Software (*cont.*)
    for event histories, 492
    for factor analysis models, 568–570
    for latent class models, 312–313
    for mean structures, 177, 241–242
Spatio-temporal contiguity, 5, 6
S-Plus, 303, 554, 570
SPSS, 177, 302
Spurious state dependence, 429
Square root linear model, 140, 141
Square tables, 270–274
Stable unit treatment value assumption (SUTVA), 18, 26
Standardized residuals
    generalized linear models and, 156
    linear regression models and, 92–93
    quasi generalized PML estimation and, 136–137
Standard nonlinear regression models, 143–146
Standard Tobit model, 410
Statistical causation, *see* Probabilistic causation
Statistics, causal inference and, 17–26
Stochastic censoring model, *see* Probit selection model
Stochastic regression imputation, 60, 61
Stochastic relaxation, 68
Stouffer–Toby data, 346
Structural equation models, 205–216, 561
    causal inference and, 28
    specification of, 235
Structural models, 500–501
Studentized residuals
    generalized linear models and, 156
    linear regression model and, 92–93
    quasi generalized PML estimation and, 136–137
Suicide, imitative, 268–270
Superexogeneity, 29
Superpopulation, 502, 508–510
Survey of Consumer Finances (SCF), 40
Survivor function
    continuous state space and, 485
    continuous-time formulations and, 460, 461–462
    dependent variable observability and, 477
    discrete state space and, 482
    proof of, 513
    sampling plans and, 503, 504, 507
    time-dependent covariates and, 471, 473–474, 475
    time-independent covariates and, 467–468
Swamy random coefficients formulation, 382, 384
Sweep inverse, 85
Symmetry + independence (SI) model, 271, 288–289
Symmetry + quasi-independence (SQI) model, 271
SYSTAT, 177

Taylor expansions
    heterogeneity test and, 419
    latent curve models and, 217
    maximum likelihood estimation and, 110

Taylor expansions (*cont.*)
pseudo-maximum likelihood estimation and, 124–125, 128
quasi generalized PML estimation and, 136
univariate nonlinear regression models and, 140
T class latent class models (LCM), 318, 326, 343
Temporal priority, 5, 6, 9, 14, 30
Granger causation and, 16
manipulative accounts and, 25–26
Tetrachoric correlation coefficient, 252
T-facets, 219–220
Three-parameter logistic model, 351
Three-way tables
models for, 282–291
odds ratios for, 265–266
Threshold models
dichotomous outcomes and, 146–149, 160
nonmetric dependent variables and, 223–226
ordered categorical variables and, 160–161
specification of, 237
Time-aggregation bias, 495–498
Time-dependent covariates, 456, 469–476
deterministic, 469, 471
endogenous, 471, 472–473, 475
exogenous, 469–471, 472–473, 475
Time-independent covariates, 465–469
Time-series data, 361–362, 368, 392, 394
Time-series deviations, 218
Time-series models, 2, 15, 27
Tobit models, 164, 221, 224, 402
censored, 150–152, 409–411
for panel data, 435–436
Standard, 410
truncated, 409–411
Tobit regression, 228
Tobit relation, 224
Total effects, 28, 29
Traffic accident frequency, 141–143, 145–146
Transition probability, 499–500
Tree-extreme-value (TEV) model, 170
Treiman occupational prestige scale, 162
True state dependence, 429
Truncated Tobit model, 409–411
Two-by-two tables, 259–263
Two-level factor analysis models, 557–560
Two-Limit Probit Relation, 224
Two-parameter logistic model for item analysis, 351
Two-way tables
Goodman's log-bilinear models for, 257–258
latent class models for, 321–324
models for, 266–282
odds ratios for, 264

Unconditional covariance structures, 221–223
Unconditional mean imputation, 45–46, 50, 60
Unconditional mean structures, 221–223
Uniform association (UA) model, 257, 280, 281, 291
Uniform logit model with symmetry, 257
Unit effect functions, 25–26
Unit effects, 17–18, 23, 24, 25–26, 29
Unit nonresponse, 42
weighting adjustments for, 46–48
Univariate distribution models, 253–259
Univariate generalized linear models (GLM), 79
Univariate linear exponential family, 115–118, 132
Univariate linear models, 153
Univariate linear regression models, 78, 80, 177
Univariate nonlinear regression models, 79, 139–160
for censored outcomes, 150–153
for count data, 139–143
for dichotomous outcomes, *see* Dichotomous outcomes
software for, 177
Univariate nonresponse, 42
imputation and, 60, 61, 63
multivariate normal model and, 51
probit selection model and, 56
Univariate probit models, 187
Univariate quasi-likelihood models, 135
Unobserved heterogeneity
event histories and, 492–495
latent class models and, 328
panel analysis, metric data and, 363–366
Unordered categorical data, 401
Unordered categorical variables, 78, 166–171
Unscaled deviance, 159

VARCL, 556, 569–570
Variable intercept models, 364–365, 375–376
Variance component analysis, 530
Variance function, 132
Variance standardization, 230–231
Varimax method, 205
Vector of fundamental parameters, 222
Visher's specification, 486

Wage inflation, 521, 562–568
Wald statistics, 230
maximum likelihood estimation and, 79, 88, 100, 104–107
pseudo-maximum likelihood estimation and, 128–129
quasi generalized PML estimation and, 135–136
Weibull model, 488, 489, 491, 499
continuous-time formulations and, 461–462
discrete state space and, 483
time-dependent covariates and, 475
time-independent covariates, 467–468
Weighted least squares estimation, 298
Weight function for generalized linear models, 553
Weighting adjustments for unit nonresponse, 46–48
White noise, 218

Whittemore's test, 287
Wishart distribution, 191–192, 560
Wishart generalized least squares (WGLS), 199–200
Wishart maximum likelihood (WML), 199, 200, 212, 215
Within-area regressions, 548
Within-group coefficients, 524
Within-group covariance, 534–535
Within-group estimator, *see* Least squares dummy variable (LSDV) estimator
Within-group regressions, 548

Yule's Q, 260, 332